Toxicology of Coal Conversion Processing

Toxicology of Coal Conversion Processing

Edited by:

ROBERT H. GRAY
Battelle, Pacific Northwest Laboratory
Richland, Washington

HARVEY DRUCKER
Argonne National Laboratory
Argonne, Illinois

MICHAEL J. MASSEY
REI Technologies, Inc.
Concord, Massachusetts

WILEY

A Wiley-Interscience Publication

JOHN WILEY & SONS

New York • Chichester • Brisbane • Toronto • Singapore

Library of Congress Cataloging in Publication Data:

Toxicology of coal conversion processing.

"A Wiley-Interscience publication."
1. Coal—Toxicology. 2. Coal gasification—Hygienic aspects. 3. Coal liquefaction—Hygienic aspects. 4. Coal gasification—Environmental aspects. 5. Coal liquefaction—Environmental aspects. I. Gray, Robert H. II. Drucker, Harvey. III. Massey, Michael J.

RA1242.C69T69 1988 662.6′2 87-21546
ISBN 0-471-80264-6

Printed in the United States of America

10 9 8 7 6 5 4 3 2 1

To A. Paul Duhamel, whose persistent and patient vision gave birth and direction, first to the Department of Energy's massive interdisciplinary synfuels characterization programs, and second to the assembly and integrated analysis of the results reported here

Contributors

John C. Craun*
ERT (A Resource Engineering Company)
Pittsburgh, Pennsylvania

Harvey Drucker
Argonne National Laboratory
Argonne, Illinois

John P. Fillo
ERT (A Resource Engineering Company)
Pittsburgh, Pennsylvania

Robert H. Gray
Battelle, Pacific Northwest Laboratory
Richland, Washington

David A. Haugen
Argonne National Laboratory
Argonne, Illinois

Walter Hubis
4031 S. Magnolia Way
Denver, Colorado

Douglas W. Later
Lee Scientific
Salt Lake City, Utah

Michael J. Massey
REI Technologies, Inc.
Concord, Massachusetts

Christopher A. Reilly, Jr.
Argonne National Laboratory
Argonne, Illinois

Roger A. Renne
Battelle, Pacific Northwest Laboratory
Richland, Washington

Barry W. Wilson
Battelle, Pacific Northwest Laboratory
Richland, Washington

**Present affiliation:* Reilly Tar & Chemical Corp., Indianapolis, Indiana.

Foreword

For those with an interest in the conversion of coal to alternate forms of fuel or chemical raw materials, or with an interest in the health and environmental impacts of complex mixtures of organic chemicals, *Toxicology of Coal Conversion Processing* is a useful and valuable book. It is the definitive review and summation of recent data on the effects on health and the environment produced by the complex mixtures formed as products or by-products in the conversion of coal. It provides a knowledgeable interpretation of a very large and complex set of data and ready accessibility to a voluminous literature, some of which might not be easily uncovered or retrieved.

The work reported here was a response to the petroleum price crisis of the early 1970s. While research and development on coal conversion were not new, the high level of attention given to the potential health and environmental consequences of the technology was. It had been learned the hard way, by seeing workers fall victims to cancer, that coal soot and the fumes from coke ovens could be hazardous to health. This time those responsible for the development of the technology, the U.S. Department of Energy (DOE) and private industry, added health and environmental safety to the list of developmental criteria in addition to such traditional ones as economics and operability.

The multiplicity of technological approaches to coal conversion and the fact that the products and by-products of concern were mainly complex mixtures of uncertain composition presented the toxicologists and analytical chemists with an enormous challenge. This book is a record of the response to that challenge. It lays out the knowledge and data that make possible the design and construction of a coal conversion industry capable of achieving high levels of health and environmental safety. In addition, it reports the significant progress that was made at unraveling the mysteries of the relationships between the composition of complex mixtures and their toxicological activity.

As of this writing in 1987, the price of oil is low, and interest in coal conversion has waned. But activity persists, and if history is any guide, the need for the information in this book will grow. Thanks to the foresight and the efforts of the DOE managers, the authors, and the editors, this vast compilation of data and knowledge is readily available not only to those with an interest in coal conversion but also to those with a concern for the biological effects of complex mixtures of chemicals.

G. K. Vick
Formerly with
Exxon Research & Engineering Company
Annandale, New Jersey

Preface

Studies of chimney sweeps in the late 1700s demonstrated that selected products of coal-derived materials can produce adverse human health effects. Two hundred years later, as the United States launched a major thrust to develop a synthetic fuels industry, there was neither a basic understanding of the causes leading to potential health effects anticipated from converting coal to alternate fuels nor a suitable database from which to derive engineering strategies to produce products with minimal health risks. As a result, the U.S. effort to develop a synthetic fuels industry included an unprecedented program for the characterization and ultimate management of potential health and environmental effects resulting from such an industry. Thus, to support process development and ultimate commercialization of an environmentally acceptable synfuels industry, the Department of Energy (DOE) implemented a comprehensive health and environmental effects research program. Important goals of the program were to determine the causes of toxicological activity observed in major synfuel products and by-products and, where possible, to identify mitigative processes to reduce or eliminate such activity. At the same time, agencies interested in regulation, primarily the Environmental Protection Agency (EPA) and the National Institute for Occupational Safety and Health (NIOSH), established complementary research programs focused on developing a database suitable for potential future regulation of this new industry.

This book emphasizes the database developed by DOE over a 10-year period on the toxicological and chemical properties of coal-derived liquids and tars and how those properties relate to engineering variables affecting process design and operation. Data from other sources that are useful for understanding that relationship are also presented. The book focuses specifically on health-effects studies associated with coal-based synthetic fuel processes; that is, coal liquefaction and gasification. Liquefaction materials selected for detailed investigation were believed to represent viable commercial products, for example, boiler fuels and transportation liquids. Gasification materials studied were for the most part secondary oils and tars which, if produced, could pose health risks in the workplace. Although health and environmental research was also performed on emissions, effluents, and solid wastes from coal-based and other synthetic fuels technologies, these studies are not addressed here.

Finally, the book is intended to serve as a primary reference for the work that was performed and for the database generated. The book is structured to serve both the casual reader and the specialist interested in detailed information on the subject. An introduction provides a brief overview of the programs conducted and the major conclusions derived from evaluation of the database. Individual chapters address, in depth, the technical work performed in the areas of toxicology, engineering, chemistry, and industrial hygiene. Appendixes document all samples and measurements referenced. A complete reference list is given at the end of each technical chapter; supplemental references are given in the appendixes.

Robert H. Gray
Harvey Drucker
Michael J. Massey

Richland, Washington
Argonne, Illinois
Concord, Massachusetts
August 1987

Acknowledgments

In efforts of the scale represented here, a few people necessarily report on the efforts of many. The research summarized in this book is the product of a large and dedicated group of people working in numerous organizations (e.g., national laboratories, industrial facilities, and universities) and spanning a variety of disciplines. The integration of their work was essential to meaningful sample collection, characterization of process operations, chemical analyses, toxicological evaluations, and ultimately interpretation of the results reported. We acknowledge particularly the U.S. Department of Energy (DOE) for its vision and persistent 10-year commitment to support multidisciplinary studies linking toxicology, chemistry, and engineering. We are especially grateful for the guidance and support provided by A. P. Duhamel, DOE Office of Health and Environmental Research, and P. H. Buhl, Office of the Assistant Secretary for Fossil Energy.

Several organizations were involved in studying the health, environmental, and safety aspects of coal liquefaction and gasification, and the data they collected are represented here. For example, the Pacific Northwest Laboratory evaluated solvent-refined coal (SRC) and other materials and initiated comprehensive health and environmental effects research programs. Similar programs were initiated by the Oak Ridge National Laboratory (ORNL), the International Coal Refining Company (ICRC), The Pittsburg & Midway Coal Mining Company, and Exxon Research & Engineering Company on H-coal, SRC-I, SRC-II, and EDS processes, respectively. Programs for coal-gasification processes were initiated by Argonne National Laboratory, the Inhalation Toxicology Research Institute, and ORNL. Environmental Research & Technology, Inc., coordinated the process-engineering aspects of the DOE's health and environmental programs for both coal liquefaction and gasification. Industrial hygiene, occupational health, and workplace monitoring programs were also planned and/or initiated by private industry for each of the major processes being developed. Finally, to achieve program objectives, working relationships were established between health and environmental research laboratories and process developers: Gulf Mineral

Resources Company, The Pittsburg & Midway Coal Mining Company, ICRC, Exxon Research & Engineering Company, the C. E. Lummus Company, Hydrocarbon Research, Inc., Ashland Synthetic Fuels, Inc., the Institute of Gas Technology, and others.

This book was reviewed in two phases. First, individual chapters were submitted for peer review by experts in each technical area (engineering, toxicology, chemistry, industrial hygiene). Second, the entire book was submitted to organizations capable of providing holistic review of the subject. Thus, we thank the following individuals and organizations for their critical review of parts or all of the various drafts of this document: S. B. Alpert, Electric Power Research Institute; W. E. Bechtold and C. H. Hobbs, Lovelace Inhalation Toxicology Research Institute; R. Biljetina, Institute of Gas Technology; N. E. Bolton, Union Carbide Corp.; M. T. Brandt, H. J. Ettinger, K. M. Hargis, J. O. Jackson, and R. C. Scripsick, Los Alamos National Laboratory; W. H. Calkins, E. I. du Pont de Nemours & Company, Inc.; W. G. Dreibelbis, J. C. Epler, M. R. Guerin, and C.-H. Ho, Oak Ridge National Laboratory; J. E. Duddy, Hydrocarbon Research, Inc.; J. M. Evans and K. E. Woodcock, Gas Research Institute; J. M. Holland, The Upjohn Company; C. M. Kelly, S. E. Morris, and A. F. Yen, Air Products and Chemicals, Inc.; M. L. Lee, Brigham Young University; G. C. Page, Radian Corp.; R. A. Pelroy, Pacific Northwest Laboratory; D. K. Schmalzer, Gulf Mineral Resources Company; R. F. Sullivan, Chevron Research Company; R. W. Tennant, National Institute of Environmental Health Sciences; G. K. Vick, Exxon Research & Engineering Company; C. M. White, Pittsburgh Energy Technology Center; and K. E. Wilzbach and V. C. Stamoudis, Argonne National Laboratory.

Finally, we are indebted to D. L. Felton for her invaluable assistance in editing and proofing the endless drafts of this book, preparation of which was supported by DOE under Contract DE-AC06-76RL0-1830 with Battelle Memorial Institute and Contract W-31-109-ENG-38 with the University of Chicago.

Glossary

AA	=	aminoanthracene
AC	=	aminochrysene
AHs	=	aliphatic hydrocarbons
AHH	=	aryl hydrocarbon hydroxylase
Amino-PAHs	=	amino-polycyclic aromatic hydrocarbons
ANL	=	Argonne National Laboratory
API	=	American Petroleum Institute
ASDA	=	antisolvent deashing
ASTM	=	American Society for Testing and Materials
ATP	=	adenosine triphosphate
BaP	=	benzo[a]pyrene
BETC	=	Bartlesville Energy Technology Center
bp	=	boiling point
BrdUrd	=	bromodeoxyuridine
BR	=	bottoms recycle
BSF	=	benzene-soluble fraction
C	=	carbon
CH_4	=	methane
CHO	=	Chinese hamster ovary
CI	=	chemical ionization
CLPP	=	coal-liquefaction pilot plant (EDS process)
CO	=	carbon monoxide

CO_2	=	carbon dioxide
COED	=	Char-Oil Energy Development
CRC	=	Chevron Research Company
CTPV	=	coal-tar pitch volatiles
CTSL	=	catalytic two-stage liquefaction
dg	=	day(s) gestation
DHF	=	deashed hydrotreater feed
DMBA	=	dimethyl benzanthracene
DMSO	=	dimethyl sulfoxide
DNA	=	deoxyribonucleic acid
DOE	=	Department of Energy
DUVAS	=	derivative ultraviolet adsorption spectrometer
EC_{50}	=	effective concentration (concentration required to cause abnormalities in 50% of survivors)
ECD	=	electron-capture detection
ECLP	=	EDS coal-liquefaction plant
EDS	=	Exxon donor solvent
EI	=	electron impact
EIMS	=	electron impact mass spectrometry
EPA	=	Environmental Protection Agency
EPR	=	electron paramagnetic resonance
EPRI	=	Electric Power Research Institute
ESP	=	electrostatic precipitator
ESR	=	electron spin resonance
FAB	=	fast-atom bombardment
FD	=	field desorption
FDMS	=	field desorption mass spectrometry
FI	=	field ionization
FID	=	flame ionization detection
FIMS	=	field ionization mass spectrometry
FOB	=	fuel-oil blend
FT	=	Fourier transform
FTIR	=	Fourier transform infrared
GC	=	gas chromatography
GC-MS	=	gas chromatography—mass spectrometry
GFETC	=	Grand Forks Energy Technology Center
H	=	hydrogen
H:C	=	hydrogen: carbon ratio
HCl	=	hydrogen chloride
HD	=	heavy distillate
He	=	helium
HGPRT	=	hypoxanthine guanine phosphoribosyl transferase

HPLC	=	high-performance liquid chromatography
HRFIMS	=	high-resolution field ionization mass spectroscopy
HRGC	=	high-resolution gas chromatography
HRMS	=	high-resolution mass spectrometry
H_2S	=	hydrogen sulfide
HTR	=	hydrotreater
ICRC	=	International Coal Refining Company
I/P	=	initiation/promotion
IP	=	intraperitoneal
IR	=	infrared
ITRI	=	Inhalation Toxicology Research Institute
ITSL	=	integrated two-stage liquefaction
LC_{50}	=	lethal concentration (concentration required for 50% mortality)
LCF	=	LC Finer
LCFF	=	LC Finer feed
LCFTLP	=	LC Finer total liquid product
LD_{50}	=	lethal dose (dose required to kill 50% of the exposed population)
LDH	=	lactate dehydrogenase
LEC	=	lung epithelial cells
LO	=	light oil
LSE	=	liquid solvent extraction
LTR	=	low-temperature reactor
LVMS	=	low-voltage mass spectrometry
MAF	=	moisture- and ash-free
MD	=	middle distillate
METC	=	Morgantown Energy Technology Center
MI	=	Mellon Institute
MRDC	=	Mobil Research & Development Corporation
MS	=	mass spectrometry
MTD	=	maximum tolerated dose
N	=	nitrogen
NBS	=	National Bureau of Standards
NCB	=	National Coal Board (British)
NH_3	=	ammonia
NMR	=	nuclear magnetic resonance
NIOSH	=	National Institute for Occupational Safety and Health
N-PACs	=	nitrogen-containing polycyclic aromatic compounds
NPD	=	nitrogen/phosphorus detector
NTSL	=	nonintegrated two-stage liquefaction
NVO	=	nonvolatile organic
O	=	oxygen
ODC	=	ornithine decarboxylase

O-PACs	=	oxygen-containing polycyclic aromatic compounds
ORNL	=	Oak Ridge National Laboratory
OSHA	=	Occupational Safety and Health Administration
OT	=	once-through
PAAs	=	primary aromatic amines
PACs	=	polycyclic aromatic compounds
PAHs	=	polycyclic aromatic hydrocarbons
PAKs	=	polycyclic aromatic ketones
PANHs	=	polycyclic aromatic nitrogen heterocycles
PAQs	=	polycyclic aromatic quinones
PASHs	=	polycyclic aromatic sulfur heterocycles
PBS	=	phosphate-buffered saline
PD	=	plasma desorption
PDMS	=	plasma desorption mass spectrometry
PDU	=	process development unit
PETC	=	Pittsburgh Energy Technology Center
PFP	=	pentafluoropropionic
pfu	=	plague-forming unit
PID	=	photoionization detection
P&M Co.	=	Pittsburg & Midway Coal Mining Company
PNL	=	Pacific Northwest Laboratory
ppm	=	parts per million
PS	=	process solvent
RAMs	=	rabbit alveolar macrophages
RBCs	=	red blood cells
rev	=	revertants
RNA	=	ribonucleic acid
RPE_{50}	=	relative plating efficiency (50%)
RTI	=	Research Triangle Institute
S	=	sulfur
SARA	=	saturates–aromatics–resins–asphaltenes
SCE	=	sister chromatid exchange
SCE_{200}	=	200% increase in SCE
SCT	=	short contact time
SE	=	standard error
SFC	=	supercritical fluid chromatography
SGC	=	simulated gas chromatography distillation
SHE	=	Syrian hamster embyro
SI	=	survival index
S-PACs	=	sulfur-containing polycyclic aromatic compounds
SRC	=	solvent-refined coal
SSMS	=	spark source mass spectrometry

TBP	=	true boiling point
THF	=	tetrahydrofuran
TFA	=	trifluoroacetic
TFPA	=	trifluoroperoxyacetic
TFT	=	trifluorothymidine
TG	=	thioguanine
TI	=	teratogenic index (LC_{50}/EC_{50})
TK	=	thymidine kinase
TLC	=	thin-layer chromatography
TLP	=	total liquid product
TLV	=	threshold limit value
TPA	=	tetradecanoylphorbol acetate
TSL	=	two-stage coal liquefaction
UDS	=	unscheduled DNA synthesis
UMD	=	University of Minnesota–Duluth
UNDERC	=	University of North Dakota Energy Research Center
UOP	=	Universal Oil Products
UV	=	ultraviolet
VB	=	vacuum bottoms
WS	=	wash solvent
wt %	=	weight percent

Contents

Toxicology of Coal Conversion Processing

CHAPTER 1

Introduction and Summary

R. H. Gray
Battelle, Pacific Northwest Laboratory, Richland, Washington

H. Drucker
Argonne National Laboratory, Argonne, Illinois

M. J. Massey
REI Technologies, Inc., Concord, Massachusetts

In the mid-1970's, the U.S. Department of Energy (DOE) initiated major research and development programs to support the commercialization of synthetic fuels technologies. Simultaneously, the DOE initiated major health and environmental research programs to ensure that commercialization would occur in an environmentally acceptable manner. The programs were unique in that, for the first time, major commitments were made to determine the potential health and environmental effects of a technology prior to commercialization.

This book focuses on results of research efforts directed at determining potential health effects from various products of coal-conversion processing, particularly coal liquids and coal tars. Major program objectives were the following:

1. Determination of the specific chemicals or chemical classes responsible for observed toxicological activity (primarily mutagenicity and carcinogenicity) in materials from various coal-conversion processes.
2. Determination of similarities and differences among toxicologically active chemicals within and between different types of coal conversion processes.
3. Determination of relations between measured toxicological activity and the physical properties and process history of coal-conversion materials.

4. Determination of the relative effectiveness of alternatives for controlling the toxicological activity of coal-conversion materials.

To attain program objectives and ensure meaningful interpretation of data, toxicologists, chemists, and process engineers worked in closely coordinated teams. To cope with the unique problems associated with the complex toxicology and chemistry of coal-conversion materials, a special strategy called *biodirected chemical characterization* was devised and systematically applied (see Section 1.2 and Chapter 5).

This book briefly reviews the chemical and toxicological database for coal conversion prior to the mid-1970s, compiles and evaluates results of the work described above, and identifies the key properties of coal-conversion materials that relate to toxicological activity. The book is organized into six chapters. This chapter provides an introduction and overall summary of the effort. Chapter 2 describes the coal-conversion processes evaluated and the materials studied. Chapter 3 discusses the toxicological methods developed and presents and evaluates the data available for coal-derived materials, while the appendix provides related information, including complete descriptions of all materials assayed for toxicological response. Chapter 4 analyzes the toxicological data from a process engineering perspective. An important feature of this book is that most materials described have been assigned identifying numbers that allow the interested reader to locate, trace, and cross-reference specific samples discussed in each chapter. Consult the appendix at the end of Chapter 4 for additional information, including an example of how to use the identifying sample numbers. Chapter 5 provides the chemical basis for toxicological response. Chapter 6 summarizes our current understanding of human health effects, based on monitoring workers at small-scale (bench and pilot plant) coal-conversion facilities. A glossary provides the reader with definitions of the many abbreviations and acronyms used throughout this book.

1.1 SUMMARY OF MAJOR CONCLUSIONS

In spite of the complexity of coal-conversion materials and the diversity of processing options, there is remarkable consistency in the observed toxicological activity of coal-derived materials:

1. Although some quantitative differences exist, the chemical classes responsible for toxicological activity in coal-conversion materials are qualitatively the same regardless of the coal-conversion process used to produce the material.
2. The chemical classes primarily responsible for mutagenic and carcinogenic activity of coal liquids and tars are high-molecular-weight, polycyclic aromatic hydrocarbons (PAHs). Amino-PAHs are also prevalent and are a major source of mutagenic activity.
3. Toxicological activity of coal-conversion materials can be predicted from their 95% distillation temperatures (temperature at which 95% of the feed material has been distilled) and their aromaticity. Toxicological activity is concentrated in those highly aromatic components boiling above 650°F and having hydrogen : carbon (H : C) ratios below 1.5.
4. Selective distillation, heavy ends process recycle, and catalytic hydrotreating, singly and in combination, can effectively reduce toxicological activity through control of product boiling point and aromaticity.

1.2 RESEARCH STRATEGY

Although program objectives focused on assessing potential health and environmental effects of a large-scale synfuels industry, there were no large-scale (i.e., demonstration- or commercial-scale) coal-conversion facilities in the United States from which to obtain study materials. Therefore, study materials, produced from a variety of coal feedstocks over a range of process conditions, were obtained from process facilities representing bench-scale units to pilot plants (processing from pounds to tons of coal per day). Because these materials proved to be essentially toxicologically identical, materials from a commercial facility would likely have similar toxicological characteristics. Thus, the question of scale was addressed under the conditions existing at the time.

Research on coal-derived materials was performed as follows. Initially, materials were screened using a battery of short-term toxicological assays. Results of these screening assays provided a preliminary indication of the potential health effects of various coal-conversion materials, identified materials that required special handling or additional processing to minimize health risks prior to widespread distribution, and aided in the selection of materials for and design of longer-term, more extensive process engineering and toxicological studies and chemical analyses. The amount of time, effort, and expense invested in performing detailed chemical analyses was dictated by the toxicological properties of the materials studied. To identify and quantify specific compounds or chemical classes responsible for observed toxicological effects, chemical separations, iterative bioassays of derived fractions, and detailed chemical analyses were performed only on those materials showing significant toxicological activity. This strategy has been termed the biodirected chemical characterization for the coupled chemical and toxicological assay of complex mixtures containing potential toxicants.

Longer-term research was limited to evaluation of those toxicologically active materials that most closely approximated materials anticipated from a large-scale industry and to which human exposure was likely. Some of these materials were subjected to *in vivo* (whole-animal) studies to evaluate potential carcinogenicity, teratogenicity, and/or other toxicological effects. Again, chemical fractionation and detailed chemical analyses were used to identify components of those materials that were toxicologically active. Results of longer-term studies were compared to those of shorter-term studies to provide a firmer basis for assessing health and environmental risks.

The influences of process or operational modifications and control technology options on potential health and environmental effects of coal-conversion materials were also evaluated. Short-term assays coupled with chemical fractionation were again applied to identify changes in the chemistry of coal-conversion materials that might account for modifications in toxicological activity.

1.3 COAL-CONVERSION DATABASE

The chemical, toxicological, and engineering database for coal-conversion oils and tars comprises thousands of separately reported analyses. Almost 400 coal-conversion materials were obtained from nearly 40 coal-conversion facilities, providing over 500 individual samples that were assayed for toxicological response. Toxicological endpoints included *in vitro* cytotoxicity, mutagenicity, and cellular transformation and in vivo acute toxicity (e.g., ocular and dermal irritation), embryotoxicity, teratogenicity, and carcinogenicity (see Chapter 3). Most materials were studied in vitro, with the Ames *Salmonella*/mammalian microsome mutagenicity assay. Because of time and cost considerations, considerably fewer (59) materials were evaluated for dermal carcinogenicity in vivo during chronic (2-year) exposures or in shorter-term (6-month) tumor initiation/promotion assays. Data on process conditions, physical properties, and chemical characteristics of samples assayed for toxicological response are available for many coal-liquefaction materials, especially those representing solvent refined coal (SRC), EDS, and H-Coal processes. More limited data, especially for physical properties and chemical characteristics, are available for gasification materials.

Four major direct coal-liquefaction processes were initially evaluated for large-scale use. These included two single-stage, SRC processes (SRC-I and SRC-II) and two catalytic processes (EDS and H-Coal). These four processes all produce liquid fuel products. The SRC-I process also yields a low-sulfur, low-ash, solid fuel. Both SRC process configurations were evaluated at a 30–50-ton/day pilot plant at Ft. Lewis, Washington. The SRC-I process and two-stage liquefaction (TSL) variations of the SRC-I process were also evaluated at a 6-ton/day pilot plant at Wilsonville, Alabama. Other TSL options were studied at smaller-scale (lb/day) process development units (PDU) in New Brunswick and Lawrenceville, New Jersey. The H-Coal process was evaluated at a 200–600-ton/day facility at Catlettsburg, Kentucky, while the EDS process was studied at a 250-ton/day pilot plant at Baytown, Texas. The SRC-II, H-Coal, and EDS processes were also studied at smaller-scale facilities in Harmarville, Pennsylvania, and Merriam, Kansas; in Trenton, New Jersey; and in Baytown, Texas, respectively.

Coal-gasification processes were evaluated at facilities representing a range of process scales. Development of new or modified gasification processes was conducted at the Institute of Gas Technology (Hygas), Chicago, Illinois and the University of North Dakota Energy Research Center, formerly the Grand Forks Energy Technology Center (slagging, fixed-bed gasifier), Grand Forks, North Dakota. In addition, fixed-bed coal-gasification processes were tested and/or used for industrial purposes at the Morgantown Energy Technology Center, Morgantown, West Virginia; the University of Minnesota–Duluth, Duluth, Minnesota; the Holston Army Ammunition Plant, Kingsport, Tennessee; and elsewhere. Studies were also conducted at other facilities; for example, Research Triangle Institute (fixed-bed reactor), Research Triangle Park, North Carolina and Mellon Institute (entrained-flow riser reactor), Pittsburgh, Pennsylvania.

1.3.1 Toxicological Results

Certain coal-derived materials contain components with high specific mutagenic activities as determined in the Ames assay. The concentration of high-molecular-weight components is related to mutagenic response; that is, higher-boiling-range (higher-molecular-weight) components show higher mutagenic activities (Fig. 1.1). The Ames assay is particularly sensitive to nitrogen-containing polycyclic aromatic compounds (N-PAC), principally amino-PAHs in these materials. The amine-rich fraction constitutes only a small portion of the weight of most crude materials but is often responsible for most of the material's mutagenic activity. Other microbial assays were used on a limited number of materials, and results were qualitatively similar to those of the Ames assay. Mutagenic crude materials were also tumorigenic to mouse skin.

Fifty-nine materials were assayed for dermal tumorigenicity in mice. Experimental conditions were either chronic (exposure to multiple doses of material over 1–2 years; 24 materials) or involved exposure to a single dose of material (initiation; 35 materials, of which 3 were also assayed chronically) followed by exposure over a period of 6 months to an

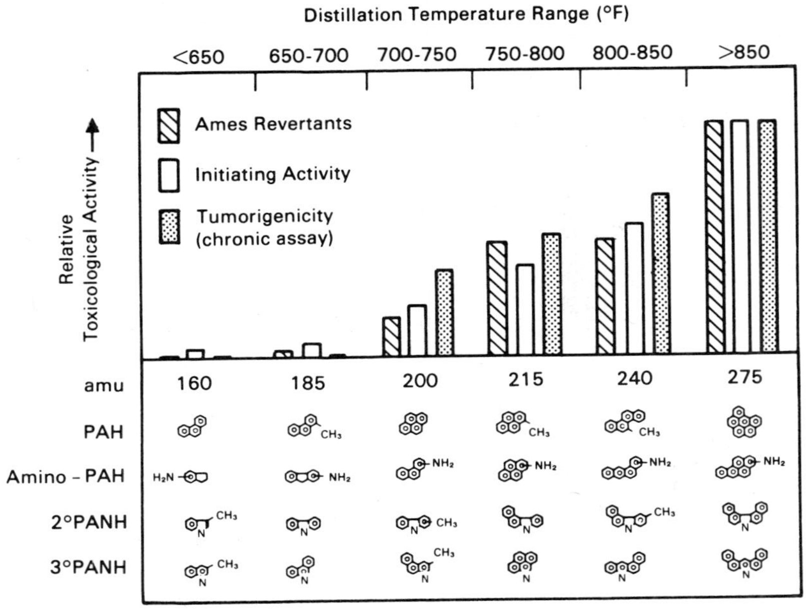

Figure 1.1 Toxicological activity and chemical composition versus boiling range of a coal-derived liquid. Chemical structures represent those in highest concentration at each boiling range for four chemical classes. PAHs = polycyclic aromatic hydrocarbons; 2° PANHs = secondary polycyclic aromatic nitrogen heterocycles; 3° PANHs = tertiary PANHs. Average molecular weight given in atomic mass units (amu), rounded to nearest 5.

agent known to promote the growth of tumors initiated by chemical carcinogens. Carcinogenic activity of crude coal-conversion materials was generally greater than that observed with materials from shale oil or petroleum technologies. PAH compounds were the dominant dermal carcinogens. Carcinogenic components of coal-derived liquids, like mutagenic components, were found in higher-boiling-range distillates (Fig. 1.1).

Expression of specific PAH mutagens and carcinogens is determined by the total composition of the material in which they are found. Antagonism or inhibition of the mutagenic and carcinogenic activity of these complex mixtures has been observed. Carcinogenic activity of fractions of some materials was greater than that of unfractionated material. Synergistic effects (those greater than the sum of the toxicological activities of specific components) were only found in the mutagenic response to aromatic amines and in the teratogenic response to one coal-derived material.

Mutagenic and carcinogenic activity of coal-derived liquids varied with the nature and severity of coal processing. Toxicological activity could be reduced through any processing that effectively cracked and hydrogenated toxicologically active components. For example, hydrotreatment, a method of upgrading hydrocarbon liquids prior to further refining or use, resulted in decreased mutagenesis in cellular assays and reduced carcinogenic activity in whole-animal systems (Fig. 1.2).

In vitro mammalian cell assays provided information on the genotoxicity and cytotoxic effects of a limited number of coal-derived materials. Because of their inherent cytotoxicity,

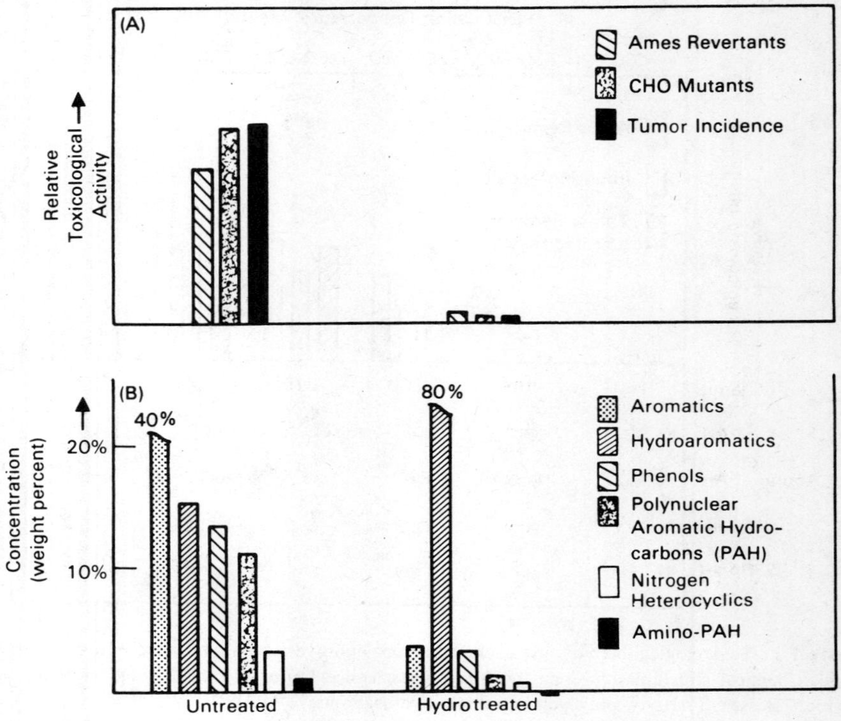

Figure 1.2 Effect of hydrotreatment on: (A) toxicological activity in Ames, Chinese hamster ovary (CHO), and mouse skin-painting assays, and (B) chemical composition of a coal-derived liquid.

most materials had to be chemically fractionated to obtain reproducible results in these assays. Test materials caused various cytotoxic responses, for example, cell death, alterations in cell growth, and loss of cell function. Cytotoxicity was generally greater for materials with low average molecular weight.

Most materials showed relatively low acute toxicity in whole animals. Oral and intraperitoneal exposure to coal-conversion materials produced a range of responses from relatively harmless to slightly toxic. Hydrotreatment reduced acute toxicity. Acute ocular exposure of rabbits to most materials tested produced slight temporary irritations; a few materials induced permanent damage to the cornea. Although most materials tested were not dermal sensitizers, several induced a substantial allergic response. Chronic dermal irritation was variable and increased with exposure concentration. Chronic exposure to a few materials induced neural and renal toxicity. Several coal liquids and a gasifier tar also induced embryotoxic and teratogenic effects in amphibians and/or rodents.

1.3.2 Engineering Analysis

The availability of physical and chemical data varied among coal-conversion materials evaluated for toxicological activity. In some cases, little was known other than the process and generic name of the stream from which the sample was obtained. Bulk properties, like specific gravity and elemental analyses, were often available because they were routinely determined to characterize process performance. Sample distillation curves were available for many coal-liquefaction materials and a few gasification tars and oils. In some cases, complete analytical data, including analysis of hydrocarbon types, basic nitrogen, and sulfur, were also available.

Sampling and handling procedures varied substantially among materials. Some samples were stored months to years under conditions that were not precisely controlled, and substantial changes could have occurred in their toxicological and chemical properties. However, many samples were refrigerated and stored in the dark under inert atmospheres and underwent little degradation. Handling of most coal-conversion materials was intermediate to these extremes. A systematic analysis suggested that, while chemical and toxicological results can reflect differences in handling and storage regimes, qualitative toxicity (i.e., active versus inactive) of materials reported here remained unaffected.

The database was evaluated to determine which engineering variables (e.g., process configurations or operational modes) and/or physical and chemical characteristics of coal-conversion materials were most often associated with toxicological response. Toxicological endpoints included in this analysis emphasized mutagenic and carcinogenic responses. The availability of critical engineering data on process operations and the small number of samples tested for carcinogenic response in longer-term assays were limiting factors in attempting to relate engineering variables to physical, chemical, and toxicological properties of coal-conversion materials.

Over the range of coal-conversion materials analyzed, boiling point and extent of aromaticity (and thus hydrogenation) were found to be the most reliable predictors of mutagenic and carcinogenic response (see Chapter 4). In particular, coal-liquefaction materials with 95% distillation temperatures above 650–700°F and having hydrogen : carbon (H : C) ratios less than 1.5 (a measure of aromaticity) were consistently mutagenic and carcinogenic, regardless of the process from which they were derived. Materials boiling below 650°F and having H : C ratios greater than 1.5 were generally inactive (Figs. 1.3 and 1.4). Factors such as coal type and handling, specific process design, and other engineering considerations had little influence on toxicological activity beyond that reflected in their effect on the 95% distillation temperature and H : C ratio.

Hydrotreating reduces overall aromaticity and heteroatom content of the feedstock and lowers, although to a lesser extent, the 95% distillation temperature. Because of reduced aromaticity, hydrotreated products may have greatly reduced mutagenic or carcinogenic activity even though their 95% distillation temperatures exceed 650°F. Reduced aromaticity after hydrotreatment reflects destruction of higher-molecular-weight PAHs, N-PACs, and related compounds that are responsible for mutagenic and carcinogenic activity.

Insufficient data (process conditions, physical properties, chemical and toxicological properties) were available to explain the presence or magnitude of mutagenic or carcinogenic activity in gasification oils and tars. Process variables controlling production of toxicologically active materials are numerous, interrelated, and complex. Nonetheless, several trends relative to the toxicological properties of tars and oils that are by-products of coal gasification were identified; they suggest that principles similar to those found for coal-liquefaction materials may apply.

1.3.3 Chemical Basis for Toxicological Response

With minor exceptions, the same chemical classes, mainly aromatic and hydroaromatic compounds and their alkylated homologs (some with heteroatomic functional groups), are

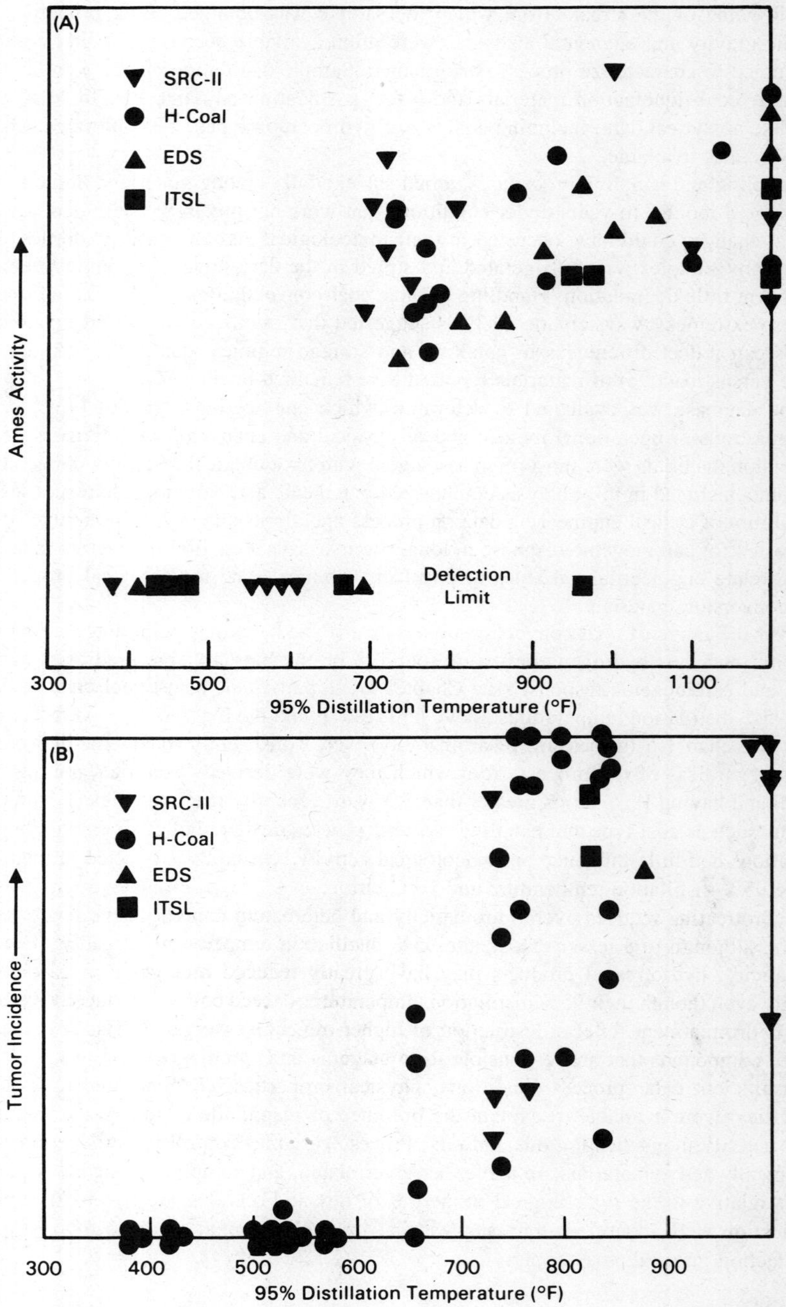

Figure 1.3 (A) Mutagenic and (B) tumorigenic activity of coal-derived liquids as a function of the 95% distillation temperature. Mutagenic activity determined in the Ames assay; tumorigenic activity determined at different doses (doses not shown) in mouse skin-painting assays.

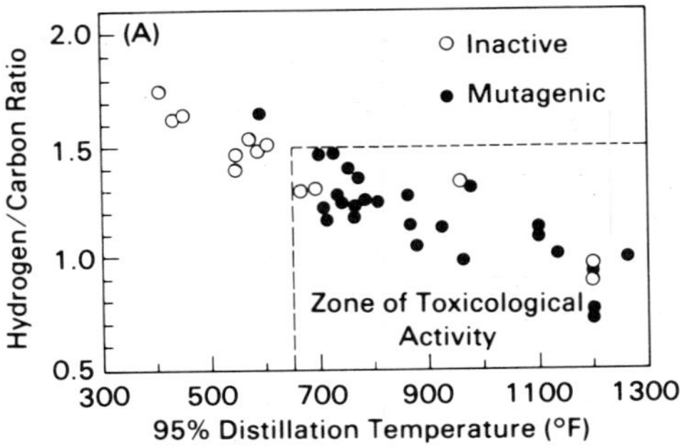

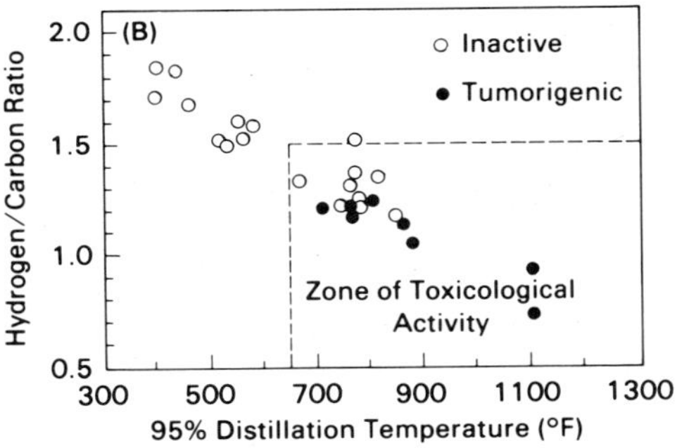

Figure 1.4 Relationships among toxicological activity, hydrogen : carbon ratio and the 95% distillation temperature for coal-derived liquids tested in (A) Ames and (B) mouse skin-painting assays.

present in all coal-liquefaction and gasification materials (see Chapter 5). Gasification materials tend to contain more oxidative reaction products, such as polycyclic aromatic compounds (PACs) with carbonyl and cyano functionalities. Heteroatomic groups found in both liquefaction and gasification materials include primary, secondary, and tertiary amino and phenolic functionalities, and heterocyclic nitrogen, oxygen, and sulfur moieties. Similarities also exist in PAH compounds found in these materials prior to upgrading. However, there may be substantial quantitative differences in the relative concentrations of the various compounds and chemical classes present.

Catacondensed PAHs (those having no adjacent condensed ring faces), cata/pericondensed PAHs (those having adjacent ring faces), and certain alkylated (C_1–C_3) catacondensed PAHs with four to six aromatic rings are responsible for most carcinogenic activity in coal-derived materials. Generally, isolates of coal-derived materials containing four- to six-ring, alkylated PAHs with three or more methylene units show decreased mutagenic and tumorigenic activity compared with PAH fractions with four- to six-ring C_0–C_3 PAHs. Other N-PAC classes, including compounds such as benzacridines, dibenzcarbazoles, dimethylbenzacridines, and dibenzacridines, also contribute to the mutagenic and carcinogenic activity of coal-conversion materials.

Coal-derived materials with significant levels of aromatic PAH analogs generally show reduced mutagenic and tumorigenic activity after hydrotreating. Mild to moderate hydrotreatment decreases the mutagenic activity of coal-conversion materials by reducing the concentration of primary amino-PAHs and, to a lesser extent, by dearomatization of PAH structures through saturation and/or carbon–carbon bond cleavage. More severe hydrotreatment significantly decreases dermal carcinogenicity, primarily from dearomatization of four- to six-ring PAH structures. Process variables other than those that determine upper distillation temperature and H : C ratios do not appear to significantly affect dermal tumorigenicity of coal-conversion materials.

The presence of amino-PAHs in coal-conversion materials, including single- and two-stage liquefaction products and gasification condensates, is readily detected by a positive response in the microbial mutagenicity assay in most cases, regardless of matrix composition. Three- and four-ring amino-PAHs are more active than those having less than three or more than four aromatic rings, with activity being a function of amino substituent position. Amino-PAHs are not as active as PAHs in skin-tumorigenicity assays.

Alterations in process conditions (i.e., within a specific process), such as catalyst quality, reaction severity, and residence time, can substantially affect the chemical composition of coal-derived liquids, particularly the toxicologically active constituents such as amino-PAHs and PAHs. However, the effects of process conditions on mutagenic or carcinogenic activity are expressed predominantly through their influence on boiling point and aromaticity. Coal type may also affect the chemical composition of coal liquids. However, because coal-conversion process conditions are optimized for each coal used, direct relations between coal type and toxicological activity cannot be determined.

1.3.4 Human Experience: Worker Exposure

Coal conversion, whether to coke, gas, liquids, or simply heat, results in products and by-products containing materials or compounds that, under certain conditions, may be deleterious to human health. Thus, exposure of workers to materials in a production facility is a major concern. The degree of hazard depends to some extent on the specific process and process conditions. However, one chemical class, PAHs, is common to and of particular concern in all processes. The PAHs can be present in liquid products, as aerosol emissions, in by-products, or associated with particulate matter. Many PAHs are potent carcinogens and, more recently, amino-PAHs have also been implicated as problematic in certain coal-conversion materials (see Section 1.3.3 and Chapter 5).

Extensive industrial hygiene monitoring and medical surveillance programs have accompanied the operation of developmental coal-conversion facilities (see Chapter 6). These invariably included engineering controls, personal hygiene, protective clothing, prescribed work practices, and health education. Anticipated health concerns were based on known hazards for coking operations and the occurrence of skin cancer among workers at a coal-liquefaction facility that operated in the early 1950s in Institute, West Virginia. Coal coking has been shown epidemiologically to pose serious occupational hazards. Health problems were associated with direct exposures to the coking operation and to the by-products, coal, tar, coal tar pitch, and creosote. Standards and industrial hygiene and medical surveillance requirements have been mandated by the Occupational Health and Safety Administration to protect coke-oven workers.

Although available only for short-term exposures to coal-derived materials, data from coal-liquefaction pilot-scale facilities are encouraging. Compared to the experience in coke plants, results suggest that if worker exposures in coal-conversion facilities are monitored and sound industrial hygiene procedures are applied, occupational health problems can be minimized.

1.4 IMPLICATIONS

Potential health and environmental risks associated with development and deployment of a large-scale coal-conversion industry were assessed during the early stages of technology development. Results of these studies, coupled with engineering and economic factors, can be used to guide further process design so that developers of new technologies can avoid the all too familiar chain of events: design the process, construct a facility, find potential health/environmental effects associated with commercialization, take legal action and/or conduct environmental research, regulate, retrofit, and/or abandon the technology. This chain of events can have economic and human costs far beyond those associated with the prospective analysis described here.

The research efforts described provided guidance for identifying commercial products with minimal health risks (e.g., those boiling below 650°F) and for selecting process modifications (e.g., selective distillation and hydrotreatment) to minimize adverse human health and environmental risks. Because of these efforts, coal liquids currently projected for commercial use are generally not expected to have higher toxicological activity than shale oil or petroleum, thus demonstrating that combining health-effects research with process development can result in safer and more environmentally responsible technology development. Although applied to developing a coal-conversion industry, the research strategy used is equally applicable to development of any energy-related or other process that produces complex chemical mixtures (e.g., petroleum, shale oil, fuels from biomass, and hazardous waste). In fact, the linking of engineering and life sciences research can be applied to the development of any technology or the environmentally safe exploitation of any resource.

CHAPTER 2

Coal Conversion Processes and Database Description

J. P. Fillo *and* J. C. Craun*

ERT (A Resource Engineering Company), Pittsburgh, Pennsylvania

Use of coal as a fuel dates back to before the ancient Greeks, in the third century B.C. Since that time, coal has been used to produce coke, smelt iron, generate heat, produce electricity, and produce both gaseous and liquid by-products, some of which constitute alternate or synthetic fuels. For instance, tar was recovered from the pile-coking process as early as 1600, and Berthelot hydrogenated coal to produce oil containing aromatics and naphthenes in 1869. In 1792, coal gas was produced for use in lighting, and by the mid-1920s, both slagging and pressure gasification of coal had been demonstrated separately. The *refining* of coal liquids dates back to the first commercial distillation of coal tar in 1818 and the large-scale catalytic hydrogenation of tar from brown coal in 1925. Coal-conversion processing to synthetic fuels and chemicals currently is practiced commercially in Europe, South Africa, and, to a limited extent, the United States.

This chapter reviews coal-conversion processes that generated process streams, products, and by-products for toxicological testing and the associated database since the early 1970s. This review begins with a historical overview of the development of coal-conversion processing. Descriptions of each technology and the associated nomenclature of unit operations, products, and by-products are presented. The specific facilities generating materials for toxicological testing are also summarized. The scope of the database discussed in this book is extensive. Almost 400 process materials were sampled from nearly 40 individual liquefaction, gasification, and hydrotreating facilities, providing over 500 individual samples for toxicological study. An overview of the types of data that comprise the database is presented. Toxicity, chemistry, and property data that are most commonly available and the associated nomenclature are briefly discussed. Finally, the chemical stability of coal-

**Present affiliation:* Reilly Tar & Chemical Corp., Indianapolis, Indiana.

conversion materials and its effect on the toxicological and physical properties of these materials are also discussed.

2.1 HISTORICAL PROCESS DEVELOPMENTS

The conversion of coal to gas and/or liquid products has a long history of development and practice. This section briefly reviews this history as perspective for more recent developments in coal-conversion technologies since the 1960s. More in-depth reviews, which provide the basis for this overview, can be found in Donath (1963), Rhodes (1966), Wu and Storch (1968), Nelson (1969), Elliott and Yohe (1981), and Probstein and Hicks (1982).

2.1.1 Coal Liquefaction

The first process for hydrogenating coal to form liquid products was patented by Bergius in 1913. In this process, coal was hydrogenated under pressure at 570–660°F, without a catalyst, to form oils. The process was converted to industrial operation in 1927, by I. G. Farben of Germany under the direction of M. Pier, and consisted of the liquid-phase hydrogenation of coal in tar using a molybdenum oxide catalyst. Heavy tar from the hydrogenation process itself was used to slurry the feed coal, while middle-boiling oil was further hydrogenated to gasoline. In 1934, iron replaced molybdenum oxide and became the standard hydrogenation catalyst. During the 1930s and 1940s, 12 commercial plants were commissioned in Germany, and by 1943 total annual oil production reached about 4×10^6 metric tons, which is equivalent to approximately 70,000 barrels/day of liquid products. Before the end of World War II, the only other commercial-scale coal hydrogenation facility was located in Billingham, Great Britain, with an annual production capacity of 150,000 metric tons of liquid products, or approximately 2600 barrels/day.

Although commercial production of coal-derived liquids continued after World War II, most facilities were converted to petroleum or coke-oven product refining, or they were dismantled. During the 1950's, synthetic-fuel production continued in East Germany and about 1.7×10^6 metric tons/year of gasoline plus diesel fuel were produced in 1956. Two dismantled plants were transported to Siberia and operated at an annual capacity of 1×10^6 metric tons/year. Although coal research has been conducted throughout Asia, Europe, Canada, Australia, and the United States, the German experience remains the single largest commercial application of coal-liquefaction technology.

Following World War II, two demonstration-scale, coal- and tar-hydrogenation facilities were constructed in the United States. Although these facilities were designated as demonstration plants, they were comparable in capacity to the major liquefaction pilot plants developed during the 1960s and 1970s (see Section 2.5). The first was operated from 1949 to 1954 by the U.S. Bureau of Mines in Louisiana, Missouri. The facility was designed to convert about 75 tons/day of coal into 200–300 barrels/day of gasoline. Liquefaction conditions were 860–900°F and 500–700-atm pressure using ferrous sulfate ($FeSO_4$) catalysts. Coal-derived oil boiling below 625°F was converted to gasoline via vapor-phase hydrogenation at 700 atm and 900°F in the presence of a clay-based catalyst containing zinc, chromium, and molybdenum (Kastens et al., 1949; Chaffee and Hirst, 1953).

The second facility was constructed at Institute, West Virginia by the Carbide and Carbon Chemical Company, and it was designed to convert about 300 tons/day of coal into over 1000 barrels/day of chemical products. Liquefaction was conducted at about 400-atm pressure and 840–1020°F in the presence of a catalyst and hydrogen gas. The reactor residence time was less than 5 min. The facility was operated from 1952 to 1956, after which a small, 12-ton/day facility was operated until 1963 (Donath, 1963; Young, 1979).

Further development of coal-liquefaction processes occurred on small, bench-scale units and process development units (PDUs) through most of the 1960s. Research and development for major pilot-plant and commercial development programs increased in the early 1970s when a variety of concepts were tested over many scales of process development. Acceleration of coal-liquefaction process development in the 1970s, prompted by a national goal of energy independence, resulted in implementation of major efforts in the United States to evaluate potential health effects of coal-derived materials from these facilities (Gray and Cowser, 1982; Wilzbach and Reilly, 1982).

2.1.2 Coal Gasification

Coal gasification was first demonstrated in 1792, when bituminous coal was distilled in an iron retort to produce gas for home lighting. Gas use proliferated quickly in England, mainly to illuminate factories. By 1812, the London and Westminster Chartered Gas, Light and Coke Company was founded to illuminate the streets of London. Over the next 15 years, city gas plants were installed in Great Britain, France, Germany, and the United States in which coal typically was carbonized in small, cast-iron retorts to produce gas.

Use of manufactured gas increased in the United States in the late nineteenth and early twentieth centuries. In 1870, gas containing carbon monoxide and hydrogen was produced by passing steam over hot coke or coal. The gas was eventually called blue gas or water gas and was first produced commercially in 1875. Gas-production facilities were typically centered in large metropolitan areas until the 1950s when the advent of large, inexpensive supplies of crude oil and natural gas resulted in their ultimate extinction.

Coal-derived gas also was used industrially to provide heat. Investigations in the early twentieth century focused on combustion of fuel in beds (1916), development of slagging gas producers (1912), and pressurized gasification of coal (1926). From 1930 to 1950, the advent of fluidization, hydrogasification, and the first application of these concepts to pressurized coal gasification further enhanced the art. Many fixed- and fluidized-bed gasifiers (more appropriately termed gas producers) were developed, and some are used today in various industrial applications throughout the world.

The gasifier used predominantly to produce commercial synthetic-fuel products was developed by Lurgi Kohle and Mineralotechnik GmbH. The first fixed-bed, pressurized Lurgi gasifier was built in 1936. It was first used to produce commercial liquid products from coal (in conjunction with Fischer–Tropsch synthesis) in 1954 at Sasolburg, South Africa. The Sasol complex has been expanded since then from its original 10,000-barrel/day capacity and now includes two additional facilities producing over 100,000 barrels/day of liquid products. In addition, the Lurgi Mark IV gasifier is used at the Great Plains Coal Gasification Plant in Beulah, North Dakota, to produce substitute natural gas (SNG) from coal.

2.1.3 Coal-Liquid Hydrotreating

Hydrogenation of coal liquids closely paralleled development of coal hydrogenation in Germany. By the mid-1920s, petroleum oils and coal tar were catalytically hydrogenated to gasoline on a large scale. Conversion occurred in the vapor phase, over fixed beds of molybdenum oxide catalyst. The process produced gasoline from the middle oil derived from catalytic coal hydrogenation. Catalyst development in the 1930s was extensive, and the use of various molybdenum, magnesium, tungsten, zinc, chromium, and nickel oxides were examined. Specific catalysts were used to catalyze different types of reaction, such as saturation of olefinic and aromatic double bonds, removal of nitrogenous bases, sulfur-bearing and oxygenated aromatics, and catalytic cracking, isomerization, and aromatization to produce acceptable gasoline yields. Coal tars, including low-temperature, brown-coal tar and high-temperature, bituminous-coal tar and pitch, were also hydrogenated in catalytic, liquid-phase processes that were similar to those used to hydrogenate coal. Products ranged from gasoline and middle-boiling oils to fuel oils, depending on the feedstock and process used. Coal hydrogenation, liquid-phase coal-tar, and vapor-phase middle-oil processes were all components of fuel-production facilities in Germany during World War II.

As with coal-hydrogenation facilities, a gradual change in feedstocks and production patterns occurred over time in coal hydrotreating. Use of coal decreased, and facilities were converted to process coke-oven light oils and petroleum crude. In 1959, the annual refining capacity of West German hydrogenation plants was 10, 1.6, 1.2, and 1.3 $\times$ 10^6 metric tons for distillation, hydrogenation, cracking, and reforming operations, respectively. The demonstration-scale facilities in Louisiana, Missouri and Institute, West Virginia also used high-pressure catalytic processes, extraction, and distillation to convert coal-derived liquids into a full range of products, including gasoline, fuel oils, and various by-product chemicals.

Further efforts to develop hydrotreatment processes to refine coal-derived liquids lagged behind the development of coal-conversion processes. Production of coal-tar chemicals resulted from the independent development of the coking industry, where distillation and extraction processes were used to recover organic by-products from coal. Although chemicals were extensively produced from coal tar, the economics of the practice were questionable. During the 1950s and 1960s, catalytic cracking and hydrotreating processes were developed and have been used in the petroleum industry. Understanding these processes has provided an important basis for studying the production and refining of coal-derived liquids.

2.2 COAL-LIQUEFACTION PROCESSING

Coal liquefaction converts solid coal into primarily liquid and solid fuels by one of three general processing schemes (Fig. 2.1): (1) direct liquefaction, or reaction of coal in the presence of a coal-derived solvent and molecular hydrogen; (2) indirect liquefaction, or gas production followed by reaction of synthesis gas containing carbon monoxide and hydrogen over a catalyst to form liquid products; and (3) pyrolysis, or coal devolatilization in either a hydrogen-rich or inert environment to recover tars and oils. The following discussion focuses primarily on direct coal liquefaction because of its extensive toxicological database. There is a limited published toxicological database for materials from indirect liquefaction processes. Coal pyrolysis/hydropyrolysis is more similar to coal gasification in process configuration and operating conditions (i.e., temperature, pressure, and residence time during coal devolatilization) and is discussed in Section 2.2.3. Comprehensive reviews of coal-liquefaction reactions and technologies are provided by Alpert and Wolk (1981), Gorin (1981), and Probstein and Hicks (1982).

2.2.1 Direct Coal Liquefaction

Direct coal liquefaction involves the addition of hydrogen to coal from a hydrogen-rich donor solvent and from the gas phase to the process- or coal-derived solvent in the presence of a catalyst. The reaction sequence includes the following types of distinct reaction:

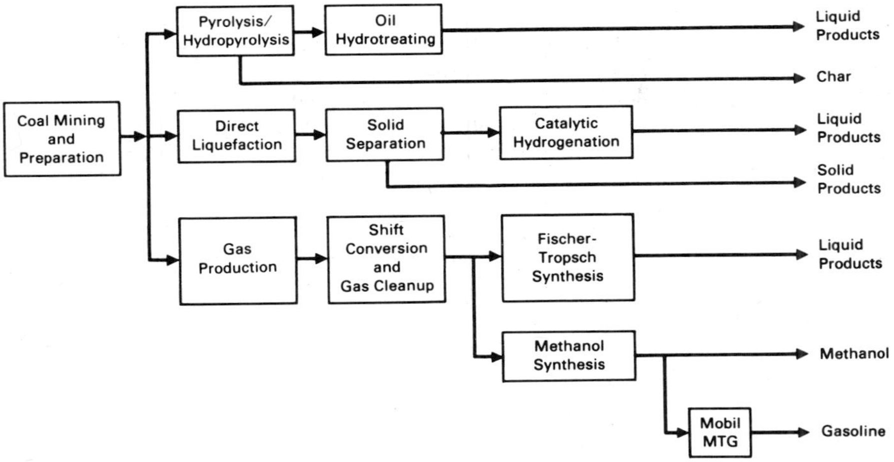

Figure 2.1 General processing schemes for coal liquefaction (adapted from Probstein and Hicks, 1982).

1. Coal dissolution—initial reaction of the coal ''molecule'' to produce primarily a heavy, nondistillable residuum.
2. Hydrogen transfer—reactions that stabilize dissolved, coal-derived residuum.
3. Hydrogen attack—reaction with hydrogen to form distillable hydrocarbon molecules from dissolved residuum and to remove heteroatoms from hydrocarbon molecules.
4. Rehydrogenation—addition of hydrogen to donor solvent molecules.

Dissolution of coal in a process-derived recycle solvent is rapid and proceeds by thermal cleavage of the most labile bonds in coal. Residuum stabilization and conversion reactions are slower. Stabilization of free radicals occurs by reaction with molecular hydrogen and/or hydrogen from hydroaromatic compounds (often referred to as donor solvents) in the solvent vehicle. Hydrocracking and hydrogenation reactions result in production of most of the distillate products. The above reactions all occur via thermal mechanisms, to some extent. Reactions 2 through 4 can be promoted by the addition of a catalyst either directly to the primary liquefaction reactor or in a second stage outside the primary reactor.

The primary process operations required to achieve direct coal liquefaction include liquefaction and fractionation. During liquefaction, coal is mixed with recycle process solvent (PS) and hydrogen, preheated, and reacted at high temperature and pressure in the liquefaction reactor or dissolver. Much of the initial coal dissolution reaction actually occurs in the slurry preheater, and the remaining reactions occur primarily in the dissolver, or second-stage reactor, as appropriate.

Fractionation occurs in two ways. First, the high-pressure, multiphase reactor product is reduced in pressure and temperature in a series of flash separators to separate compounds crudely, by differences in volatility, into vapor, liquid, and slurry phases at the conditions of the flash. Distillation provides a more definitive separation of the complex, coal-derived liquid mixtures (which can contain hundreds of individual chemicals), into wide-boiling-point cuts, also based on the volatility of individual compounds comprising the mixture. Distillation columns, containing a solid medium (e.g., trays, packing) that enhances vapor–liquid contacting and separation efficiency, are used to separate a liquid feedstock into a low-boiling overhead and a high-boiling bottoms product or to separate additional side-stream products, which are intermediate in boiling-point range between the overhead and bottoms materials. Distillation can be conducted at atmospheric pressure or under vacuum, depending on the volatility of the specific feedstock. Subcommercial-scale liquefaction facilities typically use several flash vessels and distillation columns as a means to separate various process streams and products.

Direct coal liquefaction also requires a number of additional processing steps, including hydrogen generation, solids processing, product upgrading, and waste handling, depending on the nature of the process and the scale of the particular facility. Hydrogen is required as a reactant and usually is provided through use of a large, recycle stream. Hydrocarbon gases and gaseous contaminants (e.g., H_2S, NH_3) produced during the liquefaction reactions are removed from the recycle hydrogen, and hydrogen is added to the system. Small facilities typically use an on-site captive source of hydrogen gas, whereas a commercial facility requires that large volumes of hydrogen be generated.

Unreacted coal mineral matter, combined with unreacted carbon and residuum, must be removed from the process. Typically, vacuum bottoms are obtained first from vacuum distillation and are cooled prior to recovery (e.g., on flakers or Sandvik belts) in pilot-scale facilities. Mineral residue or ash concentrate, as it is sometimes called, can be recovered either through filtration of vacuum bottoms followed by solvent washing of the filter cake or by processing vacuum bottoms through various solvent deashing processes (e.g., Kerr–McGee critical-solvent deashing, C. E. Lummus antisolvent deashing). Some processes also use hydroclones as a means of separating unreacted solids and/or catalyst. A commercial-scale facility would separate either vacuum bottoms or the mineral residue and use it to produce the hydrogen required for coal liquefaction via processes such as partial oxidation or gasification.

Various processes can be used to upgrade the value of petroleum- and coal-derived products. Pilot-scale facilities typically do not use these processes for postproduction upgrading, except possibly hydrotreating to stabilize distillate products. Product upgrading processes are discussed later, in Section 2.4.

Finally, various gaseous, aqueous, and solid residuals are produced in a coal-liquefaction facility and require processing prior to discharge or disposal. The nature and extent of the processing is a function of the facility's size, location, and the specific process used.

The product slate for a coal-liquefaction process typically includes gaseous hydrocarbons (C_1–C_5) and liquid distillate products. Although the product slate differs from process to process and across process scale within a given process, some typical product designations and approximate boiling-point ranges are as follows: naphtha (boiling from C_6 to about 400°F), middle distillate (boiling from about 400 to 600°F), heavy distillate (boiling from about 600 to over 900°F), and residuum (boiling about 900°F). Middle and heavy distillates can be blended or produced directly as a full-boiling-range fuel oil. Depending on the specific process, recycle or *pasting* solvents consist of a blend of middle and heavy distillates or heavy distillate and residuum. Most liquid products and/or product blends from subcommercial-scale coal-liquefaction facilities have been characterized chemically and toxicologically.

Other by-products and raw effluents are produced during direct coal liquefaction, and many of these have also been characterized chemically and toxicologically. Depending on the type of liquefaction process and associated operating conditions, these by-products or effluents may include:

- vacuum bottoms, containing the high-boiling hydrocarbon residuum, and unreacted carbon and mineral residue from coal;
- ash concentrate, containing the unreacted carbon and mineral matter from coal that has been separated from the hydrocarbon residuum in vacuum bottoms;
- aqueous condensate, containing dissolved organic compounds (e.g., phenols, trace aromatic hydrocarbons), inorganic compounds (e.g., dissolved and ionized acid gases, carbonate salts), and trace metals;
- spent catalyst, which must be replaced periodically with fresh catalyst; and
- slop oils, consisting of waste oils from process vents, equipment pads, and auxiliary waste-treatment operations.

The direct liquefaction of coal can be performed with and without catalysts and in single- or two-stage process operations. A description of these major options and their major differences is presented later.

2.2.1.1 Noncatalytic Coal Liquefaction

In noncatalytic liquefaction (Fig. 2.2), coal is extracted with a process-derived solvent at elevated temperatures and pressures, typically in the presence of molecular hydrogen and without the addition of a specific catalyst. Coal is mixed into a slurry with process-derived solvents at about 300–400°F. The slurry is mixed with hydrogen-rich recycle gas, is preheated at elevated pressures, and enters the dissolver reactor. Liquefaction temperatures range from about 800 to 930°F, at reactor residence times of about 30–100 min, and pressures of 100–200 atm. Reaction severity is expressed conveniently in terms of hydrogen consumption (expressed as the amount of hydrogen chemically consumed per unit of coal fed) and ranges from about 2 to 5 wt% of the coal processed. Hydrogen transfer occurs mainly from hydroaromatic compounds in the solvent. Solvent, depleted of hydrogen, is

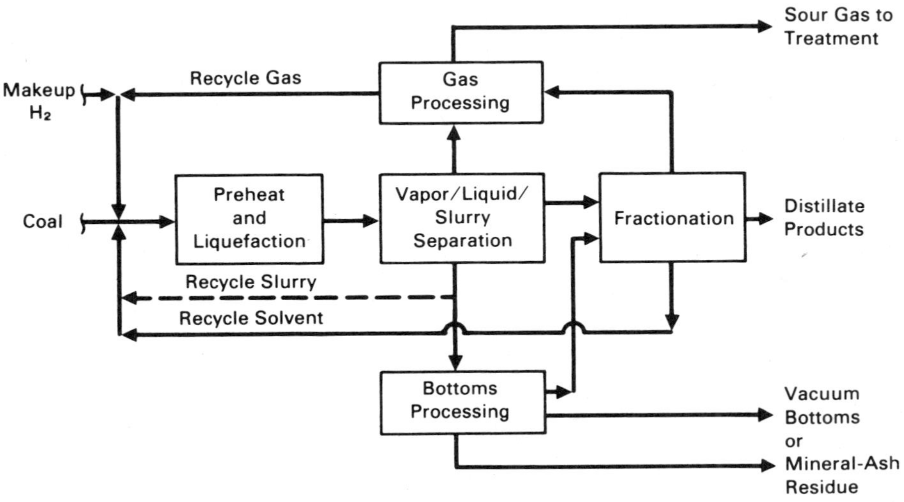

Figure 2.2 Simplified flow diagram for noncatalytic coal-liquefaction processes.

replenished simultaneously in the same reactor with the addition of high-pressure molecular hydrogen.

Following liquefaction, gaseous, liquid, and solid products are separated in downstream flash vessels and via distillation. Gases, consisting primarily of unreacted hydrogen, are recovered from flash vessels, processed to remove light hydrocarbons and acid gases, recycled to the front of the process and combined with make-up hydrogen. Liquids collected from various points in the process are fractionated to recover different boiling-range products, as previously described. A portion of the hydrocarbon liquids is recycled to slurry the feed coal and provide donor solvent. In some cases, internal process streams containing residuum and coal-derived mineral residue also comprise a portion of the process-derived recycle solvent. Mineral-bearing residuum is typically subjected to vacuum distillation to recover heavy distillates. The residuum-bearing vacuum bottoms either exit the process as a by-product to be used as fuel to generate hydrogen in larger-scale operations or are processed via filtration or solvent deashing to recover residuum or to produce SRC. In the latter case, carbon-bearing mineral residue can be recovered as mineral-ash residue for use in generating hydrogen in larger-scale operations. Hence, the product slate from noncatalytic coal liquefaction can range from primarily SRC at lower reaction severities to a full range of naphtha and distillate products at higher reaction severities.

Two noncatalytic, direct coal-liquefaction processes (SRC-I and SRC-II) have been sampled for toxicological analysis. The processes are distinguished mainly by their product slates, which reflect differences in reaction severity. The SRC-I process produces primarily a solid product and is operated at lower reactor temperature, slurry residence time, and overall pressure than the SRC-II process, which produces primarily naphtha and distillate products. In the SRC-II process, a portion of the mineral-bearing residuum is recycled to slurry the feed coal. The iron- and sulfur-bearing minerals serve to catalyze the coal-liquefaction reactions.

2.2.1.2 *Catalytic Coal Liquefaction*

The basic operating principles and conditions for catalytic liquefaction processes are similar to those for noncatalytic processes. The major difference is the addition of a specific catalyst to enhance the liquefaction reactions (see Fig. 2.3). Catalytic liquefaction can occur in two different modes:

1. Direct addition of catalyst to the liquefaction reactor to enhance reactions of coal-derived liquids with molecular hydrogen.
2. Catalytic hydrogenation of recycle solvent to increase the fraction of hydroaromatic or donor-solvent compounds, thus facilitating the liquid-phase transfer of hydrogen between recycle solvent and dissolved coal during liquefaction.

Both catalytic processing modes produce mainly naphtha and distillate products and optimally operate at conditions that deemphasize production of gases and residuum. Spent catalyst is either generated continuously or must be replaced periodically, dictating a periodic need to add fresh catalyst to the liquefaction reactor or the hydrotreater.

Four catalytic liquefaction processes (EDS, H-Coal, Synthoil, and Veba Oel/Ruhrkohle AG) have been sampled for toxicological analysis. Differences among the processes primarily reflect catalyst type and operating mode. All processes except EDS use a catalyst in the main liquefaction reactor. The Veba Oel/Ruhrkohle AG process adds a red-mud catalyst, containing iron oxides, to the feed-coal slurry. The Synthoil process uses a fixed-bed,

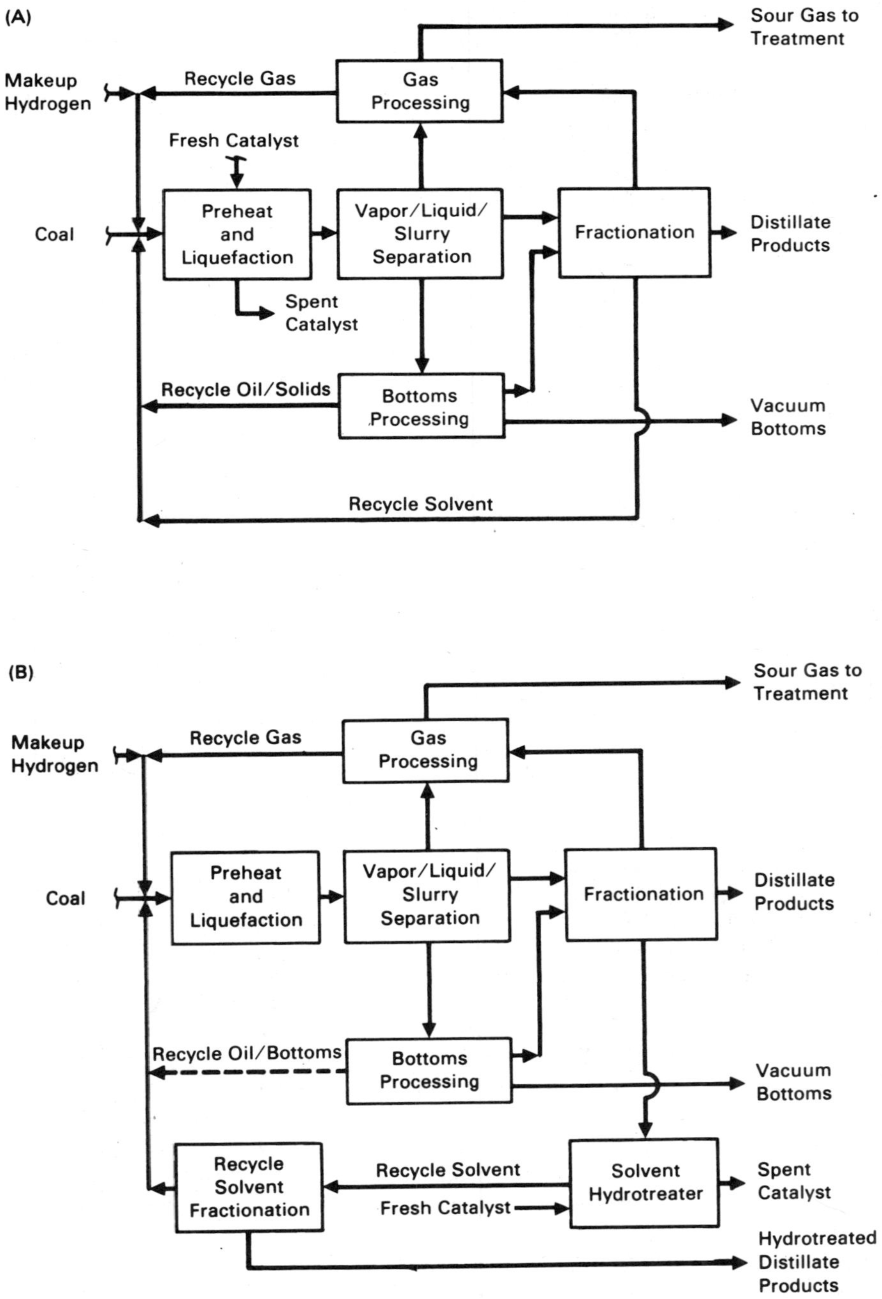

Figure 2.3 Simplified flow diagram for catalytic coal-liquefaction processes: (A) direct catalyst addition and (B) recycle solvent hydrogenation.

liquefaction reactor filled with cobalt–molybdenum catalyst on an alumina support. The H-Coal process uses an ebullated-bed reactor and typical hydrodesulfurization catalysts containing nickel–molybdenum or cobalt–molybdenum on an alumina support. Spent catalyst is periodically replaced in the H-Coal reactor to maintain constant activity in the reactor. The EDS process hydrotreats the spent-solvent vehicle in a conventional, fixed-bed hydrotreater, thus replenishing donor-solvent constituents in the recycle solvent.

2.2.1.3 Two-Stage Coal Liquefaction

Two-stage liquefaction (TSL) is a recent development designed to optimize the liquefaction process by physically separating the thermal dissolution stage from the residuum conversion stage and using a catalyst to facilitate residuum conversion to final product. Thermal dissolution is conducted under relatively mild conditions (i.e., 750–840°F and residence times on the order of 5–30 min) to produce primarily a residuum product. After solids separation, the residuum is hydrogenated catalytically to produce final products. Separation of reaction stages allows for efficient hydrogen use by permitting the optimization of the reactions in each stage.

Two-stage liquefaction can be configured in nonintegrated (NTSL), integrated (ITSL), or catalytic (CTSL) modes. In NTSL, PSs (i.e., heavy distillates) from each stage are used to prepare the feed for that stage (Fig. 2.4). For example, coal fed into stage 1 is slurried with solvent from stage 1. Likewise, the residuum fed into stage 2 is diluted with PS from stage 2.

In ITSL, PS from the second stage is used as slurry or pasting solvent for coal fed to the first stage (see Fig. 2.5). Thus, a higher-quality slurry solvent containing more hydroaromatics and fewer alkylated and heteroatom-substituted compounds is used in the stage 1 coal-dissolution step. In this mode of operation, hydrogen consumption in the first stage is optimized to stabilize only primary coal-reaction products. The second stage hydrotreats/hydrocracks the liquefied coal into final products. In CTSL, first-stage reaction products are not cooled prior to processing in the second stage. This procedure maintains free-radical concentrations in the first-stage product and facilitates hydrotreatment achieved in the second stage.

Essentially, the same wastes and by-products are produced from TSL processes as from catalytic and noncatalytic processes. However, unreacted carbon and ash are routinely produced as ash concentrate, essentially devoid of high-boiling hydrocarbons. All four TSL processes/configurations have been sampled for toxicological studies.

2.2.2 Indirect Liquefaction

Indirect liquefaction produces liquids from synthesis gas, containing varying quantities of carbon monoxide and hydrogen. The process includes components normally associated with both coal gasification and liquefaction. Coal is first gasified with steam and oxygen to produce a medium-Btu gas (Fig. 2.1). The gas is processed to remove contaminants such as H_2S, NH_3, and HCN, and the relative amounts of CO and H_2 are adjusted using CO-shift reactors. The extent of the shift reaction needed depends on the synthesis gas composition and requirements of the liquefaction process.

Cleaned synthesis gas can be converted catalytically to liquid fuels using three different processes:

1. Fischer–Tropsch synthesis—converts CO, CO_2, and H_2 into primarily straight-chain, gasoline-range hydrocarbons and produces C_4 gases, C_5–C_7 liquid hydrocarbons, diesel-

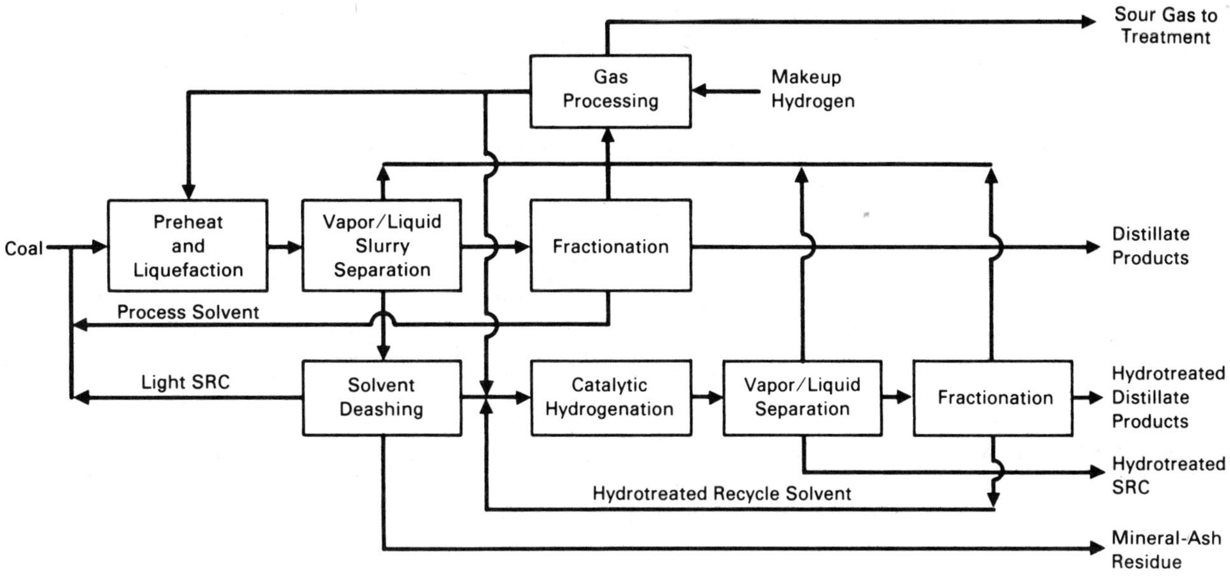

Figure 2.4 Simplified flow diagram for nonintegrated, two-stage coal-liquefaction (NTSL) processes.

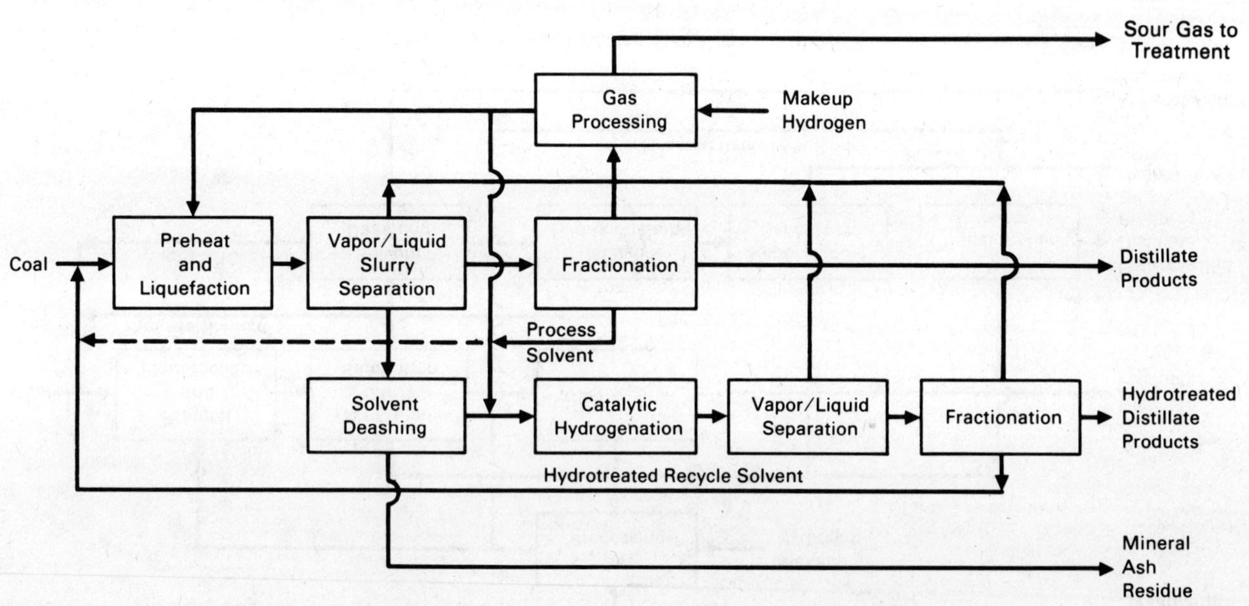

Figure 2.5 Simplified flow diagram for integrated, two-stage coal-liquefaction (ITSL) processes.

range fuel oils, and oxygen-bearing alcohols and acids. High-molecular-weight waxes are also produced as a by-product.

2. Methanol synthesis—converts CO, CO_2, and H_2 into primarily methanol and water. Smaller quantities of higher alcohols are produced and enter the undistilled, "wild methanol" product, which can be distilled to high-grade methanol.
3. Methanol-to-gasoline (MTG) synthesis—methanol can be converted into gasoline-range hydrocarbons via the Mobil MTG process. The product spectrum is limited to hydrocarbons with 10 carbon atoms or less. Water is produced as a by-product, and it contains small amounts of dissolved alcohols, aldehydes, and ketones.

Only limited toxicological data are available for materials from laboratory-scale Fischer–Tropsch facilities. Consequently, indirect liquefaction processes and the characteristics of the associated liquid products are not considered further.

2.2.3 Pyrolysis/Hydropyrolysis

In pyrolysis/hydropyrolysis, coal is reacted in either an inert or a hydrogen-rich environment to produce liquids. Process conditions vary from 1- to 150-atm pressure, 1100 to 1650°F, and residence times of less than 1 to 20 s. Liquid products range from light aromatic hydrocarbons to heavier, full-boiling-range tars and oils that can be produced at 30–40 wt% yield of the moisture- and ash-free (MAF) coal. Appreciable quantities of gaseous hydrocarbons and carbon-bearing char are also produced.

Pyrolysis/hydropyrolysis is more analogous to the coal devolatilization stage of coal gasification, based on the temperatures, pressures, residence times, and modes of processing. It can be conducted in fixed-bed, fluidized-bed, or entrained-flow reactors and produces tars and oils that are similar to gasification tars and oils. Therefore, pyrolysis/hydropyrolysis reaction configurations and conditions are discussed in the next section.

2.3 COAL GASIFICATION

Coal-gasification processes (Fig. 2.6) convert solid coal to a combustible gas by reacting it with steam and air or oxygen. The gas product can be used as fuel directly or converted to SNG or liquid products such as methanol, gasoline, or fuel oils. Comprehensive reviews of coal-gasification reactions and technologies are provided by von Fredersdorff and Elliott (1963), Johnson (1981), Hebden and Stroud (1981), Probstein and Hicks (1982), and Parekh (1982).

Coal gasification involves three distinct yet interrelated reactions:

1. Devolatilization—initial reactions of coal that produce condensable organics (e.g., tars, oils, and phenolic compounds), carbon oxides and water, and light hydrocarbon gases (e.g., methane, ethane).
2. Gasification—endothermic reaction of carbon in coal with gases such as carbon dioxide, water, and/or hydrogen to produce the primary gas products (e.g., carbon monoxide, hydrogen, and methane).
3. Combustion—exothermic reaction of carbon with oxygen to produce carbon oxides and provide heat for the endothermic gasification and devolatilization reactions.

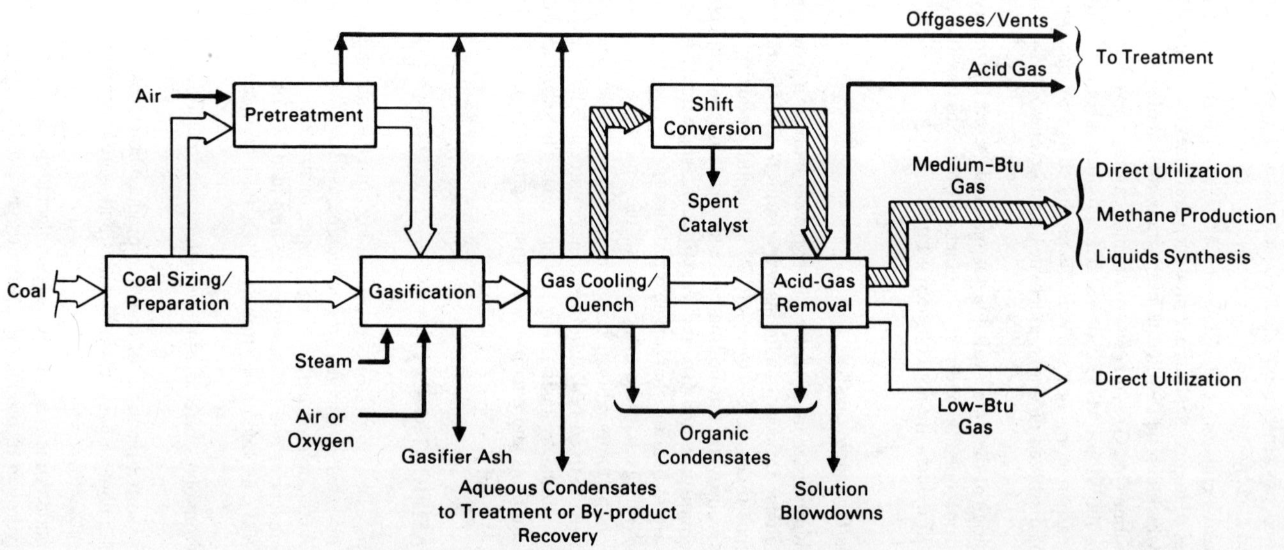

Figure 2.6 Simplified flow diagram for coal-gasification processes.

The reactions that result in production of the primary gaseous products are shown in Table 2.1. Devolatilization is not shown because a simple stoichiometric equation cannot be written that adequately represents the complexity of the devolatilization process. Considerable research has been conducted on modeling the reactions associated with the coal devolatilization process; it is summarized elsewhere (Howard, 1981).

Reactions that produce the primary gaseous products can be conducted over a wide range of conditions. Temperatures can range from 1200 to over 2900°F, pressures from 1 to 70 atm, and gas and solid residence times from seconds to hours. Fixed-bed, fluidized-bed, and entrained-bed/flow reactors can be used to contact gaseous reactants and solid coal. The primary product of gasification is a low-Btu (i.e., 100–200 Btu/SCF*) or medium-Btu (about 350 Btu/SCF) gas, depending on whether the gasifier is air- or oxygen-blown, respectively, and hence whether the product gas is diluted with nitrogen.

A gasification facility can include the following major process operations:

- coal preparation to size and/or pulverize the feed coal;
- pretreatment to oxidize thermally the surface of the coal and thus reduce the tendency of bituminous coals to cake or agglomerate;
- gas quench/cooling to recover sensible heat, remove particulates, and remove aqueous and hydrocarbon condensates from the gasifier raw process gas;
- shift conversion to adjust the relative quantities of H_2 and CO in the product gas;
- acid-gas removal to clean the product gas of reduced-sulfur and -nitrogen compounds and CO_2; and
- product synthesis to convert the product gas to fuel products such as methane, methanol, and/or distillate fuels.

Gaseous, aqueous, and solid effluents also must be processed to remove potential environmental contaminants and/or recover marketable by-products (e.g., sulfur, ammonia, phen-

Table 2.1 Summary of Principal Reactions in Coal Gasification[a]

Reaction	ΔH°_{298} (kcal/gmol)[b]
Combustion	
$C + O_2 \rightarrow CO_2$	−94.05
$C + \frac{1}{2}O_2 \rightarrow CO$	−26.42
$H_2 + \frac{1}{2}O_2 \rightarrow H_2O$	−57.80
Gasification	
$C + CO_2 \rightarrow 2CO$	41.22
$C + H_2O \rightarrow CO + H_2$	31.38
$C + 2H_2 \rightarrow CH_4$	−17.89
Water–Gas Shift	
$CO + H_2O \rightarrow CO_2 + H_2$	−9.84

[a] Data from von Fredersdorff and Elliott (1963).
[b] Negative heats of reaction indicate exothermic reactions.

*SCF = standard cubic foot.

ols). The specific process operations used are a function of the scale of the facility and the desired product and end use.

Major by-products and raw effluents of toxicological interest that are produced during gasification include:

- coal/char fines that leave the gasifier with the raw process gas;
- tars and oils produced during coal devolatilization;
- aqueous condensate consisting of coal moisture and unreacted steam and containing dissolved organic compounds (e.g., phenols, trace aromatic hydrocarbons), inorganic compounds (e.g., carbon dioxide, hydrogen sulfide, ammonia, salts), and trace metals; and
- ash or slag, consisting of unreacted carbon and mineral matter, and residual trace metals.

Of all these materials, it is the tar and oils that have been the primary focus of toxicological characterization studies. Interest in the other by-products results almost solely from the extent to which they have been contaminated with these hydrocarbons. For example, coal/char fines can be contaminated with condensed tar/oil on the surface, or aqueous condensates can contain tar/oil in suspension or in solution.

The characteristics and quantities of by-product hydrocarbons are dictated primarily by the gasifier design and its associated operating conditions. However, the ultimate distribution of hydrocarbons between the coal-char fines, the aqueous condensates, and the hydrocarbon condensates known as tars and oils is governed by the gas cooling or quench system used to remove the hydrocarbons and excess steam from the raw process gas. Because of these process considerations, toxicological studies on materials downstream of the quench system are not considered here. The importance of the individual gasifier designs and operations on hydrocarbon production and characteristics and a brief discussion of the effect of quench system configurations on the separation of oil and tar by-products are given subsequently.

2.3.1 Gasifier Configurations

2.3.1.1 Fixed-Bed Gasifiers

In conventional, fixed-bed gasifiers (Fig. 2.7), sized coal ranging from 5 to 50 mm in diameter enters the top of the gasifier and contacts reactant gases that enter from the bottom. Coal descends from top to bottom in the reactor, proceeding sequentially through drying, devolatilization, gasification, and combustion zones. Unreacted coal mineral matter is removed as a dry ash or a molten slag at the base of the gasifier, depending on whether excess steam is used to moderate combustion reactions and thus control bottom temperatures. Gasifier temperatures range from 300–1100°F at the top of the bed to 1800 to over 2900°F at the bottom.

Fixed-bed gasifiers produce substantial quantities of tars, oils, and phenolic compounds because the relatively low temperatures at the top of the bed are inadequate to crack the oils and tars to lighter gaseous components. Coal/char fines are swept through the fuel bed and are removed dry or with the hydrocarbon or aqueous condensates. Condensate production and quality vary with feedstock, reactant feed rates, and quench-system configuration. Water also can be used to quench the molten slag and sluice slag or ash from the gasifier bottom. Materials sampled for toxicological studies from fixed-bed gasification facilities include fly ash, bottom ash, tars, oils, raw gas, naphtha, wastewater, ash sluice water, and ash leachate.

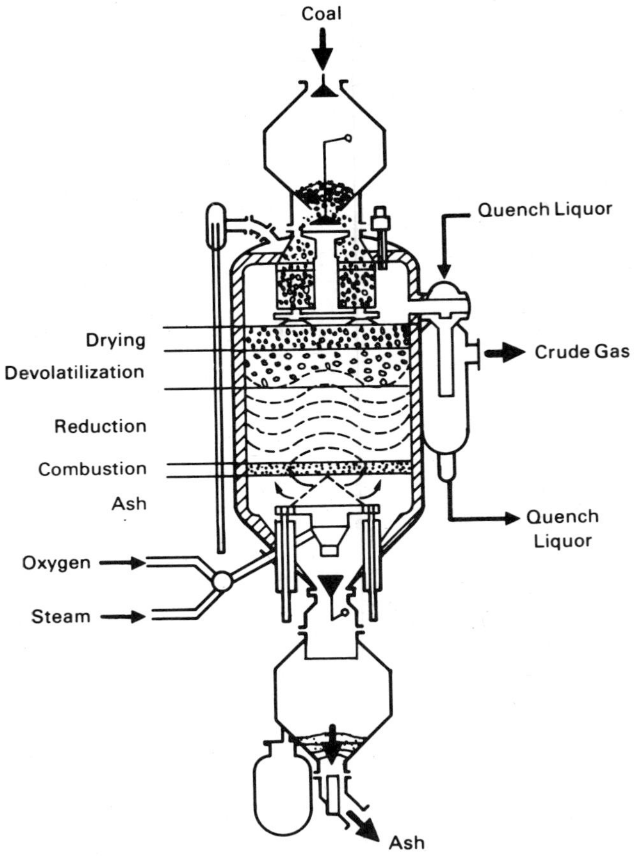

Figure 2.7 Schematic diagram of a typical fixed-bed gasifier (adapted from Hebden and Stroud, 1981).

A wide variety of fixed-bed gasification processes have been sampled for toxicological analysis. Many industrial facilities, both within and outside the United States, have been sampled and include pressurized gasifiers such as the dry-ash Lurgi, and atmospheric-pressure gasifiers such as the Foster–Wheeler Stoic, Wellman–Galusha, and Chapman. Smaller-scale, pressurized gasifiers such as the slagging fixed-bed and again the Wellman–Galusha have also been tested. Except for the dry-ash Lurgi and slagging fixed-bed gasifiers, these gasifiers are suited for producing primarily a low-Btu process gas for direct use as a fuel.

2.3.1.2 Fluidized-Bed Gasifiers

In fluidized-bed systems (Fig. 2.8), coal is ground to a maximum of 10-mm diameter and fluidized in a vessel by reactant/product gases. Reactant gases enter the gasifier near the bottom, at sufficiently high flow rates to suspend the solids in a well-mixed bed without carrying them out the top. Swirling or bubbling action creates uniform temperatures throughout the bed, and reaction rates are enhanced through intimate contact between gas and high-surface-area coal particles.

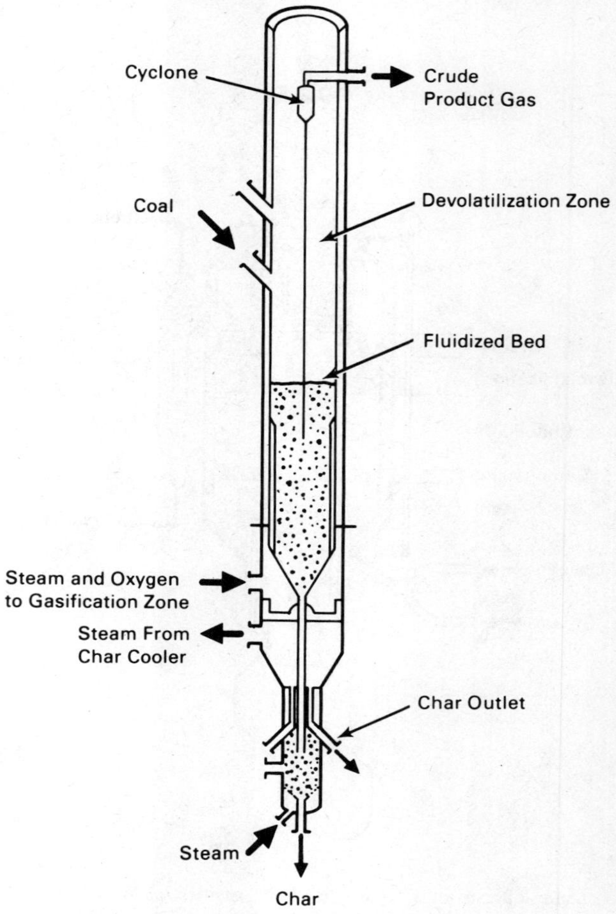

Figure 2.8 Schematic diagram of a typical fluidized-bed gasifier (adapted from Hebden and Stroud, 1981).

Fluidized-bed gasifiers can produce the full range of by-products and raw effluents. Oils, tars, and phenols are produced in negligible to substantial quantities, depending on bed temperatures and gas and solid residence times. For example, negligible quantities of condensable organics are produced if coal enters at the bottom of the bed. Organics released during coal devolatilization undergo thermal and catalytic cracking in the presence of hot char solids in the fluidized bed. This behavior is demonstrated by operating data for both the CO_2-Acceptor (CCDC, 1978) and KRW (Mann et al., 1985) gasifiers, which operate in the temperature range of 1500–1900°F. (See Chapter 4, Table 4.11 for representative data.) In contrast, coal fed at the top of the bed results in insufficient residence times for complete cracking of oils, tars, and phenols (Nakles et al., 1975; Fillo, 1979). In certain cases (e.g., Hygas process), organic solvents are used to slurry feed coal and therefore constitute a major process stream that can accumulate organic and inorganic compounds. Fine char particles leave the gasifier bed and typically are recovered by cyclones within and/or external to the gasifier. Trace quantities of heavy hydrocarbons can condense on the surface of char fines. In bottom-fed gasifiers, aqueous condensates contain mainly inorganic contaminants and trace metals. Ash is removed from the gasifier base through lockhoppers, either dry or as a water slurry. Materials sampled for toxicological studies from fluidized-bed gasifiers include char, raw gas, tars and oils, and quench water.

Fluidized-bed gasifiers can be used in pyrolysis/hydropyrolysis processes to produce significant quantities of tars and oils. These processes typically consist of a number of fluidized beds in series to carry out separately the various gasification reactions. The pyrolysis or hydropyrolysis reactor is controlled at process conditions that are optimal for production of large quantities of tars and oils, and thus the gasification of the coal and cracking of hydrocarbon products is minimized. Typical operating temperatures are in the range of 1000–1200°F.

Only two subcommercial-scale, fluidized-bed coal reactors have been sampled for toxicological analysis, the Hygas process gasifier and the Char-Oil Energy Development (COED) process, which is a pyrolysis process.

2.3.1.3 Entrained-Flow Gasification

Entrained-flow gasifiers use finer-sized coal (i.e., typically less than 0.1-mm diameter) than fixed- or fluidized-bed gasifiers. The fine coal is entrained in a gas stream in either an upflow or a downflow configuration and reacted at high temperatures (2000–2900°F) and short residence times (see Fig. 2.9). Devolatilization, gasification, and combustion occur rapidly to produce a gas high in carbon monoxide and virtually devoid of hydrocarbons, including methane. Volatilization of trace metals from coal is typically higher for entrained-flow gasifiers because of higher temperatures.

All by-products and raw effluents except for hydrocarbons are produced in entrained-flow gasifiers. Aqueous condensates contain inorganic contaminants and trace metals. Unreacted ash from coal is recovered as molten slag and fly ash or fly dust, depending on the specific gasification process. Sluice systems typically are used to cool and recover ash and slag.

Entrained-flow systems can be used at short gas/particle residence times to devolatilize coal rapidly at typical pyrolysis/hydropyrolysis temperatures. Under these conditions, significant quantities of hydrocarbon tars, oils, and phenols can be produced. Only tars and oils from experimental-scale pyrolysis/hydropyrolysis studies have been characterized toxicologically, and only these data are included in this investigation.

2.3.2 Quench Configurations

An understanding of quench system designs and operation is important in defining the nature of oils and tars produced during coal gasification. For the most part, the majority of processes use some form of a deluge quench, where the hot process gas is contacted directly with large volumes of water in venturi contactors or spray towers. This results in a multiphase mixture of hydrocarbon and aqueous condensates that are separated, with the hydrocarbons lighter than water designated as oil and those heavier than water designated as tar. Quench and condensate separation can be achieved in a single vessel or in multiple vessels, depending on the type and scale of process.

Some processes use a partial *dry* quench by cooling the process gas to just above the dew point by indirect cooling or spray injection of limited quantities of water. A dry quench

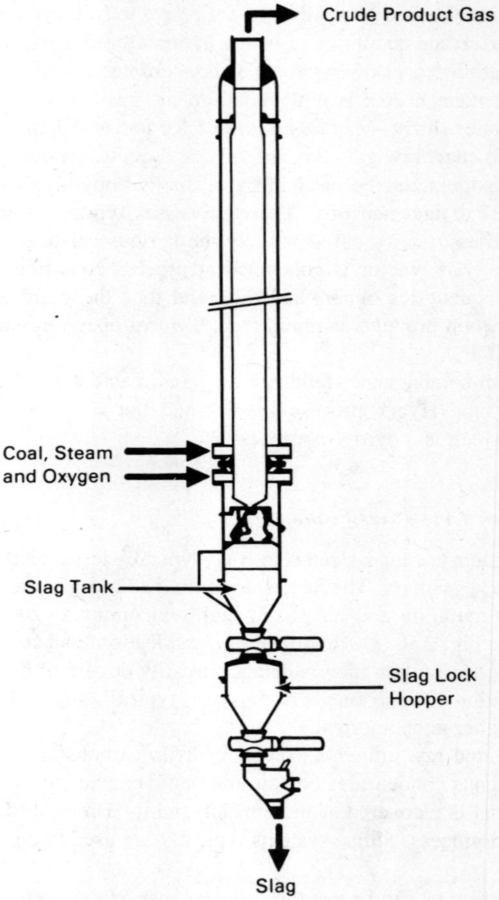

Figure 2.9 Schematic diagram of a typical entrained-flow gasifier (adapted from Hebden and Stroud, 1981).

usually is followed by a direct, deluge quench. Thus, quench configuration results in condensation of a relatively dry, heavy tar followed by the more conventional separation of oils and tars in the deluge quench condensate.

At the other extreme from the deluge quench is the more recent consideration of *hot-gas clean-up* techniques that do not require the cooling or quenching of the gas at all. These processes must address the difficulties of treating high-temperature, particulate- and hydrocarbon-laden gas streams. Their success is achieved most often by using these techniques on gasifiers that produce hydrocarbon-free product gases.

The quantities and characteristics of the aqueous and hydrocarbon condensates are a direct function of the type, conditions, and configuration of the quench system. The physical and chemical characteristics of tars and oils are affected directly, and consequently all tars and oils are not necessarily alike. Quench operation and configuration also affect the distribution of organic and inorganic contaminants among gaseous and liquid phases, and thus the composition of the aqueous condensate can vary substantially. In contrast to coal-liquefaction materials, quench systems are notoriously efficient at capturing the high-boiling, high-molecular-weight compounds that can be important in determining certain toxicological properties. Volatile hydrocarbons such as benzene, toluene, and xylenes (i.e., BTX) are not captured very effectively. Quench system characteristics not only affect the understanding of toxicological properties for gasification condensates as produced but make the extrapolation of characteristics for materials obtained at subcommercial facilities very difficult.

2.4 COAL-LIQUID HYDROTREATING

Coal liquids can be refined by a variety of chemical or physical process operations to improve specific properties of the raw coal liquid. Although refining is generally considered for coal-liquefaction products, refining coal-gasification tars and oils has been investigated to a lesser extent. Research to date on coal-liquid refining has focused on standard petroleum upgrading processes including distillation, catalytic reforming, catalytic cracking, and hydrotreating. All available toxicological data on upgraded coal liquids are on hydrotreated materials. Thus, only hydrotreating is discussed. Some refining is likely to be part of commercial, coal-liquefaction facilities. For discussions of hydrotreating and other upgrading processes, see Horne and McAfee (1960), Gates et al. (1979), Crynes (1981), Sullivan (1981), and Satterfield (1983).

Hydrotreating involves the hydrogenation of olefinic and aromatic bonds and includes reactions such as dealkylation, desulfurization, denitrogenation, deoxygenation, and demetalization. These reactions occur in the presence of a catalyst at elevated temperature and pressure. Hydrotreating can be performed over a range of severities, depending on catalyst type and reactor conditions. Relatively mild conditions are used to remove heteroatoms (e.g., hydrodesulfurization or hydrodenitrification) that are not part of the ring structure. More severe conditions result in substantial cracking of the feedstock as well as heteroatom removal from within the ring structure. Such severe processing is often referred to as hydrocracking.

A range of feedstocks from light naphtha to heavy residuum can be hydrotreated. Lighter feedstocks are generally hydrotreated to remove heteroatoms that may poison catalysts in subsequent processing units or to meet product specifications. Heavier feedstocks are also hydrotreated to remove heteroatoms or in more severe processing to improve overall yield of lighter products.

Important process conditions for hydrotreating are catalyst type, temperature, pressure, and residence time. Catalysts are usually small pellets of an inert support such as alumina, with cobalt, molybdenum, nickel, and/or tungsten impregnated on the surface. Residence time is usually expressed in terms of liquid or weight hourly space velocity. Space velocity is defined as the volume or weight of feed passing over an equivalent volume or weight of catalyst in 1 h and is inversely proportional to the residence time. Other important hydrotreating variables are catalyst age, hydrogen partial pressure, and the gas : oil feed ratio.

Typical reactor conditions for hydrotreating heavier feedstocks, which have boiling-point ranges from as low as 390–575°F to over 900°F, are pressures of 70–200 atm, reactor temperatures of 575–800°F, and liquid hourly space velocities of 0.5–2.0 h^{-1}. Reaction severity is often expressed in terms of hydrogen consumption, which is the amount of hydrogen chemically consumed per unit of oil processed. Hydrogen consumption can range from a few hundred to several thousand standard cubic feet of hydrogen per barrel of oil processed (i.e., 0.5–5 wt % of oil processed).

2.4.1 Postproduction Hydrotreatment

Most of the toxicological data for hydrotreated coal liquids are available for materials hydrotreated following their production during liquefaction. A typical system reacts the liquids with hydrogen over a fixed bed of catalyst. Unreacted hydrogen is recovered and recycled and reactor products are distilled to remove gaseous products (see Fig. 2.10). Feedstocks are usually more difficult to hydrotreat as the relative amount of high-boiling-point material increases and the required quality of the desired product increases. Both raw and hydrotreated liquid products are then characterized to determine the extent of hydrotreating achieved.

2.4.2 In-Process Hydrotreatment

Hydrotreatment is used as an integral part of a TSL process to improve its efficiency and to improve the overall quality of the final product (see Section 2.2.1.3). Typically, a mixture of residuum and heavy distillate, which is produced in the thermal dissolution stage, is hydrotreated as part of a continuous liquefaction process. Reaction severity can be minimized to produce a solid residuum product or it can be increased to convert the residuum to lighter, distillate products.

In-process hydrotreatment uses hydrogen more efficiently than postproduction hydrotreatment. Free radicals available for hydrogen stabilization are not quenched prior to treatment as they typically are during cooling in postproduction processing. Hydrogen consumption is also optimized because, in the first reaction stage, it can be limited only to that

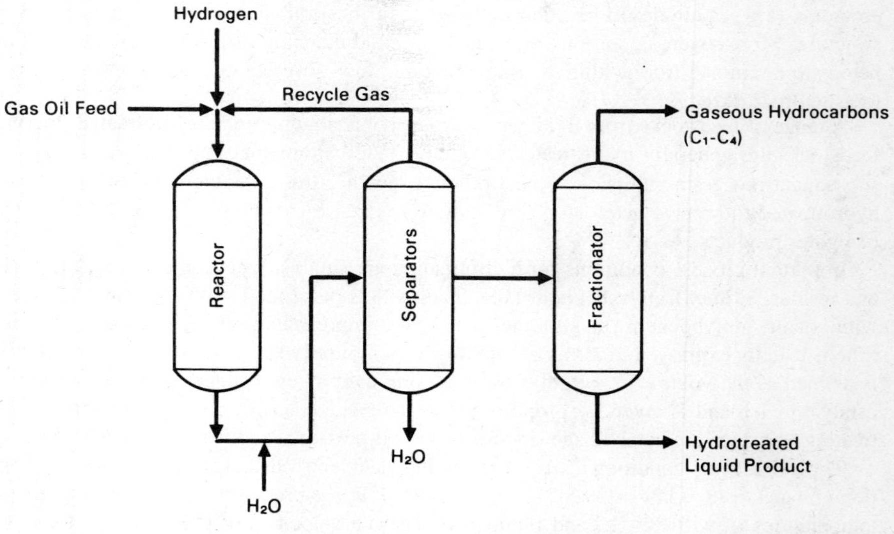

Figure 2.10 Flow diagram for a typical hydrotreating reactor system (adapted from Riedl and de-Rosset, 1979).

required to stabilize free radicals formed during coal dissolution. Hydrogen is used more efficiently in the second, catalytic stage to convert the primary coal products to the desired final products. Thus, a material of comparable properties (i.e., boiling-point range, heteroatom level, chemical composition) can be hydrotreated at a lower severity to achieve specified product properties if hydrotreatment is done as an integral part of the liquefaction process.

2.5 OVERVIEW OF COAL-CONVERSION PROCESSES AND THE TOXICOLOGICAL DATABASE

This section identifies the specific facilities that provided samples, describes the types of sample obtained for testing, and provides an overview of the types of data available. Published data for these samples can include toxicity, chemical composition, bulk physical/chemical properties (e.g., distillation, elemental analyses), and/or process conditions associated with the production of these materials. The toxicological testing database is summarized in Table 2.2.

More detailed descriptions of the database are presented elsewhere. The appendix at the end of Chapter 4 provides a summary of the process source, name, available process engineering, chemistry, and toxicology data, and primary references for samples discussed in Chapter 4. Each sample is identified by number, and the numbers are used throughout the book. The origins of each sample assayed for toxicological activity are provided in the appendix to Chapter 3. The reader interested in tracing samples through the various chapters should consult the appendix at the end of Chapter 4 for examples. More detailed process descriptions and flow diagrams and descriptions of facility-specific materials are provided by Fillo and Craun (1985).

2.5.1 Coal-Conversion Facilities

The database for coal liquefaction includes about 175 materials sampled from 19 facilities. Samples were obtained from over 60 process runs conducted in 11 processes or process

Table 2.2 Number of Coal-Conversion Process Materials Included in the Toxicological Testing Database

	Number of Individual Runs that Produced Materials and Number of Individual Materials that Were Sampled for Toxicological Tesing							
	Experimental Scale[a]		Pilot Plant		Industrial/ Commercial		Totals	
Technology	Runs	Materials	Runs	Materials	Runs	Materials	Runs	Materials
Coal gasification	52	107	12	50	8	37	72	194
Coal liquefaction	30	70	35	105	—	—	65	175
Hydrotreated coal liquids	39	41	14	22	—	—	53	63
Boiling-point cuts	6	28	6	45	—	—	12	73[b]
	127	246	67	222	8	37	202	505

[a]Includes bench- and PDU-scale operations.

[b]Includes 10 materials with boiling ranges in excess of 100°F that are not strictly considered narrow boiling-point cuts.

concepts. Tables 2.3, 2.4, and 2.5 indicate the noncatalytic, catalytic, and two-stage processes, respectively, that generated test materials. The facilities include bench-scale units operating at 1–4 lb/h, PDUs operating at 1–3 tons/day, and large pilot plants processing as much as 250 tons/day of coal. Three of the processes sampled (i.e., SRC-II, H-Coal, and EDS) were operated at all three scales of process development (i.e., bench, PDU, and pilot plant).

The database for gasification is the largest, and it includes over 200 materials sampled from 12 facilities. Samples were obtained from over 70 process runs on various fixed- and fluidized-bed gasifiers, an entrained-flow pyrolysis/hydropyrolysis reactor, and an underground coal gasification site. Eight facilities have fixed-bed gasifiers, although most toxicological data were developed on materials from the fluidized-bed Hygas gasifier and the entrained-flow Mellon Institute reactor (see Table 2.6). The facilities range in scale from the 2-lb/h bench-scale, fixed-bed gasifier at Research Triangle Institute (RTI) to the 180,000-lb/h industrial gasification facility in Kosovo, Yugoslavia. Over half the materials were obtained from bench- and PDU-scale facilities.

Toxicological characterization of hydrotreated coal-conversion materials has been conducted solely on samples from coal-liquefaction facilities. SRC-I, SRC-II, H-Coal, and EDS materials have been hydrotreated by Universal Oil Products (UOP), Bartlesville Energy Technology Center (BETC), Chevron Research Company (CRC), and Mobil Research and Development Company (MRDC; see Table 2.7). Postproduction hydrotreating studies have been conducted in bench-scale facilities that are substantially smaller than the main coal-processing units, processing from about 0.1 to 1.0 lb/h of coal liquids. Samples of raw and hydrotreated materials also have been obtained from the three TSL facilities shown in Table 2.5.

2.5.2 Coal-Conversion Materials

The materials sampled from various technologies and processes are identified using a varied nomenclature. This variability results from differences in technology, process, product slate, unit operations, and process scale, as well as developer-specific definitions of internal process streams and ultimate commercial products. Although the nomenclature for the various technologies, processes, and materials associated with these processes has been presented gradually throughout the discussion in this chapter, a brief discussion and intercomparison of the types of material sampled and tested are appropriate.

Four basic categories of materials have been sampled and tested as follows:

1. Full-boiling-range liquefaction materials
2. Narrow-boiling-range liquefaction materials
3. Gasification by-product oils and tars
4. Hydrotreated coal-liquefaction materials.

The latter two categories include primarily full-boiling-range materials. Other materials (e.g., chars, residues, wastewaters, particulates, and aerosols) also have been sampled, but they comprise a small fraction of the database and they are not the main focus of discussion in this book.

The nomenclature for full-boiling-range coal-liquefaction materials is both similar and different across the various processes and scales tested. Most of the materials tested can be classified as light oils, distillates, or residua. A summary of standard nomenclature is pre-

Table 2.3 Noncatalytic Coal-Liquefaction Processes and Facilities Providing Materials for Toxicological Characterization

Process Developer	Facility Scale	Location	Facility Operator	Operations: Mode[a]	Operations: Coal Capacity (lb/h)
Solvent-refined coal I (SRC-I) (International Coal Refining Company)	Pilot plant	Fort Lewis, Washington[b]	Pittsburg & Midway Coal Mining Company	Integrated	4000
	Pilot plant	Wilsonville, Alabama	Southern Company Services, Inc. and Catalytic, Inc.	Integrated	500
Solvent-refined coal II (SRC-II) (Pittsburg & Midway Coal Mining Company)	Pilot plant	Fort Lewis, Washington[b]	Pittsburg & Midway Coal Mining Company	Integrated	2000
	P99 process development unit	Harmarville, Pennsylvania	Gulf Research and Development Company	Nonintegrated	80
	Merriam bench unit	Shawnee Mission, Kansas	Gulf Research and Development Company	Nonintegrated	0.8

[a] Operating mode indicates whether the facility operates in a continuous or integrated mode, or whether certain operations, such as solvent recovery and slurry preparation, are conducted in a semicontinuous/batch or nonintegrated mode.

[b] The Fort Lewis, Washington pilot plant was operated in both SRC-I and SRC-II modes.

Table 2.4 Catalytic Coal-Liquefaction Processes and Facilities Providing Materials for Toxicological Characterization

Process Developer	Facility Scale	Location	Facility Operator	Operations	
				Mode[a]	Coal Capacity (lb/h)
EDS process (Exxon Research & Engineering Company)	EDS coal-liquefaction plant (ECLP)	Baytown, Texas	Exxon Research & Engineering (ER&E)	Integrated	17,000
	Coal-liquefaction pilot plant (CLPP)	Baytown, Texas	Exxon Research & Engineering (ER&E)	Integrated	80
	Recycle coal-liquefaction unit (RCLU)	Baytown, Texas	Exxon Research & Engineering (ER&E)	Nonintegrated	3
H-Coal (Hydrocarbon Research, Inc.)	Pilot plant	Catlettsburg, Kentucky	Ashland Synthetic Fuels, Inc.	Integrated	17,000[b]
	Process development unit	Trenton, New Jersey	Hydrocarbon Research, Inc.	Integrated	250
	Bench unit	Trenton, New Jersey	Hydrocarbon Research, Inc.	Nonintegrated	1
Synthoil (Pittsburgh Energy Research Center)	Process development unit	Pittsburgh, Pennsylvania	Pittsburgh Energy Research Center	Nonintegrated	40
Ruhrkohle-Veba (Veba Oel/Ruhrkohle AG)	Pilot plant	Bottrop, Federal Republic of Germany	Veba Oel AG	Integrated	17,000

[a] Operating mode indicates whether the facility operates in a continuous or integrated mode, or whether certain operations, such as solvent recovery and slurry preparation, are conducted in a semicontinuous/batch or nonintegrated mode.

[b] Capacity for the H-Coal pilot plant operating in syncrude mode.

Table 2.5 Two-Stage Coal-Liquefaction Processes and Facilities Providing Materials for Toxicological Characterization

				Operations	
Process Developer	Facility Scale	Location	Facility Operator	Mode[a]	Coal Capacity (lb/h)
Integrated two-stage liquefaction (ITSL); (U.S. DOE/EPRI)	Pilot plant	Wilsonville, Alabama[b]	Southern Company Services, Inc. and Catalytic, Inc.	Integrated	500
Nonintegrated two-stage liquefaction (NTSL) (U.S. DOE/EPRI)	Pilot plant	Wilsonville, Alabama[b]	Southern Company Services, Inc. and Catalytic, Inc.	Integrated	500
ITSL (C. E. Lummus Company.)	Process development unit	New Brunswick, New Jersey	C. E. Lummus Company	Nonintegrated	10–20
Catalytic two-stage liquefaction (Hydrocarbon Research, Inc.)	Bench unit	Trenton, New Jersey	Hydrocarbon Research, Inc.	Integrated	1–4

[a]Operating mode indicates whether the facility operates in a continuous or integrated mode, or whether certain operations, such as solvent recovery and slurry preparation, are conducted in a semicontinuous/batch or nonintegrated mode.

[b]The Wilsonville, Alabama pilot plant has operated in SRC-I, NTSL, and ITSL modes.

Table 2.6 Gasification Processes and Facilities Providing Materials for Toxicological Characterization

Process	Facility scale	Location	Facility Operator	Operations: Mode[a]	Operations: Coal Capacity (lb/h)
Fixed Bed					
Slagging	Pilot plant	Grand Forks, North Dakota	Grand Forks Energy Technology Center	Nonintegrated	2,000
McDowell–Wellman	Pilot plant	Morgantown, West Virginia	Morgantown Energy Technology Center	Nonintegrated	1,000
Foster–Wheeler Stoic	Industrial	Duluth, Minnesota	University of Minnesota	Integrated	6,000
Wellman incandescent	Industrial	York, Pennsylvania	Glen-Gery Brick Company	Integrated	2,000
Wellman	Industrial	Fort Snelling, Minnesota	Bureau of Mines	Integrated	2,600
Chapman	Industrial	Kingston, Tennessee	Holston Defense Corp.	Integrated	2,400
Research Triangle Institute	Bench	Research Triangle Park, North Carolina	Research Triangle Institute	Nonintegrated	2–3
Lurgi	Industrial	Kosovo, South Yugoslavia	Government of Yugoslavia	Integrated	180,000
Fluidized Bed					
Hygas	Pilot plant	Chicago, Illinois	Institute of Gas Technology	Integrated	6,000
COED pyrolysis	Pilot plant	Princeton, New Jersey	FMC Corp.	Nonintegrated	3,000
Entrained Flow					
Mellon Institute	Process development unit	Pittsburgh, Pennsylvania	Mellon Institute	Nonintegrated	2

[a] Operating mode indicates whether the facility operates in a continuous or integrated mode, or whether certain operations, such as coal or slurry preparation, are conducted in a semicontinuous/batch or nonintegrated mode.

Table 2.7 Hydrotreating Facilities Providing Materials for Toxicological Characterization

Operating Facility (Testing Laboratory)	Scale	Location	Capacity (lb/h)	Feed Materials
Universal Oil Products, Inc., UOP (Pacific Northwest Laboratory, PNL)	Laboratory	Des Plaines, Illinois	NA	SRC-II fuel-oil blend
Bartlesville Energy Technology Center, BETC (Inhalation Toxicology Research Institute, ITRI; and PNL)	Bench	Bartlesville, Oklahoma	0.1–0.2	SRC-II fuel-oil blend, EDS recycle solvent and fuel-oil-blend, H-Coal ASB and VSOH[a]
Mobil Research and Development Company, MRDC (Oak Ridge National Laboratory, ORNL)	Laboratory	Paulsboro, New Jersey	0.1–1.0	H-Coal oil blend, SRC-I light organic liquid, SRC-I wash solvent, and SRC-I recycle solvent
Chevron Research Company, CRC (ORNL, PNL)	Pilot plant	Richmond, Virginia	0.15–0.5	EDS oil blend, H-Coal oil blend, SRC-II oil blend

[a] ASB = atmospheric-still bottoms; VSOH = vacuum-still overhead.

sented in Fig. 2.11, along with typical boiling-point ranges for each material. Note that the boiling-point ranges are only approximations, and they can vary substantially among and between specific materials.

Samples of full-boiling-range materials can be tested in two ways. First, samples can be taken from a specific process stream and tested directly without any prior processing. Thus, results from these evaluations provide data applicable to the specific material tested. Second, samples from various process streams can be blended and/or redistilled to provide a material that approximates a product from a commercial-scale facility. For example, various samples of middle and heavy distillates from the SRC-II process have been blended to form full-boiling-range fuel oils, and they have been tested directly or hydrotreated and tested. Distillate blends from EDS, H-Coal, and SRC-I also have been formulated and tested in this manner.

Narrow-boiling-range liquefaction materials are prepared by redistilling a full-boiling-range coal liquid into well-defined boiling-point *cuts*. Boiling-point cuts typically are prepared to separate coal-derived chemicals into discrete boiling ranges and further study the characteristics or properties of the cuts that comprise the original whole material. Specific to toxicological testing, these cuts permit the investigator to elucidate further those chemicals or chemical compound classes responsible for observed toxicity. Narrow-boiling-range cuts from the SRC-I, H-Coal, EDS, SRC-II, and C. E. Lummus ITSL processes have been characterized toxicologically.

Gasification by-product tar and oils are considered as full-boiling-range materials. The specific properties of these organic condensates are determined to a large extent by the specific unit operations used to quench and/or collect gasifier products (see Section 2.3.2). Organic condensates that float on water usually are called oils and those that sink are termed tars. The closest analogy to coal-liquefaction materials might be the differences seen between middle and heavy distillates or fuel oils. Oil and tar can be blended to form a combined oil or full-boiling-range tar/oil by-product, but this usually is done only for samples from experimental-scale systems. The majority of tar and oil materials from pilot-plant and industrial-scale facilities were tested directly as sampled.

Hydrotreated coal liquids are also full-boiling-range materials. Light oil and distillate samples are hydrotreated as sampled, or they can be blended prior to hydrotreatment. The raw feedstock and hydrotreated liquid product typically are tested toxicologically and/or chemically.

2.5.3 Database for Coal-Conversion Materials

The database discussed in this book encompasses a wide diversity of disciplines. Data available for the various coal-conversion materials include toxicity by definition, as well as chemical composition, bulk physical/chemical properties, and/or process conditions associated with the specific samples tested. Some of the more common terminology referred to in the remaining chapters is presented briefly here. More detailed discussions of concepts and methods are provided in the appropriate chapters.

2.5.3.1 Toxicity Data

The coal-conversion database is composed of toxicology data that address two primary endpoints, mutagenicity and carcinogenicity. Mutagenicity data were typically developed using the Ames *Salmonella*/mammalian microsomal mutagenicity assay. In this assay, *Salmonella typhimurium* is exposed to various doses of test material and the mutagenicity is

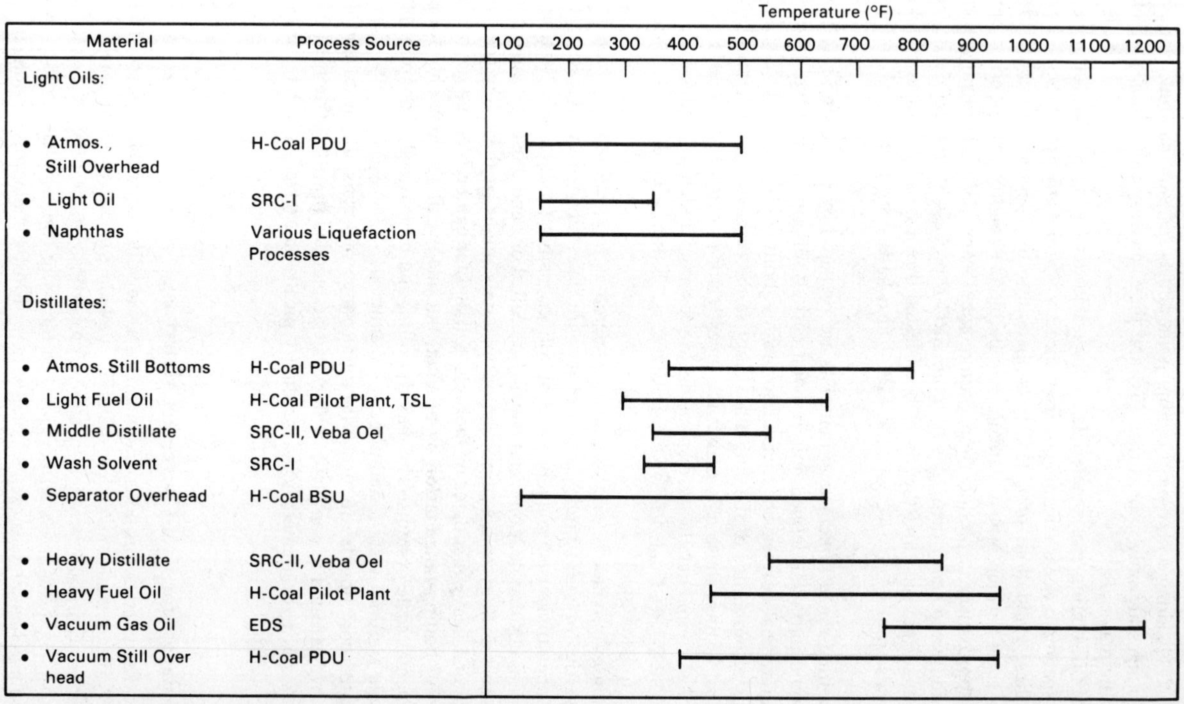
Temperature (°F)
Material
Process Source
100 200 300 400 500 600 700 800 900 1000 1100 1200
Light Oils:
• Atmos., Still Overhead
H-Coal PDU
• Light Oil
SRC-I
• Naphthas
Various Liquefaction Processes
Distillates:
• Atmos. Still Bottoms
H-Coal PDU
• Light Fuel Oil
H-Coal Pilot Plant, TSL
• Middle Distillate
SRC-II, Veba Oel
• Wash Solvent
SRC-I
• Separator Overhead
H-Coal BSU
• Heavy Distillate
SRC-II, Veba Oel
• Heavy Fuel Oil
H-Coal Pilot Plant
• Vacuum Gas Oil
EDS
• Vacuum Still Over head
H-Coal PDU

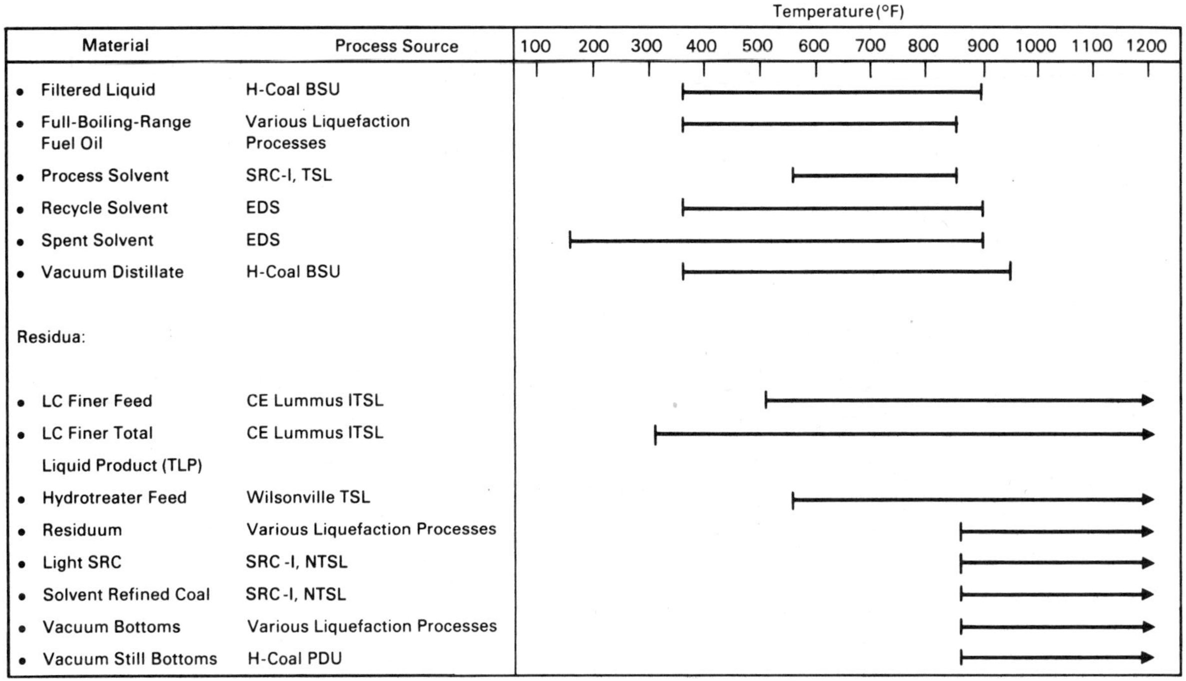

Figure 2.11 Typical boiling-point ranges for coal-liquefaction materials.

expressed in terms of the number of histidine revertants produced at a given dose of test material. Various strains of *S. typhimurium* have been used, but the most sensitive and most frequently used strains are TA98 and TA100. Testing is performed either with or without metabolic activation using Aroclor-induced, S9 rat-liver enzyme. Test results are typically expressed as revertants/microgram, which is derived by calculating a best-fit linear regression of the results over a number of doses within the linear portion of the dose–response curve. Other mutagenicity assays have been used less frequently and are discussed in Chapter 3.

Carcinogenicity of test materials is reported from two types of assay, chronic dermal carcinogenesis and initiation/promotion. Chronic dermal carcinogenesis studies are conducted by applying a specified dose of test material to the skin of laboratory mice for a period up to the life span of the mouse. Dermal application of a carcinogenic material results in growth of a tumor. The incidence of tumors in the test population is charted over the course of the test. Data are typically reported as percentage of test animals with tumors. Sophisticated mathematical models and statistical techniques also can be used to evaluate dose-related effects, tumor latency (i.e., time from first exposure to appearance of tumors), and biological behavior of tumors. Carcinogenic response also can be reported as potency, relative to carcinogenic pure compounds such as benzo[a]pyrene (BaP) or dimethylbenzanthracene (DMBA).

Initiation/promotion assays in mouse skin are conducted by application of a single dose of test material (i.e., initiation), followed by repeated application of a noncarcinogenic chemical that is known to promote the induction of tumors. Tumorigenic response can be reported as number of tumors per test animal, total number of tumors per exposure group, and percentage of test animals with tumors.

Other toxicity endpoints have been investigated for coal-conversion materials. These include studies of dermal and ocular irritation, teratogenicity and embryotoxicity to determine physical malformations in exposed cells or fetuses, genetic damage, and cell growth. Data from these test procedures comprise a small portion of the coal-conversion database and are discussed in Chapter 3.

2.5.3.2 Chemistry Data

The chemical composition of coal-conversion materials has been under investigation for decades. Many methods have been developed and applied to identify the specific chemicals and/or chemical classes that comprise coal-conversion materials. These methods are described in detail in Chapter 5.

In the context of the reported studies, chemistry methods were typically used to identify the specific chemicals and/or chemical classes responsible for observed toxicity in coal-conversion materials. Often, whole materials were fractionated using pH-fractionation and/or high-performance liquid chromatography (HPLC) techniques to isolate the major class(es) of chemicals responsible for observed sample toxicity. These fractions, as well as the whole materials, could be analyzed using gas chromatography (GC), mass spectrometry (MS), combined GC-MS, and other methods to (1) determine more specifically which compounds were responsible for observed sample toxicity and (2) determine the precise chemical composition of the sampled materials. The following discussion summarizes the nomenclature used throughout this book to describe the specific chemicals found in coal-conversion materials.

The chemical nomenclature used in this chapter for polycyclic aromatic compounds (PAC) is conventional, although the system suggested by Bartle et al. (1981) is also used in some instances. The general classes of PACs that are of interest are designated as polycyclic aromatic hydrocarbons (PAHs), oxygen-containing PACs (O-PACs), sulfur-containing PAC (S-PACs), and nitrogen-containing PACs (N-PACs). The specific functionality of heteroatoms outside the aromatic ring system is prefixed to designate PAH type: hydroxy-PAHs, cyano-PAHs, and amino-PAHs. For example, the name 2-aminonaphthalene is used instead of β-naphthylamine. Conventional names may also be used for the heterocyclic compounds, such as thiophenes, azaarenes, and furans. This approach allows unambiguous designation of mixed-function heteroatomic species, such as aminocarbazoles and hydroxythiophenes, often found in coal-derived products. The conventional designation, primary aromatic amine (PAA), is often used in the literature but is generally avoided in favor of the more specific amino-PAH. PAA is a catch-all that denotes members of several classes, and many are industrial carcinogens that have only an amino functionality in common with compounds found in some coal liquids. Names and representative structures for specific compound classes are shown in Table 2.8.

Table 2.8 Heterocyclic Polycyclic Aromatic Compounds (PACs) in Coal-Derived Materials

Representative Structure	Class	Acronym
	N-PACs[a]	
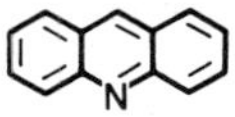	3°-Polycyclic aromatic nitrogen heterocycles	3°-PANHs
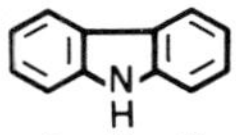	2°-Polycyclic aromatic nitrogen heterocycles	2°-PANHs
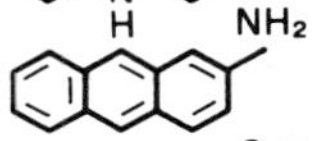	Amino polycyclic aromatic hydrocarbons	Amino-PAHs
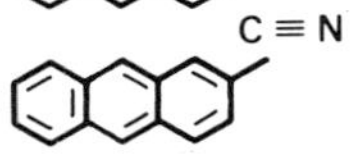	Cyano polycyclic aromatic hydrocarbons	Cyano-PAHs
	O-PACs[b]	
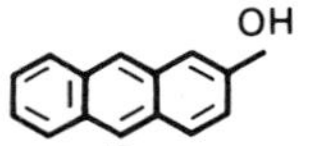	Hydroxy polycyclic aromatic hydrocarbons	Hydroxy-PAHs
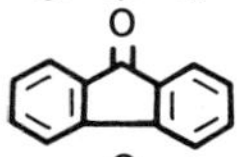	Polycyclic aromatic ketones	PAKs
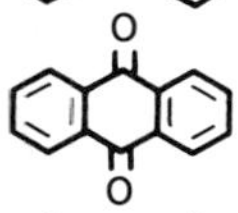	Polycyclic aromatic quinones	PAQs
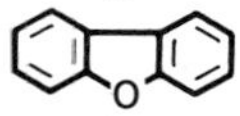	Polycyclic aromatic oxygen heterocycles	PAOHs
	S-PACs[c]	
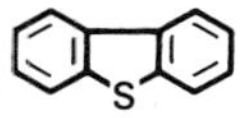	Polycyclic aromatic sulfur heterocycles	PASHs
	Multiple Heteroatom PACs	
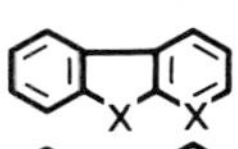	X = X ≡ N, O, or S	X, X-PACs
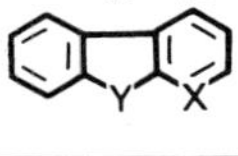	X = Y ≡ N, O, or S	X, Y-PACs

[a] Nitrogen-containing PACs.
[b] Oxygen-containing PACs.
[c] Sulfur-containing PACs.

The term *z number* used in the text is a measure of aromaticity defined by the expression C_nH_{2n-z}. Benzene (C_6H_6) therefore has a z number of 6. For a given number of carbon atoms, increasing z number indicates increasing ring condensation or aromaticity. Rank ordering of compound elemental composition by z number is often a convenient way of expressing mass spectrometric data. In addition, the heteroatomic element designation is sometimes included [e.g., carbazole ($C_{12}H_9N$) is designated as a z = 15N compound].

2.5.3.3 Sample Properties

Independent of the toxicological and chemical characterization data reported in this book, process developers frequently characterize samples of coal-conversion materials to evaluate process performance and to determine the physical and chemical properties of specific products. Two measures of these *bulk* properties that relate to the toxicity of coal-conversion materials are distillation data and elemental analysis data, and they are described here briefly. These property measurements are discussed further in Chapter 4, and the application and limitations of these methods have been reviewed by Fillo and Craun (1985).

Distillation data are available mostly for samples of coal-liquefaction and hydrotreated materials. Many distillation methods have been used to generate these data, including atmospheric and vacuum ASTM methods (e.g., D-86, D-1160), true-boiling-point (TBP) methods, and simulated GS (SGC) methods. For the purposes of this work, distillation data from all methods have been corrected where necessary to approximate a TBP distillation curve using standard American Petroleum Institute (API) corrections for petroleum materials (API, 1976) and modified corrections for coal-derived liquids (Gray, 1981). It is assumed that SGC distillations approximate the TBP curves. It is noted that few distillation data are available for gasification tars and oils.

Aromaticity is an important property for determining toxicity of petroleum- and coal-derived materials. The most available measure of aromaticity for all coal-derived materials is the atomic H:C ratio. This parameter is derived from elemental analysis data, which are usually developed to perform material balances around the various coal-conversion processes tested. These data are available for virtually all coal-conversion materials tested for toxicological activity.

2.6 CHEMICAL STABILITY OF COAL-DERIVED LIQUID SAMPLES

Chemical stability is an important consideration in the characterization of coal-derived liquid fuels. Chemical changes may occur over time in complex, coal-derived liquids that tend to be reactive because of their high heteroatom content. Degradation reactions, including precipitate or sludge formation, can adversely affect combustion characteristics and/or useful properties of these fuels. Changes in chemical composition resulting from degradation between the time samples are taken and the time they are tested may also affect the toxicological characteristics of the materials. The storage stability and aging characteristics of several coal-, shale-, and petroleum-derived materials have been studied, and Table 2.9 presents a synopsis of the relevant literature.

2.6.1 Stability of Chemical Composition and Toxicity

Wright and Weimer (1984) and Wright et al. (1984) monitored mutagenicity and chemical composition of various SRC materials as a function of storage time, temperature, dilution, cover atmosphere, and ambient lighting. Mutagenic activity was measured with the Ames

Table 2.9 Selected Literature on Stability of Petroleum and Synthetic Fuels During Storage

Investigator	Objectives	Degradation Characterization
Weimer and Wright (1985)	Analysis of weathered residual coal-conversion materials	Chemical composition of class fractions, Ames mutagenicity, initiation/promotion tumorigenicity
Kalkwarf et al. (1984)	Photodegradation of mutagens in SRC liquids	Ames Mutagenicity, primary aromatic amines
Wright and Weimer (1984), Wright et al. (1984)	Chemical/biological stability of SRC liquids	Ames mutagenicity, chemical class fractions, overall chemical composition, PAHs, N-PACs
Jones and Li (1983)	Aging of SRC-II middle distillate with CuO_2	Viscosity, hydroxyl concentrations, elemental analysis
White et al. (1983)	Phenolic-oxidative coupling in SRC-II middle distillate	Phenolics composition
Epperly (1982, 1983)	Stability of EDS liquids	Viscosity, chemical compound classes, overall chemical composition
Drozdowicz and Kelly (1983)	Chemical stability of SRC-I materials	Chemical composition
Hazlett et al. (1983)	Shale-derived diesel fuel degradation mechanism	Total insolubles, nitrogen compounds
Brinkman and Bowden (1982)	Stability of tar sands, EDS, SRC-II liquids	Precipitates and total acid number, viscosity, peroxide number, elemental analysis, dissolved O_2
Frankenfeld and Taylor (1982)	Fundamental stability study of petroleum- and shale-derived fuels	Total sediment; nitrogen, sulfur, and oxygen compounds
Frankenfeld and Taylor (1981)	Fundamental stability study of petroleum-derived fuels	Total sediment, nitrogen compounds
Bowden and Brinkman (1980)	Stability of petroleum-, shale-, tar-sand-, and coal-derived gasolines and jet fuels	Gum formation, elemental analysis, viscosity, acid number
Brinkman et al. (1980)	Annotated review—crude oil and finished fuel storage stability	—
Brinkman et al. (1979)	Stability of petroleum-, shale-, tar-sand-, and coal-derived gasolines and jet fuels	Gum formation, elemental analysis, viscosity, acid number
Brown and Karn (1979)	Stability of Synthoil liquid product	Viscosity, O_2 consumption, chemical class fractions
Karn and Brown (1976)	Aging of Synthoil product	Viscosity, chemical class fractions

assay against *S. typhimurium* strain TA98. Chemical composition was monitored with high-resolution GC (HRGC) and HRGC-MS. Data were analyzed by computer-based, multivariate analysis.

Two samples of naphtha, four of middle distillate, three of heavy distillate, and four of fuel-oil blend (2.9 : 1, middle : heavy distillate) from the SRC-II process were monitored while stored at 4°C (39°F), under nitrogen, in inert containers, in the dark. In addition, one light oil, one wash solvent, and two PS samples from the SRC-I process were monitored. After 4 years under these conditions, no significant changes in the microbial mutagenicity of any of the materials were detected; nor were there significant changes in the major components detectable by GC under these storage conditions (see Fig. 2.12).

Fuel-oil-blend (FOB) samples were also stored at −20°C and at room temperature under various atmospheric conditions (air versus nitrogen), dilution (undiluted versus dilution in methylene chloride), and ambient lighting (light versus darkness). After 1 year under these conditions, microbial mutagenicity decreased significantly or was totally eliminated in the

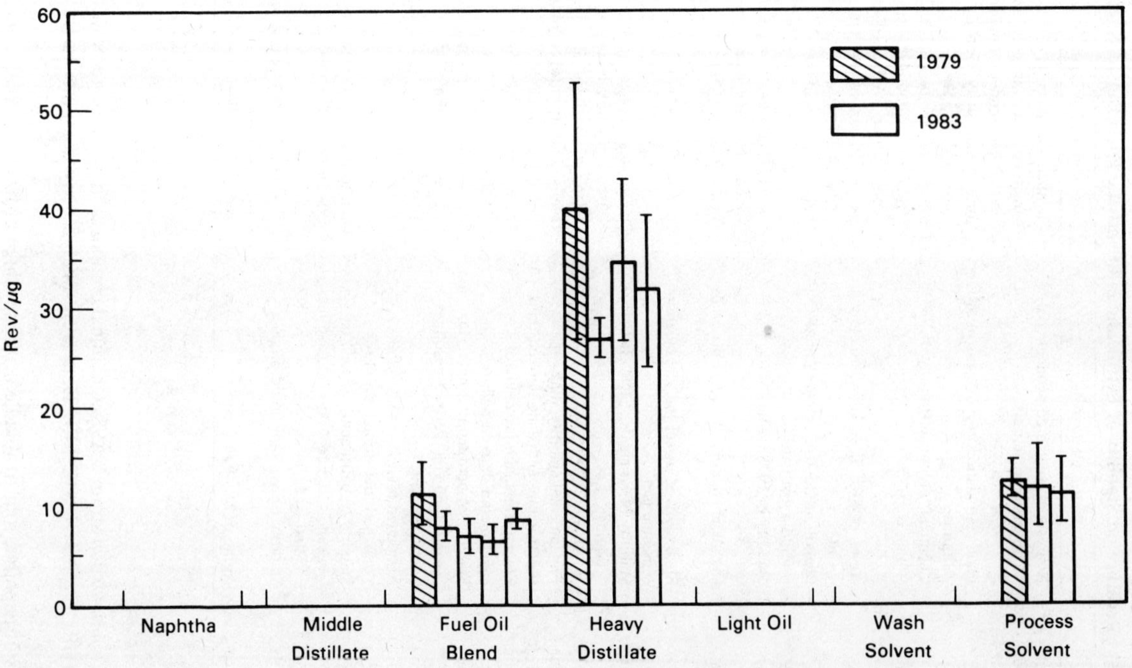

Figure 2.12 Microbial mutagenicity of solvent-refined coal (SRC) materials tested in 1979 versus those stored in a repository at 4°C (39°F), in the dark, under nitrogen, and tested in 1983 (Wright and Weimer, 1984).

samples stored diluted in methylene chloride as well as in the samples stored at room temperature, in light, under air. There were no significant changes in the major chemical constituency, as monitored by HRGC, with the exception of the samples stored diluted in methylene chloride in which the concentrations of cresols, indanols, and tetralins decreased.

Fuel-oil blend samples were also stored under controlled temperatures, in the dark, and under air. Mutagenic activity was completely eliminated after 32 weeks of storage at 60°C (140°F) and after 16 weeks at 100°C (212°F). Mutagenicity was reduced by about 75% following storage for 16 weeks at 100°C (212°F). After storage under these harsh conditions, the percentage of gas-chromatographable FOB decreased; weight percent in N-PAC and phenolic-PAC fractions decreased; and the color of stored samples darkened. Concentrations of phenols, cresols, indanols, and tetralins were consistently lower after degradation. Concentrations of amino-PAHs and indoles in the FOB also decreased dramatically.

Kalkwarf et al. (1984) studied the effects of simulated sunlight during storage on an SRC-II FOB and an SRC-I PS. Liquids were exposed to darkness and sunlight in the presence of air, then they were chemically fractionated, and chemical composition and mutagenicity were monitored for the various fractions. Mutagenicity of the amino-PAH fraction from the SRC-II fuel-oil blend decreased by 65 and 90% after it was exposed for 36 h in the dark or in simulated sunlight, respectively (see Fig. 2.13). Similarly, a decrease in the summed concentration of 14 aromatic amines in the amino-PAH subfraction was also measured (see Fig. 2.14). Similar results were also observed with the SRC-I process solvent.

Studies were also conducted on SRC-I and TSL materials from the Wilsonville, Alabama, pilot plant to determine their stability during storage (Drozdowicz and Kelly, 1983). Aliquots of SRC-I middle distillate were placed in clear, colorless, glass vials and allowed to age while exposed to light at room temperature. Samples were not in an inert atmosphere.

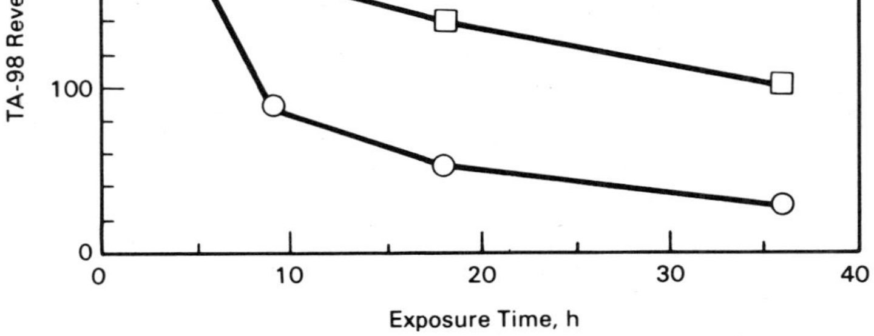

Figure 2.13 Reduction of mutagenicity in the amino-polycyclic aromatic hydrocarbon-rich subfraction of solvent-refined coal II (SRC-II) fuel-oil blend during exposure to air in darkness and in simulated sunlight (Kalkwarf et al., 1984).

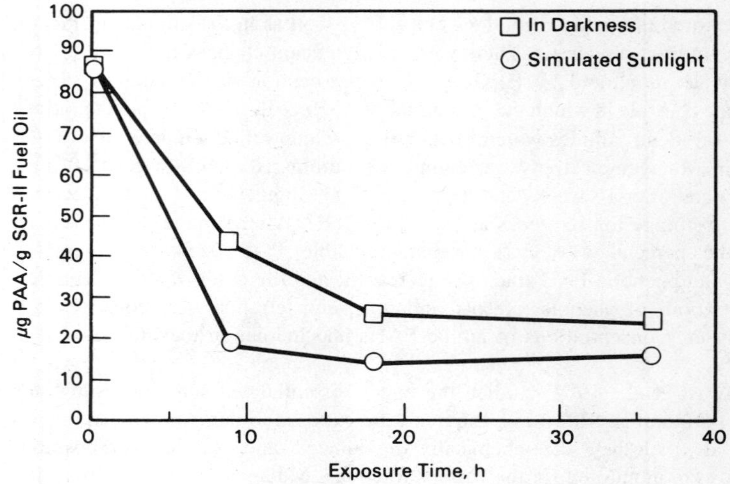

Figure 2.14 Reduction of primary aromatic amines (PAA) in the amino-polycyclic aromatic hydrocarbon subfraction of solvent-refined coal II (SRC-II) fuel oil exposed to air in darkness and in simulated sunlight (Kalkwarf et al., 1984).

After 70 days, aged samples were analyzed by capillary GC and compared with freshly prepared standards. Total areas and the relative areas of individual peaks were compared with those of corresponding peaks for standards. Relative responses of aged sample peaks were compared with those of standards and expressed as percentages (aged/standard × 100). In addition, various materials were stored in DMSO at 5°C (41°F) in an inert atmosphere for 2 months. Stabilities were expressed as the percent response for aged material relative to a standard, measured via ultraviolet (UV) spectrophotometry. Stability in all cases ranged from 90 to 99%. Average stability of aged middle distillate was 92.6%, with peak-specific values ranging from 77 to 137%. Although there was some chemical degradation, it should not affect the qualitative results of toxicity tests (i.e., active or inactive). Similar changes were cited to occur during routine handling and storage.

Finally, White et al. (1983) studied the degradation of phenolic compounds in SRC-II middle distillate. Degradation was accelerated by exposure of the sample to temperatures of 635°F (335°C), oxygen, and copper shavings. Infrared (IR) analysis showed that more quinone-type compounds were present in aged as compared to fresh materials. HRMS was used to determine elemental compositions of homologous series of multioxygenated, multibenzenoid species thought to be products of oxidative phenolics coupling in middle distillate. Although no specific structures were assigned because no standards were available, several possible structural types were postulated.

2.6.2 Stability of Physical Properties

The chemical degradation of hydrocarbon liquids can also be observed through monitoring of bulk physical properties. Physical degradation of hydrocarbon liquids results in both increased viscosity and the formation of gums and residues during storage. The combined effect of temperature and oxygen on stability of a Synthoil liquid product from Pittsburgh seam coal was studied by Karn and Brown (1976) and Brown and Karn (1979). The sample was stored under nitrogen, air, and oxygen at 30, 45, and 61°C (86, 113, and 142°F) and viscosity was measured periodically for 60 days. Viscosity (Fig. 2.15) increased over two orders of magnitude in less than 30 days at 142°F (61°C) in an oxygen atmosphere. In contrast, at 86°F (30°C), viscosity increased only 50% in nitrogen and about five- to sixfold in air.

Bowden and Brinkman (1980) used gum formation as the indicator for degradation in studies with gasoline-range liquids from Sasol, Athabasca tar-sands bitumen, and petroleum, and with jet fuels from petroleum and a Kentucky-coal-derived COED syncrude. Materials were stored at 43°C (110°F) for 32 weeks. Coal- and tar-sands-derived fuels showed the largest increases in gum formation.

Stability studies were also conducted on EDS materials (Epperly, 1982, 1983). Raw naphthas were generally unstable; spent solvent did not store well, with sediment increases exceeding 500 mg/100 mL after 100 days at 110°F, in air. Storage stability of recycle solvent was similar to that of petroleum fuels, maintaining its integrity over the course of the study. Bulk measurements of EDS FOB indicated increased viscosity with storage time, particularly at elevated temperatures; however, after 1 month at 210°F, viscosity levels were below those of typical petroleum fuels. Wall deposits were the most significant stability concern in the EDS materials and they were generally associated with increased viscosity or sediment formation. Wall deposits were not observed in petroleum fuels used for comparisons in these studies.

Stability of EDS materials decreased with increased storage temperature. Autooxidation appeared to be the primary cause of instability. Nitrogen blanketing effectively eliminated aging, even under severe storage conditions. Antioxidant additives had no substantial effect on stability of EDS fuels in storage. Increasing the volume : surface area ratio of the storage container appeared to improve stability, and steel was more effective than glass.

Weimer and Wright (1985) studied the effects of weathering on residual fugitive emission and process materials collected from both interior and exterior surfaces of the Ft. Lewis SRC pilot plant prior to and during dismantling and salvage operations. The molecular weight range of the residual materials, including those collected from handrails, pipes, and interior and exterior valve surfaces, as well as from the concrete pad directly beneath the valves, was more similar to that of HD than to that of MD or LO. Overall, lighter components appear to have volatilized or leached out of the residual materials. Residual materials appeared to have more oxygen incorporated into their structure than was present in HD itself. Excess oxygen was present in addition to that contributed by the presence of relatively high concentrations of diphenyl ether, which is the major component of Dowtherm, used in the plant as a heat-transfer fluid. Photoinduced oxidation reactions were suspected to account for much of the excess oxygen found in materials collected from external surfaces.

Selected residual materials were relatively rich in four-, five-, and six-ring PAHs. Most of the materials collected were mutagenic in the Ames assay. Those tested for tumorigenicity in the initiation/promotion assay were found to be similar to HD in their activity (D. D. Mahlum, Pacific Northwest Laboratory personal communication).

Nitrogen compounds in coal-derived liquids were also shown to participate in degradation reactions by Hazlett et al. (1983). Pyrroles accelerated formation of deposits in shale oil in 30-day stability tests at 43°C (109°F) and 80°C (176°F) (Fig. 2.16), where 3,5-dimethylpyrrole was added at 45–450 ppm total nitrogen. Deposit formation was marginally higher at 80°C (176°F), particularly for nitrogen concentrations above 135 ppm. Similarly,

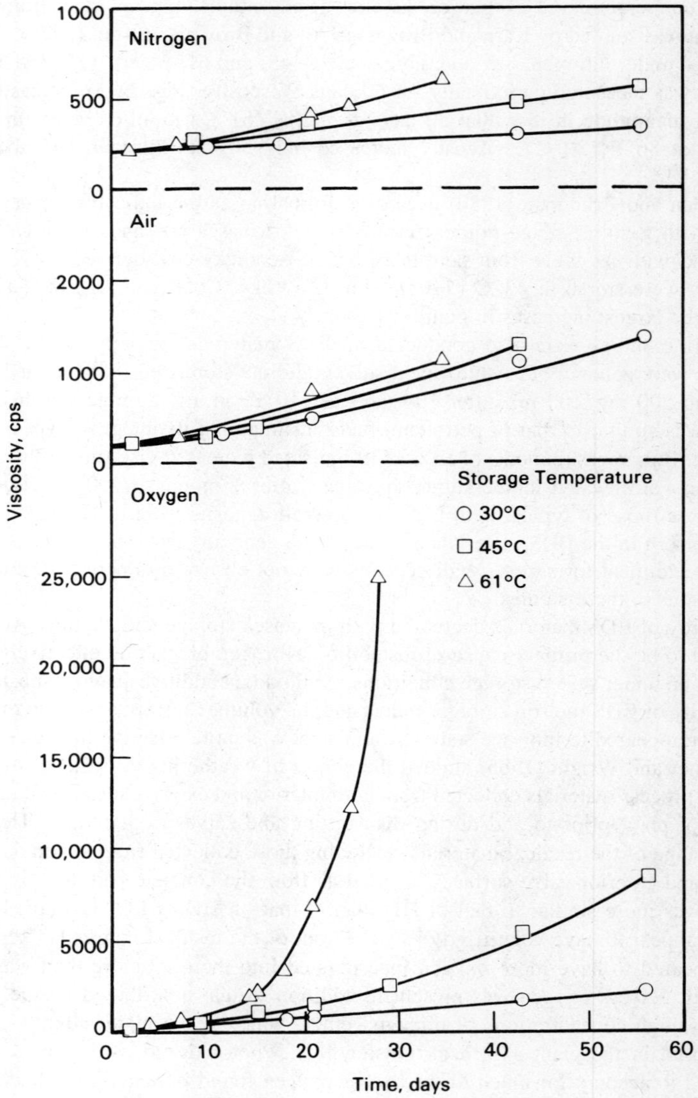

Figure 2.15 Effects of storage atmosphere and temperature on the viscosity of Synthoil coal liquid (Karn and Brown, 1976).

Frankenfeld and Taylor (1981, 1982) reported that compounds with heterocyclic nitrogen and adjacent alkyl groups promoted sediment and gum formation in petroleum and synthetic fuels.

2.6.3 Significance of Sample Stability

Three primary variables influence the degradation of petroleum- and coal-derived liquids: temperature, light, and the presence of oxygen. Generally, reaction rates increase with

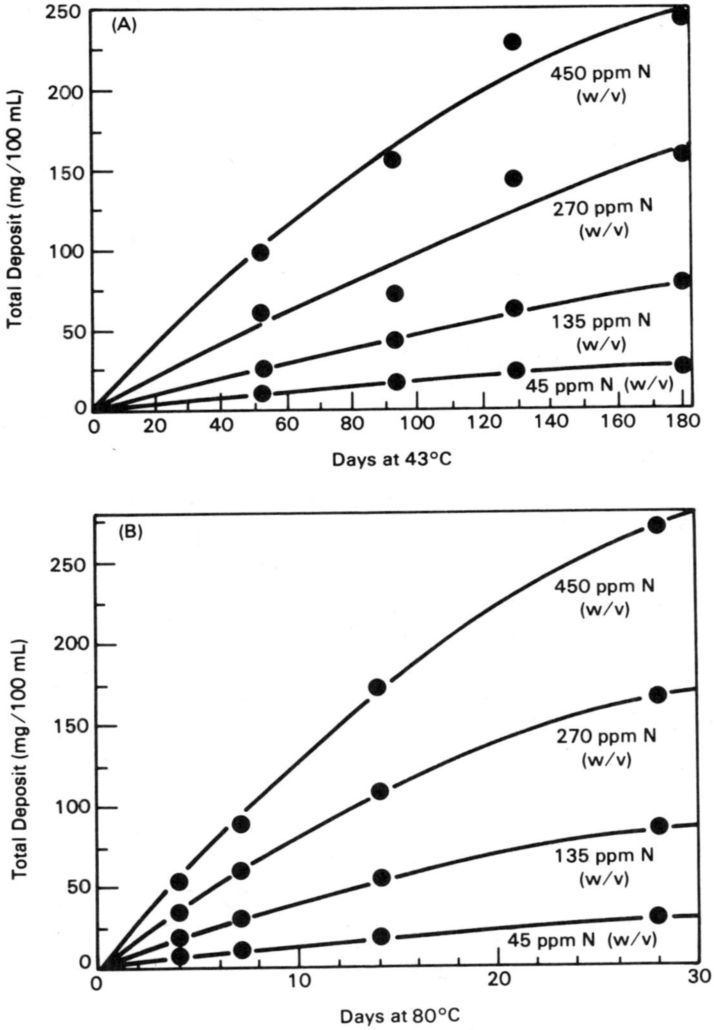

Figure 2.16 Effect of adding 2,5-dimethylpyrrole on the formation of sediment in shale oil at (A) 43°C (109°F) and (B) 80°C (192°F) (Hazlett et al., 1983).

increasing temperature, and sample degradation increases accordingly. Light also accelerates sample degradation through the formation of free radicals that participate in degradation reactions. Oxygen in the presence of light or at higher storage temperatures substantially increases degradation. Because sample degradation basically results from free-radical condensation reactions, oxygen accelerates the process. Additionally, oxygen-containing and nitrogen- and sulfur-containing compounds can act similarly.

The type and magnitude of expected changes to the chemical composition and toxicity of coal-conversion materials are a function of the specific material tested and the conditions during storage and handling between sampling and testing. Test materials are either sampled directly from process streams or taken from long-term storage. In the first case, samples are typically collected and placed immediately in dark-glass, stainless-steel, or lined containers, blanketed with inert gas, and shipped to the testing laboratory. Shipment is accomplished within days to weeks, and often the materials are cold-packaged. In the other case, materials are stored in closed containers (typically, steel drums) and may have been kept at ambient conditions or refrigerated under an inert atmosphere for periods of months to years. Once materials are received by the testing laboratory, samples are placed in long-term storage and are typically inert-blanketed, refrigerated, and stored in the dark to maintain sample integrity.

Given the nature of procedures used to store coal-conversion materials prior to testing, the following conclusions can be drawn:

1. Materials maintained in long-term storage prior to testing undergo little if any change in chemical composition and thus toxicological activity.
2. Most samples can be expected to undergo some level of chemical degradation prior to testing, and the type and magnitude of the degradation are functions of the specific handling and storage conditions.

Thus, the toxicological activity of coal-derived liquids qualitatively remains stable when materials are stored under controlled conditions. All samples may undergo minor degradation during handling, with any decrease in activity reflecting the level of stress (i.e., heat, light, and/or oxygen) applied during storage. Only under the most severe storage conditions might the complete disappearance of biological activity be expected.

Stability of coal-conversion materials has not been evaluated in initiation/promotion or long-term skin-painting assays. Thus, the effect of sample stability on the results of whole-animal testing is unknown. Different tests respond differently to different chemical classes and sample properties.

REFERENCES

Alpert, S. B., and R. H. Wolk, Liquefaction Processes, in M. A. Elliott (ed.), *Chemistry of Coal Utilization*, Second Supplementary Volume, Wiley, New York, 1981, pp. 1919–1990.

American Petroleum Institute, *Technical Data Book—Petroleum Refining*, Vol. 1, Chaps. 3 and 5, Washington, DC, 1976.

Bartle, K. D., M. L. Lee, and S. A. Wise, Modern Analytical Methods for Environmental Polycyclic Aromatic Compounds. *Chem. Soc. Rev.*, **10,** 113–158 (1981).

Bowden, J. N., and D. W. Brinkman, Stability of Alternate Fuels. *Hydrocarbon Process.*, 77–82 (1980).

Brinkman, D. W., and J. N. Bowden, Stability of Some Coal and Tar Sands Syncrude Fractions. *Fuel*, **61,** 1141–1148 (1982).

Brinkman, D. W., J. N. Bowden, and H. N. Giles, *Crude Oil and Finished Fuel Storage Stability: An Annotated Review*, DOE/BETC/RI-79/13, National Technical Information Service, Springfield, VA, 1980.

Brinkman, D. W., M. L. Whisman, and J. N. Bowden, *Stability Characteristics of Hydrocarbon Fuels from Alternative Sources*, BETC/RI-78/23, National Technical Information Service, Springfield, VA, 1979.

Brown, F. R., and F. S. Karn, *Stability Studies of Coal Liquids*, PERC/TR-79/5, National Technical Information Service, Springfield, VA, 1979.

CCDC, *Environmental Characterization: CO_2-Acceptor Process Gasification Pilot Plant*, Final Report, Vol. II, FE-1734-44. Conoco Coal Development Company, Library, PA, National Technical Information Service, Springfield, VA, 1978.

Chaffee, C. C., and L. L. Hirst, Liquid Fuel From Coal—Program Report on Hydrogenation Demonstration Plant, *Ind. Eng. Chem.*, **45,** 822–838 (1953).

Crynes, B. L., Processing Coal Liquefaction Products, in M. A. Elliott (ed.), *Chemistry of Coal Utilization*, Second Supplementary Volume, Wiley, New York, 1981, pp. 1991–2061.

Donath, E. E., Hydrogenation of Coal and Tar, in H. H. Lowry (ed.), *Chemistry of Coal Utilization*, Supplementary Volume, Wiley, New York, 1963, pp. 1041–1080.

Drozdowicz, B. Z., and C. M. Kelly, *Interim Report on the Genetic and Animal Toxicity Testing of SRC-I Products, Intermediates, and Waste Materials*, DOE/OR/03054-50, National Technical Information Service, Springfield, VA, 1983.

Elliott, M. A., and G. R. Yohe, The Coal Industry and Coal Research and Development in Perspective, in M. A. Elliott (ed.), *Chemistry of Coal Utilization*, Second Supplementary Volume, Wiley, New York, 1981, pp. 1–54.

Epperly, W. R., *EDS Coal Liquefaction Process Development Phase V, EDS Product Quality Final Report*, FE-2893-97, Exxon Research & Engineering Company, National Technical Information Service, Springfield, VA, 1982.

Epperly, W. R., *EDS Coal Liquefaction Process Development, Phase V—Quarterly Technical Progress Report for the Period October 1–December 31, 1982*, FE-2893-103, National Technical Information Service, Springfield, VA, 1983.

Fillo, J. P., *An Understanding of Phenolic Compound Production in Coal Gasification Processing*, DOE/ET/10249-101, National Technical Information Service, Springfield, VA, 1979.

Fillo, J. P., and J. C. Craun, *Engineering Analysis of the Toxicological Characterization Database for Coal Conversion Materials*, DOE/NBM-1079, National Technical Information Service, Springfield, VA, 1985.

Frankenfeld, J. W., and W. F. Taylor, *Fundamental Synthetic Fuel Stability Study, First Annual Report*, DOE/BC/10045-12, National Technical Information Service, Springfield, VA, 1981.

Frankenfeld, J. W., and W. F. Taylor, *Fundamental Synthetic Fuel Stability Study, Final Report*, DOE/BC/10045-23, National Technical Information Service, Springfield, VA, 1982.

Gates, B. C., J. R. Katzer, and G. C. A. Schuit, *Chemistry of Catalytic Processes*, McGraw-Hill, New York, 1979.

Gorin, E., Fundamentals of Coal Liquefaction, in M. A. Elliott (ed.), *Chemistry of Coal Utilization*, Second Supplementary Volume, Wiley, New York, 1981, pp. 1845–1918.

Gray, J. A., *Selected Physical, Chemical and Thermodynamic Properties of Narrow Boiling Coal Liquids from the SRC-II Process*, DOE/ET/10104-7, National Technical Information Service, Springfield, VA, 1981.

Gray, R. H., and K. E. Cowser (eds.), *Status of Health and Environmental Research Relative to Coal Liquefaction*, DOE/NBM-1016, National Technical Information Service, Springfield, VA, 1982.

Hazlett, R. N., J. V. Cooney, and E. J. Beal, *Mechanisms of Syncrude/Synfuel Degradation, First Annual Report*, DOE/BC/10525-4, National Technical Information Service, Springfield, VA, 1983.

Hebden, D., and H. J. F. Stroud, Coal Gasification Processes, in M. A. Elliott (ed.), *Chemistry of Coal Utilization*, Second Supplementary Volume, Wiley, New York, 1981, pp. 1599–1752.

Horne, W. A., and J. McAfee, Hydrogenation of Petroleum and Its Fractions, in J. J. McKetta (ed.), *Advances in Petroleum Chemistry and Refining*, Vol. III, Interscience, New York, 1960, pp. 192–277.

Howard, J. B., Fundamentals of Coal Pyrolysis and Hydropyrolysis, in M. A. Elliott (ed.), *Chemistry of Coal Utilization*, Second Supplementary Volume, Wiley, New York, 1981, pp. 665–784.

Johnson, J. L., Fundamentals of Coal Gasification, in M. A. Elliott (ed.), *Chemistry of Coal Utilization*, Second Supplementary Volume, Wiley, New York, 1981, pp. 1491–1598.

Jones, L., and N. C. Li, Aging of SRC-II Middle Distillate from Illinois No. 6 Coal. 1. In Presence of Copper and Oxygen, *Fuel*, **62,** 1156–1160 (1983).

Kalkwarf, D. R., D. L. Stewart, R. A. Pelroy, and W. C. Weimer, *Photodegradation of Mutagens in Solvent-Refined Coal Liquids*, PNL-4982, National Technical Information Service, Springfield, VA, 1984.

Karn, F. S., and F. R. Brown, *Aging Studies of the Synthoil Product, Part II: Effects of Time, Temperature, Atmosphere*, PERC/TPR-76/2, National Technical Information Service, Springfield, VA, 1976.

Kastens, M. L., L. L. Hirst, and C. C. Chaffee, Liquid Fuel from Coal, *Ind. Eng. Chem.*, **41,** 870–885 (1949).

Mann, R. M., G. E. Harris, W. R. Menzies, A. J. Simon, and W. A. Williams, *Environmental Health and Safety Database for the KRW Coal Gasification Process Development Unit*, GRI-85/0123, Gas Research Institute, Chicago, IL, 1985.

Nakles, D. V., M. J. Massey, A. J. Forney, and W. P. Haynes, *Influence of Synthane Gasifier Conditions on Effluent and Product Gas Production*, PERC/RI-75/6, National Technical Information Service, Springfield, VA, 1975.

Nelson, W. L., *Petroleum Refinery Engineering*, 4th ed., McGraw-Hill, New York, 1969.

Parekh, R. D., *Handbook of Gasifiers and Gas Treatment Systems*, WD-TR-82/008-010, National Technical Information Service, Springfield, VA, 1982.

Probstein, R. F., and R. E. Hicks, *Synthetic Fuels*, McGraw-Hill, New York, 1982.

Rhodes, F. O., The History of Coal Tar and Light Oil, in A. J. Hoiberg (ed.), *Bituminous Materials: Asphalt, Tars, and Pitches*, Vol. III, *Coal Tars and Pitches*, Interscience, New York, 1966, pp. 1–33.

Riedl, F. J., and A. J. deRosset, *Upgrading of Coal Liquids, Interim Report, Hydrotreating and Fluid Catalytic Cracking of EDS Process Derived Gas Oil*, FE-2566-30, National Technical Information Service, Springfield, VA, 1979.

Satterfield, N., *Heterogeneous Catalysis in Practice*, McGraw-Hill, New York, 1983.

Sullivan, R. F., *Upgrading Coal Liquids*, ACS Symposium Series 156, American Chemical Society, Washington, DC, 1981.

von Fredersdorff, C. G., and M. A. Elliott, Coal Gasification, in H. H. Lowry (ed.), *Chemistry of Coal Utilization*, Supplementary Volume, Wiley, New York, 1963, pp. 892–1022.

Weimer, W. C., and C. W. Wright, *Decommissioning Samples from the Ft. Lewis, WA Solvent Refined Coal Plant: Chemical Analysis and Biological Testing*, PNL-5556, Pacific Northwest Laboratory, Richland, WA. National Technical Information Service, Springfield, VA, 1985.

White, C. M., L. Jones, N. C. Li, and K. C. Tewari, Aging of SRC-II Middle Distillate from Illinois No. 6 Coal. 2. Further Evidence of Phenolic Oxidative Coupling in Coal-Derived Liquids, *Fuel*, **62,** 1397–1403 (1983).

Wilzbach, K. E., and C. A. Reilly (eds.), *Status of Health and Environmental Research Relative to Coal Gasification: 1976 to the Present*, DOE/ER-0149, National Technical Information Service, Springfield, VA, 1982.

Wright, C. W., and W. C. Weimer, *Chemical and Biological Stability of Solvent Refined Coal Liquids*, PNL-4962, National Technical Information Service, Springfield, VA, 1984.

Wright, C. W., W. C. Weimer, and D. S. Daly, Stability Studies on the Chemical Composition and Biological Activity of Coal-Derived Liquids, in M. Cook and A. J. Dennis (eds.), *Polycyclic Aromatic Hydrocarbons: Mechanisms, Methods and Metabolism*, Battelle Press, Columbus, OH, 1984, pp. 1441–1455.

Wu, W. R. K., and H. H. Storch, *Hydrogenation of Coal and Tar*, Bureau of Mines Bulletin 633, U.S. Department of the Interior, Washington, DC, 1968.

Young, R. J., Early U.S. Coal Hydrogenation Projects: The Technology and Carcinogenic Risks, in W. N. Rom and V. E. Archer (eds.), *Health Implications of New Energy Technologies*, Ann Arbor Science, Ann Arbor, MI, 1979, pp. 319–339.

CHAPTER 3

Toxicological Effects of Coal-Based Synfuels

C. A. Reilly, Jr.
Argonne National Laboratory, Argonne, Illinois

R. A. Renne
Battelle, Pacific Northwest Laboratory, Richland, Washington

A national goal in the mid-1970s was to become energy independent by 1990. Consequently, research was strongly stimulated on the development of synthetic fuels from raw materials indigenous to the United States (coal, shale, and biomass). Substantial effort has been devoted in the last two decades to investigating the use of coal-liquefaction and coal-gasification processes to produce petroleum substitutes. Concurrently with engineering development, the Energy Research and Development Administration and its descendant, the Department of Energy (DOE), and the Environmental Protection Agency (EPA) sponsored programs to evaluate potential toxicological and environmental effects associated with developing a synthetic fuels industry. The National Institute of Environmental Health Sciences and the National Institute of Occupational Safety and Health also sponsored health-related programs in this area. In addition, substantial toxicological and industrial hygiene programs were conducted by private industry (e.g., Gulf, Exxon, Texaco, and Institute of Gas Technology) and academia.

This chapter describes the toxicological database on coal-conversion materials obtained from these sources, including information on genotoxicity, carcinogenicity, teratology and embryotoxicity, acute organ toxicity, and metabolism, dosimetry, and biological transport. It also presents results from tiered approaches used to evaluate genotoxicity, wherein the initial use of rapid *in vitro* screening assays to identify materials with significant genotoxicity is followed by additional tiers of more sophisticated assays to define better the genotoxicity and potential for carcinogenicity. Attempts were made to determine the chemical classes responsible for the genotoxicity of some of the genotoxic raw materials, and results of these studies are reported here. Toxicologically active components were identified through iterative chemical class fractionation of the raw materials and toxicological assay of the resulting fractions. Selection of materials to be further fractionated and fractionation methods to be used were based on results of the toxicological studies. A more in-depth discussion of this *biodirected* approach for identification of toxic components within complex chemical mixtures is found in Chapter 5. Results of various *in vitro* and *in vivo* assays of selected materials are compared and discussed with respect to variability of results among assays, limitations, and usefulness of each assay in predicting human health effects. Data gaps or weaknesses are also identified to serve as a basis for determining future research directions. Results and conclusions are summarized at the end of each section.

3.1 PURPOSE AND BACKGROUND

Processes that convert coal to other energy forms have been used for many years. Gasification of coal was a prime energy source for utility gas at the turn of the century and the Bergius process for the direct hydrogenation of coal was used in Germany to produce bunker and transportation fuel during World War II. However, in more recent times, replacement of coal by petroleum-based fuels has greatly reduced industrial use of coal-conversion processes.

Even though there had been substantial historical experience with coal-based fuels, when there was a resurgence of interest in synthetic fuels during the 1970s, there was a paucity of data to predict associated toxic effects. For example, no reliable information is available regarding health risks associated with the German coal-liquefaction industry (Young, 1980). Some information exists for early U.S. operations, which started in the late 1940s and continued until the early 1960s in Bruceton, Pennsylvania; Louisiana, Missouri; and Institute, West Virginia. Hueper conducted skin-painting studies with mice on materials from the Bruceton, Pennsylvania (Hueper, 1953) and Louisiana, Missouri (Hueper, 1956) pilot plants. Three materials were obtained from the Bruceton plant: *centrifuge residue*, a semisolid material; a *heavy oil* of moderate viscosity; and a *light oil* of low viscosity. All three materials were carcinogenic; centrifuge residue and heavy oil were more active than light oil. Nine materials were obtained from the plant in Louisiana, Missouri. These materials, ranging from *finished gasoline* to highly viscous *centrifugation residue*, had a wide range of boiling points. All materials except finished gasoline were carcinogenic. Some correlation was evident between boiling point (bp) and relative carcinogenic potency.

Industrial hygiene studies at the Institute, West Virginia pilot plant identified benzo[a]pyrene (BaP) and other polycyclic aromatic hydrocarbons (PAH) as workplace risks (Ketcham and Norton, 1960). Epidemiological (Sexton, 1960; Palmer, 1979; Young, 1980) and experimental animal (Weil and Condra, 1960) data are also available from the Institute pilot plant. Among 359 exposed workers examined over a period of approximately 5 years, precancerous skin lesions were observed on 42 people, and 11 cancerous lesions (five confirmed histopathologically) were observed on 10 additional people (Sexton, 1960). Weil and Condra (1960) evaluated the carcinogenic potential in mice of 15 products or process materials from the Institute facility. Four materials were carcinogenic: middle oil stream (500–600°F bp range), light oil stream residue (500–715°F bp range), pasting oil (600–850°F bp range), and pitch product (>850°F bp range). In terms of tumor incidence and latency, pasting oil was the most potent carcinogen of the materials tested.

No reliable information on potential health effects associated with coal gasification is available from the early industrial uses of coal. Gasification of coal was used for fuel in the steel industry into the 1950s, and epidemiological information on health risks is available for this industry. Initial evidence for an increased risk of cancer associated with coal gasification came from studies of gas workers in the Japanese steel industry in the 1930s (Kuroda and Kawahata, 1936). Subsequent epidemiological studies conducted in Japan, England, and the United States demonstrated that gas workers experienced increased incidences of lung and bladder cancers (Kawai et al., 1967; Lloyd, 1971; Doll et al., 1972).

Many toxicological studies were conducted over the last decade to determine the potential health effects of the complex chemical mixtures found in coal-conversion processes. These studies often included substantial efforts in chemical characterization of the mixtures (see Chapter 5) to identify components responsible for the toxicity. Because of the chemical complexity of the materials being studied, a variety of chemical and toxicological methods and approaches were applied. Results of these studies have been reported in a range of sources, for example, peer-reviewed open literature, symposium proceedings, and contractor status reports. This chapter attempts to bring together the results of all available toxicological studies, so that the information is available from a single source and comparison of materials as well as methods is facilitated.

3.2 APPROACH AND RATIONALE

The chemical composition of coal-conversion materials and historical inferences from epidemiology and toxicity studies of earlier coal-conversion technologies focused the emphasis of recent toxicity studies on determining the carcinogenicity of appropriate materials. Chronic exposure of experimental animals, usually rodents, via the most appropriate route (e.g., dermal application, feeding, inhalation) has been the classical method of assaying for carcinogenicity of chemicals (Clayson, 1962). However, time and cost factors made it impractical to use chronic-exposure animal studies for the assay of all coal-conversion materials with any potential for human exposure. Therefore, a series of short-term, in vitro assays were used to determine the genotoxicity of test materials, and from the results of these assays, materials were chosen for carcinogenicity assay in in vivo (whole-animal) systems. This tiered approach (Fig. 3.1) included an initial screening with a rapid (days) and relatively inexpensive microbial in vitro assay designed to identify those materials, or fractions thereof, that were most likely to be toxic. This was followed by a second tier of longer-term (weeks) in vitro assays using cultured mammalian cells. The chemical complexity of the test materials and the broad spectrum of potential toxic effects required a variety of assays with different sensitivities and endpoints for confident identification of toxic materials. A final tier of assays was then used to evaluate the toxicity of the most relevant materials in whole-animal systems (months/years). Because of the uncertainty of the in vivo predictive capability of the first- and second-tier in vitro assays, a number of materials were also evaluated for toxicity in whole animals solely for purposes of validating in vitro results.

The in vitro assay most widely employed as the first-tier screening procedure was the Ames *Salmonella*/mammalian microsome mutagenicity assay (Ames et al., 1975). Many fewer samples were assayed in tier two (mammalian cell) systems, and there was no prevalence of data from any one assay; several mammalian cell assays were used. In addition to measurements of toxicity, mammalian cell assay systems were used to detect and measure cytotoxicity (cell death or dysfunction), an indicator for potential acute toxicity. Tier three (whole-animal) assays used for detection of carcinogenicity included repeated chronic dermal application and initiation/promotion studies in mice and inhalation studies in rats. In addition, a variety of acute and subchronic toxicity studies were performed. These included LD_{50} (the statistically derived dose that is lethal to 50% of exposed animals) studies, dermal

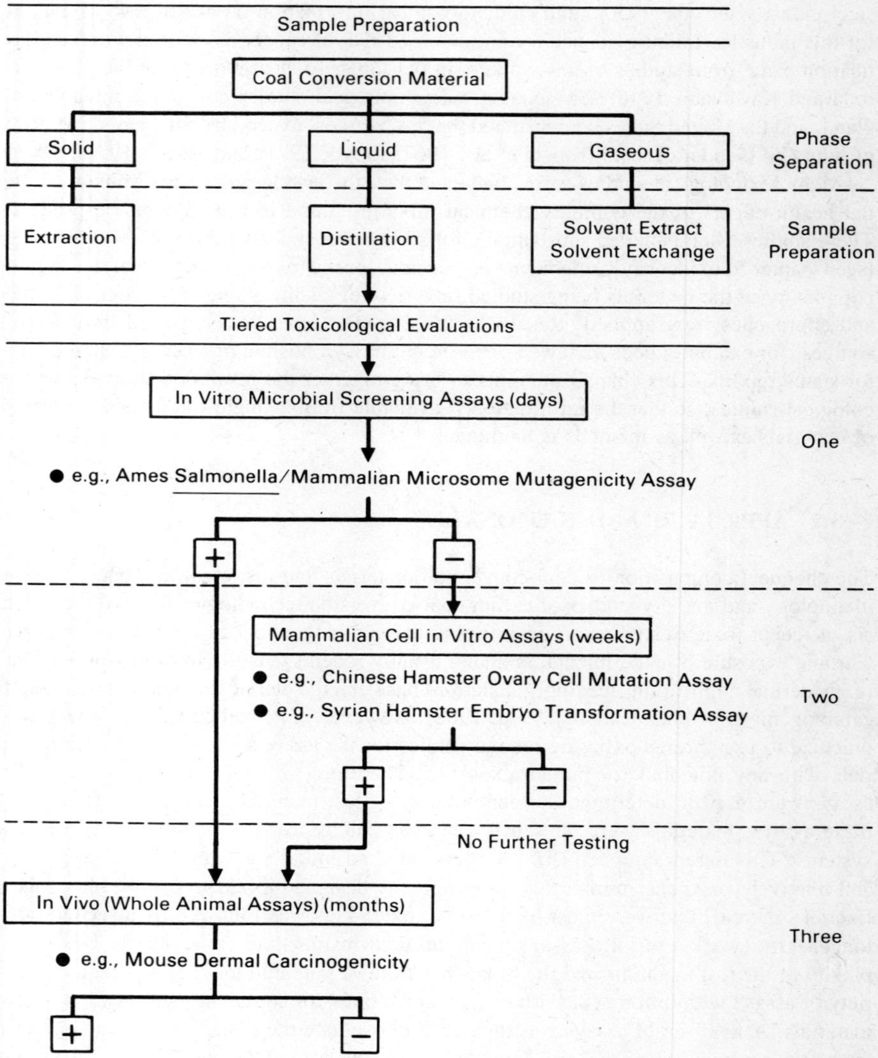

Figure 3.1 Schematic of sample preparation and tiered approach to toxicological evaluation of coal-conversion materials.

and ocular irritation, dermal sensitization studies, and oral and inhalation studies for organ toxicity other than carcinogenesis.

Of the coal-conversion materials evaluated for toxicity during the last decade and discussed in this chapter, most came from bench-scale (lb coal/h) or pilot plant (25–250 tons coal/day) facilities. Only a few materials came from industrial or commercial-scale facilities. The materials studied were chemically complex; many had solid, liquid, and/or gaseous phases. Thus, some form of phase separation was often required before toxicology studies could be performed (Fig. 3.1). After the initial separation, individual phases sometimes required additional manipulations to make them suitable for a given test procedure. For example, a strong organic solvent, such as methylene chloride, was required to extract organic components from overhead char particles collected from coal-gasification process streams. However, the severe cytotoxicity of methylene chloride required its removal and replacement with a more biocompatible solvent, such as dimethyl sulfoxide (DMSO), for the toxicity assay.

After materials of concern (i.e., significant toxicological effects, high potential for human exposure, possible bioaccumulation) were identified in screening assays, a more limited number of materials were evaluated in depth. Generally, these evaluations involved biodirected chemical characterization, that is, iterative chemical fractionations coupled with toxicological monitoring (short-term tests) to identify components responsible for the toxic effect.

3.3 DATABASE

3.3.1 Materials

Materials discussed in this chapter include major coal-conversion products and by-products; waste materials from coal conversion, such as ashes, slags, cyclone fines, and quench waters, are not discussed here, although they may contain toxic components. Human exposure to these waste materials would generally be indirect (e.g., leaching of toxic components from waste materials and subsequent aqueous passage to humans). Handling potentially hazardous wastes has been the focus of numerous control technology studies, and recent reviews are available (DOE, 1982a).

The number of products and by-products from coal gasification is more limited than that from coal liquefaction. Coal gasification results in a variable product of mixed synthesis gases (e.g., CO, CO_2, N_2, H_2, NH_3, CH_4, H_2S). Acute toxicity of these gases is well known. The ill-defined health hazards associated with gasifiers principally reflect occupational exposures to higher-molecular-weight organic constituents in fugitive aerosol emissions and condensation by-products from the cleanup of product gas streams. Coal-liquefaction processes present a somewhat broader exposure potential for the work force and general population because of the broad range of liquids produced. Products included materials similar to those already in commerce (e.g., naphtha and light gas oils) and some that are unique [e.g., solvent-refined coal I (SRC-I) solid products].

Materials were selected for study on the basis of criteria such as likelihood of human exposure, relevance to large-scale operation, quantities produced, and chemical composition. Fractions of some of the materials found to be toxic were also studied to identify chemical classes responsible for the observed toxicity. The actual materials studied are identified by a common name and a five-character code (see appendix at end of Chapter 4). Use of the code allows simplified identification and cross reference of materials discussed in other chapters. Additional information on materials can be found in the appendix at the end of this chapter.

Four types of coal-liquefaction material were characterized toxicologically: process streams, by-products, crude products, and upgraded products. Process streams consist primarily of recycle solvents that are used to slurry coal or largely unconverted material into a process unit. SRC-I process solvent (PS) and the nonintegrated two-stage liquefaction (NTSL) hydrotreater recycle solvent are examples. By-products are formed incidentally to the crude product or during product upgrading (e.g., vacuum still bottoms). Crude products, as defined here, represent the total mix of nonupgraded materials, such as H-Coal syncrude product, prior to any upgrading steps such as distillation or hydrotreatment. Upgraded products are the commercial materials derived from the crude product by procedures such as selective distillations or hydrotreatment. Table 3.1 lists the 176 coal-liquefaction materials and 24 hydrotreated materials on which one or more toxicity assays were performed.

Table 3.1 Coal-Liquefaction Materials Evaluated for Toxicity and Identification of the Assays Used for Each Material

		Toxicity Assay[a]		
Material	Sample No.	Tier One	Tier Two	Tier Three
Solvent-Refined Coal I				
Light oil	L2152	1		
	L2211	1	9	16, 19
Wash solvent	L2142	1		
	L2212	1	9	16, 19
Process solvent	L2111	1		
	L2121[b]	1–3, 5	8, 9	17
	L2132	1		
	L2213	1	9, 14	16, 18–20
		1		
Recycle coke condensate oil	L2161	1		
Solid product	L2221	1		
Solvent-Refined Coal II				
Light distillate	L1241	1	9, 14	16, 18, 19
Naphtha	L1231	1		17
Middle distillate	L1242	1	9, 14	16, 19, 20
	L1264	6	11	19, 20, 22–24
Heavy distillate	L1243	1	9, 14	16–20
Blends	L1111[b]	1–3, 5	8, 9	16, 17, 18
	L1121[b]	1, 5	9	
	L1131[b]	1, 5	9	
	L1211	1		18
	L1221	1		
	L1251[b]	1	8, 9	17
	L1261			16, 19
	L1262	1		16, 18, 19
	L1311	1		
	L1321	1		
	L1331	1		
	L1341	1		
	L1351	1		
Fuel oil	L1232	1		17
Heavy gas/oil	L1233	1		17
Stripper tower bottoms	L1234	1		17
	L1263	6	11	19, 22–24
Vacuum tower bottoms	L1235	1		17
	L1271	1		
	L1272	1		
	L1281	1	15	
Residue	L1236	1		17
Coal slurry	L1265			19, 22–24

Table 3.1 (*Continued*)

Material	Sample No.	Toxicity Assay[a]		
		Tier One	Tier Two	Tier Three
H-Coal				
Atmospheric still overhead	L3121	1		17
	L3131	1		17
	L3141			18
	L3151	1		18, 19, 22
Atmospheric still bottoms	L3122	1		17
	L3134	1		17
	L3142	1		18, 19
	L3152	1		18, 19
	L3161	1		
Vacuum still overhead	L3123	1		17
	L3135	1		17
	L3143	1	8	18, 19
	L3153	1		18, 19
	L3162	1		
	L3314	1		
	L3324	1		
Vacuum still bottoms	L3124	1		17
	L3136	1		17
	L3144	1		
	L3154	1		
	L3315	1		
	L3323	1		
	L3325	1		
Naphtha/light oil	L3132	1		17
	L3133	1		17
	L3211			18
	L3241	1		16, 18
	L3251	1		16
	L3311	1		
	L3321	1		
Hydroclone overhead	L3214	1		22
	L3247			22
	L3255	1		19, 22, 24
Light slop oil	L3218	1		19, 22–24
	L3246			19, 22–24
	L3256			19, 22–24
Heavy slop oil	L3219	1		22
	L3257			19, 22, 24
Separator bottoms	L3215	1		
	L3312	1		
	L3313	1		
	L3322	1		
Atmospheric stripper	L3254	1		
Recycle oil	L3217	1		
Waste-oil residual	L3210	1		
Light fuel oil	L3212	1		18, 19, 22–24
	L3242	1		16, 18, 19, 22–24
	L3252	1		16, 19, 22–24

Table 3.1 (*Continued*)

Material	Sample No.	Toxicity Assay[a]		
		Tier One	Tier Two	Tier Three
Heavy fuel oil	L3213	1		18, 19, 22–24
	L3234[b]	1		
	L3243	1		16, 18, 19, 22–24
	L3253	1		16, 19, 22–24
Blend	L3112	1		18, 19, 22
	L3231	1	8	16, 19
	L3232	1		16, 19, 25
	L3244[b]	1		
Vacuum flaked product	L3216	1		
	L3245	1		
EDS				
Naphtha	L4114			19, 20
	L4115			19, 20, 22, 23
	L4131	1		
	L4135			20
	L4211	1		17
Recycle solvent	L4111	1		17, 19, 20, 22, 23
	L4121			22, 23
	L4132	1		19
	L4212	1		17
	L4242	1		
	L4243	1		
	L4311			
Vacuum gas oil	L4112	1		
	L4213	1		17
	L4241	1		
Heavy gas oil	L4214	1		17
Vacuum bottoms/residue	L4133	1		
	L4215	1		17
Blend	L4113			19, 22, 23
	L4116	1		16, 19
	L4134[b]	1, 2		17
	L4221	1		
	L4231	1		
Integrated Two-Stage Liquefaction (ITSL)				
LC Finer feed	L5111	1	8	17
	L5121	1		17
	L5131	1		
LC Finer feed product	H0611	1	8	17
	H0621[b]	1		17
	H0631	1		
	H0641	1		
Short-Contact-time (SCT) heavy oil	L5132	1		
Recycle solvent	H0632	1		
Solvent bottoms	L5133	1		
Gasifier feed	L5134	1		

Table 3.1 (*Continued*)

Material	Sample No.	Toxicity Assay[a]		
		Tier One	Tier Two	Tier Three
Light fuel oil	L5311	1		
	L5321	1		
	L5331	1		
	L5341	1		
	L5351	1		
	L5361	1		
Distillate process solvent	L5315	1		
	L5325	1		
	L5335	1		
	L5345	1		
	L5355	1		
	L5365	1		
First-stage distillate	L5312	1		
	L5342	1		
Deashed product	L5314	1		
	L5324	1		
	L5334	1		
	L5344	1		
	L5354	1		
	L5364	1		
Critical solvent deashing (CSD) product	L5313	1		
	L5323	1		
	L5333	1		
	L5343	1		
	L5353	1		
	L5363	1		
Hydrotreater (HTR) feed	L5316	1		
	L5326	1		
	L5336	1		
	L5346	1		17
	L5356	1		
HTR light fuel oil	H0572	1		
	H0582	1		
	H0592	1		
HTR distillate bottoms	H0573	1		
	H0583	1		17
	H0593	1		
HTR recycle pasting solvent	H0544	1		
	H0554	1		
	H0564	1		
	H0574	1		
	H0584	1		
Total liquid product	H0633			17
Nonintegrated Two-Stage Liquefaction (NTSL)				
Light fuel oil	L5211	1		
First-stage product	L5212	1		
distillate	L5241[b]	1	10	16, 19, 22–24
Deashed product	L5214	1		

Table 3.1 (*Continued*)

		Toxicity Assay[a]		
Material	Sample No.	Tier One	Tier Two	Tier Three
CSD product	L5213	1		
Solid product	L5232	1	10	20, 22–24
HTR feed	L5216	1		17
HTR light fuel oil	H0512	1		
	H0521	1		22–24
Second-stage blend	H0524[b]	1	10	16, 19, 22–24
Residual fuel	H0522	1		
HTR distillate bottoms	H0513	1		17
HTR vacuum bottoms	H0514	1		
HTR solid product	H0523	1	10	20, 22–24
Catalytic Two-Stage Liquefaction (CTSL)				
First-stage product	L5411	1		
	L5412	1		
	L5413	1		
	L5414	1		
Second-stage product	L5415	1		
	L5416	1		
	L5417	1		
	L5418	1		
Veba Oil				
Middle distillate	L6211	1		
Heavy distillate	L6212	1–3, 5		
Synthoil				
Product oil	L6111	1	8	18, 22, 23
Zinc Halide Hydrocracking				
Distillate	L6311	1		
Hydrocarbonization				
Product oil	L6411	1		

[a] 1 = Ames *Salmonella*/mammalian microsome mutagenicity assay.
2 = *Salmonella* fluctuation assay.
3 = *Salmonella* forward mutation assay.
5 = Prophage induction assay.
6 = Yeast (*Saccharomyces*) forward mutation assay.
8 = Chinese hamster ovary (CHO) cell mutation assay.
9 = Syrian hamster embryo (SHE) cell transformation assay.
10 = Mouse lymphoma mutagenesis assay.
11 = Unscheduled DNA synthesis (UDS).
14 = Mammalian cell cytotoxicity.
15 = Alveolar macrophage assay.
16 = Teratogenicity/embryotoxicity.
17 = Dermal carcinogenicity (initiation/promotion).
18 = Dermal carcinogenicity (chronic).
19 = Oral or intraperitoneal (IP) LD_{50}.
20 = Inhalation.
22 = Dermal irritation.
23 = Ocular irritation.
24 = Hypersensitivity.
25 = Mouse spot test.

[b] Some toxicological determinations made on discrete boiling-point cuts.

Three types of coal-gasification material were studied: tars, oils, and product gases. Gasification tars are high-molecular-weight materials collected from gas cleanup devices [e.g., electrostatic precipitators (ESPs), tar traps, gravity separators] and heavier fractions of raw product gas condensates. Oils include organic by-products of intermediate molecular weight (e.g., the light layer from gravity separators, venturi scrubber condensates, and intermediate fractions of raw product gas condensates). Product gas is a low-molecular-weight, cleaned gas, either methane or a more-complex synthesis gas.

Although the total number of gasification materials (194) assayed for toxicity was similar to the number of liquefaction materials, over half of the materials were gas stream condensates from a single bench-scale experimental gasifier (Table 3.2). Detailed descriptions of specific liquefaction and gasification samples for which toxicological data are available are presented in the appendix at the end of this chapter.

Table 3.2 Coal-Gasification Materials Evaluated for Toxicity and Identification of the Assays Used for Each Material

		Toxicity Assay[a]		
Material	Sample No.	Tier One	Tier Two	Tier Three
Morgantown Energy Technology Center Gasifier				
Tar	G1211	1		
	G1221	1	15	
	G1227	1		
	G1231	1		
Gas condensate	G1212	1	8, 12	
	G1213	1	8, 12	
	G1222	1	15	
	G1223	1	15	
	G1224	1	14, 15	
	G1225	1	15	
	G1226	1		
	G1232	1		
	G1233	1		
	G1234	1		
	G1235	1		
	G1236	1		
	G1242	1		
University of Minnesota–Duluth (UMD) Gasifier				
Electrostatic precipitator tar	G1301	1, 6		19, 22–24, 26
	G1302	1		
	G1303	1		
	G1304	1		
	G1305	1		
	G1306	1, 6		19, 21–24
	G1307	1		
	G1308	1		
	G1310	1		
	G1311	1		
	G1312	1		
	G1313	1		

Table 3.2 (*Continued*)

Material	Sample No.	Toxicity Assay[a] Tier One	Tier Two	Tier Three
	G1314	1		
	G1315	1		
	G1316	1		
	G1317	1		
	G1318	1		16, 18, 19,
	G1319	1, 6		21–26
	G1320	1		
	G1321	1		
	G1322	1		
	G1323	1		
	G1327			16
Holding tank tar	G1324	1		
Vent gas cyclone tar	G1325	1		
Dirty top gas tar	G1326	1		
Kosovo Gasifier				
By-product tar	G1412	1	14	
Heavy tar and dust	G1411	1	14	
Medium oil	G1413	1	14	
Light oil	G1414	1	14	
University of North Dakota Energy Research Center Gasifier				
Tar	G1121	1	8, 12, 15	17, 18, 22, 23
	G1122	1	12, 15	
	G1123	1	12, 15	
	G1124	1	12, 15	
Oil	G1125	1	12, 15	
	G1126	1		
	G1127	1	12	
Gas condensate	G1111	1	12, 15	
Chapman Low-Btu Gasifier (Holston)				
Tar/oil	G1811	1		
	G1821	1		
Glen-Gery Brick Company Low-Btu Gasifier				
Gas condensate	G1611	1		
	G1612	1		
Ft. Snelling Low-Btu Gasifier				
Gas condensate	G1711	1	14	
	G1712	1	14	
Hygas Gasifier				
Recycle oil	G2111	1		23, 24
	G2112	1	12	

Table 3.2 (*Continued*)

Material	Sample No.	Toxicity Assay[a]		
		Tier One	Tier Two	Tier Three
	G2113	1	12, 15	19, 22
	G2122	1	12, 15	
	G2123	1		
	G2141	1		
	G2151	1		18
Char/oil slurry	G2114	1	15	22
	G2115	1		
	G2116	1	12	
	G2124	1	15	
	G2125	1	12, 15	23
	G2131	1	15	
	G2142	1		
Light oil recovery system (LORS) oil	G2143	1		
Low-temperature reactor (LTR) condensate	G2117	1		
	G2118	1	12, 15	
	G2119	1	12, 15	23, 24
	G2126	1	12, 15	
	G2127	1		
	G2132	1	12, 15	
	G2144	1	15	
Hoe Creek (In Situ Gasification)				
Tar	G4111	1		
	G4121	1	8, 12, 13	
Research Triangle Institute Gasifier				
Tar	G1511	1	14	
	G1521	1		
	G1531	1	14	
	G1541	1	14	
	G1551	1		
	G1561	1		
	G1571	1		
	G1581	1		
COED Pyrolysis Oil				
Oil	G2211	1		18, 19, 22, 23
	G2221	1, 4, 7	10	
Mellon Institute				
Gas condensate	G3144HT	1		
	G3145HT	1		
	G3146HT	1		
	G3147HT	1		

Table 3.2 (*Continued*)

Material	Sample No.	Toxicity Assay[a] Tier One	Tier Two	Tier Three
	G3148HT	1		
	G3153HT	1	12, 15	
	G3154HT	1		
	G3155HT	1		
	G3156HT	1		
	G3159HT	1		
	G3160HT	1		
	G3161HT	1		
	G3162HT	1		
	G3163HT	1		
	G3164HT	1		
	G3165HT	1		
	G3166HT	1		
	G3167HT	1		
	G3168HT	1		
	G3169HT	1		
	G3170HT	1	12, 15	
	G3171HT	1		
	G3172HT	1		
	G3173HT	1		
	G3174HT	1		
	G3175HT	1		
	G3176HT	1		
	G3178HT	1		
	G3179HT	1		
	G3180HT	1		
	G3181HT	1		
	G3182HT	1		
	G3186HT	1		
	G3187HT	1		
	G3188HT	1		
	G3189HT	1		
	G3190HT	1		
	G3191HT	1		
	G3192HT	1		
	G3193HT	1		
	G3194HT	1		
	G3195HT	1		
	G3144RT	1		
	G3145RT	1		
	G3146RT	1		
	G3147RT	1		
	G3148RT	1		
	G3153RT	1	12, 15	
	G3154RT	1		
	G3155RT	1		

Table 3.2 (*Continued*)

Material	Sample No.	Toxicity Assay[a] Tier One	Tier Two	Tier Three
	G3156RT	1		
	G3159RT	1		
	G3160RT	1		
	G3161RT	1		
	G3162RT	1		
	G3163RT	1		
	G3164RT	1		
	G3165RT	1		
	G3166RT	1		
	G3167RT	1		
	G3168RT	1		
	G3169RT	1		
	G3170RT	1	12, 15	
	G3171RT	1		
	G3172RT	1		
	G3173RT	1		
	G3174RT	1		
	G3175RT	1		
	G3176RT	1		
	G3178RT	1		
	G3180RT	1		
	G3181RT	1		
	G3182RT	1		
	G3184RT	1		
	G3186RT	1		
	G3187RT	1		
	G3188RT	1		
	G3189RT	1		
	G3190RT	1		
	G3191RT	1		
	G3192RT	1		
	G3193RT	1		
	G3194RT	1		
	G3195RT	1		
	G3146LO	1		
	G3147LO	1		
	G3148LO	1		
	G3153LO	1		
	G3154LO	1		
	G3155LO	1		
	G3156LO	1		
	G3159LO	1		
	G3160LO	1		
	G3161LO	1		
	G3163LO	1		
	G3164LO	1		

Table 3.2 (*Continued*)

Material	Sample No.	Toxicity Assay[a] Tier One	Tier Two	Tier Three
	G3165LO	1		
	G3166LO	1		
	G3167LO	1		

[a] 1 = Ames *Salmonella*/mammalian microsome mutagenicity assay.
4 = Bacterial DNA-repair assay.
6 = Yeast (*Saccharomyces*) forward mutation assay.
7 = Yeast (*Saccharomyces*) mitotic recombination.
8 = Chinese hamster ovary (CHO) cell mutation assay.
10 = Mouse lymphoma mutagenesis assay.
12 = Sister chromatid exchange (SCE).
13 = Cytogenetics.
14 = Mammalian cell cytotoxicity.
15 = Alveolar macrophage assay.
16 = Teratogenicity/embryotoxicity.
17 = Dermal carcinogenicity (initiation/promotion).
18 = Dermal carcinogenicity (chronic).
19 = Oral or intraperitoneal LD_{50}.
21 = Mouse lung adenoma assay.
22 = Dermal irritation.
23 = Ocular irritation.
24 = Hypersensitivity.
25 = Mouse spot test.
26 = Mouse dominant lethal mutation.

3.3.2 Toxicity Assays

Of the nearly 400 coal-conversion materials evaluated for toxicity, 96% were studied with a tier one assay (Table 3.3). The Ames *Salmonella*/mammalian microsome mutagenicity assay was used on all but two of the materials. In contrast, only 12 materials were studied with any other tier one assay.

Many fewer materials were studied using tier two assays, with no single assay being used with a majority of the 65 materials tested. The number of materials used with any of the nine different assays varied from one material for which cytogenetic effects were evaluated to 27 materials for which cytotoxicity in alveolar macrophages was determined.

The number of materials studied with tier three assays was similar to tier two. Of the 98 materials evaluated in tier three assays, 52 were studied with multiple assays; one material was evaluated with nine different assays. Materials in this group generally had assays that measured short-term acute effects (e.g., oral LD_{50}, dermal sensitivity), although a number of the materials were studied in dermal carcinogenicity assays. Thirty-seven materials were studied only in a dermal carcinogenicity assay. A total of 59% of the materials were evaluated for dermal carcinogenicity using either a chronic or initiation/promotion protocol. Because of the potential relation between mutagenicity, in vitro cell transformation, and carcinogenesis, and the relative size of their respective databases, the best possibility for correlations between in vitro and in vivo assays should be with the Ames assay (tier one), the Chinese hamster ovary (CHO) cell mutation and Syrian hamster embryo (SHE) cell transformation assays (tier two), and dermal carcinogenicity (tier three).

Table 3.3 Assays Used for Toxicological Evaluation of Coal-Conversion Materials and the Number of Materials Studied with Each Assay

Assay	No. Samples Tested
Tier One—Total Number of Samples with One or More Tier One Assays	379
Ames *Salmonella*/mammalian microsome mutagenicity assay	377
Salmonella fluctuation assay	4
Salmonella forward mutation assay	3
Bacterial DNA-repair assay	1
Prophage induction assay	5
Yeast (*Saccharomyces*) forward mutation assay	5
Yeast (*Saccharomyces*) mitotic recombination	1
Tier Two—Total Number of Samples with One or More Tier Two Assays	65
Chinese hamster ovary (CHO) cell mutation assay	12
Syrian hamster embryo (SHE) cell transformation assay	11
Mouse lymphoma mutagenesis assay	5
Unscheduled DNA synthesis (UDS)	2
Sister chromatid exchange (SCE)	23
Cytogenetics	1
Mammalian cell cytotoxicity	14
Alveolar macrophage assay	27
Tier Three—Total Number of Samples with One or More Tier Three Assays	98
Teratogenicity/embryotoxicity	22
Dermal carcinogenicity (initiation/promotion)	37
Dermal carcinogenicity (chronic)	24
Oral or intraperitoneal LD_{50}	43
Inhalation toxicity	10
Mouse lung adenoma	2
Dermal irritation	36
Ocular irritation	30
Hypersensitivity	24
Mouse spot test	2
Mouse dominant lethal mutation	2

3.3.3 Database Evaluation and Conclusions

The available toxicity data on nearly 400 coal-conversion materials, including data for some materials obtained with assays from all three tiers (microbial, mammalian cell, and intact animal assays), make comparisons and evaluations possible. A discussion on correlations among assays is presented in Section 3.4.3. Such comparisons are hampered, however, by specific differences among materials and the limitations of the various bioassay procedures. Notwithstanding these constraints, the greatest potential for comparison of the available toxicity data resides in the mutation assays in microbial and mammalian cell systems (tiers one and two) and whole-animal carcinogenicity assays (tier three). Some general conclusions are warranted:

- The Ames assay was found to be a reasonably good predictor of carcinogenicity of crude materials; those crude materials mutagenic in the Ames assay were for the most part also carcinogenic for mouse skin. However, when crude materials were separated into chemical class fractions, the Ames assay did not correlate well with carcinogenicity.

- The concentration of high-molecular-weight components of a given material was highly correlated with the magnitude of the genotoxic response, with higher genotoxic activities found in materials with greater concentrations of higher-molecular-weight components (i.e., bp >700°F).
- Mutagenic activities of nonfractionated materials were not exceptionally high, with materials generally containing a few components with high specific mutagenic activity.
- The observed genotoxic response was the net effect of the various components of the material, and thus correlations observed between assays may be fortuitous. This is particularly true for the Ames assay, which is particularly sensitive to aromatic amines. Thus, the amine-rich basic fraction, a small proportion of the weight of most materials, was often responsible for the bulk of mutagenic activity in a crude material. Consequently, when crude materials were separated into chemical class fractions, quantitative Ames data for these fractions did not correlate well with carcinogenicity data. The dermal carcinogenicity responses in tier three mouse assays were higher with PAH-enriched fractions than with nitrogen-containing polycyclic aromatic compound (N-PAC) fractions.
- Chemical class fractionation was required to obtain reproducible results with mammalian cell assays (tier two assays) because of the cytotoxicity of most crude materials (e.g., cell death, alterations in cell growth, loss of cell function). Of the materials studied, cytotoxicity was generally greater for materials with lower average molecular weight.
- In contrast to the Ames assay, tier two mutation assays (e.g., CHO cell mutation assay) did not appear to be highly sensitive to aromatic amines. The responsiveness of mammalian cell assays to PAH fractions indicated that these assays were potentially better predictors of dermal carcinogenicity than microbial assays.
- Biodirected chemical characterization, as described and discussed in Chapter 5 of this book, was found to be a cost-effective and very useful technique for evaluating biological activity in complex chemical mixtures. Synergistic and antagonistic biological responses were observed in the complex chemical mixtures evaluated for toxicity. For example, the mutagenicity of known PAH mutagens and carcinogens has been shown to be inhibited in a number of coal-derived materials. Thus, the toxicities of mixtures did not appear to be a simple sum of the toxicities of individual components in these mixtures.
- Because the causes for these synergistic and antagonistic responses are currently not well understood, quantitative predictions of biological effects in complex chemical mixtures based on their chemical composition are not possible at this time.
- Whole-animal acute toxicity assays (tier three) each measure a specific toxic endpoint, and thus more generalized cytotoxic endpoints (e.g., cell death, loss of function) of tier two mammalian cell cytotoxicity assays were of little predictive value.
- Hydrotreatment or fractional distillation of higher-molecular-weight hydrocarbons effectively decreased both mutagenesis in microbial and mammalian cell assays and carcinogenesis in animal studies.

The following subsections contain more detailed descriptions of the various assay procedures and results for each test material. A summary of the results and general conclusions is presented at the end of each subsection.

3.4 GENOTOXICITY AND CARCINOGENESIS

The chemical composition of the majority of fossil-based synthetic fuel products and by-products indicates a potential for some of these materials to be carcinogenic. The epidemiological data from worker experiences with certain industrial coal processes (e.g., coking operations within the steel industry) or previous synthetic fuel operations (e.g., Union Carbide's Institute, West Virginia liquefaction facility) also indicate the potential carcinogenicity of coal-conversion materials (see Chapter 6). Thus, in the toxicological evaluations of materials produced by synthetic fuel technologies, a major focus has been on the determination of the genotoxicity and carcinogenicity of these materials.

Genotoxicity has been measured in a variety of bacterial and mammalian cell assays designed to identify DNA damage. Because of the high degree of correlation between mutagenicity and carcinogenicity of pure compounds, induction of mutation has been the principal endpoint measured. The Ames *Salmonella*/mammalian microsome mutagenicity assay was the assay most commonly used.

Selected materials found to have a high degree of genotoxicity in a cellular assay or to have a high potential for human exposure were assayed directly for carcinogenic activity. Several whole-animal assays were used, with dermal carcinogenicity resulting from topical application of the test material being the most common result assessed.

3.4.1 In Vitro Assays for Genotoxicity

Most carcinogenic agents are believed to induce some heritable change in the cellular genome. Thus, screening unknown materials for mutagenic activity is a rational initial approach to evaluating the potential of a material to cause long-term health effects of genetic origin (e.g., cancer, reproductive effects). Numerous mutagenicity assays have been developed for this purpose; foremost among these is the bacterial mutagenesis assay developed by Bruce Ames (Ames et al., 1975). In addition to the *Salmonella* strains used in Ames assays, other cell types from different species have been used in mutagenesis screening, including bacteria [e.g., *Escherichia coli* (Green and Muriel, 1975)], yeast [e.g., *Saccharomyces* (Zimmerman, 1975)], and a number of mammalian cells. The mammalian cell assay most often used for detecting genotoxicity of coal-conversion materials was the CHO cell mutation assay (Hsie et al., 1978). In a multitiered testing approach, mammalian cell assays fall between the Ames assay and in vivo assays for carcinogenesis in terms of cost and duration. Mammalian cell mutagenesis may be more relevant than bacterial mutagenesis to human exposure because mammalian cells possess a cellular architecture that includes nucleosomes (DNA organized in chromatin structures), a nuclear envelope, mitochondria, and endoplasmic reticulum, all of which bacterial cells lack.

Another indicator of genotoxicity commonly measured in screening assays is induction of sister chromatid exchange (SCE, Perry and Wolff, 1974). Various SCE assays measure reciprocal exchange of DNA fragments between pairs of chromosomes. Agents that cause DNA strand breaks result in DNA from one chromosome replacing the DNA in its sister chromosome and vice versa. The exact mechanism of SCE is not understood; however, agents that induce DNA damage by covalent binding of the bases and mutation (i.e., changes in DNA base pairs) are usually capable of inducing SCE. Moreover, SCE are often induced by carcinogens/mutagens that do not cause cytological changes in chromosome structure.

Some of the tests, such as the Ames, CHO, and SCE assays, have been widely used to identify mutagenic and carcinogenic agents. Other assays, such as the prophage induction test (Moreau et al., 1976), were used on a limited basis.

Most genotoxicity assays used to screen complex mixtures were developed for analyses of pure compounds. Precautions must be taken when these assays are used to analyze synfuel materials. Chief concerns are toxicity and interactions among mixture components. The potential toxicity of a mixture to humans may be undetected in a cellular assay because of the inherent cytotoxicity of the mixture. For example, test cells may be killed by cytocidal components of the mixture, leaving no viable target cells to be affected by mutagenic components. Interactions (synergistic and/or antagonistic) among components of the mixture may also modify mutagenic response. The extent of these interactions and their effect on mutagenesis are not completely understood and limit quantitative evaluations. There is considerable evidence, however, that PAH mutagens have suppressed activity in most coal-conversion materials.

3.4.1.1 *Microbial Assays*

Numerous microbial assays have been used to evaluate genotoxic potential of coal-conversion materials, including the standard Ames *Salmonella*/mammalian microsome mutagenicity assay, a *Salmonella* fluctuation assay, a *Salmonella* forward mutation assay, *Salmonella typhimurium* and *E. coli* DNA-repair assays, the prophage induction assay, a forward mutation assay in yeast (*Saccharomyces cerevisiae*), and the mitotic recombination assay in yeast (*Saccharomyces cerevisiae*). A presentation and discussion of the results are provided in this section, preceded by a brief description of each assay.

3.4.1.1.1 Ames *Salmonella*/Mammalian Microsome Mutagenicity Assay. The Ames assay is generally performed using the plate-incorporation method (Fig. 3.2) as described by Ames (Ames et al., 1975). Mutagenicity is determined in one or more *S. typhimurium* tester strains having a mutation in the histidine operon (*his*) resulting in a histidine requirement for growth (auxotrophic strain). Test samples are dissolved in a biocompatible solvent, generally dimethyl sulfoxide (DMSO), and added to melted agar containing the test organisms, a small amount of histidine and biotin, and, where called for, metabolic activation enzymes. This top agar mixture is poured onto a petri dish containing histidine-free nutrient bottom agar. Cultures are incubated for 48 h and the revertant colonies counted. The mutation measured in this test is the ability to grow in the absence of histidine, that is, a reversion to the wild-type bacteria. In contrast to the standard method, where everything in the top agar is added to the plates at once, in a limited number of cases, the test material is preincubated for 30–60 min with the test organism prior to addition to the top agar according to the methods of Yahagi et al. (1975) and Ames et al. (1975).

Tester strains, in addition to their histidine requirement, have other alterations from the wild-type bacteria that increase their sensitivity to mutation. Strains TA1535, TA1537, and TA1538 have undergone mutations that alter the ability of the bacteria for excision repair of damaged DNA (uvrB) and that increase the permeability of the bacterial cell wall to large molecules through an alteration of the cell's lipopolysaccharide coat (rfa). Strain TA1535 reverts by a base-pair substitution mechanism (point mutation) and thus is considered a missense mutant. Strains TA1537 and TA1538 are frameshift mutants. Strains TA98 and TA100 are essentially the same strains, respectively, as TA1538 and TA1535, except that strains TA98 and TA100 contain the R-factor plasmid pKM101. The R-factor plasmid increases mutagenesis by amplifying the activity of an error-prone DNA repair system present in these organisms. Strain TA98 was consistently the most sensitive to mutation from unfractionated coal-conversion materials; thus, to provide some comparisons of data, only results for strain TA98 are given in the tables.

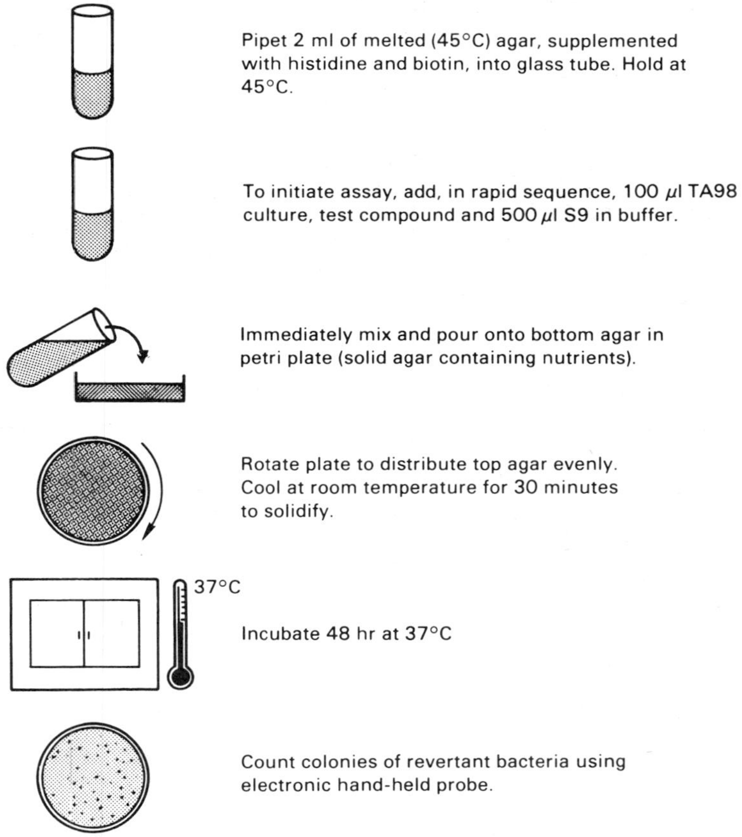

Figure 3.2 Description of the Ames *Salmonella*/mammalian microsome mutagenicity assay.

Most materials were generally studied with and without metabolic activation. The Aroclor-1254-induced rat-liver S9 system was used for activation unless specifically stated otherwise. Sample toxicity was either qualitatively determined (visual observation of cell killing or abnormal growth) or experimentally measured (cells grown on histidine-containing medium).

Samples were generally tested over a concentration range. Where possible, specific mutagenic activity (revertants per microgram of test material, rev/μg) was determined by linear regression analysis (method of least squares), ideally with results from three to five concentrations within the linear region of a dose–response curve (Draper and Smith, 1966). The slope coefficient from such an analysis was considered to represent the specific mutagenic activity. The test procedure was usually as follows: Samples were initially assayed to establish appropriate test concentrations (scaling) and to optimize metabolic activation by varying the concentration of the sample and the concentration of the S9 activating system. The area of optimal activity was graphically determined from the results of this assay. In the example shown (Fig. 3.3), the optimal S9 concentration was 10 μL and the steepest

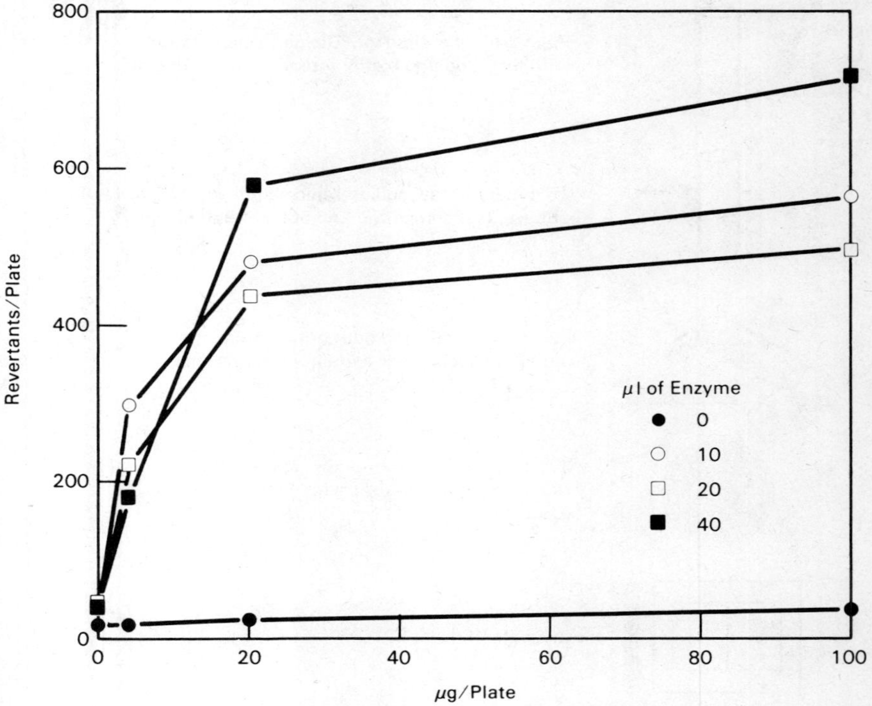

Figure 3.3 Illustration of scaling studies to determine concentrations of test material and S9 to be used in Ames *Salmonella*/mammalian microsome mutagenicity assays.

portion of the dose–response curve occurred at sample concentrations less than 5 μg/plate. Methods of calculating specific activity for each study are described in table footnotes.

A dose–response assay (with the optimal S9 concentration) was then performed, generally with five or six concentrations of the test material, all chosen from the steepest portion of the scaling curve, with two to four determinations at each concentration. Figure 3.4 shows an example of raw data for a coal-tar base fraction and a graph showing a good linear response of the S9-optimized dose–response assay. The specific mutagenic activity determination for the example with a correlation coefficient of 0.99 was calculated to be 700 rev/μg. Where specific mutagenicity was not reported by this method but data were presented that allow such a calculation, the calculation was made and used in this section.

The materials assayed, even though they were quite complex chemical mixtures differing at least qualitatively (and often quantitatively), were usually treated similarly in the assay. For the most part, such variables as different degrees of solubilization, bioavailability, and volatility were not taken into consideration in the assay procedure. In a few instances, consideration was given to these differences. For example, special attention was paid to volatile components in Kosovo gasifier samples. To evaluate the contribution of volatile components to overall mutagenicity, Ames test plates were incubated in sealable plastic bags (one bag for each set of plates for each test concentration). This treatment resulted in slightly elevated toxicity and mutant bacterial colonies that were smaller than normal, probably reflecting a decrease in available oxygen under these conditions.

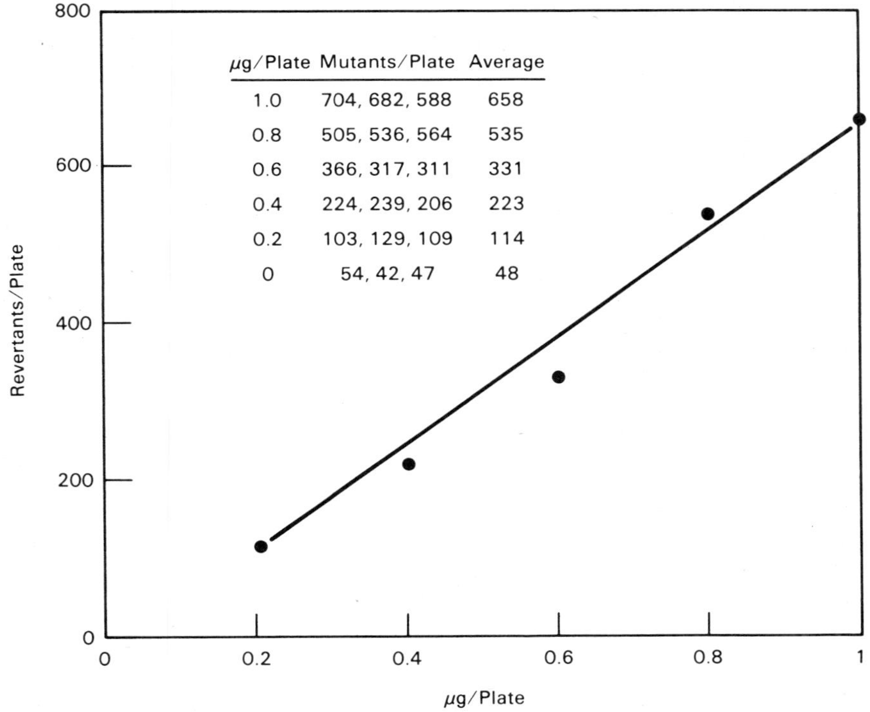

Figure 3.4 Illustration of results from Ames *Salmonella*/mammalian microsome mutagenicity assay.

The most extensive mutagenicity data sets for coal-conversion materials are the results from Ames assays; 377 materials were evaluated, including 199 liquefaction materials ranging from crude fuels through fuel blends to distillation products. Materials studied represented a range of stages in process development (i.e., scale) including bench-scale, pilot plant, process development unit (PDU), and commercial operation. At present, there are no commercial liquefaction facilities in the United States; thus, this process stage was evaluated only for some types of gasification. The majority of liquefaction materials studied represented the four major processes being developed in the United States: SRC-I, SRC-II, H-Coal, and EDS. Materials from facilities covering at least three different stages of process development were studied for each of these processes (only two stages for SRC-I). Fewer materials were studied from integrated (ITSL) and nonintegrated (NTSL) two-stage liquefaction, Synthoil, the zinc halide process (Conoco), and hydrocarbonization processes.

Ames data are available for 194 coal-gasification tars, oils, or gas condensates. With the exception of 26 tar materials from the University of Minnesota–Duluth (UMD) gasifier, only 10 materials were from industrial and commercial-scale facilities (Kosovo, Chapman/Holston, Ft. Snelling, and Glen-Gery). Considerably more data are available for pilot-plant materials. Fifty materials from the Hygas, Morgantown Energy Technology Center (METC), and University of North Dakota Energy Research Center (UNDERC) facilities were evaluated in the Ames assay. The most extensive Ames data on gasification samples came from bench-scale coal reactors at Research Triangle Institute (RTI) and the Mellon Institute (MI). Nearly 100 materials, representing 40 test runs, were assayed. These reactors provided an opportunity to evaluate the influence of process variables on toxicant formation. In addition, material from two in situ coal-gasification experiments in Wyoming and from the COED process (considered to be representative of COGAS materials) were also studied.

3.4.1.1.1.1. Liquefaction—SRC. Materials representing the SRC-I process were collected from the Ft. Lewis, Washington and Wilsonville, Alabama (now known as the Advanced Coal Liquefaction Research and Development Facility) pilot plants. Three SRC-I liquid streams from the Ft. Lewis pilot plant were evaluated for mutagenicity in the Ames assay with strain TA98 (Mahlum, 1981; Pelroy and Petersen, 1981). Light oil (LO; Sample L2211) and wash solvent (WS; Sample L2212) were nonmutagenic, while PS (Sample L2213) following metabolic activation produced 15 rev/μg (Table 3.4). (Sample numbers correspond with the more-detailed sample descriptions in the appendix at the end of this chapter.) PS also caused mutations in strains TA100 and TA1538 but not in strain TA1535. PS was fractionated by isooctane solvent–solvent separation, and fractions were assayed with *S. typhimurium* strains TA98, TA100, and TA1535, and TA1538. Basic components (isooctane-soluble and extractable with HCl), base-induced tar, and neutral (isooctane-insoluble) tar fractions were mutagenic in all strains but TA1535. Although the basic components had the highest specific activity for strain TA98 (153 rev/μg), the basic tar (25 rev/μg) and neutral tar (38 rev/μg) had greater total activity because of their larger mass (Pelroy and Petersen, 1981).

Samples of SRC-I solid product (Sample L2221) from the Ft. Lewis pilot plant and the PS (Sample L2111) from the Wilsonville pilot plant were also evaluated for Ames mutagenicity (Pelroy, 1981). Process solvent and the organic solvent-extractable components of solid product and their fractions (partition chromatography/solvent elution from Sephadex LH-20 column) were assayed with and without S9 activation in *S. typhimurium* strains TA98, TA100, TA1535, and TA1537. Both raw materials were mutagenic after metabolic activation in strain TA98 (Table 3.5) and TA100 and nonmutagenic to strains TA1535 and

Table 3.4 Results of Ames Assays on Solvent-Refined Coal I (Ft. Lewis, Washington) Materials and Fractions

Material	Relative Weight (% of total)	Specific Activity[a] (rev/μg)	Weighted Activity[b] (rev/μg)
Light oil (Sample L2211)	—	0[c]	—
Wash solvent (Sample L2212)	—	0[c]	—
Process solvent (Sample L2213)	100	15	15.0
Basic fraction	1.5	153	2.3
Basic tar	23	25	5.8
Neutral tar	27	38	10.3
Acid tar	1	2	0.02
Other fractions	48	0	0
Benzo[a]pyrene[d]	—	328 ± 46	—
2-Aminoanthracene[e]	—	1670 ± 146	—

[a]Specific activity in *Salmonella typhimurium* strain TA98 estimated by linear regression of dose-response data (i.e., slope of regression line) ± standard error (SE).

[b]Specific activity × relative weight (%).

[c]Tested at concentrations up to 1000 μg/plate.

[d]A concentration of 10 μg/plate, measured in quadruplicate.

[e]A concentration of 1 μg/plate, measured in quadruplicate.

Source: Pelroy and Petersen (1981).

Table 3.5 Results of Ames Assays on Solvent-Refined Coal I (Wilsonville, Alabama) Materials and Fractions

Material	Relative Weight (% of total)	Concentration Ranges Tested (μg/plate)	Specific Activity[a] (rev/μg)	Weighted Activity[b] (rev/μg)
Solid[c]				
Sample L2221	100	20–500	7 ± 1	7
Hexane[d]	31.9	20–500	0	0
	25.2		0	0
Toluene : hexane	7.7	2–50	17 ± 3	1
	12.7		8 ± 2	1
Methanol	16.2	2–50	32 ± 1	5
	18.3		40 ± 4	7
Toluene : methanol	38.4	20–500	0	0
	44.5		0	0
Chloroform	4.2	2–50	10 ± 1	0.4
	10.7		13 ± 2	1
Process solvent				
Sample L2111	100	20–500	0.5 ± 0.1	0.5
Hexane	75	20–500	0	0
	74		0	0
Toluene : hexane	7.1	2–50	6 ± 0.1	0.4
	8.2		4 ± 2	0.3
Methanol	16.5	2–50	18 ± 2	3
	15.4		25 ± 1	4

[a]Specific activity in *Salmonella typhimurium* strain TA98 was estimated by linear regression of dose–response data (i.e., slope of the regression line). Results are the average of three determinations ± SE. A positive response was one twice background with a correlation coefficient ≥0.8 for the regression line.

[b]Specific activity × relative weight (%).

[c]Solubilized in toluene : methanol (3 : 1).

[d]Two sets of fractions were studied from duplicate LH-20 fractionations.

Source: Pelroy (1981).

TA1537. Mutagenic activity was also present in strain TA98 following metabolic activation of the toluene/hexane (azaarene compounds) and methanol (principally primary aromatic amines and bifunctional aromatics) fractions of both materials and the chloroform fraction (highly polar compounds) of solid product. The methanol fraction accounted for about 75% of the recovered activity of solid product and 90% of PS. Unfractionated materials and fractions showing no mutagenicity in strain TA98 (the hexane fraction of the PS and the hexane and toluene/methanol fractions of the solid product) were also assayed in strains TA100, TA1535, and TA1537 with metabolic activation and found to be nonmutagenic. The hexane fraction accounted for roughly 25% of the solid product mass and 75% of the PS mass. The polycyclic aromatic hydrocarbons (PAHs), including known mutagens, occur in this material, thus mutagenic activity is apparently masked by other components.

The bp cuts of SRC-I PS (Sample L2121) and their chemical class fractions were assayed for mutagenicity with strain TA98 (Wilson et al., 1982). Seven bp cuts were tested, including an initial 100°F cut (450–550°F) and six 50°F cuts (550–850°F). Materials were assayed over one of two concentration ranges (2, 4, 8, 10, 20, and 50 μg or 0.2, 0.4, 0.8, 1.0, and 2 μg/plate) with duplicate determinations at each concentration. No mutagenicity was observed in bp cuts from 450 to 650°F. Cuts from 650 to 850°F were mutagenic (Table 3.6), with greatest response from the 750–800°F cut. Mutagenic response required metabolic activation (S9). Chemical class fractions of bp cuts were also assayed for mutagenicity. Fractions enriched in aliphatic components and essentially free of PAH or polar components were nonmutagenic. The PAH-enriched fractions were also nonmutagenic. This result is noteworthy because these fractions from bp cuts ≥700°F are enriched in five-ring PAHs, including known mutagens and carcinogens. The PAH-enriched fraction from the 800 to >850°F bp cut had a BaP concentration that exceeded 2500 ppm. The cause of the masking of mutagenicity in these fractions has not been determined. Nitrogen-containing PAH fractions in >650°F bp cuts were mutagenic, with maximum specific activity of 210 rev/μg for the 750–800°F cut. Hydroxy-PAH fractions were essentially nonmutagenic except for a weak response in the 800 to >850°F cut.

The nitrogen-containing polycyclic aromatic compound (N-PAC) fractions were separated into subfractions enriched in carbazoles, amino-PAH, and azaarenes. The carbazole subfractions of the 750–800°F and 800–850°F cuts were mutagenic (Table 3.6). Amino-PAH subfractions from cuts boiling above 700°F were mutagenic, with a maximal response of 132 rev/μg in the 750–800°F cut. Azaarene-enriched subfractions of the 750–800 and 800 to >850°F bp cuts were also mutagenic. Thus, all three classes of N-PACs contribute to the mutagenicity of bp cuts (Wilson et al., 1982).

Several distillates from the SRC-II operational mode have also been assayed for mutagenicity. Studies have been conducted on materials derived from operations of three facilities: the Ft. Lewis, Washington pilot plant, the Harmarville, Pennsylvania PDU, and the Merriam, Kansas research laboratory (bench-scale facility). Light, middle, and heavy distillates were obtained from the SRC-II pilot plant operated by the Pittsburg & Midway Coal Mining Company at Ft. Lewis, Washington (Pacific Northwest Laboratory, 1979; Pelroy and Petersen, 1981). Although light (Sample L1241) or middle distillates (Sample L1242)

Table 3.6 Results of Ames Assays on Boiling-Point Cuts and Their Chemical Class Fractions of Solvent-Refined Coal I (Wilsonville, Alabama) Process Solvent

	Specific Activity[a] (rev/μg)			
Material (wt % of cut)	650–700°F (24.5%)	700–750°F (7%)	750–800°F (6%)	800 to >850°F (12%)
Process solvent (Sample L2121)	8	15	33	4
Aliphatic	0	0	0	0
PAHs[b]	0	0	0	0
N-PACs[c]	5 (1.6)[d]	70 (21.3)	210 (63)	25 (4.8)
Carbazoles	0	0	28	15
Amino-PAHs	0	38	132	58
Azaarenes	0	0	39	20
Hydroxy-PAHs[d]	0	0	0	4 (1)

[a]Specific activity in *Salmonella typhimurium* strain TA98 estimated by linear regression of dose-response data (i.e., slope of regression line). Positive response was with a maximum response of at least 2 times background for at least one test concentration and a correlation coefficient ≥0.8 for dose-response data.

[b]Polycyclic aromatic hydrocarbons.

[c]Nitrogen-containing polycyclic aromatic compounds.

[d]Numbers in parentheses are weighted activities, that is, specific activity × relative weight (%).

Source: Wilson et al. (1982).

were nonmutagenic in TA98, heavy distillate (Sample L1243) had a specific mutagenic activity of 38 rev/μg (Table 3.7). Heavy distillate was also mutagenic in strain TA100 and TA1538 but not in strain TA1535. Solvent separation of heavy distillate produced two nonmutagenic fractions [acidic and neutral (PAH) fractions] and three fractions with significant mutagenicity. The fraction containing soluble basic components was most mutagenic, followed by the basic and neutral tar fractions. Based on mass, the neutral tar contained the most recovered mutagenicity. Thin-layer chromatography (TLC) of active fractions showed most activity was due to nitrogen-containing components (see Chapter 5). When exogenous BaP was added to Sample L1243, its mutagenic expression was inhibited (Mahlum, 1984). This suggests that the lack of mutagenic activity in the neutral fraction might be attributed to some form of masking of mutagenic expression.

A blend (2.9:1) of SRC-II middle and heavy distillate (Sample L1251) from the Ft. Lewis pilot plant was distilled selectively. Ames mutagenicity assays of the whole material and its distillate cuts, using *S. typhimurium* strain TA98, showed that mutagenicity was due to high-molecular-weight components. While the whole distillate blend had a specific activity of 2.5 rev/μg, significant mutagenic activity was found only in the two cuts above 500°F (about 40% of the blend), materials principally containing compounds of three aromatic rings or greater. The 500–510°F cut had a specific activity of 10 rev/μg and the >510°F residue induced 40 rev/μg (Weimer et al., 1980).

An SRC-II fuel oil (Sample L1211) was assayed for mutagenicity in strain TA98 as part of a study of EPA/DOE Synfuel Repository Research Materials (Kimball and Munro, 1981). An estimate of the sample's mutagenicity (1 rev/μg) was made by summing the weighted activities of chemical class fractions (Swain–Stedman separations).

An additional set of SRC-II materials (Samples L1231-L1236) representing crude oils and distillate products derived from materials produced at the Ft. Lewis pilot plant were assayed for mutagenicity in the Ames assay for comparison with similar materials from

Table 3.7 Results of Ames Assays on Solvent-Refined Coal II Distillates (Ft. Lewis, Washington)

Material	Relative Weight (% of total)	Specific Activity[a] (rev/μg)	Weighted Activity[b] (Rev/μg)
Light distillate (Sample L1241)	—	0[c]	—
Middle distillate (Sample L1242)	—	0[c]	—
Heavy distillate (Sample L1243)	100	38	38
Base fraction	1	360	3.6
Basic tar	24	52	12.5
Neutral tar	26	69	17.9
Acid tar	1	2	0.02
Other fractions	48	0	—
Benzo[a]pyrene[d]	—	185 ± 36	—
2-Aminoanthracene[e]	—	1243 ± 186	—

[a]Specific activity in *Salmonella typhimurium* strain TA98 was estimated by linear regression of dose–response data (i.e., slope of the regression line).

[b]Specific activity × relative weight (%).

[c]Tested at concentrations up to 1000 μg/plate.

[d]10 μg/plate, measured in quadruplicate.

[e]1 μg/plate, measured in quadruplicate.

Source: Pelroy and Petersen (1981).

other liquefaction processes (Calkins et al., 1983). Results are presented later in this section (Table 3.27) and show that, as with other SRC-II materials, the higher-molecular-weight materials have the highest specific mutagenic activity.

The SRC-II vacuum still bottoms (Samples L1271 and L1272) from the Ft. Lewis pilot plant were evaluated for mutagenicity using strain TA98, TA100, and TA1538 (Keleti et al., 1982). Two crude samples and their solvent extracted fractions (sequential extraction with hexane, toluene, methylene chloride, and acetonitrile) were assayed. Crude samples were mutagenic after metabolic activation in strains TA98 and TA1538. Specific activities in TA98 were 2.9 and 11.1 rev/μg. Only one crude sample was mutagenic (slightly above twice background) to strain TA100. Solvent fractions showed no single class of soluble components to be consistently responsible for the mutagenic response.

A sample of SRC-II vacuum bottoms (Sample L1281) was evaluated under EPA level 1 methodology (Brusick and Young, 1981). According to that methodology, this material had high mutagenicity (positive response at a test concentration less than 0.05 mg/plate) at a minimum effective concentration of 0.025 mg/plate (Kim et al., 1981). The *Salmonella* tester strain used was not reported.

Two sets of bp cuts from full-bp-range SRC-II process materials from the Gulf Science and Technology Company, Harmarville, Pennsylvania PDU were assayed for mutagenicity in *S. typhimurium* strains TA98, TA100, TA1535, and TA1537 (Pelroy and Wilson, 1981). Samples were assayed at concentrations of 2, 4, 8, 10, 20, and 50 μg or 0.2, 0.4, 0.8, 1, and 2 μg/plate. The first sample set (Samples L1121 and L1131) included 18 bp cuts ranging from 138°F to 1055°F. Strain TA1535 was resistant to mutation by all distillate cuts. While bp cuts below 689°F were nonmutagenic, mutagenicity was detected with strains TA98, TA100, and TA1537 in all bp cuts above 705°F, with greatest activity in cuts above 782°F. Table 3.8 shows mutagenicity results with strain TA98.

A modified Ames assay (preincubation of target cells) was also conducted on these bp cuts, using strains TA98 and TA100. Results were not significantly different from those from the standard Ames plate incorporation assay.

Five bp cuts were fractionated by high-performance liquid chromatography (HPLC) solvent elution to yield nonpolar (neutral PAH), moderately polar (enriched in amino-PAH, azaarenes, and carbazoles), transition (bifunctional polycyclics), and strongly polar (polyaromatic phenolic compounds) fractions. Three bp cuts were transitional cuts just below mutagenic cuts (565–582°F, 639–689°F, and 710–730°F); the other two were a mutagenic cut (739–761°F) and the maximally mutagenic cut (867–967°F). The bp cuts above 700°F yielded mutagenic fractions with major activity (up to 95% of the total) found in the amino-PAH-enriched fractions. In the 867–967°F cut, in addition to amino-PAH mutagens (82% of the activity), the highly polar region was also active (18% of the activity). Total activity recovered in the fractions was generally greater than that observed in unfractionated bp cuts. An amino-PAH-enriched fraction from the 639–689°F cut (nonmutagenic as an unfractionated material) was mutagenic. No activity was found following fractionation of the 565–582°F cut. Nitrous acid treatment of amino-PAH-enriched fractions under conditions known to remove amines chemically (Haugen et al., 1981; Pelroy and Stewart, 1981) reduced mutagenicity by about 90% in all bp cuts except for the 639–689°F cut (increased mutagenicity). Reduced mutagenicity after nitrosation suggests that most activity was due to amino-PAH. Although high concentrations of high-molecular-weight PAHs occurred in these materials, particularly the 739–761°F and 867–967°F cuts, their mutagenicity was not expressed in HPLC fractions (Pelroy and Wilson, 1981).

Similar results were obtained with the second sample set (Sample L1111). This sample set was assayed with strains TA98 and TA100 only. No mutagenicity was detected with the low-boiling-range material (300–700°F), while all materials above 700°F (about 20%

Table 3.8 Results of Ames Assays on Boiling-Point Cuts of Solvent-Refined Coal II Process Materials (Harmarville, Pennsylvania)

Material	Boiling-Point (°F) Cut	Specific Activity[a] (rev/μg)
Process Blend		
Samples L1121 and L1131	138–689[b]	0
	705–825	5 ± 2
	710–730	0.5 ± 0.4
	739–761	30 ± 11
	765–782	15 ± 4
	782–878	77 ± 22
	867–967	99 ± 23
	914–1055	85 ± 19
Sample L1111	300–700	0
	700–750	29 ± 12
	750–800	36 ± 20
	800–850	124 ± 19
	>850	89 ± 23

[a]Specific activity in *Salmonella typhimurium* strain TA98 was estimated by linear regression of dose–response data (i.e., slope of the regression line). A positive response was one twice background with a correlation coefficient ≥0.8 for the regression line. Results shown are the average of 3 assays ±SE.

[b]Eleven bp cuts were assayed from this range.

Source: Pelroy and Wilson (1981).

by weight of the unfractionated materials) were mutagenic in both strains. Highest specific activity was observed in the >850°F cut (4.7 wt %).

The bp cuts of SRC-II process materials (Sample L1111) were reblended according to their weight % contributions to the whole material (Pelroy and Wilson, 1981). The low-boiling-range cut (300–700°F) was blended sequentially with each of the four > 700°F cuts and each blend was assayed for mutagenicity. Weighted specific activities of the blends were compared to the sum of the weighted activities of each bp cut (Table 3.9). Experimentally derived weighted activities were somewhat higher than the sum of the activities of the bp cuts.

The 300–700°F, 800–850°F, and >850°F cuts from the second sample set were fractionated (alumina adsorption-column chromatography) into broad chemical classes (aliphatic, PAH, N-PAC, and hydroxy-PAH) and the bp cuts and their fractions were assayed for mutagenicity in strain TA98 (Mahlum et al., 1983). Highest specific mutagenic activity (179 rev/μg) was found in the >850°F cut (Table 3.10). Aliphatic and PAH-enriched fractions of the three cuts were nonmutagenic. Of the hydroxy-PAH fractions, only the fraction from the >850°F material had detectable mutagenic activity. The N-PAC fractions from all three bp cuts were mutagenic.

The possibility that mutagenic components of SRC-II materials could be selectively distilled from the net product and recycled to extinction within the process was tested using the Gulf Research and Development Company's Merriam, Kansas bench-scale unit. Ames mutagenicity of liquid product (Sample L1311) from conventional operation (on-line distillation with minimal recycling of high-molecular-weight materials) with subbituminous

Table 3.9 Results of Ames Assays of Reblended Boiling-Point Cuts of Solvent-Refined Coal II Process Materials (Harmarville, Pennsylvania)

Material	Relative Weight (% of total)	Specific Activity[a] (rev/μg)	Weighted Activity[b] (rev/μg)
	Process Blend (Sample L1111)		
Individual cuts			
300–700°F	78.9	0	0
700–750°F	5.5	42 ± 7	2.3
750–800°F	7.4	61 ± 7	4.5
800–850°F	3.5	69 ± 24	2.4
>850°F	4.7	93 ± 10	4.4
			13.6
Reblended cuts			
300–750°F	84.4	5 ± 1	4.2
300–800°F	91.8	9 ± 3	8.3
300–850°F	95.3	12 ± 3	11.4
300 to >850°F	100	17 ± 4	17.0

[a]Specific activity in *Salmonella typhimurium* strain TA98 was estimated by linear regression of dose–response data (i.e., slope of the regression line); average of five experiments ±SE. A positive response was one twice background with a correlation coefficient ≥0.8 for the regression line.

[b]Specific activity × relative weight (%).

Source: Pelroy and Wilson (1981) and Pelroy et al. (1985).

Table 3.10 Results of Ames Assays on Chemical Class Fractions of Boiling-Range Cuts of Solvent-Refined Coal II Process Materials (Harmarville, Pennsylvania)

Material	Relative Weight (% of total)	Specific Activity[a] (rev/μg)	Weighted Activity[b] (rev/μg)
	Process Blend (Sample L1111)		
800–850°F	100	57	57
Aliphatic	4	0	0
PAHs[c]	47	0	0
Nitrogen-PACs[d]	31	275	85.3
Hydroxy-PAHs[c]	16	0	0
>850°F	100	179	179
Aliphatic	2	0	0
PAHs[c]	44	0	0
Nitrogen-PACs[d]	35	150	52.5
Hydroxy-PAHs[c]	19	20	3.8

[a]Specific activity in *Salmonella typhimurium* strain TA98 was estimated by linear regression of dose–response data (i.e., slope of the regression line). A positive response was one twice background with a correlation coefficient ≥0.8 for the regression line.

[b]Specific activity × relative weight (%).

[c]Polycyclic aromatic hydrocarbons.

[d]Polycyclic aromatic compounds.

Source: Mahlum et al. (1984).

coal was compared to that of samples (L1331, L1341, and L1351) distilled to specific temperature cutoffs with the bottoms recycled to extinction (Wilson et al., 1983b). Although the conventional material had a specific mutagenic activity of 5.5 rev/μg, the three recycle samples (bp cutoff of <518°F, <554°F, and <590°F) were nonmutagenic. Similarly, a <518°F bp cut recycle sample (Sample L1321) collected while processing bituminous coal was nonmutagenic. The components which are distilled at temperatures greater than 590°F appear to encompass all the mutagenic materials and comprise only 6% of the total sample.

3.4.1.1.1.2 Liquefaction—H-Coal. H-Coal materials from Hydrocarbon Research, Inc.'s Trenton, New Jersey PDU and the Catlettsburg, Kentucky pilot plant were evaluated for mutagenicity. H-Coal PDU materials (Samples L3142–L3144 and L3151–L3154) from both syncrude and fuel-oil operations were assayed for Ames bacterial mutagenicity along with a number of other synfuel materials in a reference repository established by EPA/DOE at Oak Ridge National Laboratory (ORNL; Kimball and Munro, 1981). Mutagenicity results are shown in Table 3.11. Mutagenic activity increased with increasing average molecular weight of the materials. For example, the atmospheric still overhead (fuel-oil mode) sample, a low-molecular-weight material, was nonmutagenic, whereas a high-molecular-weight material from the same operation mode, vaccum still bottoms, had an estimated mutagenic activity of 6.3 rev/μg. In contrast to studies with SRC-II materials, neutral components were shown to be mutagenic.

Two other sample sets from the H-Coal PDU (L3311–L3315 and L3321–L3325) were also assayed for Ames mutagenicity (Schoeny et al., 1981; Warshawsky et al., 1981). Each set of liquefaction materials included light oils (vapor condensates from atmospheric distilled separator overheads), separator bottoms (heavy liquid material prior to distillation; both crude and filtered material were studied), vacuum overheads (heavy distillate oils), and vacuum bottoms (residue remaining after vacuum distillation of the separator bottoms).

Table 3.11 Results of Ames Assays on H-Coal Process Development Unit Materials

Material	Specific Activity[a] (rev/μg) Total	Acids (% contribution)	Bases (% contribution)	Neutrals (% contribution)
H-Coal (Syncrude)				
Atmospheric still bottoms (Sample L3142)	1.0	1	46	53
Vacuum still overhead (Sample L3143)	2.4	0	42	58
Vacuum still bottoms (Sample L3144)	2.0	0	66	17
H-Coal (Fuel Oil)				
Atmospheric still overhead (Sample L3151)	0	0	0	0
Atmospheric still bottoms (Sample L3152)	0.1	0	100	0
Vacuum still overhead (Sample L3153)	4.2	0	25	75
Vacuum still bottoms (Sample L3154)	6.3	0	89	9

[a]Specific activity in *Salmonella typhimurium* strain TA98 was estimated by linear regression of dose–response data (i.e., slope of the regression line). A positive response was one twice background with a correlation coefficient ≥0.8 for the regression line. Activity of material is expressed as the sum of activities of individual chemical class fractions.

Source: Kimball and Munro (1981).

Samples were prepared as DMSO extracts (10-mg/mL solutions, corrected for undissolved material) and were assayed for mutagenicity against *S. typhimurium* strains TA98, TA100, TA1535, TA1537, and TA1538. To identify chemical classes responsible for mutagenicity, an aliquot of each sample was sequentially extracted with four organic solvents of increasing polarity (hexane, toluene, methylene chloride, and acetonitrile). Extracts and the remaining residue were also assayed for mutagenicity over a threefold concentration range (100, 50, 5, 0.5, and 0.05% or 100, 50, 10, 5, and 1%) of the stock solution.

Samples from the first sample set were more mutagenic than those from the second set (Table 3.12). There were no definitive reasons reported for the difference. It was pointed out that although both sets were produced from Kentucky No. 11 coal, coal used during the period from which the second set was collected did contain a much higher ash content (18% compared to 6.4%). None of the samples contained direct-acting mutagens, all requiring metabolic activation to express mutagenicity. TA98 was the most sensitive strain and no sample induced mutation in TA1535.

Mutagenic components with the highest specific activity (>20 rev/μg) were found in toluene, methylene chloride, and acetonitrile extracts (Table 3.13). Although little material was soluble in the most polar solvent (acetonitrile), the components in this extract were often highly mutagenic (e.g., specific activity for the acetonitrile fraction of set 1 filtered separator bottoms was 108.4 rev/μg, yet was only 0.3% of the original material). In contrast, hexane-soluble materials, while having the greatest weight contribution (up to 80% of the set 1 filtered separator bottoms), had relatively low specific mutagenic activities. The decreased mutagenic activity in the second sample set was confirmed in their extracted fractions. The sum of activities of extracted fractions never equaled the observed activity of the whole material, but only in the case of the vacuum bottoms were they greatly different (Schoeny et al., 1981).

An additional set of H-Coal materials (Samples L3132–L3136 and L3121–L3124) representing crude oils and distillate products derived from materials produced at Hydrocarbon Research Inc.'s Trenton, New Jersey PDU under both syncrude and fuel-oil operation modes were assayed for mutagenicity in the Ames assay in a comparative study with similar materials from other liquefaction processes. Results are reported later in this section, in Table 3.27.

Table 3.12 Results of Ames Assays on H-Coal Process Development Unit Materials (Trenton, New Jersey)

Material	Specific Activity[a] (rev/μg) Set 1	Set 2
Light oils (Samples L3311 and L3321)	0	0.1
Separator bottoms		
Crude (Samples L3312 and L3322)	17.2	3.7
Filtered (Samples L3313 and L3323)	11.4	2.6
Vacuum overhead (Samples L3314 and L3324)	10.9	1.5
Vacuum bottoms (Samples L3315 and L3325)	27.0	14.0

[a]Specific activity in *Salmonella typhimurium* strain TA98 was estimated by linear regression of dose–response data (i.e., slope of the regression line). A positive response was one twice background with a correlation coefficient ≥ 0.8 for the regression line.

Source: Warshawsky et al. (1981).

Table 3.13 Results of Ames Assays on Fractions (Solvent Extracts) of H-Coal Process Development Unit Materials (Trenton, New Jersey)

	Specific Activity[a] (rev/μg)				
Material	Hexane	Toluene	Methylene Chloride	Acetonitrile	Residue
Separator Bottoms (Crude)					
Sample L3312	5.9 (3.1)	20.6 (5.2)	13.1 (1.9)	4.4 (0.1)	0.4 (0.02)
Sample L3322	2.2 (1.4)	3.4 (0.5)	7.3 (0.4)	27.2 (0.04)	0.05 (0.01)
Separator Bottoms (Filtered)					
Sample L3313	7.7 (4.1)	27.2 (8.1)	21.1 (2.9)	108.4 (0.3)	0.8 (0.02)
Sample L3323	1.0 (0.8)	4.1 (0.5)	6.5 (0.4)	5.4 (0.01)	0.2 (0.01)
Vacuum Overhead					
Sample L3314	9.5 (9.2)	271.4 (2.7)	63.9 (0.7)	63.5 (0.08)	13.3 (0.06)
Sample L3324	1.1 (1.1)	4.6 (0.04)	—	—	—
Vacuum Bottoms					
Sample L3315	15.2 (0.1)	24.7 (10.5)	12 (2.3)	33.3 (0.3)	2.5 (0.9)
Sample L3325	4.2 (0.08)	6.2 (1.4)	271.8 (31)	8.3 (0.06)	4.7 (3.0)

[a]Specific activity in *Salmonella typhimurium* strain TA98 was estimated by linear regression of dose–response data (i.e., slope of the regression line). A positive response was one twice background with a correlation coefficient ≥ 0.8 for the regression line. Numbers in parentheses are weighted activities, that is, specific activity × relative weight (%).

Source: Schoeny et al. (1981).

Samples were collected from the Catlettsburg, Kentucky, H-Coal pilot plant on four separate occasions, all during operations in the syncrude mode, and assayed for mutagenicity in *S. typhimurium* strain TA98 (Munro et al., 1985). The first set of samples was collected during operation with Kentucky No. 9 coal. Hydroclone overhead (Sample L3214), waste oil residual (Sample L3210), heavy slop oil (Sample L3219), and light fuel oil (Sample L3212) were assayed without fractionation and found to be mutagenic (Table 3.14). Other materials were fractionated prior to assay (Swain–Stedman acid–base procedure with Fluorisil column separation of the neutral fraction) and an estimation of the mutagenicity of the crude material was made by summing the weighted activity of the fractions. Light slop oil (Sample L3218) and recycle oil (Sample L3217) were nonmutagenic. Following fractionation, some weak evidence of mutagenicity (0.02 rev/μg) was detected with the light fuel-oil sample. All other samples, heavy fuel oil (Sample L3213), vacuum bottoms (Sample L3215), and vacuum flaked product (Sample L3216), showed significant activity, with most of the mutagenicity found in the basic and neutral fractions.

The second set of samples was derived from the process while operating on Illinois No. 6 coal and the third on Wyodak coal (Table 3.15). No obvious effects of coal type were found with respect to mutagenicity of the various sample types (Tables 3.14 and 3.15). Stabilizer naphtha (Samples L3241 and L3251) and light fuel oil (Samples L3242 and L3252) from both sets were nonmutagenic over a concentration range of 2.5–100 μg/plate. Hydroclone overhead (Sample L3255) and atmospheric stripper (Sample L3254) from the Wyodak coal-derived materials were mutagenic, as were the heavy fuel oils (Samples L3243 and L3253) from both sets.

Table 3.14 Results of Ames Assays on H-Coal Pilot-Plant Materials from Kentucky Coal No. 9 (Catlettsburg, Kentucky)

Material	Sample No.	Specific Activity[a] (rev/μg)
Hydroclone overhead	L3214	16[b]
Light slop oil	L3218	0[c]
Heavy slop oil	L3219	12.5[c]
Recycle oil	L3217	0[b]
Waste-oil residual	L3210	17.6[c]
Heavy fuel oil	L3213	3.2[d], 5.4[c]
Light fuel oil	L3212	0.02[d], 0[c]
Vacuum bottoms	L3215	5.2[d], 4.6[d], 9.8[c]
Vacuum flaked product	L3216	2.1[d], 21[c]

[a]Specific activity in *Salmonella typhimurium* strain TA98 was estimated by linear regression of dose–response data (i.e., slope of the regression line). A positive response was one twice background with a correlation coefficient ≥0.8 for the regression line.

[b]DMSO-soluble components of the material were assayed.

[c]A DMSO slurry was assayed (Ma et al., 1983).

[d]Activity of crude material estimated by summation of weighted specific activity of chemical class fractions of the whole material determined as above.

Source: Munro et al. (1985).

A sample of heavy fuel oil (Sample L3234) and a heavy oil/light oil blend (Sample L3244) from other runs using Illinois No. 6 coal were distilled into discrete bp cuts and the whole material and each cut were assayed for mutagenicity in *S. typhimurium* strain TA98 (Table 3.16). Materials from bp cuts below about 650°F were nonmutagenic while higher boiling materials and the distillation residues were mutagenic (Cowser, 1982b; Wilson et al., 1985). With the oil blend, mutagenic activity was found for the most part only in the N-PAC fractions of the bp cuts. The hydroxy-PAH fractions of the three highest cuts and the PAH fraction of the 850°F material had low but detectable activity.

Table 3.15 Results of Ames Assays on H-Coal Pilot-Plant Materials (Catlettsburg, Kentucky)

Material	Sample No.	Specific Activity[a] (rev/μg)	
		Illinois No. 6	Wyodak
Stabilizer naphtha	L3241 and L3251	0	0
Hydroclone overhead	L3255	—	3.2
Atmospheric stripper	L3254	—	1.4
Light fuel oil	L3242 and L3252	0	0
Heavy fuel oil	L3243 and L3253	39[b]	1.3[c]
Vacuum flaked product	L3245	15[d]	—

[a]Specific activity in *Salmonella typhimurium* strain TA98 was estimated by linear regression of dose–response data (i.e., slope of the regression line) when assaying a DMSO slurry of the test material.

[b]Average of 3 determinations.

[c]Average of 2 determinations.

[d]DMSO-soluble components of the material were assayed.

Source: Munro et al. (1985).

Table 3.16 Results of Ames Assays on H-Coal Heavy Fuel Oil and Its Distillate Cuts (Catlettsburg, Kentucky)

Material	Relative Weight (% of total)	Specific Activity[a] (rev/μg)	Weighted Activity[b] (rev/μg)
Heavy fuel oil (Sample L3234)	100	5.5[c]	5.5
380–475°F	9.2	0	0
465–515°F	19.2	0	0
520–590°F	19.3	0	0
600–645°F	20.3	0	0
635–700°F	17.3	12	2.1
>700°F	11.1	230	25.5
Fuel oil blend (Sample L3244)	100	11	11
650–700°F	6.8	0	0
700–750°F	1.4	65	0.9
750–800°F	4.1	45	1.9
800–850°F	1.1	104	1.1
>850°F	3.1	70	2.2

[a]Specific activity in *Salmonella typhimurium* strain TA98 was estimated by linear regression of dose–response data (i.e., slope of the regression line).

[b]Specific activity × relative weight (%).

[c]A value of 129 rev/μg (average of three determinations) was obtained by the DMSO slurry method (Munro et al., 1985).

Source: Cowser (1982b) and Wilson et al. (1985).

3.4.1.1.1.3 Liquefaction—EDS. A set of EDS samples were assayed for mutagenicity as part of a study of EPA/DOE Synfuel Repository Research Materials (Kimball and Munro, 1981). Chemical class fractions (Swain–Stedman acid–base extraction with Sephadex LH20 or Fluorisil separation of the neutral fraction) of vacuum gas oil (Sample L4241), spent solvent (Sample L4242), and hydrotreated recycle solvent (Samples L4243 and L4311) were assayed with TA98 (Table 3.17). Metabolic activation was required to demonstrate mutagenicity in the fractions and an estimate of the activity in the whole material was calculated from the sum of the weighted activities of the fractions. Only the vacuum gas oil was found to contain significant activity (9.8 rev/μg).

Table 3.17 Results of Ames Assays on EDS Materials

Material	Sample No.	Specific Activity[a] (rev/μg)	Contribution (%)		
			Acids	Bases	Neutrals
Vacuum gas oil	L4241	9.8	0	98	2
Spent solvent	L4242	0.1	2	45	50
Recycle solvent	L4243	0	0	0	0
	L4311	0.03	4	21	71

[a]Specific activity in *Salmonella typhimurium* strain TA98 was estimated by linear regression of dose–response data (i.e., slope of the regression line). A positive response was one twice background with a correlation coefficient ≥0.8 for the regression line. Activity of material is expressed as the sum of weighted activities of individual chemical class fractions.

Source: Kimball and Munro (1981).

An additional set of EDS materials (Samples L4211–L4215) representing crude oils and distillate products derived from materials produced at the Baytown, Texas PDU were assayed for mutagenicity in the Ames assay in a comparative study with similar materials from other liquefaction processes. Results are reported in Table 3.27.

Process materials from the EDS pilot plant at Baytown, Texas were also assayed for Ames mutagenicity (Later et al., 1984). Materials represented two operational modes, once-through (only hydrogenated PS recycled) and bottoms recycle (vacuum bottoms and PS recycled). Raw naphtha (Sample L4131), the volatile hydrocarbons from atmospheric distillation (bp < 400°F), recycle solvent (once-through Sample L4111 and bottoms recycle Sample L4132), hydrocarbons boiling nominally in the range 400–1000°F, vacuum gas oil (Sample L4112), heavy hydrocarbons boiling in the 700–1000°F range, vacuum bottoms (Sample L4133), the ash-bearing residuum containing hydrocarbons boiling above 1000°F, and a full-range process blend (60% naphtha, 40% PS, Sample L4134) and its high-molecular-weight distillation cuts (>650°F) were assayed in *S. typhimurium* strains TA98 and TA100 (Table 3.18). In addition to the whole materials, broad chemical class fractions (alumina adsorption-column chromatography) were also assayed. The samples with predominantly low-boiling materials had little or no mutagenicity. Vacuum gas oil had the highest specific activity (20 rev/μg). In the individual process materials with significant mutagenicity, the N-PAC fractions had the highest specific activity. The only other fractions contributing to the mutagenicity of these materials were the hydroxy-PAH fractions of the vacuum gas oil and vacuum bottoms. Mutagenic activity of the process blend was low and was limited to the >800°F bp cut (3.6% of the total material). N-PAC fractions of the >700°F cuts were the dominant mutagenic fractions, with some activity in two PAH-enriched fractions (650–700°F and >800°F).

3.4.1.1.1.4 Two-Stage Liquefaction (TSL) Technologies. TSL separates initial solvation and subsequent upgrading (hydrogenation) of coal liquids to optimize each process

Table 3.18 Results of Ames Assays on EDS Materials and Distillate Cuts (Baytown, Texas)

	Specific Activity[a] (rev/μg)			
Material (Sample No.)	Whole	PAHs	Nitrogen-PACs	Hydroxy-PAHs
Raw naphtha (L4131)	0	0	0	0
Recycle solvent				
(once through) (L4111)	<1	0	1	0
Recycle solvent				
(bottoms recycle) (L4132)	1	0	39	0
Vacuum gas oil (L4112)	20	0	18	12
Vacuum bottoms (L4133)	11	0	33	24
Process blend (L4134)				
600–650°F (3.9)[b]	0	0	0	0
650–700°F (3.2)	0	0.1	0	0
700–750°F (2.8)	0	0	13	0
750–800°F (2.2)	0	0	34	0
>800°F (3.6)	34	0.2	61	0

[a]Specific activity in *Salmonella typhimurium* strain TA98 was estimated by linear regression of dose–response data (i.e., slope of the regression line). A positive response was one twice background with a correlation coefficient ≥0.8 for the regression line.

[b]Values in parentheses are relative weight (% of total).

Source: Later et al. (1984).

step. Four types of TSL material were studied with the Ames bacterial mutagenicity assay. One type came from the C. E. Lummus Company PDU at its New Brunswick, New Jersey, Engineering Development Facility. In this PDU, the solvation and upgrading steps were separated, yet the process was an ITSL process. The upgrading unit (LC Finer) employed a proprietary catalytic hydrogeneration process called LC Fining. Materials from normal, recycle, and off-normal conditions were evaluated (Wilson et al., 1983a). Assays were performed with *S. typhimurium* strain TA98 and metabolic activation. Mutagenicities of solvent-extracted and deashed coal [LC Finer feed (LCFF)], LC Finer total liquid product, recycle solvent [>650°F material from the distillates of the LC Finer total liquid product (LCF TLP) and the deashed LC Finer], antisolvent deashing bottoms (the short-contact-time heavy oil portion and the mineral ash and insoluble organic matter used as a gasifier feed for the production of hydrogen), and the distillate cuts of these materials are shown in Table 3.19. Although mutagenic activity was detected in the LCFF collected under normal operations (Sample L5111), it was detected only in the selectively distilled material boiling

Table 3.19 Results of Ames Assays on Integrated Two-Stage Liquefaction Materials and Distillate Cuts (New Brunswick, New Jersey)

Material	Sample No.	Boiling-Point Cuts (°F)	Operating Conditions	Specific Activity[a] (rev/μg)
LC Finer feed	L5111		Normal	0 (2)
		>850		10.2
	L5121		Off-normal[b]	8.5 (12)
	L5131		Off-normal[c]	0 (10)
LC Finer total liquid product	H0611		Normal	0 (7.5)
		>850		0
	H0641		Normal	0.1 (1)
	H0621		Off-normal[b]	18 (52)
		Initial boiling point (IBP), 390		0
		390–500		0
		500–650		0
		650–850		48.8[d]
		>850		49.7[d]
	H0631		Off-normal[c]	6.5 (43)
		<650		0
		>650		4.7
Short-contact-time heavy oil	L5132			0.6[e]
Recycle solvent	H0632			17.7
Antisolvent recovery bottoms	L5133			2.9
Gasifier feed	L5134			12.5

[a]Specific activity in *Salmonella typhimurium* strain TA98 was estimated by linear regression of dose–response data (i.e., slope of the regression line). A positive response was one twice background with a correlation coefficient ≥0.8 for the regression line. Results in parentheses are for preincubation procedure.

[b]Deashed initial-stage liquefaction material recycled once prior to entering the LC Finer.

[c]Significant temperature excursion in LC Finer catalyst bed.

[d]Average of two independent determinations.

[e]Single point.

Source: Wilson et al. (1983a).

above 850°F. Low mutagenicity (0.1 rev/μg) or no mutagenicity was detected in the two normal operation LCF TLP samples (H0611 and H0641), including the >850°F material. Off-normal operation LCF TLP from deashed short-contact-time (low-sensitivity) liquids recycled once before entering the LC Finer (Sample H0621) had significant mutagenic activity (18 rev/μg). Almost all this activity occurred in the N-PAC-enriched fraction of materials boiling above 500°F, with some contribution from the hydroxy-PAH fraction.

Likewise, LCF TLP collected just after a significant temperature increase in the LC Finer catalyst beds (Sample H0631) showed significant activity (6.9 rev/μg). During this run, mutagenic activity was present in the short-contact-time reactor product, the recycle solvent, and the antisolvent deashing bottoms and its distillate gasifier feed. In the short-contact-time heavy oil sample, mutagenic activity was found in the N-PAC-enriched fraction, with a slight contribution from the PAH and hydroxy-PAH fractions.

ITSL materials evaluated in an Ames assay modified to allow preincubation of bacteria with test agents [60 min, 37°C (99°F)] had higher specific mutagenic activity (results in parentheses, Table 3.21) than when tested in the standard assay. The relative activities (LCFF > LCF TLP) were in accord with results of the standard Ames assay.

Another set of ITSL samples came from the Wilsonville Coal Liquefaction Research and Development Facility (Wilsonville, Alabama). Sample sets came from an initial start-up period, four subsequent "line-out" periods when the temperature of the second-stage catalytic hydrotreated unit was incrementally increased (680°F, 700°F, 720°F, and 740–750°F), and a temperature transition period (nonequilibrium operating period) from 700 to 720°F (Later, 1985). *S. typhimurium* strain TA98 and metabolic activation of the test material were used for the assays. No activity or only low mutagenic activity was found with most of the test materials (Table 3.20). Highest mutagenic activities were observed with materials from the lowest hydrotreater operating temperature (680°F) assayed. Chemical class fractionation (alumina adsorption-column chromatography) of two sample sets indicated that highest specific activity was found in the N-PAC fractions.

Another set of samples from the Wilsonville facility was collected during operation in the NTSL configuration (Later, 1985). With the exception of the first-stage fuel oil and

Table 3.20 Results of Ames Assays on Integrated Two-Stage Liquefaction (ITSL) Materials (Wilsonville, Alabama)

	Specific Activity[a] (rev/μg)					
Material	Start Up	680°F	700°F	720°F	720°F[b]	740°F
Light fuel oil	0(L5361)[c]	0(L5311)	0(L5321)	2(L5331)	0(L5341)	1(L5351)
Distillate process solvent	0(L5365)	2(L5315)	0(L5325)	3(L5335)	0(L5345)	2(L5355)
First-stage product distillate	—	>1(L5312)	—	—	0(L5342)	—
Deashed product	1(L5364)	7(L5314)	1(L5324)	0(L5334)	2(L5344)	5(L5354)
Critical solvent deasher product	0(L5363)	2(L5313)	2(L5323)	3(L5333)	4(L5343)	0(L5353)
Hydrotreater feed	—	2(L5316)	1(L5326)	0(L5336)	0(L5346)	0(L5356)
Hydrotreater light fuel oil	—	—	—	0(H0572)	0(H0582)	0(H0592)
Hydrotreater distillate bottoms	—	—	—	0(H0573)	0(H0583)	0(H0593)
Hydrotreater recycle pasting solvent	0(H0544)	4(H0554)	1(H0564)	0(H0574)	0(H0584)	—

[a]Specific activity in *Salmonella typhimurium* strain TA98 was estimated by linear regression of dose–response data (i.e., slope of the regression line). A positive response was one twice background with a correlation coefficient ≥0.8 for the regression line.

[b]Results from nonequilibrium transition period.

[c]Sample numbers in parentheses.

Source: Later (1985).

product distillate (Samples L5211 and L5212) and the second-stage light fuel oil (Sample H0512), which were nonmutagenic, the materials assayed had low but significant mutagenic activity (Table 3.21). Mutagenic activity was greater than that of materials sampled at the same locations in the ITSL configuration (Table 3.20) for all but the deashed product (Sample L5214). Chemical class fractionation (alumina adsorption-column chromatography) of the materials showed that, as with other coal-conversion materials, the bulk of the mutagenic material was present in the N-PAC fraction.

Another set of NTSL materials from the SRC-I pilot plant at Wilsonville were assayed in *S. typhimurium* strains TA98, TA100, TA1535, TA1537, and TA1538 with and without metabolic activation (Drozdowicz and Kelly, 1983a–c). A preliminary assay was performed with strain TA100 and six concentrations of test material (covering a threefold range) to determine the appropriate concentration range for testing. Two assays (six concentrations with two determinations for each) were then made on separate days for each material. Materials determined to be mutagenic showed a reproducible (both assays) dose-related (at least three concentrations) increase in number of revertants in at least one of the five test strains.

Solid products (Samples L5232 and H0523) from first- and second-stage low conversion operations were assayed as aqueous and saline suspensions and aqueous, saline, and DMSO extracts. A positive response in DMSO extracts demonstrates the presence of mutagens soluble in strong organic solvents, whereas that in saline and aqueous suspensions and/or extracts shows mutagenic potential in a medium more similar to physiological fluids. Saline and concentrated aqueous extracts of both materials were nonmutagenic. Saline and water suspensions of both materials were very weakly mutagenic (a black precipitate formed at high concentrations on the plates requiring hand-counting in tests with suspended material) following metabolic activation in strains TA98 (Table 3.22), TA100, TA1537, and TA1538. The DMSO extracts had high specific mutagenic activities. None of the materials was mutagenic to strain TA1535.

A blend representing NTSL low conversion first-stage process stream liquids, prepared by combining light oil (57%) and PS (43%), was distilled into nine discrete bp cuts (Sample L5241). With the exception of the initial cut (ambient to 350°F), the cuts were made at 50°F intervals. A broad middle distillate (composed of materials distilled between 400 and 650°F) and each of the nine smaller cuts were assayed for mutagenicity. A similar com-

Table 3.21 Results of Ames Assays on Nonintegrated Two-Stage Liquefaction (NTSL) Materials (Wilsonville, Alabama)

Material	Sample No.	Specific Activity[a] (rev/μg)
Light fuel oil	L5211	0
First-stage product distillate	L5212	0
Critical solvent deasher product	L5213	6
Deashed product	L5214	4
Hydrotreater feed	L5216	9
Hydrotreater light fuel oil	H0512	0
Hydrotreater distillate bottoms	H0513	6
Hydrotreater vacuum flash bottoms	H0514	5

[a]Specific activity in *Salmonella typhimurium* strain TA98 was estimated by linear regression of dose–response data (i.e., slope of the regression line). A positive response was one twice background with a correlation coefficient ≥0.8 for the regression line.

Source: Later (1985).

Table 3.22 Results of Ames Assays on Nonintegrated Two-Stage Liquefaction (NTSL) Materials (Wilsonville, Alabama)

Material	Concentration Ranges Tested (μg/plate)	Specific Activity[a] (rev/μg)
First-stage solid (Sample L5232)		
Aqueous suspension	500–60,000	0.03 (3)
Aqueous extract (concentrate)	0.2–50	0 (2)
Saline suspension	500–60,000	0.04 (4)
Saline extract	0.1–30	0 (2)
DMSO extract	0.2–50	137 (2)
Second-stage solid (Sample H0523)		
Aqueous suspension	500–60,000	0.08 (3)
Aqueous extract (concentrate)	0.5–50	0 (2)
Saline suspension	500–60,000	0.02 (4)
Saline extract	0.02–4	0 (2)
DMSO extract	0.2–40	195 (2)
2-Aminoanthracene	1	206 (15)

[a]Specific activity in *Salmonella typhimurium* strain TA98 was estimated by linear regression of dose–response data (i.e., slope of the regression line) with correlation coefficient ≥0.8. A positive response showed a reproducible dose-related (at least three concentrations) increase in number of revertants. Results shown are the average of (n) assays.

Source: Drozdowicz and Kelly (1983a, b).

plement of distillate cuts (Sample H0524) was prepared from another blend of NTSL representing low conversion second-stage process stream liquids. The initial blend contained hydrotreater unit light oil (34%) and PS (66%). Mutagenicity results for both sets of distillate cuts are presented in Table 3.23.

Some cuts induced a toxic response (smaller pinpoint colonies, diminution of the bacterial growth or lawn, and cell death) at higher test concentrations (>500 μg/plate). Only the highest bp cut (the 700–750°F heavy oil) of the first-stage blend was mutagenic. Strains TA98, TA100, and TA1538 showed a positive response following metabolic activation. In contrast, >650°F bp cuts from the second-stage blend were mutagenic. The middle oil cut No. 6 (550–600°F) was the only cut mutagenic to strain TA100. Higher bp cuts were mutagenic to strains TA98, TA100, and TA1538. The heavy oil (700–750°F) was also mutagenic to strain TA1537.

Three other NTSL materials were assayed for mutagenicity: a second-stage hydrotreater unit light oil (naphtha, Sample H0521), a residual fuel (Sample H0522), and a recycle coke condensate oil (Sample L2161). The hydrotreater unit light oil and residual fuel samples were found to be mutagenic for strains TA98, TA100, TA1537, and TA1538 following metabolic activation. The recycle coke condensate oil was mutagenic to strain TA1535 as well as the other four tester strains. Specific mutagenic activities of these materials to strain TA98 are shown in Table 3.23.

A suite of samples from a catalytic two-stage liquefaction (CTSL) process (Lawrenceville, New Jersey) was also assayed to determine mutagenicity, using metabolic activation and *S. typhimurium* strain TA98 (Later and Wilson, 1985). Tetrahydrofuran extracts of filtrates of test materials collected from operations at different temperatures were nonmutagenic or weakly mutagenic (Table 3.24). Where mutagenicity was detected, the dominant chemical class (fractionation by alumina adsorption-column chromatography) was the N-PAC.

Table 3.23 Results of Ames Assays on Nonintegrated Two-Stage Liquefaction Materials (NTSL) (Wilsonville, Alabama)

Material		Relative Weight (% of total)	Concentration Ranges Tested (μg/plate)	Specific Activity[a] (rev/μg)
	First-stage blend (Sample L5241)			
Light oil cut No. 1	(ambient to 350°F)	27.7	5-5000	0
Light oil cut No. 2	(350-400°F)	13.9	10-2500	0
Middle distillate	(400-650°F)	—	1-1000	0
Middle oil cut No. 3	(400-450°F)	15.7	5-2500	0
Middle oil cut No. 4	(450-500°F)	8.0	5-1000	0
Middle oil cut No. 5	(500-550°F)	6.6	1-1000	0
Middle oil cut No. 6	(550-600°F)	8.1	1-1000	0
Middle oil cut No. 7	(600-650°F)	6.1	1-1000	0
Heavy oil cut No. 8	(650-700°F)	4.2	1-1000	0
Heavy oil cut No. 9	(700-750°F)	3.9	5-1000	0.7
	Second-stage blend (Sample H0524)			
Light oil cut No. 1	(ambient to 350°F)	19.8	10-5000	0
Light oil cut No. 2	(350-400°F)	4.9	10-2500	0
Middle distillate	(400-650°F)	—	10-2500	0.2 (4)
Middle oil cut No. 3	(400-450°F)	4.1	5-2500	0
Middle oil cut No. 4	(450-500°F)	5.6	5-2500	0
Middle oil cut No. 5	(500-550°F)	8.2	10-2500	0
Middle oil cut No. 6	(550-600°F)	9.6	10-2500	0[b]
Middle oil cut No. 7	(600-650°F)	10.2	10-2500	0.02 (2)
Heavy oil cut No. 8	(650-700°F)	6.0	10-2500	0.2 (2)
Heavy oil cut No. 9	(700-750°F)	5.8	10-5000	2.4 (2)
Hydrotreater unit light oil (Naphtha, Sample H0521)			1-5000	2.5 (2)
Residual fuel (Sample H0522)			1-5000	5.7 (2)
Recycle coke condensate oil (Sample L2161)			1-500	10.9 (2)
2-Aminoanthracene			1	185 (12)

[a]Specific activity in *Salmonella typhimurium* strain TA98 was estimated by a linear regression of dose-response data (i.e., slope of the regression line) with correlation coefficient ≥0.8. A positive response showed a reproducible dose-related (at least three concentrations) increase in number of revertants. Results shown are the average of (n) assays.

[b]Slightly mutagenic in TA100 (0.1 rev/μg).

Source: Drozdowicz and Kelly (1983a-c).

3.4.1.1.1.5 Two-Stage Liquefaction—Other Materials. Mutagenic activity of coal-derived middle (390-610°F) and heavy (>610°F) oils (Samples L6211 and L6212) from a coal-liquefaction facility operated by Veba Oel/Ruhrkohle AG in the Federal Republic of Germany was also evaluated (Wilson and Pelroy, 1982). Crude materials and their chemical class fractions (alumina adsorption-column chromatography) were assayed in duplicate in *S. typhimurium* strain TA98 with and without metabolic activation at concentrations of 2, 4, 8, 10, 20, and 50 or 0.2, 0.5, 0.8, 1.0, and 2.0 μg/plate.

Middle oil and its fractions were nonmutagenic. Heavy oil was mutagenic (Table 3.25). No activity was detected in the aliphatic or PAH fractions. Although there was some activity in the hydroxy-PAH fraction, most activity occurred in the N-PAC fraction. All the nitrogen-containing chemical classes contributed to the mutagenicity, with the azaarenes having the highest specific mutagenic activity.

Table 3.24 Results of Ames Assays on Catalytic Two-Stage Liquefaction (CTSL) Materials (Lawrenceville, New Jersey)

Material	Sample No.	Temp. (°F)	Specific Activity[a] (rev/μg)
First-stage material	L5411	650	0
	L5412	550	0
	L5413	750	1
	L5414	750	2
Second-stage material	L5415	650	0
	L5416	550	<1
	L5417	750	<1
	L5418	750	3

[a]Specific activity in *Salmonella typhimurium* strain TA98 was estimated by linear regression of dose–response data (i.e., slope of the regression line). A positive response was one twice background with a correlation coefficient ≥0.8 for the regression line.

Source: Later and Wilson (1985).

A repository of synfuels research materials was established by the EPA and DOE. Chemical class fractions of these materials were assayed for mutagenicity in *S. typhimurium* strain TA98 with metabolic activation (Kimball and Munro, 1981). Whole materials were not tested directly. Acidic and basic components were separated (Swain–Stedman) first; then neutral components were separated by column chromatography (Fluorisil or Sephadex LH-20 columns). Weighted activities of each fraction were summed to estimate mutagenic activity of unfractionated materials.

Table 3.25 Results of Ames Assays on Coal-Liquefaction Distillates and Fractions (Veba Oel)

Material	Sample No.	Relative Weight (% of total)	Specific Activity[a] (rev/μg)	Weighted Activity[b] (rev/μg)
Middle oil	L6211	—	0	—
Heavy oil	L6212	100	11	11
Aliphatic		17.1	0	0
PAHs[c]		58	0	0
Nitrogen-PACs[d]		11.4	137	15.6
Carbazoles		4.2	22	0.9
Amino-PAH[c]		2.0	50	1
Azaarenes		4.3	108	4.6
Hydroxy-PAHs[c]		6	4	0.2
Benzo[a]pyrene		—	140	—
2-Aminoanthracene		—	875	—

[a]Specific activity in *Salmonella typhimurium* strain TA98 was estimated by linear regression of dose–response data (i.e., slope of the regression line). A positive response was one twice background with a correlation coefficient ≥ 0.8 for the regression line.

[b]Specific activity × relative weight (%).

[c]Polycyclic aromatic hydrocarbons.

[d]Polycyclic aromatic compounds.

Source: Wilson and Pelroy (1982).

Table 3.26 Results of Ames Assays on EPA/DOE Synfuel Repository Research Materials

Material	Sample No.	Specific Activity[a] (rev/μg)	Contribution (%)		
			Acids	Bases	Neutrals
Synthoil	L6111	5.2[b]	2	80	9
Zinc halide distillate	L6311	0.5	0	0	100
Hydrocarbonization product oil	L6411	5.1	0	19	9

[a] Specific activity in *Salmonella typhimurium* strain TA98 was estimated by linear regression of dose–response data (i.e., slope of the regression line). Activity of the material is expressed as the sum of activities of individual chemical class fractions corrected for their weight % contribution to the whole.

[b] Average of three reported values.

Source: Kimball and Munro (1981).

Materials representing six processes (SRC, H-Coal, EDS, Synthoil, Conoco zinc halide, and hydrocarbonization) were evaluated. Results of SRC, H-Coal, and EDS materials have been presented earlier, along with other materials from the same process. Results of a Synthoil material (Sample L6111), a zinc halide coal hydrocracking oil (Sample L6311), and hydrocarbonization product oil (Sample L6411) are shown in Table 3.26. Significant mutagenic activity occurred in the Synthoil (a refined vehicle oil) and hydrocarbonization oil. The zinc halide distillate had very little mutagenicity. Eight petroleum crude oils similarly assayed had little or no mutagenic activity (0–0.08 rev/μg). Although activity of coal-derived materials was principally confined to basic and neutral components, there was no consistent pattern; for example, 100% of the activity occurred in neutral components of the zinc halide distillate and 80% in basic components of the Synthoil sample (Kimball and Munro, 1981). The Synthoil material (Sample L6111) was also studied for mutagenicity as a DMSO extract of the whole material in strain TA98 and under these conditions was found to have a specific activity of 18.5 rev/μg (Warshawsky et al., 1981).

3.4.1.1.1.6 Two-Stage Liquefaction—Distillate Products. Four suites of crude oils and their distillate fractions representing SRC-II (Samples L1231–L1236), H-Coal syncrude (Samples L3131–L3136), H-Coal fuel oil (Samples L3121–L3124), and EDS coal liquids (Samples L4211–L4215) were evaluated for mutagenicity with *S. typhimurium* strains TA98, TA100, TA1535, TA1537, and TA1538 with and without metabolic activation (Calkins et al., 1983). Test concentrations were established after determining cytotoxicity by plating 10^3 previously exposed *S. typhimurium* strain TA1535 with excess L-histidine. Nontoxic and/or slightly toxic concentrations were assayed. Specific activity was determined by linear regression of the linear portion of the dose–response curve. Statistical analyses were used to compare mutagenic responses at each test concentration with the negative control.

Metabolic activation was required to demonstrate mutagenicity of most samples. Low levels of direct-acting mutagens (0.01–0.05 rev/μg) were detected in H-Coal (syncrude mode) and SRC-II bottoms and residue materials. Greatest activity occurred in strains TA98, TA1537, and TA1538. Coal liquids were several times more mutagenic than crude oil (southwest Louisiana petroleum).

Assay results with strain TA98 are shown in Table 3.27. Mutagenic activity increased with density of the distillate. For example, the H-Coal (syncrude mode) atmospheric still overheads were nonmutagenic, whereas vacuum still bottoms induced 8.2 rev/μg (Calkins

Table 3.27 Results of Ames Assays on Crude Coal Oils and Their Distillate Fractions

Material (Sample No.)	Specific Activity[a] (rev/μg)
SRC-II	
Naphtha (L1231)	0
Fuel oil (L1232)	1.2
Vacuum tower bottoms (L1235)	1.3
Stripper tower bottoms (L1234)	4.6
Heavy gas oil (L1233)	5.6
Residue (L1236)	4.5
H-Coal Syncrude Mode	
Atmospheric still overhead (L3131)	0
Naphtha (L3132)	0
Light gas oil (L3133)	0
Atmospheric still bottoms (L3134)	0.6
Vacuum still overhead (L3135)	1.1
Vacuum still bottoms (L3136)	2.7
H-Coal Fuel Oil Mode	
Atmospheric still overhead (L3121)	0
Atmospheric still bottoms (L3122)	1.4
Vacuum still overhead (L3123)	1.8
Vacuum still bottoms (L3124)	4.0
EDS	
Naphtha (L4211)	0
Solvent (L4212)	1.0
Vacuum gas oil (L4213)	5.8
Heavy gas oil (L4214)	2.8
Residue (L4215)	4.6
South Louisiana Petroleum Crude Oil	
Naphtha	0
Light gas oil	0
Medium gas oil	0.03
Residue	0.02

[a] Specific activity in *Salmonella typhimurium* strain TA98 was estimated by linear regression of dose–response data (i.e., slope of the regression line).

Source: Calkins et al. (1983)

et al., 1983). No major differences were found in the distillates from the three different processes.

3.4.1.1.1.7 Two-Stage Liquefaction—Hydrotreatment. In theory, a decrease in the aromaticity of coal liquids by the addition of hydrogen, that is, hydrotreatment, should decrease mutagenicity. Thus, several studies were initiated to assess the efficacy of hydrotreatment. These studies employed materials from a number of hydrotreatment processes and with varying degrees of severity (generally measured by resultant nitrogen content).

An SRC blend (Sample L1251), produced at the Ft. Lewis, Washington pilot plant, was catalytically hydrotreated (Weimer et al., 1980). Two hydrotreated samples (mild and severe hydrotreatment) and their isooctane fractions and similar materials from the untreated blend were assayed for mutagenicity in *S. typhimurium* strain TA98 with metabolic activation. Mutagenicity was detected in the untreated blend but not in hydrotreated samples. Acid and neutral fractions of the three materials were also nonmutagenic. However, basic and tar fractions (acid- and base-induced and isooctane insoluble) from the untreated blend were mutagenic (Table 3.28); the basic fraction had the highest specific activity. Both levels of hydrotreatment severity reduced mutagenicity of fractions below the level of detection (<0.05 rev/μg) except for the basic fraction. Severe hydrotreatment reduced mutagenicity of the basic fraction more than 90%.

Another SRC-II blend from the Ft. Lewis, Washington pilot plant (Sample L1221) was hydrotreated at varying severities in a bench-scale hydrogenation unit (Sutterfield et al., 1980). The untreated blend was assayed for mutagenicity with Ames *S. typhimurium* strains TA98 and TA100 using five test concentrations. In strain TA98, the specific activity was 10.9 rev/μg. Eight hydrotreated samples were assayed at one concentration (100 μg). Activity decreased at least one order of magnitude in hydrotreated samples from the 1122 rev/100 μg observed in untreated material. Activity decreased to below detectable limits when nitrogen and oxygen were below 0.04 and 0.6 wt %, respectively, and hydrogen increased from 8.4 to 10.8 wt %.

Two H-Coal materials, an atmospheric still bottoms (Sample L3161) and a vacuum overhead (Sample L3162), and two EDS materials, a 400–700°F distillate (Sample L4221)

Table 3.28 Results of Ames Assays on Solvent-Refined Coal II Distillate Blend With or Without Hydrotreatment (Ft. Lewis, Washington)

Material[a]	Nitrogen Content (ppm)	Relative Weight (% of total)	Specific Activity[b] (rev/μg)	Weighted Activity[c] (rev/μg)
SRC-II blend (Sample L1251)	8570	100	3	3
Basic		5	16	1
Acid-induced tar		6	3	<1
Base-induced tar		1	6	<1
Isooctane-insoluble tar		1	9	<1
Moderate hydrotreatment	5510	100	0	0
Basic		1	2	<1
Acid-induced tar		<1	0	0
Base-induced tar		<1	0	0
Isooctane-insoluble tar		<1	0	0
Severe hydrotreatment	1900	100	0	0
Basic		1	1	<1
Acid-induced tar		1	0	0
Base-induced tar		<1	0	0
Isooctane-insoluble tar		<1	0	0

[a] Only mutagenic fractions are shown.

[b] Specific activity in *Salmonella typhimurium* strain TA98 estimated by linear regression of dose–response data (i.e., slope of regression line). Positive response was with a maximum response of at least twice background for at least one test concentration and a correlation coefficient ≥ 0.8 for dose–response data.

[c] Specific activity × relative weight (%).

Source: Weimer et al. (1980).

and a solvent blend (L4231), were studied for mutagenic activity before and after hydrotreatment (Sutterfield, 1983). Ames assays were conducted with *S. typhimurium* strain TA98, using only one fixed test concentration (50 μg). Materials that induced a response twice background were considered to be mutagenic. The EDS solvent blend and the H-Coal atmospheric still material were mutagenic prior to hydrotreatment; the other two materials were nonmutagenic. Increasing hydrogen weight % between 1 and 3% made the mutagenic materials nonmutagenic. The EDS materials and their hydrotreated products were solvent fractionated (three fractions), and the fractions were tested for mutagenicity (100-μg test concentration). Hydrotreatment reduced the mutagenicity of the fractions.

Blended materials from SRC-I (Wilsonville, Alabama) and SRC-II (Ft. Lewis, Washington), EDS (Baytown, Texas), and H-Coal (Trenton, New Jersey PDU and Catlettsburg, Kentucky pilot plant) processes were hydrotreated to varying levels of severity (low, medium, and high based on nitrogen content) and tested for mutagenicity (Epler et al., 1981; Griest et al., 1983). Materials were either extracted with DMSO and assayed for mutagenicity in *S. typhimurium* strain TA98 with metabolic activation or separated into chemical class fractions and the fractions assayed.

In one study, the effect of hydrotreatment on the mutagenicity of recycle solvent, light organic liquid, and wash solvent from the SRC-I pilot plant and H-Coal PDU distillate was determined (Epler et al., 1981). The assay was conducted with *S. typhimurium* strain TA98 with metabolic activation on chemically fractionated materials and the activity of the unfractionated material was estimated by a summation of the activity found in the fractions (Table 3.29). The SRC-I light organic liquid (Sample L2152) and wash solvent (Sample

Table 3.29 Results of Ames Assays on Coal-Liquefaction Materials With and Without Hydrotreatment

Material	Nitrogen Content (ppm)	Relative Weight (% of total)	Specific Activity[a] (rev/μg)	Weighted Activity[b] (rev/μg)
SRC-I Recycle Solvent (Sample L2132)				
Crude	6100	100	2.7	2.7
Aliphatic	—	2.9	29.9	0.9
Neutral polyaromatic	—	19.4	9.4	1.8
Residue	—	0.2	12.0	0.02
Hydrotreatment: low severity	3700	—	NT	—
medium severity	1000	—	0	0
high severity	300	—	0	0
H-Coal Distillate (Sample L3112)				
Crude	3900	100	0.2	0.2
Ether-soluble bases	—	1.5	3.9	0.06
Neutral polyaromatic	—	5.3	3.3	0.2
Hydrotreatment: low severity	2000	—	0.4	—
Neutral polyaromatic	—	2.5	15.6	0.4
Hydrotreatment: medium severity	1300	—	0.1	—
Neutral polyaromatic	—	1.6	5.4	0.1
Hydrotreatment: high severity	900	—	0	—

[a] Specific activity in *Salmonella typhimurium* strain TA98 was estimated by linear regression of dose–response data (i.e., slope of the regression line).

[b] Specific activity × relative weight.

Source: Epler et al. (1981).

L2142) were nonmutagenic, as were their hydrotreated materials. The recycle solvent (Sample L2132) had mutagenic activity present in the aliphatic, neutral polyaromatic, and residue fractions (total weighted activity of 2.7 rev/μg) before hydrotreatment, but no activity could be detected in any of the hydrotreated materials. Some activity was detected in the H-Coal distillate (Sample L3112), with the bulk of the activity (0.2 rev/μg) in the neutral polyaromatic fraction, and minor activity in the ether-soluble base fraction. Low severity of hydrotreatment had no effect, but more severe treatment reduced activity in the neutral polyaromatic fraction, with complete elimination at high severity.

In a second study, where blended materials were evaluated for mutagenicity before and after hydrotreatment, both DMSO extracts of whole materials and fractionated materials were assayed (Griest et al., 1983). Results (Table 3.30) of activity found in the crude material before and after hydrotreatment were compared, as were the chemical classes responsible for the activity (Cowser, 1982b). A crude SRC-II blend (Sample L1262) had mutagenic activity (1.7 rev/μg), but no activity was detected in its unfractionated hydrotreated samples. Activity in the crude blend was contained in two fractions, the PAH-containing neutral polyaromatic fraction (0.2 rev/μg) and the ether-soluble base fraction (0.7 rev/μg). In this case, the estimated activities compared only qualitatively with the actual determinations. The neutral polyaromatic was the only fraction showing measurable activity after hydrotreatment. Whereas the polyaromatic fraction of the crude blend (5.9%

Table 3.30 Results of Ames Assays on Coal-Liquefaction Materials With and Without Hydrotreatment

Material	Nitrogen Content (ppm)	Relative Weight (% of total)	Specific Activity[a] (rev/μg)	Weighted Activity[b] (rev/μg)
SRC-II Blend (Sample L1262)				
Crude	—	100	1.7	1.7
Ether-soluble bases	—	5.9	0.7	0.04
Neutral polyaromatic	—	5.9	0.2	0.01
Hydrotreatment: low severity	5300	—	0	—
Neutral polyaromatic	—	2.1	0.1	<0.01
Hydrotreatment: medium severity	1710	—	0	—
Neutral polyaromatic	—	0.6	0.02	<0.01
Hydrotreatment: high severity	505	—	0	—
Neutral polyaromatic	—	0.3	0.01	<0.01
EDS Blend (Sample L4116)				
Crude	—	—	0	—
Hydrotreatment: high severity	0.2	—	0	—
H-Coal Distillate (Sample L3231)				
Crude	—	100	5.4	5.4
Crude, washed (Sample L3232)	—	100	4.9	4.9
Hydrotreatment: low severity	2650	—	0	—
medium severity	505	—	0	—
high severity	0.2	—	0	—

[a] Specific activity in *Salmonella Typhimurium* strain TA98 was estimated by linear regression of dose-response data (i.e., slope of the regression line).

[b] Specific activity × relative weight (%).

Source: Griest et al. (1983).

of total material) had a specific activity of 0.2 rev/μg, only 0.01 rev/μg were found in the high-severity hydrotreatment polyaromatic fraction (0.3% of the sample mass). An EDS blend (Sample L4116) was nonmutagenic before and after hydrotreatment. An H-Coal distillate blend (Sample L3231) had a specific mutagenic activity of 5.4 rev/μg. All hydrotreated materials were nonmutagenic.

3.4.1.1.1.8 Coal Gasification—Industrial/Commercial. In contrast to liquefaction studies, where materials from a number of stages of a given process development were evaluated, gasification materials assayed generally represented a single stage of process development. Thus, Ames mutagenicity results for coal-gasification materials are discussed according to facility scale (industrial/commercial, pilot-plant, and experimental reactors). A study on materials resulting from *in situ* gasification is also discussed. It should be noted that coal-gasification materials used in these studies were not planned for use as products or feedstocks but rather were plant residues or by-products.

Ames mutagenicity data for materials from operating industrial gasifiers (Ft. Snelling, Glen-Gery, Holston, Kosovo, and UMD) were generally obtained using *S. typhimurium* strains TA98, TA100, TA1535, TA1537, and TA1538 with and without metabolic activation. Strain TA98 was consistently the most sensitive to mutation with coal-derived materials. Assays on all except UMD materials were performed on arbitrarily fixed concentrations (4–5) of unfractionated materials, usually covering a 100-fold range (Brusick and Young, 1981). UMD materials were assayed in *S. typhimurium* strain TA98 only after fractionation. Mutagenicity of crude materials was estimated from the sum of specific activities of fractions after corrections for their relative mass.

Composite product gas condensate (Sample G1711) and test burner effluent gas particulate and organic condensate (Sample G1712) were collected from a low-Btu gasifier at the U.S. Bureau of Mines iron-ore-pelletizing pilot plant (Ft. Snelling, Minnesota). Samples were derived during operations with Indianhead lignite coal, dissolved or suspended in DMSO, and initially tested at 100, 150, 200, and 250 μL/plate (duplicate plates at each concentration). Because product gas composite was highly cytotoxic to all test strains, tests with this material were repeated at lower concentrations (10, 2, 0.4, and 0.08 μL/plate). At 10 μL/plate, the material was still cytotoxic to strains TA98 and TA1538 without metabolic activation and to TA100 with or without activation. No mutagenic activity was detected in any strain with either material (Kilpatrick et al., 1980).

Ames data on product gas particulates (>3-μm fraction and a <3-μm fraction) and volatile organic condensate (Sample G1611) and test burner flue gas particulate and organic condensate (Sample G1612) from a Wellman–Galusha gasifier (Glen-Gery Brickyard, York, Pennsylvania) were also nonmutagenic. Concentrations of each material tested were not reported. Test materials were derived from anthracite coal (Thomas et al., 1979).

Two gasification tars were collected during different operational periods at the Chapman low-Btu gasifier (combustion fuel for process heaters), Holston Munitions plant in Tennessee. A tar derived during operations with Virginia bituminous coal (Sample G1811) was mutagenic, but specific activities were not reported (Page, 1978). The other tar (Sample G1821), which had two phases (a solid tar with liquid on its surface), was also mutagenic (Warshawsky, 1979). The liquid top phase was nonmutagenic to *S. typhimurium* strains TA98 and TA100. The tar dissolved in DMSO was mutagenic to strains TA98, TA100, and TA1538 and nonmutagenic to TA1535. Mutagenesis required metabolic activation and was considered positive when the number of revertants per plate exceeded twice the spontaneous rate. Twenty-four tar concentrations (0.237–980 μg/plate) were tested in six separate experiments over a 6-month period with strains TA98 and TA100. Specific activities with each strain, calculated by linear regression of all concentrations, were 6.8 and 6.5

rev/μg, respectively (Schoeny et al., 1981). In the four individual experiments where three or more concentrations were within the linear portion of the dose–response curve, specific activities for strain TA98 were 18, 11, 6, and 3 rev/μg, respectively (Warshawsky, 1979).

Tar solutions were prepared in two additional ways. First, tar was extracted with solvents of increasing polarity (hexane, toluene, methylene chloride, and acetonitrile). The largest extract (56 wt %) was methylene chloride soluble. Five concentrations of each extract were assayed on three occasions with strains TA98 and TA100. All fractions were mutagenic, including nonextractable material (19% of sample). The weighted sum of the specific activities of fractions was significantly greater than the whole-sample activity, indicating some components in the whole tar were antagonistic to mutagenic expression (Schoeny et al., 1981). Second, because many coal-derived materials are dissolved in acetone for skin tumorigenicity studies, acetone-soluble components of the tar (acetone was not used as a solvent in the first extraction procedure) were extracted and assayed for mutagenic activity after removing the acetone. Although the amount of tar dissolved in acetone was similar to that in DMSO, mutagenicity was less. No specific activity could be calculated from the concentration range assayed (Warshawsky, 1979).

Organic by-products from the commercial gasifier (Lurgi) in Kosovo, Yugoslavia were evaluated for mutagenicity in *S. typhimurium* strains TA98, TA100, TA1535, TA1537, and TA1538 (Lee et al., 1981). Samples included tar (Sample G1412), combined heavy tar and dust (Sample G1411), medium oil (Sample G1413), and light oil/naphtha (Sample G1414). Both tar samples were mutagenic after metabolic activation in TA98 and TA1538 (Table 3.31). Medium and light oils were nonmutagenic at the concentrations tested (0.01–0.025 μL/plate). The highest concentration tested was slightly toxic (visual observation) to all tester strains.

Mutagenicity was also evaluated for by-product tars produced by the UMD Foster–Wheeler Stoic gasifier, where the gasifier provided low-Btu gas to fuel a steam boiler (Epler et al., 1983). Mutagenicity of fractions (Swain–Stedman fractionation) of 22 electrostatic precipitator tars (Samples G1301–G1308 and G1310–G1323) collected over three sampling periods (2 years) was determined in *S. typhimurium* strain TA98 (Table 3.32). Estimates of whole-tar mutagenicity were calculated by summing the specific activities of the fractions corrected for their weight percentage contribution to the whole tar. Metabolic activation was required for mutagenicity. In all but two samples, the greatest activity (%) occurred in base fractions. Of 22 samples, 17 had low mutagenicities (average, 0.3 rev/μg; range, 0–

Table 3.31 Results of Ames Assays on Kosovo, Yugoslavia Gasification By-Products

Material	Sample No.	Specific Activity[a] (rev/μg)
Tar	G1412	8.6
Heavy tar and dust	G1411	5.1
Medium oil	G1413	0
Light oil	G1414	0

[a] Specific activity in *Salmonella typhimurium* strain TA98 was estimated by linear regression of dose–response data (i.e., slope of the regression line). A positive response was once twice background with a correlation coefficient $\geq$ 0.8 for the regression line; four concentrations (0.01, 0.015, 0.02, and 0.025 μL/plate) were tested; postive control included two determinations for 2-aminoanthracene, 256 $\pm$ 17 and 373 $\pm$ 40 rev/10 μg, respectively.

Source: Lee et al. (1981).

Table 3.32 Results of Ames Assays on Tars Produced in the University of Minnesota–Duluth Gasifier

Material	Sample No.	Specific Activity[a] (rev/μg)
Electrostatic precipitator tars	G1301	0.8
	G1302	0
	G1303	0.1
	G1304	0.1
	G1305	0.1
	G1306	0.7
	G1307	0.5
	G1308	0.2
	G1310	0.1
	G1311	1.9
	G1312	0.7
	G1313	41.3
	G1314	11.8
	G1315	9.5
	G1316	0.2
	G1317	0.2
	G1318	0.2
	G1319	0.2
	G1320	8.2
	G1321	0.3
	G1322	0.3
	G1323	0.3
Holding tank tar	G1324	0
Vent gas cyclone tar	G1325	0.04
Dirty top gas tar	G1326	0.9

[a] Specific activity in *Salmonella typhimurium* strain TA98 was estimated by linear regression of dose–response data (i.e., slope of the regression line) in chemical class fractions (Swain–Stedman fractionation). Activity of the material is expressed as the sum of activities of individual chemical class fractions corrected for their weight % contribution to the whole.

Source: Epler et al. (1983).

0.8 rev/μg). The other five samples had mutagenicity values greater than 1 rev/μg (average, 12.5 rev/μg; range, 1.9–41.3 rev/μg). A sample of vent gas cyclone tar (Sample G1325) was less active (0.04 rev/μg) than most of the electrostatic precipitator tars, while dirty top gas tar (Sample G1326) taken on the same day was slightly more active (0.9 rev/μg). Holding-tank tar (Sample G1324), representing 5 days of operation, showed no mutagenic activity.

3.4.1.1.1.9 Coal Gasification—Pilot Plants. Extensive studies were conducted on by-products from three pilot-plant gasifiers. Two of the gasifiers were at DOE facilities, the McDowell–Wellman (Wellman–Galusha) atmospheric pressure stirred-bed gasifier at METC, Morgantown, West Virginia and the slagging fixed-bed gasifier at UNDERC (formerly the Grand Forks Energy Technology Center), Grand Forks, North Dakota. The third was the Hygas advanced fluidized-bed gasifier at the Institute of Gas Technology in Chicago, Illinois.

Ames assays were performed on process stream organic condensates and tars produced by the METC gasifier. Samples were collected during four runs with bituminous coal, covering a 4-year period. Metabolic activation was required to show mutagenicity of METC gasifier samples. Results with process stream condensates and the various by-product tars indicated the efficacy of the product gas cleanup system in eliminating mutagenic components.

Raw gas was collected during all four test runs. Raw gas (Sample G1242) from the first run sampled was cooled and diluted with nitrogen to simulate atmospheric conditions; process stream organics were collected on a prefilter attached to a Tenax column (Newton, 1979). A methylene chloride extract of the filter condensate was mutagenic to *S. typhimurium* strains TA98 and TA100 at 250 μg/plate (Table 3.33). Because only one concentration was assayed, specific mutagenic activities could not be determined.

Product gas samples from the other three runs included both aerosol droplets and condensates collected at several points along the gas cleanup system. A set of size-fractionated aerosol droplet samples (G1222–G1226) collected in a multicyclone train showed mutagenicity to *S. typhimurium* strain TA100. A single concentration of 50 μg/plate was assayed. While specific mutagenic activity was not significantly altered with product gas stream cleanup, removal of most organic materials (1000-fold reduction) effectively decreased total activity. There was no apparent relation between aerosol size (five effective cutoff diameters, from 5.4 to 0.3 μm) and mutagenic activity, but because only one concentration of each aerosol condensate was tested, the effect of aerosol size remains uncertain (Benson et al., 1982a).

Raw gas condensates (Samples G1226, G1232, and G1212) collected with a condensor train (different temperatures) were mutagenic to strains TA98 and TA100. Condensates were assayed over a range of concentrations, and specific mutagenic activities from 4 to 24

Table 3.33 Results of Ames Assays on the Morgantown Energy Technology Center Gasifier Product Gas

	Specific Activity[a] (rev/μg)				
Concentrations Tested (μg/plate)	Raw Gas	After Tar Trap	After Venturi Scrubber	Clean Gas	Gas Combustion Products
250	0.4[b] (G1242)	—	—	—	—
50	4.5[c] (G1222)	2.4[c] (G1223)	3.4[c] (G1224)	1.5[c] (G1225)	—
5–250	4.1[d] (G1226)	—	—	—	—
[e]	6.7 (G1232)	3.7 (G1233)	4.1 (G1234)	0 (G1235)	1.1 (G1236)
[e]	23.8 (G1212)	—	—	0 (G1213)	—

[a] Specific activity in *Salmonella typhimurium* strain TA98 was estimated by linear regression of dose–response data (i.e., slope of the regression line) unless indicated otherwise. Materials were considered to be mutagenic if they exhibited a dose–response relation and the number of revertants was at least twice the number of spontaneous revertants.

[b] Average of two plates, mutagenic at dose tested on basis of twice background (Newton, 1979).

[c] Mutagenic activity of aerosol droplets (>0.3 μm) in *Salmonella typhimurium* strain TA100, average of three plates less background. Underline indicates a response greater than twice background (Benson et al., 1982a).

[d] The 0°C condenser sample, methylene chloride soluble (Benson et al., 1982a).

[e] Concentration range (five concentrations) predetermined to be below cytotoxic levels and at a revertant induction frequency within the linear portion of the dose response (Benson et al., 1984; Shimizu et al., 1984).

rev/μg were detected in strain TA98 (Table 3.33). The mutagenicity of raw gas condensate (Sample G1232) was compared with that of condensate (Sample G1233) collected after removing relatively high-molecular-weight tars (after the tar trap) and condensate (Sample G1234) collected after removing low-molecular-weight tars (after venturi scrubber). Although there was no major difference in specific mutagenic activities (strain TA98), organic mass loadings and thus overall mutagenicity of the gas stream decreased significantly (81% mass reduction after the tar trap with an additional 78% cleanup after the venturi scrubber, or a total of 96% reduction in condensable organics). Two clean gas samples (Samples G1235 and G1213) contained little organic material other than product gas (e.g., Sample G1235 contained only 63 ng organic material/L). When assayed with TA98 or TA100, clean gas samples were nonmutagenic. Gas combustion condensate (Sample G1236) had 3.6 μg organic material/L and a specific mutagenic activity of 1.1 rev/μg (Benson et al., 1984; Shimizu et al., 1984).

Higher-molecular-weight fractions (3, 4, and 5) from chromatographic fractionation (Sephadex LH-20) of the product gas condensate (e.g., Sample G1232) and from the tars (discussed below) contributed significantly to the mutagenic activity. Mutagenicity of these materials can generally be attributed to neutral and polar PACs (Benson et al., 1984). Occasionally, fractions 1, 2, and 6 were also mutagenic. Specific compounds in each fraction are discussed in Chapter 5.

Tars had mutagenic activities consistent with those determined for product gas stream condensates. There were two sources of tar: relatively high-molecular-weight tars condensed from the raw gas stream into a humidifier tar trap by temperature reduction (1200 to 400°F) and a low-molecular-weight tar removed by the venturi scrubber (400 to 300°F).

Tar-trap tars from three separate runs were assayed (Table 3.34). One (Sample G1211) was assayed at a single concentration (250 μg/plate) with *S. typhimurium* strains TA98, TA100, TA1537, and TA1538. Crude tar was mutagenic (number of revertants exceeded twice background) following metabolic activation in all but TA100. Fractions of the tar (Sephadex LH-20 chromatography) were mutagenic in strain TA100 as well as the other tester strains (Newton, 1979).

Table 3.34 Results of Ames Assays on Morgantown Energy Technology Center Tars

Material	Concentrations Tested (μg/plate)	Specific Activity[a] (rev/μg)
Tar-trap tars		
Sample G1211	250	1.2[b]
Sample G1221	0.5–25	8.4 (0.6)
Sample G1231	—[c]	21.6 (1.8)
Venturi scrubber water (Sample G1227)		
Sediment	5–50	3.1 (0.8)
Lyophilized residue	5–50	0.4 (0.2)
Methylene chloride solubles	5–50	0.7 (0.2)
Benzo[a]pyrene	2.5	153[d]

[a] Specific activity in *Salmonella typhimurium* strain TA98 was estimated by linear regression of dose–response data (i.e., slope of the regression line) unless indicated otherwise. Values in parentheses are standard deviations (Benson et al., 1982b).

[b] Average of two plates, mutagenic at dose tested based on a mutation response of twice background (Newton, 1979).

[c] Five concentration test ranges predetermined to be below cytotoxic levels and at a revertant induction frequency within the linear portion of the dose response (Henderson, 1982).

[d] Activity at 2.5-μg concentration, average of three determinations.

Mutagenicity assays of other METC tars were conducted only with strain TA98. Tar samples G1221 and G1231 had specific mutagenic activities of 8.4 and 21.6 rev/μg, respectively (Table 3.34). Sample G1221 was assayed over a fixed range of concentrations, whereas G1231 was assayed over a concentration range experimentally predetermined to be within the linear dose–response region (Benson et al., 1982b; Henderson, 1982).

Sedimented tar, lyophilized water residue, and methylene chloride extract of organics in the venturi scrubber water (Sample G1227) were assayed for mutagenicity (Table 3.34). Tar sediment had a specific mutagenicity of 3.1 rev/μg (fixed test concentration range). Lyophilized residue or methylene-chloride-soluble organics in the water (0.17% of the sample mass) had specific mutagenic activities of 0.4 and 0.7 rev/μg, respectively (Benson et al., 1982b).

Additional assays were performed on bulk tars and oils from the UNDERC slagging fixed-bed gasifier (Wilzbach et al., 1982). All materials were from the same test run and were studied with *S. typhimurium* strain TA98. Metabolic activation was required to demonstrate mutagenicity. The three oil samples (Samples G1125–G1127) were not significantly mutagenic (all less than 1 rev/μg). The four tars (Samples G1121–G1124) were mutagenic; average specific activity was 3.8 rev/μg with a range of 1–8 rev/μg (Table 3.35). Both basic and neutral components contributed to mutagenicity, with neutral components (because of their large mass: 65%) accounting for about two-thirds of the weighted activity.

A raw gas condensate sample (Sample G1111), collected during a different test run using a side stream sampler with five temperature/pressure traps, was also assayed (Wilzbach et al., 1982). Aqueous-phase organics from all traps were not significantly mutagenic. However, oil-phase nonvolatile organic (NVO) condensates (materials with bp greater than

Table 3.35 Results of Ames Assays of University of North Dakota Energy Research Center Tars and Oils

Material	Relative Weight (% of total)	Specific Activity[a] (rev/μg)	Weighted Activity[b] (rev/μg)
Oils			
Sample G1125	—	<1	—
Sample G1126	—	<1	—
Sample G1127	—	<1	—
Tars			
Sample G1122	—	1	—
Sample G1123	—	8	—
Sample G1124	—	4	—
Sample G1121	100	2	2
Acid	30	0	0
Base	2.4	58	1.4
Neutral	65.2	4	2.6
Benzo[a]pyrene	—	255	—

[a]Specific activity in *Salmonella typhimurium* strain TA98 was estimated by linear regression of dose–response data (i.e., slope of the regression line). A positive response was considered significant as discussed by Carnes et al. (1985).

[b]Specific activity × relative weight (%).

Source: Wilzbach et al. (1982).

390°F) from all traps were mutagenic (Table 3.36). Most mutagenic activity occurred in the high-pressure cyclone (the first condensation step), where half the organic was condensed. Mutagenicity of the stream condensate calculated from the activity in each trap would be about 10 rev/μg. Thus, calculated mutagenicity of the composite gas condensate was consistent with measured mutagenicities of bulk tars (Table 3.35).

Ames assay of Hygas by-products primarily involved *S. typhimurium* strain TA98 because early investigations showed it to be consistently the most sensitive (Reilly et al., 1981, 1986). Occasionally, fractions of crude materials were also evaluated in strain TA100. Metabolic activation was required to demonstrate mutagenic activity. The Hygas process requires slurrying of feed coal with process-generated oils. In the pilot plant, slurry oil was substantially augmented with commercial-grade toluene. Samples were generally freed of volatile low-molecular-weight materials (including exogenous toluene) prior to assay. These highly volatile materials were nonmutagenic.

Greatest mutagenicity (Table 3.37) in Hygas by-products occurred in raw gas condensates from the low-temperature reactor (LTR). Specific mutagenic activity of NVO components of condensates collected from four different test runs averaged 61 rev/μg (range, 10–134 rev/μg). Samples with activities greater than 100 rev/μg all came from the same test run. No explanation was determined for the 10-fold lower values detected in the other three runs. NVO components had considerable mass, averaging 15% of the raw gas stream (up to 36% in one instance). After leaving the LTR, most condensable materials in the gas stream appear to condense on the relatively cool incoming chars in the slurry-bed drier and fall back into the gasifier, because mass contributions of NVO components of materials collected outside the gasifier are significantly lower (e.g., recycle oil; average, 2.9% mass).

Hygas recycle oil NVO had lower specific mutagenic activities than those of the LTR condensates. Specific activity of recycle oil averaged 10 rev/μg (range, 2–23 rev/μg). The NVO of a composited recycle oil (Sample G2151) used in whole-animal toxicology tests had a specific activity of 12.7 rev/μg. Char slurry oil is essentially recycle oil that has contacted pretreated char. Thus, NVO specific mutagenic activities of the samples assayed were similar to those of recycle oil; average specific mutagenicity was 9 rev/μg (range, 3–22 rev/μg). The mass weight % (average, 2.6%) of these materials is also similar. As with

Table 3.36 Results of Ames Assays of University of North Dakota Energy Research Center Product Gas Stream

Material	Relative Weight (% of total)	Specific Activity[a] (rev/μg)	Weighted Activity[b] (rev/μg)
Gas stream condensate (Sample G1111)			
High-pressure cyclone	50	19	9.5
Low-pressure cyclone	7	5	0.4
First heat exchanger	31	1	0.3
Second heat exchanger and cryogenic trap	10	<1	<0.1
Benzo[a]pyrene	—	255	—

[a]Specific activity in *Salmonella typhimurium* strain TA98 was estimated by linear regression of dose–response data (i.e., slope of the regression line). A positive response was considered significant as discussed by Carnes et al. (1985).

[b]Specific activity × relative weight (%).

Source: Wilzbach et al. (1982).

Table 3.37 Results of Ames Assays on Hygas Materials

Material	Sample No.	Relative Weight (% of total)	Specific Activity[a] (rev/μg)	Weighted Activity[b] (rev/μg)
Low-temperature reactor gas condensate, organic-phase nonvolatile organic (NVO)	G2117	17.2	134	23
	G2118	32.4	127	41.1
	G2119	35.5	104	36.9
	G2126	8.2	12	1
	G2127	3.4	21	0.7
	G2132	0.3	13	0.04
	G2144	6.2	10	0.6
Recycle oil NVO	G2111	4.3	22	0.9
	G2112	4.0	6	0.2
	G2113	5.7	23	1.3
	G2122	1.1	2	0.02
	G2123	1.0	3	0.03
	G2141	1.2	4	0.05
Char slurry oil NVO	G2114	4.0	12	0.5
	G2115	4.3	13	0.6
	G2116	3.9	22	0.9
	G2124	1.1	3	0.3
	G2125	1.8	4	0.7
	G2131	1.8	3	0.05
	G2142	1.2	6	0.7
Light oil recovery system oil NVO	G2143	0.04	5	0.002
Benzo[a]pyrene	—	—	260	—

[a]Specific activity in *Salmonella typhimurium* strain TA98 was estimated by linear regression of dose–response data (i.e., slope of the regression line). A positive response was considered significant as discussed by Carnes et al. (1985).

[b]Specific activity × relative weight (%).

Source: Reilly et al. (1983, 1986).

the LTR condensates, samples of recycled oil and char slurry oil from the initial run tested had substantially higher specific mutagenic activities than samples from the other three test runs. Oil (Sample G2143) from the light oil recovery system (oils condensed on entrained char or fines and later stripped from cyclone slurry water) had few high-molecular-weight organic components (NVO weight, 0.04%). Specific mutagenic activity of the NVO was 5 rev/μg (Reilly et al., 1986).

Several Hygas materials were fractionated into broad chemical classes (acidic, basic, and neutral components) by organic/aqueous partitioning on the basis of pH (Bourne et al., 1981) to identify the classes responsible for mutagenic activity (Stamoudis et al., 1981; Reilly et al., 1986). Mutagenicity results in *S. typhimurium* TA98 for fractions of LTR gas condensate (sample G2119), recycle oil (Sample G2111), and char slurry oil (Sample G2115) are presented in Table 3.38. The acidic fraction accounts for up to 30% of a given material but was not significantly mutagenic. With all four materials, highest specific mutagenic activity is found in the basic fraction, the smallest fraction by weight. The neutral fraction is the largest weight fraction and even though its specific mutagenic activity is significantly lower than the basic fraction, weighted activity often exceeds the basic fraction. Chemical analysis of the neutral fraction indicated that the mutagenic activity of the identifiable mu-

Table 3.38 Results of Ames Assays on Chemical Fractions of Hygas Materials

Material	Relative Weight (% of total)	Specific Activity[a] (rev/μg)	Weighted Activity[b] (rev/μg)
Low-Temperature Reactor Gas Condensate (Sample G2119)			
Organic phase NVO[c]	100	104	104
Acids	27	<1	<1
Bases	5	325	16
Neutrals	63	36	23
Recycle Oil (Sample G2111)			
NVO	100	22	22
Acids	30	<1	<1
Bases	4	83	3
Neutrals	50	3	2
Char Slurry Oil (Sample G2115)			
NVO	100	13	13
Acids	20	<1	<1
Bases	21	93	20
Neutrals	46	4	2
Char Slurry Oil (Sample G2124)			
NVO	100	4	4
Acids	26	<1	<1
Bases	3	14	<1
Neutrals	50	3	2
Benzo[a]pyrene	—	260	—

[a] Specific activity in *Salmonella typhimurium* strain TA98 was estimated by linear regression of dose–response data (i.e., slope of the regression line). A positive response was considered significant as discussed by Carnes et al. (1985).

[b] Specific activity × relative weight (%).

[c] Nonvolatile organics.

Source: Stamoudis et al. (1981) and Reilly et al. (1986).

tagens and carcinogens was not totally expressed. When BaP was added to the neutral fraction, expression of mutagenic activity was indeed inhibited (Haugen and Peak, 1983). Competition for metabolic activation was determined to be the chief contributor to the inhibition.

Char-Oil Energy Development (COED) pyrolysis oil (Sample G2221), considered similar to that which would be produced by the COGAS process, was distilled into representative commercial products (naphtha, No. 2 fuel oil, and No. 6 fuel oil). These materials and similar products prepared from two composited crude petroleums were evaluated for Ames bacterial mutagenicity (Peirce and Simmon, 1981). One crude oil was more aromatic than the COED oil (aromatic oil), the other less aromatic (paraffinic oil).

The Ames mutagenicity assay was modified to allow preincubation of *S. typhimurium* strains TA98, TA100, TA1535, TA1537, and TA1538 with test material (room temperature, 30 min). A preliminary assay with TA100 (eight test concentrations, highest being the maximum soluble concentration in DMSO with serial half-log dilutions from the high concentration) established toxic concentrations. Thereafter, five concentrations were selected based on sample solubility, toxicity, and mutagenicity. Samples were considered mutagenic if they induced a reproducible dose-related response in any tester strain.

Table 3.39 Results of Ames Assays on Char-Oil Energy Development (COED) Distillate Cuts

	Specific Activity[a] (rev/μg)		
Material	Naphtha	No. 2 Fuel Oil	No. 6 Fuel Oil
COED oil (Sample G2221)	0	0	1
Aromatic crude oil	0	0.06	0.2
Paraffinic crude oil	0	0	NDR[b]

[a]Specific activity in *Salmonella typhimurium* strain TA98 was estimated by linear regression of dose–response data (i.e., slope of the regression line). A positive response was twice background with a correlation coefficient $\geq$ 0.8 for the regression line; positive control, 2-aminoanthracene 78 rev/μg at a 2.5-μg/plate test dose.

[b]No linear dose response.

Source: Peirce and Simmon (1981).

Metabolic activation was required to detect mutagenicity. Results for strain TA98 are shown in Table 3.39. The naphtha from all three materials and No. 2 fuel oil from COED oil and paraffinic crude were nonmutagenic. The No. 2 fuel oil from aromatic crude had measurable mutagenic activity in strains TA98, TA100, and TA1538. All No. 6 fuel-oil distillates were mutagenic to these three strains. COED oil was also mutagenic to TA1537. Strain TA1535 was not responsive to any test material.

The mutagenicity of a second crude COED pyrolysis oil (Sample G2211) was found to have a specific mutagenic activity of 0.4 rev/μg (Kimball and Munro, 1981).

3.4.1.1.1.10 Coal Gasification—Experimental Reactors. Bench-scale reactors at Research Triangle Institute (RTI) and Mellon Institute (MI) provided the opportunity to evaluate the influence of process conditions on chemical and toxicological properties of resultant materials. Ames bacterial mutagenicity assays of materials from RTI were performed with *S. typhimurium* strains TA98, TA100, TA1535, TA1537, and TA1538 with and without metabolic activation. Because TA98 and TA100 were the most sensitive, the bulk of the database was gathered with these two strains. Tars and their fractions were assayed over a fixed 3-log dose concentration range (10, 100, 250, 500, and 1000 μg/plate). Minor deviations in concentration ranges assayed were necessary with tar fractions.

Mutagenicity data are available on tars for three runs with Wyoming subbituminous coal and one run each with Illinois No. 6 bituminous, western Kentucky No. 9 bituminous, and North Dakota lignite (Table 3.40). Tars derived from bituminous coals were cytotoxic; over 50% of the bacteria were killed at concentrations below 100 μg/plate with Illinois No. 6 and below 250 μg/plate with western Kentucky No. 9. Tars produced during gasification of North Dakota lignite or Wyoming subbituminous coals were not cytotoxic. Bituminous coals also produced the greatest mutagenic response according to the Brusick notation (based on lowest concentration that produces a ratio of maximum revertants/spontaneous revertants, >3, at a dose with viability >50%—see footnote *a*, Table 3.41; Nichols et al., 1979). Highest specific mutagenic activity in tar fractions occurred in the base fraction regardless of coal type.

Some activity was observed in the polynuclear aromatic and polar neutral (oxygenated components) fractions. The very high specific activity in the polar neutral fraction of Sample G1541 was considered to be artifactual, resulting from fractionation errors in the base/polar neutral separation. No mutagenic activity was found in the acid or nonpolar neutral fractions (aliphatic and one- to two-ring aromatic compounds). Metabolic activation was required to demonstrate mutagenicity of fractions and activity was expressed only with strain TA98 (Nichols et al., 1979).

Table 3.40 Results of Ames Assays on Research Triangle Institute Tars and Tar Fractions

Material	Sample No.	Brusick Notation[b]	Specific Activity[a] (rev/μg) Whole Tar	Base	Polynuclear Aromatic	Polar Neutral
Wyoming subbituminous	G1511	M (3.6)	1.2	8.5 (0.3)	1.6 (0.6)	1.4 (0.1)
Wyoming subbituminous	G1521	M (4.6)	0.4	15.6 (0.5)	0.7 (0.2)	0 (0)
Wyoming subbituminous	G1551	M (3.1)	0.3	5.8 (0.2)	0.4 (0.07)	0.3 (0.06)
Western Kentucky No. 9	G1531	H (19.2)	8.7	33.8 (1.8)	7.7 (4.8)	1.8 (0.1)
Illinois No. 6	G1541	H (14.6)	11.3	6.2 (0.4)	2 (1.3)	37.5[c] (1.4)
North Dakota lignite	G1561	M (8.4)	1.9	17.9 (0.8)	0.9 (0.3)	1.9 (0.1)

[a] Specific activity in *Salmonella typhimurium* strain TA98 was estimated by linear regression of dose–response data (i.e., slope of the regression line). A positive response was one twice background with a correlation coefficient ≥ 0.8 for the regression line; four concentrations (10, 100, 250, and 500 μg/plate) were tested. Postive control: two determinations for 2-aminoanthracene were 256 ± 17 and 373 ± 40 rev/10 μg. Values in parentheses are weighted activity (rev/μg) determined by specific activity × relative weight.

[b] H—High mutagenicity; occurs at a dose of ≤ 50 μg/plate. M—Medium mutagenicity; occurs at a dose of 50–499 μg/plate. L—Low mutagenicity; occurs at a dose of 500–5000 μg/plate. Values of parentheses are maximum revertants/spontaneous revertants at dose with viability $\geq 50\%$.

[c] Value considered to be artifactual because of error in fractionation procedure.

Source: Nichols et al. (1979).

Mode of operation (coal feed conditions) of the RTI gasifier was shown to influence the mutagenicity of specific fractions of the resultant tars but not the overall mutagenicity of the raw tar (Table 3.41). Increased specific activity was seen in the base fractions of both a Wyoming subbituminous (Sample G1571) and North Dakota lignite (Sample G1581) coal-derived tar when the gasifier was operated in a batch mode (coal is added a batch at a time) rather than a continuous-feed mode. In contrast, the specific mutagenic activity of the polynuclear aromatic fraction decreased in the batch mode from levels present under continuous-feed mode (Mixon et al., 1984).

The MI experimental riser-reactor pyrolysis unit provided a unique opportunity to evaluate mutagenic properties of coal-derived materials produced under a limited set of definable operating conditions. Three different types of gas stream condensate were collected and assayed for mutagenicity in *S. typhimurium* strain TA98 (Tables 3.42 and 3.43). Materials included tar like high-temperature (210–340°F) trap condensates, room-temperature trap-oil condensates, and cryogenic (−176°F) trap condensates. Metabolic activation was required to demonstrate mutagenicity. Condensates from test runs at 1110°F under all test conditions had low (2 rev/μg for Sample G3171HT was the highest) or statistically insignificant (<1 rev/μg) specific mutagenic activity (Fillo et al., 1983). Highest specific mutagenic activities were found in the high-temperature trap condensates from test runs at 1470 and 1830°F. Specific activities of room-temperature condensates were generally lower (up to 5- to 10-fold) than the high-temperature condensates. Fewer low-temperature condensates were assayed and they had specific mutagenic activities of 1 rev/μg or less. Crude aqueous-phase condensates from the room-temperature and low-temperature traps were not significantly mutagenic, although organic extracts of these waters were not assayed. Of the operating conditions employed (residence time, pressure, carrier gas, and temperature), temperature correlated best with mutagenic activity. Lignite coal-derived samples from the 1470°F runs had the highest average specific mutagenicities. Oil/tar yields were also highest for 1470°F runs. The 16 samples of the 1470°F runs had an average specific mutagenicity of 26 rev/μg (range, 3–80 rev/μg). The 13 samples of the 1830°F runs had somewhat lower average specific mutagenicities, averaging 18 rev/μg (range, 4–33 rev/μg).

In situ coal gasification produces synthetic gas without an industrial facility. Product gas condensate tars from the in situ gasification of coal at Hoe Creek in northeastern Wy-

Table 3.41 Effects of Coal Feed Conditions on Results of Ames Assays on Research Triangle Institute Tars and Tar Fractions

Material	Coal Feed Conditions	Specific Activity[a] (rev/μg) Whole Tar	Base	Polynuclear Aromatic
Sample G1571, Wyoming subbituminous	Batch	0.3	6.6	0.8
	Continuous	0.3	1	1.2
Sample G1581, North Dakota lignite	Batch	1.5	8	0.5
	Continuous	1.5	0.5	1.4

[a] Specific activity in *Salmonella typhimurium* strain TA98 was estimated by linear regression of dose-response data (i.e., slope of the regression line). A positive response was one twice background with a correlation coefficient ≥ 0.8 for the regression line; four concentrations (10, 100, 250, and 500 μg/plate) were tested. Positive control: two determinations for 2-aminoanthracene were 256 ± 17 and 373 ± 40 rev/10 μg.

Source: Mixon et al. (1984).

Table 3.42 Results of Ames Assays on Condensate Tars from the Mellon Institute Riser Reactor (Lignite Coal)

Carrier Gas	Pressure (atm)	Residence Time (s)	Specific Activity[a] (rev/μg) High-Temp. Trap	Room-Temp. Tap	Low-Temp. Trap
1110°F					
Hydrogen	3	1.3	2 (G3171HT)	<1 (G3171RT)	—
Helium	3	1.3	1 (G3174HT)	<1 (G3174RT)	—
Helium/water	3	1.3	<1 (G3180HT0	<1 (G3180RT)	—
Carbon dioxide	3	1.3	<1 (G3181HT)	<1 (G3181RT)	—
1470°F					
Hydrogen	3	0.2	25 (G3144HT)	2 (G3144RT)	—
	3	0.2	32 (G3145HT)	5 (G3145RT)	—
	3	0.2	13 (G3146HT)	1 (G3146RT)	1 (G3146LO)
	3	0.2	25 (G3147HT)	3 (G3147RT)	1 (G3147LO)
	3	1.3	35 (G3169HT)	3 (G3169RT)	
	10	0.2	80 (G3153HT)	3 (G3153RT)	1 (G3153LO)
Helium	3	0.2	3 (G3159HT)	2 (G3159RT)	<1 (G3159LO)
	3	1.3	32 (G3172HT)	5 (G3172RT)	
	10	0.2	47 (G3154HT)	6 (G3154RT)	4 (G3154LO)
Helium/water	3	0.2	7 (G3164HT)	4 (G3164RT)	<1 (G3164LO)
	3	0.2	13 (G3167HT)	3 (G3167RT)	2 (G3167LO)
	3	1.3	22 (G3178HT)	4 (G3178RT)	—
	3	1.3	8 (G3182HT)	4 (G3182RT)	—
Carbon dioxide	3	0.2	11 (G3160HT)	2 (G3160RT)	<1 (G3160LO)
	3	1.3	46 (G3175HT)	5 (G3175RT)	—
	10	0.2	9 (G3163HT)	4 (G3163RT)	1 (G3163LO)
1830°F					
Hydrogen	3	0.2	15 (G3148HT)	2 (G3148RT)	<1 (G3148LO)
	3	1.3	22 (G3170HT)	<1 (G3170RT)	—
	10	0.2	8 (G3166HT)	3 (G3166RT)	<1 (G3166LO)
Helium	3	0.2	15 (G3156HT)	2 (G3156RT)	1 (G3156LO)
	3	1.3	22 (G3173HT)	11 (G3173RT)	—
	10	0.2	18 (G3155HT)	2 (G3155RT)	<1 (G3155LO)
Helium/water	3	0.2	16 (G3165HT)	5 (G3165RT)	1 (G3165LO)
	3	?	12 (G3168HT)	5 (G3168RT)	—
	3	1.3	33 (G3179HT)	—	—
Carbon dioxide	3	0.2	17 (G3161HT)	3 (G3161RT)	<1 (G3161LO)
	3	0.2	32 (G3186HT)	4(G3186RT)	—
	3	1.3	4 (G3176HT)	6 (G3176RT)	—
	3	1.3	—	<1 (G3184RT)	—
	10	0.2	24 (G3162HT)	5 (G3162RT)	—.

[a] Specific activity in *Salmonella typhimurium* strain TA98 was estimated by linear regression of dose–response data (i.e., slope of the regression line). A positive response was considered significant as discussed by Carnes et al. (1985).

Source: Fillo et al. (1983).

Table 3.43 Results of Ames Assays on Condensate Tars from the Mellon Institute Riser Reactor (Subbituminous Coal)

			Specific Activity[a] (rev/μg)	
Carrier Gas	Pressure (atm)	Residence Time (s)	High-Temp. Trap	Room-Temp. Trap
1110°F				
Hydrogen	3	0.2	<1 (G3189HT)	<1 (G3189RT)
		1.3	1 (G3192HT)	<1 (G3192RT)
Carbon dioxide	3	1.3	<1 (G3195HT)	<1 (G3195RT)
1470°F				
Hydrogen	3	0.2	14 (G3187HT)	3 (G3187RT)
		1.3	32 (G3190HT)	5 (G3190RT)
Carbon dioxide	3	1.3	34 (G3193HT)	2 (G3193RT)
1830°F				
Hydrogen	3	0.2	6 (G3188HT)	1 (G3188RT)
		1.3	11 (G3191HT)	2 (G3191RT)
Carbon dioxide	3	1.3	33 (G3194HT)	6 (G3194RT)

[a] Specific activity in *Salmonella typhimurium* strain TA98 was estimated by linear regression of dose–response data (i.e., slope of the regression line). A positive response was considered significant as discussed by Carnes et al. (1985).

Source: Fillo et al. (1983).

oming were studied for mutagenicity (Timourian et al., 1982). Mutagenicity was assayed in *S. typhimurium* strains TA98 and TA100 over a fixed concentration range, 10–600 μg/plate (duplicate plates at each test dose). Specific mutagenic activity was calculated from the linear portion of the dose–response curve. Tar Sample G4111 was assayed as a crude (unfractionated) tar and was mutagenic to both strains. Specific activities were 1.1 rev/μg for TA98 and 1.5 rev/μg for TA100. Tar Sample G4121 was not assayed as a crude material. Base components of both tars and high specific activities in strain TA98 (0.8 rev/μg for G4111 and 11.9 rev/μg for G4121) although this fraction accounted for only 1% of the sample mass. Neutral components composed most of the two samples (85 and 97% of the mass) but had significantly lower specific mutagenic activities to strain TA98 than base fractions (0.3 rev/μg for G4111 and 1.1 rev/μg for G4121). Results with TA100 were similar, although this strain was slightly more sensitive to mutagenesis by neutral components.

3.4.1.1.2 Fluctuation Assay. A modification of the classical fluctuation assay (Luria and Delbruck, 1943), using a microtiter technique (Gatehouse, 1978; Gatehouse and Delow, 1979), was adapted to study bacterial mutagenesis of complex mixtures (Wilson et al., 1982). *Salmonella typhimurium* strain TA98 cells were exposed to the test material for 4 h, after which 10^6–10^7 cells were placed in microtiter wells with histidine-deficient medium for selection of mutants. Increases in turbidity and pH changes due to growth of mutants (histidine prototrophs) indicated a positive mutagenic response. Significance of the response (number of the 96 test wells positive for mutation) was determined by χ^2 statistical analysis of the responses. The fluctation assay differs from other *Salmonella* mutation assays

(Ames and forward mutation) in its greater sensitivity to weak or moderately weak mutagens.

Only four liquefaction materials were assayed with the fluctuation assay, all within the same laboratory. They included SRC-I PS (Wilsonville, Alabama), SRC-II distillate blend (Harmarville, Pennsylvania), EDS distillate blend (Baytown, Texas), and distillate material from a German process (Veba Oel). Tests were generally conducted at two concentrations per material (e.g., 1 and 10 μg/well). Mutations were detected in this assay only after metabolic activation of the test materials, with the exception of the >800°F distillate cut of the EDS material.

The bp cuts (50°F cuts from 450°F to >850°F) of SRC-I PS (Sample L2121) and their chemical fractions were assayed at concentrations of 1 and 10 μg or 10 and 100 μg (Wilson et al., 1982). Mutagenicity was detected in the three >700°F bp cuts (Table 3.44). Activity was statistically significant ($P < 0.001$) at the 10-μg/well concentration for the 700–750°F cut and at 1 and 10 μg/well in the 750–800°F and 800 to >850°F cuts. Greatest activity occurred with the 10-μg/well concentration of the 800 to >850°F cut (83% positive wells). Aliphatic and hydroxy-PAH fractions of all bp cuts showed no mutagenic activity. PAH fractions were also nonmutagenic except for the 800 to >850°F bp cut, in which a statistically significant ($P < 0.001$) mutagenic response was observed at the 10-μg/well concentration (over 30% positive wells). The N-PAC fractions of >700°F cuts were mutagenic; statistically significant ($P < 0.001$) responses occurred at the 10-μg/well concentration with the 700–750°F cut, and at 1- and 10-μg/well concentrations with the 750–800°F and 800°F to >850°F cuts (maximum responses of over 75% of the wells positive at 10 μg/well for 700–750°F and 800°F cuts). Further fractionation of N-PAC

Table 3.44 Results of Fluctuation Assays on Coal-Liquefaction Materials[a]

Material	Concentration Tested (μg/well)	Mutagenic Response (% positive wells)
SRC-I process solvent (Sample L2121)		
700–750°F	10	36
750–800°F	1	34
	10	21
800–850°F	1	78
	10	83
SRC-II distillate blend (Sample L1111)		
300 to >850°F	1	23[b]
300–700°F	10	44
700–750°F	0.1	29
	1	35
750–800°F	1	81
800–850°F	0.1	41
	1	94
>850°F	1	88
	10	100
Veba heavy oil (Sample L6212)	10	100

[a] Results are presented only for those materials shown to produce a statistically significant ($P < 0.001$) mutagenic response in *Salmonella typhimurium* strain TA98.

[b] $P < 0.01$.

Source: Wilson et al. (1982), Mahlum et al. (1983), and Wilson and Pelroy (1983).

fractions yielded subfractions enriched in carbazoles, amino-PAHs, and azaarenes. The carbazole subfractions from cuts boiling above 700°F were significantly mutagenic ($P < 0.001$) at both 1 and 10 μg/well (maximum response of 100% of wells positive at 10 μg for 800 to >850°F cut). Similarly, amino-PAH subfractions of materials boiling above 700°F were mutagenic; only 1 μg of test material/well was needed to induce a 100% positive response with the 800 to >850°F cut. Statistically significant ($P < 0.01$) activity was also observed at 10 μg/well in the amino-PAH subfractions from the material boiling between 550 and 600°F. Azaarene-enriched subfractions from 750–800°F (>30% positive at 1 μg/well) and 800 to >850°F cuts (>75% positive at 10 μg/well) were also significantly mutagenic ($P < 0.001$).

A full-boiling-range blend (Sample L1111) of SRC-II liquid (300 to > 850°F) and five distillation cuts of the blend (300–700°F, 700–750°F, 750–800°F, 800–850°F, and >850°F) were assayed at concentrations of 0.1 and 1 μg/well or 1 and 10 μg/well (Mahlum et al., 1983). The blend was mutagenic at the 1-μg test concentration (23% positive wells, $P < 0.01$; see Table 3.44). Statistically significant ($P < 0.001$) mutagenesis also occurred with all the bp cuts, with the greatest responses (>80% positive wells) occurring in cuts boiling above 750°F. All wells were positive with 10 μg of the >850°F cut. Three bp cuts (300–700°F, 800–850°F, and >850°F) were fractionated into broad chemical classes and the PAH and N-PAC fractions assayed for mutagenicity. The PAH fraction of the 300–700°F cut was nonmutagenic, but the PAH fractions of the 800–850°F and >850°F cuts were mutagenic. Greatest response (30% positive wells with $P < 0.001$) occurred with 10 μg of the PAH fraction of the 800–850°F cut. The N-PAC fractions of all three bp cuts were significantly mutagenic ($P < 0.001$). The higher bp cuts (800–850°F and >850°F) induced 100% responses at the 1-μg/well concentration and the 300–700°F cut induced about a 30% response.

A coal-derived heavy oil distillate (>620°F) from a pilot plant in West Germany (Sample L6212) and its chemical class fractions were assayed at an exposure concentration of 10 μg/well (Wilson and Pelroy, 1983). With the exception of the aliphatic fraction, heavy oil and all its fractions and subfractions were demonstrated to have a statistically significant mutagenic response ($P < 0.01$ to <0.001; Table 3.44). The heavy oil, the N-PAC fraction, and all its subfractions (carbazoles, amino-PAHs, and azaarenes) induced mutations in greater than 90% of the test wells. Fifty wells (52%) were positive with the PAH fraction and 18 wells (19%) with the hydroxy-PAH fraction.

Aliphatic, PAH, and hydroxy-PAH fractions of four 50°F bp cuts (650°F to >800°F) of an EDS distillate blend (Sample L4134) were assayed (generally at a 100 μg/well concentration) with or without metabolic activation (Later et al., 1984). The N-PAC fraction was not assayed. Aliphatic and hydroxy-PAH fractions of the four bp cuts were nonmutagenic with activation. Statistically significant (P at least < 0.01) mutagenic responses were observed with each of the PAH fractions of the bp cuts following activation. All 96 wells were positive with activation following exposure to the 750–800°F cut. Direct-acting mutagens were found in the PAH and hydroxy-PAH fraction of the >800°F material. Without activation, the >800°F cut induced mutation in roughly 40% of the wells at 10 μg/well with the PAH fraction and at 100 μg/well with the hydroxy-PAH fraction.

3.4.1.1.3 Forward Mutation Assay. *Salmonella typhimurium* strain TM677 was used to measure induction of forward mutations through resistance to the purine analog 8-azaguanine. Forward mutation assay should be capable of detecting mutagens not easily screened in the standard Ames assay. An entire gene is the target site, and all types of mutation (e.g., frameshift, point, deletion, mutuation caused by cross-linking of DNA bases) are detectable. Through use of the Skopek method, (Skopek et al., 1979), the length of exposure was found to increase from 2 to 4 h when Aroclor-induced rat liver S9 homogenates were used for metabolic activation. After exposure of bacteria (10^8), the cells were grown overnight in broth without 8-azaguanine to allow expression of induced mutations. Cultures were then plated onto Minimal E Medium with 8-azaguanine (50 μg/mL). Mutant clones (i.e., 8-azaguanine-resistant bacteria that grew into visible colonies and were presumed to arise from a single cell) were counted about 36 h after plating. Viability measurements performed on bacteria obtained immediately following the exposure period (dilution plate counts on Minimal E Medium without 8-azaguanine) allowed determination of specific mutagenic activity (mutants/clonable cell) according to the formula

$$\text{Specific mutagenic activity} = \frac{\text{Number of mutant clones (8-azaguanine-containing media)}}{\text{Number of clones (no 8-azaguanine)}} \times \text{dilution factor}$$

Statistical evaluations were based on the χ^2 statistic with level of significance based on the method of Green et al. (1976). A value greater than 1.4×10^{-4} mutants/clonable cell indicated a statistically significant response ($P < 0.05$) based on historical negative controls.

Three liquefaction materials (SRC-I PS, SRC-II distillate blend, and a distillate from Veba Oel) were assayed using this procedure. All assays were performed in the same laboratory and generally at three concentrations (e.g., 5, 10, and 20 μg/mL). Metabolic activation was required to demonstrate activity with all the materials.

Crude bp cuts (50°F cuts from 450°F to >850°F) of an SRC-I PS (Sample L2121) from the Wilsonville, Alabama pilot plant and their chemical class fractions were assayed for forward mutation induction at concentrations of 0.5, 1, and 2 μg/mL, 5, 10, and 20 μg/mL, or 50, 100, and 200 μg/mL exposure medium (Wilson et al., 1982). In contrast to the Standard Ames assay of this material (i.e., reverse mutation), where mutagenic activity occurred only in bp cuts above 650°F, statistically significant forward mutations occurred in the 600–650°F and 650–700°F bp cuts (Table 3.45). No activity could be demonstrated in the bp cuts >700°F. No forward mutations at statistically significant levels were observed with the aliphatic or hydroxy-PAH fractions of PS bp cuts. Forward mutations were observed in PAH-enriched fractions of the 650–700°F, 700–750°F, and 800°F to >850°F cuts, although higher test concentrations (up to 200 μg/mL) were needed to show positive response. It is noteworthy that the standard Ames assay did not detect any mutagenic activity in PAH fractions despite the presence of significant concentrations of known mutagens and carcinogens in these materials. N-PAC-containing fractions of all bp cuts above 600°F induced forward mutations. Concentrations tested were the same as for the crude material (5, 10, and 20 μg/mL). Maximum response (6.4 mutants/clonable cell $\times 10^{-4}$) was observed at a concentration of 20 μg/mL of the 750–800°F cut.

The N-PAC fractions were further fractionated to study chemical classes responsible for activity. Subfractions enriched in carbazoles, amino-PAHs, and azaarenes were assayed at 50, 100, and 200 μg/mL. Low but significant activity occurred in carbazole subfractions of cuts boiling above 650°F. Maximum response was 2 mutants/clonable cell $\times 10^{-4}$ for material from the 750–800°F cut. Test concentrations above 50 μg/mL of 600–650°F and 650–700°F cuts were cytotoxic (high enough to obscure response). The amino-PAH subfractions of all cuts from 550°F to >850°F, except for the 700–750°F cut, were mutagenic. Highest responses occurred in materials boiling from 550–600°F and 600–650°F (4–5 mutants/clonable cell $\times 10^{-4}$). No forward mutations occurred with azaarene-enriched subfractions from cuts boiling above 700°F. Statistically significant activity (2–3 mutants/clonable cell $\times 10^{-4}$) was present in the 650–700°F cut only.

Table 3.45 Results of Forward Mutation Assays on Coal-Liquefaction Materials[a]

Material	Concentration Tested (μg/mL)	Mutants/Clonable Cell $\times 10^{-4}$
SRC-I process solvent (Sample L2121)		
600–650°F	5	2.3
	10	2
	20	3.2
650–700°F	5	2.8
	10	3
SRC-II distillate blend (Sample L1111)		
700–750°F	5	6.4
	10	6.5
	20	4.8
750–800°F	5	11.2
	10	10
	20	9.8
800–850°F	0.5	2.2
	1	6.2
Veba heavy oil (Sample L6212)	10	2.2

[a]Results are presented only for those bp cuts shown to produce a statistically significant ($P < 0.05$ based on historical controls at more than 1.4×10^{-4} mutants/clonable cell) mutagenic response in *Salmonella typhimurium* strain TM677.

Source: Wilson et al. (1982), Mahlum et al. (1983), and Wilson and Pelroy (1983).

A full range (300 to >850°F) SRC-II blend (Sample L1111) from the Harmarville, Pennsylvania PDU and five distillate cuts (300–700°F, 700–750°F, 750–800°F, 800–850°F, and >850°F) were assayed for forward mutation (Mahlum et al., 1983). Test concentrations of 5, 10, and 20 μg/mL or 0.5, 1, and 2 μg/mL were used. The blend was nonmutagenic but some of its bp cuts were shown to induce forward mutations (Table 3.45). Greatest activity (11.2 mutants/clonable cell $\times 10^{-4}$ at a 10-μg/mL concentration) occurred in the 750–800°F cut with statistically significant activity also observed in the 700–750°F and 800–850°F cuts. The 300–700°F, 800–850°F, and >850°F cuts were fractionated into broad chemical classes and the PAH and N-PAC fractions were assayed for forward mutation induction. PAH fractions were mutagenic; the 800–850°F cut showed a statistically significant and dose-related response (5, 10, and 20 μg/mL were all positive, with 20 μg/mL having the greatest response, 4.4 mutants/clonable cell $\times 10^{-4}$). Only the N-PAC fraction from the 800–850°F cut was mutagenic. Mutagenicity of this cut appeared to be dose related (0.5, 1, and 2 μg/mL tested) and was stronger than that for comparable PAH fraction (15 mutants/clonable cell $\times 10^{-4}$ at the 2-μg/mL dose).

A coal-derived distillate heavy oil (Sample L6212) produced at a pilot plant in West Germany (Veba Oel) and its chemical class fractions were assayed for induction of forward mutation at a single test concentration of 10 μg/mL (Wilson and Pelroy, 1983). Heavy oil (>610°F) and its nitrogen- and hydroxy-PAH fractions were mutagenic ($P \leqslant 0.01$). Highest activity occurred in the N-PAC fraction (12.9 mutants/clonable cell $\times 10^{-4}$) and its azaarene subfraction (10.5 mutants/clonable cell $\times 10^{-4}$). The aliphatic and PAH fractions were nonmutagenic.

3.4.1.1.4 Microbial DNA Repair Assay. Another way to assess the ability of a material to damage DNA is its relative toxicity to a pair of bacterial strains that differ in their ability to repair DNA damage. An example of such a pair is *E. coli* strains WP2 uvrA$^+$ recA$^+$ (DNA repair-proficient) and WP100 uvrA$^-$ recA$^-$ (repair-deficient; Green and Muriel, 1975). Genotoxicity is expressed in this assay by a preferential inhibition of growth in the repair-deficient strain. In principle, materials that are toxic to cells through non-DNA-damaging mechanisms would have similar effects on growth (e.g., equally inhibitory to growth). A survival index can be determined by dividing the percentage of survival of repair-deficient strain by the percentage of survival of repair-proficient strain.

To perform the assay, overnight stock cultures are diluted in fresh nutrient broth and grown with shaking for 2 h at 37°C (98.6°F). The resulting log-phase cultures are diluted in fresh medium and 10^4 cells are then exposed to the test material or appropriate controls in the presence of a metabolic activation system (Aroclor-induced rat liver S9) for 90 min. Aliquots of the incubation mixture are then added to top agar and poured onto minimal medium plates. Plates are incubated for 48 h and the number of resultant colonies counted.

Representative commercial products (naphtha, No. 2 fuel oil, and No. 6 fuel oil) were distilled from a COED pyrolysis oil (Sample G2221) considered to be similar to that which would be produced by the COGAS process. The ability of these products and comparable distillates from two crude petroleums (one more aromatic, the other less aromatic and less paraffinic than the COED oil) were assessed for their ability to damage DNA (Peirce and Fung, 1981a). Two pairs of bacterial strains were used: *E. coli* WP2 uvrA$^+$ recA$^+$ and WP100 uvrA$^-$ recA$^-$ and *S. typhimurium* TA1979 uvrB$^+$ (DNA repair-proficient) and TA1538 uvrB$^-$ (deficient in uvrB DNA repair). A preliminary experiment was run with each material covering eight doses, the highest being 10 mg/mL, with the other seven being serial half-log dilutions. For the actual test, four doses were selected, with the highest dose being one that caused a growth reduction in strain WP100 of about 80%. DMSO was used as the solvent. The negative control was penicillin (1.5 μg/plate) and the positive control was 2-aminofluorene (20-μg/mL plate). Data from three plates at each test concentration were averaged and percentage survival was calculated relative to the solvent control.

Some genotoxicity was detected in all the materials with the exception of the COED No. 6 fuel-oil material. The paraffinic naphtha material was genotoxic in a reproducible and dose-related fashion in both sets of tester strains. Genotoxicity was demonstrated for three materials (aromatic and paraffinic No. 6 fuel oils and COED naphtha) only in *S. typhimurium*. The repair-deficient strain, TA1538, is less DNA-repair-deficient than *E. coli* WP100, and it was suggested that a response only in *Salmonella* was a reflection of TA1538 having an additional mutation (rfa) that increased the permeability of the bacterial cell wall to large organic molecules. Of the three materials, the COED naphtha was considered only weakly genotoxic because the decreases in survival index were not always clearly dose related. The aromatic naphtha material was genotoxic only in *E. coli* and thus was considered weakly genotoxic because it was not genotoxic in the less DNA-repair-deficient *Salmonella*. All three No. 2 fuel-oil materials were considered weakly genotoxic because decreases in the index were either not dose related and/or reproducible.

In this assay system, the use of a metabolic activation system (S9) reduced survival effects and necessitated evaluating the effects on growth inhibition at different concentrations, depending on whether the assay was being run with or without S9. Genotoxicity could be demonstrated without activation in most cases and sometimes with activation also.

3.4.1.1.5 Prophage Induction Assay. The induction of phage (i.e., interactions with a nonreplicating, temperate prophage resulting in initiation of replication of the virus) was measured by modification of the procedure described by Moreau et al. (1976). This assay, also called the Inductest, is based on the observation that induction of phage results from damage induced by the test material in DNA. Interactions with DNA can induce heritable changes and the assay serves as a screen to indicate that DNA interactions have occurred. *Escherichia coli* cells (10^7), strain K12 (8177), containing the temperate (nonactive) lambda virus (phage K12) are exposed with or without metabolic activation (Aroclor-induced rat liver S9) to test sample for 6 h. Then 10^3–10^4 cells are removed and killed with chloroform to release any induced phage (active virus capable of killing sensitive bacteria). The number of induced phage is then determined by titration of plaque formation (i.e., clear areas in a solid layer of bacteria resulting from virus-killed cells) in phage-sensitive *E. coli*, strain K12 (6340). The response is expressed as plaque-forming units per plate (pfu/plate).

This assay was used to evaluate DNA interactions with five liquefaction materials (SRC-I PS, SRC-II distillate blends, and a Veba Oel distillate). Metabolic activation (Aroclor-induced rat liver S9) was required to demonstrate phage induction with all the materials assayed.

SRC-I PS bp cuts (Sample L2121) from the Wilsonville, Alabama pilot plant were assayed for their ability to induce phage lytic for *E. coli* (Wilson et al., 1982). Nearly all materials assayed (50°F cuts, 450°F to >850°F) produced a low (<200 pfu/plate) but statistically significant ($P \leqslant 0.05$, when compared with the average response of negative controls, about 130 pfu/plate) induction of phage at concentrations of 5, 10, and 20 μg/mL (Table 3.46). There were no apparent differences related to temperature range between any of the bp cuts. Certain of the aliphatic, PAH, N-PAC, and hydroxy-PAH fractions of bp cuts also induced phage. Greatest responses occurred with N-PAC fractions of cuts boiling above 700°F, with a maximum response of 400 pfu/plate at a 5-μg/mL concentration of the 750–800°F cut. Activity when present in a given bp cut was generally not dose dependent. The aliphatic fraction had significant activity only in the 450–500°F cut (>250 pfu/plate). The PAH and hydroxy-PAH fractions had significant activity in all bp cuts with the exception of the hydroxy-PAH fraction of the 650–700°F cut. However, greater test concentrations than those tested with the unfractionated bp cuts (50, 100, and 200 μg/mL) had to be used to produce the activity. Three subfractions of N-PAC fractions enriched in

Table 3.46 Results of Phage-Induction Assays on Coal-Liquefaction Materials[a]

Material	Concentration Tested (μg/mL)	Plaque-Forming Units/Plate
SRC-I process solvent (Sample L2121)		
450–550°F	5	180
600–650°F	5	180
650–700°F	10	185
700–750°F	10	200
750–800°F	10	210
800 to >850°F	10	200
SRC-II distillate blend (Sample L1111)		
300 to >850°F	5	200
300–700°F	10	170
700–750°F	5	180
750–800°F	5	400
800–850°F	10	310
>850°F	10	380

[a]Results reported are for the concentration (generally three tested for each material) that demonstrated the strongest statistically significant phage-induction response in *Escherichia coli* strain K12.

Source: Wilson et al. (1982) and Mahlum et al. (1983).

carbazoles, amino-PAH, and azaarenes were also assayed for phage induction. Low to moderate increases in phage occurred with all three fractions of nearly all bp cuts with no clear relation of the activity to bp.

The prophage induction assay was also performed on distillate cuts of full-range SRC-II material (Harmarville, Pennsylvania). Two sets of distillate cuts were assayed (Pelroy and Wilson, 1981). The first set (Samples L1121 and L1131) included 18 bp cuts between 138 and 1055°F. The second set (Sample L1111) included only five bp cuts between 300°F and >850°F. In the first set, only materials boiling above 782°F showed significant phage induction (ratio of pfu exposed/pfu unexposed greater than 2); the maximum response (a ratio of 4.5) occurred with the highest-boiling material (914–1055°F cut). Similarly, only the highest-boiling cut from the second set (>850°F cut) showed a positive response (6.2 ratio). Standard Ames assay of these materials (Table 3.10) appears to be a more sensitive indicator of genetic damage.

The SRC-II blend (Sample L1111) distillate cuts were assayed for phage induction a second time along with PAH and N-PAC fractions of the bp cuts (Mahlum et al., 1983). Concentrations of 5, 10, and 20 μg/mL were used for all test materials. Statistically significant responses ($P < 0.05$) occurred with all bp cuts with the exception of the 300–700°F cut. The responses were not always dose related. Highest responses (>300 pfu/plate) occurred with 10-μg/mL concentrations of the three cuts boiling above 750°F. The 300–700°F, 800–850°F, and >850°F cuts were fractionated into broad chemical classes and the PAH and N-PAC fractions assayed. The PAH fraction of the 300–700°F cut was not active. PAH fractions of the 800–850°F and >850°F cuts were positive for phage induction; a dose-related response occurred with the >850°F cut. The 20-μg/mL concentration of this fraction induced >225 pfu/plate. N-PAC fractions were active for all three bp cuts. Although responses were larger (>650 pfu/plate with 5μg/mL of the 800–850°F cut), dose response did not occur with the fractions. Less activity (a maximum of 200 pfu/plate at the 20-μg/mL concentration) occurred with the 300–700°F bp cut.

Heavy (>610°F) distillate oil (Sample L6212) from a pilot plant in West Germany (Veba Oel) and its chemical class fractions were assayed at 25 μg/mL in the Inductest (Wilson and Pelroy, 1983). Heavy oil or its fractions produced statistically significant phage induction ($P \leq 0.05$) only in the N-PAC fraction ($P < 0.01$) and its amino-PAH and azaarene subfractions. Metabolic activation was required to demonstrate activity.

3.4.1.1.6 Yeast Mutagenesis Assay. Yeast cell DNA is organized in a fashion more related to higher organisms than bacteria having several eukaryotic chromosomes surrounded by a nuclear membrane. Zimmermann (1975) developed mutation assays in yeast that were applied to complex mixtures (Epler, 1980). The assay uses a haploid histidine-requiring, antibiotic-sensitive (canavanine) strain of *Saccharomyces cerevisiae* (XL7-10B) and can measure both forward and reverse mutations, that is, forward mutations to canavanine-resistant and reversion to non-histidine-requiring.

Yeast cells are exposed in liquid suspension for several hours with and without S9 activating enzymes. The cells are then plated onto agar growth plates containing: (1) no histidine and no canavanine (measures reversion to non-histidine-requiring), (2) histidine and canavanine (measures forward mutation to canavanine resistance), or (3) histidine and no canavanine (measures viability). Results are expressed as mutants/viable cells.

Three fractionated (acid, base, and neutral components) UMD gasifier ESP tars (Samples G1301, G1306, and G1319) were mutagenic to *S. cerevisiae*. In contrast to Ames bacterial mutagenicity assays with these materials, all ether-soluble base fractions were direct-acting mutagens. Some mutagenicity occurred in the S9-activated neutral fraction of Sample G1301 but not in the other two neutral fractions (Epler et al., 1983).

Samples of SRC-II stripper tower bottoms (L1263) and middle distillate (L1264) were also evaluated in a *Saccharomyces* mutagenicity assay. Test concentrations ranged from nontoxic to cytotoxic, with and without exogenous metabolic activation. These materials were nonmutagenic (Hubis, 1982).

3.4.1.1.7 Mitotic Recombination Assay. *Saccharomyces cerevisiae* (strain D3) is a diploid yeast heterozygous for mutations in the ade2 and his8 loci of chromosome XV. In the presence of adenine, homozygous cells produce a red pigment. Homozygosity at the ade2 locus can be induced in the D3 cells by mitotic recombination. Agents that induce mitotic recombination are considered potential DNA *clastogens* (capable of breaking DNA). Most mutagens and many carcinogens induce mitotic recombination. The utility of the yeast assay as a prescreen for mutagens and carcinogens was demonstrated by Simmon (1979). Recombinogenic potential can be determined by incubating test samples with *S. cerevisiae* D3 cells for 4 h (roller drum) and then diluting and plating exposed cells on adenine-containing medium at 27–30°C (80–86°F) for 2 days. Cells are then grown an additional 2 days at 4°C (37°F) to enhance development of the red pigment characteristic of homozygosity at the ade2 locus. The number of recombinant colonies is then microscopically determined.

Recombinogenic potential of a COED pyrolysis oil, representative of material produced in the COGAS process, was assayed with *S. cerevisiae* strain D3 (Peirce and Fung, 1981b). The COED oil (Sample G2221) was distilled into representative commercial products (naphtha, No. 2 fuel oil, and No. 6 fuel oil) and their recombinogenic potential was compared to that of similar crude oil distillates (one set slightly more aromatic, one less aromatic). A material was considered active if it induced a reproducible dose-related increase in mitotic recombination (absolute number of mitotic recombinants/plate or relative number of recombinants/10^5 survivors).

Only aromatic crude oil No. 2 fuel-oil distillate was clearly recombinogenic. Maximum response was 100 recombinants/10^5 survivors at 0.01% (w/v) test concentration. Metabolic activation was not required to demonstrate recombinogenic activity. All other materials were termed clearly not recombinogenic (crude oil, aromatic naphtha, and COED No. 6 fuel oil) or possibly weakly recombinogenic (COED naphtha, No. 2 fuel oil, and crude oil aromatic No. 6 fuel oil). Naphtha samples were most toxic to yeast cells, while No. 6 fuel-oil samples were least toxic.

3.4.1.1.8 Summary and Conclusions—Microbial Assays. The greatest number of toxicologic evaluations of coal-conversion materials were from microbial assays. Although information is available from seven different assays, only with the Ames *Salmonella* mutagenicity assay are there enough results from different processes to consider generalizations about the genetic toxicity (in this case, mutagenicity) of coal-conversion materials. Data from the other six microbial assays are available for only a few materials and the data for each assay were obtained in a single laboratory. While specific results of these assays varied from results of Ames assays of the same materials, no significant novel insights into a material's mutagenicity were obtained.

Results of the Ames assay indicate that mutagens are present in coal-derived materials and that for any given material, mutagenic activity increases with the increasing concentration of high-molecular-weight hydrocarbons. On a process stream basis, the mutagenic activity is generally low, but a limited number of components are present that have high specific mutagenic activities. The data show that, where tested, crude materials determined to be mutagenic have subsequently been shown to be carcinogenic.

Of the various *Salmonella* tester strains used, strain TA98, mutated by agents that induce frameshift mutations, was universally the most sensitive strain for unfractionated materials. Strain TA100 (base-pair substitution) was shown in some instances to be more sensitive than strain TA98 to PAH-containing fractions. The other tester strains provided no insights into a material's mutagenic potential that were not provided with strains TA98 and TA100. Coal-conversion materials required metabolic activation for mutation induction with the exception of some chemical class fractions from a few materials.

The fact that Ames test results are available for almost every type of coal-conversion material for which there is any toxicologic characterization, makes direct comparison of these results most appealing. However, such comparisons should be made judiciously, keeping in mind the generally qualitative nature of the test. Responses to standards and mixtures usually have day-to-day variations (up to 30%) within the same laboratory and, despite rigorous attempts at precision, undefined factors contribute to significantly different responses for the same bacterial strains when used in different laboratories.

The inability to make quantitative comparisons within the large data set for Ames mutagenicity test results should not detract from their value. Ames tests are extremely useful as efficient and rapid screens of complex mixtures to determine qualitatively the materials for which additional study is required.

When crude materials are separated into chemical class fractions, results of Ames mutagenicity assays were not as valuable in predicting carcinogenicity. The *Salmonella* tester strains are very sensitive to basic nitrogen-containing mutagens. Thus, a coal-conversion material may have a high specific mutagenic activity, most of which can be attributed to components in the base fraction, which is a small proportion of the total mass. The bulk of the mutagenic activity from nitrogen-containing components of coal-conversion materials can be ascribed to amino-PAHs. Because carcinogenicity potential of nitrogen-containing chemical class fractions was measured in mouse skin, an organ not particularly sensitive to tumor induction by aromatic amines, the relation between the high mutagenicity seen in the base fraction and carcinogenicity is at present undefined. Thus, correlations between the Ames assay and whole-animal carcinogenicity assays may be somewhat fortuitous.

In a few cases, mutagenicity of fractions appeared to be additive. For example, bp cuts of an SRC-II material sequentially reblended and tested for mutagenicity after the addition of each bp cut suggested additivity of the mutagenic response. However, interactions among the components of crude materials that result in masking of mutagenicity are more common and limit the value of Ames mutagenicity tests in predicting carcinogenic potency. Very often, the sum of the weighted mutagenic activities of chemical class fractions (the calculated mutagenic activity that a given fraction should contribute to the unfractionated material) exceeds the experimentally determined specific mutagenic activity of the whole material. The addition of known mutagens to fractions of coal-conversion materials has also been shown to result in inhibition of the mutagenicity of the added mutagen. The reasons for this masking of mutagenic activity in crude materials are not well defined but competition for metabolic activation has been demonstrated to play a role.

Liquefaction materials from nine different processes were assayed, including chemical class fractions or distillate cuts of many of the materials. With the SRC-I, SRC-II, H-Coal, and EDS processes, materials from a range of process development operations (bench scale, PDU, and pilot plant) were studied. While there were qualitative and quantitative chemical differences in the end products, by-products, and process streams among the various processes, the mutagenicity of a given material reflected for the most part its relative concentration of high-molecular-weight (high-bp) PAHs and N-PACs. Highest specific mutagenic activity was in the N-PAC fractions (principally ascribed to amino-PAHs), with the neutral fractions (PAH compounds) having low specific activities in spite of the presence in this fraction of many known mutagens and carcinogens.

Hydrotreatment of coal liquids was demonstrated to be effective in decreasing their mutagenicity. Because the mutagenic activity resides in the higher-boiling portions of coal liquids, fractional distillation or recycling of higher-boiling materials was also shown to be effective in decreasing mutagenicity. Very little is known about coal-derived products or by-products from off-normal operation conditions, because samples are generally collected under normal, steady-state operations. Information in this area will not be available until there is significant experience with commercial operation.

Whether or not the Ames assay database for coal gasification is truly predictive of materials from a commercial industry remains a moot point. With the exception of the UMD gasifier tars, gasification products and by-products have not been studied extensively at routinely operated facilities. The database is larger for pilot-plant facilities, but only for bench-scale coal reactors could the data sets for coal gasification be considered truly extensive.

From the number of tars, oils, and product gas condensates studied with the Ames assay, it is clear that mutagens are present in nearly all materials that are nonvolatile. Tars and higher-molecular-weight constituents of oils and product gas condensates are responsible for the bulk of mutagenic response produced in the effluents of gasification processes. In general, however, the levels of mutagenicity are relatively low, especially when considered on a process stream basis. While the Ames mutagenicity data come from materials generated from a spectrum of coal types (lignite, subbituminous, bituminous, and anthracite), evaluations of the relation of coal type and toxicant formation are limited because of the qualitative nature of Ames data.

3.4.1.2 *Mammalian Cell Assays*

Mammalian cells are employed in a number of toxicological assays. Cells used are derived from normal (e.g., rabbit alveolar macrophage) or neoplastic (e.g., mouse lymphoma) tissues. Several cellular responses indicative of direct interactions between mammalian cell DNA and coal-derived materials or their metabolites have been measured to evaluate potential genotoxicity. Measured responses included gene mutations (forward and reverse), chromosomal effects (SCE), and morphological transformation. Cytotoxicity (cell death or dysfunction) of a material can be used to assay for potential toxicological effects not necessarily involving DNA interactions. Assays used to measure cytotoxicity included two general types: (1) those requiring cell division and growth to measure the cytotoxic endpoint (e.g., cloning efficiency assays) and (2) those requiring only cell maintenance (e.g., alveolar macrophage function assays). A brief description of the individual assay procedures used to evaluate the toxicity of coal-liquefaction and coal-gasification materials and the results obtained are presented below.

3.4.1.2.1 Chinese Hamster Ovary (CHO) Cell Mutation Assay. Hsie et al. (1978) developed a mammalian cell assay system that detects forward mutation in an established CHO cell line, CHO-K1. The assay measures mutations at the hypoxanthine guanosine phosphoribosyl transferase (HGPRT) gene locus. The enzyme product of this locus is nonessential for cell survival and functions to salvage purine nucleotides. In wild-type cells, HGPRT converts hypoxanthine to inosinic acid, which can be metabolized to adenylic acid and, after modification, incorporated into DNA. In the presence of the purine analog 6-thioguanine, HGPRT converts 6-thioguanine to thioguanine ribonucleotide, which is cytotoxic to wild-type cells. The assay measures mutations resulting in loss of HGPRT to wild-type cells. The assay measures mutations resulting in loss of HGPRT enzyme activity, that is, cells capable of growth in the presence of 6-thioguanine.

The HGPRT locus is suitable for mutation experiments because: (1) loss of the enzyme does not reduce cell viability; (2) it is sensitive to any material producing an altered enzyme, therefore both point and frameshift mutagens show as positives in the system; (3) the gene for HGPRT is in single copy, thus one-hit kinetics apply; and (4) the gene for HGPRT is on the X chromosome and is heritable.

In practice, exponentially growing cells ($\sim 10^6$) are exposed for up to 5 h to graded concentrations of test material with or without an exogenous metabolic activation system (generally Aroclor-induced rat liver S9). Cells are then washed and, after additional 18–24-h incubation, removed by trypsin from their culture dishes. A small number of the exposed cells are used to determine cell viability. Viability is generally measured with the vita dye trypan blue. Dead cells take up trypan blue, which is excluded from metabolizing cells. The remaining cells are used for determining either cloning efficiency or mutation induction. Cloning efficiency measures the ability of an exposed cell to undergo division and growth, resulting in a visible colony of cells (150–300 cells grown 7 days in hypoxanthine-free growth medium). For measurement of mutation, cells are cultured in growth medium ($\sim 10^6$ cells/plate) and subcultured at 2-day intervals for an 8- or 9-day period to allow expression of mutations. Cells are then grown for mutant selection in medium containing 6-thioguanine (5×10^5 cells grown on multiple plates with up to 1×10^6 cells studied) or replated for cloning efficiency. Plates are then fixed and stained (dilute crystal violet in methanol), colonies are counted, and a mutation frequency is determined by dividing the number of mutant colonies by the number of cells plated, corrected for cloning efficiency. For a positive test, a material must exhibit a concentration-related increase over at least three concentrations, and these results must be reproducible (Hsie et al., 1981).

For one study, a DNA-repair-deficient line of CHO cells (UV-5) was used for the assay under standard assay conditions except for mutation expression, where spinner flask cultures rather than petri plates were used during a 6-day expression period (Timourian et al., 1982). Cells were treated as above for mutant selection and cloning efficiency. Mutants were selected for resistance to 6-thioguanine and azaadenine in studies using the UV-5 cell line.

Specific mutagenic activities in the CHO mutation assay of an SRC-I process solvent (Sample L2121) and an SRC-II middle and heavy distillate blend (Sample L1251) were 0.6 and 0.7 mutants/μg per 10^6 survivors, respectively (Table 3.47). Hydrotreatment lowered but did not eliminate mutagenicity of the SRC-II material (0.3 mutants/μg per 10^6 survivors; Frazier and Mahlum, 1984). Higher-boiling distillate cuts of a similar SRC-II blend (Sample L1111) produced at the Harmarville, Pennsylvania PDU were also assayed. The $>850°$F cut showed more than twice the mutagenic activity of the 800–850°F cut. Assays of chemical class fractions of the $>850°$F cut showed that PAH compounds were the greatest contributors to mutagenicity with a weighted mutagenicity of 1.8 mutants/μg per 10^6 survivors (Mahlum et al., 1983).

LCFF (Sample L5111) and LCF TLP (Sample HO611) collected during normal operation of the C. E. Lummus ITSL PDU were also assayed for mutagenicity in the CHO mutation assay (Wilson et al., 1983a). Detectable mutagenicity was found in both materials and with the PAH and N-PAC fractions of the LCF TLP (Table 3.48). The weighted activity of the two fractions was more mutagenic than the whole material.

A sample of Synthoil (L6111) was too toxic to be evaluated directly, but subfractions of its basic fraction (basic alumina and Sephadex LH chromatography) were tested for mutagenicity in the CHO assay system (Hsie et al., 1980). Benzene and isopropanol subfractions were inactive but mutagenicity was detected in the acetone subfraction following metabolic activation with mouse liver S9.

An H-Coal syncrude mode PDU vacuum still overhead material (Sample L3143) and five broad chemical class fractions of it, as well as a blended fuel oil (Sample L3231) from

Table 3.47 Results of Chinese Hamster Ovary (CHO) Cell Mutation Assay on Solvent-Refined Coal (SRC) Materials (Wilsonville, Alabama; Ft. Lewis, Washington; and Harmarville, Pennsylvania)

Material	Relative Weight (% of total)	Specific Activity[a]	Weighted Activity[b]
SRC-I process solvent (Sample L2121)		0.6	
300–700°F		0.1	
750–800°F		0.3	
800 to >850°F		0.7	
SRC-II distillate blend (Sample L1251)		0.7[c]	
Hydrotreated		0.3[c]	
SRC-II distillate blend (Sample L1111)		—	
300–700°F		NA[d]	
750–800°F		0.1	
800–850°F		0.6	
>850°F	100	1.4	1.4
Aliphatic	2	0.9	0.02
PAH[e]	44	4	1.8
Nitrogen-PACs[f]	35	1.8	0.6
Hydroxy-PAHs	19	1.7	0.3
Benzo[a]pyrene	—	10	—
6-Aminochrysene	—	2.2	—

[a] Specific activity (mutants/μg · 10^6 survivors) was estimated by linear regression of dose–response data (i.e., slope of the regression line) using the method of least squares. Mutagenic responses were corrected by subtracting the background mutation frequency ($<2 \times 10^{-6}$) (Frazier and Samuel, 1985).

[b] Specific activity (mutants/μg · 10^6 survivors) × relative weight (%).

[c] Average of two independent determinations.

[d] No detectable activity.

[e] Polycyclic aromatic hydrocarbons.

[f] Polycyclic aromatic compounds.

Source: Mahlum et al. (1983), Frazier and Mahlum (1984), and Frazier and Samuel (1985).

the Catlettsburg, Kentucky pilot plant, were assayed using the CHO mutation assay (De Marini et al., 1984). All materials were highly cytotoxic to CHO cells, but none of the materials induced a clear mutagenic response. Whereas the presence of an S9 metabolic activation system in the assay reduced the cytotoxicity, the crude vacuum still overhead material still resulted in 50% relative survival at concentrations of approximately 25 μg/mL in the presence of S9, and the blended fuel oil induced a similar level of survival at a concentration of only 0.025 μg/mL.

Materials from two gasifiers (METC and UNDERC) were assayed in the CHO mutation assay. METC raw (Sample G1212) and clean (Sample G1213) product gas condensates were cytotoxic but not very mutagenic (Shimizu et al., 1984). Metabolic activation (rat liver S9) slightly increased the cytotoxicity (cloning efficiency) of clean gas condensate while decreasing that of raw gas condensate. However, even after metabolic activation, clean gas condensate was significantly less cytotoxic [LC_{50} (lethal concentration; concentration required for 50% mortality) > 300 μg/mL] than raw gas condensate (LC_{50} < 50 μg/mL). A soft agar technique (Li and Shimizu, 1983) was used for mutant selection. Metabolic activation was required to demonstrate mutagenicity. Raw gas condensate induced 0.1 mutant/μg per 10^6 survivors whereas clean gas condensate induced only 0.003

Table 3.48 Results of Chinese Hamster Ovary (CHO) Cell Mutation Assay on Integrated Two-Stage Liquefaction Materials and Their Distillate Fractions (New Brunswick, New Jersey) and an H-Coal Distillate (Catlettsburg, Kentucky)

Material	Relative Weight (% of total)	Specific Activity[a]	Weighted Activity[b]
LC Finer feed (Sample L5111)		0.5	
>850°F		0.3	
LC Finer total liquid product (Sample HO611)		0.2	
PAHs[c]	35.0	0.9	0.3
Nitrogen-PACs[d]	6.5	2.4	0.2
>850°F		0.2	
H-Coal distillate (Sample L3244)		0.5	
Benzo[a]pyrene		10	
6-Aminochrysene		2.1	

[a] Specific activity (mutants/μg · 10^6 survivors) was estimated by linear regression of dose–response data (i.e., slope of the regression line) using the method of least squares. Mutagenic responses were corrected by subtracting the background mutation frequency ($<2 \times 10^{-6}$) (Frazier and Samuel, 1985).

[b] Specific activity (mutants/μg · 10^6 survivors) × relative weight (%).

[c] Polycyclic aromatic hydrocarbons.

[d] Polycyclic aromatic compounds.

Source: Wilson et al. (1983a) and Frazier and Samuel (1985).

mutants/μg per 10^6 survivors. It is important to note that not only is the activity lower in the clean gas condensate, but the amount of condensable organic material is considerably lower (about 50 g/m^3 in the raw gas and only 100 mg/m^3 in the clean gas).

The UNDERC tar (Sample G1121) had an LC_{50} (trypan blue exclusion) in CHO cells of 180 μg/mL after metabolic activation (Cunningham et al., 1985). The LC_{50} values for acidic, basic, and neutral fractions of the tar were 70, 126, and 72 μg/mL, respectively. While a clear dose response was not reported, the tar after metabolic activation induced 8 and 22 mutants/10^6 survivors at concentrations of 100 and 200 μg/mL, respectively. Each of the fractions of the tar contributed to the mutagenicity. At a concentration of 100 μg/mL, the acidic fraction induced 4 mutants/10^6 survivors, the basic fraction 13, and the neutral fraction 16. At 200 μg/mL, 10, 62, and 77 mutants/10^6 survivors were induced by the acidic, basic, and neutral fractions, respectively.

Basic and neutral fractions from an in-situ-derived coal-gasification tar (Sample G4121) were assayed for mutagenicity in CHO (UV-5) cells (Timourian et al., 1982). Four test concentrations of the basic fraction from 70 to 280 μg/mL were evaluated for cytotoxicity and mutagenicity. The basic fraction was not appreciably cytotoxic up to a test concentration of 210 μg/mL (>90% survival). At 280 μg/mL, 60% of the cells survived a 4-h exposure. The maximum mutagenic response occurred at 140 μg/ml, four 6-thioguanine-resistant mutants and 10 azaadenine-resistant mutants/10^5 survivors. Assays of the neutral fraction of the tar were conducted at four concentrations from 25 to 100 μg/mL. The neutral fraction was more cytotoxic (about 20% cell death at 100 μg/mL). Highest mutagenicity for 6-thioguanine (7 mutants/10^5 survivors) and for azaadenine (12 mutants/10^5 survivors) occurred at 50 μg/mL.

3.4.1.2.2 Mouse Lymphoma Mutagenesis Assay. A mammalian cell mutagenesis assay was developed by Clive (Clive and Spector, 1975; Clive et al., 1979), using a mouse lymphoma cell line (L5178Y) heterozygous at the thymidine kinase (TK) locus. These cells ($TK^+/^-$) produce TK and do not grow in the presence of the toxic thymidine analog trifluorothymidine. The assay measures the effect of chemicals on frequency of forward mutation from $TK^+/^-$ to $TK^-/^-$, a cell that grows in the presence of trifluorothymidine.

$TK^+/^-$ cells are grown in suspension culture in Fischer's medium and selectively purged weekly of $TK^-/^-$ cells by adding methotrexate (0.1 μg/mL), which blocks folate metabolism, essential for survival of $TK^-/^-$ cells. $TK^+/^-$ cells (6×10^6) are exposed for 4 h to test materials with or without S9 metabolic activation (rat liver S9) in a roller drum incubator. Test material is removed by low-speed centrifugation (1000*g*). Cells are washed (three times) and then plated in growth media at a cell density of 4×10^5 cells/mL and grown 2 days to express mutations.

After the expression period, 3×10^6 cells are plated in soft agar growth medium supplemented with trifluorothymidine. Six hundred cells are also plated in nonselective media (no trifluorothymidine) to determine cloning efficiency. Cells are grown for 11–12 days and then colonies are counted. Mutation frequency is determined by dividing the total number of trifluorothymidine mutant colonies by the number of cells plated, times the plating efficiency (number of colonies with no trifluorothymidine divided by number plated $\times$ 100).

Several NTSL materials from the Wilsonville pilot plant were evaluated in the mouse lymphoma assay (Drozdowicz and Kelly, 1983d, e). Three criteria were required for a definitive experiment: (1) plating efficiency greater than 50%, (2) the average spontaneous frequency of solvent controls less than 110×10^{-6}, and (3) mutation frequency of positive control at least twice that of solvent controls. An assay is considered positive if at least one concentration with a relative total growth greater than 10% has a mutation frequency at least twice that of solvent controls and the observed response increases with concentration. Seven test concentrations over a 10–100-fold range and determined to be below cytotoxic levels were used for each experiment. Three determinations were made with each concentration. Experiments were run with and without metabolic activation.

Saline suspension and water extract of an NTSL first-stage solid (Sample L5232) were nonmutagenic as was a saline suspension of second-stage solid (Sample H0523). Distillate cuts No. 6 and No. 8 from first-stage liquid blend (Sample L5241) and naphtha from the second-stage blend (Sample H0524) were also nonmutagenic. All other materials assayed were found to be mutagenic but not all materials required metabolic activation. Results of positive assays (maximum response values) are shown in Table 3.49. An aqueous suspension of first-stage solid product was mutagenic. Activity was probably due to suspended particulate matter; a filtered and then concentrated aqueous suspension was nonmutagenic. DMSO extracts of first- and second-stage (hydrotreated) solid products were mutagenic. A first-stage naphtha sample was mutagenic, but hydrotreatment (second stage) reduced the mutagenicity to below detectable levels. Two first-stage light distillate bp cuts and first- and second-stage middle distillates and their bp cuts were mutagenic. Five materials (the aqueous solid product suspension, the two DMSO solid product extracts, and the second-stage middle distillate and its high-bp cut, No. 9) contained direct-acting mutagens not requiring metabolic activation.

Distillates (naphtha, No. 2 fuel oil, and No. 6 fuel oil) representative of commercial products were prepared from a COED oil (Sample G2221). These distillates, considered to be similar to COGAS oil, were also assayed in the mouse lymphoma mutagenesis assay and results were compared to those for similar materials distilled from two crude petroleums, one slightly more aromatic than the COED oil and the other slightly less aromatic

Table 3.49 Results[a] of Mouse Lymphoma Mutagenesis Assays on Nonintegrated Two-Stage Liquefaction Materials (Wilsonville, Alabama)

		No Metabolic Activation (−S9)		Metabolic Activation (+S9)	
Material	Distillate Cut (°F)	Concentration Ranges Tested (μL/mL)	Mutation Frequency[b] (mutants × 10^{-6})	Concentration Ranges Tested (μL/mL)	Mutation Frequency[b] (mutants × 10^{-6})
First-stage solid (Sample L5232)					
Aqueous suspension		1–10	87 (8)	1–10	349 (1)
DMSO extract		0.4–4[c]	133 (4)	0.2–5[c]	192 (0.8)
Second-stage solid (Sample H0523)					
DMSO extract		2–10	101 (10)	0.4–4	113 (1.5)
First-stage blend (Sample L5241)					
Cut No. 1 Naphtha	—	0.07–0.2	0	0.01–0.15	169 (0.07)
Cut No. 3	400–450	0.04–0.2	0	0.009–0.08	201 (0.03)
Cut No. 4	450–500	0.01–0.1	0	0.2–0.8	433 (0.05)
Middle distillate	—	0.02–0.1	0	0.005–0.06	159 (0.04)
Cut No. 5	500–550	0.03–0.2	0	0.1–0.8	87 (0.03)
Cut No. 7	600–650	0	0	0.006–0.3	208 (0.015)
Cut No. 9	700–750	0.02–0.2	0	0.01–0.06	413 (0.03)
Second-stage blend (Sample H0524)					
Middle Distillate	—	0.01–0.1	67 (0.08)	0.02–0.2	117 (0.08)
Cut No. 5	500–550	0.03–0.2	0	0.03–0.2	306 (0.08)
Cut No. 6	550–600	0.06–0.2	0	0.03–0.2	331 (0.08)
Cut No. 7	600–650	0.01–0.1	0	0.01–0.1	123 (0.04)
Cut No. 8	650–700	0.02–0.2	0	0.005–0.08	275 (0.04)
Cut No. 9	700–750	0.01–0.1	40 (0.08)	0.005–0.08	110 (0.08)
Benzo[a]pyrene	—	—		3–5[c]	565 (5)
Solvent Control (1% DMSO)	—	—	34[d]	—	62[d]

[a] Data reported only for materials found to be mutagenic; that is, the mutation frequency for at least one test concentration, where relative total growth is greater than 10%, was at least twice that of the solvent control (1% DMSO).

[b] Maximum response, average of three determinations corrected for solvent control with concentration giving maximum response in parentheses.

[c] This value is in units of μg/mL.

[d] Average of results from studies where materials were found to be positive; (−S9) $n = 6$ and (+S9) $n = 19$.

Source: Drozdowicz and Kelly (1983d, e).

(paraffinic; Robinson and Simmon, 1981). Materials were tested over a 2-log dose range (determined in a preliminary cytotoxicity assay) with and without metabolic activation. The three naphtha products and COED-derived No. 2 fuel oil were nonmutagenic. All other materials were mutagenic after metabolic activation (Table 3.50). The results for the aromatic No. 2 fuel oil were not available.

3.4.1.2.3 Sister Chromatid Exchange (SCE). SCE is the induced transfer of genetic information between chromatids of the same chromosome without loss or damage. The exchanged DNA originates from both chromatids in the chromosome. When properly stained, one of the chromatid stains dark and one light (Perry and Wolff, 1974). The induction of SCE results in the appearance of alternating dark and light patches on the chromatid. The dark patches originate from one chromatid, the light patches from the other. As a measure of genotoxicity, the induction of SCE can be determined in all dividing mammalian cells and indicates the ability of a test material to reach the DNA and produce alterations in the chromosomes. Not all agents that produce mutations or chromosome aberrations produce SCE. Some weak mutagens produce many SCE, while some strong mutagens produce few SCE. Genotoxicity is indicated by determining the sample concentration causing an increase in SCE in exposed compared to unexposed cells. Studies with coal-derived materials principally used a mouse myeloma cell line (MPC11) grown in suspension culture, although there were limited studies with secondary cultures of Chinese hamster lung cells and an established CHO cell line (UV-5).

A variety of Hygas materials were evaluated for SCE induction in a mouse myeloma suspension cell assay (Reilly et al., 1986). A measure of cytotoxicity was first obtained by determining inhibition in cell growth, and an EC_{50} (effective concentration; the concentration of material that is needed to effect a 50% reduction in growth rate as compared to unexposed cells) was calculated (MacCoss et al., 1978). Measurement of SCE induction

Table 3.50 Results of Mouse Lymphoma Mutagenesis Assays on Crude Petroleum and COED-Derived Distillates[a]

Material	Concentration Ranges Tested (μL/mL)	Mutation Frequency[b] (mutants $\times 10^{-6}$)
Paraffinic No. 2 fuel oil	0.01–0.5	114 (0.09)
Aromatic No. 6 fuel oil	7–500	172 (160)
COED No. 6 fuel oil (Sample G2221)	3–200	106 (85)
Paraffinic No. 6 fuel oil	7–500	177 (500)
Dimethylbenzanthracene	—	492 (4)[c]
Solvent control		
1% DMSO	—	54
1% Acetone : Cyclohexane (7 : 3)	—	65[d]

[a] Data reported only for metabolically activated (+S9) materials found to be mutagenic; that is the mutation frequency for at least one test concentration, where relative total growth is greater than 10%, was at least twice that of the solvent control.

[b] Maximum response (corrected for solvent) with concentration giving maximum response in parentheses.

[c] This value is in μg/mL.

[d] Average of three determinations.

Source: Robinson and Simmon (1981).

(SCE/100 chromosomes) was conducted by light microscopy and plotted as a function of concentration of test material. A regression line was calculated for the data, and the concentration at which the number of SCE/100 chromosomes was twice (200%) the untreated control value (SCE_{200}) was determined (Reilly et al., 1981). Thus, the smaller the measurement for each endpoint, the more toxic the agent. The genotoxic potential represented by SCE induction can be considered greatest when the SCE_{200} concentration is a low cytotoxic concentration. It is significant when the SCE_{200} is less than the EC_{50}.

For the most part, only the NVO fractions of the samples (components with bp > 390°F) were evaluated (Table 3.51). While composited petroleum crude oil NVO (EPA/DOE Synfuels Research Material) had no effect in this assay up to test concentrations of 200 μg/mL, some gasification materials had substantial cytotoxicity (EC_{50} concentrations below 100 μg/mL) and SCE induction (SCE_{200} concentrations below 50 μg/mL). The lowest EC_{50} value for a test material (50 μg/mL for recycle oil sample G2112) or SCE_{200} value (28 μg/mL for LTR gas condensate sample G2119) did not approach the values for the positive control, methyl methanesulfonate (EC_{50} = 25 μg/mL and SCE_{200} = 3 μg/mL).

Similar cytotoxicity and SCE measurements in the mouse myeloma assay were made on UNDERC gasifier materials (Wilzbach et al., 1982; Reilly et al., 1986). The EC_{50} values were about 100 μg/mL for most of the tars and oils tested. One of the tars (Sample G1121) and the gas stream condensates (Sample G1111) were more cytotoxic, having EC_{50} values of 50 μg/mL or less (Table 3.52). The SCE_{200} values for all UNDERC materials were substantially lower than the EC_{50} values. The most toxic sample in terms of SCE induction (SCE_{200} = 11 μg/mL) was the high-pressure cyclone raw gas condensate, where more than 50% of the organics present in the gas stream were condensed.

Table 3.51 Results of Sister Chromatid Exchange Assays on Hygas Materials

	EC_{50}[a] (μg/mL)		SCE_{200}[b] (μg/mL)	
Material	Crude	NVO[c] Fraction	Crude	NVO Fraction
LTR Gas condensate				
Sample G2117	217	—	122	—
Sample G2118	204	—	117	—
Sample G2119	138	67	79	28
Sample G2127	—	143	—	93
Recycle oil				
Sample G2112	—	50	—	35
Sample G2113	—	55	—	47
Sample G2122	—	53	—	45
Char/oil slurry oil				
Sample G2116	—	63	—	57
Sample G2125	—	89	—	79
Methyl methanesulfonate	—	25	—	3
Composited petroleum crude	—	NE[d]	—	NE[d]

[a]Concentration of material needed to effect a 50% reduction in growth rate as compared to unexposed cells.

[b]SCE_{200} is the sample concentration inducing SCE levels 200% above levels in unexposed cells (normalized to a background SCE/100 chromosomes of 10).

[c]Nonvolatile organic.

[d]No effect up to 200 μg/mL.

Source: Reilly et al. (1986).

Table 3.52 Results of Sister Chromatid Exchange (SCE) Assays on University of North Dakota Energy Research Center Materials

Material	EC_{50}[a] ($\mu g/mL$)	SCE_{200}[b] ($\mu g/mL$)
Tar		
Sample G1121	52	15
Sample G1122	95	32
Sample G1123	90	28
Sample G1124	93	58
Oil		
Sample G1125	95	64
Sample G1127	102	47
Gas stream		
Sample G1111[c]		
High-pressure cyclone	20	11
Low-pressure cyclone	68	NE[d]
First heat exchanger	30	42
Second heat exchanger	38	69
Methyl methanesulfonate	25	3

[a] Concentration of material needed to effect a 50% reduction in growth rate as compared to unexposed cells.

[b] SCE_{200} is the sample concentration inducing SCE levels 200% above levels in unexposed cells (normalized to a background SCE/100 chromosomes of 10).

[c] Nonvolatile organics tested.

[d] No effect.

Source: Wilzbach et al. (1982) and Reilly et al. (1986).

Pyrolysis gas condensates from two test runs (one at 1470°F, the other at 1830°F) at the Mellon riser reactor were assayed in mouse myeloma cells for SCE (Fillo et al., 1983). High-temperature trap (tar) and room-temperature trap (oil) organic condensates from each run were evaluated. Concentrations of 41 $\mu g/mL$ (1470°F run, Sample G3153HT) and 29 $\mu g/mL$ (1830°F run, Sample G3170HT) tar increased SCE to twice background (SCE_{200}). Oils were somewhat less toxic, inducing an SCE_{200} of 45 $\mu g/mL$ (1470°F run, Sample G3153RT) and 79 $\mu g/mL$ (1830°F run, Sample G3170RT), respectively.

METC gasifier raw (Sample G1212) and clean (Sample G1213) gas condensates were evaluated for SCE in Chinese hamster lung cultures (Shimizu et al., 1984). While the gas cleanup system effectively removed condensable organic materials from the gas stream (raw gas having about 50 g/m^3 condensable organic materials and the clean gas only about 100 mg/m^3), the materials condensed were capable of inducing SCE. In addition to its lower mass, the clean gas condensate was also much less active than the raw gas condensate. Statistically significant ($P < 0.05$) increases in SCE were induced by the clean gas condensate at concentrations above 400 $\mu g/mL$ (>0.5 SCE/chromosome). In contrast, the frequency of SCE/chromosome was greater than 1 at concentrations of raw gas condensate below 50 $\mu g/mL$.

A DNA-repair-deficient CHO cell line (UV-5) was used for study of SCE induction by a basic tar fraction (0.6 wt % of the tar mass) from an in situ coal-gasification tar (Sample G4121). SCE were induced over a concentration range of 70–210 $\mu g/mL$ (cell survival at these concentrations was above 90%) with some indication of a dose response (1 SCE/cell at 70 $\mu g/mL$, rising to a maximum response of 5.5 SCE/cell at 210 $\mu g/mL$; Timourian et al., 1982).

3.4.1.2.4 Cytogenetic Measurements. Measurements of genetic damage included not only direct measurements of mutation but cytogenetic evidence such as chromosomal aberration, micronuclei, and polyploidy. Only limited studies of this nature were performed with coal-derived materials. For these studies, cells were grown in the presence of bromodeoxyuridine (BrdUrd) and the test agent for up to 30 h with colcimid added to the growth media for the final 5 h (holds dividing cells in metaphase to allow evaluation of chromosomal damage). A number of first-division cells (200–500), those that do not show differential chromatid staining by fluorescence plus Giemsa technique (Carrano et al., 1979), were then scored for chromosome damage. Chromosomes were evaluated for exchange and deletion-type aberrations. For polyploidy, 100 cells from both 24- and 48-h cultures (5 h colcimid arrest) were scored following Giemsa staining. Micronuclei formation was evaluated in 24- and 48-h cultures without BrdUrd, according to the method of Countryman and Heddle (1976).

The base fraction of an in-situ-derived coal-gasification tar (Sample G4121) was studied for induction of chromosome aberrations, polyploidy, and micronuclei in a DNA-repair-deficient CHO cell line (UV-5). Over a concentration range of 70–280 μg/mL, there was no statistically significant increase in aberrations, micronuclei formation, or polyploidy (Carrano et al., 1979). There was a net increase in aberrations at the three highest concentrations, thus not ruling out a significant response at higher concentrations, although at the highest concentration (280 μg/mL), there was a net survival of only 60% of the exposed cells.

3.4.1.2.5 Unscheduled DNA Synthesis. One method for detection of a material's potential for damaging DNA is to measure the DNA synthesis that accompanies repair of damage, so-called unscheduled DNA synthesis (UDS). In the UDS assay (Stich et al., 1971), the incorporation of radiolabeled thymidine, following exposure to test materials, into the DNA of cultured cells is measured by autoradiography. Cell cultures, for example, primary rat hepatocytes, are grown on coverslips in Leighton tubes for a few hours. Then appropriate concentrations of test materials are added, along with fresh medium and tritiated thymidine, and held for up to a 4-h exposure period. All cells are then washed and the coverslips are processed for autoradiography. After 2 weeks, grain counts over cell nuclei are used to determine if UDS occurred. A positive response is one in which the incorporation of tritiated thymidine is significantly higher than that of the solvent control (generally DMSO).

Measurement of UDS is most sensitive at test concentrations slightly lower than cytotoxic levels. Five concentrations covering a 2-log range, with the lower end being nontoxic and the upper end inducing 50–90% cell killing, are generally used for the assay. The appropriate concentrations of test material for the assay are first determined by measuring cytotoxicity (trypan blue exclusion) produced in cells following a 4-h exposure period to a wide range of test concentrations.

SRC-II middle distillate (Sample L1264) and stripper tower bottoms (Sample L1263) from the Ft. Lewis, Washington pilot plant were assayed for UDS (DOE, 1982c). The stripper tower bottom sample was tested over a concentration range of 0.3–40 μg/mL and found to induce statistically significant UDS at concentrations above 0.9 μg/mL. In contrast, the middle distillate was not found to induce UDS over a test concentration range of 0.04–10 μg/mL.

3.4.1.2.6 Syrian Hamster Embryo (SHE) Cell Transformation Assay. Cells exposed to carcinogens in culture undergo morphological changes characterized as transformation. Transformation is based on observed colony morphology as described by DiPaolo et al. (1971). Such transformants have been demonstrated to develop into malignant tumors upon injection into susceptible animals. In this assay, primary SHE cell cultures (3×10^7 cells in 150-cm^2 flasks) prepared by trypsinization of eviscerated and decapitated 13–14-day embryos are grown for 2 or 3 days and subcultured into 60-mm plates. Approximately 3×10^3 cells are put in each plate with culture medium (growth medium supplemented with fetal calf serum). After 24-h incubation, cells (generally 12 plates) are exposed to graded concentrations of test materials (dissolved in DMSO) in the presence or absence of a metabolic activation system (Aroclor-induced rat liver homogenate). Cells are exposed 24 h, the medium is removed, and then cells are washed and fed with fresh medium. Cells are cultured at least 8 days to allow colony formation. Colonies are fixed (10% aqueous formalin) and stained with dilute crystal violet solution. The number of morphologically transformed cells are then visually determined. Transformation frequency is calculated by dividing total number of transformed colonies by total number of colonies; there are no spontaneous transformants in this assay (Frazier and Andrews, 1983).

SRC-I light oil (Sample L2211), wash solvent (Sample 2212), and process solvent (Sample L2213) from the Ft. Lewis, Washington pilot plant were evaluated for transforming potential in the SHE assay (Frazier and Andrews, 1983). Light oil or wash solvent produced no transformants over a concentration range of 5–100 and 5–200 μg/mL, respectively. Following metabolic activation, process solvent induced transformation at concentrations between 10 and 50 μg/mL with a maximum response of 8% cell transformation with the 50-μg/mL concentration. A 20-μg/mL sample of BaP under similar conditions induced 9% cell transformation, while 200 μg/mL of Prudhoe Bay or Wilmington crude oil induced 0.5 and 0.2% transformation, respectively.

Basic and PAH components of the process solvent derived by acid–base–neutral solvent–solvent fractionation induced cell transformation. Basic components were more active than neutral components. Maximum responses of 4% and 8% transformation were seen at 50 μg/mL with the PAH and basic fractions, respectively. Because the range of test concentrations giving a linear transformation response is small, good dose–response data were not obtained. The phenomenon, common with this assay, is presently unexplained (Frazier and Andrews, 1983).

A few cell lines derived from transformed colonies (tested after a minimum of 13 passages) induced by process solvent and its basic and neutral fractions were shown to be neoplastic by their growth in athymic (nude) mice and subsequent histologic evaluation (undifferentiated sarcomas at the site of injection; Frazier and Andrews, 1983).

SRC-II light, middle, and heavy distillates (Samples L1241–L1243) from the Ft. Lewis, Washington pilot plant were also assayed in SHE cells. The light and middle distillates showed no transforming potential over test concentrations ranging from 5 to 200 μg/mL and 5 to 100 μg/mL, respectively. The heavy distillate induced about 6% transformation at all concentrations tested (5, 20, and 50 μg/mL; Frazier and Andrews, 1983). Assay of acid–base–neutral solvent-separated fractions of the heavy distillate showed transforming potential in all five fractions studied. At 10 μg/mL, the percent age of transformation was 4.4 for basic fraction, 5.4 for PAH fraction (isooctane insoluble), 1.7 for neutral fraction, 4.5 for basic tar, and 3.1 for neutral tar (Mahlum et al., 1983).

Distillate cuts of SRC-I and SRC-II material blends were also assayed (Mahlum et al., 1983). SRC-I material (Sample L2121) was generated at the Wilsonville, Alabama pilot plant. The SRC-II products came from the Ft. Lewis, Washington (Sample L1251) and Harmarville, Pennsylvania plants (Samples L1121, L1131, and L1111). Generally, results (Table 3.53) show that most transforming activity is associated with materials boiling above 750°F. All fractions (aliphatic, PAH, N-PAC, and hydroxy-PAH) of SRC-II 800–850°F and >850°F cuts derived by absorption column chromatography were active. Further anal-

Table 3.53 Assay of Solvent-Refined Coal Materials

Material	Distillate Cut (°F)	Transformation[a] (%)
SRC-I (Sample L2121)	700–750	0.6
	800–850	1.5
SRC-II (Sample L1251)	300–700	0.4
	800–850	3.3
	>850	5.8
SRC-II (Samples L1121 and L1131)	639–689	0.2
	705–825	0.5
	710–730	0.6
	739–761	0.7
	645–782	1.4
	782–878	1.5
	867–967	5.5
	914–1055	8.5
SRC-II (Sample L1111)	300–700	0.2
	700–750	0.3
	750–800	3.7
	800–850	6.8
	>850	8.9
SRC-II distillate blend (Sample L1251)	—	3.6[b]
Hydrotreated:		
moderate	—	1.2[b]
severe	—	2.0[b]
Benzo[a]pyrene	—	7.5[c]
2-Aminoanthracene	—	2.7[c]
DMSO	—	0

[a] All materials were tested at a 10-μg/mL concentration and the transformation frequency was determined by dividing the total number of transformed colonies by the total number of colonies in 12 plates.

[b] Average of two determinations.

[c] Average of three determinations.

Source: Mahlum et al. (1983) and Frazier and Mahlum (1984).

ysis of chemical class fractions (HPLC fractionation) of the distillate cuts of SRC-II (Sample L1251) showed that most activity was in PAH compounds with higher activity in the fractions of the >850°F cut. Distillate cuts of Sample L1111 were sequentially reblended and tested for transformation potential (Pelroy et al., 1985). While the summed individual weighted specific transforming activity was 1%, the specific activity of the reblended material was 6.3%.

Hydrotreatment of an SRC-II distillate blend (Sample L1251) reduced but did not eliminate transforming activity (Frazier and Mahlum, 1983; Mahlum et al., 1983). Moderate hydrotreatment (5510 ppm N) appeared to be more effective in reducing transformation than severe treatment (1900 ppm N) (Table 3.53).

3.4.1.2.7 Cytotoxicity Assays. In addition to the cytotoxicity measurements made as part of the test protocol for other mammalian cell assays, a number of mammalian cells have been used in independent assays to evaluate the cytotoxicity of coal-derived materials. Conceptually, the assay procedures are similar. Methodologies and results are presented for three cell systems: CHO, WI-38 (human lung), and VERO (monkey kidney) cells.

CHO. CHO cells have been used at several laboratories to estimate cytotoxicity. However, there were several variations in test procedures. One procedure commonly used was cloning efficiency (Brusick and Young, 1981). For this test, cells (up to 10^3) are put in culture dishes containing growth medium and allowed to adhere (6–24 h) in a humidified 5% CO_2 atmosphere at 37°C. The medium is aspirated, test material appropriately diluted in growth medium is added, and the cells exposed for 24 h. Cells are then washed three times with phosphate-buffered saline, growth medium is added, and the cells incubated (6–10 days) to allow cells capable of growth to develop into visible colonies. After this incubation, the dishes are drained, washed with phosphate-buffered saline, fixed with methanol, and the colonies Giemsa stained to facilitate identification. Surviving colonies are counted and EC_{50} (effective concentration: concentration that causes a 50% decrease in survival compared to untreated control cells) values are calculated or percentage of inhibition is determined (reduction in colonies induced by test material compared to the number of colonies developing from unexposed cells).

CHO cell cytotoxicity has also been measured as alterations in growth kinetics (ability of cells to grow and divide as members of a large population; Nichols et al., 1979). For growth kinetics assays, 10^5 cells are put in petri dishes, allowed 24 h to attach, and then are exposed for 24 h to test material. Following exposure, fresh medium is added and cells are counted at 24-h intervals for 1 week. Cytotoxicity is measured as percentage of inhibition of growth induced by test material.

CHO cells or rat lung epithelial cells have also been grown on a hydrated collagen gel to determine the toxicity of vapors (Zamora et al., 1983). Collagen solution was put into a petri dish and allowed to gel. Approximately 10^6 cells in growth medium with fetal calf serum were placed on the gel. The medium was removed (aspiration) 24 h later and the attached cells were exposed to vapors for 1 h. EC_{50} values for cell viability (trypan blue exclusion) and cloning efficiency were determined by standard procedures after removal of the cells from the gel matrix with trypsin.

Cytotoxicity values of several gasification materials were evaluated in this assay. Materials assayed were from the Kosovo, Ft. Snelling, METC, and UNDERC gasifiers and the RTI experimental reactor. By-product tar (Sample G1412), heavy tar and dust (Sample G1411), medium oil (Sample G1413), and light oil (Sample G1414) from the Kosovo gasifier were cytotoxic in a CHO clonal toxicity assay (Lee et al., 1981). Each material was tested over a range of 0.0002–2 $\mu L/mL$ (total of seven test concentrations with three determinations for each concentration). Light oil was the least toxic, and heavy tar and dust the most cytotoxic (Table 3.54). Quantitative dose–response data were not available be-

Table 3.54 Results of Chinese Hamster Ovary Cytotoxicity Assays on Kosovo Gasifier Materials

Material	Sample No.	EC_{50}[a] ($\mu L/mL$)
Light oil	G1414	0.7
Medium oil	G1413	0.1
Tar	G1412	0.03
Heavy tar and dust	G1411	0.07
Sodium azide	—	206[b]

[a] EC_{50} = concentration of test material that reduces the number of colony-producing cells by 50% compared to controls.

[b] This value is in units of $\mu g/mL$.

Source: Lee et al. (1981).

cause cytotoxicity occurred over a sharp range. For example, 97% of the cells survived exposure to 0.2 $\mu L/mL$ of light oil, while only 9% survived at 2 $\mu L/mL$ (the next highest concentration tested).

Crude tar and its chemical class fractions from three test runs at the RTI reactor, each with a different coal type, were asssayed for cytotoxicity to CHO cells (Nichols et al., 1979). Inhibition of growth kinetics and clonal efficiency were determined for two concentrations of each test material (Table 3.55). Whole tar from Illinois No. 6 coal (Sample G1541) inhibited cloning efficiency 98% and growth kinetics 72% at 50-mg/mL exposure medium. Western Kentucky No. 9 bituminous coal tar (Sample G1531) at the same con-

Table 3.55 Results of Chinese Hamster Ovary Cytotoxicity Assay on Tars and Their Fractions Produced in the Research Triangle Institute Reactor

Material	Concentration (mg/mL)	Growth Kinetics[a] (% inhibition)	Clonal Efficiency[b] (% inhibition)
Sample G1511 (Wyoming subbituminous coal)	62.5	36	26
	25	36	5
Polynuclear aromatic	125	94	80
	50	63	40
Polar neutral	12.5	78	38
	5	64	56
Acid	12.5	83	54
	5	71	49
Base	1.25	6	12
	0.5	28	10
Sample G1531 (western Kentucky coal)	125	83	56
	50	69	45
Polynuclear aromatic	125	69	23
	50	43	12
Polar neutral	125	0	0
	50	5	12
Acid	62.5	69	12
	25	52	0
Base	187.5	90	53
	75	74	50
Nonpolar neutral	125	96	23
	50	90	5
Sample G1541 (Illinois No. 6 coal)	125	72	100
	50	72	98
Polynuclear aromatic	125	77	95
	50	60	70
Polar neutral	31.25	0	42
	12.5	0	50
Acid	62.5	0	15
	25	0	18
Base	1.25	0	12
	0.5	0	22

[a] Decrease in growth kinetics induced by test material reported as percent of unexposed response.
[b] Decrease in clonal efficiency (ability of a single cell to survive exposure to test material and grow into a colony) reported as percent of unexposed response.

Source: Nichols et al. (1979).

centration was somewhat less toxic as measured by cloning efficiency (45% inhibition) but not as measured by growth kinetics (69% inhibition). At 25 mg/mL, Wyoming subbituminous coal tar (Sample G1511) inhibited cloning efficiency only 5% and growth kinetics 36%. No one tar fraction was consistently most cytotoxic. For Illinois No. 6 coal, polar neutral and polynuclear aromatic fractions were significantly cytotoxic (>50% inhibition of cloning efficiency) at the concentrations tested; for western Kentucky No. 9 coal, only the base fraction was significantly cytotoxic; for Wyoming subbituminous coal, all but the base fraction were significantly cytotoxic.

No cytotoxicity (i.e. no decrease in cell survivors) was noted in CHO cells or lung epithelial cells grown on hydrated collagen gel and exposed to METC raw product gas vapors (Zamora et al., 1983). Vapors were produced by passing air over heated raw product gas condensate (Sample G1224), then diluting with a 95% air/5% CO_2 stream.

WI-38. A cell line, WI-38, established from human lung tissue at the Wistar Institute, has been used for cytotoxicity measurements. Cells are plated at 2×10^5 cells/well (four-well cluster dishes) in growth medium plus fetal calf serum and allowed to adhere for 24 h. Assay materials first dissolved in DMSO are diluted to appropriate concentration in complete culture medium and added to test wells (3 wells/test concentration) for 20 h. Cells are then removed by trypsinization, diluted with fresh medium, and aliquoted to determine changes in cell growth (total cells and total protein/Lowry assay), cell killing (viability/trypan blue exclusion), and cell function [adenosine triphosphate (ATP) activity]. EC_{50} values (concentrations that cause a 50% decrease in the endpoint measured compared to that of untreated control cells) are then calculated (Kilpatrick et al., 1980).

Product gas (Sample G1711) and test burner effluent gas (Sample G1712) condensate and vapor components from the Ft. Snelling gasifier were evaluated for cytotoxicity in WI-38 cells. Product gas was assayed at 1.8, 6, 18, 60, and 120 μL/mL. Test burner effluent gas was assayed at 1.2, 4, 12, 40, and 80 μL/mL. The highest test concentration of each sample, 120 and 80 μg/mL, was dictated by cytotoxicity of DMSO, the test material solvent (cytotoxic at concentrations greater than 40 μL/mL). No cell killing occurred at the highest concentration of product gas assayed. However, the high concentration of particulates in this sample made it difficult to distinguish living from dead cells. Total protein and ATP activities of exposed cells were less than those of unexposed cells; EC_{50} values were 10 and 14 μL/mL, respectively. Test burner effluent gas was nontoxic as determined by total protein and ATP activities (Kilpatrick et al., 1980).

VERO. An established monkey kidney cell line, VERO cells, has also been used to evaluate cytotoxicity in protocols similar to those used with CHO and WI-38 cells. In this assay, VERO cells are exposed to test materials in small wells and allowed to grow 4 days. The number of colonies of 10 or more cells arising from division of a single cell that survived exposure are then counted and the relative plating efficiency (RPE) determined according to the formula

$$\text{RPE} = \frac{\text{number of colonies arising from exposed cells}}{\text{number of colonies arising from unexposed cells}}$$

An RPE_{50} (test concentration required to reduce plating efficiency to 50% that of the control) is then calculated to determine relative toxicities.

Cytotoxicity values of SRC-I and SRC-II materials from the Ft. Lewis, Washington pilot plant were evaluated in VERO monkey kidney cells (Pacific Northwest Laboratory, 1979).

Table 3.56 Results of VERO Monkey Cell Cytotoxicity Assays of Solvent-Refined Coal (SRC) Materials (Ft. Lewis, Washington)

Material	RPE_{50}[a] ($\mu g/mL$)
SRC-II	
Light distillate (Sample L1241)	180
Middle distillate (Sample L1242)	200
Heavy distillate (Sample L1243)	30
SRC-I	
Process solvent (Sample L2213)	35
PAH[b] fraction[c]	30
Basic fraction	100
Neutral fraction	75
Petroleum	
Prudhoe Bay crude	350
Wilmington crude	190
Diesel oil No. 2	250

[a] Concentration required to reduce the number of colony-producing cells to 50% compared to controls.
[b] Polycylic aromatic hydrocarbon.
[c] Isooctane insoluble.

Source: Pacific Northwest Laboratory (1979).

RPE_{50} values were determined for SRC-II light, medium, and heavy distillates (Samples L1241–L1243) and SRC-I process solvent (Sample L2213) and its fractions. Results are presented in Table 3.56 and demonstrate that coal-derived materials have RPE_{50} values lower than petroleum crudes.

3.4.1.2.8 Alveloar Macrophage Assay. Primary cultures of alveolar macophages have been used to study cytotoxicity and functional impairment following in vivo and in vitro exposures to environmental pollutants (Brusick and Young, 1981). Alveolar macrophage assays use primary cell cultures, having none of the inherent problems associated with interpreting results derived from long-term cultures of transformed cells. Lethal and sublethal endpoints range from a general indication of cell viability (vital dye exclusion), to more quantitative measurements such as release of lactate dehydrogenase into the culture medium, cellular ATP content and ATP-related bioluminescence, or ingestion (phagocytosis) of bacteria or polystyrene microspheres.

Alveolar macrophages are obtained by bronchopulmonary lavage from various animals (e.g., rat, rabbit, dog). Rabbit alevolar macrophages have been used most commonly. Cells are grown in nutrient medium generally containing newborn calf serum. Cells are held 24–48 h at 37°C (98.6°F) in a humidified CO_2 chamber and then exposed 18–24 h to the test material. Viability can be determined with trypan blue and an LC_{50} (concentration that kills 50% of the cells) calculated. For sublethal effects, a variety of measurements are employed. Supernatant fluids can be collected, centrifuged, and assayed for cellular enzymes released when the cell is damaged. For example, the EC_{50} (concentration of test material that effects release of 50%, as determined by linear regression, of the enzyme recovered by sonication of control cultures) for lactate dehydrogenase in the supernatant fluid is a good indication of cell membrane damage.

Similarly, functional impairment has been evaluated by measuring the exposed macrophage's ability to phagocytize polystyrene latex spheres. After exposure to the test material, spheres (1.1 mm) are added to the macrophage cultures for 2 h; cells are then fixed and stained, and the number of phagocytizing cells (those containing at least one sphere) are determined. An EC_{50} (concentration of sample that reduces the number of actively phagocytizing macrophage cells to 50% of that in unexposed cells) can be calculated.

SRC-II vacuum bottom material (Sample L1281) from the Ft. Lewis, Washington pilot plant was assayed in a rabbit alveolar macrophage assay. Cytotoxicity was low, with EC_{50} values for lethality and ATP content between 0.1 and 1 mg/mL (Kim et al., 1981).

Gel chromatographic fractions (Sephadex LH-20) of METC tar (Sample G1221) were assayed for lethality using rabbit alveolar macrophages. Of the six fractions obtained, only fraction 5 (nitrogen heterocycles and polycyclic aromatics with polar functional groups) had significant cytotoxicity with an $LC_{50} \sim 120\ \mu g/mL$ (Newton, 1979). Tar Sample G1221 was assayed with Beagle dog alveolar macrophages. Lethality and cellular release of lactate dehydrogenase were measured. Ninety-five percent of the cells were killed at a whole-tar concentration of 25 $\mu g/10^6$ cells (Newton, 1980). The EC_{50} for whole tar was 45 $\mu g/mL$. Sephadex LH-20 fractions 3 (aromatic and aliphatic hydrocarbons), 4 (PAH), and 5 showed significant cytotoxicity. Phenols present in these fractions were primarily but not solely responsible for toxicity of the tar (Hill et al., 1983). Multicyclone train gas stream aerosol condensates (Samples G1222–G1225) from the same test run as tar Sample G1221 were assayed with Beagle dog macrophages. Only the smallest aerosol particles ($>0.3\ \mu m$) were toxic as measured by lactate dehydrogenase release (concentration of 25 $\mu g/mL$ tested). Studies with materials from subsequent runs showed this was not a common finding. Cytotoxicity decreased with gas cleanup. Absolute lactate dehydrogenase release decreased in materials from all sampling points after the raw gas position (Benson et al., 1982a). Assay of condensate fractions showed cytotoxic activity was associated mainly with neutral and polar PAC (fractions 3, 4, and 5).

A number of Hygas materials, including low-temperature-reactor gas condensates, recycle oils, and char/oil slurry oils, were evaluated for lethality and functional impairment in a rabbit alveolar macrophage assay (Reilly et al., 1986). Only the nonvolatile organic fractions (components with bp $>$ 390°F) were tested and the results are presented in Table 3.57. Results of the assay provided little discrimination between samples. These materials were not particularly lethal, with LC_{50} values ranging between 100 and 200 $\mu g/mL$. (The lower the LC_{50} or EC_{50} value, the more toxic the samples.) Effective concentrations for functional impairment, that is, the ability to phagocytize latex spheres, were generally at concentrations lower than those for lethality. In a few cases, concentrations were significantly lower, for example, low-temperature-reactor gas condensate Sample G2118 ($LC_{50} = 111\ \mu g/mL$, while $EC_{50} = 39\ \mu g/mL$).

The same rabbit alveolar macrophage assay was used to evaluate materials from the UNDERC gasifier (Table 3.58). As with the Hygas material, tars (Samples G1121–G1124), an oil (Sample G1125), and a gas stream condensate (Sample G1111) did not induce significant lethality. Tars and oils were tested as crude materials but only the nonvolatile organic fraction of the gas condensate was tested. LC_{50} values varied from 80 to $>$ 2000 $\mu g/mL$. EC_{50} values for functional impairment were determined to be at concentrations lower than the LC_{50} for all materials.

High-temperature (Samples G3153HT and G3170HT) and room-temperature (Samples G3153RT and G3170RT) condensates produced by the MI experimental riser-reactor pyrolysis unit at operating temperatures of 1470 and 1830°F, respectively, were also assayed for their potential for impairing rabbit alveolar macrophage function (Fillo et al., 1983). No effect of temperature was observed, as EC_{50} values for phagocytosis of latex beads were

Table 3.57 Results of Rabbit Alveolar Macrophage Assays on Hygas Materials

Material	Cytotoxicity: Lethality LC_{50}[a] (μg/mL)	Cytotoxicity: Functional Impairment EC_{50}[b] (μg/mL)
LTR gas condensate		
Organic-phase NVO		
Sample G2117	131	61
Sample G2118	111	39
Sample G2119	121	50
Sample G2127	147	101
Sample G2144	108	105
Recycle oil NVO		
Sample G2113	165	108
Sample G2122	213	123
Char/oil slurry oil NVO		
Sample G2114	106	82
Sample G2124	115	35
Sample G2125	249	118
Sample G2131	320	58
Vanadium (V) oxide	7	5

[a] Effective concentration of test material that kills 50% of the exposed cells.

[b] Effective concentration of test material that causes a 50% reduction in the fraction of viable exposed cells able to engulf latex particles.

Source: Reilly et al. (1986).

106 μg/mL (high-temperature condensate) and 141 μg/mL (room-temperature condensate) for the 1470°F run, and 103 and 143 μg/mL, respectively, for the 1830°F run.

3.4.1.2.9 Summary and Conclusions (Mammalian Cell Assays). Toxicological information on coal-derived materials tested in mammalian cell assays is somewhat limited relative to microbial assays. Eight different assays were used to measure mutagenicity, cytotoxicity, and transformation, but only a few materials were studied with each assay. For example, in the CHO mutation assay, five liquefaction materials and four gasification materials representing six different processes were assayed. The studies were conducted in five different laboratories and there were some variations in experimental protocol. Mutagens were detected in the coal materials, but owing to the paucity of materials tested and the variation in procedures, no generalizations are warranted.

A general problem associated with studying coal-derived materials in mammalian cells is demonstrated by the CHO mutation assay. Coal-derived materials are complex mixtures of hydrocarbons and contain highly cytotoxic components that require chemical class fractionation to obtain reproducible mutation results. As with the Ames assay results, mutagenicity and genotoxicity in mammalian cell assays were associated with the higher-molecular-weight PAH and N-PAC components of the materials tested. In some assays, the PAH component had greater activity than in the microbial assays. On the whole, activities were low; for example, materials in the CHO mutation assay generally produced less than 1 mutant/10^6 survivors per μg of test material. Hydrotreatment of one material decreased but did not eliminate the mutagenic activity.

Four other assays with genotoxic endpoints were used to evaluate the toxicity of coal-

Table 3.58 Results of Rabbit Alveolar Macrophage Assays on University of North Dakota Energy Research Center Materials

Material	Cytotoxicity	
	Lethality LC_{50}[a] (μg/mL)	Functional Impairment EC_{50}[b] (μg/mL)
Tar		
Sample G1121	84	75
Sample G1122	116	59
Sample G1123	457	81
Sample G1124	104	68
Oil		
Sample G1125	180	147
Gas stream		
Sample G1111[c]		
High-pressure cyclone	222	107
Low-pressure cyclone	165	134
First heat exchanger	150	42
Second heat exchanger	159	137

[a] Effective concentration of test material that kills 50% of the exposed cells.

[b] Effective concentration of test material that causes a 50% reduction in the fraction of viable exposed cells able to engulf latex particles.

[c] Nonvolatile organics tested.

Source: Reilly et al. (1986).

derived materials. The mouse lymphoma forward mutation assay was used to study mutation induction with a number of liquefaction materials. Extracts of NTSL solid products and first- and second-stage liquid blends and a gasification-derived No. 6 fuel oil and its counterpart, petroleum-based No. 6 fuel oil, produced mutation in this assay. In contrast to the Ames assay, where mutagenicity was found only in higher-boiling distillates, activity was found in lower-boiling materials, including a naphtha fraction. Some direct-acting mutagens were also detected with this assay. The limited number of materials studied with this assay and the fact that mutations were often scored at very high cytotoxicity levels (approaching 10% survival) leaves the significance of results that conflict with other mutation assays, such as the Ames assay, in doubt.

Materials from five different processes were studied for SCE induction in three different laboratories. However, all materials were from gasification processes. The lack of information on liquefaction processes and variations in test protocols are limiting factors in making general conclusions on the value of this assay for predicting genotoxicity in coal-derived materials. SCE were induced by all materials tested, indicating that the test materials all have components that can interact with mammalian cell DNA.

UDS was studied with two liquefaction materials. An SRC-II middle distillate did not induce unscheduled DNA synthesis, while stripper tower bottoms did. Cytogenetic measurements were conducted on a base fraction from an in situ coal-gasification material, but no statistically significant aberrations, micronuclei formation, or polyploidy were observed.

Morphological transformation of cells in culture (SHE assay) was studied with four sets of liquefaction materials. Transformed colonies were induced. The relevance of morphological transformation to in vivo carcinogenesis was underscored by the observation that some of the transformants produced undifferentiated sarcomas when inoculated into nude mice. The percentage of cells transformed by a test material was shown to increase with the bp of the material, in agreement with results from microbial assays. There was some indication that this assay was more sensitive to transformation by PAH compounds than the microbial systems were to mutation by N-PACs. Hydrotreatment was studied with one material. Transformation was decreased with hydrotreatment but not eliminated.

Cultured alveolar macrophages were used to study cytotoxic and functional impairment resulting from exposure to coal-derived materials. While only in the case of gasification materials was a reasonably large number of materials studied, the assay did demonstrate that functional impairment occurred substantially below lethal concentrations. The assay, however, provided little discrimination among materials tested.

Finally, cytotoxicity measurements were made in a variety of mammalian cells. Cytotoxic determinations were conducted as part of the mammalian cell genotoxic assays. In addition, CHO, WI-38 human fibroblasts, and VERO monkey kidney cells were used to detect cytotoxicity specifically. The assays were of qualitative significance, giving some measure of relative toxicity of test materials. The wide variety of endpoints (e.g., lethality, function, growth, cloning efficiency) and limited number of materials studied do not allow for generalizations.

Mammalian cell assays show promise for evaluation of the toxicity of coal-derived materials but selection of the most appropriate tier two assay or assays is not presently possible.

3.4.2 In Vivo Assays for Carcinogenicity

Experimental data such as that generated by Hueper (1953, 1956) on materials generated in the 1940s from the Bruceton, Pennsylvania and Lousiana, Missouri liquefaction plants and the epidemiological data from the steel industry (Kawai et al., 1967; Lloyd, 1971; Doll et al., 1972), suggesting that coal-conversion materials contained carcinogenic components, pointed out the need for evaluation of the carcinogenic potential of materials from new coal-conversion technologies. Two classical in vivo assay procedures, both examining dermal carcinogenicity in mouse skin, were used in evaluating the dermal carcinogenicity of coal-converison materials produced over the last decade. The first assay evaluates tumor induction following repeated application (e.g., 3 exposures/week for a period up to the life span of the mouse) of the material on the skin of mice. The other assay uses a limited dermal initiation exposure (one to several applications of the test material) followed by a period for tumor promotion (repeated application of a promotor, defined as an agent that promotes the induction of tumors from initiated cells but is not capable of inducing carcinogenesis by itself). In some chronic skin-painting studies, animals received gross and histopathological evaluation of organs other than skin to evaluate the potential of the test material for induction of systemic carcinogenicity resulting from dermal exposure to carcinogens. A very limited number of carcinogenesis studies on coal-conversion materials have been performed using the inhalation route of exposure.

3.4.2.1 *Chronic Skin Exposure*

Repeated application of test material to the skin of experimental rodents has been the classical method of assaying for carcinogenicity in fossil-fuel crude materials, chemical fractions, and products (Leitch, 1922; Passey, 1922; Twort and Twort, 1931; Clayson, 1962). Many of the experimental data on which modern concepts of carcinogenesis are based were derived from dermal application of highly carcinogenic PAH isolated from crude fossil-fuel materials (Clayson, 1962; Weisburger and Williams, 1980). The traditional use of the chronic mouse skin-painting assay for carcinogenicity of these materials has resulted in a large database on carcinogenic response of mouse skin not only to crude petroleum materials but to their chemically derived fractions and bp distillate cuts (Smith et al., 1951; Great Britain Research Council, 1968; Tye et al., 1966; Bingham et al., 1979; Lewis et al., 1982). The use of similar methods for chemical characterization and bioassays for potential carcinogenicity has allowed researchers to compare chemical and biological data from studies of coal-conversion materials with data from similar studies on petroleum-derived materials (Holland et al., 1981; Renne et al., 1981).

Although mouse skin painting has been used for a number of years as a means of assaying for dermal carcinogenicity, some variation still exists in test protocols used for this assay (Clayson, 1962; Bernfeld, 1983); one of the most important variables is the strain of mouse used. The tumorigenic response to dermally applied carcinogens differs considerably among various strains of mice (Poel et al., 1958; Boutwell, 1964; Stenback, 1980). This may be due, at least in part, to strain differences in skin penetration and absorption (Bock, 1963) or to metabolism (Gelboin et al., 1972) of carcinogens. The majority of chronic mouse skin-painting studies on SRC, H-Coal, COED, and Synthoil used a single strain of mice, C3H/HeBd, originating from a single breeding colony.

Another important variable in skin-painting protocols is the form in which the test material is applied to the skin. In many early studies, test materials were applied undiluted (neat) or diluted in vehicles such as liquid paraffin, with variable results on carcinogenicity (Twort and Twort, 1929; Twort and Fulton, 1930). Horton et al. (1957) demonstrated that certain aliphatic solvents, such as alkylbenzenes, alkylnaphthalenes, and saturated hydrocarbons, such as *N*-hexadecane or cyclohexyldecane, increased the carcinogenic potency of PAH carcinogens when used as vehicles in skin-painting studies. Work by Holland et al. (1980) indicated that the use of a lipotropic solvent, such as cyclohexane, as a vehicle for dermal application of BaP resulted in retention of BaP in the lipid of the sebaceous glands, thus prolonging the exposure of epidermal cells in the vicinity of sebaceous gland ducts and around hair follicles. Volatile solvents, such as cyclohexane or acetone, are often used as vehicles for skin-painting studies of hydrocarbons to decrease the viscosity of the test materials, making solutions easier to pipette and application of the material more accurate and consistent.

The dose of material applied is another important variable in mouse skin-painting studies. The use of multiple doses has the obvious advantage of providing specific dose–response information for a material, but it requires additional effort and greater amounts of sample compared with a single-dose study. The maximum tolerated dose, determined by preliminary short-term tests, is frequently used as the highest dose in chronic skin-painting studies to provide a carcinogenic response as rapidly as possible. However, care must be used to ensure that the maximum tolerated dose is not dermatotoxic over long periods of application. Such a reaction might result in decreased skin-tumor response owing to sloughing of epidermis and/or ulcerative inflammation of the skin at the application site (Holland et al., 1979b).

Frequency of application is a variable that affects the total dose applied. It may also have a significant effect on dermatotoxicity of potentially irritating materials. The most frequently used protocols call for application two or three times weekly.

The total duration of exposure is also pertinent to the question of dermatotoxicity versus carcinogenicity of applied materials. Although the continuous application of the test material for the lifetime of the animal has been the most frequently used method, data from studies in which application was stopped after 22 weeks (Holland et al., 1979a) indicate a rapid increase in skin-tumor incidence immediately following cessation of exposure. This suggests that continuous application of materials with dermatotoxic potential may actually inhibit tumor formation. Nevertheless, the majority of protocols used for chronic bioassay of fossil-fuel materials in mouse skin call for continuous exposure during the lifetime of the animal.

The number of animals per experimental group must be determined on the basis of requirements for statistical significance and reliability. The natural life span of the strain of mouse used and the number of non-tumor-related deaths will profoundly impact the statistical validity of chronic mouse skin-painting studies. Most studies, including the majority of those described in this chapter, use equal numbers of male and female mice to detect sex-related differences in carcinogenic or dermatotoxic responses. Some authors (Bernfeld, 1983) recommend using only female mice to avoid losses from fighting. Individual housing is a much better way of solving this problem and also prevents mutilation of tumors by cagemates as well as ingestion of applied material by licking cagemates. However, individual caging in large-scale studies is often not practical or economically feasible.

Methods for expression, analysis, and comparison of carcinogenicity data from mouse skin-painting studies have evolved from statistical methods that graphically compare tumor incidence versus time of exposure. Sophisticated curve-fitting techniques have been used (Holland et al., 1981). Variables most frequently analyzed are skin-tumor incidence at the application site, tumor latency (time from first exposure to appearance of tumors), and biological behavior of the tumor (morphological evidence of malignancy and incidence of metastasis).

Tumor incidence is usually expressed as the number of animals with tumors as a percentage of animals at risk. However, loss of tumor-free animals from a test group will decrease the number of animals at risk, thus increasing the percentage of animals at risk that have tumors. Therefore, various methods of dealing with the loss of animals without tumors during chronic studies have been used, including using the number of animals at risk at observation of the first tumor (Iball, 1939) or adjustment of the number of animals at risk at each death without a tumor (Blanding et al., 1951; Bingham and Falk, 1969). The current method most frequently used to express graphically tumor incidence versus duration of exposure and to take into account intercurrent mortality is that of Kaplan and Meier (1958), in which the probability of tumor-free survival versus time of exposure is compared among groups.

Multiple skin tumors are very frequently observed in one mouse in chronic skin-painting studies, especially if painting is continued for a prolonged period after observation of the initial tumor. However, the number of tumors per animal is seldom used as a parameter to quantitate tumorigenic response in long-term studies, although it is used frequently in initiation/promotion assays in mouse skin.

Skin-tumor latency parameters most frequently reported are minimum latency (time from first exposure to first tumor observed in a group), median latency (time from first exposure to 50% tumor incidence), mean or average latency (total time to tumor for all animals/number of animals with tumors), and maximum latency (time from first exposure to observation of final tumor).

Biological behavior of induced skin tumors is often reported, but analysis of these data is usually limited to comparison of percentage of malignant tumors or incidence of tumor metastasis among groups. Microscopic examination of tumor tissue is required for valid classification of malignancy or presence of metastasis. It may also be required to differentiate induced epidermal tumors from induced or spontaneous subcutaneous mesenchymal tumors (Dunn et al., 1956).

Carcinogenic potency of hydrocarbon test materials is most frequently compared with highly carcinogenic, pure PAH compounds such as BaP or 7-12-dimethylbenzanthracene, although a number of comparisons have been made with crude petroleum materials. Estimation of carcinogenic potency involves comparing rate of tumor appearance and total tumor incidence; this is often complicated by variations in protocols.

Twort and co-workers (Twort and Fulton, 1930; Twort and Twort, 1931) used tumor data from large numbers of mouse skin-painting studies of oils and tars to establish a hypothetical standard of potency. They then calculated comparative potency of test materials to their standard by comparing rates of tumor response with time. Iball (1939) proposed a mathematical formula for tumorigenic potency, using skin-painting data from pure PAH compounds, in which percentage of tumor incidence was derived by mean latency to give an "index" of tumorigenicity. Blanding et al. (1951) estimated carcinogenic potency by plotting the logarithm of time-to-tumor data versus percentage of tumor response. They then fitted a straight line through the plotted points to obtain the theoretical time to 50% tumor response from this plotted line. Tumor potency was a multiple of the reciprocal of this estimated time to 50% tumor response. Horton and co-workers (Horton and Denman, 1955; Horton et al., 1957) developed an equation for calculating potency of test materials relative to methylcholanthrene or BaP; it was specific for C3H mice but allowed for variability in frequency of application.

The Kaplan–Meier analysis (Kaplan and Meier, 1958) is a popular method of graphically expressing skin-tumor incidence versus time and adjusting for mortality caused by intercurrent disease. Relative carcinogenic potency can then be estimated by comparing latency curves with those of a known carcinogen. However, as with the methods described above, this estimate of relative potency provides little basis of comparison in assays where tumor incidence does not reach or exceed 50% of exposed animals.

The Weibull mathematical model (Peto and Lee, 1973) has been used to estimate carcinogenic potency of synthetic fossil fuels and crude petroleums relative to BaP (Holland et al., 1981; Holland and Frome, 1982; Renne et al., 1986). This analytical method provides an estimate of time to median tumor response at a given dose, which can be compared with the same value for a given dose of a known carcinogen (such as BaP). The comparative potency can then be estimated on the basis of these data.

3.4.2.1.1 Liquefaction—SRC. Of the modern U.S. coal-liquefaction technologies, the SRC and H-Coal processes have been the most extensively investigated in dermal carcinogenesis studies using laboratory mice. Chronic dermal carcinogenesis studies of SRC materials, carried out in conjunction with other in vivo and in vitro assays of the same samples, began in 1978 with chronic assays of light distillate (Sample L1241) and heavy distillate (Sample L1243) from the SRC-II pilot plant at Ft. Lewis, Washington (Renne et al., 1981). These materials, studied at three dose levels diluted in acetone, were compared with samples of crude shale oil, crude petroleum, and two known carcinogens, BaP and 2-aminoanthracene. Groups of 25 male and 25 female C3H/HeBd mice were exposed three times weekly for up to 2 years.

Kaplan–Meier estimates of skin-tumor-free survival for these initial studies are presented in Fig. 3.5. Table 3.59 gives carcinogenic potency of these materials relative to BaP, using the method of Weibull as modified by Holland et al. (1981) and Holland and Frome (1982). The results for the heavy distillate sample of these initial chronic studies (Fig. 3.5) indicated that heavy distillate was a relatively potent carcinogen for mouse skin compared with the other crude fossil-fuel materials tested. Light distillate was nontumorigenic.

Results of these chronic assays of crude materials, together with the results of in vitro

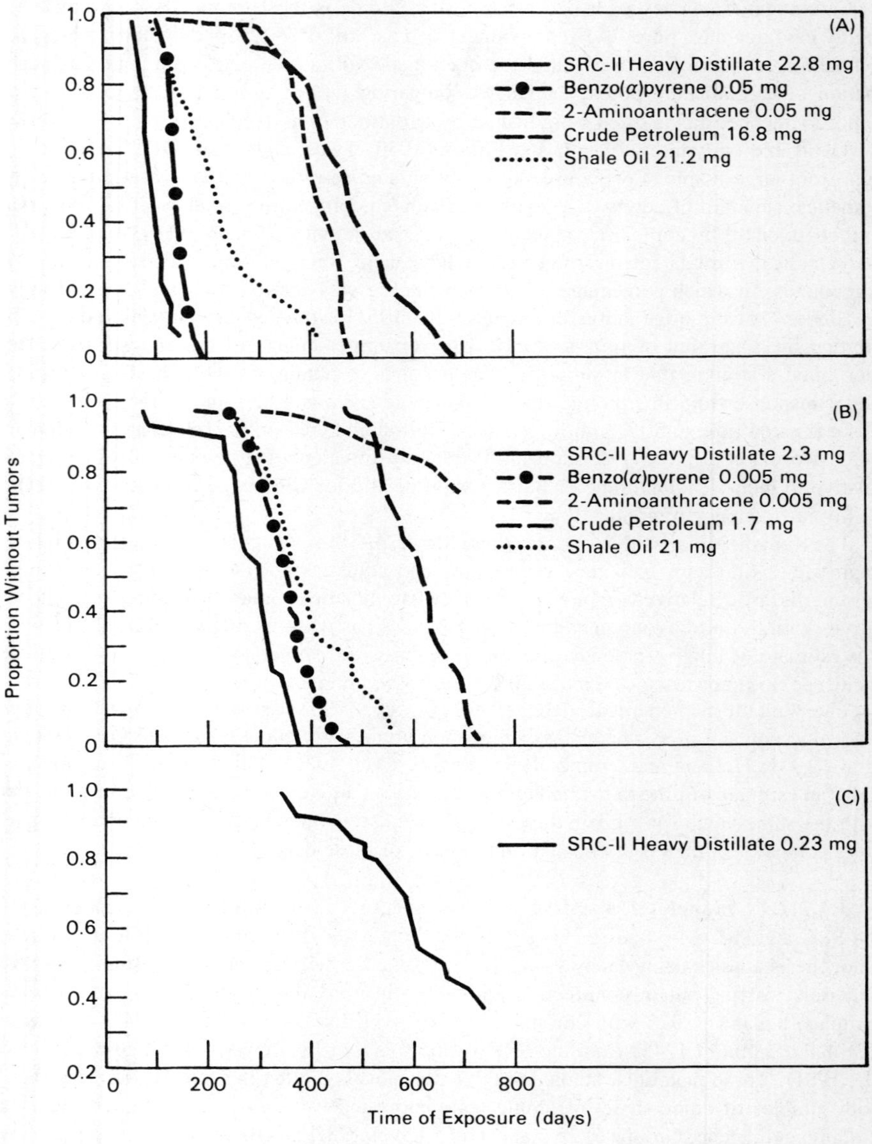

Figure 3.5 Proportion of group surviving without a skin tumor in C3H/HeBd mice exposed three times per week to crude fossil-fuel materials or known carcinogens (Kaplan–Meier estimates): (A) high dose; (B) medium dose; and (C) low dose. SRC = solvent-refined coal. (From Renne et al., 1981.)

Table 3.59 Carcinogenic Potency Relative to Benzo[a]pyrene (BaP) of Crude Fossil-Fuel Materials or Solvent-Refined Coal II Heavy Distillate (Ft. Lewis, Washington) and Its Solvent Fractions

	Potency Versus BaP[a]	
Material	%	Ratio
BaP	100	1:1
Heavy distillate (Sample L1243)	0.24	1:412
Neutral tar	0.55	1:182
Basic tar	0.50	1:200
PAH fraction[b]	0.44	1:225
Basic fraction	0.12	1:806
Livermore shale oil	0.07	1:1402
Wilmington crude petroleum	0.02	1:5000

[a] Potency data and statistical analysis method are from Holland et al. (1981) and Holland and Frome (1982).

[b] Polycyclic aromatic hydrocarbon.

studies (Pelroy and Petersen, 1981), led to chronic assays of solvent fractions of heavy distillate (Renne and Smith, 1981; Renne et al., 1986). Relative contribution (weight %) of these solvent fractions in the heavy distillate are presented in Table 3.60. Doses for all fractions except the PAH fraction were determined on the basis of relative mutagenicity in the Ames test. The PAH fraction dose was calculated on the basis of its contribution (weight %) to the unfractionated distillate. Groups of 25 male and 25 female C3H/HeBd mice were exposed three times weekly for up to 2 years. The PAH and basic fractions were diluted in acetone for application; both tar fractions were diluted in 30% methanol/70% acetone. Neutral and basic tars were the most active heavy distillate fractions at the concentrations tested (Fig. 3.6). The PAH fraction was an active dermal carcinogen having somewhat greater potency than crude heavy distillate (Table 3.59). The basic fraction, although it was more active in the Ames assay for mutagenicity, was less potent than crude heavy distillate as a mouse-skin carcinogen (Renne and Smith, 1981; Renne et al., 1986). Table 3.59 gives carcinogenic potency data relative to BaP (using the method of Weibull as modified by Holland et al., 1981) for crude SRC-II heavy distillate, other fossil-fuel crudes, and solvent fractions of SRC-II heavy distillate.

Table 3.60 Concentration of Aliphatics, Polycyclic Aromatic Hydrocarbons (PAHs), Nitrogen Polycyclic Aromatic Compounds (N-PACs), and Hydroxy-PAHs in Solvent-Refined Coal II Heavy Distillate and Its Solvent-Derived Fractions

	Relative Weight[a]				
Material	% of Total	Aliphatics	PAHs	N-PACs	Hydroxy-PAHs
Heavy distillate (Sample L1243)	100	10	42	20	28
Basic tar	24	<1	38	35	27
Neutral tar	26	<1	38	24	35
PAH fraction	41	2	81	11	5
Basic fraction	1	<1	19	68	10

[a] Weight percentage of each chemical class.

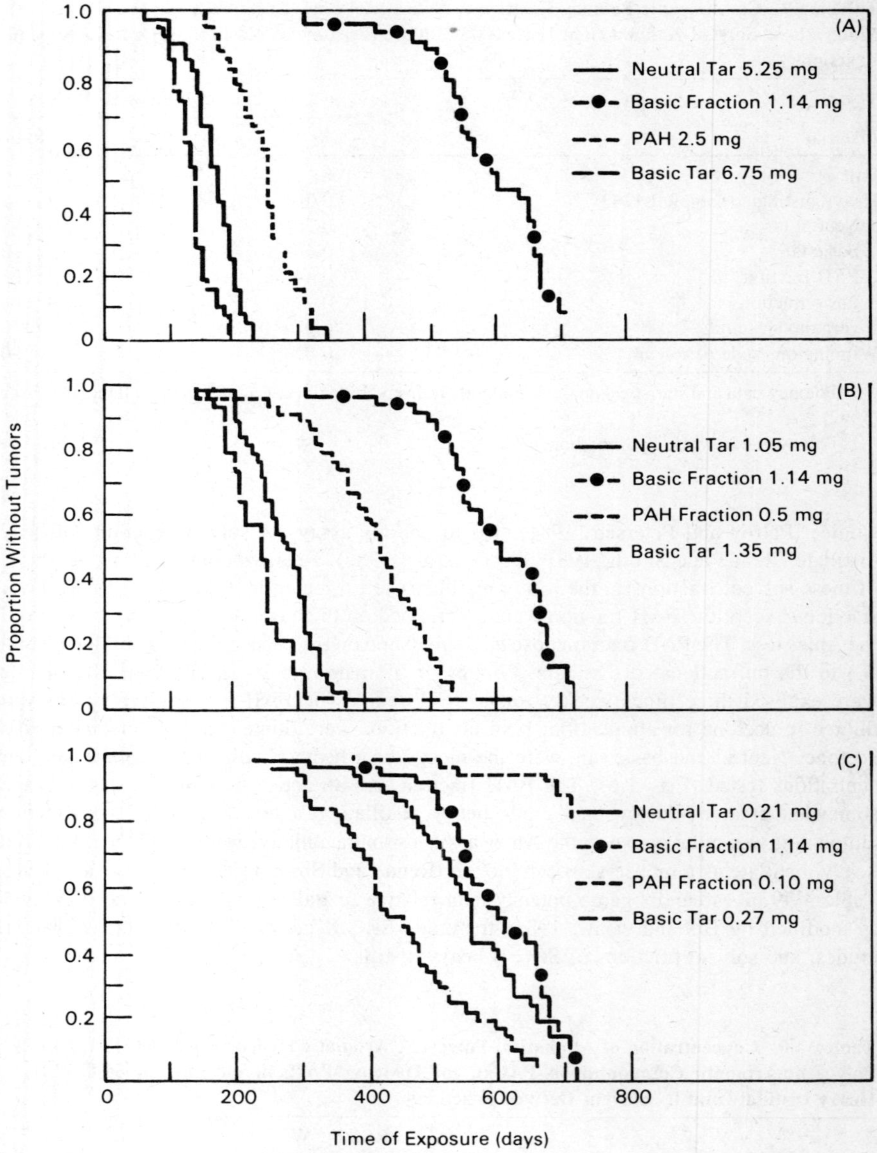

Figure 3.6 Proportion of group surviving without a skin tumor in mice exposed three times per week to solvent fractions of solvent-refined coal II heavy distillate from Ft. Lewis, Washington (Kaplan–Meier estimates): (A) high dose; (B) medium dose; and (C) low dose. PAH = polycyclic aromatic hydrocarbon.

Chemical analysis of the solvent fractions of heavy distillate (Table 3.60) indicated that the neutral and basic tars contained relatively high concentrations of PAH and N-PAC compounds. To investigate the possibility that the carcinogenic potency of crude heavy distillate was the result of synergistic or additive effects of its PAH and N-PAC compounds, chronic skin-painting assays were performed on recombined PAH and basic fractions of heavy distillate (combined at concentrations equivalent to the doses of these fractions reported in Fig. 3.6). Results of this work (Fig. 3.7) did not indicate a synergistic or additive carcinogenic effect of heavy distillate, with the tumor response to recombined PAH and basic fractions being no different from the dermal carcinogenicity of the PAH fraction (Mahlum et al., 1983; Renne and Buschbom, 1985).

Studies by Pelroy and Stewart (1981) demonstrated that destruction of amine-containing PAHs (N-PACs) by nitrosation removed most of the microbial mutagenicity of the basic fraction, basic tar fraction, and neutral tar fraction of heavy distillate. To determine whether a similar effect on carcinogenicity was present, chronic skin-painting assays were performed on nitrosated basic tar fractions of heavy distillate (Renne et al., 1986). No decrease in carcinogenic response to nitrosated basic tar was evident (Fig. 3.8), indicating that N-PACs do not play a significant role in the dermal carcinogenicity of the basic tar fraction of heavy distillate.

A somewhat similar solvent fractionation method (Fig. 3.9) to that used with the SRC-II heavy distillate was used to assay samples of an SRC-II fuel-oil blend (Sample L1211) consisting of a 6:1 mixture of middle and heavy distillates (Holland et al., 1984). Selected fractions were assayed for dermal carcinogenicity in C57Bl mice by application of 3 μg of each fraction in an acetone vehicle three times per week for 20 weeks. This was followed by promotion with 8 μg of tetradecanoylphorbol-13-acetate (TPA) in acetone for 30 weeks or until tumor formation. Tumor incidence and latency data from this study are presented in Table 3.61. Tumorigenic activity of the neutral fraction was very similar to that of the

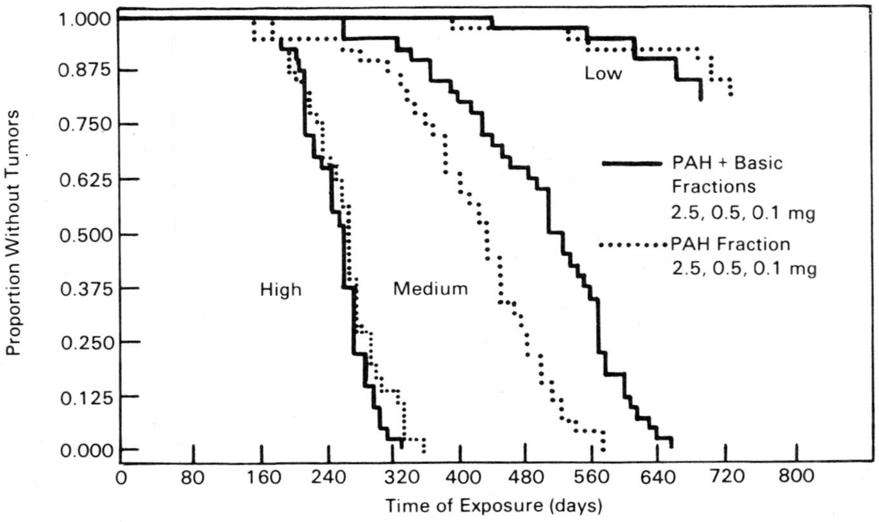

Figure 3.7 Proportion of group surviving without a skin tumor in C3H/HeBd mice exposed three times per week to recombined polycyclic aromatic hydrocarbons (PAHs) and basic solvent fractions of solvent-refined coal II heavy distillate (Kaplan–Meier estimates).

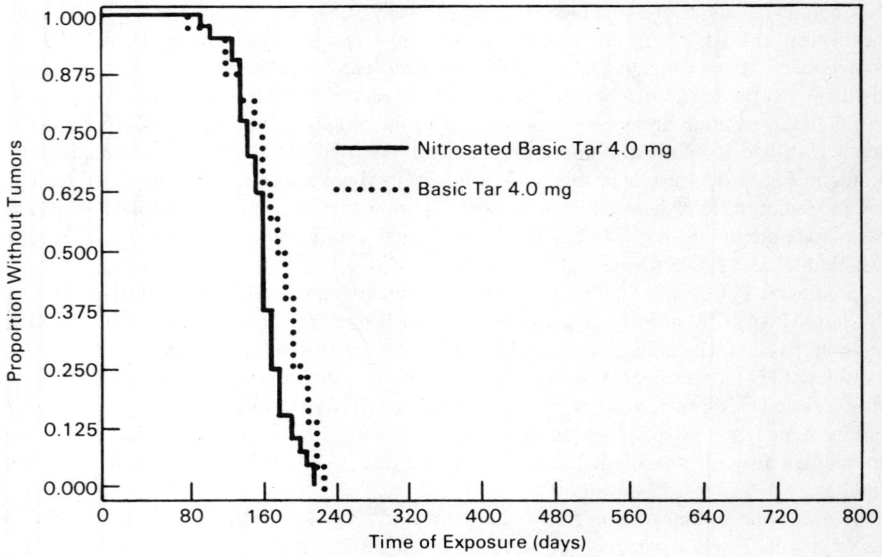

Figure 3.8 Proportion of group surviving without a skin tumor in C3H/HeBd mice exposed three times per week to nitrosated or nonnitrosated basic tar fractions of solvent-refined coal II heavy distillate (Kaplan–Meier estimates) (Renne et al., 1986).

crude fuel-oil blend. Similar assays of PAH and neutral polar subfractions of the neutral fraction indicated that the PAH subfraction was by far the most active as a dermal carcinogen, whereas the neutral polar fraction was most active as a mutagen in the Ames assay.

A series of 50°F bp distillate cuts from a whole-boiling-range material (Sample L1111) from the Harmarville, Pennsylvania SRC-II PDU were assayed for carcinogenicity by chronic skin-painting assays in groups of 25 male and 25 female C3H/HeBd mice (Renne et al., 1986). Doses of these materials varied widely because of large differences in physical properties of the bp cuts. The whole-boiling-range material and the 300–700°F bp cut were diluted in acetone; the other bp cuts were diluted in 30% cyclohexane/70% acetone. Results of this work, presented in Fig. 3.10, indicated that the dermal carcinogenicity of the crude material was due to that portion boiling above 700°F and that the dermal carcinogenicity increased directly with increasing bp above 700°F. Calculation of relative potency to BaP (Table 3.62) supports the observation of increasing potency with increasing bp.

A sample of the 750–800°F bp cut was nitrosated to destroy amino-PAH, using methods similar to those for nitrosating the heavy distillate basic tar fraction. This nitrosated material was assayed in C3H/HeBd mice, at a dose of 10.0 mg three times per week. A mixture of acetone, cyclohexane, and DMSO was used as the solvent vehicle. Another sample of nonnitrosated 750–800°F bp cut was assayed at the same dose in the same solvent vehicle for comparative purposes. Results of this assay indicated a slight but significant increase in tumor latency in the group of mice exposed to the nitrosated bp cut (Fig. 3.11). However, this difference appeared to be due to a decrease in latency in female mice exposed to the nonnitrosated material (Fig. 3.12), rather than an increase in tumor latency with the nitrosated material. Mean tumor latency was 141 $\pm$ 6 days for females painted with the nonnitrosated bp cut and 175 $\pm$ 8 days for females painted with the nitrosated bp cut. Mean

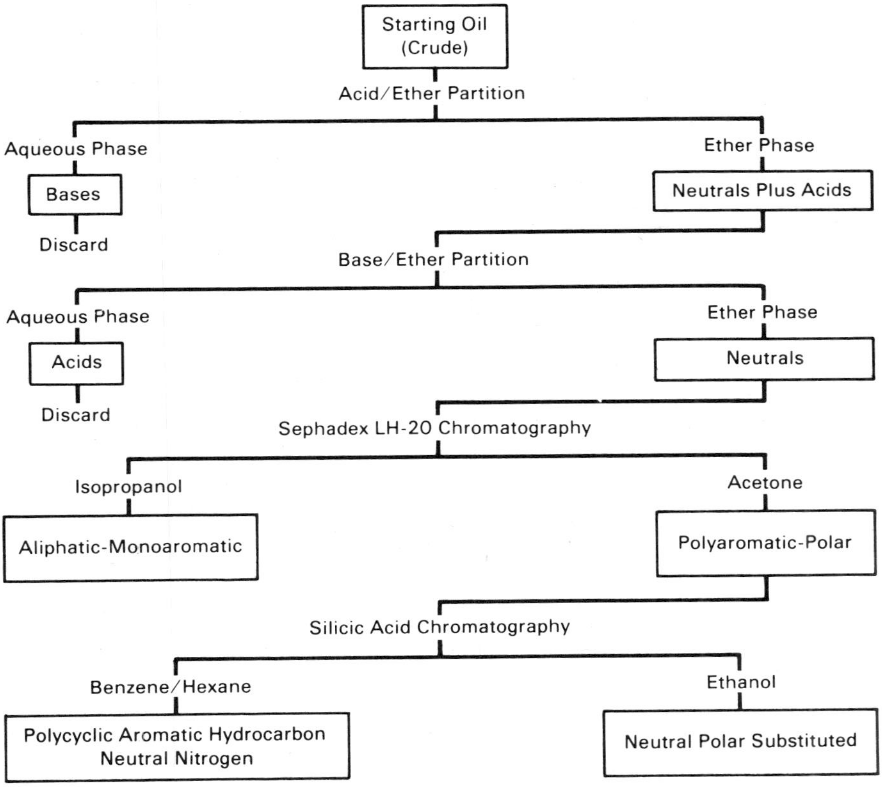

Figure 3.9 Schematic of procedures used on chemical class fractionation of H-Coal and solvent-refined coal materials (adapted from Holland et al., 1984).

Table 3.61 Skin Tumors Observed in C57B Mice Exposed to Solvent-Refined Coal II Fuel Oil Blend (Ft. Lewis, Washington)

Material	Percentage of Mice with Tumors			Average Days to Tumor (SE)
	Papilloma	Carcinoma	Total	
Fuel-oil blend (Sample L1211)	25	5	30	203 (0.6)
Neutral fraction	20	0	20	210 (0.9)
Aliphatic	0	0	0	—
Polycyclic aromatic	85	0	85	196 (0.7)
Neutral polar	5	0	5	203
Reconstituted neutral	50	0	50	210 (0.4)
Controls				
Acetone plus TPA[a]	5	0	5	294
TPA	10	0	10	154 (10)

[a]Tetradecanoylphorbol-13-acetate.

Source: Holland et al. (1984).

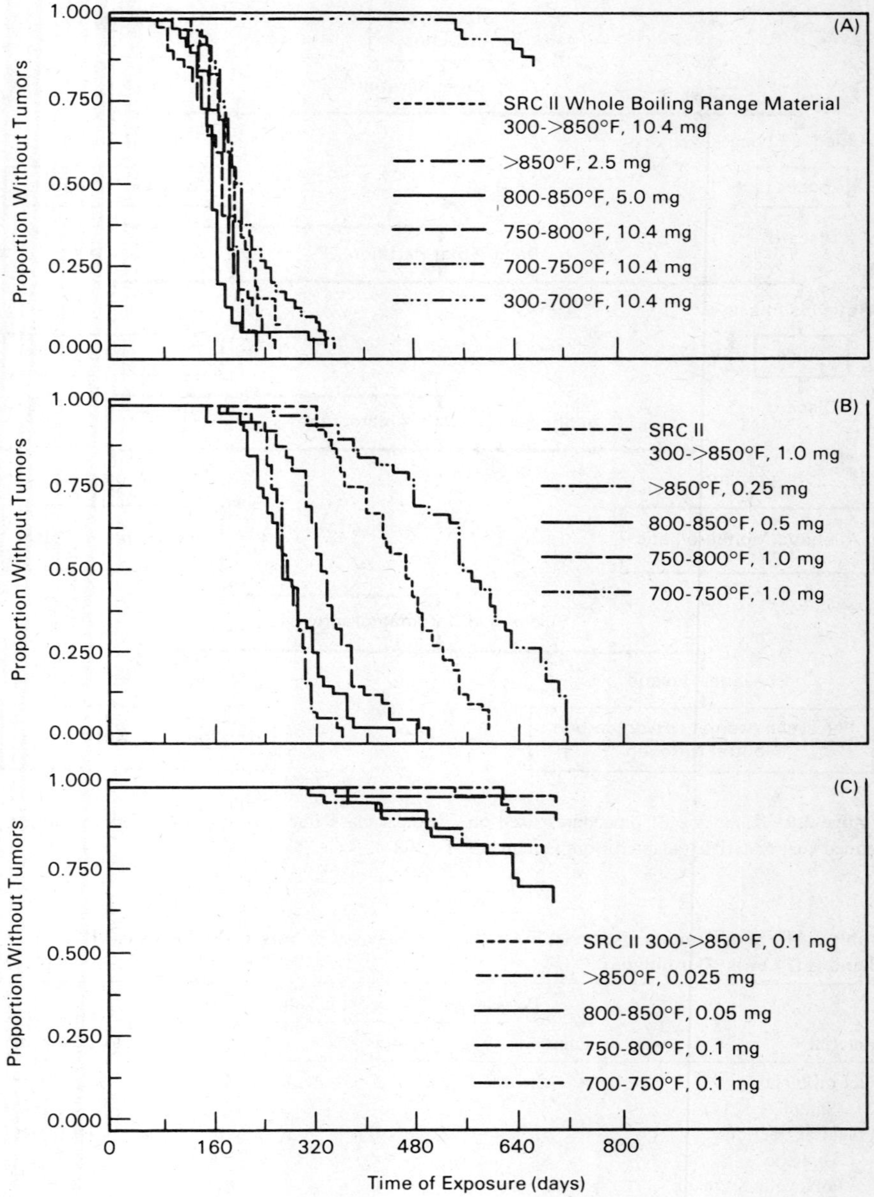

Figure 3.10 Proportion of group surviving without a skin tumor in C3H/HeBd mice exposed three times per week to a wide-boiling-range (300 to > 850°F) solvent-refined coal II (SRC-II) coal liquid from Harmarville, Pennsylvania, or to its boiling-point cuts (Kaplan–Meier estimates): (A) high dose; (B) medium dose; and (C) low dose (Renne et al., 1986).

Table 3.62 Carcinogenic Potency Relative to Benzo[a]pyrene (BaP) of Boiling-Range Cuts of Solvent-Refined Coal II Process Materials (Harmarville, Pennsylvania)

Material	Boiling-Point Cut (°F)	Potency[a] Versus BaP	
		%	Ratio
BaP	—	100.00	1:1
Process blend (Sample L1111)	—	0.18	1:556
	300–700	0.01	1:10,000
	700–750	0.15	1:667
	750–800	0.32	1:312
	800–850	1.08	1:91
	>850	2.13	1:48

[a]BaP potency data and statistical analysis method are from Holland and Frome (1982).

Source: Renne et al. (1986).

tumor latency for males was 178 ± 11 days and 182 ± 9 days, respectively, for the nonnitrosated bp cut (Fig. 3.13). Weibull analysis (Table 3.63) indicated that the nonnitrosated material was slightly more potent than the nitrosated material when data from both sexes were pooled. However, these results suggest that the amino-PAH, removed by nitrosation, do not play a major role in the dermal carcinogenicity of the 750–800°F bp cut.

To elucidate further the roles of the PAH and N-PAC chemical classes in the dermal carcinogenicity of the 750–800°F bp cut, a sample of this material was fractionated by

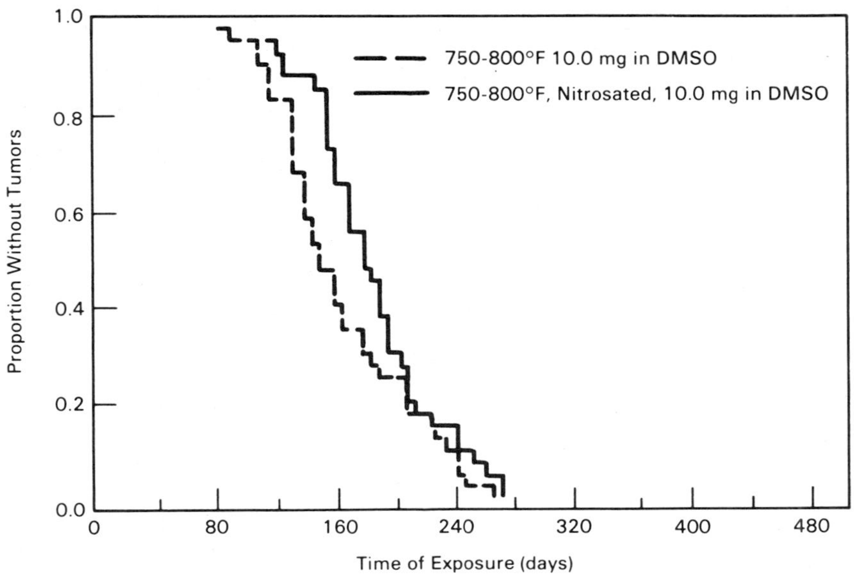

Figure 3.11 Proportion of group (both sexes) surviving without a skin tumor in mice exposed three times per week to a nitrosated or untreated boiling-point cut (750–800°F) of solvent-refined coal II heavy distillate in dimethylsulfoxide (DMSO) (Renne et al., 1986).

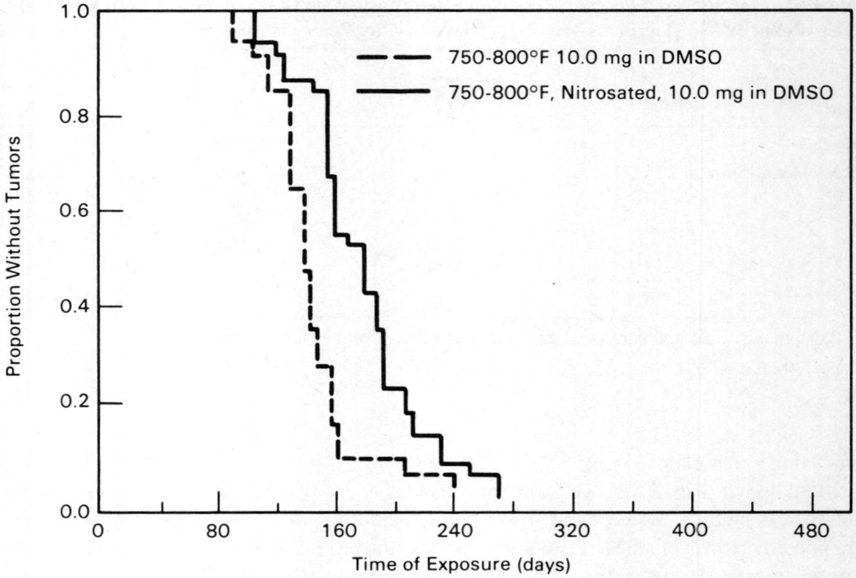

Figure 3.12 Proportion of group (females) surviving without a skin tumor in mice exposed three times per week to a nitrosated or untreated boiling-point cut (750–800°F) of solvent-refined coal II heavy distillate in dimethylsulfoxide (DMSO) (Renne et al., 1986).

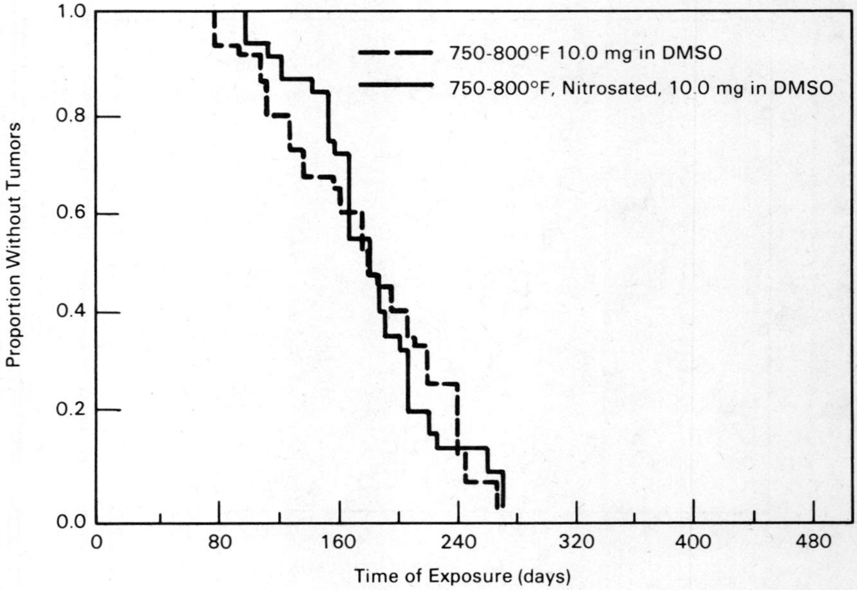

Figure 3.13 Proportion of group (males) surviving without a skin tumor in mice exposed three times per week to a nitrosated or untreated boiling-point cut (750–800°F) of solvent-refined coal II heavy distillate in dimethylsulfoxide (DMSO) (Renne et al., 1986).

Table 3.63 Carcinogenic Potency of Solvent-Refined Coal II Materials or Fractions Relative to Benzo[a]pyrene (BaP)

Material	Distillate Cut (°F)	Potency[a] Versus BaP	
		%	Ratio
BaP	—	100.00	1:1
Process blend (Sample L1111)	—	0.18	1:556
	>850	2.13	1:48
	800–850	1.08	1:91
	750–800	0.32	1:312
PAH[b]	—	0.27	1:370
Nitrogen-PACs[c]	—	0.29	1:345
	700–750	0.15	1:667
	300–700	0.01	1:10,000
In DMSO	750–800	0.23	1:435
Nitrosated, in DMSO	750–800	0.20	1:500

[a]BaP potency data and statistical analysis method are from Holland and Frome (1982).
[b]Polycyclic aromatic hydrocarbons.
[c]Polycyclic aromatic compounds

Source: Renne et al. (1986).

adsorption column chromatography using neutral alumina as an adsorbent (Later et al., 1981). The PAH and N-PAC fractions obtained were assayed in C3H/HeBd mice at doses based on the percentage composition of these fractions in the crude bp cut (Renne et al., 1986). Acetone/cyclohexane (70%/30%) was used as the vehicle solvent.

Results of this assay are presented in Fig. 3.14. Although the tumor latency periods were longer for both alumina fractions than for the unfractionated bp cut, the absolute doses of both fractions were much lower, and potency estimation using the Weibull analysis (Table 3.63) indicated that both fractions had similar potencies to the crude bp cut applied at 10.4 mg/dose in cyclohexane/acetone. Both alumina fractions were more potent than the same bp cut applied at 10.0 mg in DMSO or the whole-body range material. The Weibull analyses further indicate that the N-PAC fraction of the 750–800°F bp cut is slightly more potent than the PAH fraction of this cut. These results, considered together with the results of the nitrosation studies of this bp cut, suggest that non-amino-PAH components of the N-PAC fraction may be responsible for a large part of the carcinogenicity of this bp cut. The other major chemical classes present in the N-PAC are carbazoles and azaarenes (Later et al., 1981; Pelroy and Wilson, 1981; Pelroy et al., 1985). Certain alkylated carbazoles and azaarenes have been shown to be carcinogenic for mouse skin (Clayson, 1962).

The effects of hydrotreatment on the dermal carcinogenicity of a 1.7:1 blend of SRC-II middle and heavy distillates (Sample L1262) (Cowser, 1982b; Griest et al., 1983) from the Ft. Lewis, Washington pilot plant were studied. Materials were applied three times per week in a 50-μL volume for 50 weeks or until all 50 treated C3Hf/HeBd mice (25 of each sex) of a given treatment group developed tumors. Materials assayed, dosage, and tumor-incidence data are presented in Table 3.64. These data indicate that even the lowest-severity hydrotreatment dramatically reduced the tumorigenic response to the SRC-II distillate blend.

Chronic mouse skin-painting studies of SRC-I process materials have been performed on only one sample, process solvent (Sample L2213) from the Ft. Lewis, Washington SRC-I pilot plant (Mahlum et al., 1983). Three dose levels (0.1, 1.0, or 10.4 mg/50 μL) in acetone were applied three times weekly to groups of 25 male and 25 female C3H/HeBd

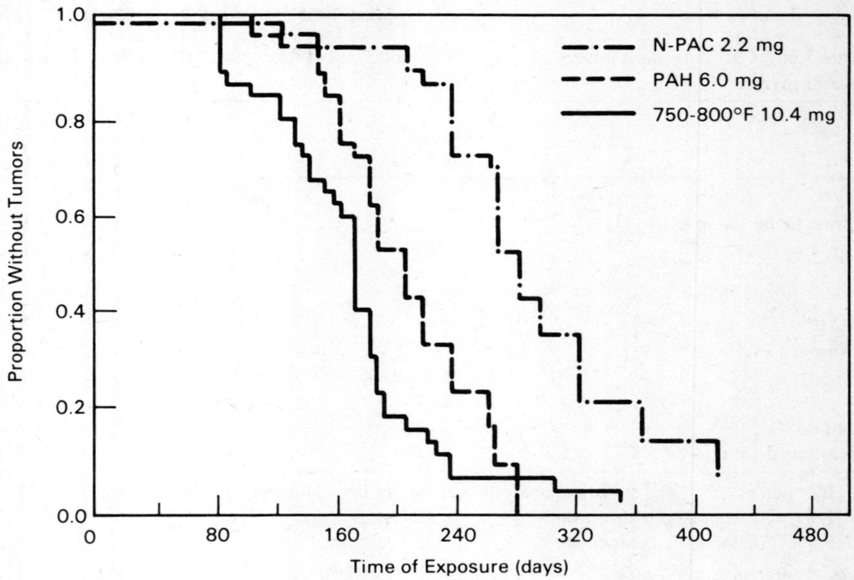

Figure 3.14 Proportion of group (both sexes) surviving without a skin tumor in mice exposed three times per week to chemical class fractions of a boiling-point cut (750–800°F) of solvent-refined coal II heavy distillate (Renne et al., 1986). N-PAC = nitrogen-containing polycyclic aromatic compounds; PAH = polycyclic aromatic hydrocarbons.

Table 3.64 Results of Mouse Skin-Painting Tests with Hydrotreated Solvent-Refined Coal II (SRC-II) Distillate Blend (Ft. Lewis, Washington)

Material	Nitrogen Content (ppm)	Dose[a] (% w/v)	Skin Neoplasms[b] (%)
SRC-II blend (Sample L1262)			
After water wash	—	100	88
		50	28
		25	20
Low-severity hydrotreatment	5300	100	4
		50	4
		25	2
Medium-severity hydrotreatment	1710	100	0
		50	0
		25	0
High-severity hydrotreatment	505	100	2
		50	2
		25	0
Benzo[a]pyrene	—	50[c]	100
Acetone	—	100	0

[a]The material was applied three times per week at 50 μL/exposure, either neat (100%) or diluted to the appropriate dose in acetone.

[b]Fifty mice (25 of each sex)/treatment group.

[c]This value is in units of μg/exposure.

Source: Griest et al. (1983).

mice. Figure 3.15 presents skin-tumor latency data from this study on a comparative basis with data from similar doses of crude SRC-II material (Sample L1111) from the Harmarville, Pennsylvania PDU. It is apparent from Fig. 3.15 that the two materials have similar carcinogenic potency for mouse skin at an individual dose of 10 mg. Carcinogenic potency of this material relative to BaP, calculated by the Weibull curve-fitting procedure (Holland et al., 1981), is 0.19% (1:526).

3.4.2.1.2 Liquefaction—H-Coal. Holland et al. (1981) have performed chronic mouse skin-painting assays on several samples from the H-Coal process in conjunction with in vitro assays of the same materials (Cowser, 1981; Holland et al., 1984; Munro et al., 1985). The first H-Coal samples assayed for dermal carcinogenicity were atmospheric overhead (Samples L3141 and L3151), vacuum overhead (Samples L3143 and L3153), and atmospheric still bottoms (Samples L3142 and L3152) collected during processing in the syncrude and fuel-oil modes from the Hydrocarbon Research, Inc. PDU in Trenton, New Jersey. BaP was used as a positive control. Fifty male and female C3Hf/HeBd mice were exposed three times per week for varying periods of time. Doses applied, duration of exposure, and skin-tumor incidence and latency data for this series of assays are presented in Table 3.65. Duration of exposure and postexposure periods varied among groups owing to wide differences in skin irritation, carcinogenic response, and quantity of test material available. Data from these studies indicated that the samples of vacuum overhead and atmospheric still bottoms from both modes of operation were carcinogenic for mouse skin, whereas atmospheric still overhead samples did not induce a carcinogenic response. Males tended to be more sensitive to hydrocarbon skin carcinogens than females. Carcinogenic potency data for each dose of these samples, calculated using the Weibull model (Holland

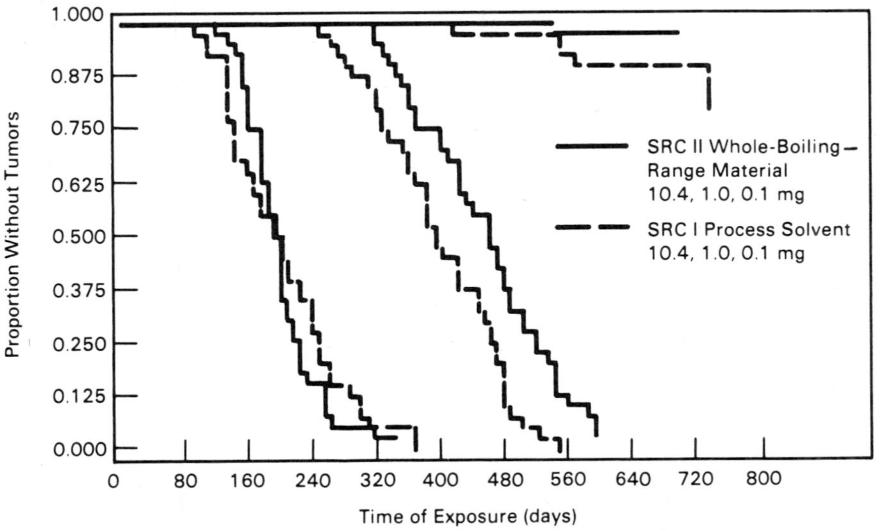

Figure 3.15 Proportion of group surviving without a skin tumor in C3H/HeBd mice exposed three times per week to solvent-refined coal I (SRC-I) process solvent or to a wide-boiling-range (300 to >850°F) SRC-II liquid from Harmarville, Pennsylvania (Kaplan–Meier estimates) (Mahlum et al., 1983).

Table 3.65 Frequency of Skin Tumors and Estimated Median Time to Tumor Following Long-Term Exposure to H-Coal Distillates

Material	Dose (mg/week)	Duration of Exposure in weeks (follow-up)	Number with Tumors[a]		Median Time to Tumor, Days (SE)	
			Female	Male	Female	Male
Syncrude Mode						
Atmospheric still	136	16 (16)	0	0	—	—
overhead	71	16 (16)	0	0	—	—
(Sample L3141)	36	16 (16)	0	0	—	—
Atmospheric still	157	28 (0)	25	25	132 (4)	104 (0)
bottoms	77	36 (0)	22	23	199 (7)	185 (8)
(Sample L3142)	38	50 (0)	16	17	324 (13)	308 (14)
Vacuum overhead	140	6 (24)	10	16	60 (3)	55 (2)
(Sample L3143)	72	28 (0)	23	24	135 (8)	106 (4)
	36	28 (0)	24	25	169 (6)	142 (5)
Fuel-Oil Model						
Atmospheric still	124	60 (0)	0	0	—	—
overhead	68	60 (0)	0	0	—	—
(Sample L3151)	34	60 (0)	0	0	—	—
Atmospheric still	157	54 (0)	21	17	291 (11)	301 (15)
bottoms	77	54 (0)	13	17	382 (18)	350 (17)
(Sample L3152)	38	54 (0)	2	5	569 (70)	503 (45)
Vacuum overhead	146	6 (24)	3	7	73 (7)	63 (3)
(Sample L3153)	74	28 (0)	25	25	129 (4)	123 (3)
	37	30 (0)	22	24	178 (5)	167 (5)
Acetone	119	60 (0)	0	0	—	—
Benzo[a]pyrene	0.015	(0)	44	48	319 (7)	316 (7)
	0.0075	88 (0)	45	50	500 (10)	465 (10)
	0.00375	106 (0)	16	26	778 (26)	711 (19)
	0.001875	106 (0)	2	2	1037 (99)	1084 (108)

[a]Ten mice/dose and sex for atmospheric still overhead sample from syncrude mode; 25 mice/dose and sex for other crude materials and acetone; and 50 mice/dose and sex for benzo[a]pyrene.

Source: Cowser (1981).

et al., 1981), are presented relative to BaP in Table 3.66. The relative carcinogenic potency of these materials correlated well with the bp data available (Cowser, 1981); 66% of the vacuum overhead sample from the syncrude mode boiled above 650°F, whereas 50% of the vacuum overhead sample from the fuel-oil mode boiled above 650°F. Values for the atmospheric still bottoms samples were 34 and 20%, respectively, for the syncrude and fuel-oil mode samples. In comparing these relative potency data with similar data from SRC materials (Tables 3.59 and 3.62 and Fig. 3.15), several differences in test protocols should be noted, including prolonged postexposure follow-up periods and sacrifice of test groups before 24 months exposure in the assays of these H-Coal materials.

In a follow-up study (Holland et al., 1984), the vacuum overheads sample from the syncrude mode (Sample L3143) was chemically fractionated using the method illustrated in Fig. 3.9. The fractions were assayed for dermal carcinogenicity in C57B1 mice by application of 3 μg of fraction in an acetone vehicle three times a week for 20 weeks, followed

Table 3.66 Estimate of Potency Relative to Benzo[a]pyrene for H-Coal Distillates

Material	Dose (mg/week)	Relative Potency (95% lower confidence limits)
Syncrude Mode		
Atmospheric still	157	0.089 (0.081–0.098)
bottoms	77	0.062 (0.055–0.068)
(Sample L3142)	38	0.044 (0.038–0.049)
Vacuum overhead	140	0.779 (0.685–0.874)
(Sample L3143)	72	0.162 (0.146–0.178)
	36	0.209 (0.189–0.230)
Fuel-Oil Mode		
Atmospheric still	157	0.012 (0.010–0.013)
bottoms	77	0.016 (0.014–0.018)
(Sample L3152)	38	0.016 (0.012–0.021)
Vacuum overhead	146	0.444 (0.352–0.541)
(Sample L3153)	74	0.186 (0.168–0.204)
	37	0.176 (0.159–0.194)

Source: Cowser (1981).

by promotion with 8 μg of TPA in acetone for 30 weeks or until tumor formation. Tumor-incidence and latency data are presented in Table 3.67. Tumorigenic activity of the neutral fraction was very similar to that of the crude vacuum overheads; similar assays of neutral subfractions indicated that both the mutagenicity and the dermal carcinogenicity were due to the neutral PAH subfraction. In contrast, the polar aromatic components, although highly mutagenic in the Ames assay, were much less active than the PAH fraction as dermal carcinogens. These findings are similar to those observed in mutagenicity and carcinogenicity assays of SRC-II heavy distillate solvent fractions.

Raw and hydrotreated samples of a blended H-Coal distillate (Sample L3112) from the PDU were assayed for carcinogenicity in C3Hf/HeBd mice to assess the effects of three degrees of hydrotreatment on carcinogenic potency (Cowser, 1981). Raw and hydrotreated samples of a crude petroleum were used as controls. Samples were diluted in a solution of

Table 3.67 Tumorigenicity in C57Bl Mice of Fractions of H-Coal Process Development Unit Materials

Material	Percentage of Mice with Tumors			Average Weeks to Tumor (SE)
	Papilloma	Carcinoma	Total	
Vacuum overhead (Sample L3143)	60	5	65	29 (0.7)
Neutral fractions	50	5	65	28 (1.6)
Aliphatic	0	0	0	—
Polycyclic aromatic	80	10	90	26 (0.9)
Neutral polar	20	5	25	31 (1.7)
Reconstituted neutral	50	10	60	29 (1.3)

Source: Modified from Holland et al. (1984).

Table 3.68 Physicochemical Properties of H-Coal Distillate Before and After Hydrotreating[a]

Property	Material and Degree of Hydrotreatment[b]			
	Raw Distillate (Sample L3112)	Low	Medium	High
Hydrotreatment temperature (°F)	—	675	675	725
Space velocity (h^{-1})	—	4	2	2
Hydrotreatment pressure (psig)	—	2500	2500	2500
Weight (%)				
Hydrogen	9.65	9.87	10.5	10.9
Oxygen	1.62	0.95	0.39	0.21
Nitrogen	0.39	0.20	0.13	0.09
Sulfur	0.107	0.002	<0.002	<0.002
Aromatic carbon (%)	57	52	45	39
Boiling range (°F)	300–700	—	—	200–650
Benzo[a]pyrene (μg/g)	57	9	3	2

[a]Mobil Research and Development Corporation.
[b]Further details unavailable.

Source: Cowser (1981).

70% acetone and 30% cyclohexane and 20 C3Hf/HeBd mice (10 of each sex/treatment group) were exposed three times per week for 65 weeks. Physicochemical properties of the samples before and after hydrotreating are presented in Table 3.68; doses applied, skin-tumor incidence, and latency data are presented in Table 3.69. Carcinogenicity was substantially reduced in the hydrotreated materials; even the lowest-severity hydrotreatment decreased the carcinogenic potency of the blended H-Coal distillate to below that of the crude petroleum.

The next series of H-Coal samples assayed for carcinogenicity in chronic skin-painting studies were obtained from the H-Coal pilot plant in Catlettsburg, Kentucky (Munro et al., 1985). Stabilizer naphtha (Samples L3211 and L3241), light fuel oil (Samples L3212 and L3242), and heavy fuel oil (Samples L3213 and L3243) from two different coal types processed in the syncrude mode were applied three times per week to groups of 25 male and 25 female C3H/HeBd mice at concentrations of 100%, 50%, or 25% w/v in acetone for periods of up to 50 weeks. Dose, skin-tumor incidence, and relative potency data from these assays are presented in Table 3.70. The heavy fuel-oil samples from both coal types were carcinogenically active; little difference between coal types was evident from tumor incidence.

Figure 3.16 (Cowser, 1982b) compares skin-tumor incidence and latency from carcinogenicity assays of the heavy fuel-oil mode from Kentucky No. 9 coal in the syncrude mode at the H-Coal pilot plant with samples of atmospheric still bottoms and vacuum overhead from the Trenton, New Jersey PDU plant. The comparison suggests that the heavy fuel oil from the pilot plant is very similar in carcinogenic potency to the atmospheric still bottoms from the PDU.

3.4.2.1.3 Liquefaction—Synthoil and COED. Crude materials and selected solvent fractions from the COED and Synthoil coal-liquefaction technologies have been assayed for carcinogenicity by chronic dermal application in mice (Holland et al., 1979a,b, 1980, 1981). Centrifuged oil from the Synthoil process (Sample L6111) and hydrotreated heavy

Table 3.69 Tumorigenicity of H-Coal Distillates in C3H/HeBd Mice

Material	Nitrogen Content (ppm)	Dose (% w/v)	Number with Skin Tumors[a] (%) Female	Male	Average Latency, Days (SE) Female	Male
H-Coal distillate (Sample L3112)	3900	50	5 (50)	4 (40)	301 (37)	342 (54)
		25	0	7 (70)	—	336 (30)
		12.5	1 (10)	1 (10)	434	390
		6.25	0	0	—	—
Low-severity hydrotreatment	2000	50	1 (10)	0	450	—
		25	0	0	—	—
		12.5	0	0	—	—
		6.25	1 (10)	0	412	—
Medium-severity hydrotreatment	1300	50	0	0	—	—
		25	0	0	—	—
		12.5	0	0	—	—
		6.25	0	0	—	—
High-severity hydrotreatment	900	50	1 (10)	0	462	—
		25	0	0	—	—
		12.5	0	0	—	—
		6.25	0	1 (10)	—	405
Recluse petroleum		50	0	3 (30)	—	381 (25.4)
		25	0	0	—	—
		12.5	0	0	—	—
		6.25	0	0	—	—
Acetone : cyclohexane		100	0	0	—	—

[a]Ten male and ten female mice/dose group.

Source: Cowser (1981).

oil (Sample G2211) from the COED process were assayed for dermal carcinogenicity in C3Hf/HeBd mice on a comparative basis with shale oil from a simulated in situ process and either a single (Wilmington, California) or composite crude petroleum (20% Wilmington, California; 20% South Swan Hills, Alberta; 20% Prudhoe Bay, Alaska; 20% Gach Sach, Iran; 10% Louisiana–Mississippi Sweet; and 10% Arabian light). Three different exposure regimens were used: (1) a single, 25-mg dose of each material applied three times per week for 22 weeks (30 mice/treatment group, 15 of each sex), followed by a postexposure examination period of 22 weeks (Table 3.71); (2) four doses (25, 12, 6, or 3 mg) applied two times per week for 30 weeks (20 mice/treatment group, 10 of each sex), followed by a 20-week postexposure period (Table 3.72); and (3) four doses (based on PAH content of the crude material and therefore different for each material assayed) applied three times per week for 24 months (50 mice/treatment group, 25 of each sex), or until death of the animal (Table 3.73). In all three studies, test materials were diluted to appropriate concentrations in a 70% acetone/30% cyclohexane mixture.

Solvent fractionation and analysis for PAH content were performed on all samples assayed (Guerin et al., 1978). In the 24-month study, samples of crude materials were diluted

Table 3.70 Skin Tumorigenicity of H-Coal Distillates (Syncrude Mode)

Material	Concentration[a] (% w/v)	Dose (mg/week)	Skin Neoplasm Incidence[b] (%)	Relative Carcinogenic Potency[c] (%)
Stabilizer naphtha	100	129	0	—
(Sample L3211)	50	69	0	—
	25	34.5	0	—
Light fuel oil	100	148	0	—
(Sample L3212)	50	75	0	—
	25	37.5	0	—
Heavy fuel oil	100	159	100	0.080
(Sample L3213)	50	77	84	0.038
	25	38.5	56	0.037
Stabilizer naphtha	100	127	0	—
(Sample L3241)	50	69	0	—
	25	34.5	0	—
Light fuel oil	100	149	0	—
(Sample L3242)	50	75	0	—
	25	37.5	0	—
Heavy fuel oil	100	160	92	0.037
(Sample L3243)	50	78	96	0.045
	25	39	66	0.042
Benzo[a]pyrene	0.1	0.037	98	—
	0.05	0.075	100	—
	0.025	0.15	100	—
Acetone	100	—	0	—

[a]Materials were diluted in acetone and applied in a total volume of 50 μL.
[b]Groups consisted of 25 male and 25 female C3H/HeBd mice.
[c]Carcinogenic potency relative to response to benzo[a]pyrene, combining tumors of both sexes.

Source: Munro et al. (1985).

so that approximately equivalent doses of total PAHs were applied in all materials assayed. To evaluate further the role of PAHs in the carcinogenicity of the crude materials, PAH fractions of crude samples were also assayed in the 24-month study (Table 3.73).

Tumor-incidence and latency data (mean time to tumor) from these studies (Tables 3.71–3.73) indicated that the Synthoil and shale oil crudes were more carcinogenic than the COED crude sample. No skin tumors were observed in the 24-month study in mice painted with the composite crude petroleum, its PAH isolate, or the PAH isolate from the crude COED sample.

The Weibull curve-fitting model was used to estimate carcinogenic potency of the synfuel materials assayed relative to BaP (Holland et al., 1981). Data from the 24-month study (Table 3.73) were used; carcinogenic potency estimates relative to BaP from these data are presented in Table 3.74. Dilution on crude materials to provide equivalent total PAH content of doses applied did not equalize tumorigenic potency, indicating that factors other than

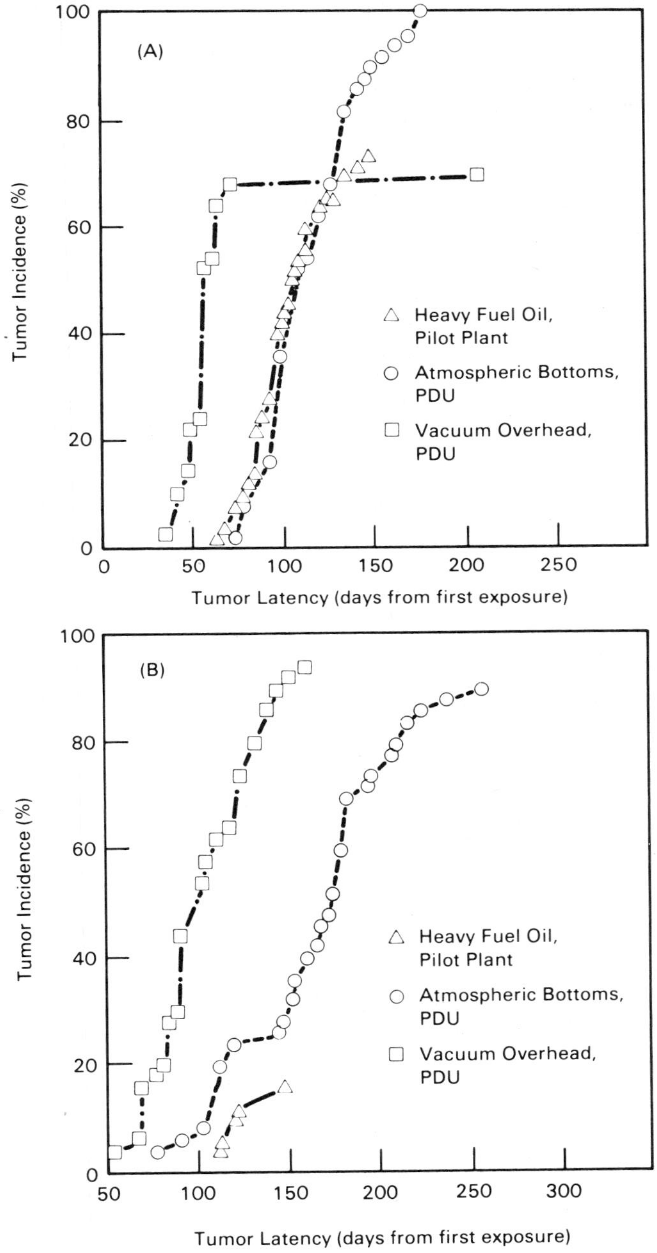

Figure 3.16 Cumulative percentage of incidence of skin tumors associated with exposure to H-Coal liquids from pilot plant or process development unit (PDU): (A) undiluted material and (B) material diluted to 50% with acetone (Cowser, 1982b).

Table 3.71 Skin Carcinogenicity of Syncrudes Assessed by Three Times Weekly Application for 22 Weeks

Material	Dose/Application (mg)	Final Number of Carcinomas[a] (%)	Mean Time to Tumor, Days (SE)
Synthoil (Sample L6111)	25	60	149 (8)
COED (Sample G2211)	25	37	191 (14)
Shale oil	25	47	154 (9)
Composite petroleum	25	0	0

[a]Thirty C3H/HeBd mice/treatment group (15 of each sex).

Source: Holland et al. (1979b).

PAHs were contributing to the carcinogenic response and that differences in composition of PAHs affected the response. The authors also pointed out that the increased potency of the PAH isolates, compared with that of the crude materials, was consistent with the view that the compounds responsible for the dermal carcinogenicity of these complex crude materials were concentrated in the PAH fraction.

With the exception of the basis for dose selection and the doses used, the study protocol for the 24-month exposures to Synthoil and COED crudes and their PAH isolates was very

Table 3.72 Skin Carcinogenicity of Syncrudes Assayed by Twice-Weekly Application for 30 Weeks

Material	Weekly Dose[a] (mg)	Number of Carcinomas[b] (%)	Mean Time to Tumor, Days (SE)
Synthoil (Sample L6111)	50	80	214 (11)
	24	35	238 (22)
	12	10	287 (50)
	6	0	—
COED (Sample G2211)	50	10	329 (17)
	24	5	253
	12	0	—
	6	0	—
Shale oil	50	35	208 (19)
	24	5	213
	12	0	—
	6	0	—
Wilmington crude petroleum	50	0	—
	24	0	—
	12	0	—
	6	0	—

[a]Applied two times/week.

[b]Twenty C3H/HeBd mice/treatment group (10 of each sex).

Source: Holland et al. (1979b).

Table 3.73 Summary of the Skin-Tumor Response in Mice Exposed Repetitively to Syncrudes

Material	Weekly Dose[a] (mg)	Number with Skin Tumors[b] (%)	Mean Time to Tumor, Days (SE)
Synthoil (Sample L6111)	3.0	92	490 (12.7)
	0.9	26	569 (17.8)
	0.6	2	596
PAH[c] fraction	0.15	8	573 (38.9)
COED (Sample G2211)	2.4	8	549 (45.8)
	0.9	2	494
	0.5	2	658
Shale oil	7.5	90	482 (14.8)
	1.5	2	315
PAH fraction	0.15	4	541 (55.5)
	0.03	2	485
	0.006	2	392
Benzo[a]pyrene	0.15	96	146 (2.6)
	0.03	98	216 (5.2)
	0.006	86	513 (14.3)
Solvent control	0.15	0	—

[a]Applied three times per week.
[b]Fifty C3H/HeBd mice/treatment group (25 of each sex).
[c]Polycyclic aromatic hydrocarbon.

Source: Holland et al. (1981).

similar to the protocols used for assay of SRC-II crudes, solvent fractions, and bp cuts. Therefore, the tumor potency data of those materials relative to BaP (Table 3.74) may be compared with similar data from Tables 3.59, 3.62, and 3.63 on SRC materials. Tumorigenic potency comparisons of PAH isolates from shale oil and Synthoil with the PAH fraction from SRC-II heavy distillate must be made with awareness that fractionation procedures differed. The shale oil and Synthoil PAH isolates were roughly four and two times as potent, respectively, as the SRC-II heavy distillate PAH fraction; however, as mentioned earlier, relatively large amounts of PAHs from the crude heavy distillate remained in the neutral tar and basic tar fractions after fractionation.

Table 3.74 Skin Carcinogenic Potency of Synthetic Petroleums and Polycyclic Aromatic Hydrocarbon Fractions Relative to Benzo[a]pyrene

Material	Relative Potency	Upper 95% Confidence Limit
Synthoil (Sample L6111)	1/503	1/128
PAH fraction	1/119	1/27
COED (Sample G2211)	1/1440	1/263
Shale oil	1/1380	1/275
PAH fraction	1/51	1/13

Source: Holland et al. (1981)

Table 3.75 Toxicity and Carcinogenicity Data from Mice Exposed Three Times Weekly to University of Minnesota–Duluth Gasifier Tar

Material	Concentration mg/week (% w/v)	Number with Skin Tumors[a] (%)		Number with Dermatitis (%)		Relative Carcinogenic Potency[b] (%)
		Female	Male	Female	Male	
ESP[c] tar	40 (25)	12 (48)	18 (72)	5	9	0.074
(Sample G1319)	20 (12.5)	25 (100)	25 (100)	4	10	0.208
	10 (6.25)	22 (88)	23 (92)	0	0	0.402

[a]Twenty-five C3H/HeBd mice/sex and dose.

[b]Relative potency is expressed as percentage of the response observed with benzo[a]pyrene [Weibull analysis, as modified by Holland and Frome (1982)].

[c]Electrostatic precipitator.

Source: Epler et al. (1983).

3.4.2.1.4 Coal Gasification. The first chronic mouse skin-painting carcinogenesis study of samples from a coal-gasification process was that of Holland (Cowser, 1981; Epler et al., 1983), who assayed a sample of tar from an ESP operating in the effluent gas stream of the UMD low-Btu coal gasifier. The tar (Sample G1319) was diluted in 70% acetone/30% cyclohexane and 50 μL were applied to groups of 25 male and 25 female C3H/HeBd mice three times per week for 32 weeks. Results of this assay, presented in Table 3.75, indicate that this tar was highly carcinogenic and dermatotoxic; the inverted dose response of carcinogenic potency was attributed to severe dermatotoxicity at the application site in mice exposed to the highest dose.

Chronic mouse skin-painting carcinogenicity studies have been performed on recycle oil (Sample G2151) from the Hygas coal-gasification pilot plant in Chicago, Illinois and on a process tar (Sample G1121) from the UNDERC coal-gasification unit (Reilly et al., 1986). The nonvolatile organic components (bp $>$ 390°F) from the Hygas recycle oil and the acid, base, and neutral fractions of the UNDERC tar were also assayed. The various materials were diluted in acetone and a total volume of 50 μL was applied three times per week to groups of 25 male and 25 female SKH (hairless) mice for 1 year or until skin-tumor incidence reached 100%. The crude process tar was also assayed in SENCAR and C3H/HeBd mice. 2-Aminoanthracene and BaP were applied as positive controls. Doses applied and results are presented in Table 3.76. It is apparent that the Hygas recycle oil tested was active as a dermal carcinogen at the doses applied. Skin-tumor incidence data from the crude UNDERC tar indicate a tumor response similar to that of the Hygas recycle oil. The neutral fraction is by far the most potent tumorigen, based on skin-tumor incidence and latency data. The differences in tumor latency and incidence in the three mouse strains painted with a 7.5-mg/week dose suggest that the SKH strain was the most sensitive of the three strains for dermal tumorigenicity with this tar.

3.4.2.1.5 Summary and Conclusions (Chronic Skin Exposure). Figure 3.17 illustrates comparative carcinogenic potency, using Holland's modification (Holland et al., 1981; Holland and Frome, 1982) of the Weibull analysis (Peto and Lee, 1973), from chronic dermal carcinogenesis assays in mice on crude process materials, tars, or residues occurring in the normal processing of materials and experimentally derived solvent or bp fractions from several coal-conversion technologies. These data are comparable because the study

Table 3.76 Incidence of Grossly Observed Skin Tumors in Mice After Exposure to Coal-Gasification By-Products

Material	Weekly Dose (mg)	Number of Mice with Skin Tumors[a] (%)	Latency[b] Days Minimum	Latency[b] Days Median	Latency[b] Days Maximum
Hygas recycle oil (Sample G2151)	150	91	70	203	441
Nonvolatile organics	105	95	14	126	231
University of North Dakota Energy Research Center tar (Sample G1121)	75	93	49	119	210
	37.5	100	70	119	266
	7.5	100	70	140	266
	7.5[d]	87	98	259	357
	7.5[e]	86	189	308	357
	0.75	74	217	368	609
	0.075	16	70	—	651
Acid fraction[c]	12.75	41	84	—	336
Base fraction[c]	1.35	89	133	322	119
Neutral fraction[c]	27	100	49	63	126
Unexposed	—	0	—	—	—
Acetone	Vehicle	0	—	—	—
2-Aminoanthracene	0.15	50	56	385	385
Benzo[a]pyrene	0.03	97	49	133	140

[a]Applied three times per week.

[b]Corrected for mice at risk excluding those mice that died before the first treated animal developed a tumor.

[c]Fractional concentrations were calculated on the treatment group receiving 25 mg crude tar/exposure.

[d]SENCAR mice.

[e]C3H/HeBd mice

Source: Reilly et al. (1986).

protocols were, in most cases, very similar (differences in protocols are noted in the text), the same strain of mice (C3Hf/HeBd) was used for the assays, and the chemical fractionation methods were similar. Potency data in Fig. 3.17 for H-Coal, UMD gasifier tar, Paraho shale oil, and recluse petroleum are from Munro et al. (1985).

It is apparent from Fig. 3.17 that crude materials from coal-conversion processes are more potent dermal carcinogens than crude shale oils or petroleums. Figure 3.17 also illustrates the increased carcinogenic potency of some tars or residues forming during coal liquefaction or gasification, when compared with crude materials. Carcinogenic potency of SRC-II bp distillates increases directly with bp; the vast majority of carcinogenic activity lies in that portion of the crude material boiling above 700°F.

Carcinogenicity of solvent fractions correlates best with PAH concentrations. However, attempts to correlate quantitatively dermal carcinogenicity in mouse skin with BaP content in materials from the H-Coal (Munro et al., 1985) and SRC (Renne et al., 1981) technologies indicated that a quantitative correlation was not present.

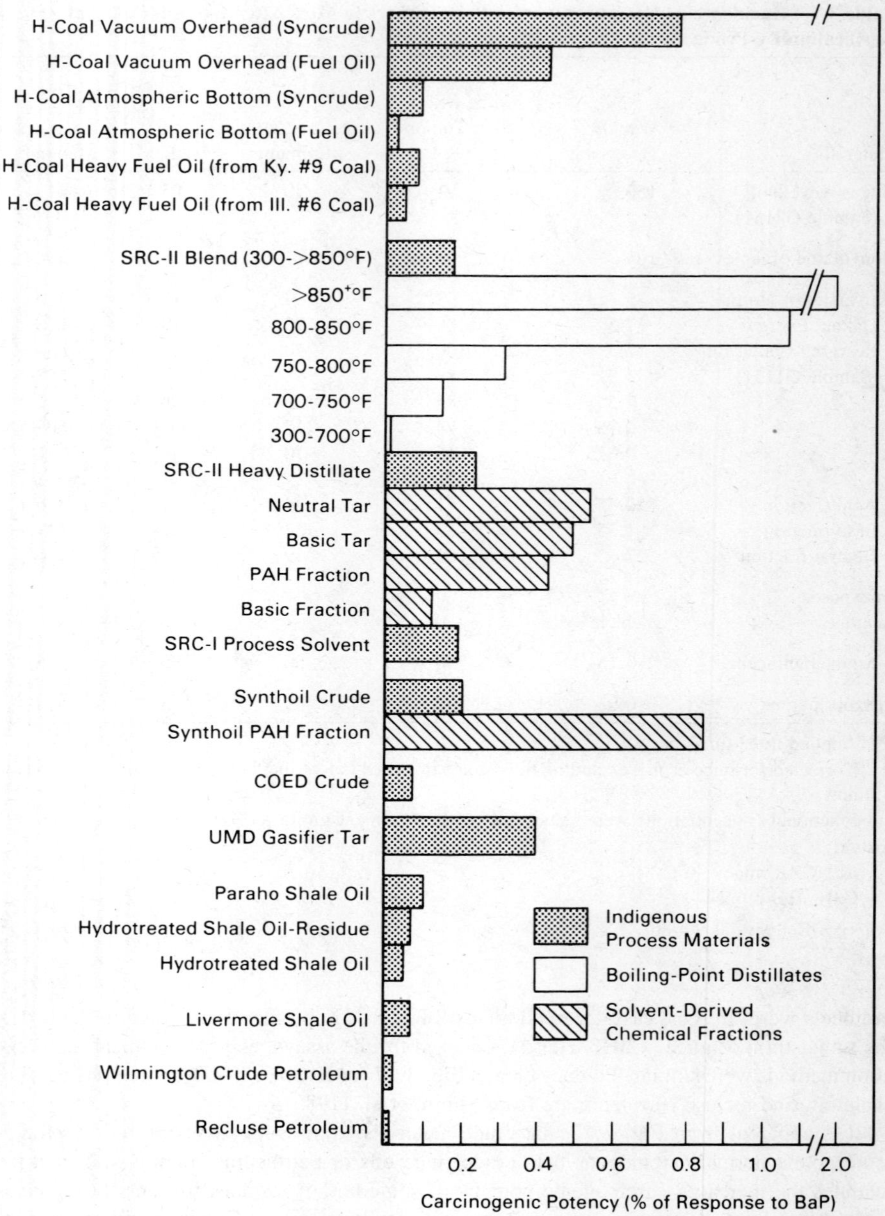

Figure 3.17 Carcinogenic potency relative to benzo[a]pyrene of various process samples, distillates, and solvent fractions of fossil-fuel materials. SRC = solvent-refined coal; PAH = polycyclic aromatic hydrocarbon; COED = Char-Oil Energy Development; UMD = University of Minnesota–Duluth.

Results from chronic studies of alumina fractions of one of the SRC-II bp cuts indicate that both PAH and N-PAC fractions are active as dermal carcinogens. Studies of a nitrosated sample of this same bp cut suggest that amino-PAHs do not play a major role in the dermal carcinogenicity of this material. These results taken together suggest that carbazoles and azaarenes, the other major chemical classes in the N-PAC alumina fraction, may play a role in the carcinogenicity of this material.

3.4.2.2 *Initiation/Promotion Assays*

The initiation/promotion assay in mouse skin has been used extensively, since it was first described in 1941, for studying the two-stage process for tumor induction (Berenblum, 1941; Van Duuren, 1969). Skin tumors are readily induced by a single application of a carcinogen (initiation), followed by repeated application of a noncarcinogenic chemical (promotion).

Many of the protocol variables previously described for chronic skin-painting studies also apply to initiation/promotion studies in mouse skin. Tumor response varies widely with the strain of mouse used in initiation/promotion assays (Ashman et al., 1982; Reiners et al., 1983; Slaga and Fischer, 1983) and with the solvent used to apply the initiator (Slaga and Fischer, 1983).

Doses used for initiation/promotion studies are, as a rule, more consistent among various experiments than those used in repeated-application studies. Although initiating doses may vary, depending on potency of the initiator being used, doses of promoter are more standardized. TPA is the most frequently used promoter for mouse skin-painting assays, usually 5–10 μg, applied twice weekly. There is ample evidence that tumor response varies with dose of initiator or promoter (Slaga and Fischer, 1983). The number of animals per experimental group is most frequently 30, and male mice are most frequently used. Most initiation/promotion studies on fossil-fuel materials use only a single dose of each material, and the duration of the studies is usually 6 months to 1 year.

Methods of expressing tumorigenic response in mouse skin initiation/promotion studies are similar to those for continuous-application studies; however, dose–response information is often not available in initiation/promotion studies since only a single initiating dose is applied. Numbers of skin tumors per animal and total number of tumors per exposure group are frequently used to express tumorigenic response. In initiation/promotion studies of dimethylbenzanthracene and BaP (Slaga et al., 1981), expression of the tumorigenic response as tumors per animal furnished a broader dose–response range than did percentage of mice with tumors. These results correlated well with percentage of carcinomas at 50 weeks of exposure. A positive correlation between tumor multiplicity and dose of initiator has also been described (Klein, 1956; Ball and McCarter, 1960). However, statistical analysis of tumorigenicity data on the basis of tumors per animal has been criticized (Peto et al., 1980) because of the tendency of a few animals in a group to develop large numbers of tumors and the difficulty of accurately counting multiple tumors that may become confluent. Another problem is the difficulty of statistically comparing tumors-per-animal data among animals that die spontaneously at various ages over long periods of time. Potency estimates from initiation/promotion data usually use BaP or other known PAH carcinogens for comparison, as in many continuous-application studies.

The biological behavior of induced skin tumors in initiation/promotion studies is reported, analyzed, and examined in depth less frequently than in chronic, continuous-application studies. Confusion of induced with spontaneous tumors is not as big a problem as in chronic studies because of the shorter duration of initiation/promotion studies.

The initiation/promotion assay in mouse skin has been used extensively for studying the carcinogenicity of coal-derived synthetic fuel materials (Calkins et al., 1983; Mahlum, 1983a,b; Mahlum et al., 1984). This assay has been used in combination with other in vivo and in vitro assays to assess the toxic potential of crude coal-liquefaction materials and their chemical or physical fractions. The small sample requirement for initiation/promotion studies allows investigators to assess the skin-tumor-initiating activity of materials obtained through complex chemical or physical separation processes, which may result in minute amounts of material.

3.4.2.2.1 Liquefaction—SRC. The first series of initiation/promotion studies with SRC materials assayed a sample of heavy distillate (Sample L1243) from the SRC-II pilot plant at Ft. Lewis, Washington and the PAH, basic, basic tar, and neutral tar solvent fractions of this material (Mahlum, 1983a). Groups of 30 strain CD-1 male mice, 3 months of age at the start of the test, were used. Initiating doses were 50 μL of each of the materials (approximately 50 mg). BaP and 7,-12-dimethylbenzanthracene (0.05 mg) diluted in 50 μL of acetone served as positive controls. TPA (5 μg in 50 μL of acetone) was applied as the promoter, twice weekly, for a period of 6 months.

Results of these assays, presented in Fig. 3.18, indicated that the tumor-initiating potency of heavy distillate was similar to that of 0.05 mg of BaP. The neutral tar and basic tar fractions of heavy distillate were nearly as active skin-tumor initiators as crude heavy distillate at the dose used; PAH and basic fractions were less active.

A crude and hydrotreated fuel-oil blend (Sample L1251) from the Ft. Lewis, Washington SRC-II plant was assayed in the mouse-skin initiation/promotion system for initiating activity, using TPA as promoter, and for promotion activity using 7,12-dimethylbenzanthracene for initiation (Frazier and Mahlum, 1984). Groups of 30 female CD-1 mice were initiated with a single 25-mg dose of crude or hydrotreated material; a similar group was initiated with 50 μg of 7,12-dimethylbenzanthracene and promoted with 50 μL of a 1:3 solution of hydrotreated material in acetone. Another group initiated with acetone only and promoted with hydrotreated material served as controls. BaP was also used as a positive control initiator, with TPA promotion.

Results of this study (Fig. 3.19) indicate significant tumor-initiating activity in the crude material; hydrotreatment reduced the tumor-initiating activity to a level comparable with petroleum-derived products. The hydrotreated material also possessed measurable skin-tumor-promoting activity with 7,12-dimethylbenzanthracene initiation, although the tumor response was much less than that observed with 7,12-dimethylbenzanthracene initiation followed by promotion with TPA (Fig. 3.20).

A full-boiling-range SRC-II material (Sample L1111) produced at the Harmarville, Pennsylvania PDU and selected bp cuts from that material were assayed for dermal tumorigenicity by the initiation/promotion protocol, using groups of 30 female CD-1 mice (Mahlum 1983a,b). Each material was diluted 1:1 with acetone, and 50 μL of the diluted material were applied in a single dose. The resulting dose was approximately 25 mg of test material per mouse. The 800–850°F bp cut was assayed twice at initiating doses of 25 mg: once in male CD-1 mice 3 months of age at the start of the study (run 1 in Fig. 3.21) and once in female CD-1 mice 6 weeks of age at the start of the study (run 2 in Fig. 3.21). Additional groups of mice initiated with 50 μg of 7-12-dimethylbenzanthracene or BaP served as positive PAH controls. Several primary aromatic amines (1-aminopyrene, 2-aminoanthracene, 6-aminochrysene, and 9-aminophenanthrene) were assayed at initiating doses of 50 μg in acetone. These materials were assayed because of data from in vitro

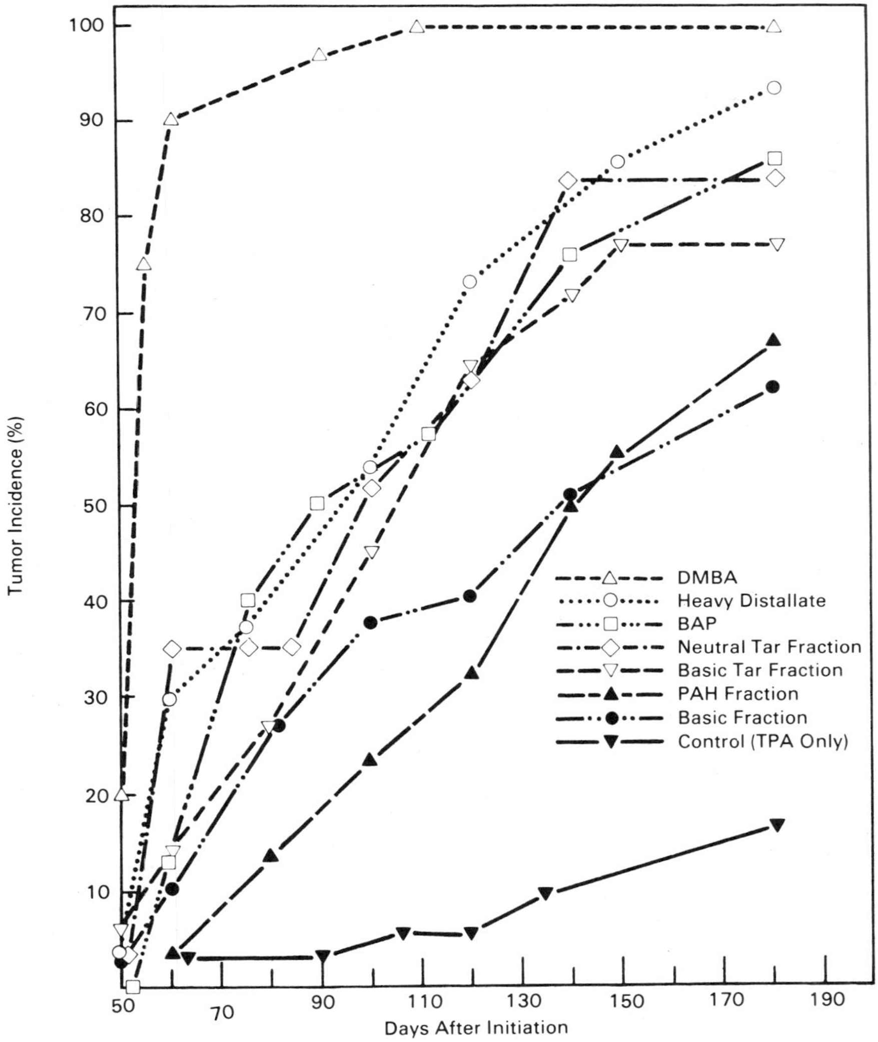

Figure 3.18 Mouse skin-tumor incidence after initiation with solvent-refined coal II heavy distillate from Ft. Lewis, Washington, or its solvent fractions followed by twice-weekly promotion with tetradecanoylphorbol-13-acetate (TPA) (Mahlum, 1983a). DMBA = 7,12-dimethylbenzanthracene; BaP = benzo[a]pyrene; PAH = polycyclic aromatic hydrocarbon.

studies (Pelroy and Stewart, 1981) suggesting that primary aromatic amines were major determinants of mutagenic activity in high-boiling coal liquids.

Results of these assays are presented in Figs. 3.21 and 3.22. The full-boiling-range material and the 800–850°F bp cut were the strongest initiators of the materials assayed, although the tumor response to the 800–850°F bp cut showed considerable variability between sexes. The 700–750°F and 750–800°F bp cuts were shown to be initiators of skin

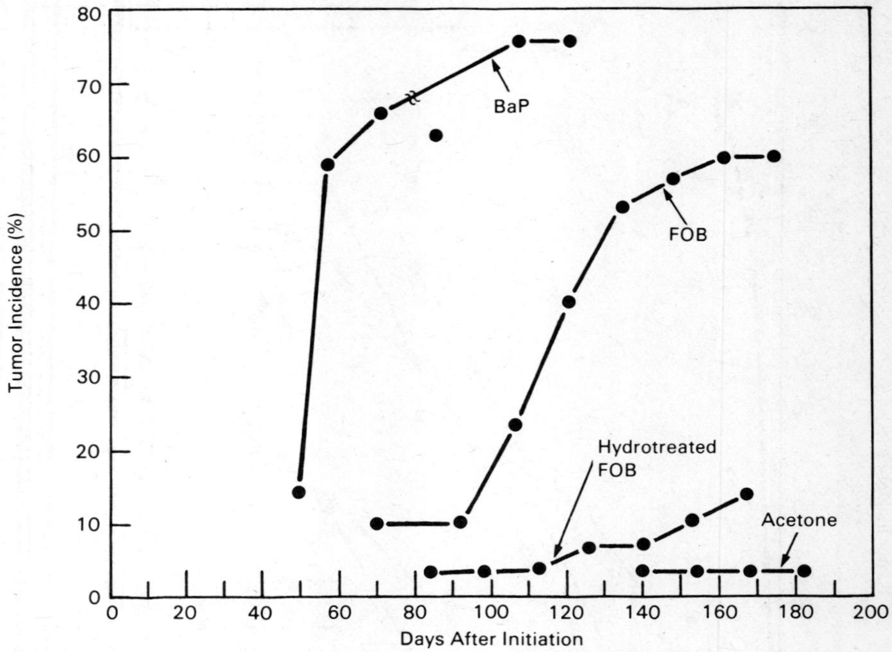

Figure 3.19 Mouse skin-tumor incidence after initiation with benzo[a]pyrene (50 μg), solvent-refined coal II (SRC-II) fuel-oil blend (FOB; 25 mg), hydrotreated SRC-II FOB (25 mg), or acetone followed by twice-weekly promotion with 5 μg of tetradecanoylphorbol-13-acetate (Frazier and Mahlum, 1984).

tumors, while the 300–700°F bp cut induced a tumor response not different from that of the TPA control group.

Of the primary aromatic amines assayed, 2-aminoanthracene and 6-aminochrysene were the most potent initiators (Fig. 3.22). Their activity was similar to that of the whole-boiling-range material, but somewhat less than that of BaP or 7,12-dimethylbenzanthracene. The initiating activity of 9-aminophenanthrene was much less, and that of 1-aminopyrene was not significantly above background.

An SRC-I process solvent (Sample L2121) produced at the Wilsonville, Alabama pilot plant and its distillate bp cuts were also assayed for dermal tumor initiation potential, using groups of 30 female CD-1 mice (Mahlum, 1983b). As in the SRC-II bp cut studies, each test material was diluted 1 : 1 with acetone and 50 μL of the diluted material (approximately 25 mg) was used as the initiation test dose.

Results of initiation/promotion assays of the SRC-I bp cuts (Fig. 3.23) were similar to those of the SRC-II bp cuts. The process solvent (550 to >850°F boiling range) was more potent as an initiator than any of its bp cuts except that cut boiling from 800°F to >850°F.

To evaluate the relative capability of different chemical classes to initiate carcinogenesis in mouse skin, fractions of the 800–850°F and >850°F bp cuts from the full-boiling-range SRC-II material (Sample L1111) and the 800 to >850°F bp cut from the SRC-I process solvent (Sample L2121) were obtained by adsorption column chromatography, using neutral alumina as an adsorbent (Later et al., 1981), and were assayed in the mouse-skin initiation/

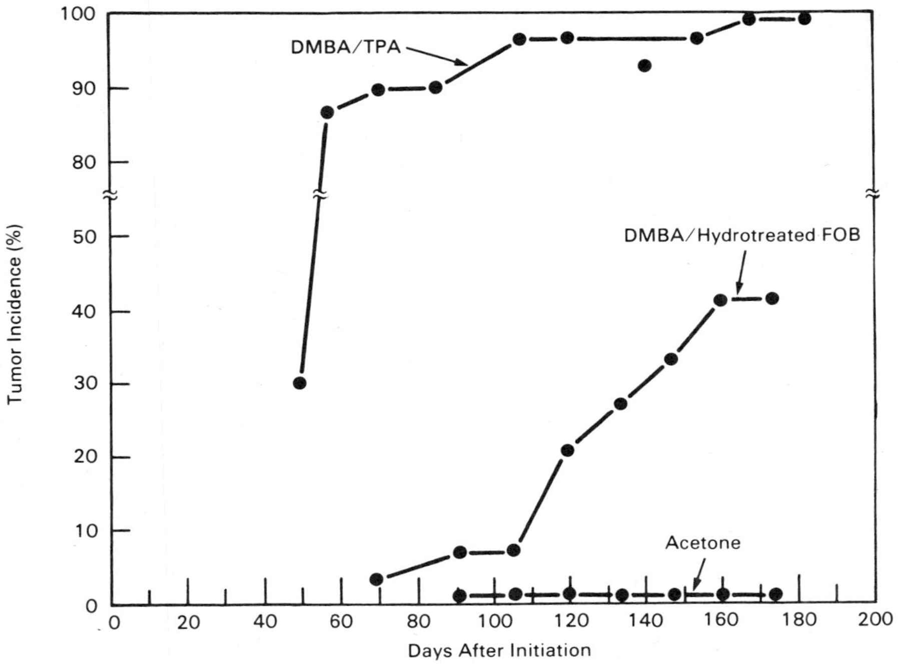

Figure 3.20 Mouse skin-tumor incidence after initiation with 7,12-dimethylbenzanthracene (DMBA) followed by twice-weekly promotion with tetradecanoylphorbol-13-acetate (TPA) or a hydrotreated solvent-refined coal II fuel-oil blend (FOB) (Frazier and Mahlum, 1984).

promotion assay (Mahlum et al., 1984). Four fractions were assayed: an aliphatic fraction (primarily aliphatic and olefinic compounds), a PAH fraction, an N-PAC fraction, and a hydroxy-PAH fraction. The unfractionated bp cuts were also assayed for comparison with the response to fractions. Groups of 30 female CD-1 mice were used. Initiating doses of the unfractionated bp cuts were 17 mg. Initiating doses of the chemical class fractions were proportional to their content in the crude bp materials at a 17-mg reference dose (Table 3.77). These materials were applied in 50 μL of acetone : methylene chloride (1 : 1) vehicle. Positive controls were initiated with 50 μg of BaP; promotion was with 5 μg of TPA, weekly for 24 weeks.

Results of these assays are presented in Figs. 3.24–3.26. The tumor-initiating activity of the SRC-II >850°F distillate cut was greater than that of the 800–850°F cut, as determined by tumor latency and incidence (Figs. 3.24 and 3.25). This suggests that the slightly increased activity of the 300–850°F material compared with that of the 800–850°F cut in the earlier assays (Fig. 3.21) is due to components boiling above 850°F.

The PAH fractions were more potent initiators of skin tumors than the crude distillate cuts in all samples assayed (Figs. 3.24–3.26). The N-PAC fractions were active, although less potent than the crude materials. Aliphatic and hydroxy-PAH fractions induced a tumorigenic response significantly above background incidence (~10%) only in the SRC-II >850°F distillate assay (Fig. 3.25). These results indicate that PAHs are the major determinants of mouse skin-tumor-initiating activity in SRC materials boiling above 800°F.

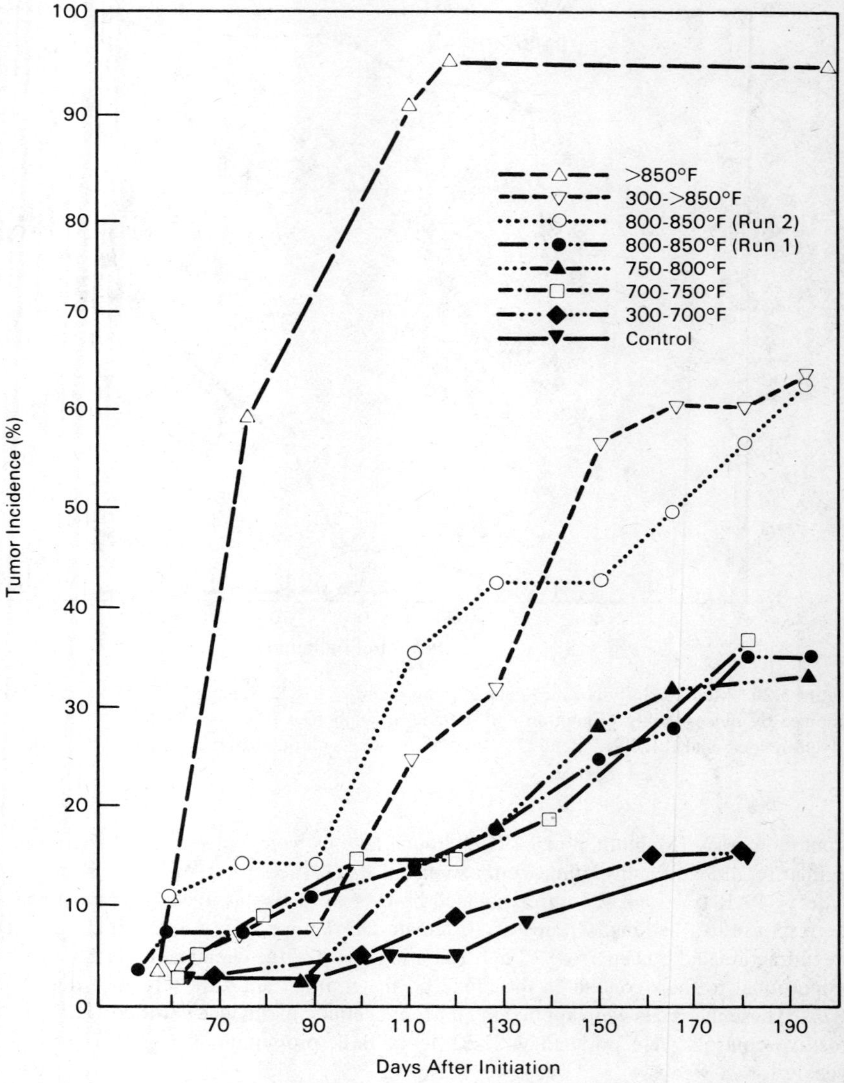

Figure 3.21 Mouse skin-tumor incidence after initiation with solvent-refined coal II distillates from Harmarville, Pennsylvania, followed by twice-weekly promotion with tetradecanoylphorbol-13-acetate (Mahlum, 1983a).

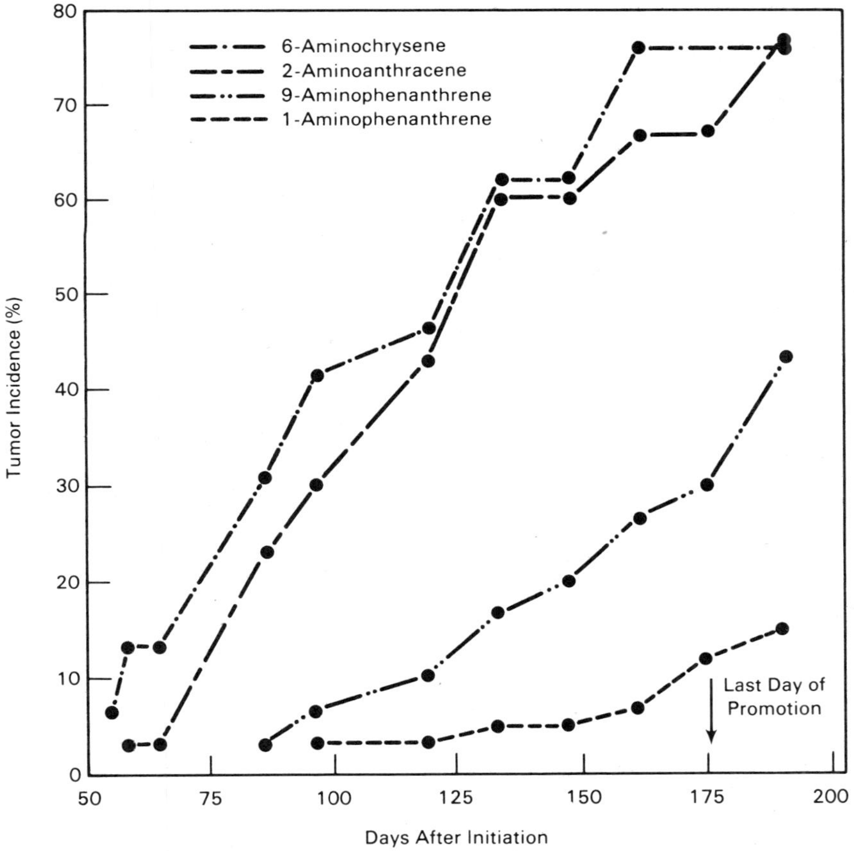

Figure 3.22 Mouse skin-tumor incidence after initiation with aromatic amines followed by twice-weekly promotion with tetradecanoylphorbol-13-acetate (Mahlum et al., 1984).

Tumor incidence rose rapidly and approached 100% by 80 days after initiation in the groups of mice initiated with the PAH fraction of the >850°F cut (Fig. 3.25). The tumor response to the crude >850°F cut was nearly this rapid. In comparison, tumor incidence rose much more slowly in response to the N-PAC fraction of this bp cut and to all the fractions and the crude material from the 800–850°F cuts. Also, 200 total tumors were observed in the group exposed to the PAH fractions of the >850°F cut. Approximately 170 total tumors were observed in the group exposed to the crude >850°F cut. No other group had more than 90 skin tumors. These results are further evidence that the high tumorigenic activity of the >850°F bp cut is due to its PAH fraction. However, the BaP contents of the 800–850°F and >850°F cuts are essentially identical (see Chapter 5), indicating either that BaP is not the determining factor in the tumor-initiating activity of these PAH fractions or that the tumor-initiating activity of BaP is suppressed in the PAH fraction of the 800–850°F cut. The tumor-initiating activity detected in the PAH and N-PAC frac-

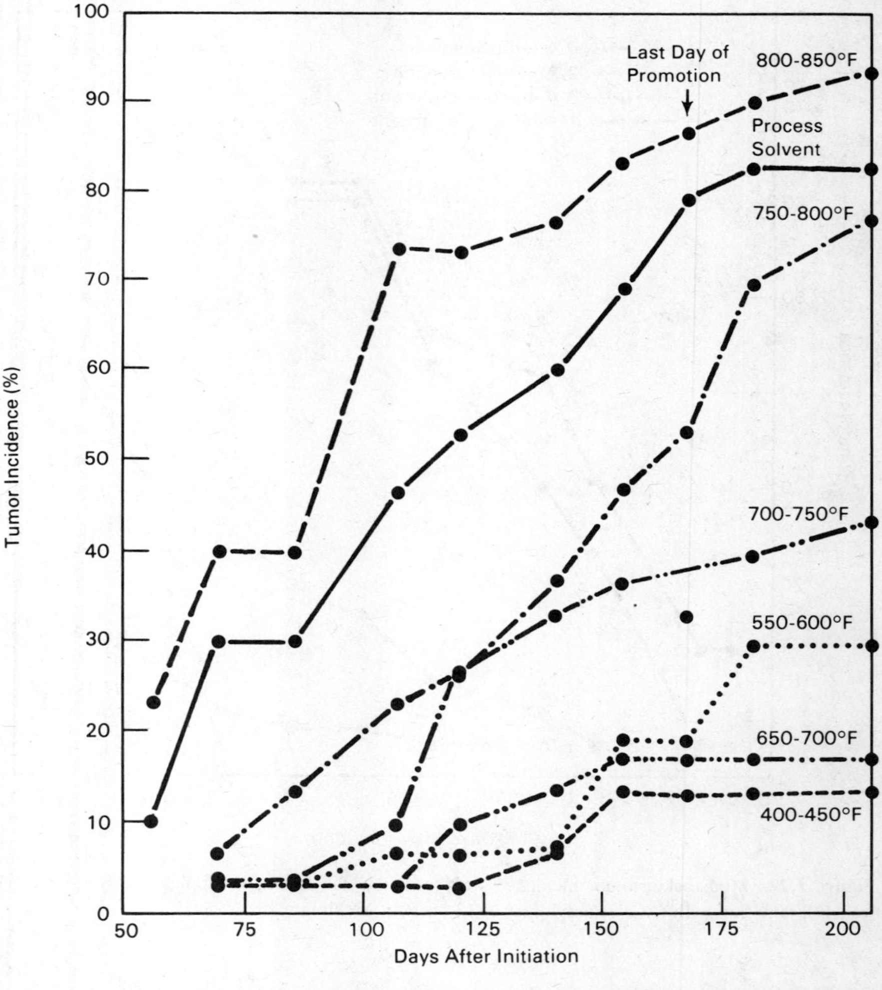

Figure 3.23 Mouse skin-tumor incidence after initiation with solvent-refined coal I distillates followed by twice-weekly promotion with tetradecanoylphorbol-13-acetate (from Mahlum et al., 1983b).

tions of the SRC-II >850°F distillate correlated well with genotoxicity measurements of these fractions in the SHE cell transformation and CHO mutation assays (Frazier and Andrews, 1983; Pelroy et al., 1984).

Calkins and co-workers (1981, 1983) examined the skin-tumor-initiating activity of standard distillate fractions of SRC-II. SRC-II naphtha (<350°F; Sample L1231), fuel oil (Sample L1232), vacuum tower bottoms (Sample L1235), and stripper tower bottoms (Sample L1234) were collected from the Ft. Lewis, Washington pilot plant during operation in the recycle slurry mode. The vacuum tower bottom sample was further distilled into a heavy gas oil, 650–850°F (Sample L1233) and a residue, >850°F (Sample L1236). Naphtha, fuel oil, and heavy gas oil were each applied as a single 50-μL dose of neat material;

Table 3.77 Weight Distribution by Percentage of Solvent-Refined Coal (SRC) Boiling-Point (bp) Distillates into Chemical Class Fractions and Initiating Doses Applied[a]

Material	Aliphatic	PAHs[b]	N-PACs[c]	Hydroxy-PAH	Total Recovery (%)
SRC-II (Sample L1111)					
>850°F	0.98 (17)	46.0 (8)	36.0 (6)	25.0 (4)	107
800 to 850°F	5.2 (1)	46.0 (8)	28.0 (5)	18.0 (3)	97
SRC-I (Sample L2121)					
800 to >850°F	4.8 (1)	19.0 (3)	36.0 (6)	40.0 (7)	99

[a]Initiating doses, indicated in parentheses, were determined by multiplying percent weight distribution by dose of unfractionated bp cut (17 mg for each bp cut).

[b]Polycyclic aromatic hydrocarbons.

[c]Nitrogen-containing polycyclic aromatic compounds.

Source: Modified from Mahlum et al. (1984).

vacuum tower bottoms, stripper tower bottoms, and residue were applied in a single 50-μL dose consisting of a 1:2 mixture of test material in toluene. Because of its acute toxicity when applied neat, naphtha was retested at a single 25-μL dose, consisting of a 66% solution of naphtha in toluene. Also, 150 μg of BaP were administered to a positive control group. Each test group consisted of 30 female Swiss ICR/Ha mice. For promotion, 2.5 μg of TPA in 0.1 mL of acetone were applied three times weekly for 6 months (unless the animal was killed earlier because it was moribund or had an advanced tumor). Results of this initiation/promotion assay of SRC-II materials are presented in Table 3.78. These data

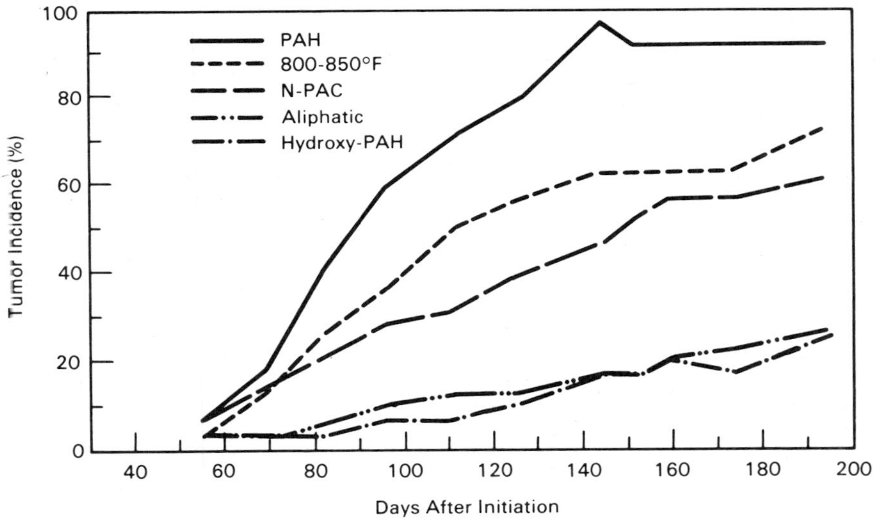

Figure 3.24 Skin-tumor incidence in mice after initiation with solvent-refined coal II 800–850°F distillate or its chemical class fractions followed by twice-weekly promotion with tetradecanoylphorbol-13-acetate (Mahlum et al., 1984). PAH = polycyclic aromatic hydrocarbons; N-PAC = nitrogen-containing polycyclic aromatic compounds.

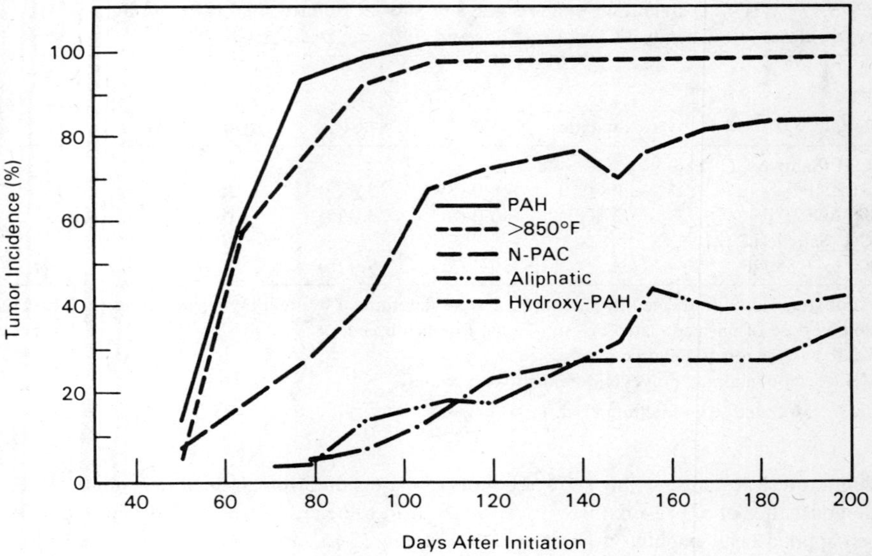

Figure 3.25 Skin-tumor incidence in mice after initiation with solvent-refined coal II >850°F distillate or its chemical class fractions followed by twice-weekly promotion with tetradecanoylphorbol-13-acetate (Mahlum et al., 1984). PAH = polycyclic aromatic hydrocarbons; N-PAC = nitrogen-containing polycyclic aromatic compounds.

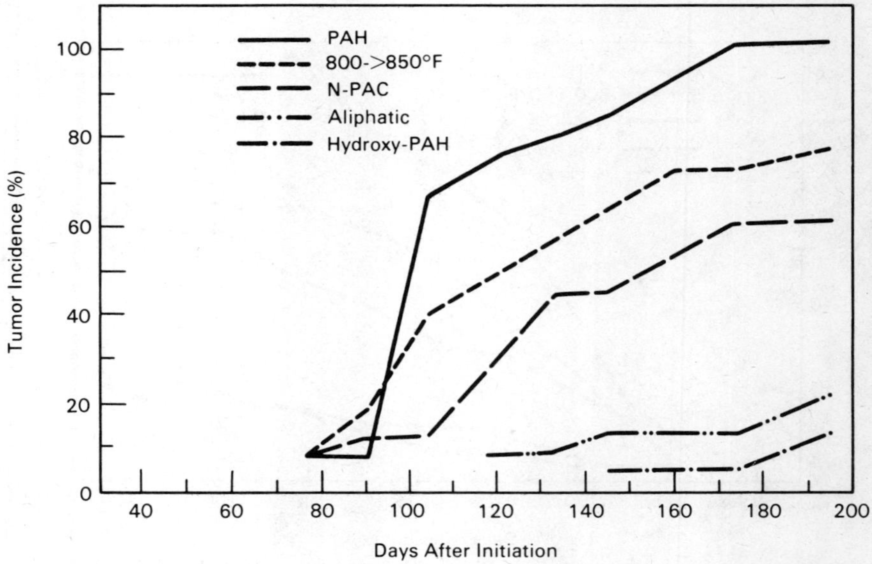

Figure 3.26 Skin-tumor incidence in mice after initiation with 800 to >850°F distillate from solvent-refined coal I process solvent or its chemical class fractions followed by twice-weekly promotion with tetradecanoylphorbol-13-acetate (Mahlum et al., 1984). PAH = polycyclic aromatic hydrocarbons; N-PAC = nitrogen-containing polycyclic aromatic compounds.

Table 3.78 Results of Initiation/Promotion Assays with Crude Petroleum and Coal-Derived Oils and Their Distillate Fractions[a]

Material	Sample No.	No. of Mice with Tumors	Total No. of Tumors	No. of Days to First Tumor
Solvent-Refined Coal II				
Naphtha	L1231	3	3	63
Fuel oil	L1232	9	21	51
Vacuum tower bottoms	L1235[b]	10	19	63
Stripper tower bottoms	L1234[b]	15	38	51
Heavy gas oil	L1233	18	56	42
Residue	L1236[b]	21	63	51
H-Coal Syncrude Mode				
Atmospheric still overhead	L3131	2	4	92
Naphtha	L3132	0	0	—
Light gas oil	L3133	1	2	76
Atmospheric still bottoms	L3134	14	48	41
Vacuum still overhead	L3135	20	103	41
Vacuum still bottoms	L3136[b]	16	50	61
H-Coal Fuel-Oil Mode				
Atmospheric still overhead	L3121	2	2	56
Atmospheric still bottoms	L3122	14	34	44
Vacuum still overhead	L3123	24	108	37
Vacuum still bottoms	L3124[b]	13	31	37
EDS				
Naphtha	L4211	0	0	—
Solvent	L4212	12	30	63
Vacuum gas oil	L4213[b]	20	66	51
Heavy gas oil	L4214	18	65	42
Residue	L4215[b]	17	41	42
South Louisiana Petroleum				
Crude oil		5	7	71
Naphtha		2	2	111
Light gas oil		3	5	71
Medium gas oil		3	3	83
Residue		5	8	62
Controls				
Benzo[a]pyrene[c]		26	192	48
Toluene and acetone		0	0	0
Toluene and TPA		4	1.5	111
Tetradecanoylphorbol-13-Acetate		3	1	92

[a]Each test group consisted of 30 female Swiss ICR/Ha mice. Only tumors lasting 30 days or more were counted.

[b]The test sample was administered in 50 μL of a 1:2 slurry in toluene.

[c]Applied at 150 μg in 50 μL of toluene.

Source: Calkins et al. (1983).

correlated well with those of Mahlum (1983a,b), indicating that the tumorigenic potency (incidence, total number of tumors, and latency) was highest in the highest-boiling materials and that the portion boiling below 350°F (naphtha) was not active.

3.4.2.2.2 Liquefaction—H-Coal and EDS. Calkins et al. (1981, 1983) also did initiation/promotion assays on syncrude-mode and fuel-oil-mode materials from the Hydrocarbon Research, Inc. PDU at Trenton, New Jersey and EDS materials from the Exxon, Inc. 1-ton/day liquefaction unit in Baytown, Texas. Doses were applied in a single 50-μL dose of neat material, except for the vacuum still bottom samples of H-Coal (Samples L3136 and L3124) and the vacuum gas oil (Sample L4213) and residue (Sample L4215) from EDS, which were diluted in toluene. The results (Table 3.78) agree with other comparative carcinogenesis studies: the coal-derived liquid materials were more potent carcinogens than shale-oil- or petroleum-derived samples, and the carcinogenicity of the coal-derived liquid materials increased with increasing bp. When the data from the coal-derived liquids are compared with regard to number of mice with tumors, total number of tumors, and time to first tumor (minimum tumor latency), the vacuum still overhead samples of H-Coal materials appear to be the most potent samples assayed.

Initiation/promotion assays have been performed on process solvent (Sample L4111) from the EDS pilot plant in Baytown, Texas, operating in the recycle mode, and from 750–800°F and >800°F distillate cuts of a 60:40 blend of raw naphtha and process solvent (Sample L4134) from the recycle mode of operation (Later et al., 1984). Twenty-five microliters of each test material (approximately 25 mg) were diluted with 25 μL methylene chloride and the diluted material was applied as the initiator dose to groups of 30 female CD-1 mice. The control group received methylene chloride vehicle only. TPA was used for promotion. Results of these assays (Fig. 3.27) suggest that the >800°F distillate cut is more active as a skin-tumor initiator (latency and incidence) than the 750–800°F cut or the process solvent, a finding similar to that observed for distillate bp cuts of SRC-I and SRC-II distillates.

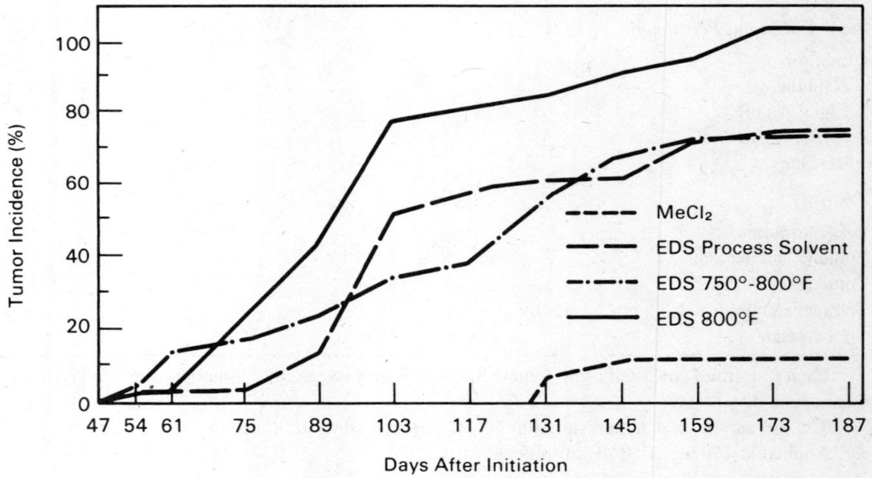

Figure 3.27 Mouse skin-tumor incidence after initiation with EDS process solvent or its distillate cuts followed by twice-weekly promotion with tetradecanoylphorbol-13-acetate (Later et al., 1984).

3.4.2.2.3 Two-Stage Liquefaction Technologies. Initiation/promotion assays in mouse skin have been performed on selected materials from several TSL technologies, concurrently with in vitro assays and chemical analyses of the same materials. Initial samples assayed were from an ITSL PDU facility operated by C. E. Lummus in New Brunswick, New Jersey (Wilson et al., 1983a). Samples assayed for tumor-initiation activity were LC Finer feed (Samples L5111 and L5121) and LCF TLP (Samples H0611 and H0621) from two separate sets of operating conditions. Samples L5111 and H0611 were collected during normal operating conditions; Sample L5121 was recycled once before entering the LC Finer, thus increasing the total residence time in the liquefaction reactor for Sample H0621. Initiating doses of 25 mg were used for all samples except Sample L5121, for which there was a difficulty with dissolution necessitating the use of only 10 mg as the initiating dose. Test materials were diluted in acetone and applied to groups of 30 female CD-1 mice. TPA was used as the promoter.

Results of the initiation/promotion assay of materials collected from the normal operation conditions are presented in Fig. 3.28. The ITSL samples taken during normal operating conditions were slightly less active than those taken during the off-normal (recycle) conditions, but these differences were not statistically significant. They were in fact less than expected from chemical analyses (see Chapter 5), in which concentrations of known PAH carcinogens, such as BaP or 5-methylchrysene, in materials from runs at normal operating conditions were lower than concentrations in materials from runs under off-normal operating conditions. Also unexpected was the relative similarity in skin-tumor-initiation activity of LCFF and TLP samples from both runs; analytical data indicated that LC Fining signifi-

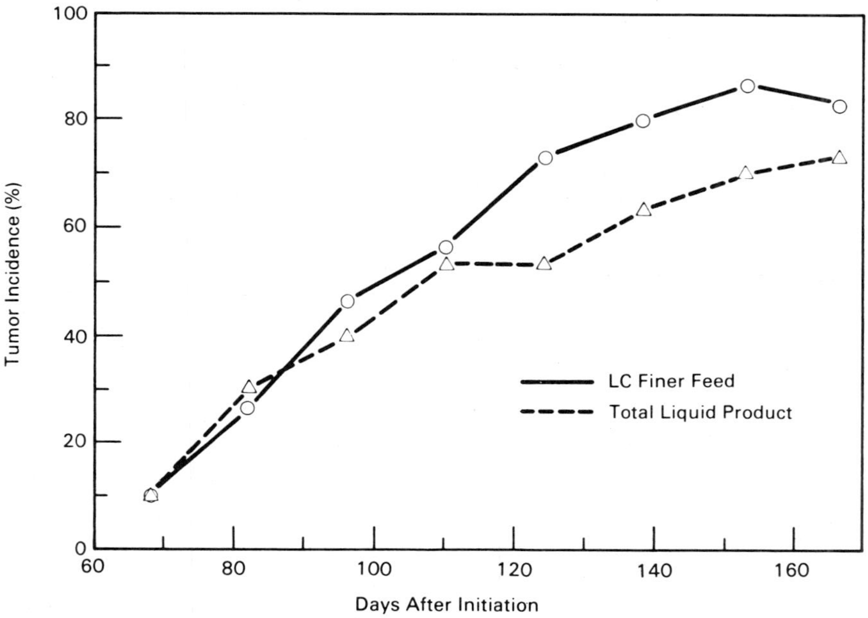

Figure 3.28 Skin-tumor incidence in mice initiated with LC Finer feed or LC Finer total liquid product from normal operation of the C. E. Lummus integrated two-stage liquefaction process development unit followed by twice-weekly promotion with tetradecanoylphorbol-13-acetate (Wilson et al., 1983a).

cantly reduced the concentrations of high-molecular-weight neutral PAHs and N-PACs in the LCF TLP compared with LCFF. The similarity in the earlier response to the two LCFF samples may be due in part to the difference in initiating dose; that is, 10 mg of the LCFF sample taken during the off-normal run versus 25 mg of the sample taken during the run at normal operating conditions.

Initiation/promotion assays in mouse skin have also been performed on process materials from the Advanced Coal Liquefaction Research and Development Facility in Wilsonville, Alabama, collected during operation of two different TSL configurations: NTSL and ITSL (Later, 1985). Materials assayed from each process made were the hydrotreater feed (Samples L5216 and L5346) and the hydrotreater distillation column bottoms (Samples H0513 and H0583). In addition, a <650°F distillate of the ITSL total liquid product (Sample H0633) from the New Brunswick, New Jersey PDU was assayed for comparison with these materials.

Results of these assays are illustrated in Figs. 3.29 and 3.30. There was some decrease in carcinogenicity with hydrotreatment in materials from both process modes; tumor incidence was decreased by 13% in the hydrotreated NTSL distillation column bottoms, compared with hydrotreater feed. The decrease with hydrotreatment was more dramatic in the ITSL materials: 50% tumor incidence in the hydrotreated distillation column bottoms versus 93% in the hydrotreater feed. It was suggested that this difference was due to the higher-quality recycle process solvent used in the ITSL mode of operation, which resulted in more effective hydrogen incorporation during both the first and second stages of the liquefaction process. The ITSL hydrotreater bottoms had skin-tumor-initiating activity approximately equal to the <650°F distillate from the ITSL PDU product, although the hydrotreater bot-

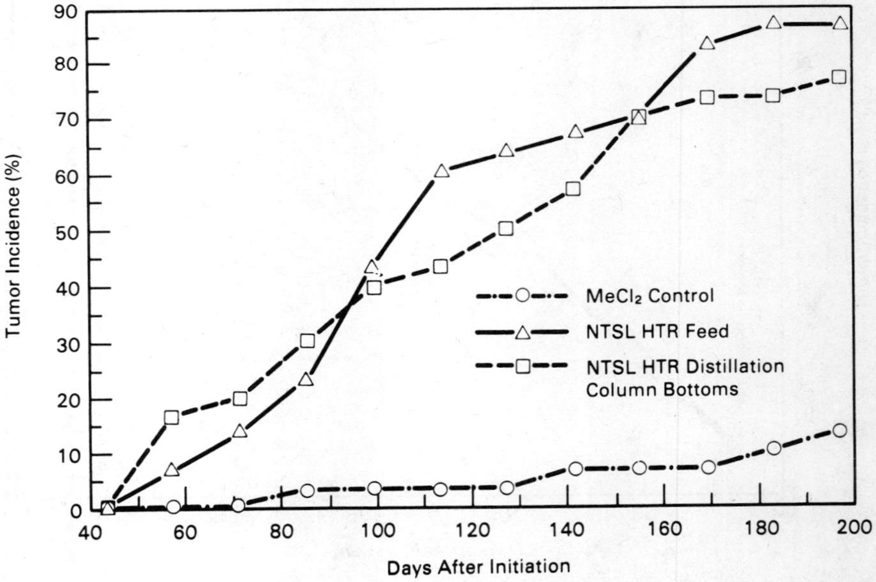

Figure 3.29 Skin-tumor incidence in mice after initiation with nonintegrated two-state liquefaction (NTSL) hydrotreater (HTR) feed and HTR distillation column bottoms followed by twice-weekly promotion with tetradecanoylphorbol-13-acetate (Later, 1985).

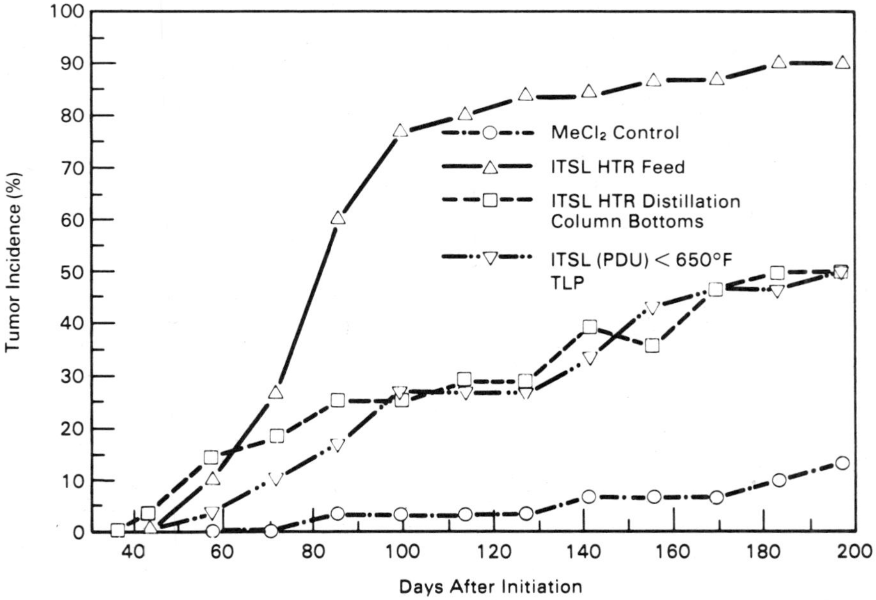

Figure 3.30 Skin-tumor incidence in mice after initiation with integrated two-stage liquefaction (ITSL) hydrotreater (HTR) feed and hydrotreater distillation column bottoms and ITSL process development unit (PDU) <650°F total liquid product, followed by twice-weekly promotion with tetradecanoylphorbol-13-acetate (Later, 1985).

toms had a somewhat higher bp range (450–850°F, approximately 60% >650°F) than did the PDU distillate.

3.4.2.2.4 Coal-Gasification Processes. An initiation/promotion assay of a crude tar (Sample G1121) from the UNDERC coal-gasification unit was performed in conjunction with chronic skin-painting studies of the same material (Reilly et al., 1986). A single initiating dose of 25 or 50 mg was applied to groups of 25 male and 25 female SKH mice. Mice were promoted twice weekly with 5 μg of TPA in 50 μL of acetone. Results are presented in Table 3.79. The test material readily initiated skin tumors, with a minimum latency shorter than the multiple-exposure assay of the same material, but with a much longer median tumor latency when applied only in a single initiating dose. Increasing the initiating dose to 50 mg decreased the latency.

3.4.2.2.5 Summary and Conclusions (Initiation/Promotion Assays). Results of initiation/promotion assays indicate that coal-conversion materials and products are more potent initiators of dermal carcinogenicity than comparable materials from shale oil or petroleum technologies. These results also indicate that the tumor-initiating potential of coal-conversion materials is due to their high-boiling components. Neutral PAHs appear to be the determinant chemical class for initiation of skin tumors in mice by high-boiling coal-conversion materials. N-PACs present in high-boiling coal liquids have a lesser amount of tumor-initiating activity.

Table 3.79 Incidence of Skin Tumors in SKH (Hairless) Mice After Initiation with Coal Gasifier Tar Followed by Twice-Weekly Promotion with Tetradecanoylphorbol-13-Acetate

Material	Initiation Dose (mg/50 μL)	Number of Mice with Tumors[a]	Total Number of Tumors[b]	Tumor Incidence (%)	Latency (days)	
					Minimal	Median
UNDERC[c] Tar (Sample G1121)	25	37	70	77	119	352
	50	37	77	76	63	274
	50[d]	24	27	48	84	—
Tetradecanoyl-phorbol-13-acetate	—	9	11	18	280	—

[a]Fifty mice per treatment group.
[b]Normalized to 50 mice.
[c]University of North Dakota Energy Research Center.
[d]No promotion.

Source: Reilly et al. (1986).

Skin-tumor response to SRC-II heavy distillate and its solvent fractions in initiation/promotion and chronic skin-painting assays are compared graphically in Fig. 3.31. Activity in the initiation/promotion assay is based on tumor incidence; activity in the chronic skin-painting assay is based on comparative potency, using the Weibull analysis. The response to heavy distillate in either assay is given a value of 1.0.

Results of initiation/promotion studies of SRC-II heavy distillate solvent fractions (Fig. 3.18) were qualitatively in accord with results of chronic skin-painting studies of the same materials; that is, the neutral tar and basic tar fractions were the most active of the solvent

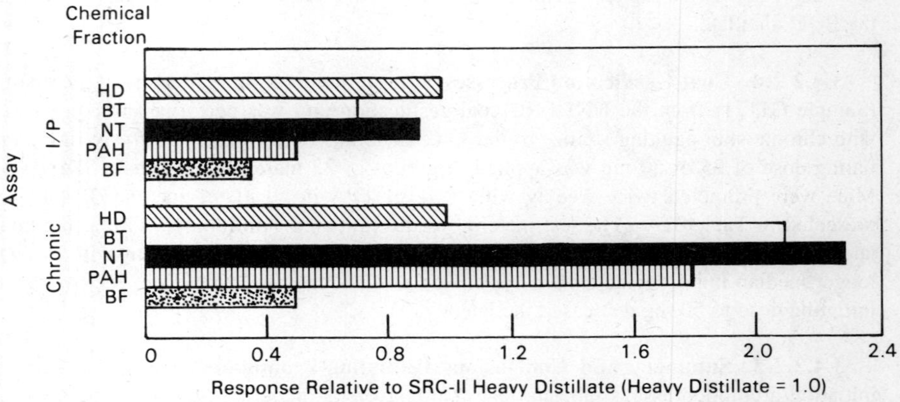

Figure 3.31 Activity of solvent-refined coal II heavy distillate (HD) and its solvent fractions in the initiation/promotion (I/P) assay in mouse skin and chronic mouse skin-painting carcinogenesis assays. Activity in the I/P assay is based on tumor incidence; activity in the chronic skin-painting studies is based on comparative potency using the Weibull analysis (see Holland et al., 1981). BT = basic tar fraction of HD; NT = neutral tar fraction of HD; PAH = polycyclic aromatic hydrocarbon fraction of HD; BF = basic fraction of HD.

fractions assayed in both assay studies, the PAH fraction was less active than the tar fractions, and the basic fraction was the least active. However, both tar fractions and the PAH fraction were more potent than heavy distillate in the chronic skin-painting studies, and the activity of the PAH fraction relative to the basic fraction was much greater in the chronic skin-painting assay than in the initiation/promotion assay.

Skin-tumor response to the whole-boiling-range SRC-II material and its bp cuts in initiation/promotion and chronic skin-painting assays are compared with the tumor response to heavy distillate in both assays in Fig. 3.32. In both assays, the skin-tumor response to the >850°F cut was much greater than to any other bp cut or to the whole-boiling-range material. The 800–850°F bp cut was next highest in activity in both assays, and the 300–700°F bp cut was least active in both assays. However, there were some noteworthy differences between initiation/promotion and chronic skin-painting study data in the bp cut assays. In the initiation/promotion assay, the 750–800°F bp cut was less active than the whole-boiling-range material; in the chronic study, it was more active. There were also large differences in the relative potency of the SRC-II heavy distillate and the higher bp cuts from the whole-boiling-range SRC-II material (Fig. 3.32). In the initiation/promotion assay, only the >850°F bp cut was more active than heavy distillate; in the chronic studies, all bp cuts over 750°F were more potent than heavy distillate, and the two bp cuts boiling over 800°F were four and nine times, respectively, as potent as heavy distillate. Figure 3.32 suggests that the potential for complete carcinogenesis is much greater than the purely tumor-initiating activity in those bp cuts boiling between 750 and 850°F. The data in Figs. 3.31 and 3.32 suggest that the initiation/promotion assay underestimates differences among active materials in relative carcinogenic potency for mouse skin.

When initiation/promotion data on H-Coal samples (Tables 3.78) are compared with the chronic skin-painting data on similar materials from the H-Coal PDU (Table 3.65), several similarities are notable, even though the samples tested in the two assays are not identical. For the syncrude-mode materials, tumor-latency data for the chronic skin painting are in the same rank order (vacuum overhead < atmospheric still bottoms), while latencies for

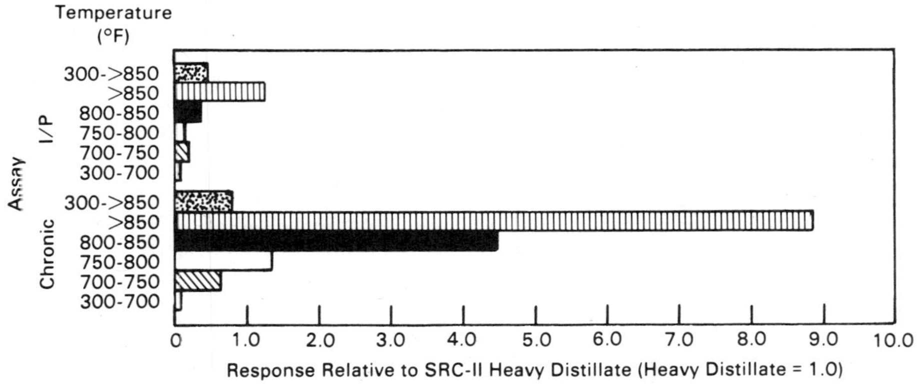

Figure 3.32 Activity of whole-boiling-range solvent-refined coal material and its boiling-point distillate cuts in the initiation/promotion (I/P) assay in mouse skin and in chronic mouse skin-painting carcinogenesis assay. Initiation/promotion activity is based on tumor incidence; chronic skin-painting activity is based on comparative potency using the Weibull analysis (see Holland et al., 1981). Responses are compared with responses to solvent-refined coal II heavy distillate.

these materials are identical in the initiation/promotion assay. However, tumor-incidence data for syncrude-mode samples differ in rank order for initiation/promotion (vacuum overhead > atmospheric still bottoms) versus chronic studies (atmospheric still bottoms > vacuum overhead). In the samples from the fuel-oil mode, both tumor-incidence and tumor-latency data for the two assays correlate rather well (tumor incidence: vacuum overhead > atmospheric still bottoms; tumor latency: vacuum overhead < atmospheric still bottoms).

3.4.2.3 *Carcinogenesis at Sites Other than Skin*

The fact that some coal-derived synfuel materials have relatively high concentrations of N-PACs as well as PAHs indicates a potential for induction of systemic carcinogenicity by these materials. The protocols for all chronic skin-painting studies of crude coal-conversion materials previously described included complete necropsy of all mice and histopathological classification of any internal and external lesions suspected of being tumors. To date, the only experimental evidence reported of systemic tumor induction from dermal application of these materials is an increased incidence of primary liver tumors in mice exposed to the basic fraction of SRC-II heavy distillate over a period of up to 2 years (R. A. Renne, unpublished data). Prevalence analysis (Peto et al., 1980) was used to test for a positive trend with respect to dose for prevalence of primary liver tumors in all groups of 25 male and 25 female C3H/HeBd mice painted with SRC-II light distillate (Sample L1241), heavy distillate (Sample L1243), heavy distillate solvent fractions, shale oil, BaP, or 2-aminoanthracene (Table 3.80). Data from untreated (shaved only) controls and acetone-painted controls were also included.

Assessment of the biological significance of these data is complicated by the relatively high spontaneous incidence of primary liver tumors in untreated C3H/He mice (Turusov and Takayama, 1979). Primary liver tumors occurred frequently in both test and control groups in which a large number of mice survived more than 1 year (Table 3.80). Prevalence analysis indicated a significant positive trend ($P = 0.02$) for liver tumor prevalence in mice painted with heavy distillate basic fraction. The main reason for this significant positive trend was a higher number of observed liver tumors than expected in the high-dose (1.14 mg) group of mice that died before 582 days on study (Fig. 3.33). This increase is suggestive of induction of heptocellular carcinomas by exposure to the heavy distillate basic fraction, which contains a high concentration of N-PACs (Table 3.60).

Inhalation is another important route of occupational exposure to genotoxic materials from coal-conversion technologies. Epidemiological (Lloyd, 1971) and experimental (Tye and Stemmer, 1967; MacEwan et al., 1976) data indicate that prolonged inhalation of coal-tar materials may induce tumors in the respiratory tract. In an industrial environment, workers may inhale aerosols or vapors of coal-conversion materials having chemical and biological properties similar to those of coal tars. BaP, a potent skin carcinogen known to be present in coal-conversion materials, readily induces tumors in the respiratory tracts of rodents when instilled intratracheally (Saffiotti et al., 1968).

Recent studies that address inhalation effects of coal-conversion materials are very limited. The effects of short-term subchronic inhalation exposure of SRC-II middle (Sample L1242) and heavy (Sample L1243) distillates collected at the Ft. Lewis, Washington pilot plant have been determined (Springer et al., 1984). Fischer-344 rats and CD-1 mice were exposed subchronically (4 weeks, 5 days/week, 6 h/day) to three concentrations of middle distillate (0.73, 0.2, and 0.04 mg/L) and the animals were sacrificed after exposure. All rats and 55% of the mice survived the exposure. Rats exposed to the high concentration failed to gain weight during the exposure. Blood samples from exposed animals had de-

Table 3.80 Incidence of Primary Liver Tumors in Mice Painted with Fossil-Fuel Materials

Material	Individual Dose[a](mg)	Mean Days on Study ± SE	Total Number of Primary Liver Tumors
SRC-II[b] light distillate	0.2	597 ± 23.9	11
(Sample L1241)	2.0	627 ± 15.5	15
	20.0	615 ± 21.0	20
RC-II heavy distillate	0.23	606 ± 15.7	15
(Sample L1243)	2.3	377 ± 14.6	5
	22.8	205 ± 4.8	4
Basic fraction	0.05	615 ± 18.7	15
	0.23	642 ± 15.5	23
	1.14	591 ± 18.6	27
PAH fraction	0.10	645 ± 18.2	13
	0.50	517 ± 10.6	12
	2.50	362 ± 7.9	2
Basic tar	0.27	527 ± 14.4	6
	1.35	348 ± 9.4	4
	6.75	247 ± 4.4	3
Neutral tar	0.21	564 ± 14.2	14
	1.05	376 ± 8.8	7
	5.25	274 ± 6.2	4
2-Aminoanthracene	0.005	623 ± 11.1	14
	0.05	413 ± 11.3	2
Benzo[a]pyrene	0.0005	666 ± 12.1	13
	0.005	461 ± 12.2	10
	0.05	248 ± 6.7	0
Untreated	0	653 ± 11.6	24
Acetone	0	628 ± 15.5	19
Shale oil	0.21	608 ± 20.5	13
	2.1	502 ± 13.4	12
	21.1	280 ± 13.6	4
Crude petroleum	0.17	624 ± 17.4	12
	1.7	618 ± 16.8	15
	16.8	564 ± 13.4	13
Acetone/methanol	0	607 ± 15.3	15

[a]This dose was applied three times weekly in acetone or acetone/methanol, a total volume of 50 μL/dose.

[b]Solvent-refined coal II.

Source: R. A. Renne (unpublished data).

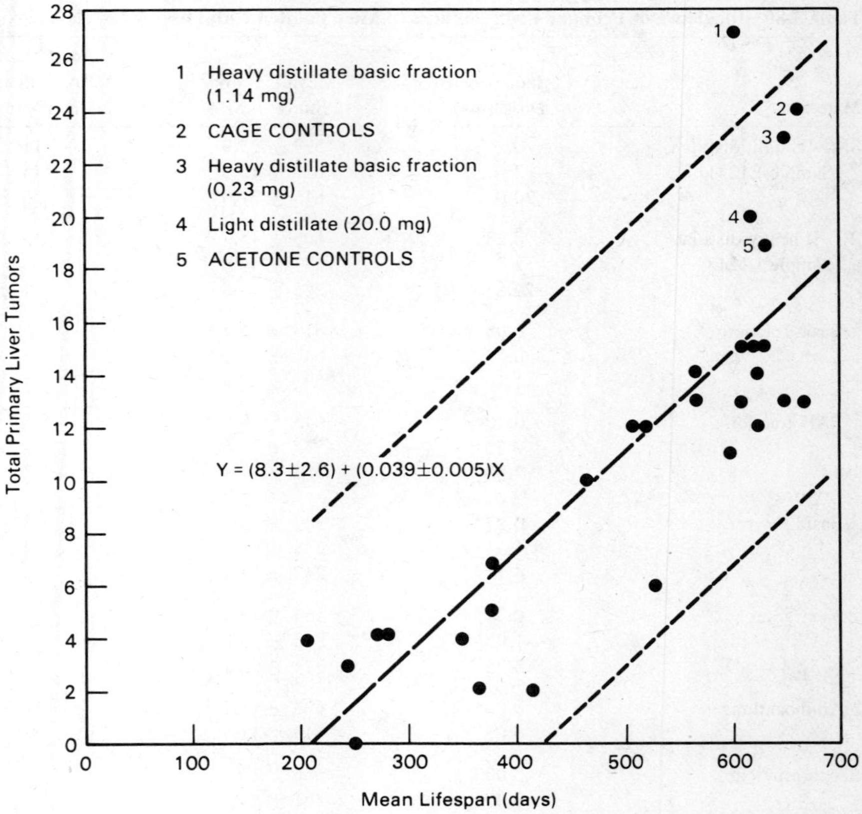

Figure 3.33 Analysis of prevalence of primary liver tumors in mice painted with crude fossil-fuel materials, solvent fraction of solvent-refined coal II heavy distillate, known dermal carcinogens, or acetone only. Fine dotted lines represent 95% prediction interval for number of liver tumors in any group at a given mean life span (R. A. Renne, unpublished data).

creased red blood cells and hemoglobin levels and increased reticulocyte and segmented neutrophils.

Twelve-week-old Fischer-344 rats were exposed to aerosolized heavy distillate at three test concentrations (0.69, 0.14 and 0.03 mg/L) for 5 or 13 weeks (6 h/day, 5 days/week). There were 64 rats (32 of each sex) in each treatment group and an unexposed control group. Ten rats of each sex for each of the five groups were sacrificed after 5 and 13 weeks of exposure and complete necropsy performed. Over 90% of the rats survived the 13 weeks of exposure. As in the middle distillate exposure, weight gain of exposed animals was less than that of controls. To date, squamous metaplasia of nasal cavity epithelium, hyperplasia of renal pelvic epithelium, and hyperplasia of mucosal epithelium of the cecum were observed in rats exposed to 0.14 or 0.69 mg/L for 13 weeks (Miller et al., 1984). The remaining animals (12 of each sex/group) were to be evaluated after an 18-month postexposure observation period.

In another in vivo study addressing carcinogenic effects of coal-derived materials occurring via nondermal exposure routes, a gasification tar was studied in a mouse lung adenoma assay (Cowser, 1981; Epler et al., 1983). Two electrostatic precipitator tar samples from the UMD gasifier were dissolved in a corn oil/propylene glycol mixture and were injected intraperitoneally, at three dose levels, into strain A/Jax mice three times per week for 8 weeks. The animals were killed 4 months after the last injection, and grossly observable lung tumors were counted at necropsy. Dosage information and results of this work are presented in Table 3.81. Sample G1319 induced a positive carcinogenic response in the lung, demonstrated by the significant increase in mean number of lung tumors and evidence of a dose response in lung-tumor incidence. A level of significance for these data was not reported. Results of assays of Sample G1306 were considered ambiguous, but criteria for a positive response in this assay have not yet been fully developed.

Complete necropsy and histopathological examinations of suspect organs were part of the protocol of chronic dermal carcinogenicity studies of coal-conversion materials. With the exception of the base fraction of an SRC-II heavy distillate material in which there was an increased incidence of primary liver tumors, there is no indication of dermal exposure resulting in systemic cancer induction. Intraperitoneal injections of a gasification tar were reported to increase the incidence of lung adenomas. A single study of cancer induction following inhalation exposure to SRC-II middle and heavy distillates is at present incomplete. Thus, while chronic dermal exposure to coal-derived materials appears to result in an extremely low incidence of systemic tumors, little is known about other routes of exposure.

Table 3.81 Results of Mouse Lung Adenoma Assay of University of Minnesota–Duluth Gasifier Tars

Material	Dose/ Injection (mg/kg)	Survivors[a] (%)	Mice with Tumors (%)	Number of Lung Tumors per Mouse ± SE
ESP[b] tar (Sample G1306)	1,000	0	0	0
	500	0	0	0
	200	43	38	0.92 ± 0.44
	100[c]	33	50	1.20 ± 0.51
	40[c]	60	39	0.44 ± 0.18
ESP tar (Sample G1319)	250	37	73	1.9 ± 0.7
	125	80	50	0.96 ± 0.30
	50	87	27	0.50 ± 0.18
Corn oil/propylene glycol	10,000	67	20	0.20 ± 0.10
	10,000[c]	53	19	0.19 ± 0.10
Urethane	1,000	85	100	17.2 ± 1.9
	1,000[c]	85	100	19.41 ± 1.55
Controls	0	100	15	0.25 ± 0.16
	0[c]	70	21	0.36 ± 0.20

[a]Thirty mice/group except for urethan-treated and untreated control groups, where only 20 mice were used.

[b]Electrostatic precipitator.

[c]Groups from low-dose study of Sample G1306.

Source: Cowser (1981a) and Epler et al. (1983).

3.4.2.4 *Metabolism, Dosimetry, and Biological Transport*

Metabolism, dosimetry, and biological transport are research areas critical to understanding the potential toxicity and carcinogenicity of coal-derived materials. However, the chemically and physically complex nature of coal-conversion materials complicates these types of study as well as those focusing on the mechanisms of toxic effects observed in response to exposure. Nevertheless, some progress has been made toward understanding the biological behavior of these materials and the mechanisms of their carcinogenicity. Most of this work to date has used a strategy of studying the distribution, metabolism, and transport of selected pure compounds considered potentially carcinogenic and/or representative of classes of hydrocarbons present in coal-conversion materials.

The observation that BaP metabolites are more rapidly removed from mouse skin when in the presence of a coal liquid (Schirmer et al., 1983) suggests that the carcinogenic response may also be altered and correlates with studies showing that the carcinogenicity of BaP and other potent PAH carcinogens is reduced by the presence of other PAHs. Pulmonary metabolism studies following inhalation exposure indicate that although BaP is rapidly cleared from the lung when administered as a pure aerosol, it is metabolized within the lung to potentially active intermediates (Mitchell, 1983). Inhalation studies with particulates indicate that pulmonary clearance of BaP is inhibited by particulates (Henry and Kaufman, 1973; Sun et al., 1982); diesel exhaust particles were especially effective in delaying pulmonary clearance of BaP (Sun et al., 1984). There is also experimental evidence indicating that prolonged clearance time increases the pulmonary carcinogenicity of BaP (Saffiotti et al., 1968; Cresia and Nettesheim, 1974). These findings are especially provocative because of the known pulmonary carcinogenicity of PAHs and the occurrence of PAH-containing particulates in many occupational environments.

Studies by O'Brien et al. (1975a,b) with pure compounds have demonstrated good correlation between mouse skin-tumor promotion and large increases in activity of ornithine decarboxylase in affected skin. Ornithine decarboxylase induction studies in mouse skin using materials from the SRC-I and SRC-II processes (Swanger et al., 1982; Springer et al., 1983) indicated an apparent qualitative correlation between ornithine decarboxylase induction and the tumorigenicity of SRC-II heavy distillate (Sample L1243). The induction of different sets of metabolic enzymes by various components of a coal-derived mixture would most likely modify the metabolism of toxic components and possibly their toxic effect. The overall effects of metabolites and transport on the dose and ultimately the toxicity of coal-derived materials remain to be defined.

3.4.3 Relation of in Vitro Mutagenesis and in Vivo Carcinogenesis Assays

The large volume of in vitro mutagenesis and in vivo carcinogenesis data described in this section, from a number of coal-conversion materials and fractions, makes it possible to examine the correlation (or lack thereof) between these sets of experimental data. This in turn permits one to evaluate the usefulness of the in vitro assays for predicting the carcinogenicity of these complex materials and their physical or chemical fractions. Comparison of the results of mutagenesis or carcinogenesis data from a number of well-characterized chemical or physical fractions of materials from different coal-conversion technologies may indicate the chemical classes or groups responsible for the observed mutagenicity and/or carcinogenicity.

Ames assays of several crude fossil-fuel materials by several different investigators have indicated that coal-liquefaction materials were much more genotoxic than shale oil or crude petroleum, correlating well with results of chronic dermal application studies (Holland et al., 1981; Renne et al., 1981) and initiation/promotion studies (Calkins et al., 1983; Mahlum et al., 1983). The work of Calkins et al. (1983), in which common distillate fractions of crude petroleum and SRC-II, EDS, and H-Coal materials were studied for Ames mutagenicity and dermal carcinogenicity (initiation/promotion), indicated a good correlation between Ames mutagenicity and skin-tumor initiation in mice. The higher-boiling distillate fractions were more mutagenic and more carcinogenic. However, Ames assays of SRC-II heavy distillate solvent-derived fractions (Pelroy and Petersen, 1981) indicated that the basic fraction was a much more potent mutagen than crude heavy distillate; neutral tar and basic tar fractions were somewhat less mutagenic than heavy distillate, while PAH fractions had little detectable mutagenic activity. These mutagenicity data do not correlate well with mouse skin carcinogenicity data on the same fractions (Fig. 3.34) and suggest that the determinant chemical class for bacterial mutagenicity in heavy distillate is the N-PAC group. Similar results of Ames assays and skin-painting studies of solvent-derived fractions of SRC and H-Coal materials were reported by Holland et al. (1984). These investigators concluded that the nitrogen-rich polar compounds (homologous to heavy distillate basic fraction) were the determinants of mutagenic activity, while the PAH-rich fractions were the determinants of skin carcinogenicity.

The large decrease in Ames mutagenicity produced by nitrosation of the Ft. Lewis, Washington SRC-II heavy distillate basic tar fraction (Pelroy and Stewart, 1981) is in sharp contrast to the results of mouse skin carcinogenicity assays (Renne et al., 1986) of this material, illustrated in Fig. 3.8. These results provide additional evidence that the N-PACs

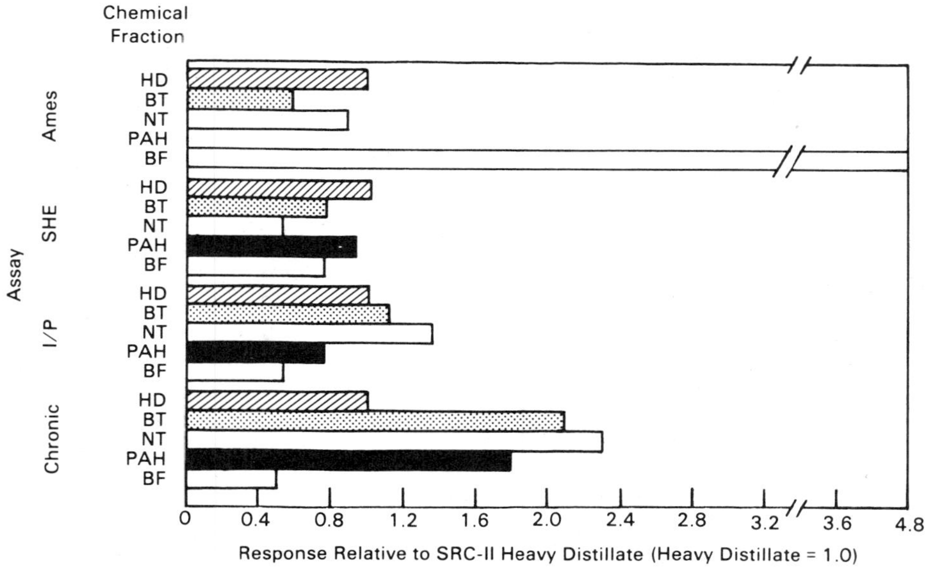

Figure 3.34 Activity of solvent-refined coal II heavy distillate (HD) and its solvent fractions in Ames assay, Syrian hamster embryo (SHE) cell transformation assay, initiation/promotion (I/P) assay in mouse skin, and chronic mouse skin-painting carcinogenesis assay. BT = basic tar fraction of HD; NT = neutral tar fraction of HD; PAH = polycyclic aromatic hydrocarbon fraction of HD; BF = basic fraction of HD.

are the determinant bacterial mutagens in this material but do not play a major role in its dermal carcinogenicity. Wilson et al. (1980) provided evidence that the mutagenicity of the basic fraction, basic tar fraction, and neutral tar fraction in the Ames assay is due in large part to N-PACs.

Ames assay data on a whole-boiling-range SRC-II material (Harmarville, Pennsylvania PDU) and its bp cuts are compared with in vivo carcinogenicity data on these materials in Fig. 3.35. In both types of assay, mutagenicity or carcinogenicity increases directly with bp, and the >850°F bp cut is much more potent in both assays than is the whole-boiling-range material or any other bp cut. However, only the >850°F bp cut has more mutagenicity than the whole-boiling-range material in the Ames assay. In the carcinogenicity studies, all bp cuts boiling above 750°F (chronic studies) or 800°F (initiation/promotion studies) have greater carcinogenic potency than does the whole-boiling-range material. These data indicate that, for the whole-boiling-range SRC-II material, potent dermal carcinogens are concentrated in that portion boiling over 750°F, whereas compounds inducing bacterial mutagenicity are present at all bp ranges. Bacterial mutagens and dermal carcinogens present in that portion boiling at >850°F are present in higher concentrations and/or are more potent mutagens or carcinogens than those present at lower bp ranges.

Removal of amino-PAH from the 750–800°F bp cut by nitrosation decreased the carcinogenicity of this material only slightly (Fig. 3.13), indicating that amino-PAHs, al-

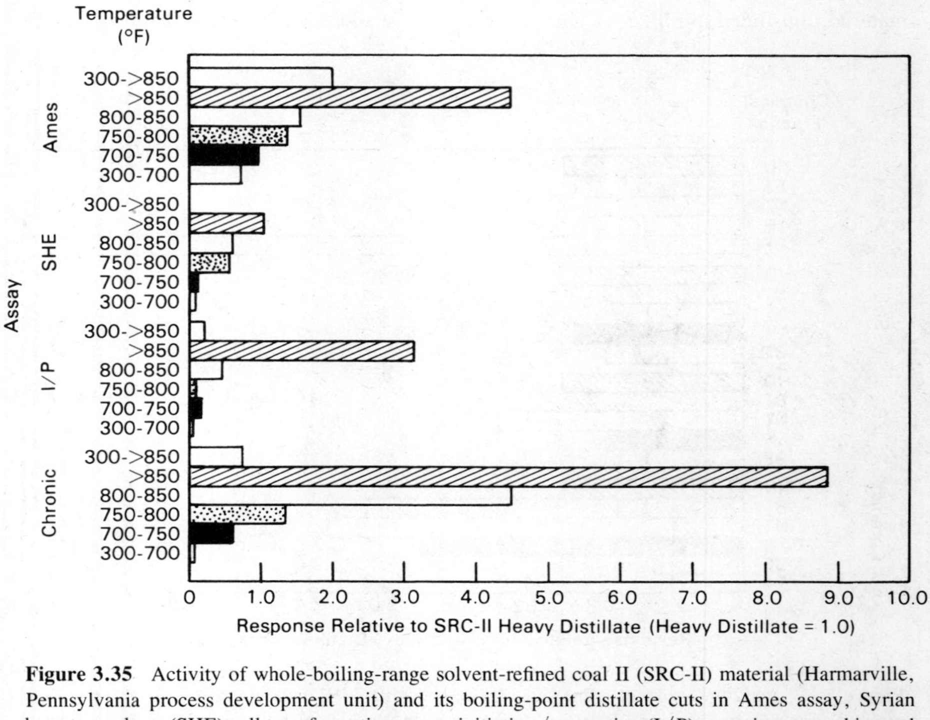

Figure 3.35 Activity of whole-boiling-range solvent-refined coal II (SRC-II) material (Harmarville, Pennsylvania process development unit) and its boiling-point distillate cuts in Ames assay, Syrian hamster embryo (SHE) cell transformation assay, initiation/promotion (I/P) assay in mouse skin, and chronic mouse skin-painting carcinogenesis assay. Responses are compared with response to SRC-II heavy distillate.

though responsible for most of the bacterial mutagenicity of SRC-II materials, are not major contributors to the dermal carcinogenicity of this bp cut.

Ames assays (Mahlum et al., 1983) of alumina-derived chemical class fractions (Later et al., 1981) of the 800 to >850°F distillates of SRC-I process solvent and the 800–850°F and >850°F distillates of SRC-II from the Harmarville, Pennsylvania PDU indicated that the microbial mutagenic activity was segregated into the alumina fraction enriched in N-PACs. Initiation/promotion data on these alumina fractions (Figs. 3.24–3.26) indicated that the N-PAC fractions, although active as mouse skin-tumor initiators, were less active than the crude bp cuts. The PAH-enriched fractions, on the other hand, were more active as tumor initiators than were the crude bp cuts. Relative potency results of chronic skin-painting studies of alumina fractions of the 750–800°F bp cut (Table 3.62) indicates that both PAH and N-PAC alumina fractions were active as dermal carcinogens. Considered together with the results of the chronic studies of nitrosated samples of this bp cut, these data suggest that the non-amino-PAH components of the N-PAC fraction (carbazoles and azaarenes) are responsible for some of the carcinogenicity of this bp cut. The results of these comparisons suggest that the Ames assay is not a good indicator of the chemical classes responsible for the majority of the dermal carcinogenic response to coal-conversion materials.

The volume of genotoxicity data on coal-conversion materials from mammalian cell culture studies is much less than that available from Ames assays. Data from SHE cell transformation assays of general crude fossil-fuel materials are in agreement with mouse skin carcinogenicity data from the same materials. SRC-II heavy distillate exposure (Sample L1243) induced a much higher rate of SHE cell transformation than did exposure to Livermore shale oil or Wilmington crude petroleum (Frazier and Andrews, 1983). However, heavy distillate also induced a higher rate of transformation than BaP or 2-aminoanthracene; these results differ from the carcinogenicity data in mouse skin (Fig. 3.5). Cell transformation activity of SRC-II heavy distillate solvent fractions in the SHE cell assay system was quite different from their carcinogenic potency in mouse skin (Fig. 3.33). Heavy distillate was more active than any of its solvent fractions; the PAHs were the most potent fraction, basic and basic tar fractions were somewhat less active, and the neutral tar was much less active than the PAH fraction.

The increased SHE cell transformation activity of SRC-II heavy distillate compared with BaP, which is approximately 400 times as potent as heavy distillate in the mouse skin-painting assay, suggests that the SHE cell system is relatively more sensitive than intact mouse skin to the genotoxic effects of complex mixtures such as heavy distillate. This is supported by the increased SHE cell-transforming activity of heavy distillate compared with its solvent fractions.

Cell transformation results on bp cuts of SRC-II whole-boiling-range material in the SHE cell assay system (Mahlum et al., 1983) are compared with in vivo carcinogenesis assay results in Fig. 3.34. As was the case in Ames assays and skin-painting assays for dermal carcinogenicity, cell transformation activity of the bp cuts increased with increased bp, with the >850°F bp cut showing considerably more activity than lower bp cuts. Further comparison of SHE cell data with skin-painting data is hindered by the lack of SHE cell data on the whole-boiling-range material.

Data from SHE cell assays of chemical class fractions obtained by HPLC from the >850°F bp cut of SRC-II blend (Sample L1111) indicated that the most SHE cell transformation activity was present in fractions high in PAHs (Mahlum et al., 1983). SHE cell activity was also detected in fractions rich in carbazoles and azaarenes, with reduced activity in the fraction highest in amino-PAHs. Although no carcinogenesis data are available on these HPLC subfractions of the >850°F bp cut, initiation/promotion data are available from alumina fractions of this bp cut (Mahlum et al., 1984). They indicate that tumor-initiating activity is greater in the PAH-rich fractions, with some activity in the fractions rich in N-PACs. Furthermore, the chronic skin-painting data from alumina fractions of the 750–800°F bp cut demonstrated dermal carcinogenic activity in the PAH and N-PAC fractions (Fig. 3.14). Certain carbazoles and azaarenes are demonstrated dermal carcinogens in mice (Clayson, 1962), and the presence of these compounds could explain the SHE cell activity of the >850°F bp cut, as well as the carcinogenicity of the N-PAC fraction of the 750–800°F and >850°F bp cuts.

The >850°F and 800–850°F bp cuts of SRC-II and HPLC fractions of these bp cuts were also assayed in a CHO cell mutation assay (Mahlum et al., 1983). This assay system measures changes in mutation rates of cultured CHO cells in response to exposure to test materials. Results of CHO assays were consistent with SHE cell transformation assay data.

The assay results discussed above suggest that there are apparently some distinct differences between in vitro and in vivo mammalian cell assay systems in their responses to the genotoxic effects of complex mixtures. However, the SHE cell transformation assay system is the most responsive in vitro mammalian cell assay to the chemical classes (PAHs, carbazoles, azaarenes) which appear to be responsible for the majority of dermal carcinogenicity and thus may be the most useful in vitro for predicting the carcinogenicity of coal-derived materials.

Comparison of the results of in vitro and in vivo assays for genotoxicity of coal-conversion materials indicates that a battery of short-term, in vitro assays can serve a useful role in choosing appropriate materials for long-term, in vivo assays and for predicting the results of the latter. The most successful strategy for future use of in vitro assay systems appears to be to use bacterial assay systems such as the Ames system together with chemical analyses to assess genotoxicity of relatively crude, complex materials. Mammalian cell assay systems should be used to evaluate less complex fractions or distillates. These results, used in concert with state-of-the-art chemical characterization techniques, should provide valuable information on the specific chemical classes or groups responsible for genotoxic effects observed.

3.4.4 Teratologic/Embryotoxic Assays

Assays for teratogenicity and embryotoxicity of drugs and chemicals are usually performed in rodents, primarily because of their short gestation period, low maintenance costs, and space requirements, relative to other mammalian species, and the large amounts of experimental data provided by large litter size. The largest drawback for using rodents as models for teratologic or embryotoxic effects in humans is the great difference in the structure and function of the rodent versus the human placenta, which results in some interspecies variation in placental transfer of chemicals to the developing fetus (Harbison, 1980). Despite this difference, the rodent is a widely accepted model for teratogenic or embryotoxic effects in humans and has been used to assay coal-conversion materials for these effects.

Development of in vitro screening assays for teratogenicity and embryotoxicity has been initiated (Wilson, 1978). However, the volume of experimental data available is insufficient to determine accurately the predictability and usefulness of these assays for teratogenicity or embryotoxicity in humans. Included among the assays being developed are those using amphibian and invertebrate embryos. One drawback of these methods is the large difference in anatomy, physiology, and embryology between egg-laying invertebrates or amphibians and placental mammals, such as rodents and humans.

Teratogenicity/embryotoxicity assays of coal-conversion materials in rats are most often performed by exposing groups of pregnant females to the test material by oral gavage or inhalation at a predetermined time during the 22–23-day gestation (dg) period, either during early (9–11 dg) or late (12–16 dg) organogenesis. Exposure over several days during organogenesis allows maximum detectability of teratogenic effects (Wilson, 1973). Longer periods of exposure have been shown to alter the teratogenic effects of some compounds by enhancing their metabolism through microsomal enzyme induction (Conney and Burns, 1972). Pregnant females are killed at 21 dg and necropsied, and the fetuses and placentas are examined for abnormalities. In postnatal studies of developmental toxicity, pregnant females are exposed to the test material and allowed to deliver offspring; dam and offspring are then examined postnatally for litter size, abnormalities in growth, and other physical and behavioral parameters of maturation.

Dominant lethal mutation assays performed in rodents involve exposing males to test material, then allowing nesting with proven fertile females over the duration of the spermatogenesis cycle. Uteri from pregnant females are then examined for presence of resorbed (dead) versus viable fetuses. The genetic basis for this assay is the induction of chromosomal damage that results in death of the embryo; this is detected as a resorbed implantation site in the uterus (Hollaender, 1973).

The *mouse spot test* is an in vivo assay that has been used for detecting somatic mutation effects of radiation (Russell and Major, 1957) and of chemicals (Russell et al., 1981). In this assay, mouse embryos, heterozygous at a number of coat-color loci, are exposed to a presumptive mutagen at a stage of development that will allow genetic events affecting the wild-type allele at these loci to occur in melanocyte-precursor cells in the epidermis. These cells will subsequently multiply to form clones that alter fur coloration. Fetuses are examined for "spots" of fur that differ in color from the rest of the coat and for other abnormalities, at birth, 12 days of age, and weaning.

Xenopus laevis (frog) embryos were used in assays for teratogenicity and embryotoxicity by Dumont et al. (1982, 1983). In this assay, groups of 50 embryos were exposed to graded concentrations of test material in water under static (nonflowing) conditions for 96 h. At selected intervals, embryos were observed, examined with a dissecting microscope, and preserved; representative tissues were later examined under light microscopy. Parameters examined included motility, pigmentation, mortality, growth, development, and the presence of anatomical abnormalities. Quantitative measurements included EC_{50} (concentration required to cause abnormalities in 50% of survivors), LC_{50} (concentration required for 50% mortality), and teratogenic index (LC_{50}/EC_{50}). Results of this assay in pure compounds have been reported to correlate well with results in mammalian teratogenicity assays (Dumont et al., 1979, 1983).

3.4.4.1 *Liquefaction—SRC*

Potential teratologic or embryotoxic effects of orally administered SRC-I and SRC-II materials have been studied in rats (Andrew et al., 1979; Pacific Northwest Laboratory, 1979, 1980). Female Wistar rats were exposed from either 7 to 11 or 12 to 16 dg to light oil (Sample L2211), wash solvent (Sample L2212), or process solvent (Sample L2213) from the SRC-I process (Ft. Lewis, Washington) or to light (Sample L1241), middle (Sample L1242), or heavy (Sample L1243) distillates from the SRC-II process (Ft. Lewis, Washington). The materials from the SRC-I materials were diluted in corn oil; SRC-II materials were administered undiluted. Rats were sacrificed at 21 dg for evaluation of embryotoxicity or were permitted to deliver offspring for postnatal monitoring of growth, physical maturation, and reflex ontogeny. Maternal mortality and embryolethality of 50% or greater were observed in some groups dosed on 7–11 dg or 12–16 dg with 3.0 g/kg body weight per day of light oil, 1.4 g/kg per day of wash solvent, or 0.7 g/kg per day of process solvent.

Malformations, including small or absent eyes, were induced in 22% of fetuses from dams given 3.01 g/kg body weight per day light oil during 7–11 dg. Cleft palate and brachydactyly (abnormal shortness of the toes) were observed in 66% of fetuses from dams dosed at 12–16 dg with 1.03 g SRC-I process solvent/kg per day. No effects on postnatal maturation were observed. The fetuses were observed for a variety of parameters and evaluated up to 35 days of age. Administration of SRC-II light distillate did not consistently induce prenatal mortality or major malformations at doses up to those that were toxic or fatal to the dam (Pacific Northwest Laboratory, 1980). Dosing with middle distillate produced some prenatal mortality but no major malformations at doses up to those that were maternally toxic. Administration of heavy distillate from 7 through 11 dg induced prenatal mortality at doses moderately toxic to the dams. Retarded intrauterine growth and fetal malformations occurred at a dose-related incidence following administration of heavy distillate at doses of 0.37, 0.73, or 1.10 g/kg per day from 12 through 16 dg. Cleft palate, hypoplastic lungs, and diaphragmatic hernias were the most frequently observed malformations.

The positive teratologic findings with SRC-II heavy distillate led to a more detailed teratology study (Pacific Northwest Laboratory, 1982), using a different strain of rat (Charles River CD). In this study, heavy distillate was diluted in milk, and somewhat lower doses of heavy distillate (Table 3.82) were administered at 12–16 dg (the period during which dosing with heavy distillate induced malformation in the initial study). Cleft palates, hypoplastic lungs, intrauterine mortality, and retarded intrauterine growth were again observed at increased frequency in litters from heavy-distillate-exposed dams. The decreases in fetal body weight and fetal lung weight appeared to be dose related; fetal lung weight expressed as a percentage of fetal body weight showed a progressive decrease with increasing dose (Fig. 3.36). Pairwise comparison showed a significant difference in this measure at the 0.14 g/kg per day dose.

A subsequent study compared the potential teratologic effects of SRC-II heavy distillate from the Ft. Lewis pilot plant with another SRC-II distillate blend (Sample L1111) from the Harmarville, Pennsylvania PDU and five bp cuts of the distillate blend (Pacific Northwest Laboratory, 1983). Pregnant CD rats were dosed on either 12–14 or 15–16 dg with 0.37 g/kg per day of heavy distillate or 0.74 g/kg per day of Harmarville distillate blend. Doses of distillate blend bp cuts, based on percentage of composition in the blend, ranged from 0.586 g/kg per day (300–700°F cut) to 0.035 g/kg per day (>850°F cut). Body and lung weights were depressed in fetuses from dams dosed with the distillate blend or heavy distillate from 12 to 14 dg. Those weights were significantly lower in fetuses from dams dosed with distillate blend from 12 through 16 dg. Fetal lung:body weight ratios were significantly lower than control values in all groups given heavy distillate or Harmarville distillate blend except those in which the dams were dosed with distillate blend on 15–16 dg. Exposure to distillate blend or heavy distillate resulted in lung hypoplasia in an average of 70% of the fetuses in all litters from exposed dams. Cleft palate occurred more frequently in fetuses from distillate-blend-dosed dams than those from heavy-distillate-dosed dams. No evidence of embryotoxicity or teratogenicity was observed in fetuses from dams dosed with any of the distillate blend bp cuts, suggesting that the malformations observed in rats dosed with distillate blend may be due to a synergistic effect of components from several bp cuts. However, the lack of response to the bp cuts may also be explained by difficulty in maintaining the bp cuts in suspension, which led to reduced bioavailability of these materials.

Teratology/embryotoxicity studies have also been performed in rats exposed to SRC-II heavy distillate (Sample L1243) by the inhalation route (Springer et al., 1982a,b). Pregnant female CD rats were exposed to an aerosol of heavy distillate at one of three concentrations

Table 3.82 Effects of Intragastric Intubation of Pregnant Rats with Solvent-Refined Coal II Heavy Distillate at 12–16 days gestation[a]

Dose (g/kg body weight)	0	0.09	0.14	0.18	0.37	0.74
Number of litters	27	20	20	26	21	3
Intrauterine mortality[b] (%)	5.1 ± 1.4	3.2 ± 1.3	9.3 ± 2.6	10.7 ± 2.3	16.6 ± 3.6	61.7 ± 31.1
Number of live pups examined	271	216	211	258	164	3
Incidence of small lungs[c]	0/9	1/1	24/4	36/9	91/18	3/2
Maternal mortality, (fraction; % in parentheses)	0/35 (0)	0/32 (0)	0/25 (0)	0/26 (0)	0.27 (0)	10/16 (67)
Maternal body weight (g at 20 dg)	334 ± 7	321 ± 8	332 ± 8	306 ± 6	281 ± 6	259 ± 8
Extragestational weight[d] (kg)	280 ± 6	273 ± 6	283 ± 4	263 ± 4	247 ± 4	251 ± 12
Maternal organ weights[e]						
Spleen	0.22 ± 0.005	0.24 ± 0.025	0.22 ± 0.006	0.23 ± 0.008	0.25 ± 0.007	0.21 ± 0.005
Thymus	0.12 ± 0.007	0.08 ± 0.004	0.07 ± 0.005	0.07 ± 0.003	0.06 ± 0.003	0.05 ± 0.007

[a]Statistically significant differences from control group are underlined.
[b]Data reported as the mean of litter means ± SE.
[c]Data are expressed as number of fetuses/number of litters.
[d]Sacrifice body weight minus weight of the uterus with products of conception.
[e]Expressed as percentage of extragestational body weight.

Source: Pacific Northwest Laboratory (1982).

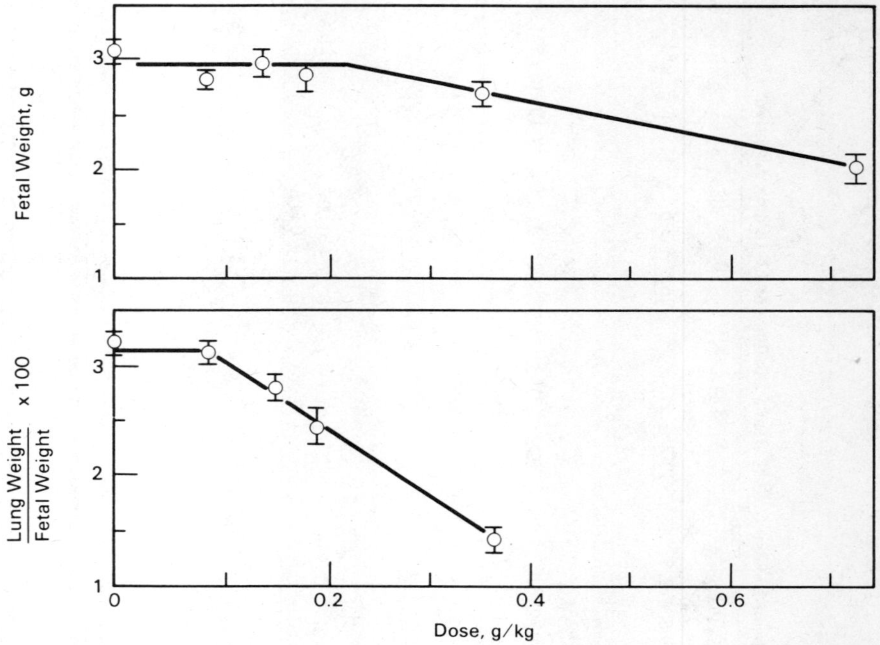

Figure 3.36 Effect of gastric intubation of solvent-refined coal II heavy distillate on fetal rat weight and relative lung (lung/body) weight (mean ± SE) (Pacific Northwest Laboratory, 1982).

for 6 h/day from 12 through 16 dg. Results of this study, summarized in Table 3.83, indicated the presence of lesions similar to those observed in the oral gavage studies of heavy distillate and Harmarville distillate blend in the high-dose group. Increased embryolethality during middle and late gestation in the group exposed to the highest concentration of heavy distillate appeared to be attributable to exposure. Fetuses from pregnant rats in the highest exposure group were smaller than fetuses from control animals, and skeletal ossification was reduced. Hypoplastic lungs and cleft palate were observed only in fetuses for the highest exposure groups, whereas dilation of the ureters, frequently observed in this strain of rat, was not present in fetuses from the highest exposure group. Body weights, at sacrifice, of dams exposed to the highest aerosol concentration of heavy distillate were significantly increased. Oral dose of heavy distillate in this inhalation exposure was estimated to be approximately 65 mg/kg per day, suggesting that the inhalation dose of heavy distillate required to produce major malformations may be lower than the oral dose. However, the authors point out that this estimated oral dose does not take into account material ingested by grooming.

Range-finding teratology studies in Charles River CD rats, using repeated dermal exposures, have been completed for two NTSL materials from the Wilsonville pilot plant (Drozdowicz and Kelly, 1983g). First-stage middle distillate (Sample L5241), at dose levels of 250, 500, 1000, 1500, and 2000 mg/kg per day, and second-stage middle distillate (Sample H0524), at dose levels of 250, 500, 750, 1000 and 2000 mg/kg per day, were applied topically, in a polyethylene glycol vehicle, to the dam's clipped dorsal skin, daily on 6–19 dg, at a constant volume of 3 mL/kg. Controls received polyethylene glycol only. Necropsy and uterine examinations were performed on all surviving females on dg 20. One

Table 3.83 Effects of Inhalation Exposure of Pregnant Rats at 12–16 days gestation to Solvent-Refined Coal II Heavy Distillate (Sample L1243)

	Exposure[a]			
Measurement	Control (0.0)	Low (0.017 ± 0.002)	Medium (0.084 ± 0.09)	High (0.66 ± 0.05)
Number of litters	23	25	25	23
Fetoplacental Measures				
Implants/litter	13 ± 2.5	13.3 ± 4.5	13.1 ± 3.1	11.6 ± 3.3
Number of resorptions[b]				
Early (7–10 dg)	24/14	22/15	26/13	15/11
Middle (11–13 dg)	0/0	3/3	0/0	8/6[f]
Late (14–18 dg)	0/0	0/0	0/0	5/4
Fetal weight (g)[c]	4.50 ± 0.49	4.33 ± 0.52	4.33 ± 0.53	3.55 ± 0.38
Crown-rump length (mm)	40 ± 2	39 ± 2	39 ± 2	36 ± 1
Placental weight (g)	0.49 ± 0.06	0.49 ± 0.10	0.46 ± 0.05	0.34 ± 0.06
Number of fetuses examined[b]	280/23	308/25	300/25	238/23
Reduced ossification	3/2	7/4	8/6	28/10
Small lungs	0/0	0/0	0/0	20/8
Cleft palate	0/0	0/0	0/0	10/2
Dilated ureter	17/11	31/18	17/9	0/0
Maternal Measures				
Body weight at sacrifice (g)	306 ± 27	312 ± 33	302 ± 29	271 ± 25
Extragestational body weight (g)[d]	231 ± 15	240 ± 22	234 ± 15	220 ± 18
Tissue weights[e]				
Lung	0.42 ± 0.02	0.43 ± 0.04	0.44 ± 0.04	0.50 ± 0.06
Spleen	0.18 ± 0.02	0.22 ± 0.03	0.20 ± 0.03	0.22 ± 0.03
Thymus	0.095 ± 0.03	0.095 ± 0.03	0.082 ± 0.03	0.040 ± 0.01

[a]Mean aerosol concentration (mg/L) ± SD.
[b]Data are expressed as number of fetuses/number of litters.
[c]Data reported as the mean of litter means ± SD.
[d]Sacrifice body weight minus weight of uterus with products of conception.
[e]Expressed as percentage of extragestational body weight.
[f]Statistically significant differences from control group are underlined.

Source: Springer et al. (1982b).

rat exposed to 2000 mg/kg per day of NTSL middle distillate and two rats exposed to 2000 mg/kg per day of second-stage middle distillate died during the study. Dermal irritation at the application site was severe at doses of 500 mg/kg, or greater, of either material. Group mean maternal body weight gain was significantly reduced in groups exposed to 1000 mg/kg per day (or higher) of NTSL middle distillate and in groups exposed to 500 mg/kg per day (or higher) of second-stage middle distillate. There were no meaningful differences in group mean maternal uterine parameters in rats exposed to NTSL middle distillate.

Effects of hydrotreatment on the embryotoxicity and teratogenicity of a blend of heavy and middle distillates of SRC-II were studied in submammalian species.

3.4.4.2 *Liquefaction—H-Coal*

With a few exceptions, embryotoxicity and teratogenicity studies of other coal-liquefaction materials have been limited to submammalian species. A series of samples from the H-Coal

Table 3.84 Teratogenic and Embryotoxic Effects on *Xenopus* Embryos of H-Coal Naphtha and Fuel Oils Derived from H-Coal, Using Various Feedstocks

Material	Sample Number	LC_{50}[a] (mg/L)	EC_{50}[b] (mg/L)	Growth (% of control)	TI[c]
Illinois No. 6 Feedstock					
Stabilizer naphtha	L3241	97.6	98.7	98	0.99
Light fuel oil	L3242	16.8	15.7	92	1.07
Heavy fuel oil	L3243	5.8	2.9	98	1.98
Wyodak Feedstock					
Stabilizer naphtha	L3251	144.8	92.2	97	1.43
Light fuel oil	L3252	21.8	37.3	104	0.59
Heavy fuel oil	L3253	8.9	12.2	102	0.75

[a] LC_{50} = concentration required to kill 50% of the embryos.
[b] EC_{50} = concentration required to cause 50% of the survivors to be abnormal.
[c] Teratogenic index (TI) = LC_{50}/EC_{50}.

Source: Munro et al. (1985).

pilot plant in Catlettsburg, Kentucky, taken during operation in the syncrude mode, were assayed for embryotoxicity and teratogenicity using frog (*Xenopus laevis*) embryos (Munro et al., 1985). Stabilizer naphtha (Sample L3241 and L3251), light fuel oil (Samples L3242 and L3252), and heavy fuel oil (Samples L3243 and L3253) samples collected during runs with two different feedstock coals were assayed. Embryolethality and teratogenicity data for these materials are summarized in Table 3.84. The greater toxicity and teratogenicity of the heavy fuel-oil samples were thought to be due to their higher PAH content. The increased embryotoxicity and teratogenicity of the fuel oils from the Illinois coal samples, compared with those of the Wyodak coal, may have been due to the concentrations of PAHs (e.g., BaP is twofold higher in concentration in the Illinois-coal-derived samples).

Other work using the *Xenopus* assay system studied the effects of several degrees of hydrotreatment on teratogenicity and embryotoxicity (Griest et al., 1983). Blended light and heavy (1 : 1.5) fuel oil (Samples L3231 and L3232) obtained from the H-Coal pilot plant during operation in the syncrude mode with Illinois No. 6 coal; EDS raw blend material (Sample L4116) consisting of No. 1 sidestream and bottoms material from the EDS Coal Liquefaction Plant (ECLP) fractionator processing recycle (hydrotreated) solvent, blended 1 : 1 by weight; and an SRC-II middle and heavy distillate blend (Sample L1261) from the Ft. Lewis pilot plant were assayed. The H-Coal and SRC-II samples were water washed prior to hydrotreatment to prevent ammonium chloride from plugging lines. Results of these assays, presented in Table 3.85, indicated that hydrotreatment reduced teratogenicity and embryolethality.

The mouse spot test for detection of mutations in somatic cells was used to assay an H-Coal 1 : 1.5 blend of light and heavy fuel oils (Sample L3232) from a run with Illinois No. 6 coal (Munro et al., 1985). The material was washed with water to remove chloride, dissolved in corn oil, and 500 or 800 mg/kg were injected intraperitoneally into 56 previously bred female C57B1 mice. A similar group of mice received corn oil alone. The frequency of animals with mutant spots was 1.6% in the treated group and 0.8% in the corn oil control group, suggesting a positive but not statistically significant effect.

Table 3.85 Teratogenicity and Embryotoxicity in *Xenopus* Embryos of Hydrotreated Synfuel Materials

Material	Nitrogen Content (ppm)	LC_{50}[a] (mg/L)	EC_{50}[b] (mg/L)	Characteristics at EC_{50} Concentration: Growth % of control	TI[c]
H-Coal blend					
Unwashed (Sample L3231)	—	6.5	11.1	93.7	0.6
Washed (Sample L3232)	—	9.0	9.9	79.4	0.9
Low hydrotreatment	2650	36.2	14.4	92.3	2.5
Medium hydrotreatment	535	58.0	46.6	103.4	1.2
High hydrotreatment	0.2	—[d]			
EDS (Sample L4116)	—	17.2	14.8	96.0	1.2
Hydrotreatment	—	—[e]	—	—	—
SRC-II distillate blend[f]					
Unwashed (Sample L1261)	—	6.9	4.3	96.9	1.6
Washed (Sample L1262)	—	7.0	4.2	88.8	1.7
Low hydrotreatment	5300	12.7	12.1	97.6	1.1
Medium hydrotreatment	1710	28.0	28.6	101.1	1.0
High hydrotreatment	505	48.7	58.8	91.4	0.8

[a]LC_{50} is the concentration required to kill 50% of the embryos.
[b]EC_{50} is the concentration required to cause 50% of the survivors to be abnormal.
[c]Teratogenic index (TI) = LC_{50}/EC_{50}.
[d]At 250 mg/L, not different from controls; insolubility problems at higher concentrations.
[e]Not different from controls at 300 mg/L; insolubility problems at higher concentrations.
[f]SRC = Solvent-refined coal II.

Source: Griest et al. (1983).

3.4.4.3 *Coal Gasification*

An aqueous extract of an ESP tar (Sample G1319) from the UMD gasifier was assayed in the *Xenopus* system at concentrations from 0.25 to 1.5% (Schultz et al., 1982; Epler et al., 1983). The acute static 96-h LC_{50} was 0.83%; the EC_{50} was 0.48%. Embryo growth, motility, and pigmentation were reduced by concentrations of 0.5% or greater. Malformations of head, eyes, viscera, and spine were frequent, and cartilage formation was abnormal. Foci of hyperplasia were present on the epidermis of exposed embryos. Severity of these abnormalities was related directly to concentration of the test material. Embryotoxic effects were attributed to phenols and the teratogenicity observed was thought to have been a synergistic effect of several compound classes or to have resulted from latent, acute, toxic effects, as suggested by the closeness of the EC_{50} and LC_{50} values.

In a subsequent study, a composite electrostatic precipitator tar (Sample G1327) from the UMD gasifier was separated into five chemical fractions and assayed in the *Xenopus* system (Cowser, 1982a; Schultz et al., 1983). Testing of each sample was also performed using newly metamorphosed frogs, in an attempt to assess toxicity to adults. Chemical class

Table 3.86 LC_{50}[a] and EC_{50}[b] Values for *Xenopus* Embryos and Frogs Exposed to Chemical Fractions of Electrostatic Precipitator Tar (Sample G1327) from the University of Minnesota–Duluth Gasifier

Material	Embryo 96-h LC_{50}[c] (mg/L)	96-h EC_{50}[c] (mg/L)	EC_{50} (mg/L)	Frog 96-h LC_{50}[d] (mg/L)	TI[e]
Tar	3.13 (2.32–4.23)	0.70 (0.31–1.57)	< 0.50	163	4.47
Acid	5.26 (4.22–6.55)	2.48 (1.49–4.11)	< 1.00	425	2.12
Base	5.34 (1.67–17.07)	1.02 (0.62–1.67)	< 0.50	299	5.24
Mono- and diaromatic	27.91 (13.20–59.00)	21.47 (15.79–29.20)	< 10.00	776	1.30
Polyaromatic	6.76 (5.21–8.77)	2.55 (1.48–4.38)	< 1.00	529	2.65
Aliphatic	—[f]	—[f]	< 10.00	—[f]	—[f]

[a]LC_{50} is the concentration required to kill 50% of the embryos.
[b]EC_{50} is the concentration required to cause 50% of the survivors to be abnormal.
[c]Calculated by least-squares methods of linear regression. All correlation coefficients are > 0.8910; 95% confidence intervals are in parentheses.
[d]Calculated by least-squares regression analyses of precentage of mortality versus concentration (mg/L).
[e]Teratogenic index (TI) = LC_{50}/EC_{50}.
[f]Effect not quantifiable at saturation level.
Source: Schultz et al. (1983).

fractions assayed and results of this work are summarized in Table 3.86. Critical examination of the data presented in Table 3.86 indicates a wide variation in embryotoxicity and teratogenicity within test groups, making it difficult to draw definite conclusions from the results presented. The data suggest that embryotoxicity and teratogenicity of the raw tar were greater than for any of its fractions. Of the five fractions, the ether-soluble base and polyaromatic fractions appeared to be the most teratogenic, and the ether-soluble base and ether-soluble acid fractions the most embryotoxic. On a weight % basis, the combined teratogenic activity of the fractions was less than that of the raw tar, suggesting that the high activity of the raw tar was due to a synergistic effect of the tar fractions. The toxic effects on frogs were one to two orders of magnitude lower than those observed for embryos.

Electrostatic precipitator tar G1319 was assayed for teratogenicity and genotoxicity in an in vivo mammalian somatic mutation assay system, the mouse spot test described earlier (Epler et al., 1983). The tar sample was diluted in corn oil and injected intraperitoneally in pregnant female mice. Doses ranged from 600 to 1000 mg/kg. In addition to decreasing survival of pups, exposure to the tar produced a slight, but apparently real, increase in the incidence of extra digits and tail kinking.

Two electrostatic precipitator tars (Samples G1301 and G1319) from the UMD gasifier were assayed in a mammalian (mouse) system for dominant lethal mutation and total reproductive capacity (Cowser, 1982a; Epler et al., 1983). With tar Sample G1301, no clear-cut dominant lethal effects were observed. There was a detectable increase in dominant-lethal mutations as determined by uterine dissection with tar Sample G1319 at all stages of spermatogenesis. In addition, average litter size was decreased in females mated within 5 days after intraperitoneal injection of 2000–2500 mg/kg of either ESP tar. It is not known whether the decrease in reproductive capacity seen with tar Sample G1319 results from dominant-lethal mutations or oocyte killing.

3.4.4.4 *Summary and Conclusions (Teratologic/Embryotoxic Assays)*

Studies in rodents suggest that some materials from coal-conversion technologies have potential for inducing embryotoxic and/or teratologic effects. Exposure to heavy distillate (Sample L1243) or distillate blend (Sample L1111) from the SRC-II process induced embryotoxicity, cleft palate, and hypoplastic lungs when administered orally to pregnant rats. Heavy distillate induced similar effects when exposure was by the inhalation route. Process solvent (Sample L2213) and light oil (Sample L2211) from the SRC-I process also induced teratologic effects when given orally to pregnant rats. A tar sample (G1319) collected from the electrostatic precipitator of the UMD gasifier induced embryotoxic and teratologic effects in mice when administered intraperitoneally.

Embryotoxicity and teratogenicity studies of other coal-conversion materials have been limited to assays in submammalian species. Exposure of frog embryos to H-Coal pilot-plant materials induced embryotoxic and teratogenic effects which were reduced when the materials were hydrotreated. Tars collected from an electrostatic precipitator at the UMD gasifier also induced malformations in frog embryos.

3.4.5 Organ Toxicity

Standard animal toxicity testing protocols usually require acute exposures by one or more routes (the choice depending in part on the anticipated route of human exposure), sensitization studies, and subchronic (90 days) or chronic exposure studies, depending on the anticipated duration of potential human exposure. Many of these studies are routinely performed in more than one animal species.

Oral gavage and/or intraperitoneal injection are the most frequent routes of exposure for acute studies, in which the chief objective is to produce data for calculating an LD_{50} (the statistically derived dose that is lethal to 50% of the exposed animals). See Weil (1952) and Litchfield and Wilcoxon (1949). These studies consist of a single exposure of groups of animals to a logarithmic range of doses of test material, diluted in a suitable vehicle such as corn oil, followed by a 7–14-day postexposure period. The LD_{50}, calculated on the basis of percentage of mortality in each dose group, is usually a hypothetical value; it may vary among species and varies widely with route of exposure. Given their many limitations, LD_{50} data should not be overinterpreted or considered highly quantitative predictors of human toxicity. Nevertheless, these values give some guidance as to the toxic response that can be expected when a material is administered by the same route of exposure as the LD_{50} assay. Table 3.87 lists categories of toxicity, ranked according to oral LD_{50} values.

Table 3.87 Categories of Toxicity from LD_{50}[a] Data Obtained by Oral Administration

Category	Oral LD_{50}
Extremely toxic	1 mg/kg or less
Highly toxic	1–50 mg/kg
Moderately toxic	50–500 mg/kg
Slightly toxic	0.5–5 g/kg
Nearly nontoxic	5–15 g/kg
Relatively harmless	> 15 g/kg

[a]The statistically derived dose that is lethal to 50% of exposed animals.

Source: Loomis (1968).

In addition to recording mortality data, clinical signs observed following exposure are often reported from LD_{50} studies. Rats and mice are most frequently used for determining the LD_{50} by oral or intraperitoneal routes, whereas rabbits are the species of choice for acute dermal toxicity studies (Loomis, 1968).

Studies of primary irritation to the skin (McCreesh and Steinberg, 1983) and eyes (McDonald et al., 1983) are frequently performed with materials that have a high potential for human exposure at these sites. Irritation studies most frequently use albino rabbits. In these studies, irritative effects of test materials at the test site are observed at several set intervals following a short exposure period and evaluated using standardized grading systems.

Sensitization studies involve a series of 8–10 epidermal applications or intradermal injections of test materials in groups of albino guinea pigs, followed by a 2–3-week waiting period, then one or several *challenge* exposures to the test material. Inflammatory response at the application site following the challenge dose is compared with the primary irritation response to determine the presence and extent of sensitization to the test material (Klecak, 1983).

Subchronic toxicity tests vary widely in duration but most frequently involve exposures for 90 days (5 days/week in parenteral routes; daily in feeding studies). Multiple doses are usually used, and larger numbers of animals per group are used than in acute studies. Histopathological examination of selected tissues is usually performed, and body weight, hematology and clinical chemistry, and food consumption data are often generated. Route of exposure is determined by the most likely anticipated human exposure route. Rats and mice are the species most often used.

Chronic non-tumor-related toxicity data are collected as part of most chronic carcinogenicity studies. Duration of exposure varies widely, but 24 months is the most frequent period of study. Usually, fewer dose groups are used than in subchronic studies, for economic reasons. Mortality and histopathology data are usually collected in detail, and clinical pathology measurements are sometimes performed.

3.4.5.1 *Oral and Intraperitoneal Exposure*

Acute oral and intraperitoneal toxicity data are available for selected process materials and/or products from several coal-conversion technologies (Table 3.88). Oral LD_{50} values ranged from 0.57 to >16 g/kg, indicating that these materials range from slightly toxic to relatively harmless in exposure by the oral route. Data in Table 3.89 indicate that hydrotreatment of crude synfuel products decreases acute oral toxicity.

Convulsion and tremors were observed in rats dosed orally with SRC-I wash solvent (Sample L2212) or light oil (Sample L2211; Pacific Northwest Laboratory, 1979), in mice exposed orally to unwashed or washed SRC-II distillate blend (Samples L1261 and L1262; Griest et al., 1983), in mice exposed orally to light fuel oil (Sample L3212) or heavy fuel oil (Sample L3213) from the H-Coal pilot plant (Munro et al., 1985), and in rats dosed orally with raw naphtha (Sample L4115) from the EDS ECLP pilot plant (McKee et al., 1984). Convulsions were also observed in mice exposed by intraperitoneal injection to heavy fuel oils from the H-Coal pilot plant.

3.4.5.2 *Dermal and Ocular Exposure*

Dermal and ocular toxicity data available for coal-conversion materials include those from skin and eye irritation studies; acute dermal toxicity studies, many of which were preliminary or dose-setting studies for chronic carcinogenesis assays; delayed hypersensitivity as-

Table 3.88 LD_{50}[a] Values for Fossil-Fuel Materials Administered to Rodents Orally and/or Intraperitoneally

Material	Oral LD_{50} (g/kg)	95% Confidence Intervals	IP LD_{50} (g/kg)	95% Confidence Inververvals	Reference
COED oil (Sample G2211)	> 5	—	—	—	Witschi (1978)
H-Coal PDU[b] (syncrude mode)					Epler et al. (1981)
Atmospheric bottoms (Sample L3142)	3.6	2.4–5.2	—	—	
Vacuum overhead (Sample L3143)	2.5	1.7–3.1	—	—	
H-Coal PDU (fuel-oil mode)					Epler et al. (1981)
Atmospheric overhead (Sample L3151)	5.8	4.7–7.2	—	—	
Atmospheric bottoms (Sample L3152)	2.3	1.9–2.6	—	—	
Vacuum overhead (Sample L3153)	2.6	2.2–3.2			
H-Coal pilot plant (syncrude mode)					Cowser (1982a)
Light fuel oil (Sample L3212)	3.2	2.5–4.0	2.3	1.6–3.3	
Heavy fuel oil (Sample L3213)	2.8	2.2–3.6	2.8	2.2–3.6	
Light slop oil (Sample L3218)	>16.0	—	>16.0	—	
Light fuel oil (Sample L3242)	2.8	1.9–4.2	3.7	3.0–4.7	
Heavy fuel oil (Sample L3243)	2.3	1.7–3.0	2.0	1.5–2.7	
Light slop oil (Sample L3246)					
Aqueous	>16.0	—	>16.0	—	
Tar	16.0	—	11.3	—	
Light fuel oil (Sample L3252)	3.7	2.6–5.4	2.8	2.0–3.9	
Heavy fuel oil (Sample L3253)	2.8	—	5.7	3.7–8.6	
Light slop oil (Sample L3256)	>16.0	—	12.7	10.1–16.0	
Hydroclone overhead (Sample L3255)	>16	—	9.0	6.7–12.1	
Heavy slop oil (Sample L3257)	>16	—	12.7	8.2–13.2	
H-Coal distillate (Sample 3112)	3.6	2.8–4.5	—	—	Epler et al. (1981)
Low-severity hydrotreatment	4.0	3.4–4.7	—	—	
High-severity hydrotreatment	5.5	3.8–7.2	—	—	
SRC-II[c]					DOE (1982b)
Coal slurry (Sample L1265)	>5.0	—	—	—	
Stripper tower bottoms (Sample L1263)	>5.0	—	—	—	

(*continued*)

Table 3.88 (*Continued*)

Material	Oral LD_{50} (g/kg)	95% Confidence Intervals	IP LD_{50} (g/kg)	95% Confidence Invervals	Reference
Middle distillate (Sample L1264)	1.8	—	—	—	
Heavy distillate (Sample L1242)	3.0	—	—	—	Mahlum (1981)
Middle distillate (Sample L1242)	3.8	—	—	—	
Light distillate (Sample L1241)	2.3	—	—	—	
SRC-I	0.6	—	—	—	Mahlum (1981)
Wash solvent (Sample L2212)					
Process solvent (Sample L2213)	2.8	—	—	—	
Light oil (Sample L2211)	2.9	—	—	—	
NTSL[d] Middle distillate				—	Drozdowicz and Kelly (1983f)
First stage (Sample L5241)	4.6	3.8–5.4	—		
Second stage (TSL) (Sample H0524)	5.3	3.9–7.0	—	—	
EDS (ECLP) pilot plant					McKee et al. (1984)
Raw naphtha (Sample L4115)	5.0	3.2–7.8	—	—	
Extracted naphtha (Sample L4114)	>5.0	—	—	—	
Recycle solvent (Sample L4111) (once through mode)	>5.0	—	—	—	
Recycle solvent (Sample L4132) (bottoms recycle mode)	>5.0	—	—	—	
Fuel oil (Sample L4113)	>5.0	—	—	—	
UMD[e] electrostatic precipitator tar					Epler et al. (1983)
Sample G1301	5.0	3.2–7.9	2.2	1.7–3.0	
Sample G1306	5.0	3.4–7.2	1.1	0.8–1.5	
Sample G1319	6.4	5.0–8.0	3.2	2.5–4.0	
Hygas recycle oil (Sample G2113)	1.3	—	—	—	Reilly et al. (1986)
Wilmington crude petroleum	>16	—	—	—	Epler et al. (1981)

[a]The statistically derived dose that is lethal to 50% of exposed animals.
[b]Process development unit.
[c]Solvent-refined coal II.
[d]Nonintegrated Two-Stage Liquefaction.
[e]University of Minnesota–Duluth.

Table 3.89 LD_{50}[a] Values for Hydrotreated Fossil-Fuel Materials Administered to Rodents Orally

Material	Oral LD_{50} (g/kg)	95% Confidence Intervals
H-Coal blend		
Unwashed (Sample L3231)	4.0	2.9–5.5
Washed (Sample L3232)	2.5	1.7–3.7
Low-severity hydrotreatment	3.6	2.7–4.8
Medium-severity hydrotreatment	11.3	
High-severity hydrotreatment	8.0	5.9–10.9
H-Coal distillate (Sample L3112)	3.6	2.8–4.5
Low-severity hydrotreatment	4.0	3.4–4.7
High-severity hydrotreatment	5.5	3.8–7.2
EDS (ECLP)[b] Blend (Sample L4116)	5.0	4.0–6.4
Hydrotreatment	14.3	10.6–19.0
SRC-II[c] distillate blend		
Unwashed (Sample L1261)	2.5	1.4–4.7
Washed (Sample L1262)	2.8	2.0–3.9
Low-severity hydrotreatment	5.7	4.1–7.8
Medium-severity hydrotreatment	5.0	4.0–6.4
High-severity hydrotreatment	7.1	5.3–9.6

[a]The statistically derived dose that is lethal to 50% of exposed animals.
[b]EDS Coal Liquefaction Plant.
[c]Solvent-refined coal II.

Source: Adapted from Epler et al. (1981) and Griest et al. (1983).

says in guinea pigs; and chronic dermal toxicity data collected during assays for dermal carcinogenesis.

Crude products from the syncrude mode of the COED (Sample G2211) and Synthoil (Sample L6111) liquefaction processes were assayed for acute dermal toxicity in rats and for dermal and ocular irritation in rabbits (Witschi, 1978). Undiluted materials were applied dermally to rabbits at a dose of 2 g/kg and to rabbits' eyes at a total volume of 0.1 mL, with no evidence of toxicity. Dermal studies in rats also indicated a low degree of dermal toxicity for these materials.

Chronic dermal exposure of mice to an atmospheric still overheads material (Sample L3151) from the H-Coal PDU during operation in the fuel-oil mode resulted in clinical signs of neurotoxicity (convulsions) after 8 months of exposure to 124 or 68 mg/week (Epler et al., 1981). Chronic dermal exposure to 50 μL of a 50% solution in acetone : cyclohexane of hydrotreated distillate (Sample L3112) from the H-Coal PDU induced systemic toxicity reflected in body weight depression and increased water consumption greater than in mice exposed to the raw distillate (Cowser, 1981).

A number of H-Coal pilot-plant materials (Catlettsburg, Kentucky) were tested for acute dermal toxicity in rats and rabbits, ocular toxicity in rabbits, and delayed hypersensitivity in guinea pigs (Munro et al., 1985). Light fuel-oil materials (Samples L3212, L3242, and L3252) from process runs with three different coal types all produced some form of slight temporary dermal irritation (e.g., reddening). Heavy fuel-oil material from two of the process runs (Samples L3213 and L3253) also had slight temporary reactions, while heavy fuel oil (Sample L3243) from the process run using Illinois No. 6 coal had no overt effect on rabbit skin. Light slop-oil materials (Samples L3218, L3246, and L3256) produced no response. Heavy slop oil (Sample L3219) and hydroclone overhead (Samples L3214 and L3247) from process runs with Kentucky No. 9 coal and Illinois No. 6 coal were also negative. Slight temporary reactions were produced with heavy slop oil (Sample L3257) and hydroclone overhead (Sample L3255) from the process run with Wyodak coal. In addition, materials from the run with Kentucky No. 9 coal (Samples L3214, L3218, L3219, L3213, and L3212) were nonirritants to rat skin (Cowser, 1982a).

Light fuel oil (Samples L3212, L3242, and L3252), heavy fuel oil (Samples L3213, L3243, and L3253), light slop oil (Samples L3218, L3246, and L3256), and hydroclone overhead (Sample L3255) materials were all negative for hypersensitivity. Only one material, heavy slop oil (Sample L3257), induced delayed hypersensitivity.

Light fuel oil (Samples L3212, L3242, and L3252), heavy fuel oil (Samples L3213, L3243, and L3253), and light slop oil (Samples L3218, L3246, and L3256) were also evaluated for acute eye irritation in rabbits (Munro et al., 1985). All three light slop oils were negative. All other materials, with the exception of the light fuel oil (Sample L3242) from the process run on Illinois No. 6 coal, induced temporary irritation ranging from slight to moderate. The light fuel oil induced moderate to severe irritation, with mild corneal opacity.

In chronic dermal carcinogenesis studies of stabilizer naphtha (Sample L3211), light fuel oil (Sample L3212), and heavy fuel oil (Sample L3243) from the H-Coal pilot plant in the syncrude mode of operation, local skin irritation at the application site was observed with chronic exposure to all test materials; it was most severe in the groups exposed to the higher concentrations (Cowser, 1982b).

A series of acute and subchronic dermal irritation, toxicity and sensitization studies, and acute ocular irritation studies were performed on several NTSL materials from the Wilsonville pilot plant (Drozdowicz and Kelly, 1983f). First-stage middle distillate (Sample L5241) was found to be moderately irritating to the rabbit eye even when washed out soon after instillation and caused severe ocular irritation, with corneal lesions, if left in the eye. In a 4-week (5 days/week) dermal toxicity study in rats, dermal irritation at the application site and reduced rate of body weight gain were observed at the lowest dose (312 mg/kg per day). Exposure-related observations at higher doses (625–5000 mg/kg per day) included the signs described above, plus scarring and subcutaneous hemorrhage at the application site. There was 100% mortality after 3 days dermal exposure to a dose of 5 g/kg per day and 40% mortality after 3 days exposure to 2.5 g/kg per day. In the guinea pig dermal sensitization assay, first-stage middle distillate was found to be a ''probable'' skin sensitizer.

Second-stage middle distillate (Sample H0524), assayed for dermal and ocular toxicity in the same manner as the first-stage middle distillate, was found to be only mildly irritating to the rabbit eye, even when not washed out. However, in the 4-week dermal toxicity study, second-stage middle distillate was also found to be irritating to rat skin, with lesions similar to those described for first-stage middle distillate. Mortality was 100% after 3 days exposure to 5 g/kg per day and 30% in the group exposed to 2.5 g/kg per day. Second-stage middle distillate was also found to be a probable skin sensitizer in the dermal sensitization study in guinea pigs.

NTSL solid fuel (Sample L5232) was minimally irritating to rabbit eyes even when washed out and mildly irritating when not washed out; second-stage solid fuel (Sample H0523) was mildly irritating in both cases. NTSL solid fuel did not cause dermal sensitization, whereas second-stage solid fuel induced dermal sensitization in the guinea pig assay and was classified as a skin sensitizer from these data.

An NTSL process intermediate light oil or naphtha (Sample H0521) was also assayed for dermal and ocular irritation in rabbits and for dermal sensitization in guinea pigs. Light oil (>1.0 g/kg body weight) had only a low dermal toxicity in rabbits but was extremely irritating to the rabbit eye, causing severe corneal injury. Light oil was also found to be a marked dermal sensitizer in the guinea pig.

Stripper tower bottoms (Sample L1263), middle distillate product (Sample L1264), and a coal slurry (Sample L1265) from the SRC-II pilot plant at Ft. Lewis, Washington were assayed for dermal and oral toxicity and acute dermal and ocular irritation in rabbits and for dermal sensitization in guinea pigs (DOE, 1982b). Systemic toxicity was low by dermal or oral exposure. Slight irritation was observed in rabbit skin following topical application of the coal slurry or stripper tower bottoms; middle distillate induced a slight to marked irritation, and the exposed skin was thickened at necropsy. Mild ocular irritation was observed in response to stripper tower bottoms. Middle distillate and coal slurry induced diffuse corneal opacity and mild to moderate conjunctival irritation. The reaction with coal slurry was greater when it was washed out of the eye shortly after exposure. Mild dermal sensitization was observed in guinea pigs exposed to all test materials.

Evidence of skin irritation was observed with repeated exposures in carcinogenesis assays of SRC-II materials in mice (Renne et al., 1981). Hair loss, diffuse thickening, and hyperpigmentation of the skin at the application site were observed, after several months, in mice exposed three times per week to 22.8 mg of SRC-II heavy distillate (Sample L1243) or 20.0 mg of SRC-II light distillate (Sample L1241) in acetone. Similar lesions were observed in mice exposed to 7.5 mg of basic tar or 5.25 mg of neutral tar fraction of heavy distillate in acetone : methanol 5 days/week for 4 weeks. These lesions were also observed in mice exposed three times per week to the basic or neutral tar fraction in the chronic carcinogenicity assay. At microscopic examination, chronic inflammation and thickening of the skin were present in mice exposed to the highest concentration of heavy distillate, light distillate, or the two tar fractions.

Naphtha (Sample L4115), recycle solvent (Sample L4111), fuel oil (Sample L4113) from the ECLP, and recycle solvent (Sample L4121) from the EDS coal-liquefaction pilot plant (CLPP) were assayed for dermal and ocular irritation (McKee et al., 1984). Mild to moderate transient ocular irritation was observed in response to all test materials; evidence of irritation was nearly absent by 14 days after exposure. The most irritating material was raw naphtha. Eye-washing reduced the severity and persistence of irritation induced by the naphtha. All EDS test materials induced considerable dermal irritation, which persisted throughout the 14-day postexposure period. The most severe dermal irritation was observed in response to raw naphtha; this was thought to be due to phenol present in this material.

Samples of three ESP tars (Samples G1301, G1306, and G1319) from the UMD gasifier were assayed for acute toxicity in rats, for dermal and ocular irritation in rabbits, and for sensitization in guinea pigs (Epler et al., 1983). No acute dermal toxicity was observed in rats. Moderate, reversible ocular lesions and slight edema of exposed skin were observed in response to all three tars in rabbits. Assays for sensitization were negative for all three tars. Chronic dermal exposure of mice to an ESP tar from the UMD gasifier in a dermal carcinogenesis assay induced dermatitis at the higher doses.

A number of Hygas coal-gasification process materials were assayed in rabbits for acute dermal and ocular toxicity and in guinea pigs for sensitization (Reilly et al., 1986). The unfractionated recycle oil (Sample G2113) and its nonvolatile organic fractions and the nonvolatile fraction of a char slurry oil (Sample G2114) induced moderate to severe erythema and edema. The unfractionated char slurry induced only a slight to moderate response. The severity of the reaction was dose dependent. Substantial to severe ocular irritation, including chronic inflammatory reaction with corneal ulceration, was induced by the nonvolatile organic fractions of recycle oil (Sample G2111) and char slurry oil (Sample G2125). A low-temperature reactor condensate nonvolatile organic sample (G2119) was a moderate eye irritant; inflammatory reactions cleared within 7 days after exposure. Recycle oil (Sample G2111) and low-temperature reactor gas condensate (Sample G2119) were markedly positive for dermal sensitization.

Acute dermal toxicity studies in mice and acute ocular irritation studies in rabbits were performed on a process tar (Sample G1121) from the UNDERC gasifier (Wilzbach et al., 1982). No dermal toxicity was noted in mice exposed dermally to a 50% concentration of tar three times per week for 2 weeks. Transient irritation resulted from a single ocular exposure to the tar.

3.4.5.3 *Inhalation Exposure*

Acute inhalation toxicity studies of process solvent (Sample L2213) from the SRC-I pilot plant at Ft. Lewis, Washington were performed in guinea pigs (Loscutoff et al., 1983). This work involved assessment of pulmonary damage from exposure to aerosols of process solvent, at several concentrations, by histamine aerosol challenge tests, pulmonary function evaluations, and morphological evaluation of lung tissue. Measurable changes in pulmonary function were produced after 8–12 days of exposure to 0.15 or 0.19 mg/L process solvent. Hypersensitivity to aerosolized histamine occurred after 12 days of exposure to process solvent aerosols; microscopic examination of lung tissues after 12 days of exposure revealed the presence of pneumonia.

Acute (single, 4-h exposure) inhalation studies were performed in rats on two test materials from the NTSL plant at Wilsonville, Alabama (Drozdowicz and Kelly, 1983f). Groups of five male and five female rats were exposed for 4 h to a dust aerosol of NTSL solid fuel (Sample L5232) at a mean concentration of 5.1 mg/L or second-stage solid fuel (Sample H0523) at a mean concentration of 5.3 mg/L. Dyspnea was observed during exposures to NTSL solids; no clinical signs were observed during exposure to Kerr–McGee ash concentration. Female rats exposed to NTSL solid fuel gained less weight than controls; no other signs were observed in any group. Lung lesions, assumed to be inhaled test material, were grossly visible at necropsy of all test animals.

Middle distillate (Sample L1264) from the SRC-II pilot plant at Ft. Lewis was assayed for acute inhalation toxicity using single, 4-h inhalation exposures in rats (DOE, 1982b). Exposure concentrations of 2.8, 3.0, 3.4, 4.5, and 5.0 mg/L were used. The calculated LC_{50} was 3.3 mg/L, with 95% confidence limits of 2.8–3.9 mg/L. Ocular lesions were grossly observed in animals surviving exposure to the four highest concentrations.

A subchronic inhalation exposure of rats and mice to aerosolized SRC-II heavy distillate (Sample L1243) was described previously (Section 3.4.2.3). Tissues from selected groups of rats and mice were examined for pathological changes at 5 and 13 weeks of exposure. Evidence of toxicity was observed in the nose, lungs, liver, spleen, bone marrow, kidneys, and intestinal tract in rats exposed for 13 weeks to 0.14 or 0.69 mg heavy distillate/L for 6 h/day, 5 days/week (Miller et al., 1984). Severity and incidence of these lesions in rats appeared to be dose related. Similar liver changes were observed in exposed mice.

Raw (Sample L4115), methanol-extracted (Sample L4114), and hydrotreated (Sample L4135) naphthas and a single sample of recycle solvent (Sample L4111) from ECLP were assayed for acute inhalation toxicity in Sprague–Dawley rats and Swiss–Webster mice (McKee et al., 1984). The animals received a single 1-h exposure to a concentration of 5 mg/L or to the maximum attainable concentration of the test material. Naphtha atmospheres were generated as vapors by heating; the recycle solvent atmosphere was generated as an aerosol. Animals were observed for 14 days following exposure. Labored breathing and evidence of eye irritation were observed during exposure to the recycle solvent at a concentration of 2.7 mg / L, and one of 10 mice died. Moderate alopecia was observed in four of nine surviving mice. No clinical signs of toxicity were observed in animals exposed to the naphthas at concentrations of approximately 5 mg / L, and no mortality occurred in these groups.

3.4.5.4 *Summary and Conclusions (Organ Toxicity)*

Data from acute oral or intraperitoneal toxicity assays of coal-conversion materials in rodents indicate a wide range of toxicity, varying from slightly toxic to relatively harmless. These data also indicate that hydrotreatment decreases acute oral toxicity of these materials.

Data from dermal and ocular toxicity and dermal sensitization studies in rodents also indicate a wide range of toxicity of coal-conversion materials. Varying degrees of dermal irritation were observed in response to repeated exposures to a number of coal-conversion materials assayed for carcinogenicity in mice. This irritation was seen especially when test materials were applied in high concentrations. Evidence of neurotoxicity and renal toxicity were observed in mice exposed chronically to some H-Coal materials.

Acute inhalation toxicity studies of SRC materials in rodents indicated adverse effects of exposure on pulmonary function and clinical and pathological effects on the lungs of exposed animals. Subchronic exposure induced pathological changes in several organ systems.

REFERENCES

Ames, B. N., J. McCann, and E. Yamasaki, Methods for Detecting Carcinogens and Mutagens with the *Salmonella*/Mammalian-Microsome Mutagenicity Test. *Mutat. Res.*, **31**, 347–364 (1975).

Andrew, F. D., D. D. Mahlum, and M. R. Petersen, Developmental Toxicity of Solvent Refined Coal-Related Hydrocarbons. *Toxicol. Appl. Pharmacol.*, **48,** A27 (1979).

Ashman, L. K., A. W. Murray, M. G. Cook, and I. Kotlarski, Two-Stage Skin Carcinogenesis in Sensitive and Resistant Mouse Strains. *Carcinogenesis*, **3,** 99–102 (1982).

Ball, J. K., and J. A. McCarter, A Study of Dose and Effect in Initiation of Skin Tumours by a Carcinogenic Hydrocarbon. *Br. J. Cancer*, **14,** 577–590 (1960).

Benson, J. M., J. O. Hill, C. E. Mitchell, G. J. Newton, and R. L. Carpenter, Toxicological Characterization of the Process Stream from an Experimental Low-Btu Coal Gasifier. *Arch. Environ. Contam. Toxicol.*, **11,** 363–371 (1982a).

Benson, J. M., C. E. Mitchell, R. E. Royer, C. R. Clark, R. L. Carpenter, and G. J. Newton, Mutagenicity of Potential Effluents from an Experimental Low-Btu Coal Gasifier. *Arch. Environ. Contam. Toxicol.*, **11,** 547–551 (1982b).

Benson, J. M., R. L. Hanson, R. E. Royer, C. R. Clark, and R. F. Henderson, Toxicological and Chemical Characterization of the Process Stream Materials and Gas Combustion Products of an Experimental Low-Btu Coal Gasifier. *Environ. Res.*, **33,** 396–412 (1984).

Berenblum, I., The Cocarcinogenic Action of Crotin Resin. *Cancer Res.*, **1,** 44–48 (1941).

Bernfeld, P., Looking Back on 20 Years of an Experimental Protocol on "Mouse-Skin-Painting Studies." *Prog. Exp. Tumor Res.*, **26,** 18–42 (1983).

Bingham, E., and H. L. Falk, Environmental Carcinogens: The Modifying Effect of Cocarcinogens on the Threshold Response. *Arch. Environ. Health*, **19,** 774–783 (1969).

Bingham, E., R. P. Trosset, and D. Warshawsky, Carcinogenic Potential of Petroleum Hydrocarbons: A Critical Review of the Literature. *J. Environ. Pathol. Toxicol.*, **3,** 483–563 (1979).

Blanding, F. H., W. H. King, W. Priestley, J. Rehner, and N. J. Lindern, Properties of High-Boiling Petroleum Products. Quantitative Analysis of Tumor-Response Data Obtained from the Application of Refinery Products to the Skin of Mice. *Arch. Ind. Hyg. Med.*, **4,** 335–345 (1951).

Bock, F. G., Species Differences in Penetration and Absorption of Chemical Carcinogens. *Natl. Cancer. Inst. Monogr.*, **10,** 361–372 (1963).

Bogovski, P., Tumors of the Skin, in V. S. Turosov (ed.), *Pathology of Tumors in Laboratory Animals,* Vol. II, *Tumors of the Mouse*, International Agency for Research on Cancer, Lyon, France, 1979, pp. 1–42.

Bourne, S., A. M. Jirka, and M. J. Peak, Chemical and Biological Characterization of High-Btu Coal Gasification (The HYGAS Process): III. Mass and Mutagenicity Balances in Chemical Fractionations, in D. D. Mahlum, R. H. Gray, and W. D. Felix (eds.), *Coal Conversion and the Environment: Chemical, Biomedical, and Ecological Considerations*, CONF-801039, National Technical Information Service, Springfield, VA, 1981, pp. 173–180.

Boutwell, R. K., Some Biological Aspects of Skin Carcinogenesis, in F. Homburger (ed.), *Progress in Experimental Tumor Research*, Vol. 4, Karger, Basel, Switzerland, 1964, pp. 207–250.

Brusick, D. J., and R. R. Young, *Procedures Manual Level 1 Environmental Assessment Biological Tests*, EPA-600/8-81-024, Environmental Protection Agency, Research Triangle Park, NC, 1981.

Calkins, W. H., J. F. Deye, R. W. Hartgrove, C. F. King, and D. F. Krahn, Synthetic Crude Oils from Coal: Mutagenicity and Tumor-Initiation Screening Tests, in D. D. Mahlum, R. H. Gray, and W. D. Felix (eds.), *Coal Conversion and the Environment: Chemical, Biomedical, and Ecological Considerations*, CONF-801039, National Technical Information Service, Springfield, VA, 1981, pp. 462–470.

Calkins, W. H., J. F. Deye, R. W. Hartgrove, C. F. King, and D. F. Krahn, Mutagenesis and Skin Tumour Initiation by Shale- and Coal-Derived Oils. *Fuel*, **62,** 857–864 (1983).

Carnes, B. A., S. S. Dornfeld, and M. J. Peak, A Quantitative Comparison of a Percentile Rule with a 2-Fold Rule for Assessing Mutagenicity in the Ames Assay. *Mutat. Res.*, **147,** 15–21 (1985).

Carrano, A. V., L. H. Thompson, D. G. Stetka, J. L. Minkler, J. A. Mazrimas, and S. Fong, DNA-Crosslinking, Sister-Chromatid Exchange and Specific-Locus Mutations. *Mutat. Res.*, **63,** 175–188 (1979).

Clayson, D. B., *Chemical Carcinogenesis*, J. and A. Churchill, Ltd., London, 1962.

Clive, D., and J. F. S. Spector, Laboratory Procedure for Assessing Specific Locus Mutations at the TK Locus in Cultured L5178Y Mouse Lymphoma Cells. *Mutat. Res.*, **31,** 17–29 (1975).

Clive, D., K. O. Johnson, J. F. S. Spector, A. G. Batson, and M. M. M. Brown, Validation and Characterization of the L5178Y/TK^{+1} Mouse Lymphoma Mutagen Assay System. *Mutat. Res.*, **59,** 61–108 (1979).

Conney, A. H., and J. J. Burns, Metabolic Interactions Among Environmental Chemicals and Drugs. *Science*, **178**, 576–586 (1972).

Countryman, P. I., and J. A. Heddle, The Production of Micronuclei from Chromosome Aberrations in Irradiated Cultures of Human Lymphocytes. *Mutat. Res.*, **41,** 321–332 (1976).

Cowser, K. E., *Life Sciences Synthetic Fuels Semiannual Progress Report for Period Ending June 30, 1981*, ORNL/TM-7926, Oak Ridge National Laboratory, Oak Ridge, TN, 1981.

Cowser, K. E., *Life Sciences Synthetic Fuels Semiannual Progress Report for the Period Ending December 31, 1981*, ORNL/TM-8229, Oak Ridge National Laboratory, Oak Ridge, TN, 1982a.

Cowser, K. E., *Life Sciences Synthetic Fuels Semiannual Progress Report for the Period Ending June 30, 1982*, ORNL/TM-8441, Oak Ridge National Laboratory, Oak Ridge, TN, 1982b.

Cresia, D. A., and P. Nettesheim, Respiratory Cocarcinogenesis Studies with Ferric Oxide: A Test Case of Current Experimental Models, in E. Karbe and J. F. Park (eds.), *Experimental Lung Cancers: Carcinogenesis and Bioassays*, Springer-Verlag, Frankfurt, Germany, 1974, pp. 234–245.

Cunningham, M. L., D. A. Haugen, F. R. Kirchner, and C. A. Reilly, Jr., Toxicologic Responses to a Complex Coal Conversion By-Product: Mammalian Cell Mutagenicity and Dermal Carcinogenicity, in M. D. Waters, S. S. Sandhu, J. Lewtas, L. Claxton, G. Strauss, and S. Nesnow (eds.), *Short-Term Bioassays in the Analysis of Complex Environmental Mixtures*, Vol. IV, Plenum Press, New York, 1985, pp. 113–123.

De Marini, D. M., P. A. Brimer, and A. W. Hsie, Cyctotoxicity and Mutagenicity of Coal Oils in the CHO/HGPRT Assay. *Environ. Mutagen*, **6,** 517–527 (1984).

Department of Energy, *Status of Health and Environmental Research Relative to Solid Wastes from Coal Conversion*, DOE/NBB-0008/1, National Technical Information Service, Springfield, VA, 1982a.

Department of Energy, *Solvent Refined Coal (SRC) Process: Health Programs*, DOE/ET/10104-56, National Technical Information Service, Springfield, VA, 1982b.

Department of Energy, *Solvent Refined Coal (SRC) Process: Health Programs*, DOE/ET/10104-57, National Technical Information Service, Springfield, VA, 1982c.

Deprospo, J. R., *Comparative Safety Assessment Studies of Three Distillate Fractions Obtained from Coal-Derived Pyrolysis and Natural Crude Oils*, M. S. Thesis, Drexel University, Philadelphia, PA, 1982.

DiPaolo, J. A., R. L. Nelson, and P. J. Donovan, Morphological, Oncogenic, and Karyological Characteristics of Syrian Hamster Embryo Cells Transformed in Vitro by Carcinogenic Polycyclic Hydrocarbons. *Cancer Res.*, **31,** 1118–1127 (1971).

Doll, R., M. P. Vessey, R. W. R. Beasley, A. R. Buckley, E. C. Fear, R. E. W. Fisher, E. J. Gammon, W. Gunn, G. O. Hughes, K. Lee, and B. Norman-Smith, Mortality of Gasworkers—Final Report of a Prospective Study. *Br. J. Ind. Med.*, **29,** 394–406 (1972).

Draper, N. R., and H. Smith, *Applied Regression Analysis*, 2nd ed., Wiley, New York, 1966.

Drozdowicz, B. Z., and C. M. Kelly, *Interim Report on the Genetic and Animal Toxicity Testing of SRC-I Products, Intermediates, and Waste Materials, Appendix A, Microbial Mutagenicity Evaluations (Ames Test)—Basic Program Reports*, DOE/OR/03054-50, National Technical Information Service, Springfield, VA, 1983a.

Drozdowicz, B. Z., and C. M. Kelly, *Interim Report on the Genetic and Animal Toxicity Testing of SRC-I Products, Intermediates, and Waste Materials, Appendix A-1, Microbial Mutagenicity Evaluations (Ames Test)—Option IA Reports*, DOE/OR/03054-50, National Technical Information Service, Springfield, VA, 1983b.

Drozdowicz, B. Z., and C. M. Kelly, *Interim Report on the Genetic and Animal Toxicity Testing of SRC-I Products, Intermediates, and Waste Materials, Appendix A-2, Microbial Mutagenicity Evaluations (Ames Test)—Option IB Reports*, DOE/OR/03054-50, National Technical Information Service, Springfield, VA, 1983c.

Drozdowicz, B. Z., and C. M. Kelly, *Interim Report on the Genetic and Animal Toxicity Testing of SRC-I Products, Intermediates, and Waste Materials, Appendix B, In Vivo Cytogenetics and In Vitro Mammalian Mutagenicity (Mouse Lymphoma Test)—Basic Program Reports*, DOE/OR/03054-50, National Technical Information Service, VA, 1983d.

Drozdowicz, B. Z., and C. M. Kelly, *Interim Report on the Genetic and Animal Toxicity Testing of SRC-I Products, Intermediates, and Waste Materials, Appendix B-1, In Vitro Mammalian Mutagenicity (Mouse Lymphoma Test) Evaluations—Option IIA Reports*, DOE/OR/03054-50, National Technical Information Service, Springfield, VA, 1983e.

Drozdowicz, B. Z., and C. M. Kelly, *Interim Report on the Genetic and Animal Toxicity Testing of SRC-I Products, Intermediates, and Waste Materials, Appendix D, Acute Animal Studies Reports*, DOE/OR/03054-50, National Technical Information Service, Springfield, VA, 1983f.

Drozdowicz, B. Z., and C. M. Kelly, *Interim Report on the Genetic and Animal Toxicity Testing of SRC-I Products, Intermediates, and Waste Materials, Appendix E, Subchronic Dermal Studies Reports*, DOE/OR/03054-50, National Technical Information Service, Springfield, VA, 1983g.

Drozdowicz, B. Z., and C. M. Kelly, *Interim Report on the Genetic and Animal Toxicity Testing of SRC-I Products, Intermediates, and Waste Materials, Appendix G, Sample History and Documentation*, DOE/OR/03054-50, National Technical Information Service, Springfield, Va, 1983h.

Dumont, J. N., T. W. Schultz, and R. D. Jones, Toxicity and Teratogenicity of Aromatic Amines to *Xenopus laevis*. *Bull. Environ. Contam. Toxicol.*, **22,** 159–166 (1979).

Dumont, J. N., T. W. Schultz, and S. M. Newman, A Frog Embryo Teratogenesis Assay: *Xenopus* (FETAX)—A Model for Teratogen Screening. *Teratology*, **25,** 37A (1982).

DuMont, J. N., T. W. Schultz, and J. L. Epler, The Response of the FETAX Model to Mammalian Teratogens. *Teratology*, **27,** 39A (1983).

Dunn, T. B., W. E. Heston, and M. K. Deringer, Spontaneous Fibrosarcomas in Strains C3H and C57Bl Female Mice, and F_1 and Backcross Hybrids of These Strains. *J. Natl. Cancer Inst.*, **17,** 639–655 (1956).

Epler, J. L., The Use of Short-Term Tests in the Isolation and Identification of Chemical Mutagens, in F. J. deSerres and A. Hollaender (eds.), *Chemical Mutagens, Principles and Methods for Their Detection*, Vol. 6, Plenum Press, New York, 1980, pp. 239–270.

Epler, J. L., F. W. Larimer, C. E. Nix, T. Ho, and T. K. Rao, Comparative Mutagenesis of Test Materials from the Synthetic Fuel Technologies, in D. Scott, B. A. Bridges, and F. H. Sobels (eds.), *Progress in Genetic Toxicology*, Elsevier/North-Holland Biomedical Press, New York, 1977, pp. 275–284.

Epler, J. L., R. J. M. Fry, F. W. Larimer, T. K. Rao, J. N. Dumont, T. W. Schultz, A. W. Hsie, H. Witschi, L. H. Smith, W. M. Haschek, and J. M. Holland, *Health Effects Research in Direct Coal Liquefaction: Studies of H-Coal Distillates: Phase I. PDU Samples—The Effects of Hydrotreatment*, ORNL/TM-8071, Oak Ridge National Laboratory, Oak Ridge, TN, 1981.

Epler, J. L., R. J. M. Fry, T. K. Rao, F. W. Larimer, J. N. Dumont, T. W. Schultz, L. B. Russell, W. M. Generoso, H. Witschi, L. H. Smith, W. M. Haschek, and J. M. Holland, *Biomedical Response to Products and Effluents from the University of Minnesota-Duluth Gasifier*, ORNL/TM-8821, Oak Ridge National Laboratory, Oak Ridge, TN, 1983.

Fillo, J. P., V. C. Stamoudis, J. R. Stetter, and S. W. Vance, *Influence of Coal Devolatilization Conditions on the Yield, Chemistry, and Toxicology of By-Product Oils and Tars*, DOE/ET/14746-11, National Technical Information Service, Springfield, VA, 1983.

Frazier, M. E., and T. K. Andrews, Jr., Transformation of Syrian Hamster Embryo Cells by Synfuel Mixtures. *J. Toxicol. Environ. Health*, **11,** 591–606 (1983).

Frazier, M. E., and D. D. Mahlum, Mutagenic and Carcinogenic Activity of a Hydrotreated Coal Liquid. *J. Toxicol. Environ. Health*, **13,** 531–543 (1984).

Frazier, M. E., and J. E. Samuel, *Genotoxicity of Complex Mixtures: CHO Cell Mutagenicity Assay*, PNL-5337, Pacific Northwest Laboratory, Richland, WA. National Technical Information Service, Springfield, VA, 1985.

Gatehouse, D. G., Detection of Mutagenic Derivatives of Cyclophosphamide and a Variety of Other Mutagens in a "Microtitre" Fluctuation Test, Without Microsomal Activation. *Mutat. Res.*, **53,** 289–296 (1978).

Gatehouse, D. G., and G. F. Delow, The Development of a "Microtitre" Fluctuation Test for the Detection of Indirect Mutagens, and its Use in the Evaluation of Mixed Enzyme Induction of the Liver. *Mutat. Res.*, **60,** 239–252 (1979).

Gelboin, H. V., N. Kinoshita, and F. J. Wiebel, Microsomal Hydroxylases: Induction and Role in Polycyclic Hydrocarbon Carcinogenesis and Toxicity. *Fed. Proc.*, **31,** 1298–1309 (1972).

Great Britain Medical Research Council. The Carcinogenic Action of Mineral Oils: A Chemical and Biological Study, Special Report. Series No. 306, Her Majesty's Stationery Office, London, 1968.

Green, M. H. L., and W. J. Muriel, Use of Repair-Deficient Strains of *Escherichia coli* and Liver Microsomes to Detect and Characterize DNA Damage Caused by the Pyrrolizidine Alkaloids Heliotrine and Monocrotaline. *Mutat. Res.*, **28,** 331–336 (1975).

Green, M. H. L., W. J. Muriel, and B. A. Bridges, Use of a Simplified Fluctuation Test to Detect Low Levels of Mutagens. *Mutat. Res.*, **38,** 33–42 (1976).

Griest, W. H., J. M. Giddings, and J. A. Klein, *Effects of Hydrotreatment on the Properties of Coal-Derived Liquid Products: A Status Report*, ORNL/TM-8836, Oak Ridge National Laboratory, Oak Ridge, TN, 1983.

Guerin, M. R., J. L. Epler, W. H. Griest, B. R. Clark, and T. K. Rao, Polycyclic Aromatic Hydrocarbons from Fossil Fuel Conversion Processes, in P. W. Jones and R. I. Freudenthal (eds.), *Carcinogenesis—A Comprehensive Survey*, Vol. 3, *Polynuclear Aromatic Hydrocarbons*, Raven Press, New York, 1978, pp. 21–33.

Harbison, R. D., Teratogens, in V. Doull, C. D. Klaassen, and M. O. Amdur (eds.), Casarett and Doulls' *Toxicology: The Basic Science of Poisons*, Chap. 8, Macmillan, New York, 1980, pp. 158–175.

Haugen, D. A., and M. J. Peak, Mixtures of Polycyclic Aromatic Compounds Inhibit Mutagenesis in the *Salmonella*/Microsome Assay by Inhibition of Metabolic Activation. *Mutat. Res.*, **116,** 257–269 (1983).

Haugen, D. A., M. J. Peak, and C. A. Reilly, Jr., Chemical and Biological Characterization of High-Btu Coal Gasification (The HYGAS Process) II. Nitrous Acid Treatment for Detection of Mutagenic Primary Aromatic Amines: Nonspecific Reactions, in D. D. Mahlum, R. H. Gray, and W. D. Felix (eds.), *Coal Conversion and the Environment: Chemical, Biomedical, and Ecological Considerations*, CONF-801039, National Technical Information Service, Springfield, VA, 1981, pp. 115–127.

Henderson, R. F., *Low Btu Gasifier Emissions Toxicology—Status Report*, J. M. Benson (ed.), LMF-95. Inhalation Toxicology Research Institute, Los Alamos National Laboratory, Albuquerque, NM, 1982.

Henry, M. C., and D. G. Kaufman, Clearance of Benzo[alpha]pyrene from Hamster Lungs After Administration on Coated Particles. *J. Natl. Cancer Inst.*, **51,** 1961–1964 (1973).

Hill, J. O., R. E. Royer, C. E. Mitchell, and J. S. Dutcher, In Vitro Cytotoxicity to Alveolar Macrophages of Tar from a Low-Btu Coal Gasifier. *Environ. Res.*, **31,** 484–492 (1983).

Hollaender, A., *Chemical Mutagens, Principles and Methods for Their Detection*, Vol. 3, Plenum Press, New York, 1973.

Holland, J. M., and E. L. Frome, Statistical Evaluation in the Carcinogenesis Bioassay of Petroleum Hydrocarbons, in H. M. McFarland, C. E. Holdsworth, J. A. MacGregor, R. W. Call, and M. L. Kane (eds.), *The Toxicology of Petroleum Hydrocarbons*, American Petroleum Institute, Washington, DC, 1982, pp. 196–207.

Holland, J. M., R. O. Rahn, L. H. Smith, B. R. Clark, S. S. Chang, and T. J. Stephens, Skin Carcinogenicity of Synthetic and Natural Petroleums. *J. Occup. Med.*, **21,** 614–618 (1979a).

Holland, J. M., M. S. Whitaker, and J. W. Wesley, Correlation of Fluorescence Intensity and Carcinogenic Potency of Synthetic and Natural Petroleums in Mouse Skin. *Am. Ind. Hyg. Assoc. J.*, **40,** 496–503 (1979b).

Holland, J. M., M. J. Whitaker, and L. C. Gipson, Chemical and Biological Factors Influencing the Skin Carcinogenicity of Fossil Liquids, in W. N. Rom and V. E. Archer (eds.), *Health Implications of New Energy Technology*, Ann Arbor Science, Ann Arbor, MI, 1980, pp. 463–480.

Holland, J. M., D. A. Wolf, and B. R. Clark, Relative Potency Estimation for Synthetic Petroleum Skin Carcinogens. *Environ. Health Perspect.*, **38,** 149–155 (1981).

Holland, J. M., F. W. Larimer, T. K. Rao, J. L. Epler, C.-H. Ho, M. V. Buchanan, and M. R. Guerin, The Distribution of Dermal Tumorigens in Coal Liquids: Relationship of Tumorigenicity and Microbial Mutagenicity, *J. Appl. Toxicol.*, **4,** 117–123 (1984).

Horton, A. W., and D. T. Denman, Carcinogenesis of the Skin—A Re-examination of Methods for the Quantitative Assessment of the Potencies of Complex Materials. *Cancer Res.*, **15,** 701–709 (1955).

Horton, A. W., D. T. Denman, and R. P. Trosset, Carcinogenesis of the Skin. II. The Accelerating Properties of Aliphatic and Related Hydrocarbons. *Cancer Res.*, **17,** 758–766 (1957).

Hsie, A. W., J. P. O'Neill, D. B. Couch, J. R. SanSebastian, P. A. Brimer, R. Machanoff, J. C. Fuscoe, J. C. Riddle, A. P. Li, N. L. Forbes, and M. H. Hsie, Quantitative Analyses of Radiation- and Chemical-Induced Lethality and Mutagenesis in Chinese Hamster Ovary Cells. *Radiat. Res.*, **76,** 471–492 (1978).

Hsie, A. W., P. A. Brimer, J. P. O'Neill, J. L. Epler, M. R. Guerin, and M. H. Hsie, Mutagenicity of Alkaline Constituents of a Coal-Liquified Crude Oil in Mammalian Cells. *Mutat. Res.*, **78,** 79–84 (1980).

Hsie, A. W., D. A. Casciano, D. B. Couch, D. F. Krahn, J. P. O'Neill, and B. L. Whitfield, The Use of Chinese Hamster Ovary Cells to Quantify Specific Locus Mutation and to Determine Mutagenicity of Chemicals. A Report of the Gene-Tox Program. *Mutat. Res.*, **86,** 193–214 (1981).

Hubis, W., *U.S. Department of Energy, Solvent Refined Coal (SRC) Process, Health Program: Final Report, Industrial Hygiene, Clinical and Toxicological Programs*, DOE/ET/10104-48, National Technical Information Service, Springfield, VA, 1982.

Hueper, W. C., Experimental Studies on *Cancerigenesis* of Synthetic Liquid Fuels and Petroleum Substitutes. *Arch. Ind. Hyg. Occup. Med.*, **8,** 307–327 (1953).

Hueper, W. C., Experimental Carcinogenic Studies on Hydrogenated Coal Oils. I: Bergius Oils. *Ind. Med. Surg.*, **25,** 51–55 (1956).

Hydrocarbon Research, Inc., H-Coal Process Development Unit Operations in Fuel Oil Mode, PDU Run 1, Hydrocarbon Research Inc., Lawrenceville, NJ. U.S. Department of Energy, Washington, DC, 1976.

Iball, J., The Relative Potency of Carcinogenic Compounds. *Am. J. Cancer*, **35,** 188–190 (1939).

Kaplan, E. L., and P. Meier, Nonparametric Estimation from Incomplete Observations. *J. Am. Stat. Assoc.*, **53,** 457–481 (1958).

Kawai, M., H. Amamoto, and K. Harada, Epidemiologic Study of Occupational Lung Cancer. *Arch. Environ. Health*, **14,** 859–864 (1967).

Keleti, G., J. Bern, M. A. Shapiro, W. P. Gulledge, and G. T. Moore, Mutagenicity of SRC-II Coal Liquefaction Wastewater Treatment Residues. *Environ. Sci. Technol.*, **16,** 826–830 (1982).

Ketcham, N. H., and R. W. Norton, The Hazards to Health in the Hydrogenation of Coal. III. The Industrial Hygiene Studies. *Arch. Environ. Health*, **1,** 194–207 (1960).

Kilpatrick, M. P., R. A. Magee, T. E. Emmel, and G. C. Page, *Environmental Assessment: Source Test and Evaluation Report—Wellman–Galusha (Ft. Snelling) Low-Btu Gasification*, EPA-600/7-80-097, Environmental Protection Agency, Research Triangle Park, NC, 1980.

Kim, J. I., M. D. Notich, H. C. Higman, and K. M. Wagner, *Environmental Assessment Source Test and Evaluation Report: Fort Lewis SRC Coal Liquefaction Pilot Plant*, H-C1016/002-8-996, Hittman Associates, Inc., Columbia, MD, 1981.

Kimball, R. F., and N. B. Munro, *A Critical Review of the Mutagenic and Other Genotoxic Effects of Direct Coal Liquefaction*, ORNL-5721, Oak Ridge National Laboratory. National Technical Information Service, Springfield, VA, 1981.

Klecak, G., Identification of Contact Allergens: Predictive Tests in Animals, in F. N. Marzulli and M. I. Maibach (eds.), *Dermatotoxicity*, Chap. 9, Hemisphere Publishing, Washington, DC, 1983, pp. 193–236.

Klein, M., Induction of Skin Tumors in the Mouse with Minute Doses of 9,10-Dimethyl-1,2-benzanthracene Alone or With Croton Oil. *Cancer Res.*, **16,** 123–127 (1956).

Kuroda, S., and K. Kawahata, Uber die Gewerbliche Entstehung des Lungenkrebses bei Generatorgasarbeitern. *Z. Krebsforch.*, **45,** 36 (1936)

Later, D. W., *Two-Stage Coal Liquefaction Process Materials from the Wilsonville Facility Operated in the Nonintegrated and Integrated Modes: Chemical Analyses and Biological Testing*, PNL-5215, Pacific Northwest Laboratory, Richland, WA. National Technical Information Service, Springfield, VA, 1985.

Later, D. W., and B. W. Wilson, *Preliminary Chemical Analysis and Biological Testing of Materials from the HRI Catalytic Two-Stage Liquefaction (CTSL) Process*, PNL-5274, Pacific Northwest Laboratory, Richland, WA. National Technical Information Service, Springfield, VA, 1985.

Later, D. W., M. L. Lee, K. D. Bartle, R. C. Kong, and D. L. Vassilaros, Chemical Class Separation and Characterization of Organic Compounds in Synthetic Fuels. *Anal. Chem.*, **53,** 1612–1620 (1981).

Later, D. W., R. A. Pelroy, and B. W. Wilson, *Chemical Analysis and Biological Testing of Materials from the EDS Coal Liquefaction Process: A Status Report*, PNL-4960, Pacific Northwest Laboratory, Richland, WA. National Technical Information Service, Springfield, VA, 1984.

Lee, K. W., W. S. Seames, R. V. Collins, K. J. Bombaugh, and G. C. Page, *Environmental Assessment: Source Test and Evaluation Report—Lurgi (Kosovo) Medium-Btu Gasification*, EPA-600/7-81-142, Environmental Protection Agency, Research Triangle Park, NC, 1981.

Leitch, A., Paraffin Cancer and Its Experimental Production. *Br. Med. J.*, **2,** 1104–1106 (1922).

Lewis, S. C., R. W. King, S. T. Cragg, and D. W. Hillman, Skin Carcinogenic Potential of Petroleum Hydrocarbons. 2. Carcinogenesis of Crude Distillate Fractions and Chemical Class Subfractions, in H. M. McFarland, C. E. Holdsworth, J. A. MacGregor, R. W. Call, and M. L. Kane (eds.), *Toxicology of Petroleum Hydrocarbons*, American Petroleum Institute, Washington, DC, 1982, pp. 185–195.

Li, A. P., and R. W. Shimizu, A Modified Agar Assay for the Quantification of Mutation at the Hypoxanthine Guanine Phosphoribosyl Transferase Gene Locus in Chinese Hamster Ovary Cells. *Mutat. Res.*, **111,** 365–370 (1983).

Litchfield, J. T., and F. Wilcoxon, A Simplified Method of Evaluating Dose–Effect Experiments. *J. Pharmacol.*, **96,** 99–108 (1949).

Lloyd, J. W., Long-Term Mortality Study of Steelworkers. V. Respiratory Cancer in Coke Plant Workers. *J. Occup. Med.*, **13,** 53–68 (1971).

Loomis, T. A., *Essentials of Toxicology*, Lea & Febiger, Philadelphia, PA, 1968.

Loscutoff, S. M., B. W. Killand, R. A. Miller, R. L. Buschbom, D. L. Springer, and D. D. Mahlum, Pulmonary Toxicity of Inhaled Coal Liquid Aerosols (Boiling Range 230–450°C). *Toxicol. Appl. Pharmacol.*, **67,** 346–356 (1983).

Luria, S. E., and M. Delbruck, Mutations of Bacteria from Virus Sensitivity to Virus Resistance. *Genetics*, **28,** 491–511 (1943).

Ma, C. Y., C.-H. Ho, R. B. Quincy, M. R. Geurin, T. K. Rao, B. E. Allen, and J. L. Epler, Preparation of Oils for Bacterial Mutagenicity Testing. *Mutat. Res.*, **118,** 15–24 (1983).

MacCoss, M., E. K. Ryu, and T. Matsushita, The Synthesis, Characterization, and Preliminary Biological Evaluation of 1-beta-D-arabinofuranosylcytosine-5′-diphosphate-L-1,2-Dipalmitin. *Biochem. Biophys. Res. Commun.*, **85,** 714–723 (1978).

MacEwan, J. D., A. Hall, and L. D. Scheel, Experimental Oncogenesis in Rats and Mice Exposed to Coal Tar Aerosols, in *Proceedings of the 7th Annual Conference on Environmental Toxicology*, Aerospace Medical Research Laboratory, Dayton, OH, 1976, pp. 66–81.

Mahlum, D. D. (ed.), *Chemical, Biomedical and Ecological Studies of SRC-I Materials from the Ft. Lewis Pilot Plant: Status Report*, PNL-3474, Pacific Northwest Laboratory, Richland, WA. National Technical Information Service, Springfield, VA, 1981.

Mahlum, D. D., Initiation/Promotion Studies with Coal-Derived Liquids. *J. Appl. Toxicol.*, **3,** 31–34 (1983a).

Mahlum, D. D., Skin-Tumor Initiation Activity of Coal Liquids with Different Boiling-Point Ranges. *J. Appl. Toxicol.*, **3,** 254–258 (1983b).

Mahlum, D. D., Expression of Benzo[a]pyrene and 6-Aminochrysene Initiating Activity in Complex Mixtures from Coal. *Toxicologist*, **4,** 52 (1984).

Mahlum, D. D., M. E. Frazier, R. A. Pelroy, and R. A. Renne, *Comparison of in Vitro and in Vivo Studies with Coal Liquids from the SRC-II Process*, PNL-4836, Pacific Northwest Laboratory, Richland, WA. National Technical Information Service, Springfield, VA, 1983.

Mahlum, D. D., C. W. Wright, E. K. Chess, and B. W. Wilson, Fractionation of Skin Tumor-Initiating Activity in Coal Liquids. *Cancer Res.*, **44,** 5176–5181 (1984).

McCreesh, A. H., and M. Steinberg, Skin Irritation Testing in Animals, in F. N. Marzulli and H. I. Maibach (eds.), *Dermatotoxicity*, Chap. 6, Hemisphere Publishing, Washington, DC, 1983, pp. 147–166.

McDonald, T. O., V. Seabaugh, J. A. Shadduck, and H. F. Edelhauser, Eye Irritation, in F. N. Marzulli and H. I. Maibach (eds.), *Dermatotoxicity*, Chap. 28, Hemisphere Publishing, Washington, DC, 1983, pp. 555–610.

McKee, R. H., R. W. Biles, R. W. Kapp, and J. P. Hinz, The Acute Toxicity of Coal-Liquefaction-Derived Material. *J. Appl. Toxicol.*, **4,** 198–205 (1984).

Miller, R. A., D. L. Springer, and D. D. Mahlum, Pathologic Response of Fischer-344 Rats and CD-1 Mice After Subchronic Inhalation Exposure to a High-Boiling Coal Liquid. *Fed. Proc.*, **43,** 574 (1984).

Mitchell, C. E., The Metabolic Fate of Benzo[a]pyrene in Rats after Inhalation. *Toxicology*, **28,** 65–73 (1983).

Mitchell, C. E., R. F. Henderson, and R. O. McClellan, Distribution, Retention, and Fate of 2-aminoanthracene in Rats after Inhalation. *Toxicol. Appl. Pharmacol.*, **75,** 52–59 (1984).

Mixon, F. O., J. G. Cleland, T. J. Hughes, D. A. Green, and S. K. Gangwal, Laboratory-Scale Simulation of Coal Gasification for Biological Studies, in K. E. Cowser (ed.), *Synthetic Fossil Fuel Technologies: Results of Health and Environmental Studies*, Butterworth Publishers, Stoneham, MA, 1984, pp. 199–227.

Moreau, P., A. Bailone, and R. Devoret, Prophage λ Induction in *Escherichia coli* K12 envA uvrB: A Highly Sensitive Test for Potential Carcinogens. *Proc. Natl. Acad. Sci. USA*, **73,** 3700–3704 (1976).

Munro, N. B., R. J. M. Fry, R. B. Gammage, W. M. Haschek, E. E. Calle, J. A. Klein, and T. W. Schultz, *Indirect Coal Liquefaction: A Review of Potential Health Effects and Worker Exposure During Gasification and Synthesis*, ORNL-5938, Oak Ridge National Laboratory, Oak Ridge, TN. National Technical Information Service, Springfield, VA, 1983.

Munro, N. B., D. M. DeMarini, J. N. Dumont, W. C. Dunn, J. L. Epler, W. N. Generoso, M. E. Goad, W. M. Haschek, J. M. Holland, A. W. Hsie, C. C. Morse, T. K. Rao, J. D. Regan, L. B. Russell, T. W. Schultz, L. H. Smith, and H. R. Witschi, *Occupational Toxicological Evaluation of Materials from the H-Coal Pilot Plant*, ORNL/TM-9197, Oak Ridge National Laboratory, Oak Ridge, TN. National Technical Information Service, Springfield, VA, 1985.

Newton, G. J., *Low-Btu Gasifier Emissions Toxicology Program*, LF-75, Inhalation Toxicology Research Institute, Los Alamos National Laboratory, Albuquerque, NM, 1979.

Newton, G. J. (ed.), *Low-Btu Gasifier Emissions Toxicology*, LMF-77, Inhalation Toxicology Research Institute, Los Alamos National Laboratory, Albuquerque, NM, 1980.

Nichols, D. G., J. G. Cleland, D. A. Green, F. O. Mixon, T. J. Hughes, and A. W. Kolber, *Pollutants from Synthetic Fuels Production: Environmental Evaluation of Coal Gasification Screening Tests*, EPA-600/7-79-202, Environmental Protection Agency, Research Triangle Park, NC, 1979.

Oak Ridge National Laboratory, *Health Effects Research in Direct Coal Liquefaction, Studies of H-Coal Distillates: Phase I. PDU Samples—The Effects of Hydrotreatment*, ORNL/TM-8071, Oak Ridge National Laboratory, Oak Ridge, TN, 1981.

O'Brien, T. G., R. C. Simsiman, and R. K. Boutwell, Induction of the Polyamine-biosynthetic Enzymes in Mouse Epidermis by Tumor-Promoting Agents. *Cancer Res.*, **35,** 1662–1670 (1975a).

O'Brien, T. G., R. C. Simsiman, and R. K. Boutwell, Induction of the Polyamine-biosynthetic Enzymes in Mouse Epidermis and Their Specificity for Tumor Promotion. *Cancer Res.*, **35,** 2426–2433 (1975b).

Pacific Northwest Laboratory, *Biomedical Studies on Solvent Refined Coal (SRC-II) Liquefaction Materials: A Status Report*, PNL-3189, Pacific Northwest Laboratory, Richland, WA. National Technical Information Service, Springfield, VA, 1979.

Pacific Northwest Laboratory, *Pacific Northwest Laboratory Annual Report for 1979 to the DOE Assistant Secretary for Environment, Part 1 Biomedical Sciences*, PNL-3300, Pacific Northwest Laboratory, Richland, WA. National Technical Information Service, Springfield, VA, 1980.

Pacific Northwest Laboratory, *Pacific Northwest Laboratory Annual Report for 1981 to the DOE Office of Energy Research, Part 1 Biomedical Sciences*, PNL-4100, Pacific Northwest Laboratory, Richland, WA. National Technical Information Service, Springfield, VA, 1982.

Pacific Northwest Laboratory, *Pacific Northwest Laboratory Annual Report for 1982 to the DOE Office of Energy Research, Part 1 Biomedical Sciences*, PNL-4600, Pacific Northwest Laboratory, Richland, WA. National Technical Information Service, Springfield, VA, 1983.

Page, G. C., *Environmental Assessment: Source Test and Evaluation Report—Chapman Low-Btu Gasification*, EPA-600/7-78-202, Environmental Protection Agency, Research Triangle Park, NC, 1978.

Palmer, A., Mortality Experience of 50 Workers with Occupational Exposures to the Products of Coal Hydrogenation Processes. *J. Occup. Med.*, **21,** 41–44 (1979).

Passey, R. D., Experimental Soot Cancer. *Br. Med. J.*, **2,** 1112–1113 (1922).

Peirce, M. V., and W. Fung, *An Evaluation of the Genotoxic Activity of COGAS Oils in a Differential DNA Repair Assay*, Reports to the COGAS Development Co. Genex Corporation, Rockville, MD, 1981a.

Peirce, M. V., and W. Fung, *An Evaluation of the Recombinogenic Activity of COGAS Oils in the Saccharomyces cerevisiae D3 Assay*, Reports to the COGAS Development Co. Genex Corporation, Rockville, MD, 1981b.

Peirce, M. V., and V. F. Simmon, *An Evaluation of the Mutagenic Activity of COGAS Oils in the Preincubation Modification of the Ames Salmonella/Microsome Assay*, Reports to the COGAS Development Co. Genex Corporation, Rockville, MD, 1981.

Pelroy, R. A. (ed.), *The Mutagenic and Chemical Properties of SRC-I Materials: A Status Report*, PNL-3604, Pacific Northwest Laboratory, Richland, WA. National Technical Information-Service, Springfield, VA, 1981.

Pelroy, R. A., and M. R. Petersen, Mutagenic Characterization of Synthetic Fuel Materials by the Ames/*Salmonella* Assay System. *Mutat. Res.*, **90,** 309–320 (1981).

Pelroy, R. A., and D. L. Stewart, The Effects of Nitrous Acid on the Mutagenicity of Two Coal Liquids and Their Genetically Active Chemical Fractions. *Mutat. Res.*, **90,** 297–308 (1981).

Pelroy, R. A., and B. W. Wilson, *Fractional Distillation as a Strategy for Reducing the Genotoxic Potential of SRC-II Coal Liquids: A Status Report*, PNL-3787, Pacific Northwest Laboratory, Richland, WA. National Technical Information Service, Springfield, VA, 1981.

Pelroy, R. A., D. D. Mahlum, M. E. Frazier, R. A. Renne, D. L. Stewart, and E. K. Chess, Comparison of in Vivo Carcinogenesis and in Vitro Genotoxicity of Complex Hydrocarbon Mixture, in K. E. Cowser (ed.), *Synthetic Fossil Fuel Technologies: Results of Health and Environmental Studies*, Butterworth Publishers, Stoneham, MA, 1984.

Pelroy, R. A., M. E. Frazier, D. W. Later, C. W. Wright, and B. W. Wilson, Effect of Process Distillation on Mutagenicity and Cell-Transformation Activity of Solvent-Refined, Coal-Derived Liquids. *Fuel*, **64,** 639–643 (1985).

Perry, P., and S. Wolff, New Giemsa Method for the Differential Staining of Sister Chromatids. *Nature*, **251,** 156–158 (1974).

Peto, R., and P. Lee, Weibull Distributions for Continuous Carcinogenesis Experiments. *Biometrics*, **29,** 457–470 (1973).

Peto, R., M. C. Pike, N. E. Day, R. G. Gray, P. N. Lee, S. Parish, J. Peto, S. Richards, and J. Wahrendorf, Guidelines for Simple, Sensitive Significance Tests for Carcinogenic Effects in Long-Term Animal Experiments, in *Long-Term and Short-Term Screening Assays for Carcinogenesis: A Critical Appraisal*, IARC Monographs, Supplement 2, IARC, Lyon, France, 1980, pp. 311–423.

Poel, W. E., D. Stanton, E. Peters, and H. O. Wade, Comparative Susceptibilities of Seven Inbred Strains of Mice to Quantified Applications of 3,4-Benzopyrene. *Proc. Am. Assoc. Cancer Res.*, **2,** 335 (1958).

Reilly, C. A., Jr., M. J. Peak, T. Matsushita, F. R. Kirchner, and D. A. Haugen, Chemical and Biological Characterization of High-Btu Coal Gasification (The HYGAS Process). IV: Biological Activity, in D. D. Mahlum, R. H. Gray, and W. D. Felix (eds.), *Coal Conversion and the Environment: Chemical, Biomedical, and Ecological Considerations*, CONF-801039, National Technical Information Service, Springfield, VA, 1981, pp. 310–324.

Reilly, C. A., Jr., K. E. Wilzbach, J. R. Stetter, D. A. Haugen, F. R. Kirchner, V. C. Stamoudis, M. J. Peak, T. Matsushita, A. S. Bopari, and R. E. Jones, *Synfuels Environmental Research Program for High-Btu Coal Gasification: Health Effects Summary Report*, ANL/SER-7, Argonne National Laboratory, Argonne, IL, 1986.

Reiners, J., K. Davidson, K. Nelson, M. Mamrack, and T. J. Slaga, Skin Tumor Promotion: A Comparative Study of Several Strains of Mice, in R. Langenbach, S. Nesnow, and J. M. Rice (eds.), *Proceedings of a Symposium: Organ and Species Specificity in Chemical Carcinogenesis*, Plenum Press, New York, 1983, pp. 173–188.

Renne, R. A., and R. L. Buschbom, Health Effects of Complex Mixtures, in *Proceedings of the First Annual DOE Contractors Meeting, Health and Environmental Effects of Complex Mixtures*, CONF-8506155, National Technical Information Service, Springfield, VA, 1985.

Renne, R. A., and L. G. Smith, Carcinogenic Effects of Coal Conversion Materials, in *Proceedings of the Workshop on Industrial Hygiene and Occupational Meetings in Coal Conversion Projects*, National Technical Information Service, Springfield, VA, 1981, pp. 9–20.

Renne, R. A., L. G. Smith, and D. D. Mahlum, Epidermal Carcinogenicity of Some Crude Fossil Fuels in Mice: A Preliminary Report, in D. D. Mahlum, R. H. Gray, and W. D. Felix (eds.), Coal Conversion and the Environment: Chemical, Biomedical, and Ecological Considerations, CONF-801039, National Technical Information Service, Springfield, VA, 1981, pp. 471–481.

Renne, R. A., C. W. Wright, L. G. Smith, and R. L. Buschbom, Epidermal Carcinogenesis Studies of Synthetic Fossil Fuel Materials in Mice. *Toxicology*, **40,** 311–325 (1986).

Robinson, D. E., and V. F. Simmon, *Evaluation of the Mutagenic Activity of COGAS Oils in the L5178Y* $TK^{+}/^{-}$ *Mouse Lymphoma Mutagenesis Assay*, Reports to the COGAS Development Co. Genex Corporation, Rockville, MD, 1981.

Royer, R. E., C. E. Mitchell, R. L. Hanson, J. S. Dutcher, and W. E. Bechtold, Fractionation, Chemical Analysis, and Mutagenicity Testing of Tar from a Low-Btu Gasifier. *Environ. Res.*, **31,** 460–471 (1983).

Russell, L. B., and M. H. Major, Radiation-Induced Presumed Somatic Mutations in the House Mouse. *Genetics*, **42,** 161–175 (1957).

Russell, L. B., P. B. Selby, E. vonHalle, W. Sheridan, and L. Valcorie, Use of the Mouse Spot Test in Chemical Mutagenesis: Interpretation of Past Data and Recommendations for Future Work. *Mutat. Res.*, **86,** 355–379 (1981).

Saffiotti, U., F. Cefis, and L. H. Kolb, A Method for the Experimental Induction of Bronchogenic Carcinoma. *Cancer Res.*, **28,** 104–124 (1968).

Schirmer, R. E., R. B. Westerberg, L. J. Felice, B. J. Kelman, and D. L. Springer, Percutaneous Absorption and Metabolism of Carcinogens Associated with Synfuel Processing. *Toxicol. Lett.*, **18,** 80 (1983).

Schoeny, R., D. Warshawsky, L. Hollingsworth, M. Hund, and G. Moore, Mutagenicity of Products from Coal Gasification and Liquefaction in the *Salmonella*/Microsome Assay. *Environ. Mutagen.*, **3,** 181–195 (1981).

Schultz, T. W., J. N. Dumont, B. R. Clark, and M. W. Buchanan, Embryotoxic and Teratogenic Effects of Aqueous Extracts of Tar from a Coal Gasification Electrostatic Precipitator. *Teratogenesis Carcinog. Mutagen.*, **2,** 1–11 (1982).

Schultz, T. W., J. N. Dumont, and M. W. Buchanan, Toxic and Teratogenic Effects of Chemical Class Fractions of a Coal-Gasification Electrostatic Precipitator Tar. *Toxicology*, **29,** 87–99 (1983).

Sexton, R. J., The Hazards to Health in the Hydrogenation of Coal. IV. The Control Program and the Clinical Effects. *Arch. Environ. Health,* **1,** 208–231 (1960).

Shimizu, R. W., J. M. Benson, A. P. Li, R. F. Henderson, and A. L. Brooks, Evaluation of the Genotoxicity of Process Stream Extracts from a Coal Gasification System. *Environ. Mutagen.*, **6,** 825–834 (1984).

Simmon, V. F., In Vitro Assays for Recombinogenic Activity of Chemical Carcinogens and Related Compounds with *Saccharomyces cerevisiae* D3. *J. Natl. Cancer Inst.*, **62,** 901–909 (1979).

Skopek, T. R., H. L. Liber, D. A. Kaden, R. A. Hites, and W. G. Thilly, Mutation of Human Cells by Kerosene Soot. *J. Natl. Cancer Inst.*, **63,** 309–312 (1979).

Slaga, T. J., and S. M. Fischer, Strain Differences and Solvent Effects in Mouse Skin Carcinogenesis Experiments Using Carcinogens, Tumor Initiators and Promoters. *Prog. Exp. Tumor Res.*, **26,** 85–109 (1983).

Slaga, T. J., S. M. Fischer, L. L. Triplett, and S. Nesnow, Comparison of Complete Carcinogenesis and Tumor Initiation and Promotion in Mouse Skin: The Induction of Papillomas by Tumor Initiation-Promotion, a Reliable Short-Term Assay. *J. Environ. Pathol. Toxicol.*, **4,** 1024–1041 (1981).

Smith, W. E., D. A. Sunderland, and K. Sugiura, Experimental Analysis of the Carcinogenic Activity of Certain Petroleum Products. *Arch. Ind. Hyg. Occup. Med.*, **4,** 299–314 (1951).

Springer, D. L., M. L. Clark, D. H. Willard, and D. D. Mahlum, Generation and Delivery of Coal Liquid Aerosols for Inhalation Studies. *Am. Ind. Hyg. Assoc. J.*, **43,** 486–491 (1982a).

Springer, D. L., K. A. Poston, D. D. Mahlum, and M. R. Sikov, Teratogenicity Following Inhalation Exposure of Rats to a High-Boiling Coal Liquid. *J. Appl. Toxicol.*, **2,** 260–264 (1982b).

Springer, D. L., D. D. Mahlum, and H. A. Ragan, Inhalation Exposure Study of a Solvent Refined Coal Material, in *Proceedings of the 13th Conference on Environmental Toxicology*, AFAMRL-TR-82-101, National Technical Information Service, Springfield, VA, 1983, pp. 317–327.

Springer, D. L., R. A. Miller, W. C. Weimer, H. A. Ragan, R. L. Buschbom, and D. D. Mahlum, *Effects of Inhalation Exposure to SRC-II Heavy and Middle Distillates*, PNL-5273, Pacific Northwest Laboratory, Richland, WA. National Technical Information Service, Springfield, VA, 1984.

Stamoudis, V. C., S. Bourne, D. A. Haugen, M. J. Peak, C. A. Reilly, J. R. Stetter, and K. Wilzbach, Chemical and Biological Characterization of High-Btu Coal Gasification (The HYGAS Process). I. Chemical Characterization of Mutagenic Fractions, in D. D. Mahlum, R. H. Gray, and W. D. Felix (eds.), *Coal Conversion and the Environment: Chemical, Biomedical, and Ecological Considerations*, CONF-801039, National Technical Information Service, Springfield, VA, 1981, pp. 67–95.

Stein, T. R., J. G. Bendoraitis, A. V. Cabal, R. B. Callen, M. J. Dabkowski, R. H. Heck, H. R. Ireland, and C. A. Simpson, *Upgrading of Coal Liquids for Use as Power Generation Fuels, Annual Report for the Period February 1976–January 1977,* EPRI AF-444, Mobil Research & Development Corporation, Paulsboro, N.J. Electric Power Research Institute, Palo Alto, CA, 1977.

Stein, T. R., A. V. Cabal, R. B. Callen, M. J. Dabkowski, R. H. Heck, C. A. Simpson, and S. S. Shih, *Upgrading of Coal Liquids for Use as Power Generation Fuels, Annual Report for the Period February 1977–January 1978*, EPRI AF-873. Mobil Research & Development Corporation, Paulsboro, N.J. Electric Power Research Institute, Palo Alto, CA, 1978.

Stenback, F., Skin Carcinogenesis as a Model System: Observations on Species, Strain and Tissue Sensitivity to 7,12-Dimethylbenz[a]anthracene With or Without Promotion from Croton Oil. *Acta Pharmacol. Toxicol.*, **46,** 89–97 (1980).

Stich, H. F., R. H. C. San, and Y. Kawazoe, DNA Repair Synthesis in Mammalian Cells Exposed to a Series of Oncogenic and Non-oncogenic Derivatives of 4-Nitroquinoline 1-Oxide. *Nature*, **229,** 416–419 (1971).

Sun, J. D., R. K. Wolff, and G. M. Kanapilly, Deposition, Retention, and Biological Fate of Inhaled Benzo[a]pyrene Adsorbed onto Ultrafine Particles and as a Pure Aerosol. *Toxicol. Appl. Pharmacol.*, **65,** 231–244 (1982).

Sun, J. D., R. K. Wolff, G. M. Kanapilly, and R. O. McClellan, Lung Retention and Metabolic Fate of Inhaled Benzo[a]pyrene Associated with Diesel Exhaust Particles. *Toxicol. Appl. Pharmacol.*, **73,** 48–59 (1984).

Sutterfield, F. D. Hydrogenation Effects on Mutagenesis of Alternate Crudes, Ph.D. Dissertation, University of Tulsa, Tulsa, OK, 1983.

Sutterfield, D., W. C. Lanning, and R. E. Royer, Screening of Hydrotreated Coal Liquids for Mutagenic Activity, in R. Clusen (ed.), *Second U.S. Department of Energy Environmental Control Symposium*, CONF-800-334/1, Vol. 1, National Technical Information Service, Springfield, VA, 1980, pp. 649–657.

Swanger, D., D. D. Malhum, and D. L. Springer, Ornithine Decarboxylase Induction by Chemically Complex Liquids from Two Solvent Refined Coal Processes. *Carcinogenesis*, **4,** 805–810 (1982).

Thomas, W. C., K. N. Trede, and G. C. Page, *Environmental Assessment: Source Test and Evaluation Report—Wellman–Galusha (Glen-Gery) Low-Btu Gasification*, EPA-600/7-79-185, Environmental Protection Agency, Research Triangle Park, NC, 1979.

Timourian, H., J. S. Felton, D. H. Stuermer, S. Healy, P. Berry, M. Tompkins, G. Battaglia, F. T. Hatch, L. H. Thompson, A. V. Carrano, J. Minkler, and E. Salazar, Mutagenic and Toxic Activity of Environmental Effluents from Underground Coal Gasification Experiments. *J. Toxicol. Environ. Health*, **9,** 975–994 (1982).

Turusov, V. S., and S. Takayama, Tumors of the Liver, in V. S. Turusov (ed.), *Pathology of Tumors in Laboratory Animals*, Vol. 2, International Agency for Research on Cancer, Lyon France, 1979, pp. 193–250.

Twort, C. C., and J. D. Fulton, Further Experiments on the Carcinogenicity of Synthetic Tars and Their Fractions. *J. Pathol. Bacteriol.*, **33,** 119–144 (1930).

Twort, C. C., and J. M. Twort, The Carcinogenic Potency of Mineral Oils. *J. Ind. Hyg. Toxicol.*, **13,** 204–226 (1931).

Twort, C. C., and J. M. Twort, Some Factors Influencing the Development of Experimental Oil and Tar Tumours in Mice. *Lancet*, **216,** 1108–1111 (1929).

Tye, R., and K. L. Stemmer, Experimental Carcinogenesis of the Lung. II. Influence of Phenols in the Production of Carcinoma. *J. Natl. Cancer Inst.*, **39,** 175–179 (1967).

Tye, R., M. J. Burton, E. Bingham, Z. Bell, and A. W. Horton, Carcinogens in a Cracked Petroleum Residuum. *Arch. Environ. Heatlh*, **13,** 202–207 (1966).

Van Duuren, B. L., Tumor-Promoting Agents in Two-Stage Carcinogenesis. *Prog. Exp. Tumor Res.*, **11,** 31–68 (1969).

Warshawsky, D., *Exploratory Research on Mutagenic Activity of Coal Related Material*, Pittsburgh Energy Technology Center, Pittsburgh, PA, 1979.

Warshawsky, D., R. Schoeny, L. Hollingsworth, M. Hund, and G. Moore, Mutagenicity of Products from Coal Liquefaction and Gasification, in D. D. Mahlum, R. H. Gray, and W. D. Felix (eds.), *Coal Conversion and the Environment: Chemical, Biomedical, and Ecological Considerations*, CONF-801039, National Technical Information Service, Springfield, VA, 1981, pp. 294–309.

Weil, C. S., Tables for Convenient Calculation of Median-Effective Dose (LD_{50} and ED_{50}) and Instructions in Their Use. *Biometrics*, **8,** 249–263 (1952).

Weil, C. S., and N. I. Condra, The Hazards to Health in the Hydrogenation of Coal. II. Carcinogenic Effect of Materials on the Skin of Mice. *Arch. Environ. Health*, **1,** 187–193 (1960).

Weimer, W. C., B. W. Wilson, R. A. Pelroy, and J. C. Craun, *Initial Chemical and Biological Characterization of Hydrotreated Solvent Refined Coal (SRC-II) Liquids: A Status Report*, PNL-3464, Pacific Northwest Laboratory, Richland, WA. National Technical Information Service, Springfield, VA, 1980.

Weisburger, J. H., and G. M. Williams, Chemical Carcinogens, in J. Doull, C. D. Klaassen, and M. O. Amdur (eds.), *Toxicology: The Basic Science of Poisons*, Chap. 26, Macmillan, New York, 1980, pp. 84–138.

Wilson, B. W., and R. A. Pelroy, *Biological and Biodirected Chemical Analysis of Veba Oel Coal Liquids*, Battelle Pacific Northwest Laboratories, Richland, WA, 1983.

Wilson, B. W., R. A. Pelroy, and J. T. Cresto, Identification of Primary Aromatic Amines in Mutagenically Active Subfractions From Coal Liquefaction Materials. *Mutat. Res.*, **79,** 193–202 (1980).

Wilson, B. W., R. A. Pelroy, and D. D. Mahlum, *Chemical Characterization and Genotoxic Potential Related to Boiling Point for Fractionally Distilled SRC-I Coal Liquids*, PNL-4277, Pacific Northwest Laboratory, Richland, WA. National Technical Information Service, Springfield, VA, 1982.

Wilson, B. W., P. Buhl, and E. C. Moroni, *Biological Testing and Chemical Analysis of Materials from an Integrated Two-Stage Coal Liquefaction Process: A Status Report*, PNL-4553, Pacific Northwest Laboratory, Richland, WA. National Technical Information Service, Springfield, VA, 1983a.

Wilson, B. W., R. A. Pelroy, R. P. Anderson, and J. Freel, *Chemical and Biological Effects of Heavy Distillate Recycle in the SRC-II Process*, PNL-4815, Pacific Northwest Laboratory, Richland, WA. National Technical Information Service, Springfield, VA, 1983b.

Wilson, B. W., D. W. Later, C. W. Wright, and D. L. Stewart, *Chemical Analysis of Distilled Oils from the H-Coal Direct Liquefaction Process: A Status Report*, PNL-5288, Pacific Northwest Laboratory, Richland, WA. National Technical Information Service, Springfield, VA, 1985.

Wilson, J. G., *Environment and Birth Defects*, Academic Press, New York, 1973.

Wilson, J. G., Survey of in Vitro Systems: Their Potential Use in Teratogenicity Screening, in J. G. Wilson and F. C. Fraser (eds.), *Handbook of Teratology*, Vol. 4, Plenum Press, NY, 1978, pp. 135–153.

Wilzbach, K. E., J. R. Stetter, C. A. Reilly, Jr., and W. G. Willson, *Environmental Research Program for Slagging Fixed-Bed Coal Gasification*, ANL/SER-1, Argonne National Laboratory, Argonne, IL, 1982.

Witschi, H. P., *Acute Toxicity Studies with Syncrudes and Natural Petroleum*, ORNL/TSCA, Oak Ridge National Laboratory, Oak Ridge, TN, 1978.

Yahagi, T., M. Tegawa, Y. Seino, T. Matsushima, T. Nagao, T. Sugimura, and Y. Hashimoto, Mutagenesis of Carcinogenic Azo Dyes and Their Derivatives. *Cancer Lett.*, **1,** 91–96 (1975).

Young, R. J., Early U.S. Coal Hydrogenation Project: The Technology and Carcinogenic Risks, in W. N. Rom and V. E. Archer (eds.), *Health Implications of New Energy Technologies*, Chap. 24, Ann Arbor Science, Ann Arbor, MI, 1980, pp. 319–339.

Zamora, P. O., J. M. Benson, T. C. Marshall, B. V. Mokler, A. P. Li, A. R. Dahl, A. L. Brooks, and R. O. McClellan, Cytotoxicity and Mutagenicity of Vapor Phase Pollutants in Rat Lung Epithelial Cells and Chinese Hamster Ovary Cells Grown on Collagen Gels. *J. Toxicol. Environ. Health*, **12,** 27–38 (1983).

Zimmerman, F. K., Procedures Used in the Induction of Mitotic Recombination and Mutation in the Yeast *Saccharomyces cerevisiae. Mutat. Res.*, **31,** 71–86 (1975).

APPENDIX TO CHAPTER 3
DESCRIPTIONS OF SAMPLES ASSAYED FOR TOXICOLOGICAL ACTIVITY AND DESCRIBED IN CHAPTER 3

Citations in the text of this appendix are listed in references for Chapter 3 or the appendix to Chapter 4.

Liquefaction Products (Liquids)

Solvent-Refined Coal I, Pilot Plant, Ft. Lewis, Washington

Sample L2211–L2213. A suite of samples, including light oil (Sample L2211), wash solvent (Sample L2212), and process solvent (Sample L2213), were collected (12/15/76) by personnel from the Pacific Northwest Laboratory (PNL) directly from the process streams during the final stages of an extended production run designed to supply materials for an SRC product burn test. Feed coal was Kentucky No. 9 and No. 14. The light oil had a density of 0.725; the wash solvent, 0.959; and the process solvent, 1.043. The process solvent was fractionated by solvent–solvent separation, based on polarity. This isooctane-based fractionation scheme resulted in standard acid, basic, and neutral fractions as well as neutral (isooctane-insoluble), acid, and basic "tar fractions." Crude materials and the process solvent fractions [stored at 4°C (39°F)] were dissolved in DMSO prior to toxicity testing (Mahlum, 1981).

Solvent-Refined Coal I, Pilot Plant, Wilsonville, Alabama

Samples L2131, L2132, L2141, L2142, L2151, and L2152. A suite of samples, including light oil (Sample L2151), wash solvent (Sample L2141), and recycle solvent (Sample L2131), were collected and sent to the Mobil Research and Development Corporation (MRDC) for hydrotreatment. Splits of these samples (Samples L2152, L2142, and L2132) were later used for toxicity testing (ORNL, 1981). Each sample was hydrotreated by MRDC at three different severity levels (Stein et al., 1977, 1978). Splits of each of the nine hydrotreated samples were also used for toxicity testing (ORNL, 1981). Designations of the hydrotreated samples are as follows:

Feedstock	Low Severity	Moderate Severity	High Severity
Light oil (L2151/L2152)	H0301/H0304	H0307/H0310	H0313/H0316
Wash solvent (L2141/L2142)	H0302/H0305	H0308/H0311	H0314/H0317
Recycle solvent (L2131/L2132)	H0303/H0306	H0309/H0312	H0315/H0318

Each sample was assigned two different numbers because physical property data measured by Oak Ridge National Laboratory (ORNL) and MRDC (see Table 4A.3) indicated that the samples may have degraded during storage and/or that the samples collected for ORNL may not have corresponded precisely to the run balance periods and samples reported by MRDC. The first sample number in these pairs (i.e., under feedstock) represents the MRDC sample and the second number represents the ORNL sample.

Sample L2121. A sample of process solvent characterized as having a boiling-point (bp) range of 450–850°F with some >850°F heavy ends was distilled into 50°F bp cuts at Air Products, Inc. (Linwood, Pennsylvania). Boiling-point cuts of 450°F and higher (seven samples) were studied for toxicity. Each of the bp cut materials was further fractionated by adsorption-column chromatography (neutral alumina column) into an aliphatic hydrocarbon fraction (fraction A-1, hexane elution), a neutral polycyclic aromatic hydrocarbon fraction (fraction A-2, benzene solution), a nitrogen-containing polycyclic aromatic compound fraction (fraction A-3, chloroform: ethanol elution), and a hydroxy polycyclic aromatic hydrocarbon fraction (fraction A-4, tetrahydrofuran: ethanol elution). Stock solutions of each material were made in DMSO (10,000 μg/mL) and stored at -196°F under nitrogen prior to testing (Wilson et al., 1982).

Sample L2111. A sample of process solvent, a material that gives rise to end products, was fractionated by solvent elution column chromatography using Sephadex LH-20. Sephadex was preswelled with methanol: water (85:15) and packed in a glass column. The process solvent sample was diluted in hexane and added to the top of the column. Three fractions were obtained for toxicological assay by sequential elution with hexane, toluene: hexane (1:9), and methanol. Each fraction was dried and redissolved in DMSO prior to toxicological assay (Pelroy, 1981).

Solvent-Refined Coal II, Pilot Plant, Ft. Lewis, Washington

Sample L1241. A sample of light distillate naphtha (134–353°F) was obtained from the raw product distillation section of the pilot plant when the plant was operating on Illinois No. 6 coal (2/8/78) and was considered to represent products of stable and upset process conditions (Pacific Northwest Laboratory, 1979).

Sample L1242. A sample of middle distillate (366–541°F) was obtained from the raw product distillation section of the pilot plant when the plant was operating on Blacksville coal (1/29/78). The material was considered to represent products of stable and upset process conditions (Pacific Northwest Laboratory, 1979).

Sample L1243. A sample of heavy distillate (570–850°F) was obtained from the raw product distillation section of the pilot plant when the plant was operating on Blacksville coal (1/29/78). The material was considered to represent products of stable and upset process conditions (Pacific Northwest Laboratory, 1979).

Sample L1251. A sample of blended middle and heavy distillates (2.9:1 blend ratio) was collected by the Pittsburg & Midway Coal Mining Company (P&M Co.) from tankage (3/23/79). The sample represented materials generated over a series of pilot-plant runs beginning in October 1978. During these test runs, Ohio/West Virginia Powhatan No. 5 coal was used. The distillate blend was fractionated by solvent–solvent separation based on polarity. This isooctane-based fractionation scheme resulted in standard acid, basic, and neutral fractions as well as neutral, acid, and basic "tar fractions." In addition, the blend was fractionally distilled and 20 bp cuts at approximately 50°F intervals (ambient to 509°F) were collected. Crude blend, its distillate bp cuts, and solvent subfractions were dissolved in DMSO prior to toxicity testing (Weimer et al., 1980).

Two drums of this distillate blend were sent up to UOP, Des Plaines, Illinois for hydrotreatment. The two drum samples (Samples L1251 and L1252) had slightly different bulk physical and chemical properties; one drum was used as the feedstock for moderate hydrotreatment (Sample L1251), the other for severe hydrotreatment (Sample L1252; Weimer et al., 1980). The moderate-severity feedstock (Sample L1251) was used for toxicological testing. Hydrotreatment was conducted in standard research-scale reactors using a commercial UOP catalyst (UOP DCB). Samples of the >400°F products from both mild [1.95% H_2 consumption (Sample H0102)] and severe [2.85% H_2 consumption (Sample H0104)] hydrotreatment were used for toxicological evaluation. Each of the materials was fractionated by solvent–solvent separation based on polarity. This isooctane-based fractionation scheme resulted in standard acid, basic, and neutral fractions as well as neutral, acid, and basic "tar fractions" (Weimer et al., 1980).

Sample L1221. A sample of SRC-II fuel blend was obtained from the P&M Co. The sample was collected from run No. 77 SR12 on Pittsburg seam coal (Consol's Blacksville No. 2 mine, West Virginia). The sample was a blend of middle (350–550°F) and heavy (550–850°F) distillates mixed in the same ratio as produced (75.5 : 24.5). The product blend was hydrotreated at the Bartlesville Energy Technology Center with a commercial catalyst in a bench-scale hydrogeneration unit at eight varying degrees of severity. The feed (Sample L1221) and hydrotreated (Samples H0401–H0408) samples were evaluated for toxicity (Sutterfield et al., 1980).

Samples L1261 and L1262. A sample of SRC-II middle and heavy distillate blend (1.7 : 1) was water washed, then both the unwashed (Sample L1261) and washed (Sample L1262) samples were used as feedstocks for hydrotreatment conducted at three levels of severity by the Chevron Research Company. The specification basis for the hydrotreatment was reduction of nitrogen content. The washed crude (Sample L1262) and hydrotreated materials (Samples H0201–H0203) were shipped to ORNL for toxicity testing (Griest et al., 1983).

Samples L1231–L1236. A suite of samples collected from the Ft. Lewis pilot plant (10/78–2/79), operating in the recycle slurry mode (Pittsburg No. 8 bituminous coal, Powhatan No. 5 mine, Ohio) was supplied by P&M Co. Samples included naphtha (Sample L1231), fuel oil (Sample L1232), stripper tower bottoms (Sample L1234), and vacuum tower bottoms (Sample L1235). All samples were dissolved in DMSO and tested as received with the exception of the stripper tower bottoms sample. This material was distilled into a heavy gas oil (Sample L1233) and residue (Sample L1236), using a Vigreux column (Calkins et al., 1983).

Samples L1271 and L1272. Two samples of the vacuum still bottoms from the Ft. Lewis pilot plant operating on Powhatan No. 6 coal in the SRC-II mode were collected for toxicological evaluation. Twenty milligrams of the solid material were suspended in 20 mL of DMSO by stirring for 24 h. The material was sterilized by autoclaving. The serial organic extraction protocol of Warshawsky (Warshawsky et al., 1981) was used on the vacuum bottoms yielding hexane, toluene, methylene chloride, and acetonitrile-soluble fractions as well as an insoluble residue (Keleti et al., 1982).

Sample L1281. A sample of vacuum still bottoms (mineral matter and organics with distillation temperature greater than 850°F) was collected (3/78) by Hittmann Associates, Inc. as part of an EPA Environment Assessment Source Test and Evaluation. The sample was used in Level 1 EPA Toxicological Testing (Kim et al., 1981).

Sample L1211. A sample of a fuel-oil blend (5.75 : 1, middle : heavy distillate) derived from the SRC-II process was obtained by P&M Co. and sent to the EPA/DOE Synfuels Repository at ORNL (Kimball and Munro, 1981).

Samples L1263 and L1264. A suite of samples, including stripper tower bottoms and middle distillate, were collected (3/80 and 4/80) and sent to SRI International for toxicological study. The stripper tower bottoms sample (Sample L1263) was a dark, tarlike substance. This is a heavy stream from which gases and light liquids have been separated. A portion of this stream constitutes the recycled slurry component of coal slurry. Samples of this material were collected at the stripper bottoms (208) sample point. The middle distillate (Sample L1264) was a dark liquid with an approximate boiling range of 350–550°F, with a kerosenelike odor (Hubis, 1982).

Solvent-Refined Coal II, Process Development Unit, Harmarville, Pennsylvania

Samples L1121 and L1131. Two full-boiling-range blends of SRC-II process materials were collected by Gulf Science and Technology Company from the various process streams during operations with Powhatan No. 5 mine coal. A set of 18 distillate cuts covering bp ranges from 138 to 1055°F (Samples N1106–N1123) were prepared for toxicological evaluations at PNL. Samples N1106–N1118 were prepared from a blend (Sample L1121) collected June 1979 during runs No. 55 and No. 56. Samples N1119–N1123 were prepared from a blend (Samples L1131) collected in December 1979 during runs No. 62 and No. 66. In addition, selected distillate cuts were fractionated by HPLC (semipreparative, normal-phase NH_2 column) solvent elution (hexane, methylene chloride, and isopropanol) to yield nonpolar, moderately polar, transition, and strongly polar fractions. Samples were prepared at concentrations of 10,000 μg/mL in DMSO and stored at -196°F until used (Pelroy and Wilson, 1981).

Sample L1111. A full-boiling-range blend of SRC-II process materials was collected during run No. 77 (2/81) by Gulf Science and Technology Company from the various process streams during operations with Powhatan No. 5 mine coal. In addition to the crude blend, a set of five distillation cuts [300–700°F, 700–750°F, 750–800°F, 800–850°F, and >850°F (Samples N1101–N1105)] were prepared for toxicological evaluation. The precision of the distillation was such that 95% of the material in any given cut had its bp within that temperature range. The distillate cuts were fractionated by adsorption-column chromatography (neutral alumina column) into an aliphatic hydrocarbon fraction (fraction A-1, hexane elution), a neutral polycyclic aromatic hydrocarbon fraction (fraction A-2, benzene elution), a nitrogen-containing polycyclic aromatic compound fraction (fraction A-3, chloroform : ethanol elution), and a hydroxy polycyclic aromatic hydrocarbon fraction (fraction A-4, tetrahydrofuran : ethanol elution). The A-3 fraction was further subfractionated by a second adsorption-column chromatography (silicic acid) to produce fractions enriched in carbazoles (S1), amino-PAHs (S2) and azaarenes (S3). Stock solutions of the distillate cuts and their various subfractions were prepared at concentrations of 10,000 μg/mL in DMSO and stored at -196°F under nitrogen until used (Pelroy and Wilson, 1981; Mahlum et al., 1938).

Solvent-Refined Coal II, Bench-Scale Unit, Merriam, Kansas

Sample L1311. A sample of distillate products from conventional mode operation, using Amax Belle Ayr Mine subbituminous coal, was collected at Gulf Research and Development Company's Merriam, Kansas laboratory. In this process mode, high-temperature, high-pressure separator bottoms contain middle and heavy distillates, SRC, insoluble organic matter, and ash. A fraction of the high-temperature, high-pressure separator bottoms is used to recycle; the remainder is distilled on-line in an atmospheric column. A sample of this material was used for toxicological evaluation (Wilson et al., 1983b).

Samples L1331, L1341, and L1351. Three samples of recycled distillate product were collected at Gulf Research and Development Company's Merriam, Kansas laboratory. All oil products from the high-temperature, high-pressure separator were glass distilled to a specified cutpoint and the heavy ends recycled in the feed slurry to extinction. Samples of net product (distillate cuts of materials with bp below 518, 554, and 590°F) were collected from an experimental run (28 days), using Amax Belle Ayr Mine (Campbell County, Wyoming) subbituminous coal, and used for toxicological testing (Wilson et al., 1983b).

Sample L1321. A sample of recycle distillate product from a run using Powhatan No. 3 bituminous coal was collected at Gulf Research and Development Company's Merriam, Kansas laboratory. All oil products from the high-temperature, high-pressure separator were glass distilled to a cutpoint of 518°F. The net product of this distillation (materials with bp < 518°F) was assayed for toxicological activity (Wilson et al., 1983b).

H-Coal, Hydrocarbon Research, Inc., PDU, Trenton, New Jersey

Samples L3141–L3144 and L3151–L3154. A suite of samples from the H-Coal PDU was provided by Hydrocarbon Research, Inc. from both syncrude (SC) and fuel-oil (FO) operating modes (9/77 and 9/78) to the EPA/DOE Synfuels Repository at ORNL. Samples included SC atmospheric still overhead (Sample L3141), SC atmospheric still bottoms (Sample L3142), SC vacuum still overhead (Sample L3143), SC vacuum still bottoms (Sample L3144), FO atmospheric still overhead (Sample L3151), FO atmospheric still bottoms (Sample L3152), FO vacuum still overhead (Sample L3153), and FO vacuum still bottoms (Sample L3154) (Kimball and Munro, 1981).

Samples L3111 and L3112. A composite distillate material from the H-Coal process development unit (PDU) was supplied to MRDC for hydrotreating at three levels of severity (Samples H0319–H0321) (Hydrocarbon Research, Inc., 1976; Stein et al., 1977). Split samples of the feedstock (Sample L3112) and hydrotreated materials (Samples H0322–H0324) were later sent to ORNL for toxicological testing (Epler et al., 1981). The ORNL and MRDC split samples were assigned different sample numbers (see Chapter 4, Table 4A.3) because physical property data measured by ORNL (Sample L3112) and MRDC (Sample L3111) for the samples indicate that they may have degraded somewhat during storage and/or that the samples collected for ORNL may not correspond precisely to the run balance periods and samples reported by MRDC.

Samples L3131–L3136. A suite of samples was produced from Illinois No. 6 coal (Burning Star Mine) in the syncrude mode run No. 5 (9/77). Samples included atmospheric still overhead (Sample L3131), naphtha (Sample L3132), light gas oil (Sample L3133), atmospheric still bottoms (Sample L3134), vacuum still overhead (Sample L3135), and vacuum still bottoms (Sample L3136). Samples were diluted in DMSO prior to toxicity testing (Calkins et al., 1983).

Samples L3161 and L3162. Atmospheric still bottoms (Sample L3161) and vacuum still overhead (Sample L3162) samples were collected from the Hydrocarbon Research, Inc. PDU for hydrotreatment tests at the Bartlesville Energy Technology Center (Sutterfield, 1983). The samples were generated during the August–September 1980 Wyodak coal run. The atmospheric still bottoms sample (Sample L3161) was hydrotreated at two different severities (Samples H0431 and H0432), and the vacuum still overhead (Sample L3162) was treated at three different severities (Samples H0441–H0443). The two feeds and five products were evaluated in the Ames assay.

Samples L3121–L3124. A suite of samples was produced from Illinois No. 6 coal in fuel-oil mode run No. 3 (4/76 and 5/76). Samples included atmospheric still overhead (Sample L3121), atmospheric still bottoms (Sample L3122), vacuum still overhead (Sample L3123), and vacuum still bottoms (Sample L3124). Samples were diluted in DMSO prior to toxicity testing (Calkins et al., 1983).

H-Coal, Hydrocarbon Research, Inc., Bench-Scale Unit, Trenton, New Jersey

Samples L3311–L3315 and L3321–L3325. Two suites of samples were produced from separate runs both using Kentucky No. 11 coal. Each suite consisted of a sample of light oil (Samples L3311 and L3321), separator bottoms (Samples L3312 and L3322), filtered liquor (Samples L3313 and L3323), vacuum overhead distillate oils (Samples L3314 and L3324), and vacuum bottoms (Samples L3315 and L3325). All samples were dissolved in DMSO and filter sterilized prior to toxicity testing (Schoeny et al., 1981).

H-Coal Pilot Plant, Catlettsburg, Kentucky

Samples L3231 and L3232. An H-Coal fuel-oil blend (Sample L3231) was created from a 1 : 1.5 blend of light and heavy oils collected (9/25/81) during operation in the syncrude mode on Illinois No. 6 coal. The fuel-oil blend was sent to Chevron Research Company for hydrotreatment at three different severities, using either the original blend

(Sample L3231) or a water-washed blend (Sample L3232) as feedstocks. Both feedstocks and the three hydrotreated products (Samples H0204–H0206) were sent to ORNL for toxicity testing (Griest et al., 1983).

Samples L3210–L3219. A suite of samples was collected (5/14/81) by Ashland Synthetic Fuels, Inc. during operation of the pilot plant in the syncrude mode on Kentucky No. 9 coal. Samples included hydroclone overhead (Sample L3214), light slop oil (Sample L3218), heavy slop oil (Sample L3219), recycle oil (Sample L3217), waste-oil residue (Sample L3210), heavy fuel oil (Sample L3213), light fuel oil (Sample L3212), stabilizer naphtha (Sample L3211), vacuum bottoms (Sample L3215), and vacuum flake product (Sample L3216) (Munro et al., 1985).

Sample L3241–L3246. A suite of samples was collected (10/29/81) during a process run (syncrude mode) using Illinois No. 6 coal. The samples included stabilizer naphtha (Sample L3241), light slop oil (Sample L3246), light fuel oil (Sample L3242), heavy fuel oil (Sample L3243), hydroclone overhead (Sample L3247), and vacuum flaked product (Sample L3245). A DMSO slurry was made for toxicity testing (Munro et al., 1985).

Sample L3234. A sample of heavy fuel oil was collected (9/24/81) during a process run (syncrude mode) using Illinois No. 6 coal. The oil was distilled into six different boiling-range fractions (Samples N3211–N3216) (Cowser, 1982b).

Sample L3244. H-Coal pilot-plant light (Sample L3242) and heavy (Sample L3243) oils collected (10/30/81) during operation in the syncrude mode were blended at a ratio of 1.0:1.4. The material was fractionally distilled into narrow (50°F) bp cuts (Wilson et al., 1985).

Sample L3251–L3257. A suite of samples was collected (7/8/82) from a process run (syncrude mode) using Wyodak coal. The samples included stabilizer naphtha (Sample L3251), hydroclone overhead (Sample L3255), atmospheric stripper (Sample L3254), light fuel oil (Sample L3252), heavy fuel oil (Sample L3253), light slop oil (Sample L3256), and heavy slop oil (Sample L3257) (Munro et al., 1985).

EDS Bench-Scale Unit, Baytown, Texas

Sample L4311. A hydrotreated recycle solvent was collected from the EDS Recycle Coal Liquefaction Unit (RCLU) and sent to the EPA/DOE Synfuels Repository (Kimball and Munro, 1981).

EDS PDU, Baytown, Texas

Samples L4211–L4215. A suite of samples collected (9/77) at the EDS Coal Liquefaction Pilot Plant (CLPP) (1 ton/day, Illinois No. 6 bituminous coal, Monterey Mine) and supplied by Exxon, Inc. Samples included naphtha (Sample L4211), solvent (Sample L4212), and vacuum gas oil (Sample L4213). Samples were dissolved in DMSO and tested as received with the exception of the vacuum gas oil, which was further distilled (Vigreux column) to yield a heavy gas oil (Sample L4214) and residue (Sample L4215) (Calkins et al., 1983).

Sample L4221. A 400–700°F distillate from a Wyodak coal run at CLPP was sent to Bartlesville Energy Technology Center for hydrotreatment tests (Sutterfield, 1983). The distillate was hydrotreated at four different severity levels, and the feed (Sample L4221) and products (Samples H0411–H0414) were evaluated in the Ames assay.

Sample L4231. A solvent blend from an Illinois No. 6 coal run at CCLP was sent to Bartlesville Energy Technology Center for hydrotreatment tests (Sutterfield, 1983). The solvent blend was composed of 2.5 parts of 400–750°F distillate and 1 part 700–1000°F vacuum gas oil. The blend was hydrotreated at four different severity levels, and the feedstock (Sample L4231) and products (Samples H0421–H0424) were evaluated in the Ames assay.

Samples L4241, L4242, and L4243. A suite of samples was obtained from CLPP and sent to the EPA/DOE Synfuel Repository. Samples included vacuum gas oil (Sample L4241), spent solvent (Sample L4242), and hydrotreated recycle solvent (Sample L4243) (Kimball and Munro, 1981).

EDS Coal Liquefaction Plant, Baytown, Texas

Samples L4111, L4112, and L4131–L4134. A suite of samples from either the "once through" (only process solvent recycled) or bottoms and vacuum gas oil (VGO) recycle (vacuum bottoms mixed with hydrogenated process solvent) operation mode using Illinois No. 6 coal were collected at the EDS Coal Liquefaction Plant (ECLP) for toxicological studies. Samples included raw naphtha (bottoms and VGO recycle operation, Sample L4131), recycle solvent (materials from both modes of operation, Samples L4111 and L4132), heavy vacuum gas oil (once through operation, Sample L4112), and vacuum bottoms (bottom and VGO recycle operation, Sample L4133). In addition, a full-boiling-range material from a bottom recycle operation (Sample L4134) was simulated by blending raw naphtha (Sample L4131) and process solvent (Sample L4132) (60 : 40). This reblended material was fractionally distilled and 50°F cuts with bp cuts greater than 650°F (Samples N4101–N4105) studied for toxicity. The process samples and distillates were fractionated by adsorption-column chromatography (neutral alumina column) into an aliphatic hydrocarbon fraction (fraction A-1, hexane elution), a neutral polycyclic aromatic hydrocarbon fraction (fraction A-2, benzene elution), a nitrogen-containing polycyclic aromatic compound fraction (fraction A-3, chloroform : ethanol elution), and hydroxy polycyclic aromatic hydrocarbon fraction (fraction A-4, tetrahydrofuran : ethanol elution). Stock solutions of the distillate cuts and their various subfractions were prepared at concentrations of 10,000 μg/mL in DMSO and stored at −196°F under nitrogen until used (Later et al., 1984).

Sample L4116. A 1 : 1 distillate blend of No. 1 sidestream (Sample L4111) and bottom (Sample L4112) material from the ECLP fractionator processing recycle (hydrotreated solvent) collected during use of Illinois No. 6 coal was hydrotreated by the Chevron Research Company for toxicity testing. A single hydrotreated sample was available for testing (Griest et al., 1983).

Samples L4111, L4113–L4115, L4121, L4132, and L4141. A suite of samples was collected from the ECLP pilot plant during operations with Illinois No. 6 coal. Samples included naphtha (Sample L4115), a methanol–water extracted naphtha (Sample L4114), recycle solvent (Sample L4111), and fuel oil (Sample L4113) from a once-through operation mode. A recycle solvent (Sample L4132) and a hydrotreated naphtha (Sample L4135) were collected during a bottoms and VGO recycle (>800°F material from the vacuum distillation tower) mode of operation. Two recycle solvents, similar to Sample L4132, except that they were collected during operations with Wyodak subbituminous coal (Sample L4121) and Texas lignite (Sample L4141), were also saved for testing. Samples were stored at 4°C (39°F) under nitrogen until tested (McKee et al., 1984).

Integrated Two-Stage Liquefaction, C. E. Lummus PDU, New Brunswick, New Jersey

Sample H0633. A total liquid product material (<650°F) from the C. E. Lummus ITSL PDU in New Brunswick, New Jersey was collected and sent to PNL for toxicity assay (Later, 1985).

Samples L5111 and H0611. A set of samples was collected during processing of Indiana No. 5 bituminous coal under normal operating conditions at the C. E. Lummus PDU in New Brunswick, New Jersey. The set included a sample of the LC Finer feed (Sample L5111) and the LC Finer total liquid product (Sample H0611). LC Finer feed is the product of hydroextraction of a coal slurry followed by solvent deashing. The LC Finer total liquid product is the result of catalytic cracking of the feed, resulting in an upgrading of the product (Wilson et al., 1983a). Each of these materials was further distilled to produce >850°F material. The crude sample and the >850°F cuts were also further fractionated by adsorption-column chromatography (neutral alumina column) into an aliphatic hydrocarbon fraction (fraction A-1, hexane elution), a neutral polcyclic aromatic hydrocarbon fraction (fraction A-2, benzene elution), a nitrogen-containing polycyclic aromatic compound fraction (fraction A-3, chloroform: ethanol elution), and a hydroxy polycyclic aromatic hydrocarbon fraction (fraction A-4, tetrahydrofuran: ethanol elution). Each fraction was subfractionated by a second adsorption-column chromatography (silicic acid adsorbent) to produce subfractions enriched in carbazoles (S1), amino-PAHs (S2), and azaarenes (S3). Stock solutions of the distillate cuts and their various subfractions were prepared at concentrations of 10,000 μg/mL in DMSO and stored at −196°F under nitrogen until used (Wilson et al., 1983a).

Sample H0641. A sample of LC Finer total liquid product was collected during processing of Illinois No. 6 coal under normal operating conditions at the C. E. Lummus PDU in New Brunswick, New Jersey. The crude sample was further fractionated by adsorption-column chromatography (neutral alumina column) into an aliphatic hydrocarbon fraction (fraction A-1, hexane elution), a neutral polcyclic aromatic hydrocarbon fraction (fraction A-2, benzene elution), a nitrogen-containing polycyclic aromatic compound fraction (fraction A-3, chloroform: ethanol elution), and a hydroxy polycyclic aromatic hydrocarbon fraction (fraction A-4, tetrahydrofuran: ethanol elution). Each fraction was subfractionated by a second adsorption-column chromatography (silicic acid adsorbent) to produce subfractions enriched in carbazoles (S1), amino-PAHs (S2), and azaarenes (S3). Stock solutions of the distillate cuts and their various subfractions were prepared at concentrations of 10,000 μg/mL in DMSO and stored at −196°F under nitrogen until used (Wilson et al., 1983a).

Samples L5121 and H0621. A set of samples was collected during processing of Indiana No. 5 bituminous coal during an experimental alternate mode of operation at the C. E. Lummus PDU, in which the LC Finer feed was recycled back to the short-contact-time reactor for a single cycle before entering the LC Finer. The sample set included the recycled LC Finer feed material (Sample L5121) and the LC Finer total liquid product (Sample H0621). The LC Finer total liquid product was selectively distilled into five bp cuts (ambient to 390°F, 390–500°F, 500–650°F, 650–850°F, and >850°F). The crude sample and the distillate cuts were fractionated further by adsorption-column chromatography (neutral alumina column) into an aliphatic hydrocarbon fraction (fraction A-1, hexane elution), a neutral polcyclic aromatic hydrocarbon fraction (fraction A-2, benzene elution), a nitrogen-containing polycyclic aromatic compound fraction (fraction A-3, chloroform: ethanol elution), and a hydroxy polycyclic aromatic hydrocarbon fraction (fraction A-4, tetrahydrofuran: ethanol elution). Each fraction was subfractionated by a second adsorption-column chromatography (silicic acid adsorbent) to produce subfractions enriched in carbazoles (S1), amino-PAHs (S2), and azaarenes (S3). Stock solutions of the distillate cuts and their various subfractions were prepared at concentrations of 10,000 μg/mL in DMSO and stored at −196°F under nitrogen until used (Wilson et al., 1983a).

Samples L5131–L5134, H0631, and H0632. A set of samples was collected during processing of Illinois No. 6 coal subsequent to a significant exothermic excursion (catalyst beds experienced temperatures up to 1000°F) at the C. E. Lummus PDU. Samples from this off-normal operation mode included short-contact-time reactor heavy oil (nondeashed LC finer feed, Sample L5132), LC finer feed (Sample L5131), LC finer total liquid product

(Sample H0631), which is composed of the heavy (>650°F) and light (<650°F) distillates of the LC Finer total liquid product, recycle solvent (>650°F distillate residue from the antisolvent deashing unit and LC Finer total liquid product, Sample H0632), and antisolvent recovery bottoms (Sample L5133) and its distillate gasifier feed fraction (mineral ash and insoluble organic matter bottoms, Sample L5134). The LC Finer total liquid product was fractionally distilled into two bp cuts: 650 and >850°F. The raw samples and the distillate cuts were fractionated further by adsorption-column chromatography (neutral alumina column) into an aliphatic hydrocarbon fraction (fraction A-1, hexane elution), a neutral polcyclic aromatic hydrocarbon fraction (fraction A-2, benzene elution), a nitrogen-containing polycyclic aromatic compound fraction (fraction A-3, chloroform : ethanol elution), and a hydroxy polycyclic aromatic hydrocarbon fraction (fraction A-4, tetrahydrofuran : ethanol elution). Each fraction was subfractionated by a second adsorption-column chromatography (silicic acid adsorbent) to produce subfractions enriched in carbazoles (S1), amino-PAHs (S2), and azaarenes (S3). Stock solutions of the distillate cuts and their various subfractions were prepared at concentrations of 10,000 μg/mL in DMSO and stored at −196°F under nitrogen until used (Wilson et al., 1983a).

Veba Oel/Ruhrkohle AG, West Germany

Samples L6211 and L6212. Samples of middle (390–610°F) and heavy (>610°F) distillate raw oils were collected (12/80) by Veba Oel personnel and sent to PNL for toxicological testing. Each of the materials was fractionated by adsorption-column chromatography (neutral alumina column) into an aliphatic hydrocarbon fraction (fraction A-1, hexane elution), a neutral polycyclic aromatic hydrocarbon fraction (fraction A-2, benzene elution), a nitrogen-containing polycyclic aromatic compound fraction (fraction A-3, chloroform : ethanol elution), and a hydroxy polycyclic aromatic hydrocarbon fraction (fraction A-4, tetrahydrofuran : ethanol elution). The A-3 fraction was further subfractionated by a second adsorption-column chromatography (silicic acid adsorbent) to produce fractions enriched in carbazoles (S1), amino-PAHs (S2), and azaarenes (S3). Stock solutions of the distillate cuts and their various subfractions were prepared at concentrations of 10,000 μg/mL in DMSO and stored at −196°F under nitrogen until used (Wilson and Pelroy, 1982).

Synthoil, Pittsburgh Energy Technology Center, Pittsburgh, Pennsylvania

Sample L6111. A sample of a final refined product intended as a vehicle oil was produced in a fixed-bed catalytic reactor fed deashed (0.1–0.2%) powdered coal and recycled product oil (Synthoil FB-60-CLP). A sample was sent to the EPA/DOE Synfuels Repository at ORNL (Kimball and Munro, 1981).

Zinc Halide, Conoco Coal Development Company, Stamford, Connecticut

Sample L6311. A 390°F distillate at the product of zinc halide hydrocracking of coal was obtained from the Conoco Coal Development Company and sent to the EPA/DOE Synfuels Repository at ORNL (Kimball and Munro, 1981).

Hydrocarbonization, Oak Ridge National Laboratory, Oak Ridge, Tennessee

Sample L6411. A sample of hydrocarbonization product oil (HC-21) was produced at ORNL and studied for toxicity (Kimball and Munro, 1981).

Nonintegrated Two-Stage Liquefaction, Wilsonville, Alabama

Samples L5211–L5214, L5216, and H0512–H0514. A suite of samples was collected by Catalytic, Inc. personnel during "lined-out" operation in the nonintegrated two-stage liquefaction (NTSL) mode. Samples included light fuel oil (Sample L5211), first-stage product distillate (Sample L5212), deashed product (Sample L5214), critical solvent deasher (CSD) product (Sample L5213), hydrotreater (HTR) feed (Sample L5216), HTR light fuel oil (Sample H0512), HTR distillate bottom (Sample H0513), and HTR vacuum flash bottoms (Sample H0514) (Later, 1985).

Samples L5221, L5231, and L5241. A blend of low-conversion, first-stage SRC liquids [57% SRC light oil (Sample L5231) from the November 1981 test run on Fies mine Kentucky No. 9 coal and 43% SRC process solvent (Sample L5221) from the February 1982 test on Illinois No. 6 coal] was prepared from materials collected by the International Coal Refining Company and then distilled (15 plate stills at reflux ratios of about 10 : 1) at Air Products and Chemicals, Inc., Linwood, Pennsylvania. This blend (Sample L5241) was distilled into nine bp cuts of defined temperature range. The first cut (Sample N5210) represented light oil materials within a bp range of ambient to 350°F. Materials with higher boiling ranges were separated into eight bp cuts of light, middle, and heavy oils in bp ranges increasing by 50°F (350–400°F, 400–450°F, etc.) up to 750°F (Samples N5211–N5218). In addition, a middle distillate sample containing materials boiling within the 400–650°F bp range was assayed (Sample N5219). Aliquots of each fraction tested were diluted in DMSO prior to testing (Drozdowicz and Kelly, 1983h).

Sample H0524. A blend of low-conversion, second-stage SRC liquids [34% hydrotreater unit light oil (Sample H0521) and 66% process solvent (Sample H0522)] was prepared from materials collected during November 1981 by the International Coal Refining Company during an operational run using Fies mine Kentucky No. 9 coal and then distilled (15 plate stills at reflux ratios of about 10 : 1) at Air Products and Chemicals, Inc., Linwood, Pennsylvania. This blend (Sample H0524) was distilled into nine bp cuts of defined temperature range. The first fraction (Sample N5220) represented light oil materials within a bp range from ambient to 350°F. Materials with higher boiling ranges were separated into eight fractions of light, middle, and heavy oils by bp ranges increasing by 50°F (350–400°F, 400–450°F, etc.) up to 750°F (Samples N5221–N5228). In addition, a middle distillate sample containing materials within the 400–650°F bp range was assayed (Sample N5229). Aliquots of each fraction tested were diluted in DMSO prior to testing (Drozdowicz and Kelly, 1983h).

Sample H0521. A sample of second-stage light oil (naphtha) was collected from the hydrotreater unit by the International Coal Refining Company during operation in November 1981 on Fies mine Kentucky No. 9 coal. The material was diluted in DMSO and tested for toxicity as collected (Drozdowicz and Kelly, 1983h).

Sample H0522. A sample of SRC residual fuel was collected from the hydrotreater unit by the International Coal Refining Company during operation of the pilot plant in November 1981 using Fies mine Kentucky No. 9 coal in the two-stage liquefaction (TSL) low-conversion mode. The material was diluted in DMSO for toxicological assay (Drozdowicz and Kelly, 1983h).

Sample L2161. A sample of recycle coke condensate oil was obtained (1982) from experimental coker run No. 55 at the ALCOA Laboratories, New Kensington, Pennsylvania. This oil was the recycled product from four test runs using SRC process solvents as the feed and included material from the International Coal Refining Company's Wilsonville runs No. 225, 235, 209, and 213. The material was diluted into DMSO prior to testing (Drozdowicz and Kelly, 1983h).

Sample L5232. A raw solid product sample was collected (11/81) by the International

Coal Refining Company during operation of the pilot plant using Kentucky No. 9 coal in a manner simulating the SRC-I demonstration plant. The sample was sent to the Air Product and Chemicals, Inc. facility at Linwood, Pennsylvania, where the material was repackaged and distributed for toxicological assay. For the Ames assay, pulverized test material was mixed with either saline, DMSO, or water. The mixture was shaken vigorously and, with the exception of the aqueous suspension used to produce the water extract concentrate, was then sonicated (4 min). Suspensions were then stirred overnight under a nitrogen blanket. Aliquots for testing were removed the next morning. The saline and aqueous suspensions were tested without further treatment. The extracts were clarified (centrifugation and filtration) prior to toxicity testing. An aqueous extract concentrate was made by lyophilization of the filtered extract. All dilutions were made with DMSO (Drozdowicz and Kelly, 1983h).

Sample H0523. A sample of SRC product was collected (11/81) from the hydrotreater unit by the International Coal Refining Company during operation of the pilot plant using Fies mines Kentucky No. 9 coal in the two-stage liquefaction low-conversion mode. The sample was sent to the Air Product and Chemicals, Inc. facility at Linwood, Pennsylvania, where the material was repackaged and distributed for toxicological assay. For the Ames assay, pulverized test material was mixed with either saline, DMSO, or water. The mixture was shaken vigorously and, with the exception of the aqueous suspension used to produce the water extract concentrate, was then sonicated (4 min). Suspensions were then stirred overnight under a nitrogen blanket. Aliquots for testing were removed the next morning. The saline and aqueous suspensions were tested without further treatment. The extracts were clarified (centrifugation and filtration) prior to toxicity testing. An aqueous extract concentrate was made by lyophilization of the filtered extract. All dilutions were made with DMSO (Drozdowicz and Kelly, 1983h).

Integrated Two-Stage Liquefaction, Wilsonville, Alabama

Samples L5311–L5316, L5321, L5323–L5326, L5331, L5333–L5336, L5341–L5346, L5351, L5353–L5356 and H0544, H0554, H0564, H0572–H0574, H0582–H0584, and H0592–H0593. Six sample suites were collected by Catalytic, Inc. during operations in the integrated two-stage liquefaction (ITSL) mode. Samples in the suites included light fuel oil, distillate process solvent, first-stage distillate, deashed product, CSD product, hydrotreater feed, hydrotreater light fuel oil, hydrotreater distillate bottoms, and hydrotreater recycle pasting solvent. The suites represented a start-up period (Samples L5361, L5363–L5365, and H0544) and four major subsequent line-out periods when the catalytic hydrotreater unit was incrementally increased to temperatures of 680°F (Samples L5311–L5316 and H0554), 700°F (Samples L5321, L5323–L5326, and H0564), 720°F (Samples L5331, L5333–L5336, and H0572–H0574), and 740°F (Samples L5351, L5353–L5356, H0592, and H0593). An additional suite of materials was collected from a nonequilibrium transition period from 700 to 720°F (Samples L5341–L5346 and H0582–H0584) (Later, 1985).

Catalytic Two-Stage Liquefaction, Lawrenceville, New Jersey

Samples L5411–L5418. A number of materials from initial runs of the CTSL process were collected by Hydrocarbon Research, Inc. First-stage product (Samples L5411–L5414) and second-stage product (Samples L5415–L5418) were collected; the first-stage product was a tetrahydrofuran extract of filtered product, and the second-stage product was pressure-filtered, topped separation bottoms. Crude and chemical class fractions were evaluated for toxicity (Later and Wilson, 1985).

Liquefaction Products (Solids)

Solvent-Refined Coal I Pilot Plant, Ft. Lewis, Washington

Sample L2221. A product sample derived from a December 1976 run was collected (3/77) from storage. The sample was first finely ground and then extracted and fractionated with organic and aqueous solvents by two different procedures. In the first, the ground product was dissolved in toluene:methanol (3:1) added to Sephadex LH-20; the liquid was then removed by gassing with nitrogen. The dry Sephadex beads containing the product were then added to the top of a glass column packed with methanol:water (85:15) and preswelled Sephadex LH-20. In the second procedure, the ground material was added dry to Sephadex. Hexane was then added to the dry mixture and this swelled material was added to the column. Test materials were then removed from the column by sequential elution with hexane, hexane:toluene (9:1), methanol, toluene:methanol (3:1), and chloroform. Fractions were dried, redissolved in DMSO, and then used for toxicological assays (Pelroy, 1981).

Gasification Products (Tars)

Morgantown Energy Technology Center Gasifier, Morgantown, West Virginia

Sample G1211. The bulk of the relatively heavy molecular weight organic tars are removed from the process stream by passing the product gas (~1200°F) through a humidification chamber. The humidification reduces the gas temperature to 400°F, which condenses the heavy tars. A sample of these tars was collected (12/77) from the tar trap, filtered, and extracted with methylene chloride prior to toxicity testing. Whole tar or Sephadex LH-20 fractionated tar was evaluated (Newton, 1979).

Sample G1221. A tar-trap tar sample was obtained (12/79) while still hot and liquid, dissolved in tetrahydrofuran, and filtered (Whatman No. 1) prior to bioassay. This crude tar extract and/or chromatographic fractions (Sephadex LH-20) were bioassayed (Benson et al., 1982b).

Sample G1231. A tar-trap tar sample was collected (9/80) as described for Sample G1221. Crude tar extract and/or chromatographic fractions (Sephadex LH-20) were bioassayed (Henderson, 1982).

University of Minnesota–Duluth Gasifier, Duluth, Minnesota

Samples G1301–G1308 and G1310–G1323. Twenty-two electrostatic precipitator (ESP) tars were collected over a 2-year period (5/79–2/80). The tars were first subfractionated by a liquid–liquid partitioning into ether-soluble and -insoluble acids and bases. The remaining neutral components were fractionated from a Sephadex LH-20 column into aliphatic, aromatic, polyaromatic, and polar (residue) components by isopropanol elution. Solvents were removed by rotary evaporation and the residue dissolved in DMSO prior to toxicity testing (Epler et al., 1983). Samples G1301 (UMD 13), G1306 (UMD 19), and G1319 (UMD 83) were assayed for toxicological effects other than bacterial mutagenicity.

Sample G1324. A bulk tar collected (11/79) from the holding tank was prepared for toxicological assay as described for Sample G1301 (Epler et al., 1983).

Sample G1325. A vent gas cyclone tar was collected (3/80) and prepared for toxicological assay as described for Sample G1301 (Epler et al., 1983).

Sample G1326. A top gas tar was collected (3/80) and prepared for toxicological assay as described for Sample G1301 (Epler et al., 1983).

Sample G1327. A composite ESP tar representing 11 ESP tar samples was collected over a 2-week operation period (3/12/81–3/26/81) using Colorado coal. The composited material and chemical class fractions were tested for embryotoxicity (Cowser, 1982a).

Kosovo Gasifier (Lurgi), Pristina, Yugoslavia

Sample G1412. An organic liquid/solid crude tar sample identified as particularly volatile was prepared for toxicological assay by extraction and solvent exchange. A 4-mL aliquot of the sample was extracted by shaking at room temperature for several hours with 36 mL of methylene chloride. The extracted organics were then exchanged with the bioassay-compatible solvent DMSO by shaking the methylene chloride phase with equal volumes of DMSO. The methylene chloride was removed under a stream of nitrogen and the sample was reconstituted to 40 mL with DMSO. The sample was checked for bacterial contamination by culturing in nutrient agar before testing (Lee et al., 1981).

Sample G1411. A heavy tar and dust sample (organic liquid/solids) was prepared for toxicological assay as described for Sample G1412 (Lee et al., 1981).

University of North Dakota Energy Research Center* Gasifier, Grand Forks, North Dakota

Samples G1121–G1124. Raw process gas (~700°F) is cooled first in a water-spray precooler, where high-boiling, heavy tars are partially condensed. This precooled gas is then freed of the majority of organic compounds by further temperature condensation (temperature drop to about 300°F) with cooled, recycled gas liquor in a spray washer. Tars, oils, and waters are then gravity separated. The four tar samples were collected (12/80) from this tar/oil/water separator. Raw samples or their pH-separated fractions (acid, base, and neutral components) were dissolved in DMSO prior to toxicological assay (Wilzbach et al., 1982).

Chapman Low-Btu Gasifier (Holston), Kingsport, Tennessee

Sample G1811. A sample of by-product tar and oil was collected from the recycle quench liquor separator tank following plant operations with Virginia bituminous coal. Tar and oil settle to the bottom of the tank and an aliquot of these materials was obtained by lowering a "bailer" to the tank floor. Upon retrieval, aqueous portions were decanted and the combined tar–oil material was assayed (Page, 1978).

Sample G1821. The Department of Energy provided the Kettering Laboratory's Department of Environmental Health (University of Cincinnati Medical College) with a variety of coal-conversion materials for determination of mutagenic activity (3/21/78). This coal-gasification tar sample (ETTM-02) is reported to be from the Holston gasifier (J. Fillo, personal communication). The sample was a brownish-black solid with a small amount of brown liquid on the surface. The surface liquid was toxicologically studied directly with no further preparation. The solid tar was studied (1) dissolved with DMSO, (2) solvent extracted, or (3) dissolved in acetone. In the first method, the tar material was weighed and mixed with DMSO to provide a presumptive concentration of 10 mg/mL. Not all the material dissolved, and the amount of insoluble material was subtracted from the total to determine the appropriate adjusted concentrations. Further dilutions for testing were made with DMSO. For solvent extraction, a solvent volume five times the weight of the sample was added to the sample and agitated vigorously in the dark for 2 h at room temperature. The material was centrifuged and the solvent was removed. The procedure was repeated with a second volume of solvent, the two solvent extracts were pooled, and the solvent was

*Formerly the Grand Forks Energy Technology Center.

removed by evaporation under nitrogen. The procedure was carried out sequentially with increasingly polar solvents (hexane, toluene, methylene chloride, and acetonitrile). In addition, acetone-soluble components were extracted from the tar in a procedure similar to the solvent-exchange procedure. All samples were filter sterilized prior to toxicological testing. In general, samples were tested over a 3 log concentration range (Warshawsky, 1979).

Hoe Creek, Wyoming

Samples G4111 and G4121. Tars were condensed out of product gas from two (1977 and 1979) underground coal-gasification experiments, Hoe Creek II and III. The samples were first extracted with methylene chloride and then fractionated into organic acids, bases, and neutrals by pH-dependent liquid–liquid extraction (Timourian et al., 1982).

Research Triangle Institute Gasifier, Research Triangle Park, North Carolina

Samples G1511, G1521, G1531, G1541, G1551, G1561, G1571, and G1581. In this experimental gasifier, effluent gas passes through a particulate trap and then into a refrigerated condenser unit, where low-volatile organic tars are trapped. Eight such samples of condensed tars were collected for bioassay. Three of the tars were produced during the gasification of Wyoming subbituminous coal in runs No. 33, No. 35, and No. 47 (Samples G1511, G1521, and G1551, respectively). The other three tars resulted from the gasification of western Kentucky No. 9 (Sample G1531), Illinois No. 6 (Sample G1541), and North Dakota lignite (Sample G1561) coals. Tar samples were derived from unspecified runs in which the gasifier was operated under two coal-feed conditions (batch or continuous), using Wyoming subbituminous coal (Sample G1571) and North Dakota lignite (Sample G1581). Because of the complexity of these tar condensates, solvent fractionation (solvent partition) was performed on the samples. The whole-tar and acid, base, nonpolar neutral (aliphatics and 1-2 ring aromatics), medium-polar neutral (polynuclear aromatic hydrocarbons), and polar neutral (oxygenated material) subfractions of whole tar were assayed for toxicity. All test samples were dissolved in DMSO prior to toxicological assay (Nichols et al., 1979).

Gasification Products (Oils/Light Tars)

Morgantown Energy Technology Center Gasifier, Morgantown, West Virginia

Sample G1227. Venturi scrubber water was collected (12/79) from the scrubber decanter. Two samples were derived from this water: sedimented tar and aqueous soluble organic species. The sedimented tar was dissolved in methylene chloride and either toxicologically tested as a crude tar extract or after chromatographic subfractionation on a Sephadex LH-20 column. The aqueous soluble organics were removed from the water by extraction (four times) with methylene chloride following removal of aqueous suspended tar particles by filtration through a Whatman No. 15 filter. Water-soluble organics (lyophilization) and methlyene-chloride-soluble organics and their acid, base, and neutral fractions were used for toxicological testing (Benson et al., 1982b).

Kosovo Gasifier (Lurgi), Pristina, Yugoslavia

Sample G1413. A medium oil (organic liquid) identified as particularly volatile was prepared for toxicological assay by extraction and solvent exchange. A 4-mL aliquot of the sample was extracted by shaking at room temperature for several hours with 36 mL of methylene chloride. The extracted organics were then exchanged with the bioassay-compatible solvent DMSO by shaking the methylene chloride phase with equal volumes of DMSO. The methylene chloride was removed under a stream of nitrogen and the sample was reconstituted to 40 mL with DMSO. The sample was checked for bacterial contamination by culturing in nutrient agar before testing (Lee et al., 1981).

Sample G1414. A light oil or naphtha sample was prepared for toxicological assay as described for Sample G1413 (Lee et al., 1981).

University of North Dakota Energy Research Center Gasifier, Grand Forks, North Dakota*

Samples G1125–G1127. Raw process gas (~700°F) is cooled in a water-spray precooler, where high-boiling, heavy tars are partially condensed. This precooled gas is then freed of the majority of organic compounds by further temperature condensation (temperature drop to about 300°F) with chilled, recycled gas liquor in a spray washer. Tars, oils, and waters are gravity separated. Three oil samples were collected (12/80) from the tar/oil/water separator concurrently with tar samples (Samples G1125–G1127). Raw samples were dissolved in DMSO prior to toxicological assay (Wilzbach et al., 1982).

Hygas Gasifier, Institute of Gas Technology, Chicago, Illinois

Samples G2111–G2113 and G2122, G2123, G2141, and G2151. Recycle oil samples were collected from the oil storage tank during four Hygas pilot-plant test-run periods (12/79–7/80), all of which treated western Kentucky bituminous coal. In this process, raw product gas is first cleaned of entrained char and fines with a cyclone, and the gas is then water quenched (100°F). The majority of condensable organics are thus removed from the gas as oils; these oils are then removed from the quench water in an oil/water separator and sent to the oil storage tank. An additional but small source of the oil in this tank comes from the light oil recovery system which removes oil from the cyclone slurry water. Oil in this tank is then used to slurry coal char to the gasifier, hence the name recycle oil. The initial slurry liquid was toluene, and toluene was used for make-up volume in the recycle oil stream. Because exogenous toluene (volatile compound) was a major constituent of the recycle oil samples, volatile components were removed from all samples by evaporative distillation (rotary evaporation at 120°F and 1 torr). The resulting residue of nonvolatile organics (NVO) contains compounds with atmospheric boiling points greater than 390°F. These residues were dissolved in DMSO prior to toxicological assay. Samples G2111–G2113 came from Hygas run No. 84 (12/6–9/79), G2122 and G2123 from run No. 85 (3/28–30/80), G2141 from run No. 87 (7/1/80). A composited sample of recycle oil and its NVO fraction (Sample G2151), made up of materials from Hygas runs No. 84 (22%) and No. 87 (78%), was used for whole-animal studies (Reilly et al., 1986).

Samples G2114–G2116, G2124, G2125, G2131, and G2142. Recycle oil is slurried with char and pumped to the gasifier. Samples of this slurried oil were collected (12/79–7/80) from the same test runs as recycle oil samples. Only the NVO portion (prepared as above) of the slurry oil was evaluated toxicologically. Samples G2114–G2116 came from Hygas run No. 84 (12/6–12/9/79) G2124 and G2125 from run No. 85 (3/28–3/30/80), G2131 from run No. 86 (5/9–5/14/80), and G2142 from run No. 87 (7/1/80) (Reilly et al., 1986).

*Formerly the Grand Forks Energy Technology Center.

Sample G2143. A cyclone is used as the principal method of removing solid materials from the Hygas product gas stream. Solids from the cyclone are collected in a water slurry. Organic oils condensed on these solids become mixed with the water; these oils are recovered and used as part of the recycle oil with a light oil recovery system. The NVO fraction of a sample of light oil from Hygas run No. 87 (7/1/80), obtained as for the recycle oil samples, was used for toxicological assay (Reilly et al., 1986).

COED, Princeton, New Jersey

Sample G2211. An oil sample from the Char-Oil Energy Development (COED) pyrolysis process was collected by FMC Corporation, partially hydrotreated, and sent to the EPA/DOE Synfuels Repository at ORNL (Sample No. 1106). The sample has also been identified as Synfuel B, B-1, and B-2 (Kimball and Munro, 1981).

Sample G2221. A sample of hydrotreated raw COED process pyrolysis oil produced from Pittsburg seam coal and deemed representative of material that would result from the COGAS coal-conversion process was evaluated for toxicity. The raw oil was fractionally distilled into three commercially usable fractions—naphtha, No. 2 fuel oil, and No. 6 fuel oil. These fractions were used for the toxicity tests (Deprospo, 1982).

Gasification Products (Gas)

Morgantown Energy Technology Center Gasifier, Morgantown, West Virginia

Sample G1242. The condensable phase of cooled and nitrogen-diluted (simulated atmospheric mixing) raw gas was collected (12/77) on a Tenax sampler prefilter. This condensate was extracted with methylene chloride, evaporated to dryness under nitrogen, and dissolved in DMSO prior to toxicity testing (Newton, 1979).

Samples G1222–G1226. Size-fractionated process stream aerosol droplets, which form upon cooling and reducing the pressure of the process gas (five size cut-off ranges, >5.4, >2.1, >1.4, >0.6, and >0.3 μm), were collected (12/79) as undiluted gas condensates using a five-stage multicyclone sampling train at sequential points along the gas cleanup system. The cleanup system is designed to reduce the organic load by condensation resulting from discrete temperature and pressure drops. Samples collected were: G1222—after the cyclone (raw gas), G1223—after the tar trap (major removal of the relatively high-molecular-weight organic tar species), G1224—after the Venturi scrubber (removal of lower-molecular-weight organic tar/oil species), and G1225—after sidestream cleanup (clean gas). Droplets in each cyclone were treated as separate samples, each extracted with methylene chloride prior to toxicity testing. A sample of raw gas (Sample G1226) was also collected in a four-stage Teflon condenser train: ambient (32–50°F), ice bath (32°F), and two dry-ice/acetone temperature traps (−170°F). The 32°F condensate had the largest sample mass and was used after methylene chloride extraction for toxicological assays (Benson et al., 1982a).

Samples G1232–G1235. Undiluted samples of product gas condensates were collected (9/80) at four locations within the gas cleanup system: G1232—after the cyclone (raw gas), G1233—after the tar trap (major removal of relatively high-molecular-weight organic tar species), G1234—after the Venturi scrubber (removal of lower-molecular-weight organic tar/oil species), and G1235—after the electrostatic precipitator, direct and indirect coolers, Holmes–Stretford unit, and alkali scrubber (clean gas). The samples were obtained with a four-stage condenser train (ambient, ~77°F, 32°F, and two −170°F temperature traps) that separates condensates according to vapor pressure. Acetone-suspended condensates from each trap were pooled from a given sampling location. The acetone was removed by rotary evaporation, and the condensates were extracted with methylene chloride. After removal of the methylene chloride, the condensates were diluted in DMSO for toxicity testing. The clean gas sample had a large aqueous component, thus both a methylene-chloride-soluble fraction (organic constituents) and water-soluble fraction (lyophilized residue) were tested for toxicity (Benson et al., 1984).

Sample G1212–G1213. Product gas was sampled (6/81) at two points within the gas cleanup stream: G1212—after the cyclone (raw gas), and G1213—after the electrostatic precipitator (relatively clean gas). The sample was obtained using a four-stage Teflon condenser (~77°F, 32°F, and two −170°F temperature traps). Materials from all four traps were pooled and evaluated as a single sample. Organic constituents of the condensates were methylene-chloride-extracted prior to toxicological testing (Shimizu et al., 1984).

Sample G1236. Gas combustion products were collected (9/80) after the corrosion/erosion unit with a four-stage condenser train (32–50°F, 32°F, and two −170°F temperature traps). The sample had a large aqueous as well as organic component, thus both the methylene-chloride-soluble fraction (organic constituents) and water-soluble fraction (lyophilization) of the pooled condensates from each trap were tested for toxicity (Benson et al., 1984).

Glen-Gery Brick Company Low-Btu Gasifier, York, Pennsylvania

Sample G1611. Clean product gas from anthracite coal was collected (4/1/78) using a source assessment sampling system (SASS) train. Particulate matter (a >3-μm fraction and a <3-μm fraction) and volatilized organics (extract from XAD resin) were studied (Thomas et al., 1979).

Sample G1612. Anthracite-derived, low-Btu gas was combusted in a test burner and the flue gas collected (4/8/78) with a high-volume SASS train. Two types of material were gathered for toxicological assay: the particulate matter and the volatilized organics (extract from XAD resin) (Thomas et al., 1979).

University of North Dakota Energy Research Center Gasifier, Grand Forks, North Dakota*

Sample G1111. Raw product stream gas condensates were collected (7/80) by a sidestream sampler located immediately downstream of the gasifier and prior to any gas cleanup. In general, the sidestream samples were an oil–water mixture, with the aqueous phase being the principal component. The sampler collected five discrete samples: a high-pressure cyclone condensate (100–400 psig, ~300°F), low-pressure cyclone condensate (18 psig, 200°F), first-stage condenser (60°F), second-stage condenser (32°F), and a cryogenic trap. The second-stage condenser condensate and cryogenic trap condensate were combined and treated as a single sample. Nonvolatile organic components (high molecular weight, bp > 390°F) were collected from the oil and aqueous phases of each condensate as a residue upon removal of the more-volatile species by evaporative distillation at 120°F (1 torr). These residues were dissolved in DMSO for toxicological assay (Wilzbach et al., 1982).

*Formerly the Grand Forks Energy Technology Center.

Ft. Snelling Low-Btu Gasifier, Ft. Snelling, Minnesota

Sample G1711. A product gas sample was obtained (12/13–12/14/78) at the start-up vent flare duct just prior to the cyclone (no gas cleanup) when the gasifier was operating with Indian Head lignite. The sample was collected using a high-volume SASS train. A hot electrostatic precipitator (285°F) was first used to remove particles and oil and tar aerosols (constituents that might plug conventional filters). A glass-fiber polish filter was used after the ESP. Organic components remaining in the gas were then removed with an organic module. Condensable organics were collected by cooling the gas and organic components present in the vapor phase by trapping in an XAD resin canister. Materials collected in the various SASS train components (ESP and organic module) were composited in ethyl ether and extracted with methylene chloride (1 : 1). Samples were solvent exchanged with DMSO for toxicological testing by mixing them with an equal volume of DMSO for 1 min, then removing the methylene chloride and ethyl ether under nitrogen gas (6 h, 86°F). The resulting solvent-exchanged sample was concentrated three-fold prior to bioassay (Kilpatrick et al., 1980).

Sample G1712. A SASS train sampler (see Sample G1711) was used to collect (12/12/78) organics from the test burner combustion effluent gas. The sample was taken downstream of the test burner at a 90° elbow leading to the scrubber. After removal of the particulates in sized cyclones (10, 3, and 1 μm) and a final glass filter, the organics were collected in an organic module as described for Sample G1711. Materials collected in the cyclones, filter, and organic module were composited in ethyl ether and extracted with methylene chloride (1 : 1). The composited sample was solvent exchanged with DMSO, as described for Sample G1711, and reconstituted two-fold prior to bioassay (Kilpatrick et al., 1980).

Hygas Gasifier, Institute of Gas Technology, Chicago, Illinois

Samples G2117–G2119, G2126, G2127, G2132, and G2144. Product gases present in the Hygas pilot-plant low-temperature reactor (1100°F) were collected from four test runs (12/79–7/80), using western Kentucky bituminous coal. Gas samples were freed of entrained char (avoids particle adsorption of organics upon cooling) by a stainless-steel mesh filter and were flash condensed. The sampling train consisted of a condenser, a series of stainless-steel sampling bombs, and a back-pressure regulator. Hot gas is cooled in the condenser, and the raw condensate is collected. Aqueous- and oil-phase condensates resulted. Organic components in aqueous phases were extracted with methylene chloride prior to testing. All oil-phase condensates were toxicologically evaluated, but only a few aqueous-phase samples were studied. The condensates were studied either in their crude form or fractionated to their nonvolatile organic (NVO) components, the condensate residue (high-molecular-weight species, atmospheric bp > 390°F) that results from removal of the more volatile species by evaporative distillation at 120°F (1 torr). In some cases, this NVO fraction was fractionated further into acidic, basic, and neutral components by pH extraction techniques. Crude and fractionated materials were dissolved in DMSO prior to toxicological assay. Samples G2117–G2119 were from Hygas run No. 84 (12/6–12/9/79), G2126 and G2127 were from run No. 85 (3/28–3/30/80), G2132 was from run No. 86 (5/9–5/14/80), and G2144 was from run No. 87 (7/1/80) (Reilly et al., 1981, 1986).

Mellon Institute Gasifier, Pittsburgh, Pennsylvania

Samples G3144HT–G3148HT, G3153HT–G3156HT, G3159HT–G3176HT, G3178HT–G3182HT, and G3186HT-G3195HT. Organic components of product gas were collected

(5/81–9/82) in a condensation sampling train. A high-temperature tar, which forms as the product gas temperature drops to ~210°F (207–338°F), is the first sample obtained. These samples represent the higher-molecular-weight organics and were the major portion of the organic species present in the gas. Thirty-three high-temperature (HT) tars were obtained from test runs with North Dakota lignite pyrolyzed under variable operating conditions covering a range of reactor residence times (0.2 and 1.3 s), coal carrier gases (H_2, He, CO_2, and He/H_2O), pressures (3 and 10 atm), and temperatures (1110, 1470, and 1830°F). The HT tar samples represented 26 different combinations of operating conditions. An additional nine HT samples, each from test runs with unique operation conditions using Montana Rosebud subbituminous coal, were also collected. All samples were organic condensates, except that aqueous phases were present in the runs using steam as a carrier dissolved in DMSO. Only organic phases were used for bioassay (Fillo et al., 1983).

Samples G3144RT–G3148RT, G3153RT–G3156RT, G3159RT–G3176RT, G3178RT, G3180RT–G3182RT, G3184RT, and G3186RT–G3195RT. A second set of samples, the room temperature (RT) trap oils, condense from the gas at temperatures between ambient and 212°F. These were collected (5/81–9/82) by further cooling the gas (water-cooled heat exchanger) to approximately ambient temperatures (97 to 45°F). Aqueous and organic condensates were always present in this trap. Thirty-three organic-phase samples were obtained from test runs with North Dakota lignite pyrolyzed under variable operating conditions covering a range of reactor residence times (0.2 and 1.3 s), coal carrier gases (H_2, He, CO_2, and He/H_2O), pressures (3 and 10 atm), and temperatures (1110, 1470, and 1830°F). An additional nine RT organic-phase samples, each from test runs with unique operation conditions using Montana Rosebud subbituminous coal, were also collected. All samples were dissolved in DMSO prior to toxicological assay (Fillo et al., 1983).

Samples G3146LO–G3148LO, G3153LO–G3156LO, G3159LO–G3161LO, and G3163LO–G3167LO. A third set of condensates were collected (5/81–9/82) by passing the ambient temperature gas stream through a low-temperature trap (refrigerated cylinder operated at −176°F). These samples, the low-temperature light oils (LOs), contained aqueous and organic phases. Seventeen LOs were obtained from test runs with North Dakota lignite pyrolyzed under variable operating conditions covering a range of carrier gases (H_2, He, CO_2, and He/H_2O), pressures (3 and 10 atm), temperature (1470°F and 1830°F), and a constant residence time (0.2 s). All samples were dissolved in DMSO prior to toxicological assay (Fillo et al., 1983).

CHAPTER **4**

Engineering Analysis of the Toxicological Database for Coal-Conversion Materials

J. C. Craun* *and* J. P. Fillo
ERT (A Resource Engineering Company), Pittsburgh, Pennsylvania

This chapter presents an analysis of the coal-conversion toxicological database that is focused on understanding the engineering factors that influence the toxicological characteristics of various coal-conversion samples. Major factors addressed are

- sample collection, handling, and storage procedures,
- physical properties,
- process type and conditions, and
- schemes for separating raw coal liquids into products.

Results provide a basis for identifying general trends in the toxicological characteristics of coal-conversion materials.

The issue of sample collection and handling is important in determining whether the results of toxicological testing are representative of the material being sampled. This issue is addressed briefly in Section 4.1 but is not covered in this chapter at length, other than to confirm that these sample concerns appear to play a relatively minor role in explaining variations in toxicological characteristics compared to other factors, such as sample properties and process conditions. As described in Section 4.1 and in Chapter 2, the toxicological characteristics of coal-conversion oil and tar samples are relatively insensitive to storage

**Present affiliation:* Reilly Tar & Chemical Corp., Indianapolis, Indiana and handling conditions as long as the samples are not stressed by significant exposures to light or high temperatures in the presence of oxygen.

The relationships between toxicological characteristics and the physical properties of coal-conversion materials are central to the data analysis in this section. This central role reflects a number of factors. First, as described in Section 4.1, there is a substantial body of historical research indicating that sample boiling points and related properties are indicative of the toxicological characteristics of petroleum- and coal-derived liquids. This is also seen for more recent coal-conversion toxicological characterization work. Second, the relationships between sample properties and toxicological characteristics need to be defined before the effects of process conditions can be explored. Finally, a clear definition of the effects of sample properties in common engineering terms should be an aid to process developers, industrial hygienists, and others concerned with controlling the potential for adverse health effects of coal-conversion materials.

Understanding how process conditions affect toxicological characteristics should also aid efforts to control potential health risks in coal-conversion facilities. While available data are not adequate to develop quantitative explanations of the reaction mechanisms that produce toxic constituents, some trends in this area are discernible. These trends help indicate the extent to which process modifications or controls may be effective in minimizing or eliminating the production of certain toxic constituents.

The basic approach to the data analysis in this chapter is first to examine the relations between sample properties and toxicological characteristics, then to look at the relations between process conditions and toxicological characteristics. This order is necessary because sample properties are strongly related to toxicological characteristics. Therefore, the effects of these sample properties must be defined before the effects of process conditions can be considered.

The toxicological data analyzed in this chapter address only mutagenic and carcinogenic activity. These effects have been the primary focus of toxicological research to date on coal-conversion materials. However, coal-conversion materials can also exhibit other toxic effects (see Chapter 3).

4.1 APPROACH

Certain basic physical properties of a complex hydrocarbon material can be indicative of the material's toxicological properties. It has been shown that the boiling range of a coal-conversion material is indicative of its mutagenic/carcinogenic activity (see Chapters 3 and 5). Thus, knowledge of the relation between sample boiling points and toxicological characteristics is required before samples from different coal-conversion processes, or from the same process but collected under different process conditions, can be compared. For example, it would be inaccurate to conclude that lack of carcinogenic activity in a solvent-refined coal II (SRC-II) middle distillate and the presence of carcinogenic activity in an H-Coal heavy distillate means that the H-Coal process produces more carcinogens or more potent carcinogens. In fact, an SRC-II middle distillate boils over a sufficiently low temperature range that carcinogenic activity is not detected. Higher-boiling SRC-II distillates, like H-Coal heavy distillates, possess carcinogenic activity. Thus, mutagenic and carcinogenic activities can only be compared meaningfully for coal-conversion materials with similar boiling ranges. In addition, when materials of similar boiling ranges are compared, the relative process yields of the materials should be recognized. Different types of coal-gasification processes and conditions, for example, can produce by-product tars with similar properties and carcinogenic activities but at yields that differ by an order of magnitude (see Section 4.4).

The example above illustrates the importance of accounting for sample properties when comparing toxicological and chemical data for different coal-conversion materials. This section briefly reviews historical toxicological data on petroleum- and coal-derived materials that indicate important sample properties. The scope of the available database that provides information to evaluate relations among these sample properties and toxicological effects of coal-conversion materials is described. Finally, the data analysis techniques used in evaluating these relations throughout the rest of this chapter are described.

4.1.1 Key Sample Properties

Numerous investigations have been conducted on the carcinogenicity of petroleum and petroleum-derived products. Work from the 1950s and 1960s provides a basis for defining bulk sample properties that should be relevant in evaluating differences in mutagenic/carcinogenic activity of coal-derived liquids. Key results of these early studies have been reviewed by Fillo and Craun (1985) in the context of potential correlations with coal-conversion sample properties. A useful general review of the carcinogenicity of petroleum-derived materials is provided by Bingham et al. (1980).

The experimental literature for petroleum-derived materials indicates two important determinants of biological activity. First, boiling-point range is important. Early work by Smith et al. (1951), for example, showed that carcinogenic activity in cracked petroleum stocks is generally confined to narrow-boiling-range materials boiling above 700°F or to full-boiling-range materials with a significant fraction boiling above 700°F (Tables 4.1 and

Table 4.1 Carcinogenicity of Boiling-Point Cuts for Tar from a Petroleum-Derived, Steam-Cracked Paraffinic Gas Oil

			Activity at 244 Days (%)		
Sample Number[a]	Boiling Range[b] (°F)	Volume (%)	Cancer Index (CI)[c]	Tumor Index (TI)[d]	CI/TI
109	IBP–1050	100.0	23[e]	46[e]	0.50
393	IBP–600	27.4	0	4	0.0
394	600–650	15.4	0	10	0.0
395	650–700	15.2	0	10	0.0
396	700–750	6.2	12	58	0.21
397	750–800	5.4	33	74	0.45
398	800–1050	6.3	41	72	0.57

[a] Samples 393–398 are narrow-boiling-range distillates obtained from Sample 109.

[b] Boiling ranges are approximate as distillation was imperfect; each cut contains higher-boiling components than cut points indicate. Initial boiling point not specified. IBP = initial boiling point.

[c] $$\text{CI} = \text{Cancer Index} = \frac{\text{total number of animals developing cancerous tumors}}{\text{original number of animals minus number dead at 90 days from causes other than cancer}}$$

[d] $$\text{TI} = \text{Tumor Index} = \frac{\text{total number of animals developing tumors}}{\text{original number of animals minus number dead at 90 days without tumors}}$$

[e] Indices reported at 250 days for Sample 109.

Source: Data from Smith et al. (1951).

Table 4.2 Carcinogenicity of Boiling-Point Cuts for Tar from a Petroleum-Derived, Catalytically Cracked Oil

			Activity at 300 Days (%)		
Sample No.[b]	Boiling Range (°F)[c]	Volume (%)	Cancer Index[a] (CI)	Tumor Index[a] (TI)	CI/TI
191	600–1010+	100.0	16	32	0.5
404	600–650	24.1	0	0	—
405	650–700	29.1	0	0	—
406	700–750	21.4	0	30	0
407	750–800	11.1	0	29	0
408	800–900	1.9	14	36	0.39
409	900–930	2.1	17	50	0.34
410	930–950	2.1	22	43	0.51
411	950–965	2.1	28	64	0.44
412	965–980	2.1	10	60	0.17
413	980–1010	1.5	19	46	0.41
415[d]	1010+	2.5	4	15	0.27

[a] See Table 4.1 for definition of indices.
[b] Samples 404–415 are narrow-boiling-range distillates obtained from Sample 191.
[c] Equivalent temperatures at atmospheric pressure for boiling-point cuts distilled at 2–10 mm Hg.
[d] Sample 415 was a 50% solution in pyridine of the bottom boiling above 1010°F.

Source: Data from Smith et al. (1951).

4.2). There is also some indication that carcinogenic activity decreases for materials boiling above 1000–1100°F.

Second, aromaticity of petroleum materials is related to carcinogenic activity. For example, thermally or catalytically cracked petroleum stocks, enriched in aromatic condensed-ring compounds, are typically more active than straight-run paraffinic or naphthenic materials (Bingham et al., 1980; Fillo and Craun, 1985). Aromatic petroleum fractions are usually more active than nonaromatic fractions. In particular, 4- to 6-ring polycyclic aromatic hydrocarbons (PAHs) are associated with carcinogenic aromatic materials and fractions.

Although not as extensive, experimental data on the carcinogenicity of materials from early coal-liquefaction pilot plants are consistent with the boiling-point effect described above for petroleum materials (Fillo and Craun, 1985). Skin-painting data reported by Hueper (1956) on oils from the direct coal-hydrogenation facility at Louisiana, Missouri, indicate marginal or no activity in samples expected to have lower boiling ranges. Weil and Condra (1960) reported skin-painting data on liquids from the direct coal-hydrogenation pilot plant at Institute, West Virginia, that showed no activity in a light oil boiling below 500°F and increasing activity in higher-boiling oils.

As described in Chapters 3 and 5, more recent toxicological testing of coal-liquefaction samples has confirmed that mutagenic/carcinogenic activity is associated with the higher-boiling range liquids. In addition, amino-PAH compounds are responsible for much of the mutagenic activity observed in coal liquids. These data are considered in detail in Section 4.2.

4.1.2 Scope of Available Data

The available data for petroleum and coal-liquefaction materials suggest that boiling point and aromaticity are likely to correlate with mutagenic and carcinogenic activity in coal-conversion materials. Given this perspective, the approach required for the data analysis in this chapter is determined by the size and scope of the database that incorporates both physical properties and toxicological effects data for coal-derived oils and tars.

Table 4.3 summarizes the scope of the database available to relate sample properties (boiling range and ultimate analysis) and toxicological characteristics (Ames *Salmonella*/mammalian microsome mutagenicity, tumor-initiating activity, and dermal carcinogenicity) of coal-conversion materials. Four basic material categories are shown: full-boiling-range liquefaction materials, narrow-boiling-range coal liquids, gasification by-product oils and

Table 4.3 Numbers of Coal-Conversion Samples for Which Both Sample Properties and Toxicological Data Are Available

	Number of Samples with Both Types of Data Available[a]			
Types of Data	Full-Boiling-Range Liquefaction Samples	Narrow-Boiling-Range Cuts	Gasification Samples	Hydrotreatment Samples
Ames assay with				
Ultimate analysis	0	0	26	3
Distillation	8	40[b]	0	6
Both	80	23	3	44
Total	88	63	29	53
Initiation/promotion assay with				
Ultimate analysis	0	0	0	0
Distillation	2	12[b]	0	0
Both	27	0	0	4
Total	29	12	0	4
Skin-painting assay with				
Ultimate analysis	0	0	0	0
Distillation	2	4[b]	0	0
Both	14	0	1	6
Total	16	4	1	6
Total number of samples with at least one bioassay and at least one property measured	91	63	29	56
Total number of samples with at least one bioassay with or without measured properties	176[c]	63	194[d]	65[e]

[a] Numbers determined from the appendix unless otherwise noted.

[b] Includes information on nominal boiling range and/or actual distillation data.

[c] From Chapter 3, Table 3.1 minus hydrotreating samples listed therein.

[d] From Chapter 3, Table 3.2.

[e] Fifty-eight samples listed in the appendix to this chapter, plus seven samples listed in Chapter 3, Table 3.1, that are not listed in the appendix.

tars, and hydrotreated coal liquids. Materials from the latter two categories generally have wide boiling ranges. Samples with various combinations of toxicological and physical properties data are shown for each category of materials. Note that a given sample may be represented more than once in the table (e.g., a sample with distillation, mutagenicity, and initiation data is counted twice).

The types of data in Table 4.3 were selected (see Fillo and Craun, 1985) to provide the largest number of coal-conversion materials with consistent data on both sample properties and mutagenic/carcinogenic activity. Distillation data are critical for examining boiling-point effects. Ultimate analyses provide the largest consistent data set to evaluate effects of aromaticity [e.g., by atomic hydrogen : carbon (H : C) ratios or hydrogen contents]. Although this is a crude measure of aromaticity, the analysis is widely employed and contributes significantly to correlation. Ames, initiation/promotion, and skin-painting data represent the largest consistent data sets on the mutagenic/carcinogenic effects of coal-conversion materials. Gas chromatography–mass spectrometry (GC-MS) chemical data are important in identifying specific mutagenic/carcinogenic compounds within these complex samples. Further details on the compilation, origin, and quality of the data represented by Table 4.3 are available in Fillo and Craun (1985).

Table 4.3 indicates that of the 498 coal-conversion materials tabulated in Chapter 2 as having had some toxicological testing performed, almost half (239) have relevant physical property data available (distillation, or ultimate analysis, or both). This proportion is highest for narrow-boiling-range and hydrotreated materials and lowest for gasification materials. Almost all coal-liquefaction materials (full-boiling-range, narrow-boiling-range, or hydrotreated materials) for which physical property data could be located included distillation data, because such data are routinely used for process characterization. It is also noteworthy that almost all samples with at least one bioassay and at least one physical property included an Ames assay (233 of 239 samples), either alone or in conjunction with other bioassays.

4.1.3 Data Analysis Techniques

The appendix at the end of this chapter provides a complete listing of all coal-conversion samples with toxicological and chemical data that are discussed in this chapter. This listing provides a unique identification number for each sample, summarizes the types of biological, chemical and engineering data available, and provides references. All the samples and data discussed in this chapter can be traced back to the original references given in the appendix, using the sample identification numbers given here and in Table 4A.5, which lists the numbers of the samples used in the various data plots. In many instances, however, some manipulation of the original data was required to develop the data analysis figures and tables presented here. The procedures used for such manipulations are described in this section, together with the basic assumptions underlying much of the data analysis.

4.1.3.1 *Sample Data Linkage*

There are very few instances in the coal-conversion toxicological database where physical properties and toxicological characteristics (chemical and/or biological) were both measured on precisely the same sample. Instead, physical properties and toxicological characteristics were often measured on different samples of a given process stream or process material that were collected at different times, ranging from a few hours to many days apart. In most cases, however, sufficient information is available on process operating conditions and process stream sample properties to assume reasonable correspondence between the different samples. In some cases, adequate process data are available to demonstrate steady-state operation with only minor variations in bulk sample properties. In others, it must be assumed that the properties of the material sampled were reasonably steady or that any variations were slight relative to the accuracy of the property measurements and bioassay performed. In any case, some assumptions must be made to establish the correspondence of different samples for physical and toxicological testing.

The important exception to the problem described above is the database for narrow-boiling-range cuts of selected coal-liquefaction products. For these samples, physical properties and toxicological characteristics were measured on the same materials as part of the experimental design, although often with significantly different lag times between generating a sample and performing the different types of test.

The strength of the basis for linking physical properties and toxicology samples is indicated by the interpretive indices assigned to the samples. As described in the appendix, the interpretive index reflects the level of available or inferable knowledge concerning sample collection and handling procedures, sample properties, and process conditions. Interpretive indices have been assigned to every sample considered in this chapter and listed in the appendix. The primary references listed in the appendix can be consulted for the full extent of process data available for linking properties and toxicology data for any given sample.

The data analysis presented in the remainder of this chapter documents general trends that reflect many samples with varying degrees of confidence in linking sample property and toxicology data. Hence, any errors introduced by the assumptions required in linking sample data are either minor relative to other sources of error, such as imprecision in testing methods, or are reflected in the relative confidence with which certain trends can be established. A few instances where potential sample linkage errors could significantly alter specific trends or conclusions drawn from this chapter's data analysis are noted in the chapter. Otherwise, the general difficulties and assumptions required in linking sample property and toxicology data should be recognized when evaluating and using the data analysis results presented in this chapter. As noted above, this caution does not apply to narrow-boiling-range cuts prepared from selected coal-liquefaction products.

4.1.3.2 *Sample Stability*

The question of sample stability is related to the linkage of sample property and toxicological data. For most samples in the coal-conversion toxicological database, the time elapsed between sample collection and analysis of physical properties is substantially different from the time between collection and analysis of toxicological characteristics, in some cases by over a year. Therefore, certain assumptions must be made regarding the extent to which a sample's physical properties and toxicological characteristics change over time.

As discussed in Chapter 2, the toxicological characteristics of coal liquids are relatively insensitive to storage and handling conditions as long as samples are not stressed by severe exposures to light and/or high temperatures in the presence of oxygen. These same stresses are also important in determining the extent of sample physical degradation, as indicated by increased viscosity and gum formation (see Fillo and Craun, 1985, for a brief review of the relevant literature). If samples are stressed sufficiently, the available data indicate that there may be some reduction in mutagenic activity and a decrease in the concentrations of related mutagenic compounds. However, in most instances mutagenic samples retain their toxicological activity, at least qualitatively, even though they may have undergone some stress during storage and handling.

For most samples in the coal-conversion toxicology database, the interval between sample collection and property testing was shorter than the corresponding interval for toxicological characterization. In addition, it is reasonable to assume that the process developers, who in most cases were directly responsible for sample property testing as part of routine monitoring of process performance, took adequate steps to prevent significant sample degradation prior to testing. Hence, the controlling factor in the data analysis presented in this chapter generally is the extent to which samples tested for toxicological characteristics were degraded. While there are very few instances where the stability of a sample tested for toxicological characteristics was actually measured, available documentation on sample handling and storage conditions can often be used, in light of the sample degradation experimental results discussed in Chapter 2, to make reasonable inferences about the likely extent or lack of significant sample degradation. In many cases, however, there is little or no documentation of sample storage conditions, thus requiring subjective judgment in assessing potential sample degradation.

The assumptions and inferences made concerning sample stability are reflected in the interpretive indices reported in the appendix for each sample. Most of the samples have sampling/handling indices of B or better, indicating that the sample is known, or can reasonably be inferred, not to have degraded significantly. However, samples with sampling/handling indices of C (suspected or inferred degradation) are included in the data analysis presented in the remainder of this chapter to maximize the scope of the available sample property/toxicological database, which would be much reduced in some key areas if C-rated data were eliminated. Fillo and Craun (1985) report that there are no biases in data analysis results similar to those presented here that can be attributed clearly to sample degradation. Moreover, the successful delineation, later in this chapter, of general trends for data sets that include C-rated samples indicates, similar to the case for sample data linkage, that any errors introduced by sample degradation are either small relative to other sources of error or are reflected in the relative confidence with which certain trends can be established. Nonetheless, the potential that sample degradation may have affected toxicological characterization results for certain samples included in this chapter's data analysis should be recognized.

4.1.3.3 *Sample Properties Data*

Sample distillation data are a critical engineering component of the coal-conversion toxicological database. Various distillation methods were used to characterize the materials considered in this chapter. Distillation methods most commonly represented in the database are distillation by simulated gas chromatography (SGC); atmospheric distillation, usually by American Society for Testing and Materials (ASTM) method D-86; and vacuum distillation for heavier distillates or bottoms, usually by ASTM method D-1160 (ASTM, 1977a–c). The applications and limitations of these methods in the context of the coal-conversion toxicological database have been reviewed by Fillo and Craun (1985).

In this chapter, the primary concern with different distillation methods is how well they represent the true boiling point (TBP) curve. The data analysis presented here assumes that SGC distillations approximate the TBP curve but that D-86 and D-1160 distillations need to be corrected to yield an approximate TBP curve. The corrections usually applied for D-1160 distillations were based on the American Petroleum Institute (API) correction curves for petroleum distillates (API, 1976). The Gulf correlation developed for SRC-II liquids (Gray, 1981), which is a modified version of the API correction method, was usually used for D-86 distillations. The estimated accuracy of the SGC or corrected ASTM distillations is about 10–30°F at a given yield. Further distillation corrections applied are given by Fillo and Craun (1985).

There is no consistency in reporting distillation yields on a weight or volume basis for samples in the coal-conversion toxicological database. Both are used about equally, which introduces additional error when comparing distillation curves. However, the difference is usually only a few percentage points in yield for a given temperature. Weight % yields are slightly less than volume % yields at a given temperature, because specific gravity increases with boiling point. Conversely, at a given yield, the temperature for a weight % yield distillation curve is higher than that for a volume % yield curve, typically by 10–20°F. No attempt was made to convert reported distillation curves to a common yield basis (weight or volume %), since the required data on specific gravity versus boiling point are usually not available for specific samples. In addition, inaccuracies introduced by not making the correction are relatively minor compared to the variability in toxicological results and the distillation database as a whole.

In some cases, standard distillation data are not available for coal-conversion materials, but distillation curves can be estimated roughly from reported distillation yield data. For example, distillation curves are not reported by Calkins et al. (1980) for the materials they studied. However, distillation curves could be estimated from gas chromatography (GC)-simulated analyses reported by Calkins and co-workers as yields of five predefined, boiling-point fractions, ranging from 150 to 330°F in width. In addition, distillation data are available from process developers for selected EDS and H-Coal liquefaction materials tested by Calkins et al. (1980).

The data analysis in this chapter frequently uses the 95% distilled temperature (also referred to as the 95% point) in correlating toxicological characteristics. Vacuum still bottoms and high-boiling residua have heavy ends boiling beyond the upper temperature limit of available distillation methods. Data for such samples were plotted with their 95% point temperatures as being greater than 1000°F. For distillate materials with only a 90% and either a 100% or other endpoint temperature reported, the 95% point was estimated as the 90% point plus one-third the difference between the 90% temperature and the final boiling point. This procedure was normally applied to corrected D-86 or D-1160 distillation data.

The data analysis in this chapter uses the average boiling point as the correlate for narrow-boiling-range cuts. In practice, this is usually a nominal average, based on the nominal initial and final boiling points for a narrow cut (e.g., 775°F is taken as the average boiling point for a nominal 750–800°F cut). In the few instances where actual distillations of the narrow cuts were performed, the average of the 10, 50, and 90% boiling points was used. In cases where the cut is reported as a nominal final end cut (e.g., greater than 800°F), the average boiling point was assumed to be 50°F higher than the nominal starting point.

Elemental analyses required to develop atomic H : C ratios are generally available. However, in certain cases carbon analyses are not reported (e.g., hydrotreated materials from Mobile Research & Development Corporation and Bartlesville Energy Technology Center studies), and carbon content was estimated by the difference between 100% and the sum of the hydrogen, nitrogen, oxygen, and sulfur analyses. Atomic H : C ratios were estimated from these data.

4.1.3.4 *Toxicological Data*

Toxicological results analyzed in this chapter are from the Ames *Salmonella*/mammalian microsome mutagenicity assay (Ames assay), initiating activity in the initiation/promotion assay, and chronic dermal exposure (skin-painting) assays. The limitations of these assays are discussed in Chapter 3. Details on how their results were used in this chapter's data analysis are given below.

Only Ames assay results for strain TA98 were considered in the engineering data analysis. This strain has been used more frequently and is generally more sensitive than other strains for coal liquids (Later et al., 1984). For consistency, Ames assay results reported as "none detected" were assigned a detection limit of 0.02 revertants/microgram (rev/μg). The latter is near the lowest reported detection limits, which generally range from 0.01 to 0.10 rev/μg.

Throughout this chapter, Ames assay data are reported as the slope of the linear portion of the dose–response curve (rev/μg). A distinction is made between Ames assay results for unfractionated samples and those representing the weighted sum of results for individual fractions (i.e., the sum of mutagenic activities measured in various chemical class fractions weighted by the relative proportions of the fractions). This distinction is required because results on fractions are not always additive. Some constituents in a crude sample may mask the mutagenicity of components. Thus, weighted-sum results are often more sensitive and give a lower effective detection limit (see Chapter 3).

While Ames assay mutagenicities measured on unfractionated samples and those calculated as a weighted average from sample fraction assays represent distinctly different tests, the two types of data give consistent results in the trends developed in this chapter's data analysis. This is consistent with results reported by Fillo and Craun (1985), who correlated unfractionated and weighted-sum strain TA98 Ames assay results for 42 coal-liquid samples where both types of analysis were available. Of 33 samples where both types gave positive responses, the two assays agreed within a factor of 10 for all but three samples. For two-thirds of the samples, the agreement was within a factor of 2, which is of the same order of magnitude as the interlaboratory accuracy of the assay. There were seven cases for which the Ames assay gave no detectable response on unfractionated samples, but where testing of fractions gave a positive weighted-sum result. There were only two cases where unfractionated samples showed activity, and fractionated samples did not.

The samples considered in this chapter generally had Ames assay results reported as either unfractionated or weighted-sum mutagenicities. In some cases, both types were reported. There are cases where unfractionated sample results were reported, but sufficient data were also reported to calculate weighted-sum results; in that case, weighted-sum mutagenicities were calculated. Samples with both unfractionated and weighted-sum results are used both ways in the data analysis in this chapter. For example, there are instances in Fig. 4.1 and 4.8 (shown later in this chapter) where one sample has both an unfractionated and a weighted-sum mutagenicity plotted.

Initiation/promotion results are expressed in this chapter as percentage of test animals with tumors, measured after 180 days. Data from different studies can be displayed in this fashion because the laboratories reporting initiation/promotion results used similar test protocols, usually with 30 mice per test (see Chapter 3). Individual data points are coded for different initiator dose ranges. A linear dose–response curve has been demonstrated for a single SRC-II heavy distillate sample (Mahlum et al., 1983), but this may not be a valid assumption for all samples.

Chronic skin-painting results are presented in this chapter as percentage of test animals with tumors, with data coded for the dose range tested. Where both sexes were tested, which is the standard procedure, the percent tumor data used is the average for males and females. However, this is a semiquantitative measure, because varying test protocols were used at different laboratories, and it ignores other measures of carcinogenic potency, such as total number of tumors or latency period. A more quantitative expression of carcinogenicity is response relative to a known carcinogen, such as benzo[a]pyrene (BaP). However, there are fewer coal-conversion samples for which responses relative to BaP can be calculated (see Chapter 3). Where relative potency data are available, they are plotted as the average value from different doses.

4.2 ENGINEERING ANALYSIS OF THE COAL-LIQUEFACTION TOXICOLOGICAL DATABASE

Coal-liquefaction processes have produced the largest number of materials with data available on both toxicological characteristics and key sample properties. The analysis of this database yields important trends for subsequent discussions of the coal-liquid hydrotreating and coal-gasification databases. In addition, the coal-liquefaction database includes the results of extensive characterization of narrow-boiling-range coal liquids. The data for these narrow-boiling materials provide important perspectives on the relative importance of sample properties and process conditions as indicators of mutagenic and carcinogenic activity of coal liquids.

4.2.1 Narrow-Boiling-Range Coal Liquids

Most of the toxicological and chemical studies of narrow-boiling-range coal liquids were performed at a single laboratory, where investigators tested materials from the SRC-II, SRC-I, EDS, and H-Coal processes (Pelroy and Wilson, 1981; Wilson et al., 1982; Later et al., 1984; Wilson et al., 1985). In addition, Drozdowicz and Kelly (1983) report results for narrow-boiling-range cuts derived from the first and second stages of the Wilsonville SRC pilot plant operated in the nonintegrated two-stage liquefaction (NTSL) mode, and Cowser (1982) reports results from H-Coal pilot-plant distillate cuts. Narrow-boiling-range cuts were prepared in laboratory distillation apparatus from full-boiling-range materials produced during nine different process runs. A wide range of reaction severities was represented during these runs, as indicated by hydrogen consumptions, which ranged from 2.3 to 6.2 % of moisture and ash-free feed coal (Table 4.4). Bituminous coals were the feeds in each run.

Full-boiling-range liquefaction products (*parent samples*) were generally distilled to produce narrow cuts boiling over a nominal range of 50°F (Table 4.4). Distillations of the narrow cuts themselves are generally not available to check the accuracy of the nominal boiling ranges. However, ASTM Method D-1160 distillations are reported by Sullivan (1983) for the nominal 50°F H-Coal cuts (Table 4.5). These data indicate that the nominal boiling ranges correspond only approximately to the corrected D-1160 distillation curves. However, ASTM D-1160 distillations, even when corrected, do not necessarily reflect true boiling-point curves for narrow cuts of high-boiling coal liquids (Gray, 1981). Table 4.5 indicates that nominal distillation ranges should not be assumed to give precise indications of actual boiling ranges for narrow cuts.

4.2.1.1 *Mutagenic/Carcinogenic Activity*

The Ames and initiation/promotion assays have been used most consistently on narrow-boiling cuts. Other mutagenesis and carcinogenesis assays have also been applied on occasion, but not to all samples.

Figures 4.1 and 4.2 show that Ames assay mutagenicity and initiation/promotion activity, respectively, are strongly affected by boiling range of the material tested. Ames assay activity is limited to material boiling above about 650°F. Above 650°F, mutagenic activity increases rapidly with temperature, up to about 800°F, at which point the activity appears to level off. Figure 4.2 shows similar trends for initiation/promotion results, although some activity may occur down to about 600°F for SRC-I materials, and there is no apparent leveling off of activity.

Chronic skin-painting data have been reported (Mahlum et al., 1983) for one set of

Table 4.4 Process Conditions for Narrow-Boiling Cut Feeds

No. of Samples	Narrow Cut Sample Nos.	Cut Boiling Range[a]	Process	Parent Sample No.	Parent Sample Name
4[d]	N1102–N1105	50°F	SRC-II P-99 PDU[e]	L1111	Process solvent
13[d]	N1106–N1118	Generally 20–40°F	SRC-II P-99 PDU	L1121	Process solvent
5[d]	N1119–N1123	100–140°F	SRC-II P-99 PDU	L1131	Process solvent
7[f]	N2103–N2109	50°F	SRC-I Ft. Lewis pilot plant	L2121	Process solvent
5[g]	N3201–N3205	50°F	H-Coal pilot plant	L3242/43	1.4 : 1.0 Light fuel oil: heavy fuel oil blend
6[g]	N3211–N3216	45–95°F	H-Coal pilot plant	L3234	Heavy fuel oil
5[h]	N4101–N4105	50°F	EDS-ECLP	L4131/32	Naphtha/process solvent blend
9[i]	N5210–N5218	50°F	NTSL first-stage	Blend of 57% L5231 and 43% L5221	Light oil[j] Process solvent
9[i]	N5220–N5228	50°F	NTSL second-stage	Blend of 34% H0521 and 66% H0522	HTU Light oil[k] HTU process solvent[k]

[a]Based on 10–90% boiling-point range for samples with distillations reported (N1106–N1123), otherwise based on nominal boiling ranges reported.
[b]Measured in lb moisture-free coal/hr · ft^3.
[c]MAF = moisture- and ash-free.
[d]Gray (1981) and McIlvried (1982).
[e]PDU = process development unit.
[f]Catalytic, Inc. (1982).
[g]Morrison et al. (1983).

narrow-boiling-range cuts and are consistent with the results shown in Figs. 4.1 and 4.2. A 300–700°F distillate (Sample N1101) from the SRC-II P99 process development unit (PDU) showed 0.01% potency relative to BaP, with a regular increase in potency for three 50°F cuts starting at 700°F and a fourth, >850°F, cut (see Chapter 3, Table 3.62).

Other cellular assays have also been used to characterize narrow-boiling-range cuts. These include microbial mutagenesis assays (fluctuation test, forward mutation assay, prophage induction test) and mammalian cell assays (Syrian hamster embryo cell transformation and Chinese hamster ovary cell mutation) (Mahlum et al., 1983; Later et al., 1984). Results confirm genotoxic activity above 700°F in SRC-II materials, although the lowest-boiling cut tested was a full-boiling-range, 300–700°F material (Mahlum et al., 1983). Thus, activity in cuts in the range of 600–700°F may have been masked. When 600–650°F and 650–700°F EDS cuts were tested, significant activity was reported in the 650–700°F

Coal Type	Coal Concentration (wt %)	Liquefaction Temperature (°F)	Residence Time (hr)	Space Velocity[b]	Pressure (psig)	H_2 Consumption (wt % MAF coal)[c]
Pittsburgh Ireland Mine	30.0	855	0.99	23.1	2000	4.70
Pittsburgh Powhatan #5	29.6	850	1.02	21.8	2000	4.65
Pittsburgh Powhatan #5	30.6	855	1.00	22.4	1870	3.63
Kentucky #9	36.4	840	—	40.0	2100	2.37
Illinois #6	—	850	—	29.7	3000	6.19
Illinois #6	—	845	—	31.2	3010	4.34
Illinois #6	31.6	840	1.53	—	2550	4.53
Kentucky #9	38.7	836	~0.7	—	2100	2.3
Illinois #6	33.3	785	~0.7	—	2000	—
Kentucky #9[l] first stage	37.5	837	~0.7	—	2100	2.2
second stage[l]	45.4	782	—	2	2400	2.6

[h]Epperly (1982, 1983).

[i]Drozdowicz and Kelly (1983) and Catalytic, Inc. (1983).

[j]Light oil L5231 was obtained on 11/7/81 during run 235 under conditions shown in first row, while process solvent Sample L5221 was obtained 2/8/82 during run 236 under conditions shown in second row.

[k]HTU = hydrotreating unit.

[l]First row gives first-stage process conditions and second row gives second-stage process conditions, both during run 235. Second-stage feed concentration and hydrogen consumption are relative to solvent-refined coal feed, and pressure shown is hydrogen partial pressure.

Table 4.5 Distillation Data for H-Coal Narrow-Boiling-Range Cuts

Nominal Boiling Range (°F)	Actual Distillation Temperature[a] (°F)				
	Start	10%	50%	90%	Endpoint
650–700	599	653	665	672	691
700–750	706	713	722	734	755
750–800	730	741	751	772	809
800–850	793	804	816	843	890
>850	—	854	961	1100	1173

[a]All are corrected ASTM D-1160 distillations, except for the >850°F cut, which was simulated by thermogravimetric analysis.

Source: Data from Sullivan (1983).

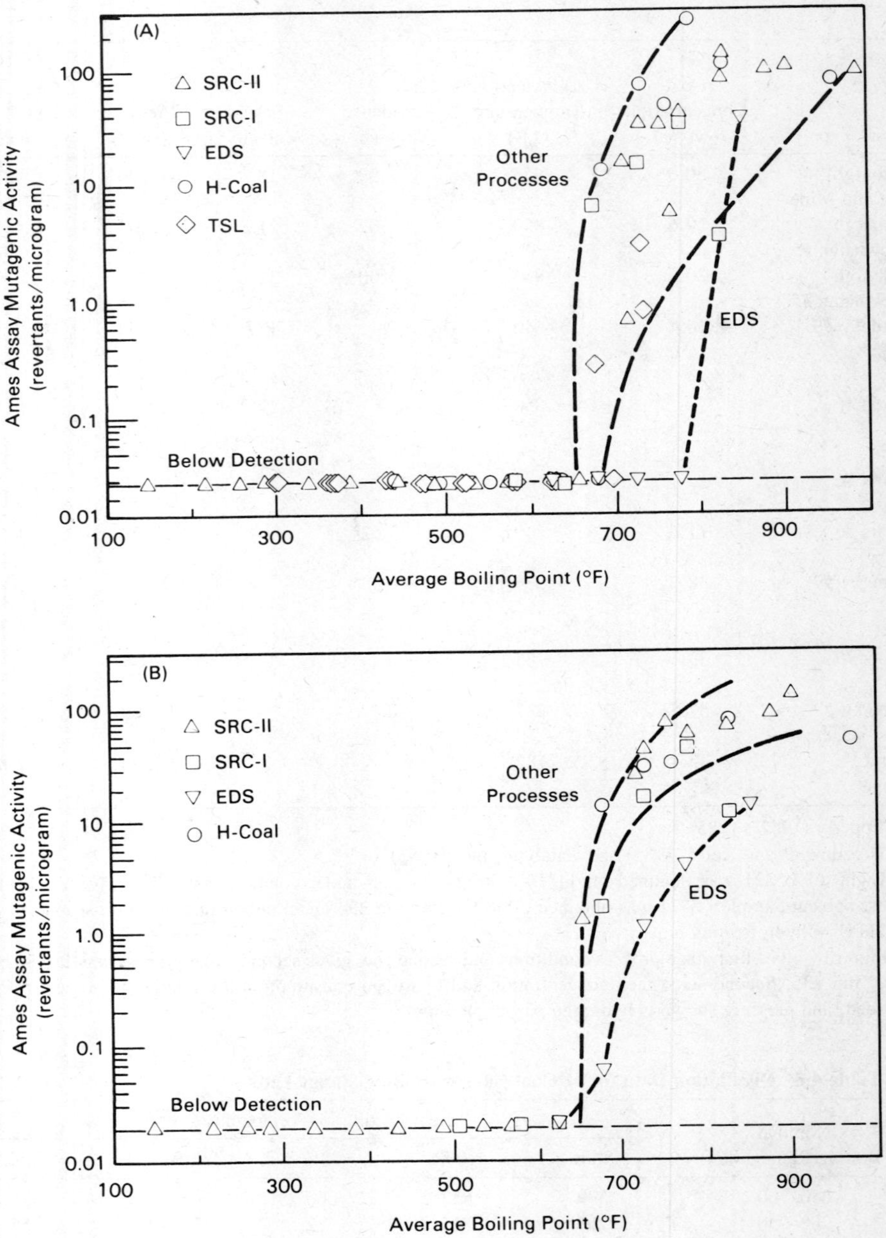

Figure 4.1 Ames assay mutagenicity versus average boiling point for narrow-boiling-range coal-liquefaction samples: (A) unfractionated samples and (B) fractionated samples (weighted-sum results). SRC = solvent-refined coal.

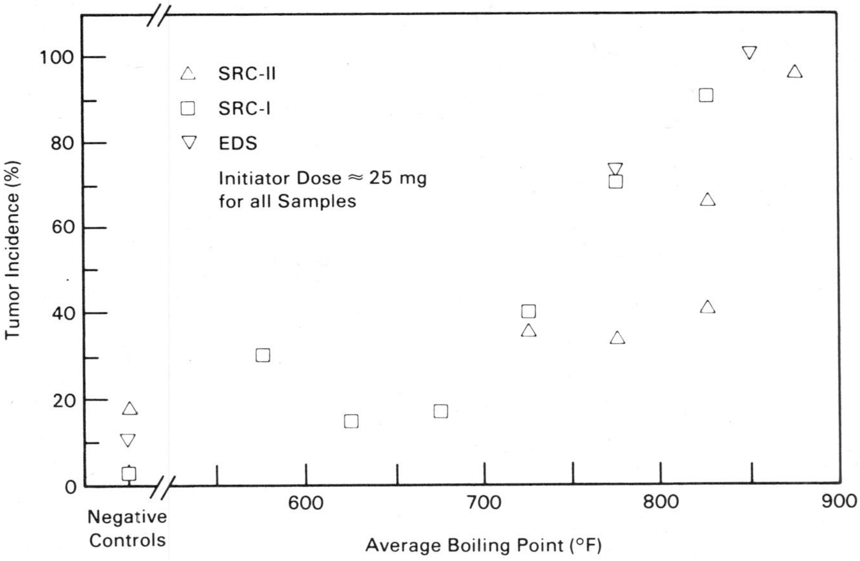

Figure 4.2 Initiating activity for narrow-boiling-range coal-liquefaction samples. SRC = solvent-refined coal.

cut for two microbial mutagenesis assays: Ames and fluctuation (Later et al., 1984). Thus, activity can be expected for materials boiling above 650°F.

4.2.1.2 *Chemical Characteristics*

The separation of active (i.e., mutagenic/carcinogenic) and inactive narrow-boiling-range cuts reflects the boiling points of the mutagenic and carcinogenic compounds present in the full-boiling-range liquefaction products. While all such compounds cannot be identified definitively, certain PAH, heterocyclic-PAH, and amino-PAH compounds in coal liquids are known cellular mutagens and animal carcinogens (Kimball and Munro, 1981). Within these chemical classes, compounds in the higher-molecular-weight range, containing four to six aromatic rings, are generally most active. Since boiling point increases with molecular weight, there is a correlation between boiling range and mutagenic/carcinogenic response.

Figures 4.3 and 4.4 show concentrations of selected PAHs and amino-PAHs, respectively, in different SRC-I, SRC-II, H-Coal, and EDS narrow-boiling-range cuts. The compounds were selected for their consistency in being reported in different samples and for their toxic properties. Benz[a]anthracene, chrysene, and BaP are animal carcinogens (IARC, 1982), and the amines shown are mutagenic in the Ames assay (Kimball and Munro, 1981). Boiling points of the compounds are also shown in the figures. Concentrations are generally at a maximum near or somewhat below the boiling point of the pure compound, except for aminonaphthalenes (Fig. 4.4A). While a given compound usually occurs in a few cuts up to 100°F above and below its true boiling point, the range between minimum and maximum concentrations is often two or three orders of magnitude. These results reflect the inherent inefficiencies of the distillation processes used to prepare the samples.

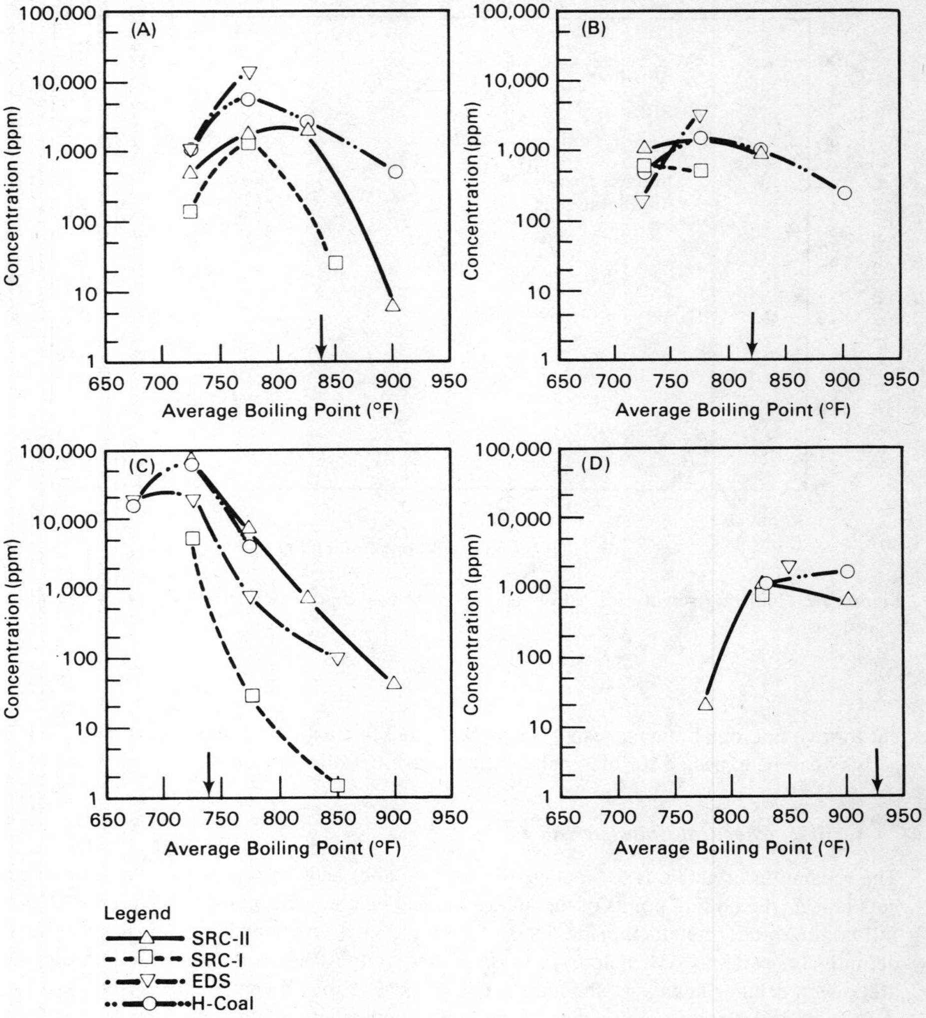

Figure 4.3 Concentrations of selected polycyclic aromatic hydrocarbon compounds in narrow-boiling-range coal liquids versus average boiling point. Boiling point of pure compounds (Anderson and Wu, 1963) indicated by arrows (↓): (A) chrysene, (B) benz[a]anthracene, (C) pyrene, and (D) benzo[a]pyrene. SRC = solvent-refined coal.

Figure 4.3 shows that concentrations of various PAH compounds are generally consistent for the SRC-I, SRC-II, H-Coal, and EDS samples within an order of magnitude. However, concentrations of amino-PAHs in EDS cuts are generally an order of magnitude lower than those in SRC-II cuts (Fig. 4.4). This likely reflects the effects of hydrotreatment in the recycle solvent loop of the EDS process, which deaminates amino-PAHs (see Section 4.3).

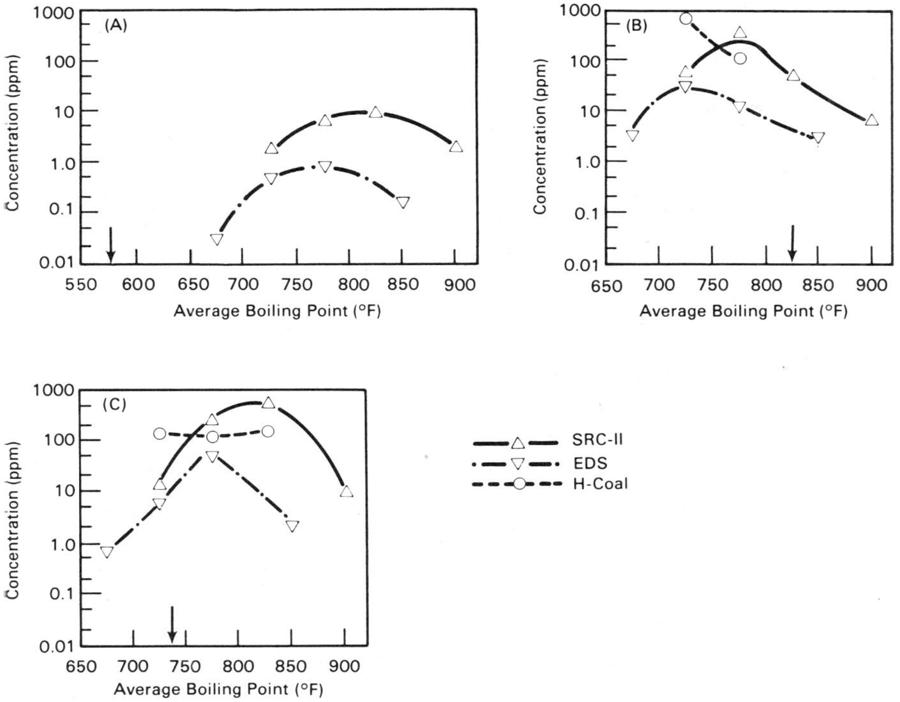

Figure 4.4 Concentrations of selected amino polycyclic aromatic hydrocarbon compounds in narrow-boiling-range coal liquids versus average boiling point. Approximate boiling point of pure compound (Later, 1985b) indicated by arrows (↓): (A) aminonaphthalenes, (B) aminophenanthrenes, and (C) aminopyrenes. SRC = solvent-refined coal.

4.2.1.3 *Other Sample Properties*

Many other sample properties (e.g., aromaticity, nitrogen content, density) are also related to boiling range. Studies by Gray (1981) on a series of 19 narrow-boiling cuts prepared from a PDU-scale SRC-II process solvent illustrate these relations. Eighteen of these samples (N1106–N1123) were also tested toxicologically (Pelroy and Wilson, 1981).

Figure 4.5 shows how hydrocarbon types varied in Gray's narrow-boiling-range SRC-II samples. Note that asphaltene content (high-molecular-weight, highly condensed aromatic compounds) increases sharply at about 700°F, which roughly corresponds to the onset of mutagenic and carcinogenic activity. Thus, one would expect mutagenic/carcinogenic activity and asphaltene content to be correlated, since boiling range affects both properties. However, the relation between asphaltene content (or other measures of aromatic ring size and structure) and mutagenic/carcinogenic activity of a given boiling range cannot be determined from the data available for narrow-boiling cuts.

Figure 4.6 shows how selected ultimate analysis results vary with boiling range for narrow-boiling-range SRC-II cuts (Gray, 1981) and H-Coal cuts (Sullivan, 1983). The atomic H:C ratio (Fig. 4.6A) is an indicator of aromaticity, and nitrogen content (Fig.

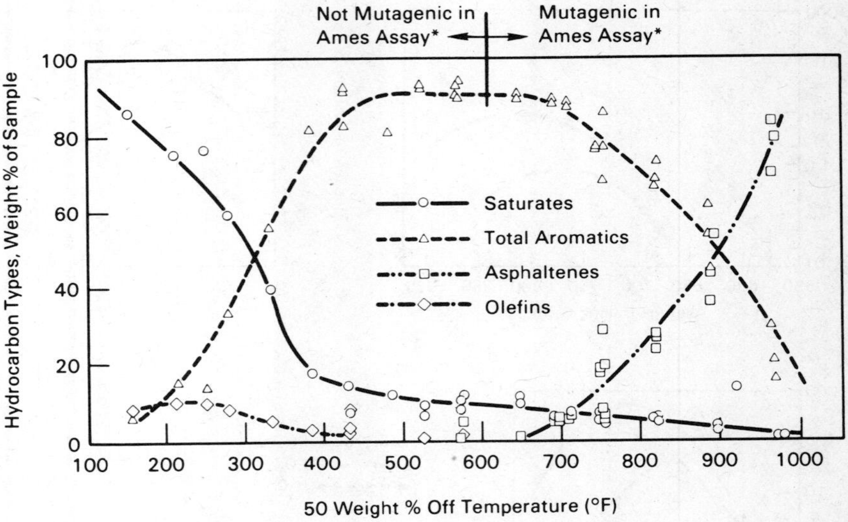

Figure 4.5 Variation in the composition of hydrocarbon types in narrow-boiling range cuts of solvent-refined coal-II coal liquid (adapted from Gray, 1981).

4.6B) is an indicator of azaarene and amine content. Certain higher-molecular-weight compounds within these classes are carcinogens and/or mutagens. In both cases, the demarcations between mutagenically active and inactive samples are shown. H-Coal and SRC-II cuts both show trends of aromaticity and nitrogen content increasing with boiling point. However, the SRC-II cuts are somewhat more aromatic with higher nitrogen content, presumably reflecting more severe hydrogenation in the H-Coal reactor (Table 4.4).

4.2.1.4 *Process Comparisons*

The data for narrow-boiling-range cuts show closely comparable results for Ames assay mutagenicity, initiation/promotion activity, and selected PAH and amino-PAH concentrations for comparable boiling-range cuts from five different coal-liquefaction processes (SRC-I, SRC-II, EDS, NTSL, and H-Coal). Noteworthy exceptions are the lower Ames assay mutagenicities and amino-PAH concentrations for EDS cuts. This apparently reflects the effect of hydrotreatment in the EDS recycle solvent loop in deaminating mutagenic amino-PAHs (Figs. 4.1 and 4.4). Hydrotreatment of the EDS recycle solvent had less effect on PAH concentrations and initiation/promotion activity, however (Figs. 4.2 and 4.3).

The comparable toxicological and chemical characterization results for narrow cuts produced from five different liquefaction processes under substantially different process conditions indicate the importance of the boiling range compared to process conditions in assessing the toxicological characteristics of coal liquids. Coal liquids produced in catalytic and noncatalytic liquefaction processes at processing severities ranging from 2.3 to 6.2 wt% hydrogen consumption on a moisture- and ash-free (MAF) basis show comparable mutagenic/carcinogenic activity for a given narrow-boiling range. Relatively slight changes

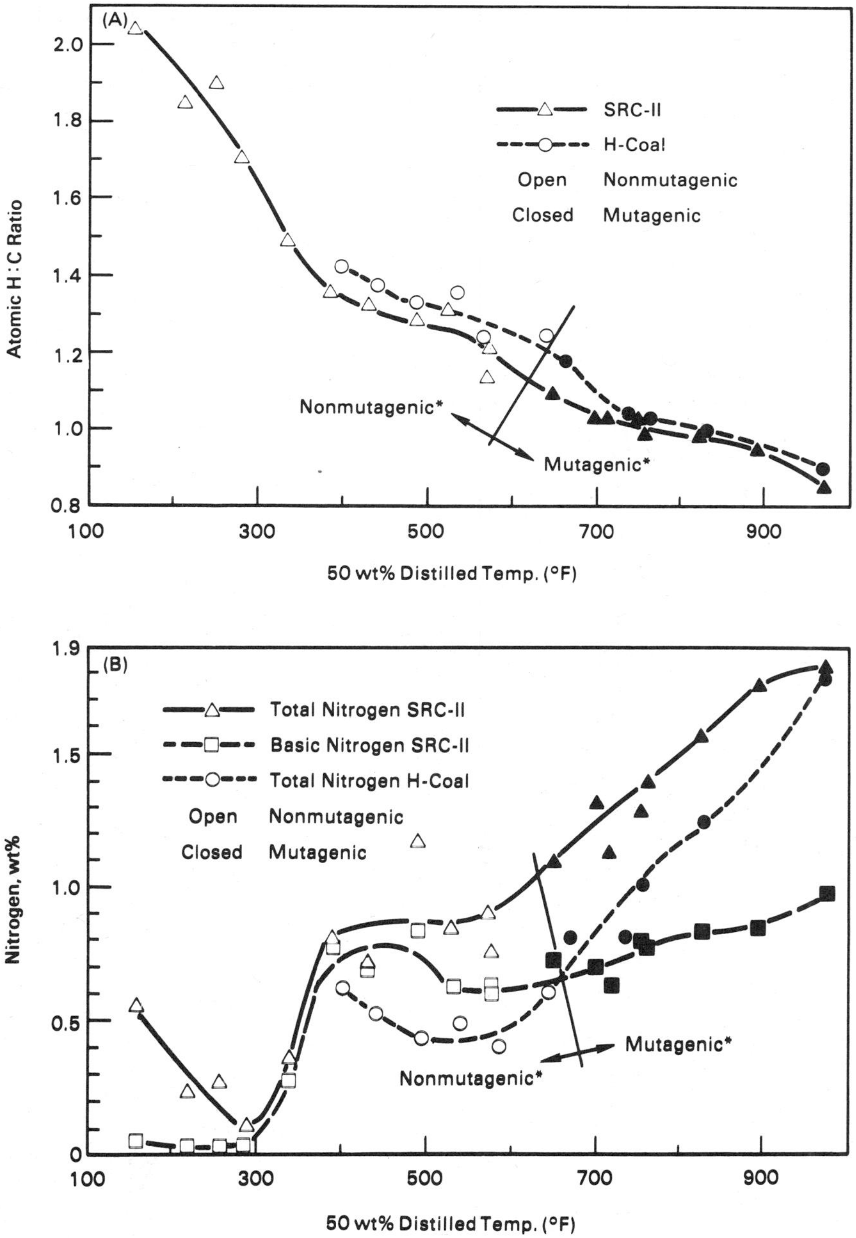

*See Figure 4.1

Figure 4.6 Variation in (A) hydrogen : carbon (H : C) ratio and (B) nitrogen content for narrow-boiling-range solvent-refined coal II (SRC-II) and H-Coal cuts (data from Gray et al., 1983; Sullivan, 1983).

(50°F) in the boiling range, however, can affect mutagenic/carcinogenic activity substantially.

4.2.1.5 *Summary*

Toxicological testing (Ames, initiation/promotion, and skin-painting assays, plus some other cellular assays) of narrow-boiling-range cuts of coal-liquefaction products shows that mutagenic/carcinogenic activity is confined to materials boiling above about 650°F. Above this point, mutagenic/carcinogenic activity increases sharply for higher-boiling cuts. Chemical analyses of narrow-boiling cuts show that known mutagenic and carcinogenic agents follow generally similar trends. Closely comparable toxicological and chemical characteristics are observed for cuts generated from five different liquefaction processes (SRC-I, SRC-II, NTSL, EDS, and H-Coal) over widely ranging processing severities (2.3–6.2 wt % MAF hydrogen consumption). This indicates that the process history of a coal-liquefaction material should be less important than the material's distillation characteristics in determining its mutagenic/carcinogenic activity.

4.2.2 Full-Boiling-Range Coal-Liquefaction Materials

Toxicological data for narrow-boiling-range cuts of coal-liquefaction products show that mutagenic/carcinogenic activity is strongly correlated with boiling point. Cuts boiling below about 650°F show no mutagenic/carcinogenic activity, while above 650°F, such activity increases rapidly. Thus, the amount of high-boiling material in a full-boiling-range coal-liquefaction material should determine the extent of mutagenic/carcinogenic activity.

4.2.2.1 *Importance of the Shape of the Distillation Curve*

Figure 4.7 shows distillation curves for four SRC-II materials: three fuel oils and a middle distillate. Although each material has a similar weighted average boiling point, the fuel oils and middle distillate have distinctly different distillation curves and toxicological properties. The three fuel oils were mutagenic in the Ames assay, while the middle distillate was not. One fuel oil (Sample L1251) was active in the initiation/promotion assay; another (Sample L1262) was carcinogenic in a chronic skin-painting assay. The third fuel oil was not tested in an animal assay.

The three mutagenic and presumably carcinogenic fuel-oil samples have significant "tails" at the heavy ends of their distillation curves (Fig. 4.7). Each fuel oil contained about 10% material boiling above 650°F, while the middle distillate boiled entirely below 600°F. Thus, some measure of the heavy end (the tail) of the material's distillation curve, rather than some overall average boiling point, is required to predict mutagenic/carcinogenic activity.

Fillo and Craun (1985) showed that the 95% distilled temperature is a convenient measure of the heavy end of a distillation curve that can be reliably used to segregate active (i.e., mutagenic and/or carcinogenic) and inactive coal-liquefaction materials based on their distillation curves. This indicator parameter for the heavy end of distillation curves is used in this analysis of the toxicological database for full-boiling-range coal-liquefaction materials.

4.2.2.2 *Mutagenic/Carcinogenic Activity*

Results of Ames, initiation/promotion, and skin-painting assays on full-boiling-range coal-liquefaction materials where distillation data are available are shown in Figs. 4.8–4.11. In

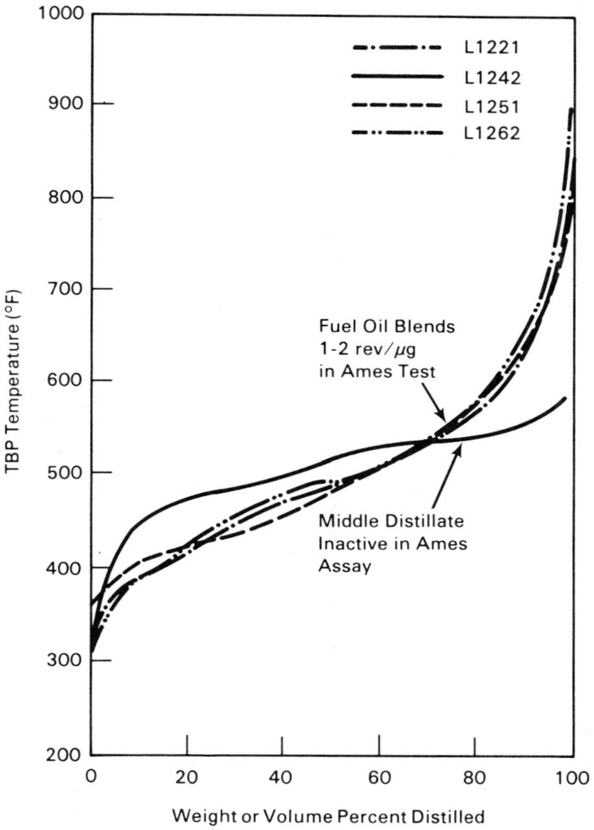

Sample No.	Distillation Characteristics: Method[(a)]	Units[(b)]	ABP1[(c)]	ABP2[(d)]	Ames Assay Strain TA98 Activity, rev/μg: Crude	Weighted Sum
L1221	SGC	wt	510	503	-	1.0
L1242	D-86CG	vol	506	508	<0.01	-
L1251	D1160CP	vol	512	502	2.5	1.2
L1262	SGC	vol	511	505	1.7	0.9

Notes:

(a) SGC = simulated by gas chromatography; D-86CG = American Society for Testing and Materials (ASTM) Method D-86 corrected to total boiling point (TBP) by Gulf procedure (Gray, 1981); D-1160 CP = ASTM Method D-1160 corrected to TBP by American Petroleum Institute procedure (API, 1976).

(b) Distillation yield units (weight or volume basis).

(c) Average of 10, 50 and 90% boiling point (ABP1) temperatures.

(d) Average of 10, 30, 50, 70 and 90% boiling point (ABP2) temperatures.

Figure 4.7 Distillation curves for four solvent-refined coal-II distillates with similar average boiling points but dissimilar mutagenic activity. TBP = true boiling point.

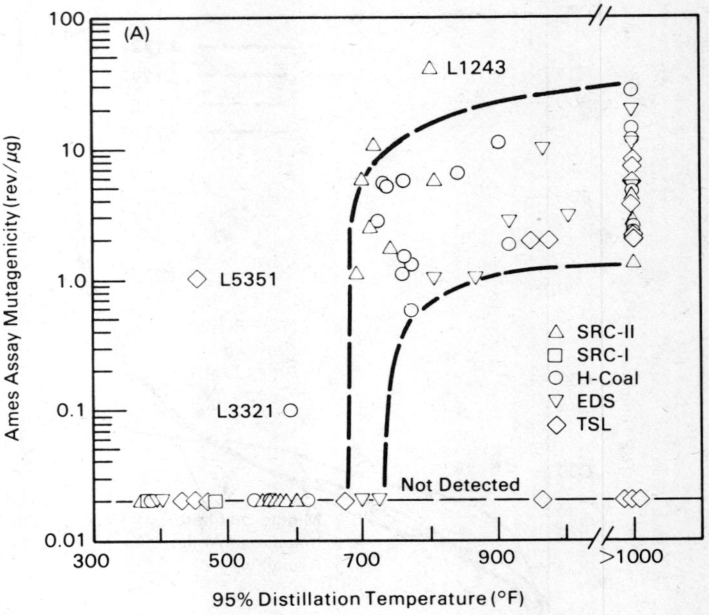

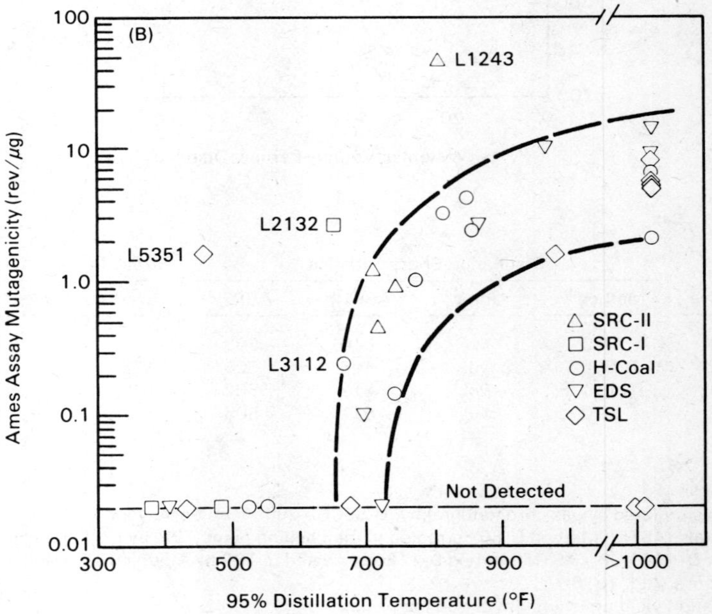

Figure 4.8 Ames assay mutagenicity versus 95% point for full-boiling-range coal-liquefaction samples: (A) unfractionated samples and (B) fractionated samples (weighted-sum results). Numbered points are discussed in text. SRC = solvent-refined coal; TSL = two-stage liquefaction.

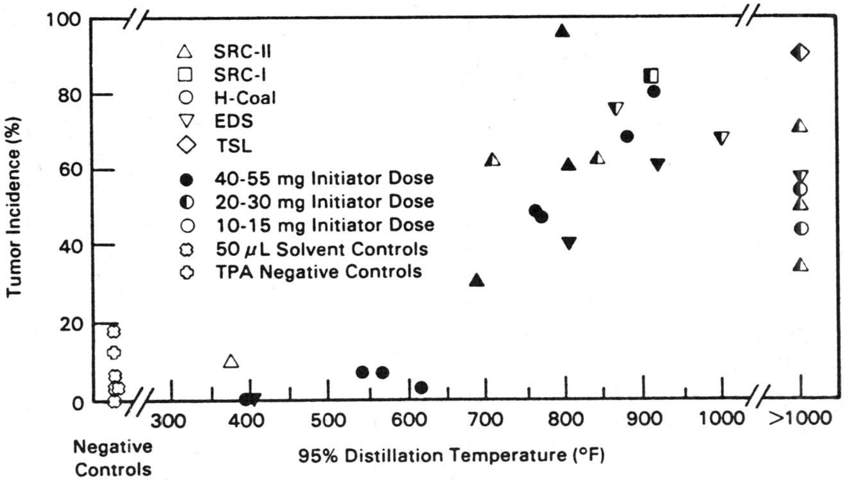

Figure 4.9 Initiating activity versus 95% point for full-boiling-range coal-liquefaction samples. TSL = two-stage liquefaction; SRC = solvent-refined coal; TPA = 12-*O*-tetradecanoylphorbol-13-acetate.

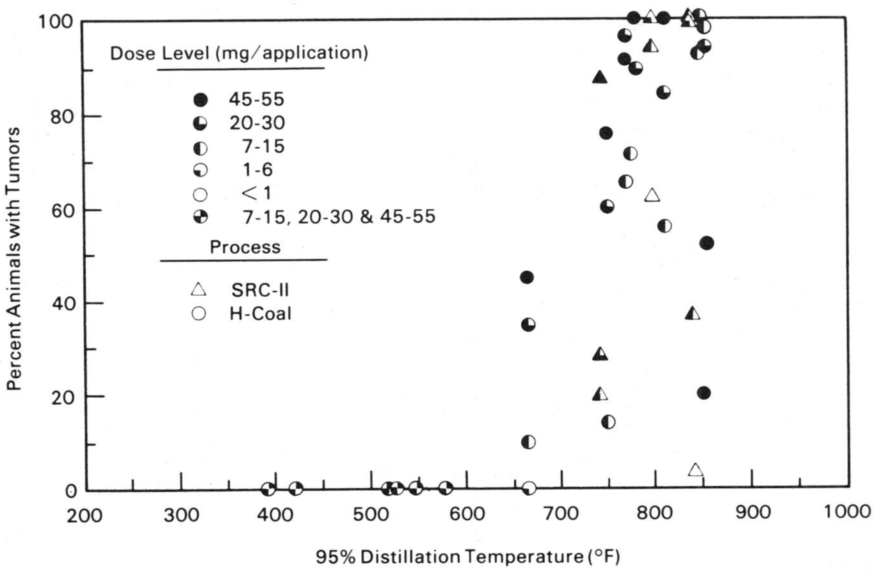

Figure 4.10 Skin-painting activity versus 95% point for full-boiling-range coal-liquefaction samples. SRC = solvent-refined coal.

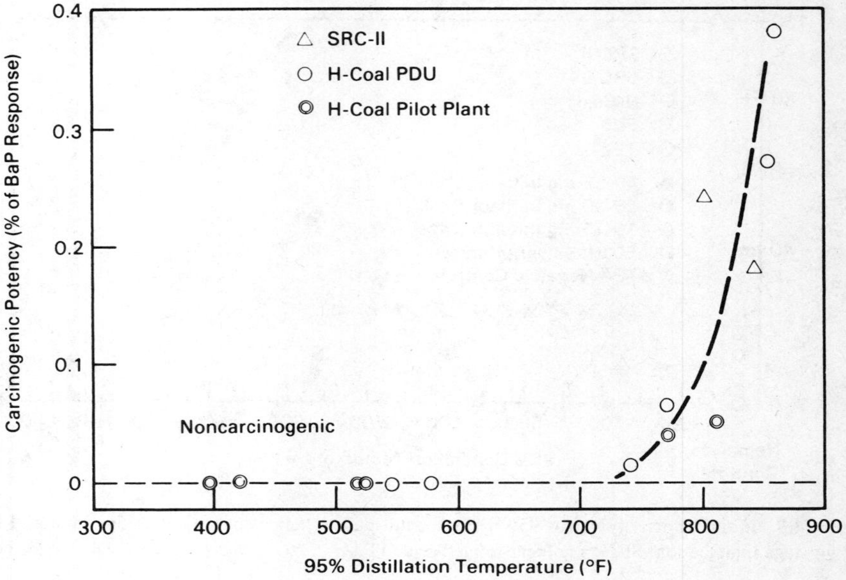

Figure 4.11 Carcinogenic potency relative to benzo[a]pyrene versus 95% point for full-boiling-range coal-liquefaction samples. Relative potency data from Chapter 3. SRC = solvent-refined coal; PDU = process development unit.

each case, the 95% point is used to express the heavy-end characteristics of samples, and the bioassay results are plotted as the dependent variable. The results of skin-painting assays are shown as percentage of the test animals with tumors in Fig. 4.10 and, for the slightly smaller number of samples with the necessary data available, as carcinogenic potency relative to BaP in Fig. 4.11. Details on approaches and assumptions used to construct these plots are given in Section 4.1.3 and by Fillo and Craun (1985).

The crucial feature common to Figs. 4.8–4.11 is that Ames assay, initiation/promotion, and chronic skin-painting results all show the same threshold temperature for mutagenic/carcinogenic effects of full-boiling-range coal-liquefaction materials. Expressed as the 95% point, this temperature is about 650–675°F, which is consistent with data for narrow-boiling cuts. This indicates that full-boiling-range samples containing over 5% of material boiling above about 650°F have enough marginally active material in the 650–700°F range and enough significantly active material in the >700°F range (probably a couple of percent) to be mutagenic/carcinogenic.

Once the threshold temperature is exceeded, all the samples tested are carcinogenic and, with few exceptions, mutagenic. There is no consistent pattern in the magnitude of the mutagenic/carcinogenic response as a function of increasing 95% point temperature, however. Ames and initiation/promotion assay activities rise sharply once the threshold temperature is exceeded, then fall within broad bands for 95% points from about 700°F to >1000°F (Figs. 4.8 and 4.9). Samples from different processes may show somewhat different patterns within these broad bands but do not always provide comparable coverage of the range of 95% points. Carcinogenic potency, relative to BaP, exhibits a steady increase with increasing 95% points (Fig. 4.11). While the carcinogenic potency data set is dominated by H-Coal samples, two SRC-II samples fit the H-Coal data trend, and H-Coal pilot-plant and PDU samples give consistent results.

The data in Figs. 4.8–4.11 show considerable variation in level of activity once the threshold 95% point is crossed. In particular, variation in the Ames data is substantial, amounting to a factor of over 10. However, variation decreases to about a factor of 3 for the initiation/promotion assay, and to roughly ±50% for chronic skin painting. These variations could reflect a number of factors.

First, there is considerable uncertainty in the variables plotted; none is precise. Ames data from multiple laboratories, collected over a period of many years, may show absolute variations of well over a factor of 2 (Chapter 3). Distillation curves are not precise, particularly at the heavy end, where 95% points may only be accurate to within 25–50°F (Fillo and Craun, 1985).

Second, samples were obtained from a variety of processes and process conditions and differ in their data quality and physical and chemical properties other than distillation. However, close inspection of the data indicates that data quality, process type, process conditions, and other physical properties do not account appreciably for the observed variability. Rather, variability appears to be inherent in the data used, particularly when results are compared for multiple samples tested by multiple laboratories over an extended period of time.

Figures 4.8–4.11 provide a simple means to predict whether a given full-boiling-range coal-liquefaction material is likely to be mutagenic or carcinogenic. All that is required is knowledge of the distillation curve for the heavy end of the sample. Although the 95% point is used here, other measures can be used to obtain similar results. The threshold of about 650°F for the 95% points is consistent with narrow-boiling-range cut data. Equally important, the various bioassays (Ames, initiation/promotion, and skin painting) all confirm this threshold.

4.2.2.3 *Chemical Characteristics*

The chemical composition of full-boiling-range coal liquids has been studied by many investigators for various reasons. This discussion is limited to a few examples illustrating how concentrations of selected carcinogenic and mutagenic PAHs are related to the toxicological endpoints considered above in the context of the 95% point.

Figure 4.12 shows concentrations of BaP and selected four-, five-, and six-ring PAH compounds versus the 95% point for full-boiling-range coal-liquefaction materials. The compounds plotted generally represent known animal carcinogens (IARC, 1982). All three figures show increasing concentrations with increasing 95% point. Moreover, concentrations generally increase sharply from 600 to 700°F, particularly for BaP. It is also noteworthy that concentrations start to increase rapidly at temperatures as much as 100–200°F below the boiling points of the pure compounds (based on pure-compound boiling points from Anderson and Wu, 1963).

Chemical interactions responsible for mutagenic/carcinogenic activity of coal-liquefaction materials are complex, and the trends shown in Fig. 4.12 are greatly simplified. However, the data do show trends consistent with more detailed efforts to identify the chemicals responsible for mutagenic and carcinogenic activity in coal liquids. Concentrations of specific chemicals implicated in mutagenic/carcinogenic activity increase at the same general threshold temperature as biological activity. These findings are consistent with boiling points of known mutagens and carcinogens in coal-liquefaction materials and the chemical characteristics of narrow-boiling cuts.

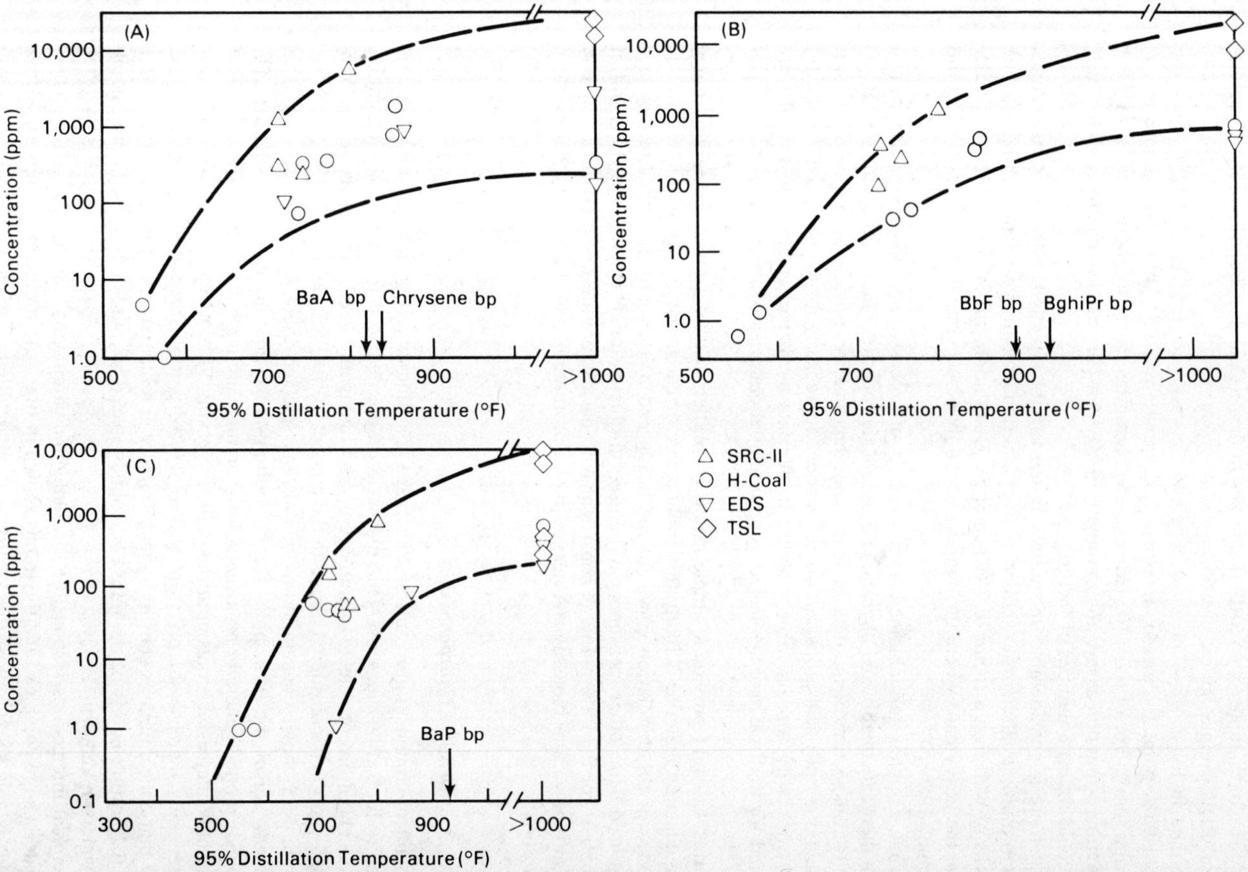

Figure 4.12 Concentrations of selected polycyclic aromatic hydrocarbon compounds versus 95% point for full-boiling-range coal-liquefaction samples: (A) benz[a]anthracene (BaA) + chrysene; (B) benzo[b&j]fluoranthenes (BbF) + dibenz[a,c & a,h]anthracenes + benzo(ghi)perylene (BghiPr); and (C) benzo[a]pyrene (BaP). SRC = solvent-refined coal II, TSL = two-stage liquefaction, bp = boiling point.

4.2.2.4 *Data Quality*

Full-boiling-range coal liquids show consistent trends in the presence or absence of mutagenic/carcinogenic activity versus 95% point, but there is considerable variation in the magnitude of activity above the threshold 95% point of about 650°F. While some variability is expected in comparing results from different laboratories, there is also the possibility that different test methods contribute to the apparent variability in the magnitude of mutagenic/carcinogenic activity. Potential sources of variability include (1) the use of different distillation methods, (2) differences in bioassay procedure, and (3) linkage of distillation and bioassay data from different laboratories testing different samples of the same material at different times.

Figure 4.13 shows initiation/promotion assay results reported in Fig. 4.9, coded by the two laboratories that generated the data: Pacific Northwest Laboratory (PNL) (Mahlum et al., 1983; Wilson et al., 1983a; Later et al., 1984) and Du Pont (Calkins et al., 1980, 1983). The figure shows a fairly clear indication of higher initiation/promotion activity for a given 95% distillation temperature for samples tested at PNL. (Note that the PNL samples in the 700–900°F 95% point range were tested at only about half of the initiator dose used by Calkins for his samples in the same range.) This difference may be due to variations in the test protocol, such as different vehicles for dilute samples (i.e., Calkins used toluene, PNL used methylene chloride or acetone), or different phorbol promotion schedules (2.5 μg/100 μL acetone thrice weekly by Calkins versus 5 μg/50 μL acetone twice weekly by PNL; see Chapter 3). In addition, the 95% distillation temperatures for Calkins' samples had to be estimated, and less is known about the storage history of these samples (Fillo and Craun, 1985). While the data from both laboratories show consistent trends, the differences noted above caution against quantitative comparisons of their results.

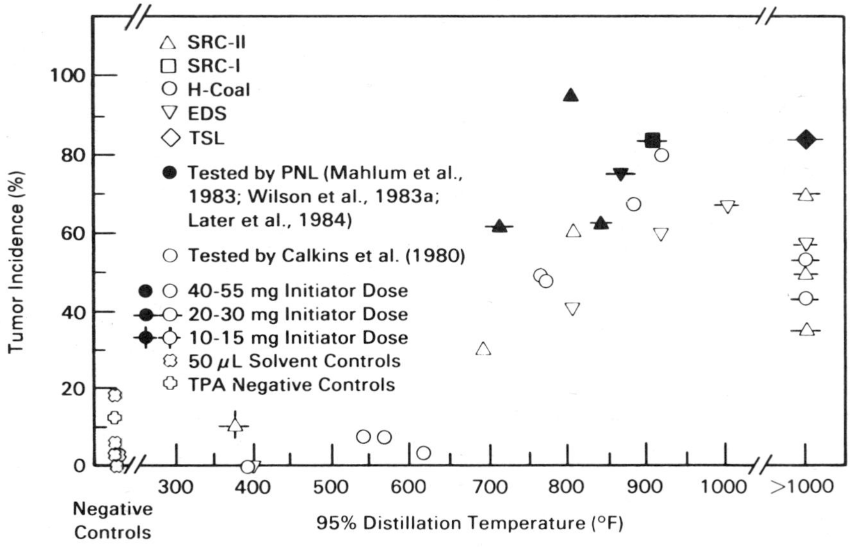

Figure 4.13 Initiating activity versus 95% point for full-boiling-range coal-liquefaction samples coded by testing laboratory. SRC = solvent-refined coal; TSL = two-stage liquefaction; TPA = 12-*O*-tetradecanoylphorbol-13-acetate.

The manner in which toxicological and distillation data are linked can also introduce variability. There are few instances where there is documentation that both toxicological and physical property data were obtained for the same sample (see Section 4.1.3). In most instances, certain assumptions are required to link biological and physical data on the same or a similar sample.

Figure 4.8 contains two examples of the problems that data linkage can pose. The figure shows an SRC-I recycle solvent sample (L2132) and an H-Coal fuel-oil sample (L3112) with positive weighted-sum Ames responses at 95% points of about 660°F. Both materials were feedstocks used in hydrotreating studies (Stein et al., 1977), samples of which were assayed for mutagenic activity (Epler, 1981). It is not clear that the samples assayed for biological activity match directly with those tested for physical properties, thus the reported 95% points may not be representative. Moreover, the distillations were simulated by GC (ASTM Method D-2887), which can be imprecise and inaccurate at the heavy end for high-boiling materials and may not correspond directly with corrected D-86 or D-1160 distillation results (Fillo and Craun, 1985). For example, the SGC distillation data reported for these and related samples (Stein et al., 1977) typically show an unusually large spread of 150–250°F between the 95% and 100% distillation temperatures. The 95% points shown for these samples are important in locating the 95% point threshold temperature for Ames activity. If these sample results are ignored, or relocated, the 95% threshold temperature for mutagenicity could be set as high as 700°F.

Figure 4.8 also shows three samples that lie outside the range observed for other samples. Samples L5351 and L3321 were both mutagenically active, even though their 95% points were below 600°F. Sample L1243 showed high activity compared to that of other samples with similar 95% points (about 800°F). Samples L5351 (Later, 1985a), and L3321 (Warshawsky and Schoeny, 1980) were both reported as being "slightly" mutagenic. Neither sample had distillation data reported for the exact material tested, although relevant distillation data for samples from the corresponding process streams are available. The samples tested may have exhibited unusual chemical characteristics or contained some higher-boiling material from poor process separations, or they may simply reflect spurious test results. The 95% point for Sample L1243 was estimated based on process knowledge and could be off by up to 100°F. Because the Ames assay activity for this sample would be unusually high for any 95% point, the result must reflect unusual chemical characteristics or a very responsive Ames assay.

The examples given above illustrate two sources of difficulty in attempting quantitative comparison of mutagenic/carcinogenic activity for samples with different distillation characteristics. Different distillation methods and sample degradation are other potential sources. Fillo and Craun (1985) report no obvious signs of bias from these sources for the coal-liquefaction database as a whole, although biases may exist in comparing discrete data sets.

4.2.2.5 *Summary*

The high-boiling, heavy end of a full-boiling-range coal-liquefaction sample's distillation curve is closely related to its mutagenic/carcinogenic activity. The 95% distilled temperature can be used to characterize a sample's heavy-end distillation relative to its genotoxic activity. Mutagenic/carcinogenic activity (Ames, initiation/promotion, and skin-painting assays) is confined to samples with 95% points above about 650°F. Above this threshold, mutagenic/carcinogenic activity increases as the 95% point increases. Chemical analyses of full-boiling-range samples indicate that concentrations of mutagenic and carcinogenic PAHs follow generally similar trends. A coal-liquefaction material containing sufficient high-boiling material can be expected to be mutagenic and carcinogenic within the detection limits of the bioassays used. This "sufficient" amount is at least 5% of the material boiling above about 650°F, which typically implies a few percent boiling above at least 700°F.

Considerable variation in mutagenic/carcinogenic activity is observed once the threshold 95% point is exceeded. The amount of variation appears to decrease as higher-tier bioassays are used (e.g., Ames results are more variable than skin-painting results). Observed variations may, in part, reflect biases among results from different laboratories and testing procedures. Hence, quantitative comparisons must be made cautiously. Nonetheless, the three bioassays considered (Ames, initiation/promotion, and skin painting) are all consistent in defining a 95% point that reliably distinguishes between active and inactive materials. This simple yes/no segregation is useful, independent of significant variations in the magnitude of activity once the threshold 95% point is exceeded.

4.2.3 Comparisons of Liquefaction Processes

Examination of Figs. 4.8–4.11 reveals no obvious, consistent patterns that distinguish the toxicological data for SRC-I, SRC-II, H-Coal, EDS, and two-stage liquefaction (TSL) materials. However, the considerable variability in the data may be attributable to differences in process conditions or physical properties other than boiling-point range. These possibilities are evaluated below.

4.2.3.1 *Other Sample Properties*

Atomic H:C ratios are the most widely available measure of aromaticity for coal-liquefaction materials with mutagenesis/carcinogenesis test data. However, Fillo and Craun (1985) indicate that atomic H:C ratios do not account satisfactorily for the variability observed in Figs. 4.8–4.11. This result is not surprising, given the limitations inherent in using atomic H:C ratios to characterize a coal liquid's aromaticity in relation to its mutagenic/carcinogenic activity. First, H:C ratios reflect a material's total composition, not just that of aromatic species. For example, a mixture composed entirely of naphthenic compounds (fully, or almost fully, saturated rings) may have the same H:C ratio as a mixture composed of fully condensed aromatics and saturated aliphatics. Second, H:C ratios vary with boiling range for coal-liquefaction and other complex hydrocarbon mixtures. As boiling range increases, ring structures, independent of alkylation or partial saturation, become larger and more fully condensed, and H:C ratios decrease. Finally, H:C ratios cannot be used to focus directly on the higher-boiling ends, as can be done with distillation data. For example, two full-boiling-range distillates could have similar distillations and H:C ratios for their heavy ends (e.g., >650°F material) and yet contain different amounts of light ends (e.g., <400°F material) with higher H:C ratios. The two whole materials would then have different H:C ratios, even though the H:C ratios of the heavy ends are similar.

Sample properties other than gross aromaticity may also influence toxicological effects. Physical properties, such as viscosity, solids content, or solubility in various solvents, may modify the effective dose of test material administered. High viscosity, poor solubility, and related properties may account for the variability in mutagenic/carcinogenic responses obtained with very-high-boiling materials. (Note results for samples with 95% points of 1000°F and above in Figs. 4.8 and 4.9.) As an example of such properties, note the sharp increase in asphaltene content for narrow-boiling-range SRC-II cuts in Fig. 4.5, where asphaltenes are defined as the toluene-soluble fraction of hexane insolubles (Gray, 1981).

Unfortunately, few data are available to evaluate the importance of these sample properties. Table 4.6 shows distillation, physical properties, and bioassay data for three SRC-II and H-Coal residual oils (Calkins et al., 1980). The samples have similar distillation

Table 4.6 Ames and Initiation/Promotion Assay Data for Three Coal-Liquefaction Residuum Samples with Similar Physical Properties

Property or Assay	SRC-II[a] Residuum (Sample L1236)	SRC-II[a] Vacuum Bottoms (Sample L1235)	H-Coal Vacuum Bottoms (Sample L3124)
Distillation (%)			
<644°F	10	10	14
644–806°F	15	17	15
806–932°F	9	13	13
>932°F	66	60	58
Hydrogen (wt %)	3.85	4.08	4.56
Atomic hydrogen:carbon ratio	0.68	0.73	0.77
Density at 25°C (g/cm^3)	1.46	1.46	1.49
Ames mutagenicity for crude samples (rev/μg)	4.5	1.3	2.3[b]
Initiation/promotion assay results[c]			
Mice with tumors (%)	70	33	43
Total number of tumors	63	19	31

[a] Solvent-refined coal II.

[b] Calkins et al. (1983) report 4.0 rev/μg for this sample.

[c] All three samples tested as a 50-μL dose of a 2:1 mix of toluene:sample.

Source: Data from Calkins et al. (1980).

curves and densities, with H:C ratios and hydrogen contents that span a relatively narrow range and increase as boiling point decreases. However, Ames and initiation/promotion data for these samples, while internally consistent (i.e., Sample L1236 is most active and Sample L1235 is least active in both assays), show no relation to boiling range or aromaticity. Test results span a large-enough range (factor of 2–4) that they probably reflect real differences within the accuracy limits of the tests. It is not clear how these differences are related to physical properties. Unmeasured sample properties, such as viscosity, may affect observed differences in test results, or subtle changes in chemistry not reflected in gross property measurements may be responsible.

4.2.3.2 *Process Operating Mode*

Chemical and toxicological differences in coal liquids may reflect differences in operating modes for a given process. A wide range of chemical and toxicological data have been developed on EDS materials obtained from the 1-ton/day coal-liquefaction pilot plant (CLPP) and the 200-ton/day EDS coal-liquefaction plant (ECLP). At ECLP, three different types of coal were used as feedstocks, and the facility was operated in the once-through mode and in modes where portions of the heavy vacuum gas oil and vacuum bottoms were recycled. Data developed on these materials (Later et al., 1984). are summarized in Table 4.7.

Although process feedstocks and operating modes varied, differences in mutagenic activities for the samples in Table 4.7 can be explained largely from the properties of the

Table 4.7 Mutagenic Activity and Chemical Class Composition of EDS Coal-Liquefaction Materials

Run/Material	Properties		Ames Activity[a] (rev/μg)		Chemical Class Fractions[a] (wt %)				A-PAH Concentration[a] (μg/g)
	Atomic H:C	95% Point (°F)	Crude	Weighted Sum	AHs	PAHs	N-PACs	H-PAHs	
ECLP, once through, Illinois No. 6[b]									
Recycle solvent (ETP-03/L4111)[c]	1.47	724	<1	0.02	69	21	2	8	14.2
Heavy vacuum gas oil (VGO; ETP-04/L4112)	0.99	>1200	20	9.0	5	31	22	41	115.4
ECLP, bottoms/VGO recycle, Illinois No. 6[d]									
Naphtha (ETP-11/L4131)	1.75	406	0	0	28	4	1	11	—
Recycle solvent (ETP-13/L4132)	1.27	866	1	2.7	46	37	7	13	48.4
Vacuum bottoms (ETP-14/L4133)	—	~1200	11	14	2	7	24	27	45.2
CLPP, once through, Illinois No. 6[e]									
Naphtha (L4211)	1.72	399	0	—	—	—	—	—	—
Recycle solvent (L4212)	1.24	804	1	—	—	—	—	—	—
Vacuum gas oil (L4213)	0.91	~1000	6	—	—	—	—	—	—
ECLP, bottoms/VGO recycle, Texas lignite[d]									
Recycle solvent (ETP-22/L4141)	1.38	800	—[f]	—	59	25	3	9	—
ECLP, bottoms recycle, Wyodak subbituminous									
Recycle solvent (ETP-08/L4121)	1.43	754	—	—	57	25	3	11	—

[a] Data from Later et al. (1984). AHs = aromatic hydrocarbons; PAHs = polycyclic aromatic hydrocarbons; N-PACs = nigrogen-containing polycyclic aromatic compounds; H-PAHs = hydroxy-PAHs; A-PAH = amino-PAH.

[b] Samples obtained from 200-ton/day EDS coal-liquefaction plant (ECLP), once-through operating mode.

[c] Sample numbers assigned for EDS Toxicology Program/this data analysis.

[d] ECLP operating with complete recycle of the heavy VGO and partial recycle of vacuum bottoms.

[e] Samples obtained from 1-ton/day coal-liquefaction pilot plant (CLPP). Data from Calkins et al. (1980) and Epperly (1978).

[f] — indicates no data available.

materials, particularly the 95% point. Mutagenic activity increases with the 95% point for various materials from any run. Naphthas are consistently inactive, and heavy vacuum gas oil and/or vacuum still bottoms are more active than recycle solvents. Concentrations of PAHs, nitrogen-containing polycyclic aromatic compounds (N-PACs) and amino polycyclic aromatic hydrocarbons (amino-PAHs) increase with higher-boiling ranges, as expected. Mutagenic activity of recycle solvent for ECLP runs with Illinois No. 6 coal tracks with the 95% point, as do chemical class fractions and amino-PAH concentrations. Recycle solvent ETP-13 (Sample L4132) had the highest levels of PAHs, N-PACs, and amino-PAHs and a 95% point of 866°F. Recycle solvent ETP-03 (Sample L4111) had the lowest levels of PAHs, N-PACs, and amino-PAHs and a 95% point of 724°F. Recycle solvents ETP-22 (Sample L4141) and ETP-08 (Sample L4121) had levels of PAHs and N-PACs and 95% points that were intermediate.

Data for materials from the Merriam, Kansas SRC-II PDU operating in a heavy distillate recycle mode (Wilson et al., 1983b) can also be explained by differences in sample properties. Of five total liquid products, one generated in the conventional SRC-II mode (Sample L1311) and four where the heavy distillate fraction was recycled (Samples L1321–L1351), only the conventional product was biologically active, with a 95% point of 703°F. Products produced in the heavy distillate recycle mode were inactive in the Ames assay and had 95% points ranging from 546 to 600°F.

4.2.3.3 *Process Conditions*

Different process conditions may also influence biological activity of resultant materials. There are few instances where documented process conditions, physical property data, and toxicological data are all available for samples derived under significantly different process conditions for a given liquefaction process. One such instance involves two H-Coal PDU runs. Comparing process conditions and properties data (HRI, 1978, 1979) and bioassay data (Epler, 1981) for samples from these two runs suggests that differences in sample properties greatly outweigh any effects that could be attributed to different process conditions. Both runs were made with Illinois No. 6 coal, but they differed in reaction severity. PDU run 5 (Table 4.8) was operated in the syncrude mode, with high reaction severity, as evidenced by a hydrogen consumption of 5.5 wt % on a dry coal basis (HRI, 1978). PDU run 7 was operated in the less-severe fuel-oil mode, with a hydrogen consumption of 3.4 wt % (HRI, 1979).

Table 4.8 shows that the products tested from these two H-Coal runs had comparable physical properties and mutagenic/carcinogenic activity. Ames assay mutagenicities and skin-painting potencies relative to BaP are consistent with sample properties. Slight differences in bioactivity can be explained by differences in physical properties (e.g., compare skin-painting potencies and 95% points for the two atmospheric still bottoms samples; see also Fig. 4.11) or limitations in testing accuracy (e.g., differences in Ames results for the vacuum still overhead samples). The major difference in these runs was in the product yield structure. The most-severe run (PDU run 5) gave 46 wt % yield of liquid boiling below 975°F (dry coal basis), while PDU run 7 gave only a 27 wt % yield of the same material. Mutagenic/carcinogenic activity and physical properties were similar for comparable distillates in the spectrum of liquid products. The primary effect of increased reaction severity was on relative products yield.

EDS data provide another example of the effects of changing process conditions on mutagenic/carcinogenic activity. The different runs with toxicological data shown in Table 4.7 had hydrogen consumptions ranging from 2.8 to 4.7 wt % (dry, ash-free coal basis).

Table 4.8 Mutagenic/Carcinogenic Activity and Properties of Materials from the H-Coal Process Development Unit (PDU) During Runs 5 and 7

		Properties			
Sample Description	Sample No.[a]	95% Point (°F)	H:C Ratio	Ames Activity, Weighted Sum (rev/μg)[b]	Carcinogenic Potency Relative to Benzo[a]pyrene[c] (%)
PDU Run 5—5.5% Hydrogen Consumption[d]					
Atmospheric still overhead	L3141	576	1.60	—[e]	0
Atmospheric still bottoms	L3142	771	1.26	1.0	0.065
Vacuum still overhead	L3143	855	1.15	2.4	0.38
Vacuum still bottoms	L3144	>1100	0.80	2.0	—
PDU Run 7—3.4% Hydrogen Consumption[f]					
Atmospheric still overhead	L3151	547	1.62	0	0
Atmospheric still bottoms	L3152	738	1.29	0.1	0.015
Vacuum still overhead	L3153	849	1.17	4.2	0.27
Vacuum still bottoms	L3154	>1100	0.78	6.3	—

[a] Oak Ridge National Laboratory sample numbers are 1308–1311 for PDU run 5 and 1312–1315 for PDU run 7 (Epler 1981).
[b] Data from Epler (1981).
[c] Potency results are average of values for three different doses for each sample. See Chapter 3, Table 3.66.
[d] Hydrogen consumption for period 5; properties data for period 4 (HRI, 1978).
[e] — indicates not tested.
[f] Hydrogen consumption for period 5; properties data for period 8B (HRI, 1979).

The different reaction severities are attributable, in part, to changes in operating mode and coal feeds. Nonetheless, differing bioassay results for various EDS materials can be explained largely by physical property differences.

4.2.3.4 *Summary*

The distillation characteristics of the heavy end of full-boiling-range coal-liquefaction samples correlate well with mutagenic and carcinogenic activity. The 95% point is an effective measure of heavy-end characteristics relative to toxicological effects. Ames, initiation/promotion, and skin-painting assay data show that mutagenic/carcinogenic activity can be expected in samples with 95% points greater than about 650°F (see Figs. 4.8–4.11). There is considerable variation in the activity observed once this threshold is crossed that cannot be explained by gross process or sample property effects. Subtle process differences, physical property changes, or chemistry changes may be responsible, in part. However, it is likely that the variation primarily reflects test results from numerous laboratories and test protocols over many years. Nonetheless, the 95% point reliably predicts the presence or absence of mutagenic/carcinogenic activity.

4.3 ENGINEERING ANALYSIS OF THE COAL-LIQUID HYDROTREATING TOXICOLOGICAL DATABASE

Hydrotreatment is closely related to liquefaction. Both processes upgrade feed materials by adding hydrogen. Liquefaction converts coal to a liquid, and hydrotreating produces a liquid of higher quality. Many liquefaction processes incorporate hydrotreatment as part of the coal-conversion process (e.g., hydrotreatment of the recycle solvent in the EDS process and hydrotreatment as the second reaction stage in the TSL processes) or to upgrade crude liquefaction products for market (e.g., SRC-II and H-Coal conceptual demonstration plant designs). However, there are important differences between the two processs. Hydrotreating uses specific active catalysts to treat liquid materials in a two-phase flow system of gas and oil, whereas liquefaction involves a three-phase system (coal, oil, and gas). The nature of the feedstocks and the processing required are obviously different, and different reaction networks are involved.

The extent and type of available data for liquefaction and hydrotreating are similar, particularly in the availability of sample distillation data (see Table 4.3). This allows hydrotreatment to be viewed in terms comparable to liquefaction processes with respect to relations between sample properties and toxicological characteristics.

4.3.1 General Process Chemistry

There is a large body of literature on process engineering and chemistry of hydrotreating petroleum and coal-derived liquids. A primary result of catalytic hydrotreatment is saturation of aromatic and olefinic compounds accompanied by ring-opening, dealkylation, and heteroatom removal reactions. For highly aromatic feedstocks, these reactions must be preceded by ring-saturation reactions, as illustrated by the reaction pathways reported by Scheppele et al. (1981) for phenanthrene and acridine (Fig. 4.14). In addition, little of the hydrogen consumed in hydrotreating coal-derived liquids is used for heteroatom removal or cracking to light gases. Instead, most hydrogen is consumed by ring-saturation reactions, where it remains in the liquid product (Fig. 4.15). These characteristic reactions are re-

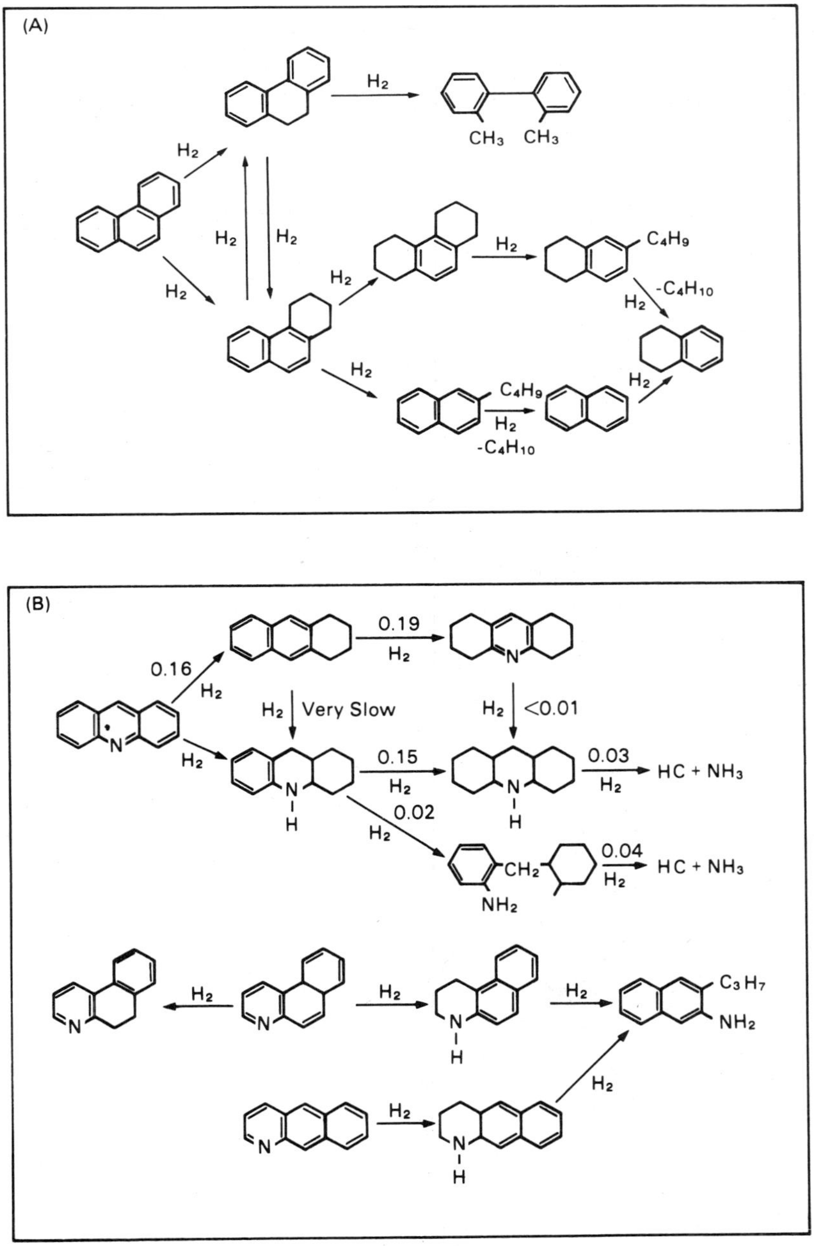

Figure 4.14 Reaction pathways for model compounds in coal-derived liquids (adapted from Scheppele et al., 1981). (A) phenanthrene; (B) acridine and other benzoquinolines.

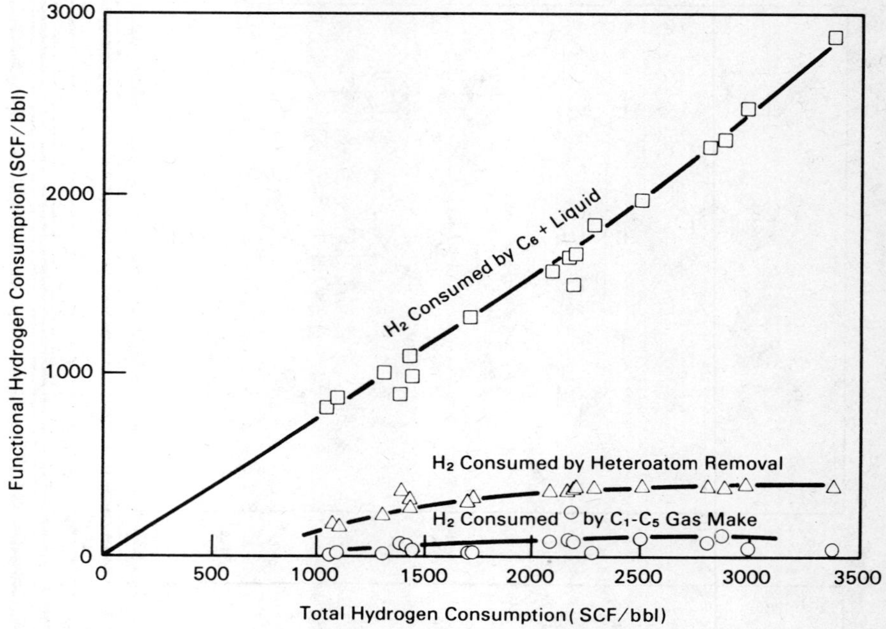

Figure 4.15 Hydrogen utilization in hydroprocessing of solvent-refined coal (SRC) recycle solvent (adapted from Stein et al., 1977). SCF/bbl = standard cubic feet per barrel.

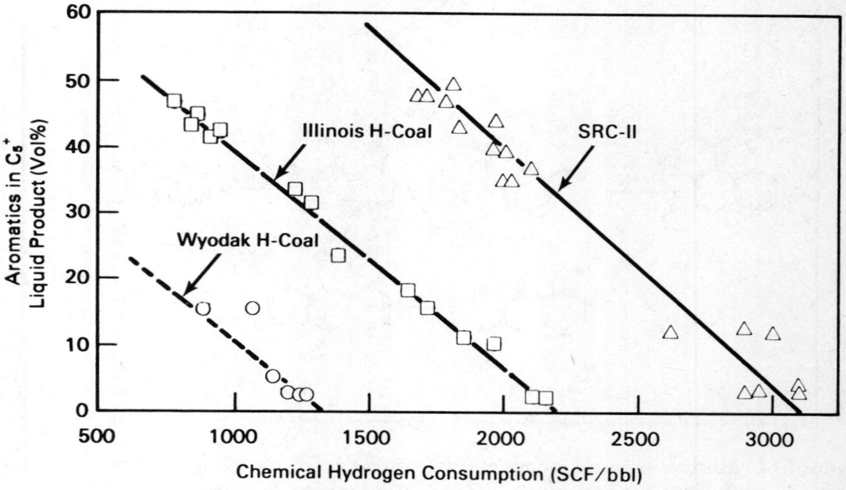

Figure 4.16 Product aromatic content versus hydrogen consumption for hydrotreated H-Coal and solvent-refined coal-II (SRC-II) fuel oils (adapted from O'Rear et al., 1981). SCF/bbl = standard cubic feet per barrel.

flected in reduced aromaticity of liquid products with increasing hydrotreatment severity (Fig. 4.16). Feedstock differences are also important, with higher-boiling and more-aromatic feeds requiring more hydrogen to achieve a given product aromatics content (compare Figs. 4.16 and 4.17).

Hydrotreating also results in removal of heteroatoms from feedstocks. More-severe processing results in more complete heteroatom removal. Desulfurization is generally achieved much more readily than denitrogenation and deoxygenation (Maruhnich et al., 1982; Sutterfield, 1983). Products with very low nitrogen content (less than 1 ppm) can be achieved with sufficiently severe hydrotreating, as illustrated by Fig. 4.18 for an SRC-II fuel oil.

Hydrogenation and heteroatom removal reactions can affect the toxicological characteristics of a coal liquid. For example, Later (1985a) reported that the initiation/promotion activity of BaP decreases with increasing degree of hydrogenation. Falk et al. (1964) reported that partially hydrogenated PAHs can inhibit the carcinogenic activity of other PAH compounds. Weimer et al. (1980) showed that coal-liquid hydrotreatment provides substantial removal of amino-PAHs, which are major mutagens in coal liquids.

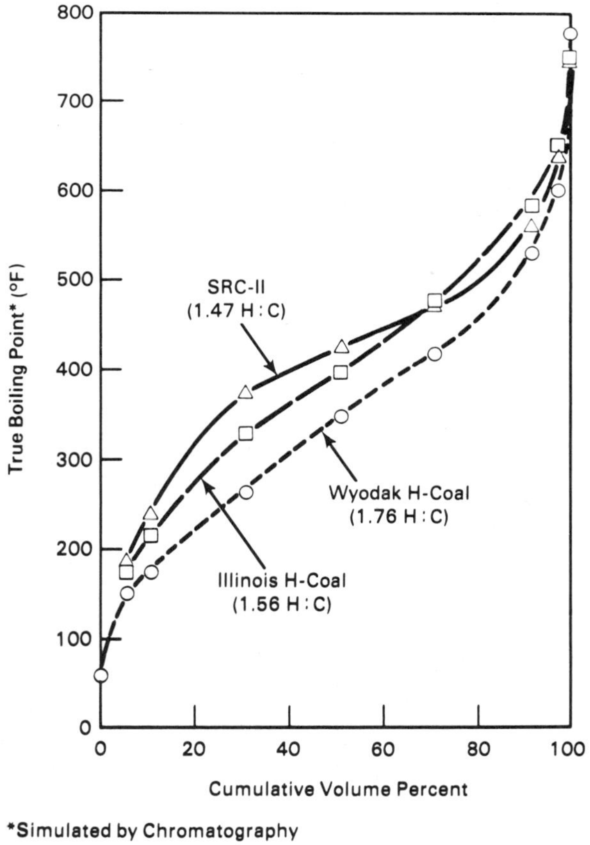

Figure 4.17 True-boiling-point distillations for H-Coal and solvent-refined coal-II (SRC-II) fuel oils used as hydrotreatment feedstocks (adapted from O'Rear et al., 1981). H : C = hydrogen : carbon ratio.

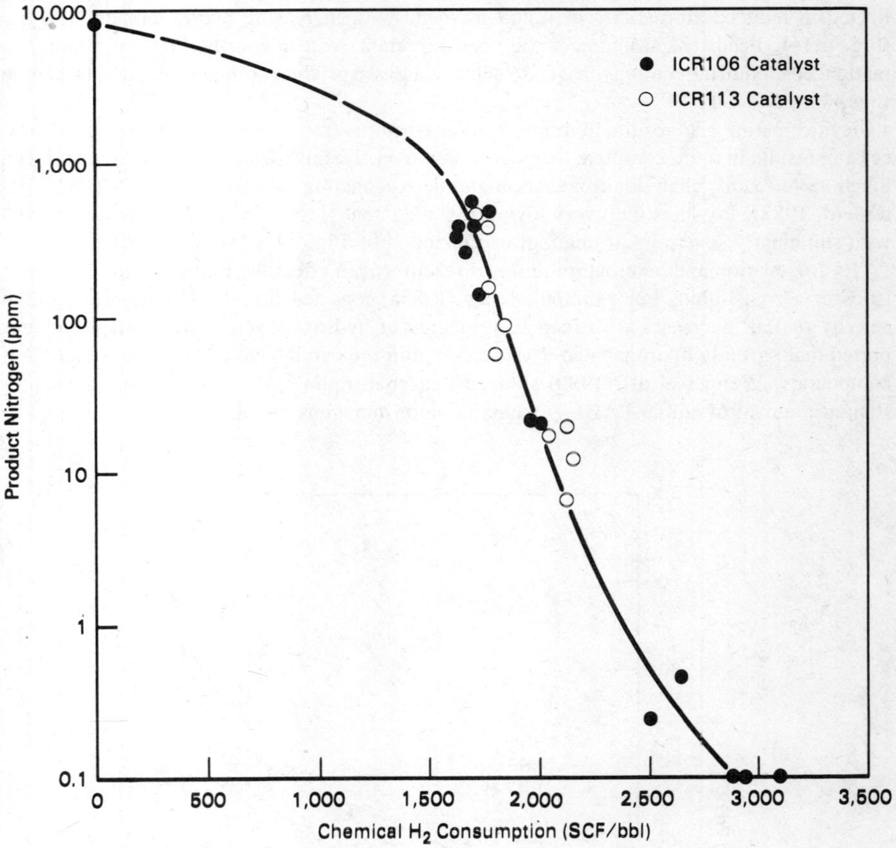

Figure 4.18 Product nitrogen versus hydrogen consumption for hydrotreatment of a solvent-refined coal-II (SRC-II) liquid at 750°F (adapted from Frumkin et al., 1981).

4.3.2 Feedstock and Process Severity Effects

During hydrotreatment, larger, mutagenic/carcinogenic aromatic ring structures are partially hydrogenated or cracked into smaller rings, and aromatic amines are deaminated. Thus, catalytic hydrotreating should reduce or eliminate the mutagenic or carcinogenic activity of feed material. However, hydrotreatment does not always remove mutagenic/carcinogenic activity. Its effectiveness depends on the properties of the feedstock and the severity of the treatment, both of which affect physical properties that correlate with toxicological effects, namely, boiling range and aromaticity.

Figure 4.19 shows the effects of feedstock properties and process severity on reducing the Ames assay activity of coal liquids by hydrotreatment. Process severity is indicated by total hydrogen consumption in weight %. The average boiling point of the feedstocks are shown in the figure. Additional selected properties for the range of feedstocks plotted in Fig. 4.19 are given in Table 4.9. Figure 4.20 shows distillation curves for selected hydro-

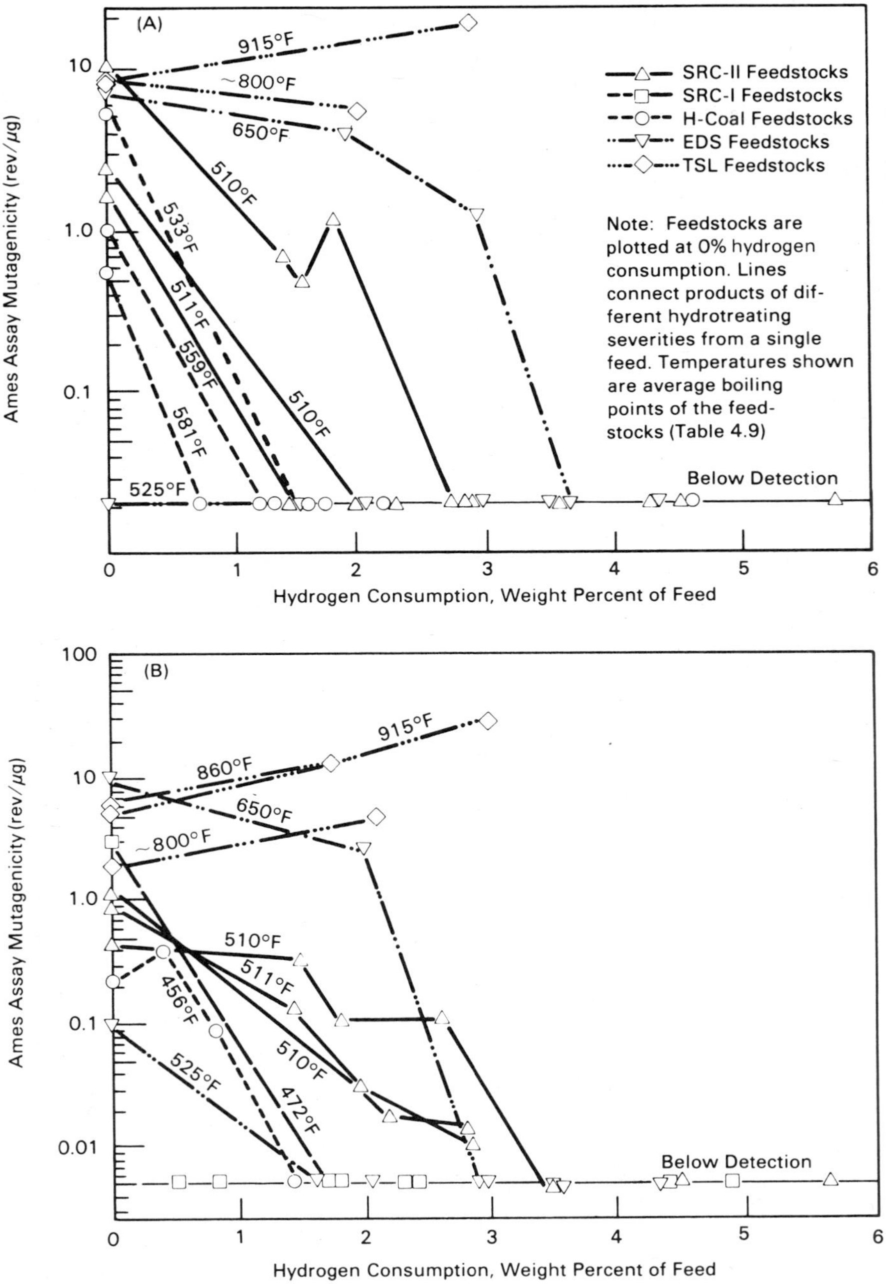

Figure 4.19 Ames assay mutagenicity versus hydrogen consumption for hydrotreated coal liquids: (A) unfractionated samples; (B) fractionated samples (weighted-sum results). SRC = Solvent-Refined Coal; TSL = Two-Stage Liquefaction.

Table 4.9 Selected Characteristics of Coal-Liquid Hydrotreatment Feedstocks and Testing Conditions

Feedstock Characteristics					Hydrotreatment Test Conditions			
Feedstock Sample No.	Description[a]	Average Boiling Point[b] (°F)	95% point (°F)	Atomic H:C Ratio	Hydrotreating Laboratory[c]	Number of Tests	Range of Hydrogen Consumptions (wt %)	Product Sample No.
L1221	SRC-II fuel oil	510	714	1.16	BETC	8	1.53–5.68	H0401-08
L1251/52[d]	SRC-II fuel oil	512/508	711/705	1.21/1.21	UOP	2	1.95–2.85	H0102, 04
L1261/62[d]	SRC-II fuel oil	511	743	1.24	CRC	3	1.44–2.81	H0201-03
L3111/12[e]	H-Coal syncrude	456	665	1.33	MRDC	3	0.42–1.43	H0319-21/22–24[e]
L3161	H-Coal PDU atmospheric still bottoms	559	754	1.39	BETC	2	1.34–1.67	H0431, 32
L3162	H-Coal PDU vacuum still overhead	581	772	1.35	BETC	3	0.72–1.82	H0441-43
L3231/32[d]	H-Coal fuel oil	532/534	731/733	1.26/1.28	CRC	3	1.44–4.61	H0204-06
L2131/32[e]	SRC-I recycle solvent	472	653	1.06	MRDC	3	1.72–4.88	H0313-15/16–18[e]
L2141/42[e]	SRC-I wash solvent	512	665	1.23	MRDC	3	0.85–4.42	H0307-09/10–12[e]
L2151/52[e]	SRC-I light organic liquid	298	420	1.62	MRDC	3	0.49–2.40	H0301-03/04–06[e]
L4221	EDS–CLPP solvent	525	694	1.30	BETC	4	1.57–3.51	H0411-14
L4231	EDS–CLPP fuel-oil blend	650	963	0.99	BETC	4	1.92–4.36	H0421-24
L5111	TSL LC Finer feed (run 7–37)	810	>1000	0.96	C. E. Lummus	1	3.62	H0611
L5121	TSL LC Finer feed (run 7–90)	915	>1100	0.74	C. E. Lummus	1	2.94	H0621
L5131	TSL LC Finer feed (run 8–52)	860	>1100	0.91	C. E. Lummus	1	1.69	H0631
L5141	TSL LC Finer feed (run 9-0524)	875	>1100	0.90	C. E. Lummus	1	2.33	H0641
L5216	Wilsonville NTSL H-Oil feed run 241)	~800[f]	>1000	1.13	Catalytic, Inc.	1	2.1	H0513

[a]SRC = solvent-refined coal; PDU = process development unit; CLPP = coal-liquefaction pilot plant; TSL = two-stage liquefaction; NTSL = nonintegrated two-stage liquefaction.

[b]Average of 10, 50, and 90% points. For LC Finer feed and H-Oil feed, 50% points are reported.

[c]BETC = Bartlesville Energy Technology Center; CRC = Chevron Research Company; MRDC = Mobil Research & Development Corporation; UOP = Universal Oil Products, Inc.

[d]Two different, but similar, feedstocks were used for multiple hydrotreatment runs (see appendix to chapter 3).

[e]First number is for samples tested for properties by MRDC; second is for samples for toxicological testing.

[f]Estimated from product distillation.

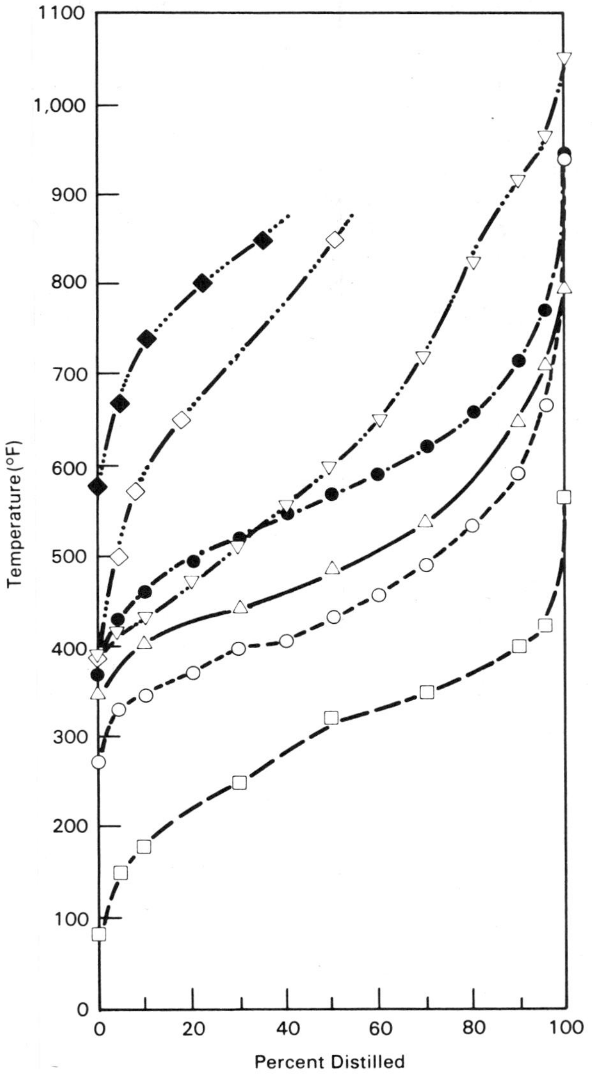

Symbol	Sample Number	Description
△	L1251	SRC-II Fuel Oil
□	L2151	SRC-I Light Organic Liquid
○	L3111	H-Coal PDU Syncrude
●	L3162	H-Coal PDU Vacuum Still Overhead
▽	L4231	EDS-CLPP Fuel Oil
◇	L5111	TSL LC Finer Feed
◆	L5121	TSL LC Finer Feed

Figure 4.20 Distillation curves for selected hydrotreatment feedstocks. SRC = solvent-refined coal; PDU = process development unit; CLPP = coal-liquefaction pilot plant; TSL = two-stage liquefaction.

treating feedstocks plotted in Fig. 4.19, which range from naphtha-range distillates (e.g., SRC-I light organic liquid, Sample L2151) to atmospheric residua (e.g., C. E. Lummus LC Finer feeds, Samples L5111–L5141), with most being full-boiling-range fuel-oil blends (e.g., SRC-II fuel oil, Sample L1251).

It is noteworthy that throughout this section, data from the second stage of TSL processes are analyzed together with data from separate product upgrading hydrotreating studies. These two different hydrotreatment applications are similar in that both involve catalytic processing of coal liquids. They differ, however, in that TSL second-stage hydrotreating generally involves much heavier feedstocks containing insoluble organic and unconverted materials. In addition, TSL second-stage hydrotreating feeds may experience different time–temperature histories than product hydrotreating feeds (e.g., hot product from a TSL first-stage reactor entering the second-stage hydrotreater versus a product hydrotreating feed that is cooled to ambient temperatures).

Figure 4.19 shows two consistent trends. First, increased hydrogen consumption results in greater reductions in Ames assay activity. Second, heavier feedstocks generally require more hydrogen addition than lighter feeds to achieve comparable reduction in Ames assay activity. These trends are demonstrated for a total of 46 hydrotreating tests on 17 different feedstocks performed by six different hydrotreating laboratories, with hydrogen consumptions ranging from 0.4 to 5.7 wt % (Table 4.9).

There are fewer data available for the effects of hydrotreatment on initiation/promotion and skin-painting activity. Figure 4.21 summarizes the dose–response data for the two experiments reported with skin-painting assays of coal liquids hydrotreated at different severities (Epler, 1981; Griest et al., 1983). The three hydrotreating severities used for both feedstocks in all cases reduced tumor incidences to 5% or less. The low tumor-incidence rates and lack of consistent dose–response data for the hydrotreated products make it difficult to draw quantitative conclusions about skin-painting activity versus hydrotreating severity or feedstock effects. It does appear, however, that the higher-boiling and more aromatic SRC-II fuel oil (Table 4.9) required more-severe hydrotreatment to achieve tumor-incidence rates comparable to products from the lighter, H-Coal syncrude. Ames assays of these materials (Fig. 4.19) gave results consistent with the skin-painting assays, although mutagenic activity was still detected in products hydrotreated at lower severities.

4.3.3 Sample Properties Effects

As more hydrogen is added to a feedstock, it gets increasingly lighter, with a lower boiling range and lower aromaticity. In addition, levels of heteroatomic compounds are reduced. At some point, the product should be light enough that it no longer contains significant amounts of the higher-boiling, aromatic and heteroatomic compounds responsible for mutagenic/carcinogenic activity. Heavy feeds should require more hydrogen than lighter feeds to become inactive. Hence, the trends seen in Fig. 4.19 for reduction of mutagenic/carcinogenic activity as a function of hydrotreating severity and feedstock characteristics should be related to changes in product properties.

Figures 4.22 and 4.23 show the 95% point versus Ames assay and skin-painting activity, respectively, for full-boiling-range coal-liquefaction and hydrotreated materials. Figure 4.22 shows that hydrotreated products generally have a lower range of mutagenic activity than crude liquefaction materials. In particular, there is a higher incidence of inactive hydrotreated products with 95% points above the mutagenic activity threshold of 650°F than for nonhydrotreated materials. Similarly, skin-painting tumor-incidence rates for the three hydrotreated SRC-II products with 95% points in the range of 667–734°F are lower than those

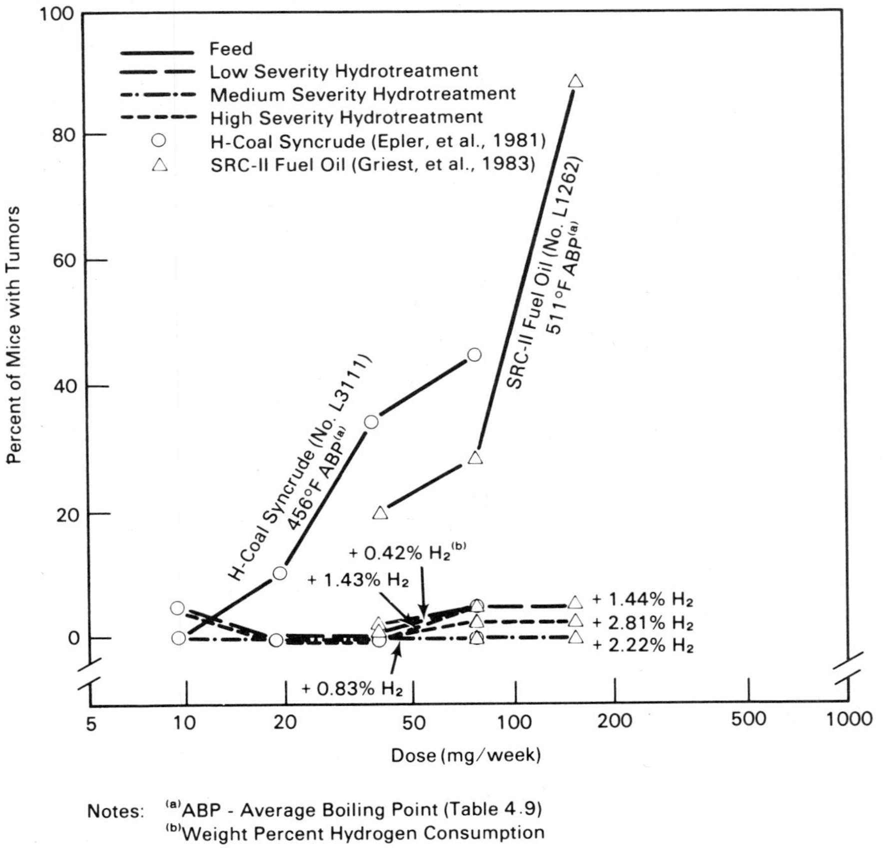

Figure 4.21 Percentage of mice with tumors versus dose for crude and hydrotreated H-Coal syncrude and solvent-refined coal II (SRC-II) fuel oil.

of liquefaction materials with comparable 95% points (Fig. 4.23). There are initiation/promotion test data for only three hydrotreated samples. Results for an SRC-II fuel oil (Sample H0104) lie near the activity threshold in Fig. 4.9 (668°F 95% point, 20% tumors, 11 total tumors for 25-mg dose; Mahlum et al., 1983). Two full-boiling-range products from the C. E. Lummus ITSL hydrotreater (Samples H0611 and H0621), with estimated 95% points well over 1200°F, yielded 76 and 77 tumors for 25-mg doses (Wilson et al., 1983a). All three sample points lie within the range of results for crude liquefaction samples (Fig. 4.9).

Figures 4.22 and 4.23 indicate that sample properties other than reduced boiling range are associated with the decreased mutagenic/carcinogenic activity of catalytically hydrotreated materials. For example, reduced aromaticity should correlate with reduced toxicological activity, based on the known effects of hydrotreatment on sample chemistry.

Figure 4.24 shows Ames assay activity for hydrotreated coal liquids as a function of the sample's 95% distillation point and atomic H : C ratio. Ames results are shown as active or

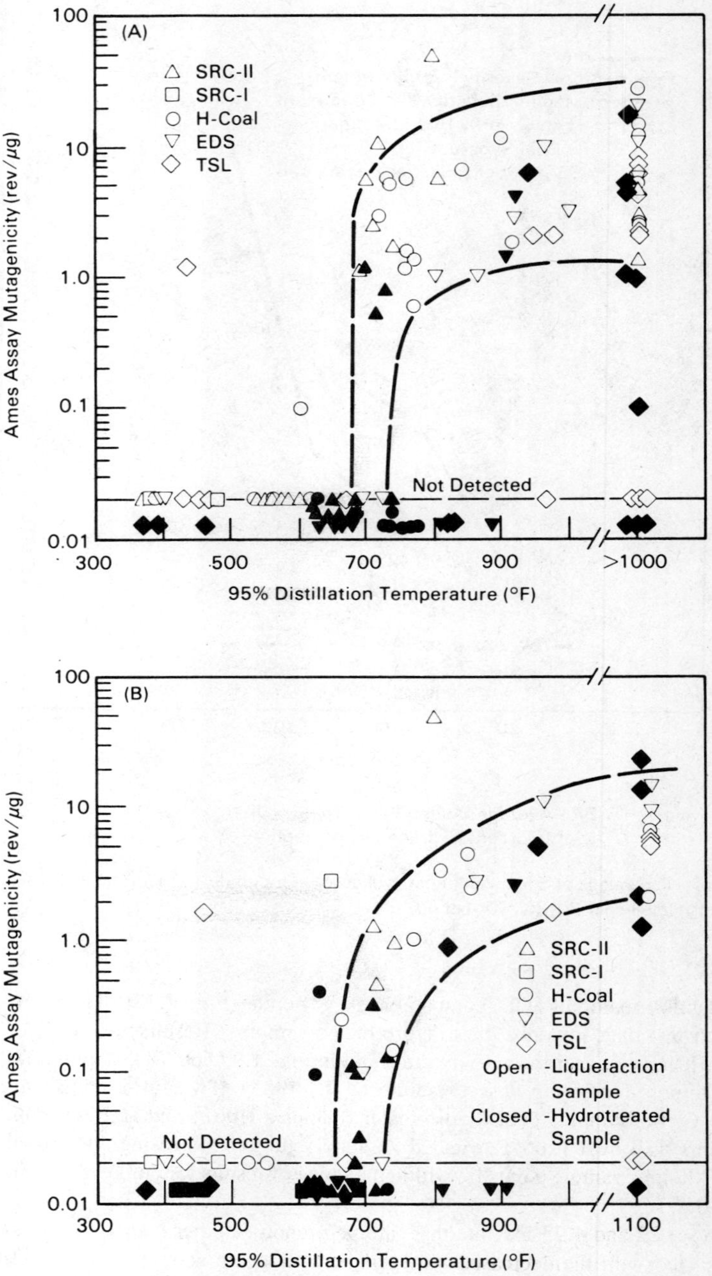

Figure 4.22 Ames assay mutagenicity versus 95% point for full-boiling-range coal-liquefaction (open white symbols) and hydrotreatment (closed black symbols) samples: (A) unfractionated samples; (B) fractionated samples (weighted-sum results). TSL = two-stage liquefaction.

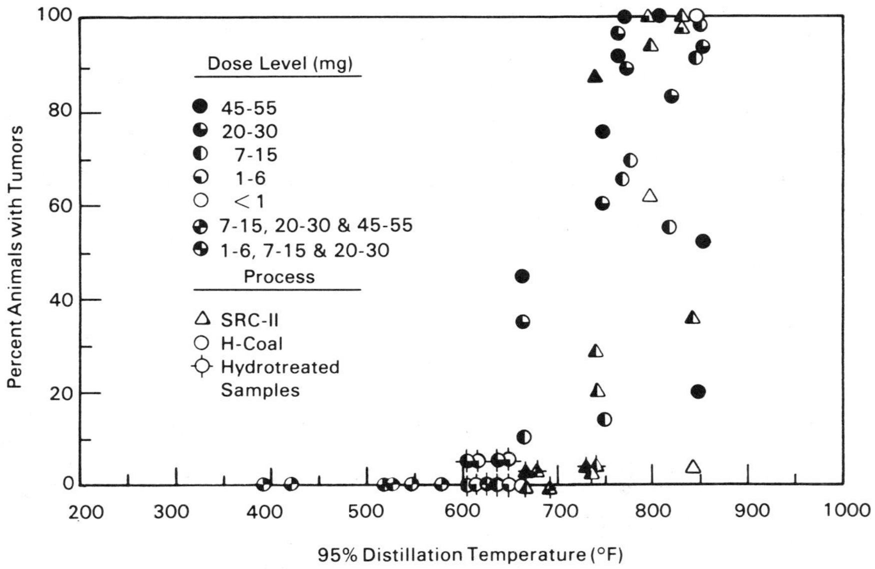

Figure 4.23 Skin-painting activity versus 95% point for full-boiling-range coal-liquefaction and hydrotreatment samples. SRC = solvent-refined coal.

inactive, according to the detection limits of the tests, which were usually in the range of 0.01–0.1 rev/μg. Data from a given hydrotreatment experiment are connected; data for the feedstock are the lowest right-hand points and data for products from increasingly severe hydrotreatment progress toward the upper left corner. As expected, increasing hydrotreatment severity reduces the 95% point (and the product distillation curve in general) and the aromaticity of products (i.e., increasing H:C ratio). Relative reductions in aromaticity are greater than those for the 95% point. For example, the H:C ratio may increase 40–50%, while the 95% point is reduced by only about 10%.

Figure 4.24 indicates that mutagenic and nonmutagenic samples can be segregated by two properties: The first is a 95% point of 625–675°F, while the second is an atomic H:C ratio of 1.4–1.5. The activity dividing lines for these two properties appear to be independent; that is, the lines are roughly vertical and horizontal. However, increasing hydrotreatment severity for a given feedstock affects both properties (i.e., the points for a given set of increasing severity hydrotreatment tests lie on a diagonal line). This explains why sufficiently severe hydrotreatment can eliminate mutagenic activity even though the product still has a high-boiling tail end (i.e., 95% point greater than 650°F). Enough hydrogen can be added to alter or destroy mutagenic/carcinogenic compounds, as reflected in increased H:C ratios, even though higher-molecular-weight materials are not totally eliminated, as reflected by 95% points above 700°F. This explains the higher incidence of nonmutagenic hydrotreated products with high 95% points compared to liquefaction products when only the 95% points are considered (Fig. 4.22).

Figure 4.24 also provides perspective on the relationship among feedstock properties, hydrotreatment severity, and mutagenic activity of products. For lighter feedstocks, the loss of Ames activity in products can be related to a decreased 95% point or to decreased aro-

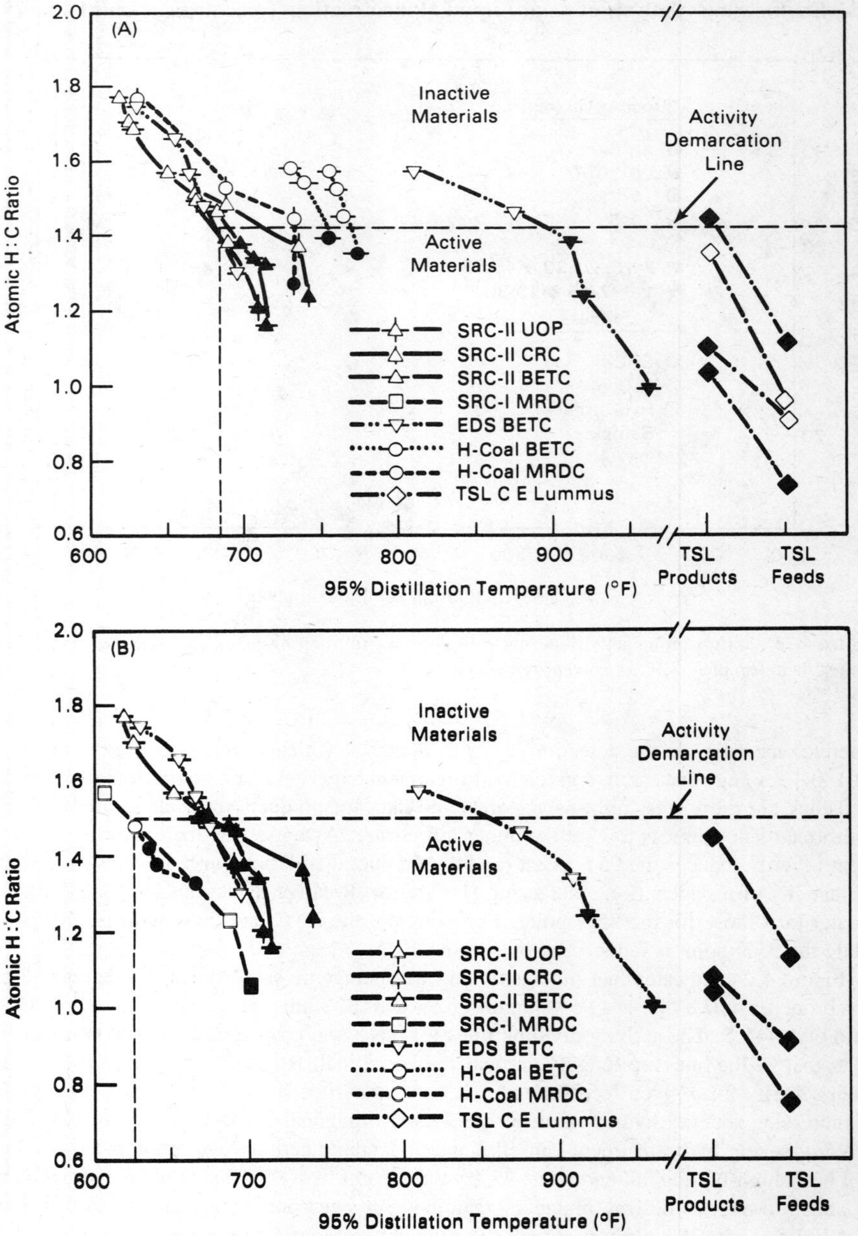

Figure 4.24 Atomic hydrogen : carbon (H : C) ratio versus 95% point for hydrotreatment feeds and products categorized as active (solid black) or inactive (open white) by the Ames assay: (A) unfractionated samples and (B) fractionated samples (weighted-sum results). The lines drawn connect products from hydrotreatment experiments (see Table 4.9) involving different severity treatment of single feeds. The feeds are shown as points at the lower-right end of the lines; products from increasingly severe hydrotreatment progress to the upper left. Two-stage liquefaction (TSL) LC Finer feeds and products are too heavy to have measurable 95% points; they have been plotted at arbitrarily high 95% points, as indicated. SRC = solvent-refined coal; UOP = Universal Oil Products; CRC = Chevron Research Company; BETC = Bartlesville Energy Technology Center; MRDC = Mobil Research & Development Corporation.

maticity, because inactive products generally lie on the inactive side of either demarcation line. For heavier feedstocks, however, the loss of Ames activity relates to decreased aromaticity, because the products still have 95% points above 650°F (e.g., the Bartlesville H-Coal and heavy EDS samples in Fig. 4.24). Finally, for very heavy feeds, corresponding to atmospheric residua or heavier materials (e.g., feeds to the C. E. Lummus hydrotreater in the ITSL process) even the addition of large amounts of hydrogen (from 2 to 3 wt % hydrogen consumption) is inadequate to reduce aromaticity sufficiently to cross the H:C ratio activity threshold. Hence, products from very heavy feedstocks are mutagenic, even after relatively severe hydrotreatment.

The few chronic skin-painting and initiation/promotion data available for hydrotreated products (Fig. 4.25) fit the pattern shown in Fig. 4.24 for Ames assay data. For the very heavy C. E. Lummus samples, both feeds and products were active in the initiation/promotion assay (Wilson et al., 1983a; Later, 1985a). For lighter feedstocks and products, initiation/promotion and chronic skin-painting data indicate that the activity dividing lines may be at the outer range of those shown (i.e., those based on weighted-sum Ames assay results). The most severely hydrotreated H-Coal (Sample H0322; Epler, 1981) and SRC-II (Sample H0203; Griest et al., 1983) products tested in the chronic skin-painting assay showed marginal activity. The severely hydrotreated SRC-II product tested in the initiation/promotion assay was active (Sample H0104; Mahlum et al., 1983). It is uncertain whether the chronic skin-painting activity is significant, because the products had tumor-incidence rates of 0–5% (Fig. 4.21), which may be within the range of experimental error compared to negative controls.

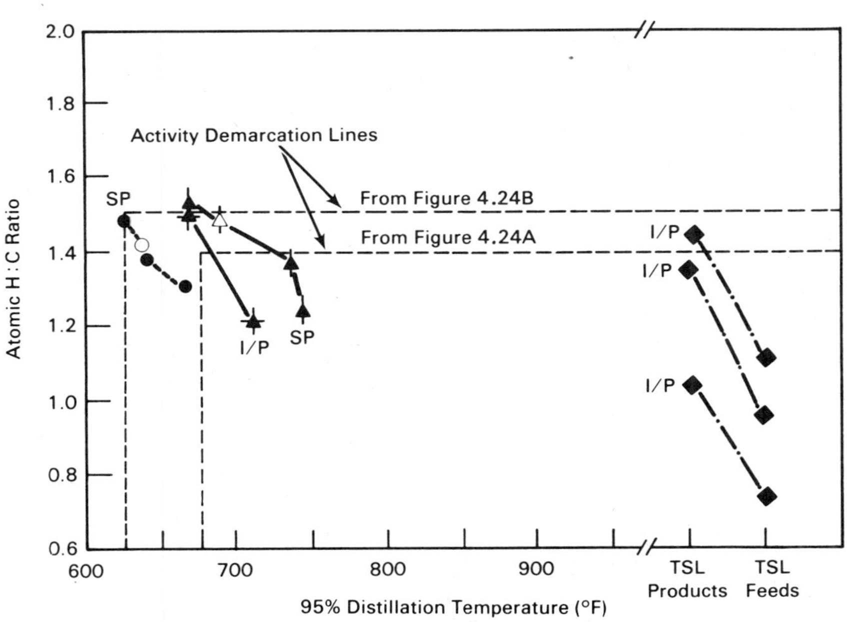

Figure 4.25 Atomic hydrogen:carbon (H:C) ratio versus 95% point for hydrotreatment feeds and products categorized as active (solid black symbols) or inactive (open white symbols) by initiation/promotion (I/P) and skin-painting (SP) assays. TSL = two-stage liquefaction. See Fig. 4.24 for symbol identifications.

It must be noted when using Figs. 4.24 and 4.25 that the H : C ratio is a very crude measure of sample chemistry, reflecting gross aromaticity and, to a lesser degree, heteroatom removal. Chemical effects that are too subtle to be reflected meaningfully by H : C ratios likely play an important role in the reduction of mutagenic/carcinogenic activity by hydrotreatment. For example, it has been shown that amino-PAH compounds are important contributors to the mutagenicity of coal liquids (Gray and Cowser, 1982) and that hydrotreatment is very effective in deaminating these compounds, with attendant decreases in mutagenicity (Weimer et al., 1980).

It should also be noted that the activity demarcation lines in Figs. 4.24 and 4.25 cannot be located precisely. The locations of the demarcation lines appear to correlate with the type of Ames assay used, being more inclusive for the weighted-sum result from testing chemical class fractions (Fig. 4.24B) than from testing unfractionated crude samples (Fig. 4.24A). Fillo and Craun (1985) discuss uncertainties associated with the data for samples located near the activity demarcation lines. These include linkage of some of the bioassay and properties data and the extent to which certain bioassay results clearly indicate the presence or absence of mutagenic/carcinogenic activity.

4.3.4 TSL Hydrotreatment

The second stage of TSL processes involves catalytic hydrogenation of liquid product from the first-stage reactor. This application of hydrotreatment illustrates the sample property concepts discussed in the previous section because the feedstocks are such high-boiling materials (essentially atmospheric distillation residua, almost all of which boiled above 600–700°F; see Fig. 4.20). The most complete toxicological data, process history, and physical property data for TSL hydrotreating is for samples from four LC Finer hydrotreating runs at the C. E. Lummus ITSL facility (Wilson et al., 1983a). The data are discussed in some detail here because of their importance in understanding the toxicological implications of hydrotreating high-boiling coal liquids.

Figure 4.26 shows the effects of hydrotreatment severity on atomic H : C ratio and 50% boiling point for the four LC Finer runs. (LC Finer feeds and products are too heavy to measure the 95% point.) While properties of the feedstocks and process mode and conditions varied, changes caused by hydrotreatment were consistent (i.e., the slopes of the feed–product lines in Fig. 4.26 are all similar). LC Finer runs 7-37 and 9-0524 were performed under normal operating conditions, although feed coals to the first-stage liquefaction reactor differed (Indiana No. 5 for run 7-37 and Illinois No. 6 for run 9-0524). Run 7-90 treated a feed that had been recycled through the liquefaction reactor once before entering the LC finer. Samples from run 8-52 were collected after a temperature excursion where bed temperatures in the LC finer were 150–250°F higher than normal. Although the detailed chemical composition of LC finer products may have been altered by these process variations, bulk properties of the products were consistent relative to severity of treatment and heaviness of the feedstock (Fig. 4.26).

Selected chemical and toxicological data for LC Finer feeds (LCFF) and products (Wilson et al., 1983a) are shown as a function of hydrogen consumption in Figs. 4.27 and 4.28. Not all feeds and products were analyzed chemically or tested with each bioassay.

Concentrations of specific carcinogenic PAHs and N-PACs were consistently higher in feed and product samples from run 7-90 compared to run 7-37 (Fig. 4.27). However, the LCFF and resulting product were heavier (i.e., more aromatic and higher boiling) in run 7-90 than in run 7-37 (Fig. 4.26). This could account in part for the differences observed in sample chemistry between these runs, because all compounds shown in Fig. 4.27 have

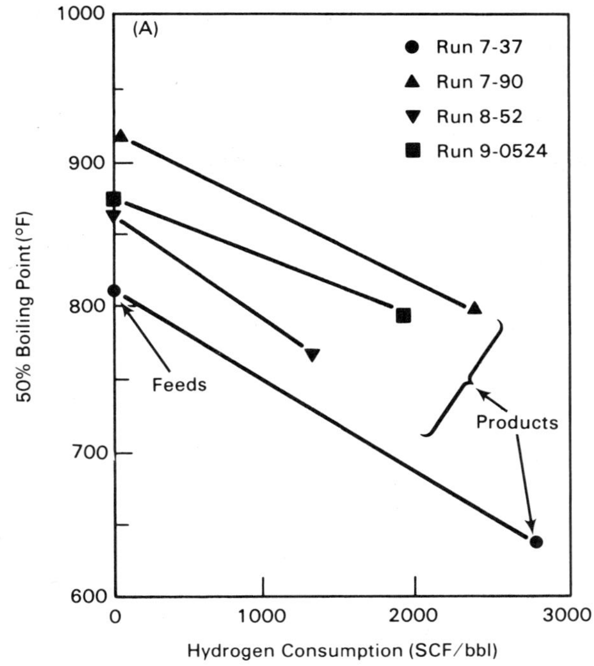

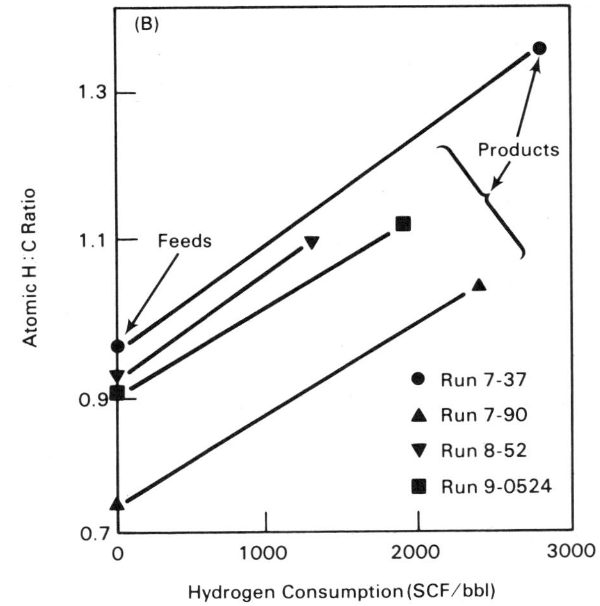

Figure 4.26 (A) Distillation midpoints and (B) atomic hydrogen : carbon (H : C) ratios of integrated two-stage liquefaction LC Finer feeds and products as a function of LC Finer hydrogen consumption. SCF/bbl = standard cubic feet per barrel.

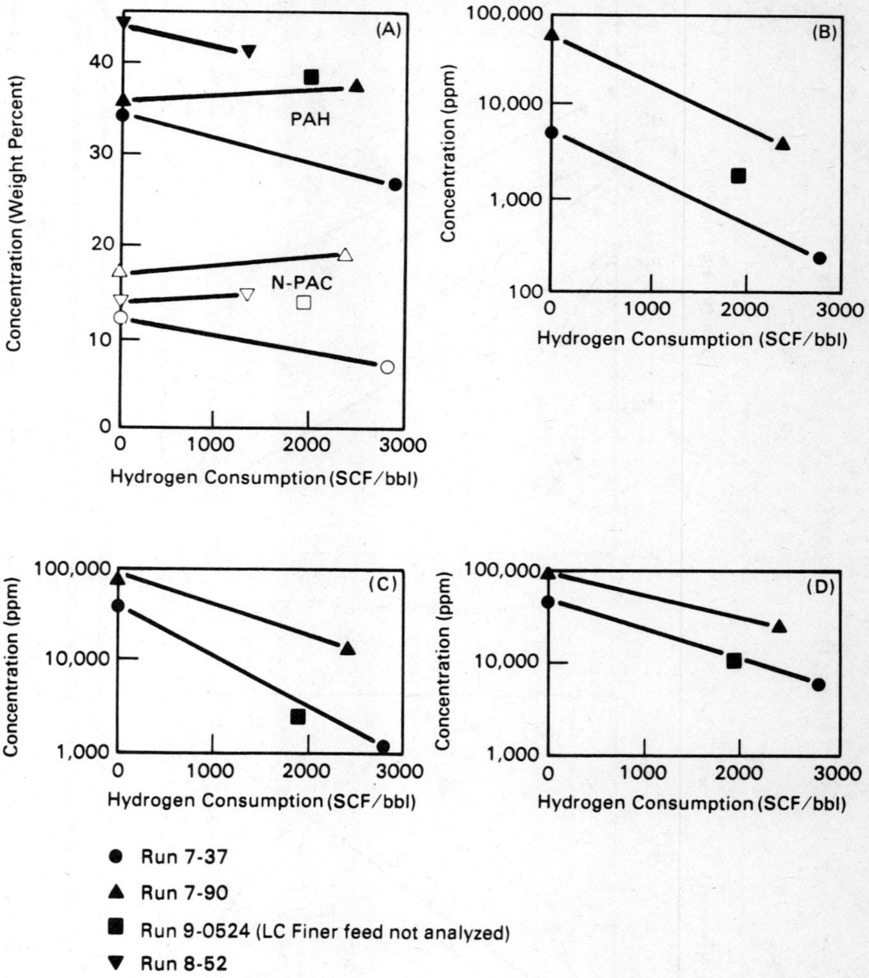

Figure 4.27 Selected chemical class and polycyclic aromatic hydrocarbon compound concentrations in integrated two-stage liquefaction LC Finer feeds and products as a function of LC Finer hydrogen consumption (chemical data from Wilson et al., 1983a). (A) Polycyclic aromatic hydrocarbons (PAH) and nitrogen-containing polycyclic aromatic compounds (N-PAC). (B) Benz[a]acridine + benz[c]acridine; (C) benz[a]anthracene + chrysene; (D) benzo[a]pyrene + dibenz[a,h]anthracene + benzo[b,j,k]fluoranthenes. SCF/bbl = standard cubic feet per barrel.

boiling points from 800 to 1000°F (Anderson and Wu, 1963). The role of hydrotreatment in removing these compounds was comparable for these two runs, as indicated by the similar feed–product slopes in Fig. 4.27. Similar conclusions can be drawn from chemical class fractionation data.

The bioassay data (Fig. 4.28) offer a different perspective on the effects of hydrotreating high-boiling feedstocks. There are no consistent trends with respect to effects of product

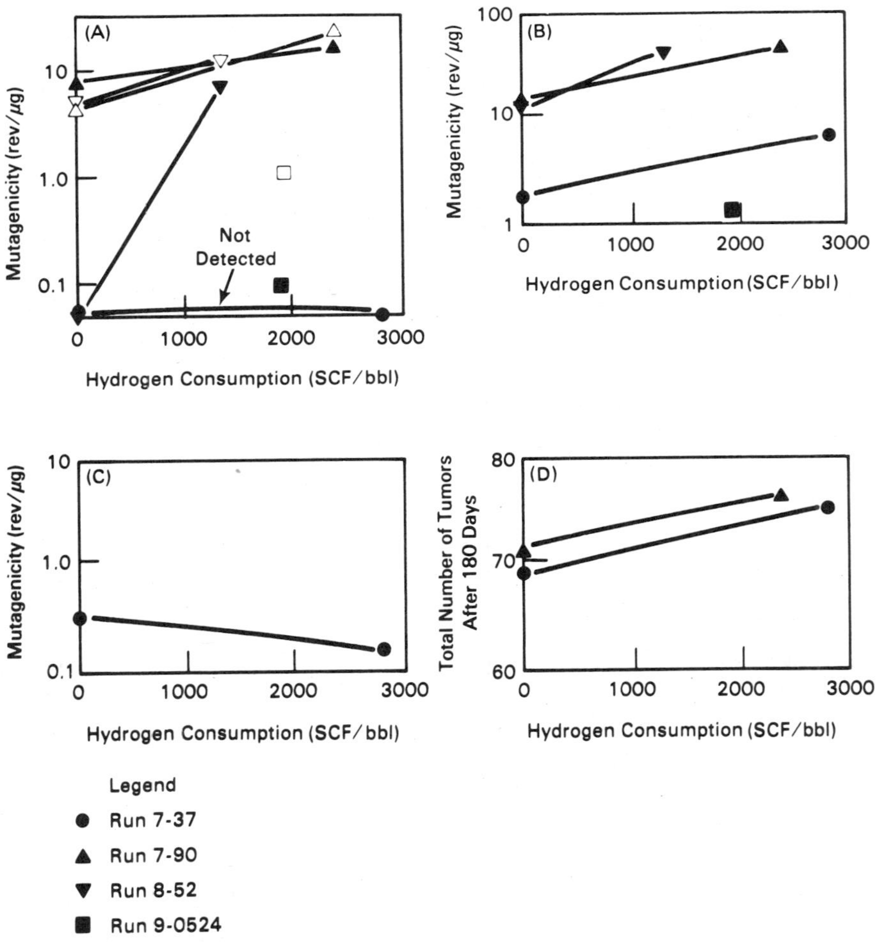

Figure 4.28 Selected bioassay results for integrated two-stage liquefaction LC Finer feeds and products as a function of LC Finer hydrogen consumption (bioassay data from Wilson, et al., 1983a): (A) Ames assay (strain TA98), (B) preincubational Ames assay, (C) Chinese hamster ovary assay, and (D) initiation/promotion assay. SCF/bbl = standard cubic feet per barrel. Lack of data for a given run and bioassay means that samples were not tested. LC Finer feed for Run 9-0524 was not tested. Closed (black) symbols represent unfractionated samples; open (white) symbols represent fractionated samples (weighted-sum results).

properties, run conditions, or hydrotreatment severity on mutagenic/carcinogenic activity. Initiation/promotion test data indicate that samples from runs 7-37 and 7-90, representing the lightest and heaviest feedstocks and products, respectively (Fig. 4.26), have comparable activity, with hydrotreatment slightly increasing the activity. In contrast, preincubation Ames assay data indicate that run 7-90 materials are significantly more active than run 7-37 materials and that hydrotreatment again increases biological activity. Standard Ames assay results indicate that run 7-37 materials are inactive (both feed and product), while run 7-90 materials are among the most active.

While it is difficult to make quantiative comparisons of bioassay results for TSL hydrotreatment, all the feeds and products tested were active in one assay or another. This finding is consistent with the bulk properties of the various samples. The results from run 7-37 (lightest feed and product) and run 7-90 (heaviest feed and product) indicate that hydrotreatment of very heavy feedstocks may actually increase mutagenic/carcinogenic activity (Fig. 4.28). This could be a result of compounds that are too large to give positive results in the bioassays used being reduced to smaller compounds that are active. For example, high-molecular-weight heterocyclic N-PACs could be partially hydrogenated to yield mutagenic amino-PAH intermediates, or high-molecular-weight PAHs could be reduced to lower-molecular-weight, possibly alkylated PAHs that are genotoxic (see Fig. 4.14 for example reaction pathways). The bioassay results for the run 7-37 feed and the run 7-90 product support this hypothesis, since the latter showed substantially greater mutagenic activity (Fig. 4.28), even though these two materials had comparable 50% points and H : C ratios (Fig. 4.26).

4.3.5 Summary

The effects of hydrotreatment on mutagenic/carcinogenic activity of crude coal-liquefaction materials can be related to changes in bulk sample properties. Hydrotreatment reduces the boiling range of feedstock heavy ends and reduces overall aromaticity and heteroatom content. Because of reduced aromaticity, hydrotreated products may have no mutagenic/carcinogenic activity, even though their 95% points are above 650°F. Reduced aromaticity reflects the destruction, during hydrotreatment, of higher-molecular-weight PAHs, amino-PAHs, and related compounds that are responsible for mutagenic/carcinogenic activity.

4.4 ENGINEERING ANALYSIS OF THE COAL-GASIFICATION TOXICOLOGICAL DATABASE

Inherent differences between gasification and liquefaction processes directly affect the analysis of toxicological data for gasification oil and tar by-products. Liquefaction processes produce gas, liquid, and solid products, but liquids are generally the primary products. Gasification processes primarily produce gas, with varying amounts of oils and tars as by-products. Of concern here is whether physical properties of coal-gasification oils and tars are related to mutagenic/carcinogenic activity, as they are for liquefaction and hydrotreated materials.

4.4.1 Physical Properties

Only four gasification tar samples have both toxicological data and distillation curves available (Table 4.10). Three samples are from the Research Triangle Institute (RTI) bench-scale, fixed-bed reactor and one is from the Grand Forks Energy Technology Center (GFETC) pilot plant. Ames assay data are available for all four samples; chronic skin-painting data are available for the GFETC tar sample. Comparing the data in Table 4.10 with Figs. 4.8 and 4.10 shows that the Ames assay and skin-painting data for these four gasification tars fall in the same range of bioactivities as do liquefaction samples with comparable 95% points.

Table 4.10 Summary of Ames Assay and Skin-Painting Data for Gasifier Tars with Distillation Curves Available

Parameter	Sample Names and Numbers			
	RTI Tar[a] (G1541)	RTI Tar[a] (G1551)	RTI Tar[a] (G1561)	GFETC Tar[b] (G1121)
Coal type	Illinois No. 6	Wyoming subbituminous	North Dakota lignite	Indianhead lignite
Tar distillation[c] (°F)				
10/30%	390/430	370/400	370/420	341/516
50/70%	480/640	430/520	460/550	640/NA
90/95%	900/1020	700/810	750/840	NA/(>1000)[d]
Atomic H:C ratio	NA[e]	1.15	1.0	0.43
Ames assay mutagenicity (rev/μg)				
Crude sample (TA98)	11.3	0.3	1.9	2.0
Weighted sum (TA98)	3.1	0.3	1.3	2.0
Skin-painting activity, % mice with tumors at weekly dose (mg) shown				
75	NA	NA	NA	93
37.5				100
7.5				100
0.75				74
0.075				16

[a]RTI = Research Triangle Institute; data from Cleland et al. (1979), Gangwal et al. (1979), Gangwal, 1981, Nichols et al. (1979), Green et al. (1981), and Mixon et al. (1984).

[b]GFETC = Grand Forks Energy Technology Center; data from Stearns-Roger (1981)) and Wilzbach et al. (1982); see Chapter 3, Table 3.35.

[c]RTI tar distillations by simulated gas chromatography; GFETC tar distillation under vacuum by a nonstandard method.

[e]NA indicates data not available.

The limited data available indicate that gasifier tars and liquefaction products with comparable distillation curves show roughly comparable mutagenic/carcinogenic activity. Although there is some threshold temperature for the occurrence of toxicological activity in gasifier oils and tars, as shown by the consistent lack of mutagenic activity in their volatile fractions (nominal endpoint, 390°F; Reilly et al., 1981), its location in the range of 400–800°F is unknown. Recycle-oil samples from the Hygas pilot plant are also inactive and typically contain 90–95% toluene (the make-up solvent for the recycle oil) and a few tenths of a percent of three-, four- and five-ring PAHs (Anastasia et al., 1982a,b). The distillation endpoint of these oils must be above 700°F, but the 95% point should be below this. Nonvolatile fractions (nominally, >400°F material) are usually active once the diluent toluene is removed (Reilly et al., 1981).

Ultimate analysis data are available for a considerable number of gasification samples (see Table 4.3). Therefore, use of atomic H:C ratios as a surrogate for boiling range is of interest because these two properties are related (Fig. 4.6). Figure 4.29 shows Ames assay data for gasification oils and tars as a function of H:C ratio. Most data are for the University of Minnesota–Duluth (UMD), low-Btu industrial gasifier; some data are for the RTI bench-scale unit and the industrial Lurgi plant in Kosovo, Yugoslavia. The extensive toxicological database for Hygas oils and tars was not considered here because there are no ultimate analysis results available. An extensive database is also available for the PDU-scale Mellon Institute (MI) entrained-flow pyrolysis/hydropyrolysis reactor. These data are shown separately in Fig. 4.30 because of their unusually large number and because they were generated by pyrolysis/hydropyrolysis reactions, as opposed to the full spectrum of gasification reactions. In addition, the MI tars may be heavier than the tars in Fig. 4.29, since the former were condensed at 100°C.

Data for gasifier tars show broad scatter compared to data for liquefaction and hydrotreated samples. Tars with H:C ratios from 0.3 to 1.6 have significant Ames assay mutagenic activity (greater than 1 rev/μg). While there is no clear overall trend (Fig. 4.29), MI tars show peak mutagenicity at an H:C ratio of about 0.7 (Fig. 4.30), which corresponds to high-molecular-weight, highly aromatic material with little alkylation or ring saturation. Some MI tar samples with very low H:C ratios also reflect carryover of char fines, which may affect the Ames assay results (Fillo et al., 1983).

Five RTI samples show a consistent pattern of increasing Ames assay activity with decreasing H:C ratios (Fig. 4.29). The H:C ratios may also track with boiling range. The sample with the highest H:C ratio (1.15) has a 95% point of 810°F; the sample with a 95% point of 840°F has an H:C ratio of 1.00 (Table 4.10). Moreover, Ames assay activities for a given H:C ratio generally agree for RTI and MI samples. (Compare Figs. 4.29 and 4.30.)

The scatter in Fig. 4.29 primarily reflects UMD samples collected under various operating modes, using different coals and often spanning standby or shutdown periods. More-

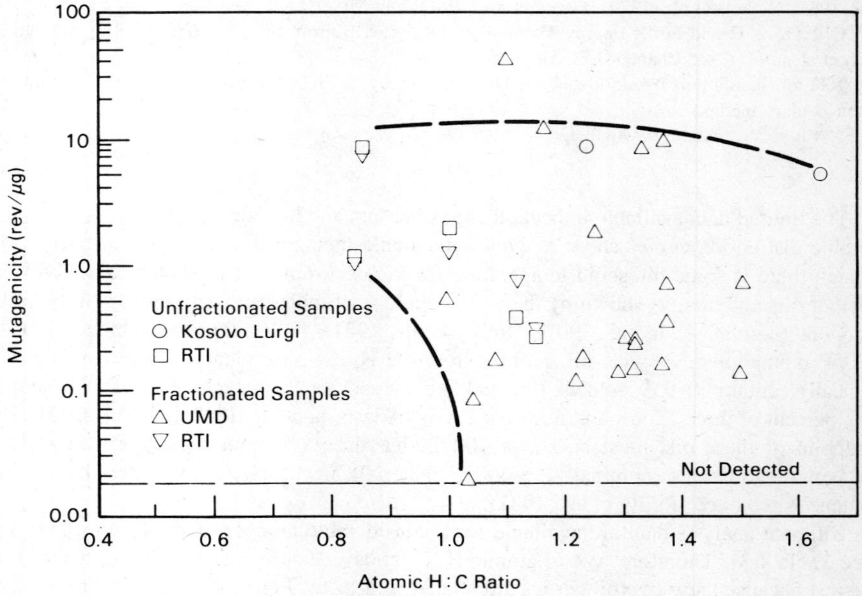

Figure 4.29 Ames assay mutagenicity versus atomic hydrogen:carbon (H:C) ratio for gasification samples. RTI = Research Triangle Institute; UMD = University of Minnesota–Duluth.

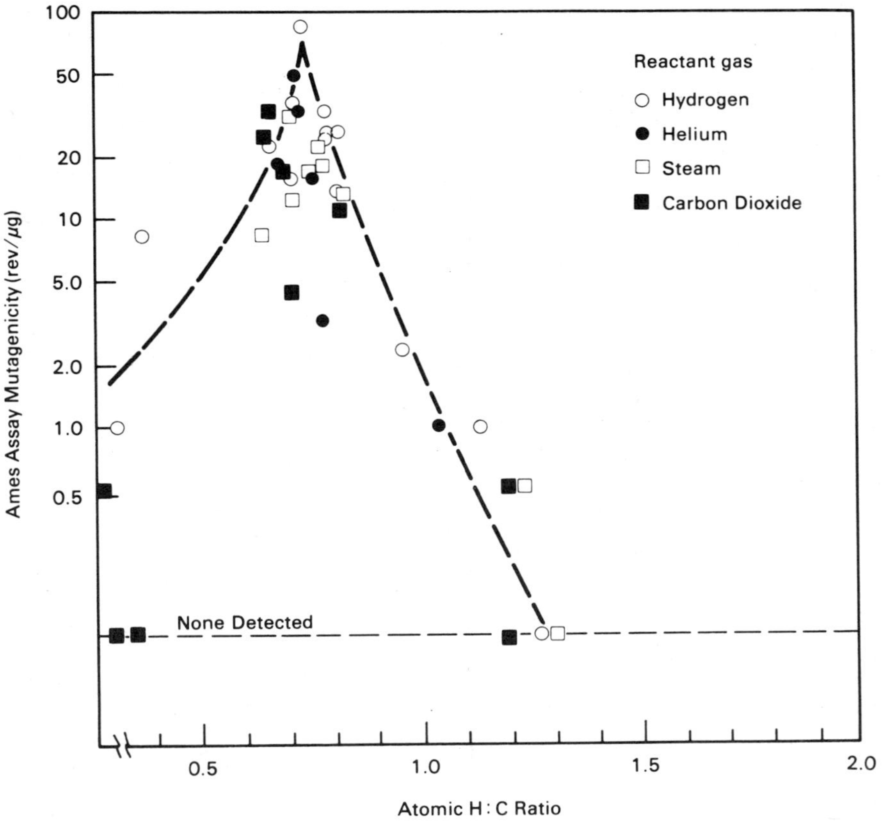

Figure 4.30 Ames assay mutagenicity versus atomic hydrogen : carbon (H : C) ratio for Mellon Institute high-temperature trap tars (data from Fillo et al., 1983).

over, by-product tar at UMD is collected in an electrostatic precipitator (ESP), and variations in ESP operating conditions may affect tar collection efficiencies.

In summary, the combined toxicology/physical properties database is small for gasification oils and tars. Results for four samples with distillation data are comparable to those for full-boiling-range liquefaction products. There are insufficient data to determine whether an activity threshold 95% point at 650°F also applies to gasification materials. Atomic H : C ratios tend to support trends seen in liquefaction samples.

4.4.2 Process Conditions

Some qualitative intra- and interprocess comparisons may explain the wide range in oil/tar production rates and toxicological effects obtained for various gasification processes. A conceptual approach for discussing gasification oil/tar production and chemistry is that of primary and secondary reaction pathways. The concept has been used by several investigators to study coal gasification and pyrolysis reactions and kinetics (Essenhigh, 1981;

Howard, 1981; Solomon, 1981; Talwalker, 1983). Coal is seen as being broken into primary fragments by thermal reactions at labile bonds in the coal structure. Primary fragments are then subjected to various secondary reactions, including thermal and catalytic cracking and condensation reactions. Secondary reactions yield a range of gas, liquid, and solid (char) products, depending on the relative intensity of the secondary reactions compared to the primary reactions.

This conceptual framework can be used to evaluate trends in oil/tar yields and chemistry, both of which affect mutagenic/carcinogenic activity and physical properties of the oil/tar products. For example, sufficient residence time at high temperatures results in negligible production of by-product oils and tars, because secondary cracking reactions are sufficiently intense to decompose all intermediate oil/tar products to gaseous constituents. This is illustrated schematically for lignite coal in Fig. 4.31A, with supporting experimental-scale pyrolysis data from Suuberg (1977) and Fillo et al. (1983) shown in Fig. 4.31B. Suuberg's data essentially show yields of primary reaction products, because negligible residence time was provided by his experimental apparatus. The MI riser-reactor data show effects of secondary reactions on cracking primary products to gaseous species. Table 4.11 shows similar trends in pilot- and commercial-scale processes.

For gasification processes that produce by-product oils and tars, the primary/secondary reaction scheme also provides a conceptual framework to explain differences in the chemical characteristics and related physical properties of the oils and tars. As residence time and temperature increase, primary products, such as multiring, high-molecular-weight aromatics, should be cracked to lighter aromatics and gaseous species. The expected profile of aromatic ring sizes in the product oil/tar is illustrated conceptually in Fig. 4.32. Actual ring-size profiles obtained from GC-MS analyses of by-product oils and tars from various gasification facilities are shown in Fig. 4.33. For the MI riser reactor, increasing reaction temperatures follow the expected trend. Other variables, such as gaseous environment, residence time, and coal type, are partly responsible for the broad range in the MI data. Composition data for the low-temperature reactor oil from the HYGAS reactor (a light oil produced under long residence time from the low-temperature reactor— the coal devolatilization stage of this multistage, fluidized-bed gasifier—prior to mixing with recycle oil) also show a broad range. Oil compositions for fixed-bed gasifier tars (GFETC and a Lurgi plant) show intermediate profiles, most likely reflecting combinations of primary and secondary reactions occurring in the top of the reactors, where there are gradual temperature and residence time profiles for gas and solid reactants.

The trends described above are general. Residence time and temperature are not the only important process variables. Gaseous environment, pressure, coal type, gas–solid contacting characteristics, and other variables can also affect oil/tar yields and chemistry. However, residence time and temperature appear to be useful first-order variables for a conceptual framework.

The chemical composition of by-product oils and tars should be reflected in their toxicological characteristics. Very light oils with negligible quantities of higher-molecular-weight aromatic compounds should not be mutagenic/carcinogenic; the opposite should be true for heavy tars. However, very heavy tars may be inactive because their constituents are not metabolized to genotoxic compounds, or they may have physical properties that are incompatible with bioassay systems (e.g., high viscosity, poor solubility).

Figure 4.34 shows the trend in Ames assay mutagenicity observed with tars produced by the MI riser reactor. Tars produced at 600°C largely consisted of high-molecular-weight, primary reaction products (see Fig. 4.33C) and showed little or no mutagenic activity. Tars produced at 800°C or 1000°C, with more secondary products in the bioactive molecular-

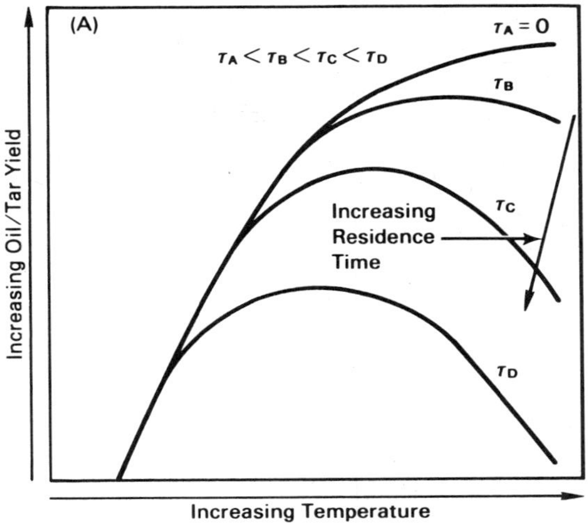

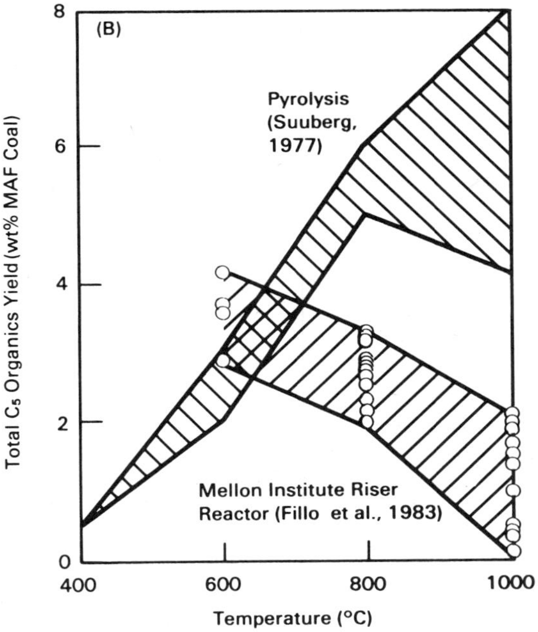

Figure 4.31 Expected trends (A) and observed results (B) for oil/tar yields in devolatilization of lignite coal (Fillo et al., 1983).

Table 4.11 Condensable Hydrocarbon Production for Selected Coal Gasification Processes

Gasification Process	Type	Temperature (°F)	Pressure (atm)	Gas-Phase Residence Time (s)	Production Rate (lb/ton MAF Coal)[a]	
					Tar and Oil	Aqueous Phenols
Lurgi, dry ash	Fixed bed	600–2100	<30	<30	80–200	5–20
Lurgi, slagging	Fixed bed	600–2600	<20	<20	50–150	10–30
Hygas	Fluidized bed	600–1900	35–70	5–20	20–140[b]	1–20
Synthane	Fluidized bed	1000–1800	<40	<1–10	0.1–180	1–12
CO_2-Acceptor	Fluidized bed	1500–1600	10–12	20	2[c]	<0.1
KRW[d]	Fluidized bed	1500–1900	16–17	10–15	Negligible	Negligible
Texaco	Entrained flow	2000–2500	<70	<10	Negligible	Negligible
Koppers–Totzek	Etrained flow	2500–3500	1	<5	Negligible	Negligible

[a] MAF = moisture- and ash-free.

[b] Primarily oils (i.e., benzene, toluene, xylene) due to coal devolatilization in the low-temperature reactor as calculated by the Hygas oil production model.

[c] Primarily benzene, as measured in gasifier raw product gas.

[d] Kellog–Rust Westinghouse (Mann et al., 1985).

Source: Nakles and Fillo (1983).

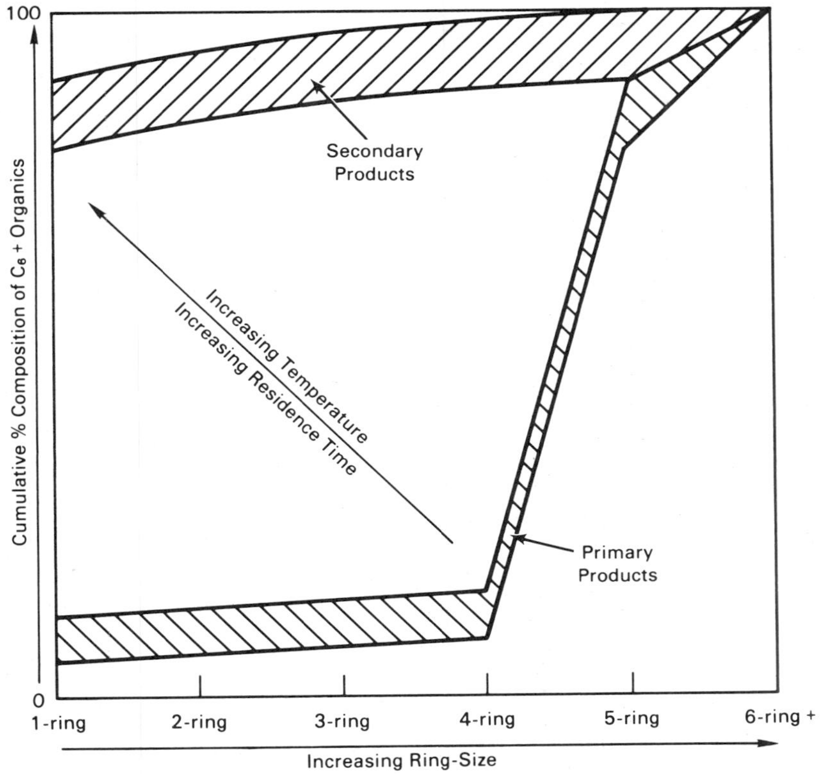

Figure 4.32 Changes in oil/tar chemistry anticipated during the devolatilization of coal (Fillo et al., 1983).

weight range, were mutagenic. Mutagenicity appeared to decrease as reaction temperatures increased from 800 to 1000°C, presumably as bioactive materials were cracked to inactive lower-molecular-weight or gaseous products, depending on the specific reactor conditions involved.

Various measures of reaction severity have been developed to represent the extent of thermal cracking of complex hydrocarbon mixtures (Hougen and Watson, 1947; Duncan et al., 1978). These measures may help explain variations in the mutagenic activity of gasification oils and tars. Duncan et al. (1978) defined a kinetic severity function expressed as the integral, over time, of the rate constant for thermal decomposition of anthracene (as reported by Virk et al., 1974) and used the function to correlate product yields for riser-reactor cracking of coal. Figure 4.35 shows that this function, defined in the figure legend, roughly correlates with Ames assay activity for a series of seven HYGAS low-temperature-reactor oil/tar samples. Results in Fig. 4.35 are for two measures of biological activity: (A) Ames testing of crude sample of the nonvolatile oil (NVO) fraction (>200°C) of low-temperature-reactor (LTR) oils weighted by the fraction of NVO in the whole LTR oil and (B) results of (A) normalized to a revertant production rate (rev/g moisture-and ash-free coal), using coal feed rate and LTR oil yield data. This correlation primarily quantifies the

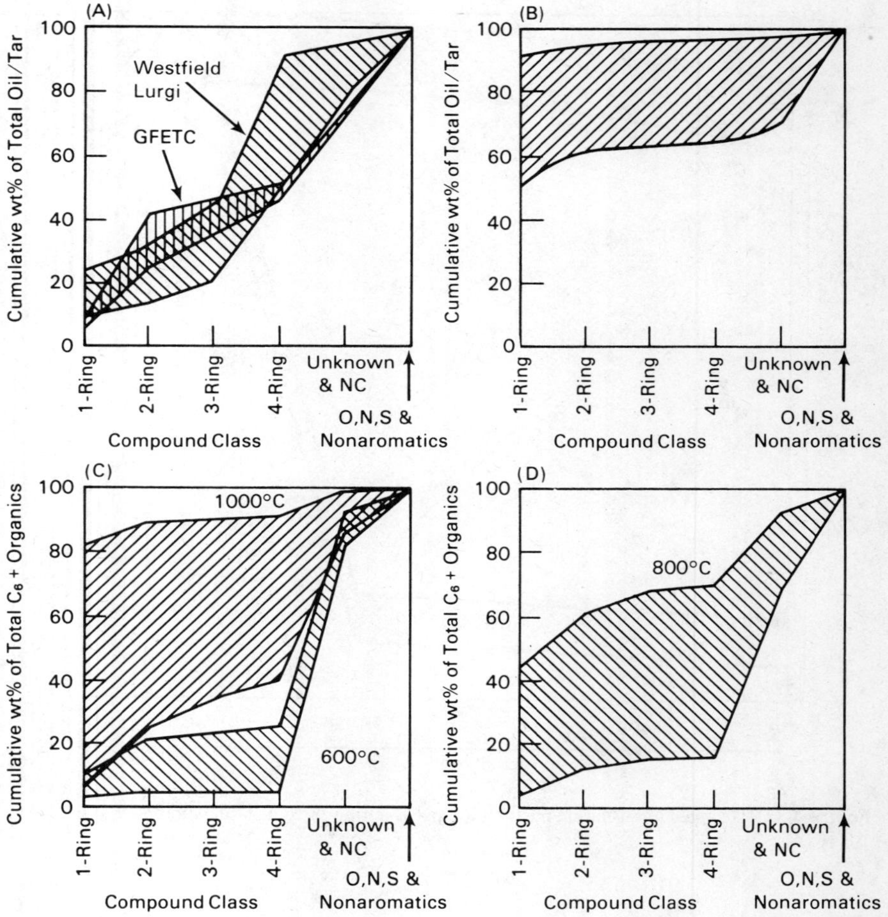

Figure 4.33 Cumulative distribution of compound classes in total C_6^+ organics from various gasifiers (Fillo et al., 1983): (A) fixed-bed gasifiers, GFETC = Grand Forks Energy Technology Center; (B) Hygas low-temperature reactor oil; and (C) and (D) Mellon Institute riser reactors at three reactor temperatures. NC = nonchromatographable; O, N, S = oxygen-, nitrogen-, or sulfur-containing compounds.

effect of secondary reactions that crack mutagenic compounds into lower-molecular-weight, inactive compounds or gaseous products. However, the approach cannot be used to compare data from different gasification reactors because of the many varying process conditions other than temperature and residence time in different gasification processes.

4.4.3 Summary

The database for coal-gasification processes is inadequate to explain the presence or magnitude of mutagenic/carcinogenic activity in gasification oil/tars. Physical property data are

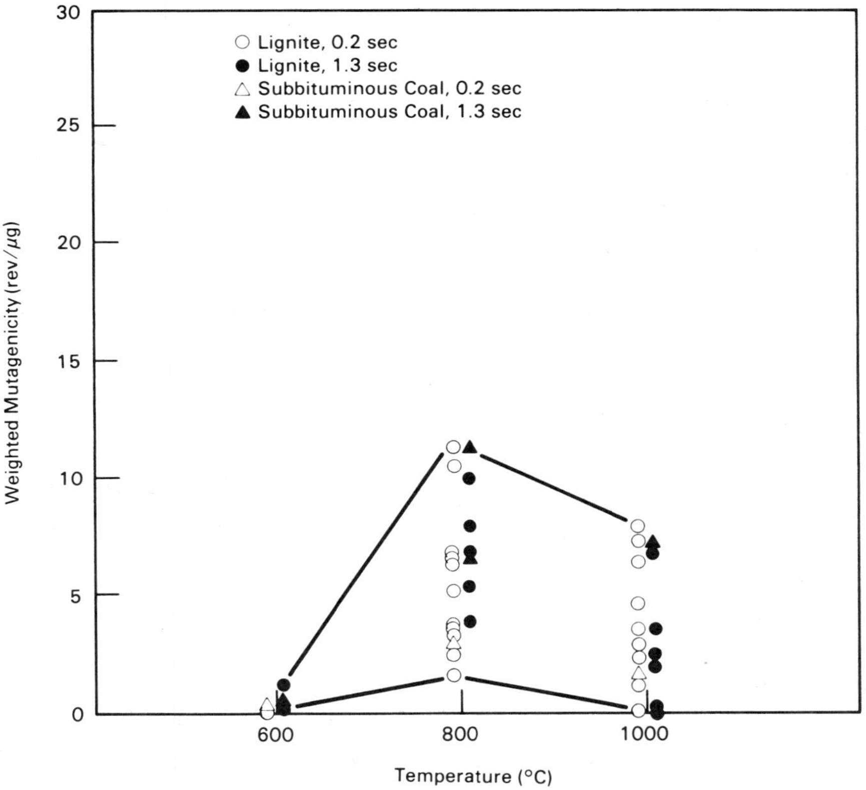

Figure 4.34 Ames assay mutagenicity of oil/tar produced by Mellon Institute reactor (adapted from Fillo et al., 1983). Legend indicates reactor conditions (i.e., coal type and residence time).

too sparse to evaluate effects of sample properties on mutagenic/carcinogenic activity (see Table 4.3). Process conditions are available for too few samples to make interprocess comparisons. Process variables controlling production of biologically active materials are likely to be numerous, with complex interrelations.

Nonetheless, there are several trends and relations in the database for gasification processes. There is some indication that sample properties are related to mutagenic/carcinogenic activity in a manner similar to that seen for liquefaction and hydrotreated materials. A model of primary/secondary reactions provides a conceptual framework to evaluate toxicological data, both within and between different gasification processes. Further refinement and definition of the model may allow for better predictive capabilities relating to the mutagenic/carcinogenic activity of coal-gasification oils and tars.

4.5 INTERTECHNOLOGY COMPARISONS

Heavy-end distillation characteristics (i.e., 95% point) and aromaticity (i.e., H:C ratio) are related to the presence or absence of mutagenic/carcinogenic activity in hydrotreating

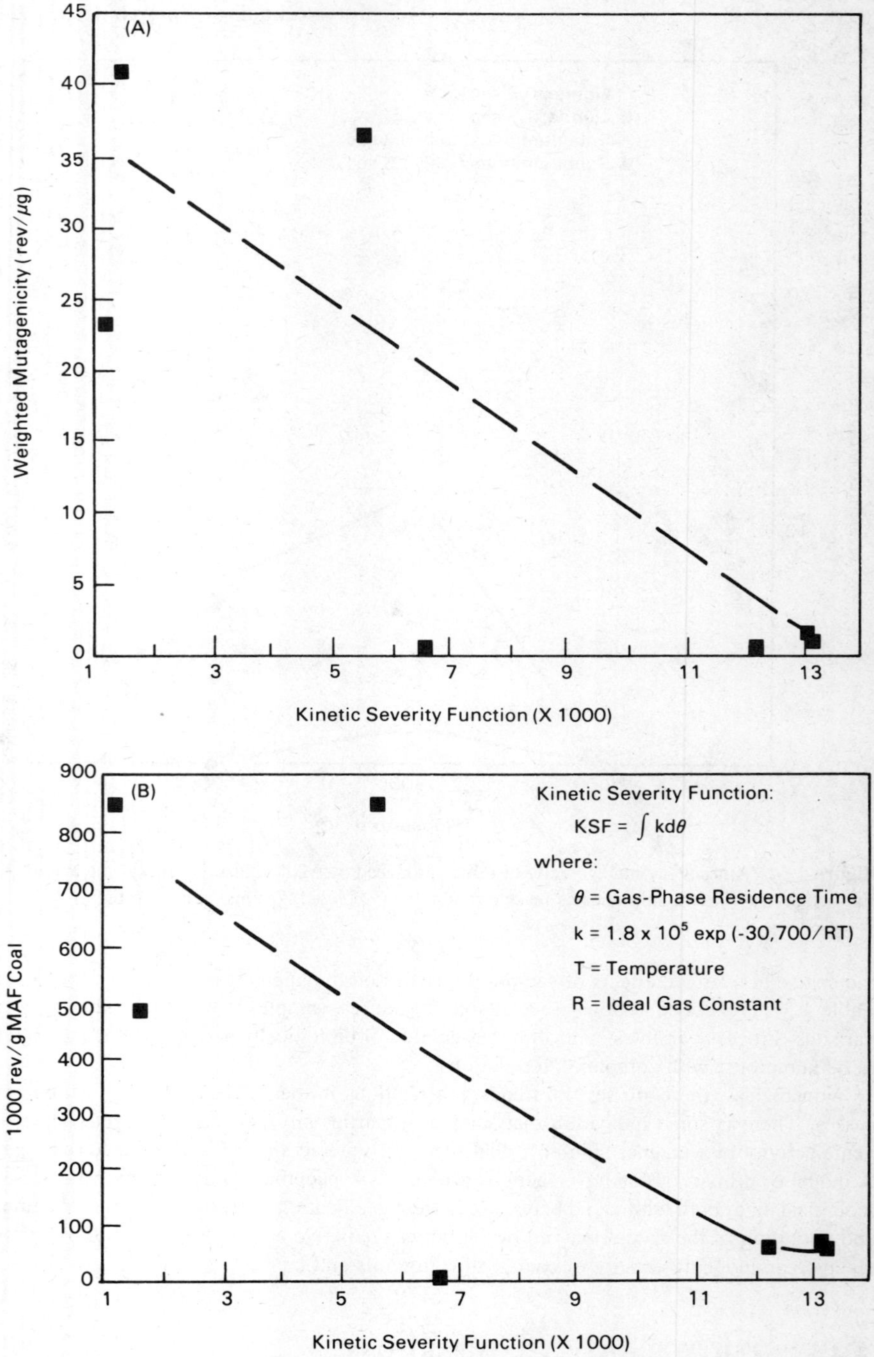

Figure 4.35 Ames mutagenicity assay versus kinetic severity function for low-temperature-reactor oils from the Hygas process: (A) weighted mutagenicity (rev/μg); (B) feed-coal-normalized mutagenicity (1000 rev/g moisture- and ash-free coal).

feedstocks and products. These sample properties are also related to such activity in crude liquefaction products and gasification oils and tars. The 95% point is the key property for liquefaction products. The H : C ratio is key for gasification materials, primarily because of the lack of distillation data, although the H : C ratio partially reflects distillation properties.

Figures 4.36–4.38 show the relationships among 95% points, H : C ratios, and toxicological activity, as determined by Ames, initiation/promotion, and chronic skin-painting dermal carcinogenicity assays, respectively, for coal-liquefaction, hydrotreatment and gasification samples. Activity generally occurs for samples with a 95% point above 650°F and an H : C ratio below 1.5. These demarcations between active and inactive samples apply for all materials in each assay. In only a few instances are samples actually active that would be expected to be inactive based on their properties. These instances are primarily for Ames data, and often involve samples with uncertain data linkage (see Section 4.2.2.4) or slight toxicological activity (e.g. active hydrotreated samples in the inactive zone in Fig. 4.38; see Section 4.3.3). Nonetheless, the correlation between sample properties (i.e. 95% point and H : C ratio) and toxicological activity is remarkably robust, considering the large number of samples, processes and testing laboratories involved.

Figures 4.36–4.38 show that the effects of heavy-end boiling range and aromaticity on the presence/absence of mutagenic/carcinogenic activity of liquefaction and hydrotreated materials are consistent. Only three gasification samples are represented, but these data are consistent with liquefaction results.

Figures 4.36–4.38 establish a conceptual basis to explain the relationships among sample properties, sample chemistry, process engineering and mutagenic/carcinogenic activity of coal-conversion liquids. Figure 4.39 schematically combines the results of Fig. 4.36–4.38 to illustrate this conceptual framework. The trends shown in Fig. 4.39 are discussed below.

There are various options to liquefy coal, including catalytic or noncatalytic processing, low-to-high reaction severity, and various schemes for solvent hydrogenation and recycling, including hydrotreatment of recycle solvent, as in the EDS process. The end result is a highly aromatic liquid product. These variations, including coal type, affect the aromaticity and chemistry of the products. However, the main effect is on product yields (i.e., gas versus liquid versus solids) and conversion efficiencies. Coal liquids appear to be generally similar in their bulk chemistry with respect to toxicological effects, are highly aromatic, and contain mutagenic/carcinogenic compounds in higher-boiling products. This reflects the basic chemistry of coal and coal-liquefaction/dissolution reactions.

In contrast, catalytic hydrotreatment yields products with more variable chemical characteristics. Hydrotreatment of a range of feedstocks reduces product aromaticity with relatively minor effects on boiling range. Reduced product aromaticity is also accompanied by significant heteroatom removal. The reductions in aromatic and heteroatom contents are controlled by hydrotreatment severity, as indicated by hydrogen consumption, via changes in operating temperature, pressure, residence time, and catalyst. In theory, liquefaction products ranging from distillates to residue can be hydrotreated at sufficient severity to destroy the compounds responsible for mutagenic/carcinogenic activity, which should be reflected in reducing product aromaticity below the threshold (i.e., H : C ratios greater than about 1.5) for mutagenic/carcinogenic activity. For liquefaction processes, product aromaticity and related chemistry appear to be affected only moderately by reaction severity, and product distillation is the only controllable property that is predictive of mutagenic/carcinogenic activity.

In summary, simple physical properties, namely, boiling range and aromaticity, can be effective predictors of whether a complex coal-derived material is mutagenic or carcino-

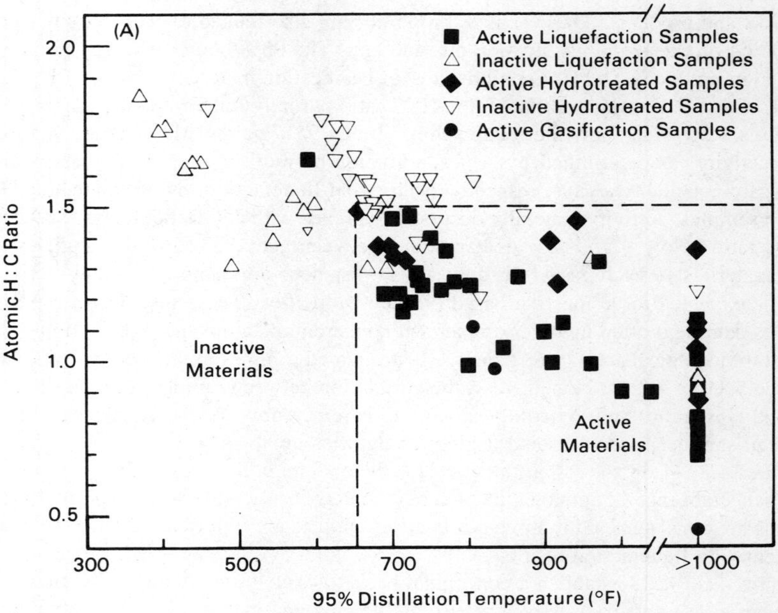

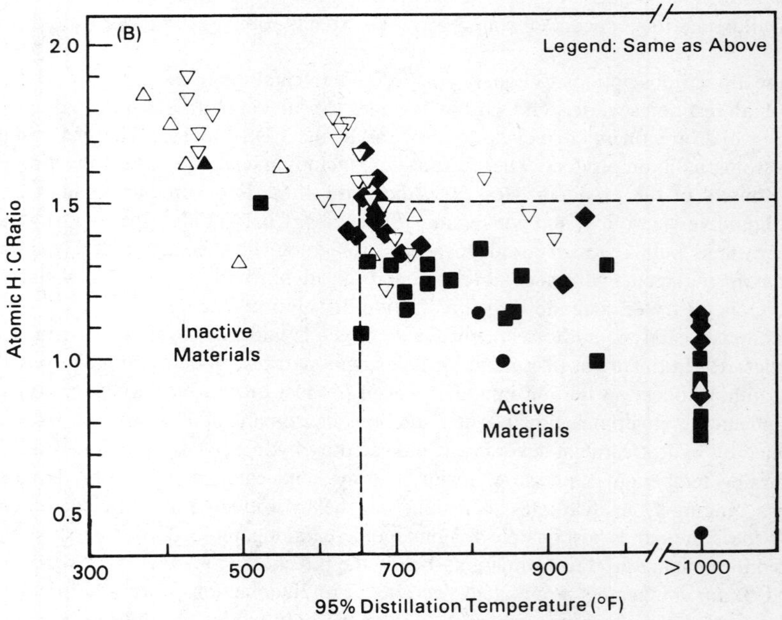

Figure 4.36 Atomic hydrogen : carbon (H : C) ratio versus 95% point for full-boiling-range coal-liquefaction, gasification, and hydrotreatment samples categorized as active or inactive by the Ames assay: (A) unfractionated samples and (B) fractionated samples (weighted-sum results).

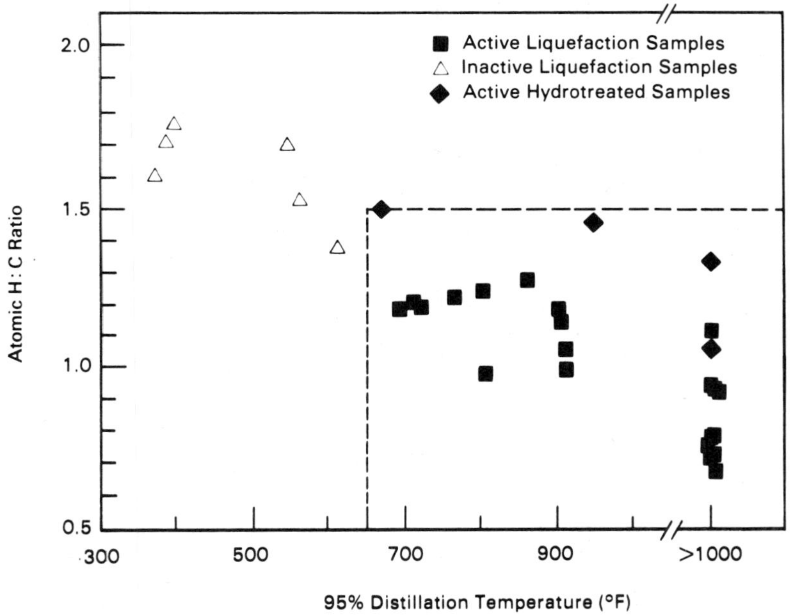

Figure 4.37 Atomic hydrogen : carbon (H : C) ratio versus 95% point for full-boiling-range coal-liquefaction and hydrotreatment samples categorized as active or inactive by the initiation/promotion assay.

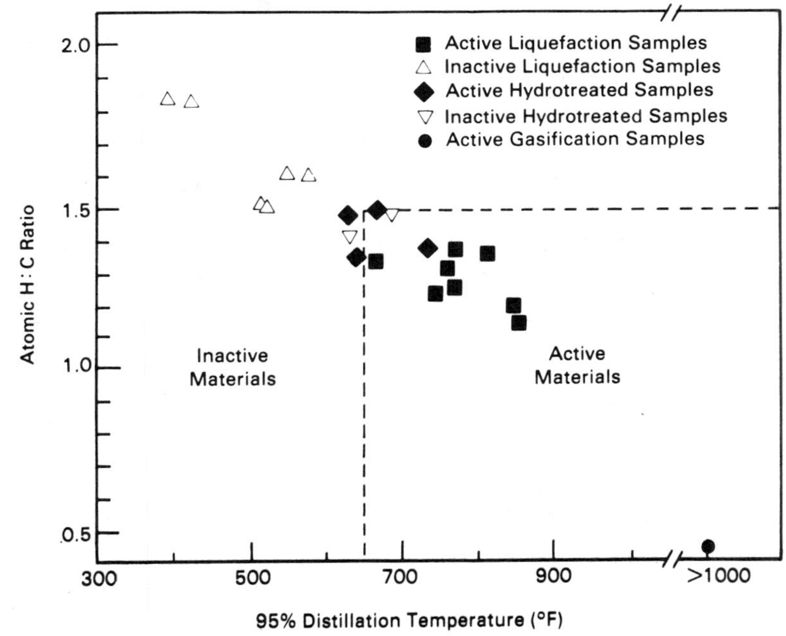

Figure 4.38 Atomic hydrogen : carbon (H : C) ratio versus 95% point for full-boiling-range coal-liquefaction, gasification and hydrotreatment samples categorized as active or inactive by chronic skin-painting carcinogenicity assays.

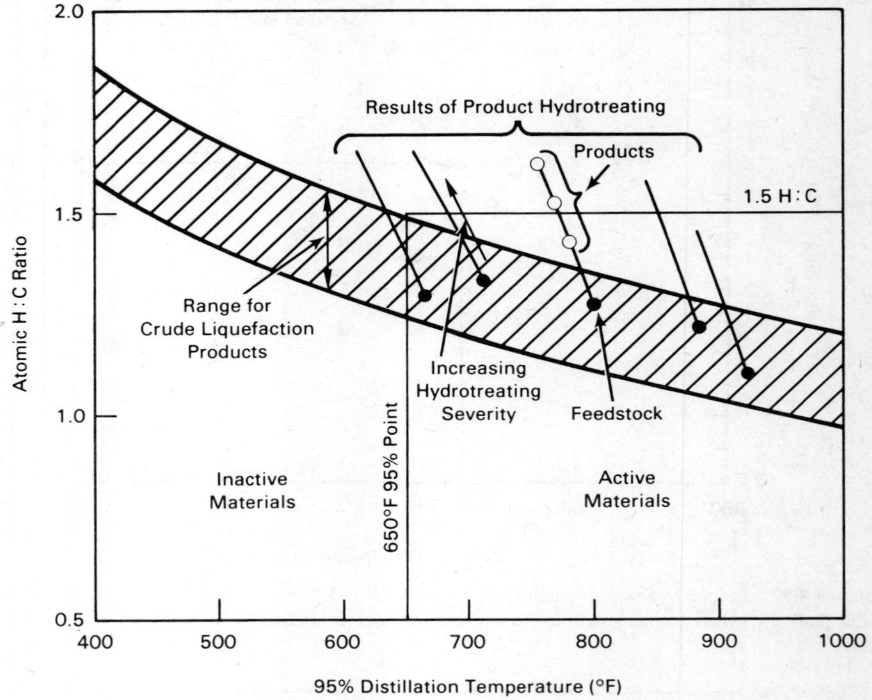

Figure 4.39 Schematic representation of the effects of coal-liquefaction processing on mutagenic/carcinogenic activity and sample properties. H:C = hydrogen:carbon ratio.

genic. In particular, a 95% point of about 650°F and an H:C ratio of about 1.5 appear to separate active and inactive materials. There is a substantial database to support this conclusion for crude and hydrotreated coal-liquefaction products. There are indications that this may also be true for gasification oils and tars.

REFERENCES

Anastasia, L. J., R. J. Evans, S. J. Bossart, R. H. Karst, and C.-S. Juan, *Environmental Assessment of the Hygas Process*, Vol. IV, *Hygas Tests 65 Through 76*, FE-2433-27, National Technical Information Service, Springfield, VA, 1982a.

Anastasia, L. J., R. J. Evans, R. H. Karst, and C.-S. Juan, *Environmental Assessment of the Hygas Process*, Vol. V, *Final Report—Hygas Tests 77 Through 87*, FE-2433-28, National Technical Information Service, Springfield, VA, 1982b.

Anderson, H. C., and W. R. K. Wu, *Properties of Compounds in Coal-Carbonization Products*, Bureau of Mines Bulletin 606, U.S. Bureau of Mines, Washington, DC, 1963.

API, *Technical Data Book—Petroleum Refining*, Vol. I, Chaps. 3 and 5, American Petroleum Institute, Washington, DC, 1976.

ASTM, Standard Method for Distillation of Petroleum Products, ANSI/ASTM D 86-77, *Annual Book of ASTM Standards, Part 23*, American Society for Testing and Materials, Philadelphia, PA, 1977a, pp. 8–20.

ASTM, Standard Method for Distillation of Petroleum Products at Reduced Pressures, ANSI/ASTM D 1160-77, *Annual Book of ASTM Standards, Part 23*, American Society for Testing and Materials, Philadelphia, PA, 1977b, pp. 577–589.

ASTM, Standard Method for Boiling Range-Distribution of Petroleum Fractions by Gas Chromatography, ANSI/ASTM D 2887-73, *Annual Book of ASTM Standards, Part 24*, American Society for Testing and Materials, Philadelphia, PA, 1977c, pp. 770–778.

Bingham, E., R. P. Trosset, and D. Warshawsky, Carcinogenic Potential of Petroleum Hydrocarbons. *J. Environ. Pathol. Toxicol.*, **3**, 483–563 (1980).

Calkins, W. H., J. F. Deye, R. W. Hartgrove, C. F. King, and D. F. Krahn, *Synthetic Crude Oils Carcinogenicity Screening Tests—Final Report*, COO-4758-4, E. I. du Pont de Nemours, Wilmington, DE, 1980.

Calkins, W. H., J. F. Deye, R. W. Hartgrove, C. F. King, and D. F. Krahn, Mutagenesis and Skin Tumor Initiation of Shale- and Coal-Derived Oils and Their Distillation Fractions. *Fuel*, **62**, 857–864 (1983).

Catalytic, Inc., *1980 Operation of SRC Pilot Plant, Wilsonville, Alabama*, EPRI AP-2235, U.S. Department of Energy, Washington, DC and Electric Power Research Institute, Palo Alto, CA, 1982.

Catalytic, Inc., *Operation of the Wilsonville Advanced Coal Liquefaction Research & Development Facility, Annual Technical Report, January to December 1981*, DOE/ET/10154-120, U.S. Department of Energy, Washington, DC, 1983.

Cleland, J. G., S. K. Gangwal, C. M. Sparacino, R. M. Zwedinger, D. G. Nichols, and F. O. Mixon, *Pollutants from Synthetic Fuels Production: Coal Gasification Screening Test Results*, EPA-600/7-79-200, U.S. Environmental Protection Agency, Research Triangle Park, NC. National Technical Information Service, Springfield, VA, 1979.

Cowser, K. E., *Life Sciences Synthetic Fuels Semiannual Progress Report for the Period Ending June 30, 1982*, ORNL/TM-8441, Oak Ridge National Laboratory, Oak Ridge, TN, 1982.

Drozdowicz, B. Z., and C. M. Kelly, *Interim Report on the Genetic and Animal Toxicity Testing of SRC-I Products, Intermediates, and Waste Materials*, DOE/OR/03054-50, International Coal Refining Company, Allentown, PA, 1983.

Duncan, D. A., J. L. Beeson, and R. D. Oberle, *Research and Development of Rapid Hydrogenation for Coal Conversion to Synthetic Motor Fuels (Riser Cracking of Coal)*, FE-2307-30, U.S. Department of Energy, Washington, DC, 1978.

Epler, J. L. (ed.), *Health Effects Research in Direct Coal Liquefaction—Studies of H-Coal Distillate: Phase I. PDU Samples—The Effect of Hydrotreatment*, ORNL/TM-8071, Oak Ridge National Laboratory, Oak Ridge, TN, 1981.

Epperly, W. R., *EDS Coal Liquefaction PDU Phase IIIB—Quarterly Technical Progress Report for the Period October 1–December 31, 1977*, FE-2893-7, National Technical Information Service, Springfield, VA, 1978.

Epperly, W. R., *EDS Coal Liquefaction Process Development Phase V—Quarterly Technical Progress Report for the Period January 1–March 31, 1982*, DOE/ET/10069-89, National Technical Information Service, Springfield, VA, 1982.

Epperly, W. R., *EDS Coal Liquefaction Process Development Phase V—Quarterly Technical Progress Report for the Period January 1–March 31, 1983*, FE-2893-110, National Technical Information Service, Springfield, VA, 1983.

Essenhigh, R. H. Fundamentals of Coal Combustion, in M. A. Elliott (ed.), *Chemistry of Coal Utilization*, Second Supplementary Volume, Wiley, New York, 1981, pp. 1153–1312.

Falk, H. L., P. Kotin, and S. Thompson, Inhibition of Carcinogenesis. *Arch. Environ. Health*, **9**, 169–179 (1964).

Fillo, J. P., and J. C. Craun, *Engineering Analysis of the Toxicological Characterization Database for Coal Conversion Materials*, DOE/NBM-1079 (DE85019553), National Technical Information Service, Springfield, VA, 1985.

Fillo, J. P., V. C. Stamoudis, J. R. Stetter, and S. W. Vance, *Influence of Coal Devolatilization Conditions on the Yield, Chemistry and Toxicology of By-Product Oils and Tars*, DOE/ET/14746-11, National Technical Information Service, Springfield, VA, 1983.

Frumkin, H. A., R. F. Sullivan, and B. E. Stangeland, Catalytic Upgrading of SRC-II Syncrude, in R. F. Sullivan (ed.), *Upgrading Coal Liquids*, ACS Symposium Series 156, American Chemical Society, Washington, DC, 1981, pp. 75–113.

Gangwal, S. K., Rapid Boiling Point Assay of Tar from Coal Gasification by Gas Chromatography. *J. High Resol. Chrom. & Commun.*, **4,** 580–581 (1981).

Gangwal, S. K., P. M. Grohse, D. E. Wagoner, D. J. Minick, C. M. Sparacino, and R. A. Zweidinger, *Pollutants from Synthetic Fuels Production: Sampling and Analysis Methods for Coal Gasification*, EPA-600/7-79-021, prepared for the U.S. Environmental Protection Agency, Research Triangle Park, NC. National Technical Information Service, Springfield, VA, 1979.

Gray, J. A., *Selected Physical, Chemical and Thermodynamic Properties of Narrow Boiling Coal Liquids from the SRC-II Process*, DOE/ET/10104-7, National Technical Information Service, Springfield, VA, 1981.

Gray, J. A., C. J. Brady, J. R. Cunningham, J. R. Freeman, and G. M. Wilson, Thermophysical Properties of Coal Liquids. 1. Selected Physical, Chemical and Thermodynamic Properties of Narrow Boiling Range Coal Liquids. *Ind. Eng. Chem. Process Design Dev.*, **22,** 410–424 (1983).

Gray, R. H., and K. E. Cowser, *Status of Health and Environmental Research Relative to Coal Liquefaction: 1976 to the Present*, DOE/NBM-1016, U.S. Department of Energy. National Technical Information Service, Springfield, VA, 1982.

Green, D. A., J. G. Cleland, D. P. Daugherty, W. J. McMichael, F. O. Mixon, and R. S. Truesdale, *Coal Gasifier Parameters Influencing Environmental Pollutant Production*, RTI/1934/00-03F, Research Triangle Institute, Research Triangle Park, NC, 1981.

Griest, W. H., J. M. Giddings, and J. A. Klein, *Effects of Hydrotreatment on the Properties of Coal-Derived Liquid Products: A Status Report*, ORNL/TM-8836, Oak Ridge National Laboratory, Oak Ridge, TN, 1983.

Hougen, O. A., and K. M. Watson, *Chemical Process Principles, Part III—Kinetics and Catalysis*, Wiley, New York, 1947, pp. 884–887.

Howard, J. B., Fundamentals of Coal Pyrolysis and Hydropyrolysis, in M. A. Elliott (ed.), *Chemistry of Coal Utilization*, Second Supplementary Volume, Wiley, New York, 1981, pp. 665–784.

Hueper, W. C., Experimental Carcinogenic Studies on Hydrogenated Coal Oils—I. Bergius Oils, *Ind. Med. Surg.*, **25**, 51–55 (1956).

HRI, *Laboratory Program to Support H-Coal Pilot Plant Operations, PDU Run 5*, FE-2547-19, Hydrocarbon Research Inc. U.S. Department of Energy, Washington, DC, 1978.

HRI, *Laboratory Program to Support H-Coal Pilot Plant Operations, PDU Run 7*, FE-2547-39, Hydrocarbon Research Inc. U.S. Department of Energy, Washington, DC, 1979.

IARC, IARC Monographs on the Evaluation of the Carcinogenic Risk of Chemicals to Humans, *Chemical Industrial Processes and Industries Associated with Cancer in Humans*. IARC Monographs, Vol. 1–29, Suppl. 4, October 1982. International Agency for Research on Cancer, Lyons, France, 1982.

Kimball, R. F. and N. B. Munro, *A Critical Review of the Mutagenic and Other Genotoxic Effects of Direct Coal Liquefaction*, ORNL-5721, Oak Ridge National Laboratory, Oak Ridge, TN, 1981.

Later, D. W., R. A. Pelroy, and B. W. Wilson, *Chemical Analysis and Biological Testing of Materials from the EDS Coal Liquefaction Process: A Status Report*, PNL-4960, Pacific Northwest Laboratory, Richland, WA, National Technical Information Service, Springfield, VA, 1984.

Later, D. W., *Two-Stage Coal Liquefication Materials from the Wilsonville Facility Operated in the Nonintegrated and Integrated Modes: Chemical Analyses and Biological Testing*, PNL-5215, Pacific Northwest Laboratory, Richland, WA. National Technical Information Service, Springfield, VA, 1985a.

Later, D. W., Nitrogen-Containing Polycyclic Aromatic Compounds in Coal-Derived Materials, in A. B. Bjorseth and T. Ramsdahl (eds.), *Handbook of Polycyclic Hydrocarbons*, Vol. 2, Marcel Dekker, New York, 1985b, pp. 265–349.

Mahlum, D. D., M. E. Frazier, R. A. Pelroy, and R. A. Renne, *Comparison of in Vitro and in Vivo Studies with Coal Liquids from the SRC-II Process*, PNL-4836, Pacific Northwest Laboratory, Richland, WA. National Technical Information Service, Springfield, VA, 1983.

Mann, R. M., G. E. Harris, W. R. Menzies, A. J. Simon, and W. A. Williams, *Environmental Health and Safety Database For KRW Coal Gasification Process Development Unit*, GAI-8510123, Gas Research Institute, Chicago, IL, 1985.

Maruhnich, E. D., R. C. Weber, J. C. Craun, and F. B. Melnick, *The Effects of Hydrotreating on the Chemical and Toxicological Properties of Coal-Derived Liquids*, DOE/ET/14746-8, U.S. Department of Energy, National Technical Information Service, Springfield, VA, 1982.

McIlvried, H. G., *Summary of Results from SRC-II Runs Made on Process Development Unit P-99, Runs P99-15 to P99-85, Interim Report for March 1977–June 1981*, DOE/ET/10104-42, Pittsburg & Midway Coal Mining Co., Denver, CO. National Technical Information Service, Springfield, VA, 1982.

Mixon, F. O., J. G. Cleland, T. J. Hughes, D. A. Green, and S. K. Gangwal. Laboratory-Scale Simulation of Coal Gasification for Biological Studies, in K. E. Cowser (ed.), *Synthetic Fossil Fuel Technologies, Results of Health and Environmental Studies*, Butterworth, Stoneham, MA, 1984, pp. 199–227.

Morrison, J. A., M. J. Dabkowski, and F. K. Young, *H-Coal Pilot Plant Topical Report—H-Coal Run No. 8 with Illinois Coal in Syncrude Mode*, FE-2260, Ashland Synthetic Fuels, Inc., Ashland, KY. U.S. Department of Energy, Washington, DC, 1983.

Nakles, D. V., and J. P. Fillo, Impact of Process Design on Effluent Control Technology Needs in Coal Gasification, in *Proceedings: Workshop on Environmental Control Technology for Coal Gasification*, AP-3006, Electric Power Research Institute, Palo Alto, CA, 1983.

Nichols, D. G., J. G. Cleland, D. A. Green, F. O. Mixon, T. J. Hughes, and A. W. Kolber, *Pollutants from Synthetic Fuels Production: Environmental Evaluation of Coal Gasification Screening Tests*, EPA-600/7-79-202, U.S. Environmental Protection Agency, Research Triangle Park, NC, 1979.

O'Rear, D. J., R. F. Sullivan, and B. E. Strangeland, Catalytic Upgrading of H-Coal Syncrudes, in R. F. Sullivan (ed.), *Upgrading Coal Liquids*, ACS Symposium Series 156, American Chemical Society, Washington, DC, 1981, pp. 115–144.

Pelroy, R. A., and B. W. Wilson, *Fractional Distillation as a Strategy for Reducing the Genotoxic Potential of SRC-II Coal Liquids: A Status Report*, PNL-3787, Pacific Northwest Laboratory, Richland, WA. National Technical Information Service, Springfield, VA, 1981.

Reilly, C. A., M. J. Peak, T. Matsushita, F. R. Kirchner, and D. A. Haugen, Chemical and Biological Characterization of High-Btu Coal Gasification (The Hygas Process). IV. Biological Activity, in D. D. Mahlum, R. H. Gray, and W. D. Felix (eds.), *Coal Conversion and the Environment: Chemical, Biological and Ecological Considerations*, CONF-801039, National Technical Information Service, Springfield, VA, 1981, pp. 310–324.

Scheppele, S. E., G. J. Greenwood, R. J. Pancirov, and T. R. Ashe, Chemical Composition of Raw and Upgraded Anthracene Oil and the Chemistry of Coal-Liquids Upgrading, in R. F. Sullivan (ed.), *Upgrading Coal Liquids*, ACS Symposium Series 156, American Chemical Society, Washington, DC, 1981, pp. 39–73.

Smith, W. E., D. A. Sunderland, and K. Sugiura, Experimental Analysis of the Carcinogenic Activity of Certain Petroleum Products, *Arch. Ind. Hyg. Occup. Med.*, **4**, 229–314 (1951).

Solomon, P. R., Coal Structure and Thermal Decomposition, in B. D. Blaustein, B. C. Bockrath, and S. Friedman (eds.), *New Approaches in Coal Chemistry*, ACS Symposium Series 169, American Chemical Society, Washington, DC, 1981, pp. 61–71.

Stearns-Roger, Inc., *Fixed-Bed Slagging Gasifier Pilot Plant, Gasifier Run RA-87, December 1-4, 1980*, Stearns-Roger, Inc., Denver, CO, 1981.

Stein, T. R., J. G. Bendoraitis, A. V. Cabal, R. B. Callen, M. J. Dabkowski, R. H. Heck, H. R. Ireland, and C. A. Simpson, *Upgrading of Coal Liquids for Use as Power Generator Fuels,—Annual Report for the Period February 1976–January 1977*, EPRI AF-444, Mobil Research & Development Corporation, Paulsboro, NJ. Electric Power Research Institute, Palo Alto, CA, 1977.

Sullivan, R. F., *Refining and Upgrading of Synfuels from Coal and Oil Shales by Advanced Catalytic Processes—Eighth Interim Report—Task 11: Distillation of H-Coal into Narrow Boiling Fractions*, FE-2315-93, Chevron Research Co., Richmond, CA. U.S. Department of Energy, Washington, DC, 1983.

Sutterfield, F. D., *Hydrogenation Effects on Mutagenesis of Alternate Crudes*, Ph.D. Thesis, University of Tulsa Graduate School, Tulsa, OK, 1983.

Suuberg, E. M., *Rapid Pyrolysis and Hydropyrolysis of Coal*, D.Sc. Thesis, Massachusetts Institute of Technology, Cambridge, MA, 1977.

Talwalker, A. T., *A Topical Report on Coal Pyrolysis*, DOE/MC/19316-1408, National Technical Information Service, Springfield, VA, 1983.

Virk, P. S., L. E. Chambers, and H. N. Woebke, Thermal Hydrogasification of Aromatic Compounds, in L. G. Massey (ed.), *Coal Gasification*, American Chemical Society, Washington, DC, 1974, pp. 237–258.

Warshawsky, D., and R. S. Schoeny, *Exploratory Research on Mutagenicity of Coal-Derived Materials*, DOE-ET-00222-4, University of Cincinnati, OH. U.S. Department of Energy, Washington, DC, 1980.

Weil, C. S., and N. I. Condra, The Hazards to Health in the Hydrogenation of Coal. II. Carcinogenic Effect of Materials on the Skin of Mice. *Arch. Environ. Health*, **1**, 187–193 (1960).

Weimer, W. C., B. W. Wilson, R. A. Pelroy, and J. C. Craun, *Initial Chemical and Biological Characterization of Hydrotreated Solvent Refined Coal (SRC-II) Liquids: A Status Report*, PNL-3464, Pacific Northwest Laboratory, Richland, WA. National Technical Information Service, Springfield, VA, 1980.

Wilson, B. W., R. A. Pelroy, and D. D. Mahlum, *Chemical Characterization and Genotoxic Potential Related to Boiling Point for Fractionally Distilled SRC-I Coal Liquids*, PNL-4277, Pacific Northwest Laboratory, Richland, WA. National Technical Information Service, Springfield, VA, 1982.

Wilson, B. W., P. Buhl, and E. C. Moroni, *Biological and Chemical Analysis of Materials from an Integrated Two Stage Coal Liquefaction Process: A Status Report*, PNL-4553, Pacific Northwest Laboratory, Richland, WA. National Technical Information Service, Springfield, VA, 1983a.

Wilson, B. W., R. A. Pelroy, R. P. Anderson, and J. Freel, *Chemical and Biological Effects of Heavy Distillate Recycle in the SRC-II Process*, PNL-4815, Pacific Northwest Laboratory, Richland, WA. National Technical Information Service, Springfield, VA, 1983b.

Wilson, B. W., D. W. Later, C. W. Wright, and D. L. Stewart, *Chemical Analysis and Mutational Assay of Distilled Oils from the H-Coal Direct Liquefaction Process: A Status Report*, PNL-5288, Pacific Northwest Laboratory, Richland, WA. National Technical Information Service, Springfield, VA, 1985.

Wilzbach, K. E., J. R. Stetter, C. A. Reilly, and W. G. Wilson, *Environmental Research Program for Slagging Fixed-Bed Coal Gasification*, ANL/SER-1, Argonne National Laboratory, Argonne, IL, 1982.

APPENDIX TO CHAPTER 4
SUMMARY OF THE TOXICOLOGICAL DATABASE FOR COAL-CONVERSION OILS AND TARS

As an aid to the reader interested in obtaining more information on any given sample, this appendix provides a listing of all samples discussed in Chapter 4. The types of data available and the appropriate references are given for each sample. Separate tables with this information are provided for liquefaction, gasification, hydrotreatment, and narrow-boiling-range samples (Tables 4A.1–4A.4, respectively).

Tables 4A.1–4A.4 provide the sample number, sample name, interpretive index (explained below), and references for the process history, physical properties, toxicological testing, and chemical analyses data used in Chapter 4. These references are listed at the end of this appendix. Tables 4A.1–4A.3 provide the 95% point, distillation method, and atomic H : C ratio for liquefaction, gasification, and hydrotreatment samples. Table 4A.3 also provides hydrogen consumption data for the hydrotreatment runs that produced the samples listed. Table 4A.5 lists identifying numbers for samples used in constructing each of the figures in Chapter 4. The data in Tables 4A.1–4A.5 enable the interested reader to identify individual data points in the figures presented in Chapter 4. Table 4A.6 provides the process operator/developer run number and the run or sampling date for various samples.

To help sort and interpret data from diverse sources, a three-letter interpretive index (Table 4A.7) was assigned to each coal-conversion material having toxicological data. This index is reported as XYZ, where

X = sampling and handling index
Y = sample properties index
Z = process history index

The alphabetical position of each letter defines the level of available or inferred knowledge in the assigned category: The levels are ranked from A to D, reflecting highest to lowest level of knowledge, respectively. Criteria for each level in each category are described in Table 4A.7. Taken as a whole, the index provides a mechanism for understanding the potential utility of each sample for various types of data analysis.

Sampling and handling procedures varied substantially among coal-conversion materials. In some cases, samples were stored for months to years under conditions that were not precisely controlled. As discussed in Chapter 2, substantial changes could have occurred in biological and chemical properties of such materials through sample degradation. However, many samples were stored in the dark under inert atmospheres at subambient temperatures, and it is unlikely that sample degradation occurred. The lack of sample degradation has been documented, in some cases, through periodic testing. The handling of most coal-conversion materials has been intermediate to these extremes, and inferences regarding sample degradation must be made.

The availability of bulk physical and chemical data for coal-conversion samples varies substantially. Bulk properties like specific gravity and elemental analyses are often available, as these are routinely determined by developers to characterize process performance. For many coal-liquefaction materials, including hydrotreated liquids, sample distillation curves obtained by standard ASTM or gas chromatography methods are available. In some cases, complete analytical data, including such analyses as hydrocarbon types, basic nitrogen, and carbon residue, are available.

It is often tedious and difficult to obtain the processing history for individual samples. Available information might include only a process name and/or a stream description. In other cases, process histories may identify specific samples and associated sampling times within steady-state operating periods and completely document feedstock properties, process conditions, and yields of the material sampled. In many cases, uncertainties exist in the availability and/or interpretation of sample-specific data.

Materials discussed in this book have been assigned sample numbers that allow comparison of information on a given sample in the various chapters and appendixes. The sample numbers are five-character codes; the first character identifies the origin of the sample: L = liquefaction, G = gasification, H = hydrotreatment, N = narrow-boiling-range cut. The next two digits identify the facility where the sample originated, using the code given in Table 4A.6. The fourth digit usually identifies the process run or sampling period, and the fifth digit distinguishes samples collected during the same run. The appendix at the end of Chapter 3 briefly describes the process and sample histories of every material discussed in Chapter 3 and listed in this appendix.

As an example of how to use the unique sample numbers, suppose we are interested in all information on SRC-II heavy distillate (HD) samples from the Ft. Lewis, Washington pilot plant. Sample descriptions in the relevant section in the appendix to Chapter 3 indicate one HD sample (Sample L1243) and a number of middle and heavy distillate blends. How do we track information for the HD sample in the book?

This appendix provides references for the process history of Sample L1243 and for some of the basic toxicological testing performed. Table 4A.1 indicates that mutagenicity (A = Ames), initiation/promotion (I/P), and chronic skin-painting (SP) assays were among those performed on Sample L1243. Results from these three assays are analyzed in Chapter 4, because that chapter relates toxicological effects and engineering data. Table 4A.5 indicates the correlations in which the data were used (Fig. 4.8, 4.9, 4.10, etc). The data points representing Sample L1243 in the figures can be determined by noting that the sample's 95% point is 800°F (Table 4A.1). This is sufficient information to locate Sample L1243 in some correlations (e.g., Fig. 4.10 and 4.11, where there is only one SRC-II data point near 800°F). However, for others (e.g., Fig. 4.9, where there are two SRC-II data points near 800°F), the value of the dependent variable for the sample must be obtained from Chapter 3, Chapter 5, or the referenced literature to locate the specific data point.

Toxicological data and additional information on testing protocols can be found in Chapter 3 by searching relevant sections of the text. The section on Ames data (Section 3.4.1.1.1) is organized by process, and the data for Sample L1243 can be found therein. Similarly, chronic skin-painting and initiation/promotion data can be found in Sections 3.4.2.1 and 3.4.2.2, respectively.

The extent of other toxicological results is indicated in Table 3.1, and the data and references can be found in appropriate sections of Chapter 3. For Sample L1243, Table 3.1 indicates that results are available for a Syrian hamster embryo (SHE) cell transformation assay (see Section 3.4.1.2.6), for mammalian cell cytotoxicity assays (see Section 3.4.1.2.7), for teratogenicity assays (see Section 3.4.4), and for various organ toxicity assays (see Section 3.4.5).

Finally, we can check the relevant process sections in Chapter 5 for available chemical characterization data. Such data for SRC-II materials, including HD, are discussed in Section 5.13.1.1. Thus, the reader can find all available information on a given sample in the book by working back and forth between the appendixes to Chapters 3 and 4, Chapter 4, Tables 3.1 and 3.2 in Chapter 3, and the relevant process sections of Chapters 3 and 5. Further details can be obtained by consulting the references given throughout the book. Tables 4A.1, 4A.2, and 4A.3 also give sample numbers used in the referenced literature by other investigators for the samples listed. Again, this should help the interested reader find the original literature for a given sample.

Table 4A.1 Liquefaction Process Samples

Sample Number	Sample Name[b]	Interpretive Index[c]	95% Point (°F)	Distillation Method[d]	H:C Ratio	Reference Number[a] Process History	Physical Properties	Toxicological Testing[e]	Chemical Analysis[f]	Other Sample Numbers
				SRC-II P-99 PDU						
L1111	Process solvent	BCA	840	Est. E	—	2262, 2036	2039, 2036	I/P = 2151 SP = 2411	—	
L1121	Process solvent	BCA	—	—	—	2262, 2036	2039, 2036	—	—	
L1131	Process solvent	BCA	—	—	—	2262, 2036	2039, 2036	—	—	
				SRC-II Ft. Lewis, Washington Pilot Plant						
L1211	FBR fuel oil	CBC	714	D86CG	1.22	2400	2400	A = 2120	2412	ORNL 1701
L1221	Fuel-oil blend	CBA	714	SGC	1.16	2111	2032	A = 2032, 2348	—	
L1231	Naphtha	CBC	375	Est. A	1.60	2040	2040	A, I/P = 2040	—	
L1232	FBR fuel oil	CBC	690	Est. A	1.20	2040	2040	A, I/P = 2040	—	
L1233	Heavy gas oil	CBC	805	Est. A	0.97	2040	2040	A, I/P = 2040	—	
L1234	Stripper bottoms	CBC	>1000	Est. A	0.77	2040	2040	A, I/P = 2040	—	
L1235	Vacuum bottoms	CBC	>1000	Est. A	0.73	2040	2040	A, I/P = 2040	—	
L1236	Resid	CBC	>1000	Est. A	0.68	2040	2040	A, I/P = 2040	—	
L1241	Naphtha	BBC	370	D86CG	—	2111	2005	A = 2005	—	
L1242	Middle distillate	BBC	564	D86CG	—	2111	2005	A = 2005	—	
L1243	Heavy distillate	BBC	800	Est. B	—	2111	2005	A = 2005 I/P, SP = 2151	2412	
L1251	Fuel oil	CAC	711	D1160CP	1.21	2075	2075	A, I/P = 2151	2412	UOP 3777-6
L1252	Fuel oil	CAC	705	D1160CP	1.21	2075	2075	—	—	UOP 3777-7
L1261	Fuel oil blend	CBB	755	SGC	—	2190	2180	—	2304	SGQ 7559,
L1262	Washed fuel oil	CBB	743	SGC	1.24	2190	2180	A, SP = 2304	2304	WOW 4451, ACD 916
				SRC-II Merriam, Kansas Bench Unit						
L1311	FBR distillate	BBB	703	SGC	1.46	2159	2159	A = 2336	—	
L1321	<270°C Distillate	BBB	546	SGC	1.39	2160	2160	A = 2336	—	

Table 4A.1 (*Continued*)

Sample Number	Sample Name[b]	Interpretive Index[c]	95% Point (°F)	Distillation Method[d]	H:C Ratio	Reference Number[a] Process History	Physical Properties	Toxicological Testing[e]	Chemical Analysis[f]	Other Sample Numbers
L1331	<270°C Distillate	BBB	573	SGC	1.53	2236	2236	A = 2336	—	
L1341	<290°C Distillate	BBB	586	SGC	1.49	2236	2236	A = 2336	—	
L1351	<310°C Distillate	BBB	600	SGC	1.51	2236	2236	A = 2336	—	
				SCR-I Wilsonville, Alabama Pilot Plant						
L2121	Process solvent	BBB	911	D1160CP	1.19	2067	2067	I/P = 2066	—	—
L2131/32[g]	Recycle solvent	CBC	653/—	SGC/—	1.06/1.08	2401	2332/2027	A = 2027	—	—/ONRL 1613
L2141/42[g]	Wash solvent	CBC	665/—	SGC/—	1.23/1.10	2401	2333-34/2027	A = 2027	—	—/ORNL 1609
L2151/52[g]	Light organic liquid	CBC	420/—	SGC/—	1.62/1.60	2401	2332/2027	A = 2027	—	—/ORNL 1605
				SRC-I Ft. Lewis, Washington Pilot Plant						
L2211	Light oil	CCC	380	Est. C	1.85	2061	2061	A = 2061	—	—
L2212	Wash solvent	CCC	480	Est. C	1.31	2061	2061	A = 2061	—	—
L2213	Process solvent	CCC	850	Est. C	1.09	2061	2061	A = 2061	—	—
					H-Coal PDU					
L3111/12[g]	Raw distillate	CBC	665/—	SGC/—	1.33/1.31	2053	2332/2027	A, SP = 2027	2027	ORNL 1601
L3121	ASOH	CBB	540	Est. A	1.68	2040, 2255	2040	A, I/P = 2040	—	
L3122	ASB	CBB	770	Est. A	1.22	2040, 2255	2040	A, I/P = 2040	—	
L3123	VSOH	CBB	915	Est. A	1.14	2040, 2255	2040	A, I/P = 2040	—	
L3124	VSB	CBB	>1000	Est. A	0.77	2040, 2255	2040	A, I/P = 2040	—	
L3131	ASOH	CBB	562	D86CG	1.53	2040, 2052	2040	A, I/P = 2040	—	
L3132	Naphtha	CBC	390	Est. A	1.77	2040, 2052	2040	A, I/P = 2040	—	
L3133	Light gas oil	CBC	615	Est. A	1.36	2040, 2052	2040	A, I/P = 2040	—	
L3134	ASB	CBB	720	D1160CG	1.18	2040, 2052	2040	A, I/P = 2040	—	
L3135	VSOH	CBB	915	D1160CG	1.05	2040, 2052	2040	A, I/P = 2040	—	
L3136	VSB	CBB	>1000	Est. D	0.75	2040, 2052	2040	A, I/P = 2040	—	
L3141	ASOH	CBB	576	D86CG	1.60	2052	2052	SP = 2027	2412	ORNL 1308

L3142	ASB	CBB	771	D1160CG	1.25	2052	2052	A, SP = 2027	2412	ORNL 1309
L3143	VSOH	CBB	855	D1160CG	1.15	2052	2052	A, SP = 2027	2412	ORNL 1310
L3144	VSB	CBB	>1000	Est. D	0.80	2052	2052	A = 2027	—	ORNL 1311
L3151	ASOH	BCC	547	D86CG	1.62	2051	2051	A, SP = 2027	2412	ORNL 1312
L3152	ASB	BCC	738	D1160CG	1.29	2051	2051	A, SP = 2027	2412	ORNL 1313
L3153	VSOH	BCC	849	D1160CG	1.17	2051	2051	A, SP = 2027	2412	ORNL 1314
L3154	VSB	BCC	>1000	Est. E	0.78	2051	2051	A = 2027	2412	ORNL 1315
L3161	ASB	BBB	754	SGC	1.39	—	2348	A = 2348	—	
L3162	VSOH	BBB	772	SGC	1.35	—	2348	A = 2348	—	
				H-Coal Catlettsburg, Kentucky Pilot Plant						
L3211	Stabilizer naphtha	BBC	421	SGC	1.82	2016	2016, 2099	A = 2099, SP = 2411	—	ORNL 587
L3212	Light fuel oil	BBC	524	SGC	1.50	2016	2016, 2099	A = 2099, SP = 2411	—	ORNL 589
L3213	Heavy fuel oil	BBC	812	SGC	1.36	2016	2016, 2099	A = 2099, SP = 2411	—	ORNL 591
L3231	Fuel oil blend	BAB	731	SGC	1.26	2047	2179	A = 2304	2304	WOW 4734
L3232	Washed fuel oil	BAB	732	SGC	1.28	2047	2179	A = 2304	2304	WOW 4747
L3234	Heavy fuel oil	BAB	760	SGC	—	2047	2179	—	—	ORNL 695
L3241	Stabilizer naphtha	BBB	393	SGC	1.84	2047	2047, 2099	A = 2099, SP = 2411	—	ORNL 886
L3242	Light fuel oil	BBB	515	SGC	1.52	2047	2047, 2099	A = 2099, SP = 2411	—	ORNL 887
L3243	Heavy fuel oil	BBB	768	SGC	1.37	2047	2047, 2099	A = 2099, SP = 2411	—	ORNL 888
				H-Coal Bench-Scale Unit						
L3313	Filtered liquid	CBB	>1000	D1160CP	1.01	2402	2402	A = 2290, 2291	—	ETTM 09
L3314	Vacuum overhead	CBB	900	D1160CP	1.12	2002	2402	A = 2290, 2291	—	ETTM 10
L3315	Vacuum bottoms	CBB	>1000	D1160CP	0.72	2402	2402	A = 2290, 2291	—	ETTM 11
L3321	Separator overhead	CBB	592	D86CG	1.64	2402	2402	A = 2290, 2291	—	ETTM 14
L3323	Filtered liquid	CBB	>1000	D1160CP	1.11	2402	2402	A = 2290, 2291	—	ETTM 16
L3324	Vacuum overhead	CBB	760	D1160CP	1.25	2402	2402	A = 2290, 2291	—	ETTM 17
L3325	Vacuum bottoms	CBB	>1000	D1160CP	0.79	2402	2402	A = 2290, 2291	—	ETTM 18

Table 4A.1 (*Continued*)

Sample Number	Sample Name[b]	Interpretive Index[c]	95% Point (°F)	Distillation Method[d]	H:C Ratio	Reference Number[a]: Process History	Physical Properties	Toxicological Testing[e]	Chemical Analysis[f]	Other Sample Numbers
				EDS ECLP Pilot Plant						
L4111	Recycle solvent	AAB	724	SGC	1.47	2127, 2337	2403	A = 2354	2412	ETP 3
L4112	Heavy VGO	ABB	>1000	SGC	0.99	2127, 2337	2403	A = 2354	2412	ETP 4
L4121	Recycle solvent	AAB	754	SGC	1.43	2127, 2337	2403	—	—	ETP 8
L4131	Naphtha	AAB	406	SGC	1.75	2127, 2337	2403	A = 2354	—	ETP 11
L4132	Recycle solvent	AAB	866	SGC	1.27	2127, 2337	2403	A, I/P = 2354	2412	ETP 13
L4133	Vacuum bottoms	ABB	>1000	Est. E	—	2127, 2337	2037	A = 2354	2412	ETP 14
L4141	Recycle solvent	AAB	800	SGC	1.38	2127, 2337	2403	—	—	ETP 22
				EDS CLPP Pilot Plant						
L4211	Naphtha	CBB	399	SGC	1.72	2040, 2331	2040, 2331	A, I/P = 2040	—	
L4212	Spent solvent	CBB	804	SGC	1.24	2040, 2331	2331	A, I/P = 2040	—	
L4213	VGO	CBB	1000	Est. A	0.91	2040, 2331	2331	A, I/P = 2040	—	
L4214	Heavy gas oil	CBB	915	Est. A	0.99	2040, 2331	2331	A, I/P = 2040	—	
L4215	Resid	CBB	1050	Est. A	0.91	2040, 2331	2331	A, I/P = 2040	—	
L4221	Solvent	BBC	694	SGC	1.30	—	2348	A = 2348	—	
L4231	Solvent/VGO blend	BBC	963	SGC	0.99	—	2348	A = 2348	—	
				C. E. Lummus ITSL PDU						
L5111	LC Finer feed	BBB	>1000	Est. D	0.96	2414	2035	A, I/P = 2414	2412	
L5121	LC Finer feed	BBB	>1000	Est. D	0.74	2414	2035	A, I/P = 2414	2412	
L5131	LC Finer feed	BBC	>1000	Est. D	0.91	2414	2320	A = 2414	2412	
L5141	LC Finer feed	BBB	>1000	Est. D	0.90	2414	2321	—	—	
				Wilsonville, Alabama NTSL Pilot Plant						
L5211	Light fuel oil	BCB	432	SGC	1.63	2115, 2237	2237	A = 2415	—	5226-018
L5212	Middle distillate	BCB	670	SGC	1.31	2115, 2237	2237	A = 2415	—	5226-019
L5214	Deashed SRC	BCB	>1000	Est. E	0.81	2115, 2237	2237	A = 2415	2412	5226-020

L5216	HTR feed	BCB	>1000	Est. E	1.13	2115, 2237	2237	A, I/P = 2415	—	5526-021
L5221	Process solvent	—	—	—	—	2410	2410	—	—	
L5231	Light oil	—	—	—	—	2410	2410	—	—	
				Wilsonville, Alabama ITSL Pilot Plant						
L5311	Light fuel oil	BBB	475	SGC	—	2239	2239	A = 2415	—	5226-076
L5314	Deashed SRC	BBB	>1000	SGC	—	2239	2239	A = 2415	—	5226-080
L5315	Process solvent	BBB	945	SGC	—	2239	2239	A = 2415	—	5226-077
L5331	Light fuel oil	BBB	450	SGC	1.63	2239	2239	A = 2415	—	5226-048
L5334	Deashed SRC	BBB	>1000	Est. D	0.92	2239	2239	A = 2415	2412	5226-050
L5335	Process solvent	BBB	964	SGC	1.33	2239	2239	A = 2415	—	5226-049
L5344	Deashed SRC	BDC	>1000	Est. E	—	2239	2239	A = 2415	—	5226-063
L5351	Light fuel oil	BBB	454	SGC	1.63	2239	2239	A = 2415	—	5226-064
L5354	Deashed SRC	BCB	>1000	Est. E	0.90	2239	2239	A = 2415	—	5226-066
L5355	Process solvent	BBB	976	SGC	1.31	2239	2239	A = 2415	—	5226-065

[a] References are listed in numerical order at the end of this appendix.

[b] FBR = full-boiling range, ASOH = atmospheric still overhead, ASB = atmospheric still bottoms, VSOH = vacuum still overhead, VSB = vacuum still bottoms, VGO = vacuum gas oil.

[c] See appendix text and Table 4A.7 for explanation.

[d] SGC = simulated by gas chromatography, D86CG = ASTM D-86 corrected by Gulf procedure (Gray, 1981), D1160CP = ASTM D-1160 corrected by API procedure (API, 1976), D1160CG = ASTM D-1160 corrected by Gulf procedure (Gray, 1981), Est. A = estimated from yields at 350, 660, and 849°F (Calkins et al., 1980), Est. B = estimated from D-86 and distillations for other SRC-II heavy distillates, Est. C = upper end of reported nominal boiling range, Est. D = from distillation method where end temperature was achieved at well below 100% overhead, Est. E = from process knowledge.

[e] A = Ames assay, I/P = initiation/promotion assay, SP = skin-painting assay.

[f] References are for data used in Figure 4.12. Chemical data for other samples listed in this table often can be found in the other references listed here and in Chapter 5.

[g] Sample numbers to left of slash represent samples with properties tested by MRDC (ref. 2332-34). Sample numbers to right of slash represent samples tested toxicologically by ORNL, which also reported ultimate analysis data (ref. 2027).

Table 4A.2 Gasification Process Samples[a]

Sample Number	Sample Name[c]	Interpretive Index[d]	95% Point (°F)	Distillation Method[e]	H:C Ratio	Reference Number[b]: Process History	Reference Number[b]: Physical Properties	Reference Number[b]: Toxicological Testing[f]	Reference Number[b]: Chemical Analysis	Other Sample Numbers
					GFETC Pilot Plant					
G1121	Crude tar	BBC	>1000	Est. A	0.43	2070, 2116	2070, 2404	A = 2116, SP = 2411	2116	
					UMD Industrial Gasifier					
G1302	ESP tar	BCC	—	—	1.03	2102, 2346	2041, 2102, 2122	A = 2122, 2041	2122	UMD-05
G1303	ESP tar	BCC	—	—	1.28	2102, 2346	2041, 2102, 2122	A = 2122, 2041	2122	UMD-16
G1304	ESP tar	BCC	—	—	1.04	2102, 2346	2041, 2102, 2122	A = 2122, 2041	2122	UMD-17
G1305	ESP tar	BCC	—	—	1.50	2102, 2346	2041, 2102, 2122	A = 2122, 2041	2122	UMD-18
G1306	ESP tar	BCC	—	—	1.50	2102, 2346	2041, 2102, 2122	A = 2122, 2041	2122	UMD-19
G1307	ESP tar	BCC	—	—	0.99	2329, 2340, 2345, 2102	2041, 2102, 2122	A = 2122, 2041	2122	UMD-23
G1308	ESP tar	BCC	—	—	1.07	2329, 2340, 2345, 2102	2041, 2102, 2122	A = 2122, 2041	2122	UMD-24
G1310	ESP tar	BCC	—	—	1.22	2329, 2340, 2345, 2102	2041, 2102, 2122	A = 2122, 2041	2122	UMD-26
G1311	ESP tar	BCC	—	—	1.24	2329, 2340, 2345, 2102	2041, 2102, 2122	A = 2122, 2041	2122	UMD-27
G1312	ESP tar	BCC	—	—	1.37	2329, 2340, 2345, 2102	2041, 2102, 2122	A = 2122, 2041	2122	UMD-28
G1313	ESP tar	BCC	—	—	1.09	2329, 2340, 2345, 2102	2041, 2102, 2122	A = 2122, 2041	2122	UMD-29
G1314	ESP tar	BCC	—	—	1.16	2329, 2340, 2345, 2102	2041, 2102, 2122	A = 2122, 2041	2122	UMD-30

G1315	ESP tar	BCC	—	—	1.36	2329, 2340, 2345, 2102	2041, 2102, 2122	A = 2122, 2041	2122	UMD-36
G1316	ESP tar	BCC	—	—	1.36	2329, 2340, 2345, 2102	2041, 2102, 2122	A = 2122, 2041	2122	UMD-37
G1317	ESP tar	BCC	—	—	1.32	2329, 2340, 2345, 2102	2041, 2102, 2122	A = 2122, 2041	2122	UMD-38
G1318	ESP tar	BCC	—	—	1.32	2329, 2340, 2345, 2102	2041, 2102, 2122	A = 2122, 2041	2122	UMD-39
G1319	ESP tar	BCC	—	—	1.23	2329, 2340, 2345, 2102	2041, 2102, 2122	A = 2122, 2041	2122	UMD-83
G1320	ESP tar	BCC	—	—	1.32	2329, 2340, 2345, 2102	2041, 2102, 2122	A = 2122, 2041	2122	UMD-89
G1321	ESP tar	BCC	—	—	1.30	2329, 2340, 2345, 2102	2041, 2102, 2122	A = 2122, 2041	2122	UMD-90
G1322	ESP tar	BCC	—	—	1.32	2102, 2329, 2345	2041, 2102, 2122	A = 2122, 2041	2122	UMD-91
G1323	ESP tar	BCC	—	—	1.37	2102, 2329, 2345	2041, 2102, 2122	A = 2122, 2041	2122	UMD-92
					Lurgi Kosovo Industrial Facility					
G1411	Heavy tar	CCC	—	—	1.63	2148	2148	A = 2148	2148	
G1412	Light tar	CCC	—	—	1.23	2148	2148	A = 2148	2148	
					RTI Bench-Scale Unit					
G1511	GF tar 33	BCB	—	—	0.83	2088, 2092, 2093, 2096	2096	A = 2096	2088, 2092, 2096	
G1521	GF tar 35	BCB	—	—	1.11	2088, 2092, 2093, 2096	2096	A = 2096	2088, 2092, 2096	
G1531	GF tar 41	BCB	—	—	0.85	2088, 2092, 2093, 2096	2096	A = 2096	2088, 2092, 2096	
G1541	GF tar 44	BCB	1020	SGC	—	2088, 2092, 2093, 2096	2090, 2096	A = 2096	2088, 2092, 2096	
G1551	GF tar 47	BBB	810	SGC	1.15	2088, 2092, 2093, 2096	2090, 2096	A = 2096	2088, 2092, 2096	

Table 4A.2 (*Continued*)

Sample Number	Sample Name[c]	Interpretive Index[d]	95% Point (°F)	Distillation Method[e]	H:C Ratio	Reference Number[b]: Process History	Physical Properties	Toxicological Testing[f]	Chemical Analysis	Other Sample Numbers
G1561	GF tar 51	BBB	840	SGC	1.00	2088, 2092, 2093, 2096	2090, 2096	A = 2096	2088, 2092, 2096	
						Hygas Pilot Plant				
G2117	LTR oil	ACA	—	—	—	2416	2416	A = 2411	2416	
G2118	LTR oil	ACA	—	—	—	2416	2416	A = 2411	2416	
G2119	LTR oil	ACA	—	—	—	2416	2416	A = 2411	2416	
G2125	LTR oil	ACA	—	—	—	2416	2416	A = 2411	2416	
G2126	LTR oil	ACA	—	—	—	2416	2416	A = 2411	2416	
G2132	LTR oil	ACA	—	—	—	2416	2416	A = 2411	2416	
G2144	LTR oil	ACA	—	—	—	2416	2416	A = 2411	2416	

[a] Samples from the Mellon Institute riser reactor (Fillo et al., 1983) are not included; they were treated as a separate sample set in Chapter 4.
[b] References are listed in numerical order at the end of this appendix.
[c] ESP = electrostatic precipitator, LTR = low-temperature reactor.
[d] See appendix text and Table 4A.7 for explanation.
[e] SGC = simulated by gas chromatography, Est. A = nonstandard vacuum distillation method.
[f] A = Ames assay, SP = skin-painting assay.

Table 4A.3 Hydrotreating Process Samples

Sample Number	Sample Name[c]	Interpretive Index[d]	Hydrogen Consumption[b] (wt. %)	95% Point (°F)	Distillation Method[e]	H:C Ratio	Reference Number[a]: Process History	Physical Properties	Toxicological Testing[f]	Chemical Analysis	Other Sample Numbers
					UOP Hydrotreatment						
H0102	Mod. Sev. SRC-II	BAA	1.95	691	D1160CP	1.38	2075	2075	A = 2075	2075	UOP 3777-34
H0104	High Sev. SRC-II	BAA	2.85	668	D1160CP	1.50	2075	2075	I/P = 2151, A = 2075	2075	UOP 3777-41
					CRC Hydrotreatment						
H0201	Low Sev. SRC-II	BAA	1.44	734	SGC	1.37	2180	2180	A, SP = 2304	2180	CRC SGQ 7779, ACD 917
H0202	Mod. Sev. SRC-II	BAA	2.22	687	SGC	1.48	2180	2180	A, SP = 2304	2180	CRC SGQ 7711, ACD 918
H0203	High Sev. SRC-II	BAA	2.81	667	SGC	1.51	2180	2180	A, SP = 2304	2180	CRC SGQ 7795, ACD 919
H0204	Low Sev. H-Coal	BAA	1.44	734	SGC	1.45	2179	2179	A = 2304	2179	CRC SGQ 8031, ACD 934
H0205	Mod. Sev. H-Coal	BAA	2.16	687	SGC	1.53	2179	2179	A = 2304	2179	CRC SGQ 7965, ACD 935
H0206	High Sev. H-Coal	BAA	4.61	628	SGC	1.75	2179	2179	A = 2304	2179	CRC SGQ 8014, ACD 936
					MRDC Hydrotreatment[g]						
H0301/4	High Sev. SCR-I LOL	CBB	2.40	425	SGC	1.88/1.91	2332	2332/2027	A = 2027	—	ORNL 1608
H0302/5	Mod. Sev. SRC-I LOL	CBB	1.80	429	SGC	1.80/1.83	2332	2332/2027	A = 2027	—	ORNL 1607
H0303/6	Low Sev. SRC-I LOL	CBB	0.49	443	SGC	1.66/1.72	2332	2332/2027	A = 2027	—	ORNL 1606
H0307/10	High Sev. SCR-I WS	CBB	4.42	460	SGC	1.73/1.78	2333, 2334	2333, 2334/2027	A = 2027	—	ORNL 1612
H0308/11	Mod. Sev. SRC-I WS	CBB	2.31	—	—	1.44/1.38	2333, 2334	2333, 2334/2027	A = 2027	—	ORNL 1611

Table 4A.3 (*Continued*)

Sample Number	Sample Name[a]	Interpretive Index[b]	Hydrogen Consumption[c] (wt. %)	95% Point (°F)	Distillation Method[d]	H:C Ratio	References: Process History	References: Physical Properties	References: Toxicological Testing[e]	References: Chemical Analysis	Other Sample Numbers
H0309/12	Low Sev. SRC-I WS	CBB	0.85	—	—	1.31/1.20	2333, 2334	2333, 334/2027	A = 2027	—	ORNL 1610
H0313/16	High Sev. SRC-I RS	CBB	4.88	607	SGC	1.57/1.50	2332	2332/2027	A = 2027	—	ORNL 1616
H0314/17	Mod. Sev. SRC-I RS	CBB	—	—	—	—/1.36	2332	2332/2027	A = 2027	—	ORNL 1615
H0315/18	Low Sev. SRC-I RS	CBB	1.72	688	SGC	1.23/1.22	2332	2332/2027	A = 2027	—	ORNL 1614
H0319/22	High Sev. H-Coal	CBB	1.43	626	SGC	1.48/1.47	2332	2332/2027	SP = 2027	—	ORNL 1604
H0320/23	Mod. Sev. H-Coal	CBB	0.83	636	SGC	1.42/1.42	2332	2332/2027	SP = 2027	—	ORNL 1603
H0321/24	Low Sev. H-Coal	CBB	0.42	640	SGC	1.38/1.33	2332	2332/2027	SP = 2027	—	ORNL 1602
						BETC Hydrotreatment					
H0401	SRC-II Prod. 4	BBB	1.53	707	SGC	1.34	2032, 2348	2032, 2348	A = 2348	—	
H0402	SRC-II Prod. 8	BBB	1.37	716	SGC	1.33	2032, 2348	2032, 2348	A = 2348	—	
H0403	SRC-II Prod. 2	BBB	1.80	696	SGC	1.38	2032, 2348	2032, 2348	A = 2348	—	
H0404	SRC-II Prod. 1	BBB	2.66	682	SGC	1.47	2032, 2348	2032, 2348	A = 2348	—	
H0405	SRC-II Prod. 3	BBB	3.52	651	SGC	1.57	2032, 2348	2032, 2348	A = 2348	—	
H0406	SRC-II Prod. 5	BBB	4.23	628	SGC	1.68	2032, 2348	2032, 2348	A = 2348	—	
H0407	SRC-II Prod. 6	BBB	4.52	624	SGC	1.70	2032, 2348	2032, 2348	A = 2348	—	
H0408	SRC-II Prod. 7	BBB	5.68	619	SGC	1.77	2032, 2348	2032, 2348	A = 2348	—	
H0411	EDS Wyodak Prod. 2	BBC	1.57	675	SGC	1.48	2348	2348	A = 2348	—	
H0412	EDS Wyodak Prod. 1	BBC	2.04	666	SGC	1.57	2348	2348	A = 2348	—	
H0413	EDS Wyodak Prod. 3	BBC	2.89	655	SGC	1.66	2348	2348	A = 2348	—	
H0414	EDS Wyodak Prod. 4	BBC	3.51	630	SGC	1.74	2348	2348	A = 2348	—	

H0421	EDS Illinois No. 6 Prod. 1	BBC	1.92	919	SGC	1.24	2348	2348	A = 2348	—	
H0422	EDS Illinois No. 6 Prod. 2	BBC	2.94	910	SGC	1.38	2348	2348	A = 2348	—	
H0423	EDS Illinois No. 6 Prod. 3	BBC	3.55	876	SGC	1.46	2348	2348	A = 2348	—	
H0424	EDS Illinois No. 6 Prod. 4	BBC	4.36	810	SGC	1.57	2348	2348	A = 2348	—	
H0431	H-Coal ASB Prod. 1	BBB	1.34	738	SGC	1.54	2348	2348	A = 2348	—	
H0432	H-Coal ASB Prod. 2	BBB	1.67	729	SGC	1.58	2348	2348	A = 2348	—	
H0441	H-Coal VSOH Prod. 1	BBB	0.72	766	SGC	1.45	2348	2348	A = 2348	—	
H0442	H-Coal VSOH Prod. 2	BBB	1.43	759	SGC	1.52	2348	2348	A = 2348	—	
H0443	H-Coal VSOH Prod. 3	BBB	1.82	754	SGC	1.57	2348	2348	A = 2348	—	
				Wilsonville NTSL Second-Stage Hydrotreater							
H0512	HTR light fuel oil	BCB	—	461	SGC	1.81	2115, 2237	2237	A = 2415	2415	5226-023
H0513	HTR bottoms	BCB	2.1	944	SGC	1.45	2115, 2237	2237	A, I/P = 2415	2415	5226-022
H0514	HTR recycle solvent	BCB	—	>1000	Est.	0.88	2115, 2237	2237	A = 2415	2415	5226-024
H0521	HTR light oil	—	2.6	—	—	—	2410	2410	—	2410	2410
H0522	HTR process solvent	—	2.6	—	—	—	2410	2410	—	2410	2410
				Wilsonville ITSL Second-Stage Hydrotreater							
H0554	HTR recycle solvent	BBB	—	>1000	SGC	—	2239	2239	A = 2415	2415	5226-082
H0564	HTR recycle solvent	BDB	—	>1000	Est.	—	2239	2239	A = 2415	2415	5226-105
H0572	HTR light fuel oil	BBB	—	388	SGC	—	2239	2239	A = 2415	2415	5226-051
H0573	HTR bottoms	BBB	—	813	SGC	—	2239	2239	A = 2415	2415	5226-052

Table 4A.3 (*Continued*)

Sample Number	Sample Name[a]	Interpretive Index[b]	Hydrogen Consumption[c] (wt. %)	95% Point (°F)	Distillation Method[d]	H:C Ratio	References: Process History	References: Physical Properties	References: Toxicological Testing[f]	References: Chemical Analysis	Other Sample Numbers
H0574	HTR recycle solvent	BBB	—	>1000	SGC	1.22	2239	2239	A = 2415	2415	5226-054
H0584	HTR recycle solvent	BDC	—	>1000	Est.	—	2239	2239	A = 2415	2415	5226-058
H0592	HTR light fuel oil	BBB	—	373	SGC	—	2239	2239	A = 2415	2415	5226-068
H0593	HTR bottoms	BBB	—	817	SGC	1.20	2239	2239	A = 2415	2415	5226-069
				C. E. Lummus ITSL Second-Stage Hydrotreater							
H0611	LCF TLP	BBB	3.62	>1000	SGC	1.35	2035	2035	A, I/P = 2414	2414	
H0621	LCF TLP	BBB	2.94	>1000	SGC	1.04	2035	2035	A, I/P = 2414	2414	
H0631	LCF TLP	BBC	1.69	>1000	SGC	1.09	2320	2035	A = 2414	2414	
H0641	LCF TLP	BBB	2.33	>1000	SGC	1.12	2321	2035	A = 2414	2414	

[a] References are listed in numerical order at the end of this appendix.

[b] Derived from values reported in SCF H_2 consumed per barrel of oil fed using feed density and assuming 60°F and 1 atm in converting SCF to lb moles.

[c] Mod. = moderate, Sev. = severity, HTR = hydrotreater, LCF = LC Finer, TLP = total liquid product, LOL = light organic liquid, WS = wash solvent, RS = recycle solvent, Prod. = product, ASB = atmospheric still bottoms, VSOH = vacuum still overhead.

[d] See appendix text and Table 4A.7 for explanation.

[e] SGC = simulated by gas chromatography. D1160CP = ASTM D-1160 corrected by API procedure (API, 1976). Est. = estimated from process knowledge.

[f] A = Ames assay, I/P = initiation/promotion assay, SP = skin-painting assays.

[g] Sample numbers to left of slash represent samples with properties tested by MRDC (2332-34). Sample numbers to right of slash represent samples tested toxicologically by ORNL, which also reported ultimate analysis data (2027).

Table 4A.4 Narrow-Boiling-Range Cuts of Liquefaction Process Samples

Sample Number	Boiling Range[b] (°F)	Interpretive Index[c]	Reference Number[a]			
			Process History	Physical Properties	Toxicological Testing[d]	Chemical Analyses
			SRC-II P-99 Process Solvent Cuts (L1111)[e]			
N1102	700–750	BCA	2036	NA[f]	A = 2039, I/P, SP = 2151	2412
N1103	750–800	BCA	2036	NA	A = 2039, I/P, SP = 2151	2412
N1104	800–850	BCA	2036	NA	A = 2039, I/P, SP = 2151	2412
N1105	>850	BCA	2036	NA	A = 2039, I/P, SP = 2151	2412
			SRC-II P-99 Process Solvent Cuts (L1121)			
N1106	123–269	BAA	2036	2036	A = 2039	2036
N1107	193–234	BAA	2036	2036	A = 2039	2036
N1108	245–262	BAA	2036	2036	A = 2039	2036
N1109	264–301	BAA	2036	2036	A = 2039	2036
N1110	315–360	BAA	2036	2036	A = 2039	2036
N1111	367–400	BAA	2036	2036	A = 2039	2036
N1112	417–449	BAA	2036	2036	A = 2039	2036
N1113	470–501	BAA	2036	2036	A = 2039	2036
N1114	521–546	BAA	2036	2036	A = 2039	2036
N1115	549–588	BAA	2036	2036	A = 2039	2036
N1116	621–691	BAA	2036	2036	A = 2039	2036
N1117	699–729	BAA	2036	2036	A = 2039	2036
N1118	739–764	BAA	2036	2036	A = 2039	2036
			SRC-II P-99 Process Solvent Cuts (L1131)			
N1119	647–781	BAA	2036	2036	A = 2039	2036
N1120	709–822	BAA	2036	2036	A = 2039	2036

Table 4A.4 (*Continued*)

Sample Number	Boiling Range[b] (°F)	Interpretive Index[c]	Reference Number[a]			
			Process History	Physical Properties	Toxicological Testing[d]	Chemical Analyses
N1121	778–879	BAA	2036	2036	A = 2039	2036
N1122	840–967	BAA	2036	2036	A = 2039	2036
N1123	915–1056	BAA	2036	2036	A = 2039	2036
			SRC-I Process Solvent Cuts (L2121)			
N2103	450–550	BCB	2066, 2067	NA	A = 2066	2412
N2104	550–600	BCB	2066, 2067	NA	A, I/P = 2066	2412
N2105	600–650	BCB	2066, 2067	NA	A, I/P = 2066	2412
N2106	650–700	BCB	2066, 2067	NA	A, I/P = 2066	2412
N2107	700–750	BCB	2066, 2067	NA	A, I/P = 2066	2412
N2108	750–800	BCB	2066, 2067	NA	A, I/P = 2066	2412
N2109	800–850+	BCB	2066, 2067	NA	A, I/P = 2066	2412
			H-Coal Fuel-Oil Blend Cuts (L3242/43)			
N3201	650–700	BAB	2047	NA	A = 2413	2413
N3202	700–750	BAB	2047	NA	A = 2413	2413
N3203	750–800	BAB	2047	NA	A = 2413	2413
N3204	800–850	BAB	2047	NA	A = 2413	2413
N3205	>850	BAB	2047	NA	A = 2413	2413
			H-Coal Heavy Fuel-Oil Cuts (L3234)			
N3211	380–475	BCB	2047, 2137	NA	A = 2411	NA
N3212	465–515	BCB	2047, 2137	NA	A = 2411	NA
N3213	520–590	BCB	2047, 2137	NA	A = 2411	NA
N3214	600–645	BCB	2047, 2137	NA	A = 2411	NA
N3215	635–700	BCB	2047, 2137	NA	A = 2411	NA
N3216	>700	BCB	2047, 2137	NA	A = 2411	NA

		EDS ECLP Naphtha/Process Solvent Blend Cuts (L4131/32)				
N4101	600–650	ACB	2354, 2127, 2337	NA	A = 2354	2354
N4102	650–700	ACB	2354, 2127, 2337	NA	A = 2354	2354
N4103	700–750	ACB	2354, 2127, 2337	NA	A = 2354	2354
N4104	750–800	ACB	2354, 2127, 2337	NA	A, I/P = 2354	2354
N4105	>800	ACB	2354, 2127, 2337	NA	A, I/P = 2354	2354
		NTSL First-Stage Product Blend Cuts (L5231/L5221)[g]				
N5210	<350	ABA	2410	2410	A = 2411	2410
N5211	350–400	ABA	2410	2410	A = 2411	2410
N5212	400–450	ABA	2410	2410	A = 2411	2410
N5213	450–500	ABA	2410	2410	A = 2411	2410
N5214	500–550	ABA	2410	2410	A = 2411	2410
N5215	550–600	ABA	2410	2410	A = 2411	2410
N5216	600–650	ABA	2410	2410	A = 2411	2410
N5217	650–700	ABA	2410	2410	A = 2411	2410
N5218	700–750	ABA	2410	2410	A = 2411	2410
		NTSL Second-Stage Product Blend Cuts (H0521/H0522)[g]				
N5220	<350	ABA	2410	2410	A = 2411	2410
N5221	350–400	ABA	2410	2410	A = 2411	2410
N5222	400–450	ABA	2410	2410	A = 2411	2410
N5223	450–500	ABA	2410	2410	A = 2411	2410
N5224	500–550	ABA	2410	2410	A = 2411	2140
N5225	550–600	ABA	2410	2410	A = 2411	2410
N5226	600–650	ABA	2410	2410	A = 2411	2410
N5227	650–700	ABA	2410	2410	A = 2411	2410
N5228	700–750	ABA	2410	2410	A = 2411	2410

[a] References are listed in numerical order at the end of this appendix.

[b] Boiling range given is nominal range, except for samples N1106–N1123, where range is 10–90% point based on cut distillations reported by Gray (1981).

[c] See appendix text and Table 4A.7 for explanation.

[d] A = Ames assay, I/P = initiation/promotion assay, SP = skin-painting assay.

[e] Sample number in parentheses is that of parent sample. See Chapter 4, Table 4.4.

[f] NA = not available.

[g] The B index for sample properties indicates that specific gravities and ultraviolet and gas chromatography spectra are available.

Table 4A.5 List of Sample Numbers Used in Chapter 4 Figures

Figure Number	Sample Numbers for Data Plotted
4.1A	N1102–05, N1106–23, N2103–09, N3201–05, N3211–16, N4101–05, N5210–18, N5220–28
4.1B	N1102–18, N1122, N2103–09, N3201–05, N4101–05
4.2	N1102–05, N2104–09, N4104–05
4.3	N1102–05, N2107–09, N3201–05, N4102–05
4.4	N1102–05, N3202–04, N4102–05
4.5	N1106–23
4.6	N1106–23, N3201–05 (plus additional cuts reported in Sullivan, 1983)
4.7	Listed with the figure
4.8A	L1221, L1232–36, L1241–43, L1251, L1262, L1311, L1321, L1331, L1341, L1351, L2211, L2212, L3121–24, L3131–36, L3161–62, L3231–32, L3313–15, L3321, L3323–25, L4111–12, L4131–33, L4211–15, L4221, L4231, L5111, L5121, L5131, L5211–12, L5214, L5216, L5311, L5314–15, L5331, L5334–35, L5344, L5351, L5354–55
4.8B	L1221, L1243, L1251, L1262, L2132, L2211–12, L3112, L3142–44, L3151–54, L3212–13, L4111–12, L4131–33, L4221, L4231, L5121, L5131, L5211–12, L5214, L5216, L5344, L5351, L5354–55
4.9	L1111, L1231–36, L1243, L1251, L2121, L3121–24, L3131–36, L4132, L4211–15, L5216
4.10	L1111, L1243, L1262, L3111, L3141–43, L3151–53, L3211–13, L3241–43
4.11	L1111, L1243, L3141–43, L3151–53, L3211–13, L3241–43
4.12A	L1211, L1243, L1251, L1261, L3141–43, L3151–54, L3244, L4111–12, L4132–33, L5111, L5121
4.12B	L1211, L1243, L1251, L1261, L3141–43, L3151–54, L4112, L4133, L5111, L5121
4.12C	L1211, L1243, L1251, L1261–62, L3111, L3141, L3151–52, L3154, L3231–32, L4111–12, L4132–33, L5111, L5121, L5214, L5334
4.13	Same as Fig. 4.9
4.14–4.18	Not applicable
4.19A	L1251/52, H0102,4; L1262, H0201–03; L1221, H0401–08; L3231/32, H0204–06; L3161, H0431,2; L3162, H0441–43; L4221, H0411–14; L4231, H0421–24; L5121, H0621; L5216, H0513
4.19B	L1251/52, H0102,4; L1262, H0201–03; L1221, H0401,3–5,7–8; L3111/12, H0322–24; L2131/32, H0316,18; H0310–12; H0304–06; L4221, H0411–14; L4231, H0421–24; L5121, H0621; L5131, H0631; L5216, H0513
4.20	Listed with the figure
4.21	L1262, H0201–03; L3111, H0322–24
4.22A	Open points same as Fig. 4.8A. Closed points are: H0102,4, H0201–03, H0204–06, H0401–08; H0411–14, H0421–24, H0431,32, H0441–43, H0611,21,31,41, H0512–14, H0554, H0564, H0572–74, H0592,93, H0584
4.22B	Open points same as Fig. 4.8B. Closed points are: H0102,4, H0201–03, H0204–06, H0304–06, H0310–12, H0316–18, H0322–24, H0401,3–5,7–8, H0411–14, H0421–24, H0621,31,41, H0512–14, H0584, H0592,93
4.23	Nonhydrotreated samples same as Fig. 4.10. Hydrotreated samples are: H0201-3 and H0322-24
4.24A	L1221, H0401–08; L1251/52, H0102,4; L1262, H0201–03; L3231/32, H0204–06; L3161, H0431,32; L3162, H0441–43; L4221, H0411–14; L4231, H0421–24; L5111, H0611; L5121, H0621; L5131, H0631; L5216, H0513
4.24B	L1221, H0401,3–5,7–8; L1251/52, H0102,4; L1262, H0201–03; L2131/32, H0313/16, H0315/18; L3111/12, H0319–21/22–24; L4221, H0411–14; L4231, H0421–24; L5121, H0621; L5131, H0631; L5216, H0513

Table 4A.5 (*Continued*)

Figure Number	Sample Numbers for Data Plotted
4.25	L1251/52, H0104; L1262, H0204–06; L3111/12, H0319–21/22–24; L5111, H0611; L5121, H0621; L5216, H0513
4.26–4.28	L5111,21,31,41; H0611,21,31,41
4.29	G1302–08, G1310–23, G1411–12, G1511,21,31,51,61
4.30–4.34	Not applicable
4.35	G2117–19, G2125–26, G2132, G2144
4.36A	L1221, L1232–36, L1251, L1262, L1311,21,31,41,51,61, L2211–12, L3122–24, L3134–36, L3161–62, L3231–32, L3313–15, L3321, L3323–25, L4111–2, L4131–33, L4211–15, L4221, L4231, L5111, L5121, L5131, L5211–12, L5214, L5216, L5331, L5334–35, L5351, L5354–55, H0102, H0104, H0201–03, H0204–06, H0401–08, H0411–14, H0421–24, H0431–32, H0441–43, H0512–14, H0574, H0611,21,31,41, G1121, G1551, G1561
4.36B	L1221, L1251, L1262, L2132, L2211–12, L3112, L3142–44, L3151–54, L3212–13, L4111–12, L4131–33, L4221, L4231, L5121, L5131, L5211–12, L5214, G5216, H0102, H0104, H0201–03, H0301–03, H0307, H0313, H0315, H0319–21, H0401,3-5,7-8, H0411–14, H0421–24, H0512–14, H0621,31,41, G1121, G1551, G1561
4.37	L1231–36, L1251, L2121, L3121–24, L3131–36, L4132, L4211–15, L5111, L5121, L5216, H0104, H0513, H0611, H0621
4.38	L1262, L3111, L3141–43, L3151–53, L3211–13, L3241–43, H0201–03, H0322–24, G1121
4.39	Not applicable

Table 4A.6 Process and Run Codes

Process Facility	Process Code	Run Code	Process Run Number	Run or Sampling Date
	Liquefaction Processes			
SRC-II P-99 PDU, Harmarville, PA	11	1	77	11/17/80
		2	55&56	6/17/79
		3	62&66	12/7/79
SRC-II pilot plant, Ft. Lewis, WA	12	1	—	5/21/78
		2	77SR12	11/77
		3	—	2/79
		4	78SR18/19	1/78
		5	78SR22-29	12/78
		6	—	4/6/80
SRC-II bench unit, Merriam, KS	13	1	454-RA	6/82
		2	466-R	10/82
		3	467-RA	11/82
		4	467-RB	11/82
		5	467-RC	11/82
SRC-I pilot plant, Wilsonville, AL	21	2	220	11/7/80
		3	46A	—
		4	—	1/6/77
		5	—	—

Table 4A.6 (*Continued*)

Process Facility	Process Code	Run Code	Process Run Number	Run or Sampling Date
SRC-I pilot plant, Ft. Lewis, WA	22	1	—	12/76
H-Coal PDU, Trenton, NJ	31	1	1	10/75
		2	3	5/76
		3	5	9/77
		4	5-Period 5	9/9/77
		5	7-Period 5	9/13/78
		6	—	8–9/80
H-Coal pilot plant, Catlettsburg, KY	32	1	7	5/14/81
		3	8	9/25/81
		4	8	10/29/81
H-Coal bench unit, Trenton, NJ	33	1	177–143	—
		2	177–144	—
EDS ECLP pilot plant, Baytown, TX	41	1	—	5/23/81
		2	—	10/23/81
		3	—	2/18/82
		4	—	8/17/82
EDS CLPP pilot plant, Baytown, TX	42	1	256–263	9/77
		2	—	1976
		3	—	1976
C. E. Lummus ITSL PDU, Bloomfield, NJ	51	1	3LCF7-37	3/1/81
		2	3LCF7-90	10/1/81
		3	3LCF8-52	2/10/82
		4	3LCF9-0524	5/24/82
NTSL pilot plant, Wilsonville, AL	52	1	241 C/D	9/22/82
		2	236	2/8/82
		3	235	11/81
ITSL pilot plant, Wilsonville, AL	53	1	242–680	12/13/82
		3	242–720A	12/20/82
		4	242–720B	1/6/83
		5	242–740	1/10/83
	Gasification Processes			
GFETC slagging fixed-bed pilot plant, Grand Forks, ND	11	2	RA-87	12/1/80
UMD Foster–Wheeler Stoic gasifier, Duluth, MN	13	02–06	5	11/27–12/12/79
		07–23	6	2/4–2/28/80
Lurgi commercial plant, Kosovo, Yugoslavia	14	1	—	1977–79
RTI bench unit, Research Triangle Park, NC	15	1	33	7/11/78
		2	35	7/25/78
		3	41	9/6/78
		4	44	10/5/78

Table 4A.6 **(*Continued*)**

Process Facility	Process Code	Run Code	Process Run Number	Run or Sampling Date
		5	47	12/21/78
		6	51	1979
Hygas pilot plant, Chicago, IL	21	1	84	12/79
		2	85	3/80–4/80
		3	86	5/80
		4	87	7/80
	Hydrotreatment Processes			
UOP (Universal Oil Products), Des Plaines, IL	01	—	—	—
Chevron Research Company, Richmond, CA	02	—	—	—
Mobile Research & Development Company, Paulsboro, NJ	03	—	—	—
BETC, Bartlesville, OK	04	—	—	—
TSL pilot plant, Wilsonville, AL	05	1	241C/D	9/22/82
		2	235	11/81
		5	242–680	12/13/82
		6	242–700	12/16/82
		7	242–720A	12/20/82
		8	242–720B	1/6/83
		9	242–740	1/10/83
C. E. Lummus TSL PDU, Bloomfield, NJ	06	1	3LCF7-37	3/1/81
		2	3LCF7-90	10/1/81
		3	3LCF8-52	2/10/82
		4	3LCF9-0524	5/24/82

Table 4A.7 Criteria Used to Define the Interpretive Index for the Engineering Analysis of the Coal-Conversion Toxicological Testing Database

Index	Sampling/Handling	Sample Properties	Process History
D	Sampling and handling procedures known to be inadequate to preserve sample integrity, and/or sample(s) known to have degraded significantly	Name is known that relates sample(s) to a process stream of known approximate characteristics	Name and description are known that relate the sample to a specific process and/or stream
C	Sampling and handling procedures not documented but suspected (inferred) to be inadequate to	Known properties include nominal boiling-point range, specific gravity or density, and/or elemental composition	Run number(s) or date(s) known; or feedstock, process conditions, and/or yields known. However, process

Table 4A.7 (*Continued*)

Index	Sampling/Handling	Sample Properties	Process History
	preserve sample integrity		operations judged as poor (e.g., substantial nonsteady operation before and/or during sampling)
B	Sampling and handling procedures not documented but suspected (inferred) to be adequate to preseve sample integrity	Boiling-point distribution is known in addition to properties for Level C	Partial or inferred feedstock, process conditions, and yields known and can be directly associated with the sample
A	Sampling and handling procedures documented and adequate to preserve sample integrity, and/or sample(s) known not to have degraded	Additional inspections (e.g., hydrocarbon types, carbon residues) are available in addition to those in Level B	Complete information available and documented on feedstock, process conditions, and yields of sampled materials

References for Tables 4A.1–4A.4

2005

Biomedical Studies on Solvent Refined Coal (SRC-II) Liquefaction Materials: A Status Report, PNL-3189, Pacific Northwest Laboratory, Richland, WA. National Technical Information Service, Springfield, VA, 1979.

2016

Morrison, J. A., and C. L. Chen, *H-Coal Pilot Plant Topical Report T-9 Operating Results for Coal Run No. 1*, FE-2260, Ashland Synthetic Fuels, Inc., Ashland, KY, 1981.

2027

Epler J. L. (ed.), *Health Effects Research in Direct Coal Liquefaction, Studies of H-Coal Distillates: Phase I PDU Samples—The Effects of Hydrotreatment*, ORNL/TM-8071, Oak Ridge National Laboratory, Oak Ridge, TN. National Technical Information Service, Springfield, VA, 1981.

2032

Sutterfield, D., W. C. Lanning, and R. E. Royer, Screening of Hydrotreated Coal Liquids for Mutagenic Activity, in *Proceedings Second U.S. DOE Environmental Control Symposium*, Reston, VA, March 17–19, 1980, CONF-800334/1. National Technical Information Service, Springfield, VA.

2035

Schindler, H. D., and J. M. Chen, *Integrated Two Stage Liquefaction Topical Technical Progress Report*, Completion of Indiana V Program, Period April 1, 1981–September 30, 1981, 14804-Q7, The Lummus Company, Bloomfield, NJ and Cities Service Research & Development Company, Tulsa, OK. National Technical Information Service, Springfield, VA, 1982.

2036

Gray, J. A., *Selected Physical, Chemical & Thermodynamic Properties of Narrow Boiling Range Coal Liquids from the SRC-II Process*, Topical Report for Period March 1980 to February 1981, Gulf Research & Development Co., Denver, CO, 1981.

2037

Epperly, W. R., *EDS Coal Liquefaction Process Development Phase V*, Annual Technical Progress Report for the Period July 1, 1981–June 30, 1982, FE-2893-93. Exxon Research & Engineering Company, Annandale, NJ. National Technical Information Service, Springfield, VA, 1982.

2039

Pelroy, R. A., and B. W. Wilson, *Fractional Distillation as a Strategy for Reducing the Genotoxic Potential of SRC-II Coal Liquids: A Status Report*, PNL-3787, Pacific Northwest Laboratory, Richland, WA. National Technical Information Service, Springfield, VA, 1981.

2040

Calkins, W. H., J. F. Deye, R. W. Hartgrove, C. F. King, and D. F. Krahn, *Synthetic Crude Oils Carcinogenicity Screening Tests*, Final Report, Covering the Period October 16, 1978 to August 30, 1980. E. I. Du Pont de Nemours and Co., Inc., Wilmington, DE, 1980.

2041

Griest, W. H., and B. R. Clark, *Sample Management and Chemical/Physical Properties of Products and Effluents from the University of Minnesota–Duluth Low-Btu Gasifier*, ORNL/TM-8427, Oak Ridge National Laboratory, Oak Ridge, TN. National Technical Information Service, Springfield, VA, 1982.

2047

Morrison J. A., M. J. Dabkowski, and F. K. Young, *H-Coal Pilot Plant Topical Report Coal Run No. 8 with Illinois Coal in Syncrude Mode T-10*, FE-2260, Ashland Synthetic Fuels, Inc., Ashland, KY, 1983.

2051

Laboratory Program to Support H-Coal Pilot Plant Operations, PDU Run 7, FE-2547-39, Hydrocarbon Research, Inc., Lawrenceville, NJ, 1979.

2052

Johnson, E. S., and A. G. Comolli, *Laboratory Program to Support H-Coal Pilot Plant Operations—PDU Run 5*, FE-2547-19, Hydrocarbon Research, Inc., Lawrenceville, NJ, 1978.

2053

H-Coal Process Development Unit Operations in Fuel Oil Mode—PDU Run 1, Hydrocarbon Research, Inc., Lawrenceville, NJ, 1976.

2061

Mahlum, D. D., *Chemical Biomedical & Ecological Studies of SRC-I Materials from the Ft. Lewis Pilot Plant: A Status Report*, PNL-3474, Pacific Northwest Laboratory, Richland, WA. National Technical Information Service, Springfield, VA, 1981.

2066

Wilson, B. W., R. A. Pelroy, and D. D. Mahlum, *Chemical Characterization and Genotoxic Potential Related to Boiling Point for Fractionally Distilled SRC-I Coal Liquids*, PNL-4277, Pacific Northwest Laboratory, Richland, WA. National Technical Information Service, Springfield, VA, 1982.

2067

1980 Operation of SRC Pilot Plant Wilsonville, Alabama, EPRI AP-2235, Catalytic, Inc., Philadelphia, PA. Electric Power Research Institute, Palo Alto, CA, 1982.

2070

Stearns-Roger, Inc., *Fixed-Bed Slagging Gasifier Pilot Plant, Gasifier Run RA-87, December 1–4, 1980*, Stearns-Roger Inc., Grand Forks, ND, 1981.

2075

Weimer, W. C., B. W. Wilson, R. A. Pelroy, and J. C. Craun, *Initial Chemical & Biological Characterization of Hydrotreated Solvent Refined Coal (SRC-II) Liquids: A Status Report*, PNL-3464, Pacific Northwest Laboratory, Richland, WA. National Technical Information Service, Springfield, VA, 1980.

2088

Mixon, F. O., J. G. Cleland, T. J. Hughes, D. A. Green, and S. K. Gangwal, Laboratory-Scale Simulation of Coal Gasification for Biological Studies, in K. E. Cowser (ed.), *Fifth ORNL Life Sciences Symposium: Synthetic Fossil Fuel Technologies, Results of Health and Environmental Studies*, Butterworth, Stoneham, MA, 1984.

2090

Gangwal, S. K., Rapid Boiling Point Assay of Tar from Coal Gasification by Gas Chromatography, *J. High Resol. Chrom. & Commun.*, **4**, 580–581 (1981).

2092

Green, D. A., J. G. Cleland, D. P. Daugherty, W. J. McMichael, F. O. Mixon, and R. S. Truesdale, *Coal Gasifier Parameters Influencing Environmental Pollutant Production*, RTI/1934/00-03F, Research Triangle Institute, Research Triangle Park, NC, 1981.

2093

Cleland, J. G., S. K. Gangwal, C. M. Sparacino, R. M. Zwedinger, D. G. Nichols, and F. O. Mixon, *Pollutants from Synthetic Fuels Production: Coal Gasification Screening*, EPA-600/7-79-200, Research Triangle Institute, Research Triangle Park, NC, 1979.

2096

Nichols, D. G., J. G. Cleland, D. A. Green, F. O. Mixon, T. J. Hughes, and A. W. Kolber, *Pollutants from Synthetic Fuels Production: Environmental Evaluation of Coal Gasification Screening Tests*, EPA-600/7-79-202, Environmental Protection Agency, Research Triangle Park, NC, 1979.

2099

Cowser, K. E., *Life Sciences Synthetic Fuels Semiannual Progress Report for the Period Ending December 31, 1982*, ORNL/TM-8700, Oak Ridge National Laboratory, Oak Ridge, TN. National Technical Information Service, Springfield, VA, 1983.

2102

Klein, J. A., M. T. Harris, and R. L. Ruenger, *Operational History of the University of Minnesota–Duluth Gasifier*, ORNL/CF-81/316, Oak Ridge National Laboratory, Oak Ridge, TN, 1982.

2111

Beskind, M. M., and R. F. Cascone, *A Data Review and Operating Experience Evaluation of the SRC-II Process*, ORNL/SUB-7240/2, Oak Ridge National Laboratory, Oak Ridge, TN. Scientific Design Company, Inc., New York, NY, 1978.

2115

The Wilsonville Advance Coal Liquefaction Research & Development Facility Topical Report No. 2 Non-Integrated Two-Stage Liquefaction, DOE/PC/50041-15, Catalytic, Inc., Philadelphia, PA. National Technical Information Service, Springfield, VA, 1983.

2116

Wilzbach, K. E., et al., *Environmental Research Program for Slagging Fixed-Bed Coal Gasification: A Status Report, November 1981*, ANL/SER-A, Argonne National Laboratory, Argonne, IL. National Technical Information Service, Springfield, VA, 1982.

2120

Kimball, R. F., and N. B. Munro, *A Critical Review of the Mutagenic and Other Genotoxic Effects of Direct Coal Liquefaction*, ORNL-5721, Oak Ridge National Laboratory, Oak Ridge, TN. National Technical Information Service, Springfield, VA, 1981.

2122

Biomedical Response to Products and Effluents from the University of Minnesota–Duluth Gasifier, ORNL/TM-8821, Oak Ridge National Laboratory, Oak Ridge, TN. National Technical Information Service, Springfield, VA.

2127

Epperly, W. R., *EDS Coal Liquefaction Process Development Phase V Quarterly Technical Progress Report for the Period January 1–March 31, 1982*, DOE/ET/10069-89, Exxon Research & Engineering Company, Annandale, NJ. National Technical Information Service, Springfield, VA, 1982.

2137

Cowser, K. E., *Life Sciences Synthetic Fuel Semiannual Progress Report for the Period Ending June 30, 1982*, ORNL/TM-8441, Oak Ridge National Laboratory, Oak Ridge, TN. National Technical Information Service, Springfield, VA.

2148

Lee, K. W., W. S. Seames, R. V. Collins, K. J. Bombaugh, and G. C. Page, *Environmental Assessment: Source Test and Evaluation Report—Lurgi (Kosovo) Medium-Btu Gasification*, Final Report, EPA-600/7-81-142, Environmental Protection Agency, Research Triangle Park, NC. National Technical Information Service, Springfield, VA, 1981.

2151

Mahlum, D. D., M. E. Frazier, R. A. Pelroy, and R. A. Renne, *Comparison of in Vitro and in Vivo Studies with Coal Liquids from the SRC-II Process*, PNL-4836, Pacific Northwest Laboratory, Richland, WA. National Technical Information Service, Springfield, VA, 1983.

2159

Alexander, B. F., and R. P. Anderson, *Merriam Laboratory Progress Report, June 1983*, Gulf Research & Development Company, Merriam, KS, 1982.

2160

Alexander, B. F., and R. P. Anderson, *Merriam Laboratory Progress Report, October 1982*, Gulf Research & Development Company, KS, 1982.

2179

Sullivan, R. F., *Refining and Upgrading of Synfuels from Coal & Oil Shales by Advanced Catalytic Processes, 7th Interim Report Task 10*, DOE/ET/10532-T11, Chevron Research Company, Richmond, CA, 1982.

2180

Sullivan, R. F., *Refining and Upgrading of Synfuels from Coal and Oil Shales by Advanced Catalyst Processes, 4th Interim Report Task 9*, DOE/ET/10532-T5, Chevron Research Company, Richmond, CA, 1981.

2190

Craun, J. C., M. J. Massey, J. P. Fillo, G. M. Wong-chong, R. W. Rittmeyer, R. B. Charna, J. E. Bratina, S. W. Vance, and A. Allen, *Environmental Assessment in the ERDA–DOE Coal Gasification Development Program, Final Technical Report for the Period July 1978–July 1980*, DOE/ET/10249-100, Vol. 2, Carnegie-Mellon Institute for Research and Environmental Research & Technology, Inc., Pittsburgh, PA. National Technical Information Service, Springfield, VA, 1981.

2236

Alexander, B. F., and R. P. Anderson, *Merriam Laboratory Progress Report*, Gulf Research & Development Company, Merriam, KS, 1982.

2237

The Wilsonville Advanced Coal Liquefaction Research & Development Facility Technical Progress Report, Run 241 with Illinois 6 Coal, DOE/PC/d50041-14, Catalytic, Inc., Philadelphia, PA. National Technical Information Service, Springfield, VA, 1983.

2239

The Wilsonville Advanced Coal Liquefaction Research & Development Facility Technical Progress Report, Run 242 with Illinois 6 Coal, DOE/PC/50041, Catalytic, Inc., Philadelphia, PA. National Technical Information Service, Springfield, VA, 1983.

2255

H-Coal Integrated Pilot Plant Phase I—Final Report, December 1973–June 1977, Volume II, RCP/T-1544/1, Vol. II, Hydrocarbon Research, Inc., Lawrenceville, NJ, 1977.

2262

McIlvried, H. G., *Summary of Results from SRC-II Runs Made on Process Development Unit P-99, Runs P99-15 to P99-85, Interim Report for March 1977–June 1981*, DOE/ET/10104-42, The Pittsburg & Midway Coal Mining Company, Denver, CO. National Technical Information Service, Springfield, VA, 1982.

2290

Warshawsky, D., and R. S. Schoeny, *Exploratory Research on Mutagenicity of Coal-Related Materials*, Final Report, DOE-ET-00222-4, National Technical Information Service, Springfield, VA, 1980.

2291

Warshawsky, D., *Exploratory Research on Mutagenicity of Coal-Related Materials*, Final Report of Coal Conversion Contract No. EW-78-5-22-0222, National Technical Information Service, Springfield, VA.

2304

Griest, W. H., J. M. Giddings, and J. A. Klein, *Effects of Hydrotreatment on the Properties of Coal-Derived Liquid Products: A Status Report*, ORNL/TM-8836, Oak Ridge National Laboratory, Oak Ridge, TN. National Technical Information Service, Springfield, VA, 1983.

2320

Schindler, H. D. and J. M. Chen, *Integrated Two-Stage Liquefaction: Topical Technical Progress Report, Status of Illinois No. 6 Program, Period October 1, 1981–March 30, 1982*, 14804-Q8, The Lummus Company, Bloomfield, NJ and Cities Service Research and Development Company, Tulsa, OK, 1982.

2321

Schindler, H. D., and J. M. Chen, *Integrated Two-Stage Liquefaction: Topical Report, Steady State Illinois No. 6 Program, Period April 1, 1982–July 6, 1982*, 14804-Q9, The Lummus Company, Bloomfield, NJ and Cities Service Research and Development Company, Tulsa, OK, 1983.

2329

Dreibelis, W. G., *Observations of the Coal Gasifier at the University of Minnesota–Duluth, Run 6*, January 31, 1980 to March 2, 1980, Lake Superior Basin Studies Center of the University of Minnesota–Duluth, Duluth, MN, 1980.

2331
Epperly, W. R., *EDS Coal Liquefaction PDU Phase IIIB, Quarterly Technical Progress Report for the Period October 1–December 31, 1977*, FE-2893-7, Exxon Research & Engineering Company, Annandale, NJ. National Technical Information Service, Springfield, VA, 1978.

2332
Stein, T. R., J. G. Bendoraitis, A. V. Cabal, R. B. Callen, M. J. Dabkowski, R. H. Heck, H. R. Ireland, and C. A. Simpson, *Upgrading of Coal Liquids for Use as Power Generation Fuels, Annual Report for the Period February 1976–January 1977*, EPRI AF-444 Project 361-2, Mobil Research & Development Corporation, Paulsboro, N.J. Electric Power Research Institute, Palo Alto, CA, 1977.

2333
Stein, T. R., A. V. Cabal, R. B. Callen, M. J. Dabkowski, R. H. Heck, C. A. Simpson, and S. S. Shih, *Upgrading of Coal Liquids for Use as Power Generation Fuels, Annual for the Period February 1977–January 1978*, EPRI AF-873, Project 361-2, Mobil Research & Development Corporation, Paulsboro, NJ. Electric Power Research Institute, Palo Alto, CA, 1978.

2334
Appendix A to reference 2333.

2336
Wilson, B. W., R. A. Pelroy, R. P. Anderson, and J. Freel, *Chemical and Biological Effects of Heavy Distillate Recycle in the SRC-II Process*, PNL-4815, Pacific Northwest Laboratory, Richland, WA. National Technical Information Service, Springfield, VA, 1983.

2337
Epperly, W. R., *EDS Coal Liquefaction Process Development Phase V Quarterly Technical Progress Report, January 1–March 31, 1983*, FE-2893-110, National Technical Information Service, Springfield, VA, 1983.

2340
Soderberg, W. E., *Coal Gasification Duluth Campus Heating Plant Monthly Technical Progress Report June 1981 (Run No. 8)*, FE-2571-26, University of Minnesota–Duluth, Duluth, MN, 1981.

2345
Soderberg, W. E., *Coal Gasification Duluth Campus Heating Plant Interim Report for the Period January 31, 1980–March 2, 1980 (Run No. 6)*, FE-2571-9, University of Minnesota–Duluth, Duluth, MN, 1980.

2346
Soderberg, W. E., *Coal Gasification Duluth Campus Heating Plant Interim Report for the Period November 13, 1979–December 13, 1979 (Run No. 5)*, FE-2571-8, University of Minnesota–Duluth, Duluth, MN, 1979.

2348
Sutterfield, F. D., *Hydrogenation Effects on Mutagenesis of Alternate Crudes*, Ph.D. Thesis, University of Tulsa, Tulsa, OK, 1983.

2354
Later, D. W., R. A. Pelroy, and B. W. Wilson, *Chemical Analysis and Biological Testing of Materials from the EDS Coal Liquefaction Process: A Status Report*, PNL-4960, Pacific Northwest Laboratory, Richland, WA. National Technical Information Service, Springfield, VA, 1984.

2400
J. P. Fillo, personal communication with W. Griest, Oak Ridge National Laboratory, Oak Ridge, TN, 1983.

2401
J. C. Craun, personal communication with P. Angevine, Mobil Research & Development Corporation, Paulsboro, NJ, 1983.

2402
J. P. Fillo, personal communication with L. Lorenzi, Pittsburgh Energy Technology Center, Pittsburgh, PA, 1983.

2403
J. P. Fillo, personal communication with J. Vick, Exxon Research & Development Co., Annandale, NJ, 1984.

2404
J. P. Fillo, personal communication with C. Reilly, Argonne National Laboratory, Argonne, IL, 1983.

2410
Drozdowicz, B. Z., and C. M. Kelly, *Interim Report on the Genetic and Animal Toxicity Testing of SRC-I Products, Intermediates and Waste Materials*, DOE/OR/03054-50, Summary and Appendixes C and G, International Coal Refining Company, Allentown, PA, 1983.

2411
Chapter 3 of this book.

2412
Chapter 5 of this book.

2413
Wilson, B. W., D. W. Later, C. W. Wright, and D. L. Stewart, *Chemical Analysis and Mutational Assay of Distilled Oils from the H-Coal Direct Liquefaction Process: A Status Report*, PNL-5288, Pacific Northwest Laboratory, Richland, WA. National Technical Information Service, Springfield, VA, 1985.

2414
Wilson, B. W., P. Buhl, and E. C. Moroni, *Biological Testing and Chemical Analysis of Materials from an Integrated Two-Stage Liquefaction Process: A Status Report*, PNL-4553, Pacific Northwest Laboratory, Richland, WA. National Technical Information Service, Springfield, VA, 1983.

2415
Later, D. W., *Two-Stage Coal Liquefaction Process Materials from the Wilsonville Facility Operated in the Nonintegrated and Integrated Modes: Chemical Analyses and Biological Testing*, PNL-5215, Pacific Northwest Laboratory, Richland, WA. National Technical Information Service, Springfield, VA, 1985.

2416
Anastasia, L. J., R. J. Evans, R. H. Karst, and C.-S. Juan, *Environmental Assessment of the Hygas Process*, Vol. V, Final Report—Hygas Tests 77 through 87, FE-2433-28, National Technical Information Service, Springfield, VA, 1982.

CHAPTER 5

Chemistry of Coal-Conversion Materials Related to Toxicology and Process

B. W. Wilson
Battelle, Pacific Northwest Laboratory, Richland, Washington

D. W. Later
Lee Scientific, Salt Lake City, Utah

D. A. Haugen
Argonne National Laboratory, Argonne, Illinois

5.1 PURPOSE

Chemical compositions of coal-derived products have been examined with increasing levels of sophistication since the turn of the century. However, only in the last two decades has a broad-based, interdisciplinary effort been made to correlate chemical compositions of these materials with toxicological and process engineering data. In the 1970s and 1980s, with the resurgence of processing coal for use as a petroleum substitute, both gasification and liquefaction process developers were cognizant of potentially adverse health effects associated with fossil-fuel conversion and considered these issues in plant design and operation. This chapter reviews the chemistry of coal-conversion materials as it relates to their toxicological activity as defined by laboratory assays for mutagenicity, carcinogenicity, and teratogenicity. To a lesser extent, composition is also related to process conditions.

Coal-conversion materials derive their chemical composition from that of coals in general. Therefore, coal chemistry and our current understanding of coal-liquefication and -gasification reactions are discussed briefly. Emphasis is placed on the formation during liquefaction or gasification of compounds that are toxicologically active. Chemical compositions of coal-liquefaction and -gasification materials are also compared to those of other coal and non-coal-derived fossil fuels or by-products. Although the composition of coal-conversion materials is complex, their toxicological activity arises from compounds or compound classes that are, in general, common to petroleum-, shale-, and coal-derived materials prior to upgrading.

The term *biodirected* denotes the iterative process of chemical separations followed by bioassay and detailed chemical analysis of toxicologically active fractions from coal-conversion materials. Methods used for biodirected chemical analysis of coal-derived materials often differ from those used in process research and engineering development. Therefore, these analytical techniques are reviewed and discussed with special reference to those that depart from standard approaches.

5.2 HISTORICAL PERSPECTIVE: 1770–1970

Kennaway (1955) and his group in England first applied the general approach of iterative separation, bioassay, and instrumental analysis to identification of coal-derived carcinogens in the 1920s. Building on observations of Pott in 1775 (Pott, 1963) and the work of Yamagiwa and Ichikawa (1915) and Bloch and Dreifuss (1921), Kennaway and his co-workers (Heiger, 1930; Cook et al., 1933) developed a scheme to separate coal-derived materials by distillation and paper chromatography. Fractions were then characterized by spectrophotometric techniques, and identification of compounds was confirmed by synthesis, when possible. During this period, much of the theoretical work on spectrophotometry and other chemical and physical properties of polycyclic aromatic hydrocarbons (PAHs) was done by Clar (1964). Bioassays for dermal carcinogenic potential were conducted with mice and rabbits. However, because the mouse strains used were not bred for sensitivity to chemical carcinogens, several hundred days were often required to test animals to develop tumors from active fractions. Tumor response was found by these early workers to be associated primarily with those isolates containing parent and methylated PAHs with four to six aromatic rings.

Biological activity was first associated with coal-derived nitrogen-containing compounds by Rehn (1895), who linked occupational exposure to anilines with increased incidence of bladder tumors. The compound 2-aminonaphthalene and other amino-PAHs, such as 4-aminobiphenyl, benzidine, and diphenylamine, were ultimately identified as important industrial carcinogens (Case and Hosker, 1954; Case et al., 1954) and within the next decade or so industrial production of these and other coal-derived amino-PAHs ceased in the West.

5.3 NOMENCLATURE AND DEFINITIONS

The chemical nomenclature used in this chapter for polycyclic aromatic compounds (PACs) was described in Chapter 2. The general classes of PACs are PAHs, oxygen-containing PACs (O-PACs), sulfur-containing PACs (S-PACs), and nitrogen-containing PACs (N-PACs). The specific functionality of heteroatoms that are substituents on the aromatic ring system is prefixed to designate PAH type: hydroxy-PAHs, cyano-PAHs, amino-PAHs, and so on. Conventional names may also be used for the heterocyclic compounds, such as thiophenes, azaarenes, and furans. This approach allows unambiguous designation of mixed-function heteroatomic species, such as aminocarbazoles and hydroxythiophenes, often found in coal-derived products. The conventional designation, primary aromatic amines (PAAs), often used in the literature, is generally avoided in favor of the more specific amino-PAHs.

The sample numbers used in this chapter begin with L (for liquefaction material), G (for gasification materials), S (for solid materials), H (for hydrotreated materials), or N (for narrow-boiling-range materials). These designations allow comparison of chemical data for specific samples with the engineering and toxicology data for these materials where discussed in other chapters. Designations used by other sample repositories are also provided in some tables.

5.4 CHEMICAL CARCINOGENESIS

Carcinogenic agents include ultraviolet (UV) light, ionizing radiation, certain inorganic chemicals, and a host of natural and manufactured organic chemicals. Exposure to organic chemicals is one of the primary occupational and environmental carcinogenic risks posed by coal-conversion technologies. This section is a brief overview of chemical carcinogenesis, in which topics most relevant to this book are emphasized. More comprehensive information is presented in early and recent reviews (Arcos and Argus, 1974; Heidelberger, 1975; Miller and Miller, 1976, 1981; Gelboin and Ts'o, 1978, 1979, 1981; Nelson, 1982; Searle, 1984).

The structures of selected organic chemical carcinogens found in coal-derived materials are shown in Fig. 5.1. The mode of action in which carcinogenic response is induced for these compounds is based on principles that have become well established during the past

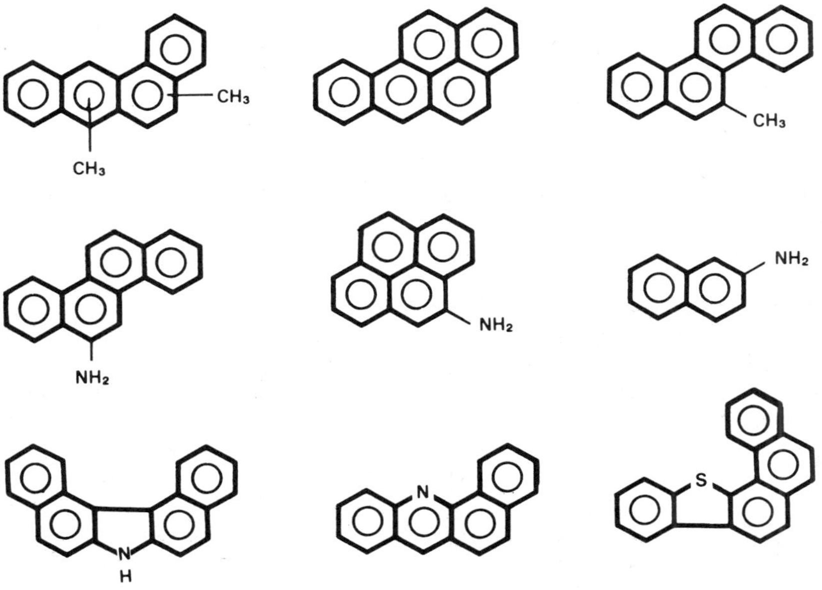

Figure 5.1 Structures of selected polycyclic aromatic compounds identified as carcinogens in coal-derived materials.

25 years. Most carcinogenic organic chemicals and their metabolites bind covalently to celluar macromolecules, including protein, ribonucleic acid (RNA), and most importantly deoxyribonucleic acid (DNA). DNA bears the information required for expressing all cellular characteristics, including regulation of growth and cellular function, processes disrupted in carcinogenesis. The chemical carcinogens thus far known to arise from coal-conversion processes do not react directly with nucleic acids and proteins but instead require metabolic transformation to form reactive chemical species.

Vertebrates, including humans, have evolved enzyme systems that facilitate the excretion of potentially toxic, lipophilic chemicals that are not glucuronic acid and the sulfhydryl-containing tripeptide glutathione, which are metabolized by enzyme systems that act on the major food components, such as proteins, carbohydrates, and fats (Gelboin and Ts'o, 1978, 1979, 1981; Jakoby, 1980, Jakoby et al., 1982; Guengerich and Liebler, 1985). The initial detoxification reaction is the addition of oxygen catalyzed by a cytochrome P-450-dependent enzyme system. This system is found in the membranous endoplasmic reticulum of mammalian cells and is most abundant in liver.

The principal fate of the oxygenated products is their enzyme-catalyzed conjugation with polar biochemical intermediates, including sulfate ion, gluconic acid, and the sulfhydryl-containing tripeptide glutathione. Reaction with sulfate and glucuronic acid requires an initial, energy-dependent activation of these cofactors to enhance their reactivity. The products of these conjugation reactions are relatively polar and are readily excreted.

A wide variety of chemicals are metabolized by this enzyme system, including constituents of edible plants, physiological compounds such as steroids and fatty acids, pharmaceutical and therapeutic agents, other synthetic chemicals, and products of incomplete combustion of carbonaceous material. Metabolism and excretion of foreign chemicals appear essential for survival. However, some oxygenated, or subsequently conjugated, metabolites are sufficiently reactive to form covalent adducts with nucleic acids and proteins. These adducts are formed by the reaction of electrophilic centers, in the activated metabolites, with nucleophilic sites in biological macromolecules. For proteins, these nucleophilic sites include amino and sulfhydryl moieties. For nucleic acids, nucleophilic sites include amino-, nitrogen- and electron-rich aromatic carbons that are constituents of the pyrimidine and purine bases. Some activated metabolites react at phosphate ester sites, causing breaks in the nucleic acid structure.

Binding of chemicals to DNA can cause alterations in the genetic information in cells. According to the most prevalent working hypothesis, these alterations, sometimes detected as mutations, are the beginning of the tumorigenic process. In some experimental models, this alteration of the genome is termed *initiation*. Representatives of several types of chemical found in coal-conversion materials are tumor initiators, as evidenced directly by their tumorigenicity or indirectly by their ability to cause mutations in cell cultures. Upon metabolic activation, some metabolites of these chemicals bind covalently to DNA.

Some constituents of coal-conversion materials, although not carcinogenic, are operationally defined as either *cocarcinogens* or *promoters* of carcinogenesis (Slaga et al., 1978; Slaga, 1983, 1984a–c). Promotors are noncarcinogenic chemicals that enchance the effect of carcinogens when administered days or even months after brief treatment with a carcinogen (the initiator). Promoters are believed to enhance the expression of genetic alterations caused by initiators. They are not effective if administered before initiation. Cocarcinogenic chemicals, while not inherently carcinogenic themselves, enhance the effect of carcinogens with which they are simultaneously administered. Biological effects of carcinogens are not as well defined experimentally as those of promoters.

Carcinogenesis may involve several phenomena. During prolonged simultaneous exposure to chemical mixtures, cocarcinogens may mimic some of the characteristics of promoters. Many chemicals, including carcinogens and cocarcinogens, alter cellular levels of enzymes responsible for their metabolism and metabolic activation. Thus, by this mechanism and others, cocarcinogens may mediate metabolic activation of carcinogens, or they may alter cellular sensitivity to chemical carcinogens by other undefined mechanisms. It is probable that all these functions have a role in cocarcinogenesis.

Within a chemical class consisting of isomeric and structurally similar members, some may be potent carcinogens, while others may be inactive. A primary factor governing bioactivity is the specificity of enzymes responsible for metabolic activation. Three-dimensional surfaces of catalytic sites on enzymes accommodate some isomers and allow formation of reactive metabolites; for other isomers, productive enzyme–substrate interaction is stearically hindered. Furthermore, for similar chemicals subject to the same metabolic fate, subtle stearic and electronic characteristics influence reactivity of electrophilic sites. Although some generalizations have been discovered regarding structure–activity relations, prediction of carcinogenicity based only on chemical structure is not yet possible.

Enzymes responsible for metabolic activation of chemical carcinogens catalyze a variety of oxidative reactions, including hydroxylation of aliphatic carbons, epoxidation and hydroxylation of aromatic chemicals, N-dealkylation of aromatic and aliphatic amines, O-dealkylation of aromatic ethers, N-hydroxylation of amines and N-oxidation at other nitrogen centers, and dehalogenation of aliphatic halogen compounds. They also catalyze the reduction of azo- and nitro-aromatic compounds. Because of their toxicological importance in coal-conversion materials, metabolic activation of PAHs and aromatic amines is described below.

The initial step in PAH metabolism involves formation of arene oxides via the addition of oxygen (Gelboin and Ts'o, 1978; Thakker et al., 1982; Dipple et al., 1984). Some epoxides are relatively stable and can be isolated; others readily undergo nonenzymatic rearrangement or further oxidation, forming phenols and quinones, respectively. Electrophilic epoxides are also subject to stereo-specific enzymatic hydrolysis, forming *trans*-dihydrodiols. Subsequent stereo-selective epoxidation of some dihydrodiols yields highly reactive vicinal diol-epoxides. Covalent adducts formed by reaction of diol-epoxides with purified DNA have been isolated and identified as being the same as those arising from metabolic activation in animal or cellular systems. These reactions are shown in Fig. 5.2 for benzo[a]pyrene (BaP), the carcinogen for which the mechanism of metabolic activation was first elucidated. For bay-region diol-epoxides, as illustrated for BaP, reactivity of electrophilic benzylic carbon is high. Studies of numerous other four- to six-ring PAHs indicate that if formation of bay-region diol-epoxides is structurally possible, the PAHs are likely to be carcinogens. However, enzymatic oxidation does not occur at bay regions of all PAHs. Futhermore, the presence of alkyl substituents may enhance or diminish reactivity of diol-epoxides via electronic or stearic effects. Alkyl substituents in bay regions of PAHs prevent formation of diol-epoxides and greatly diminish carcinogenicity as predicted by the mechanistic model.

Although epoxidation (usually detected as hydroxylation) of benzene, naphthalene, and three-ring PAHs occurs, these compounds generally have negligible carcinogenic activity. However, many four- to six-ring PAHs and their alkyl homologs are carcinogenic (Arcos and Argus, 1974; Dipple et al., 1984). The best-characterized include alkyl homologs of benz[a]anthracene and chrysene, BaP and benzofluoranthenes, and dibenzanthracenes. Some methyl and dimethyl homologs of angular catacondensed PAHs, for example, benz[a]anthracene, chrysene, and dibenzanthracene, have markedly greater carcinogenicity than the respective unsubstituted compounds. For the noncarcinogenic PAHs phenanthrene and fluoranthene, some methyl and dimethyl homologs have significant mutagenic and carcinogenic activity. Depending on the structure of the PAHs and the position of alkyl sub-

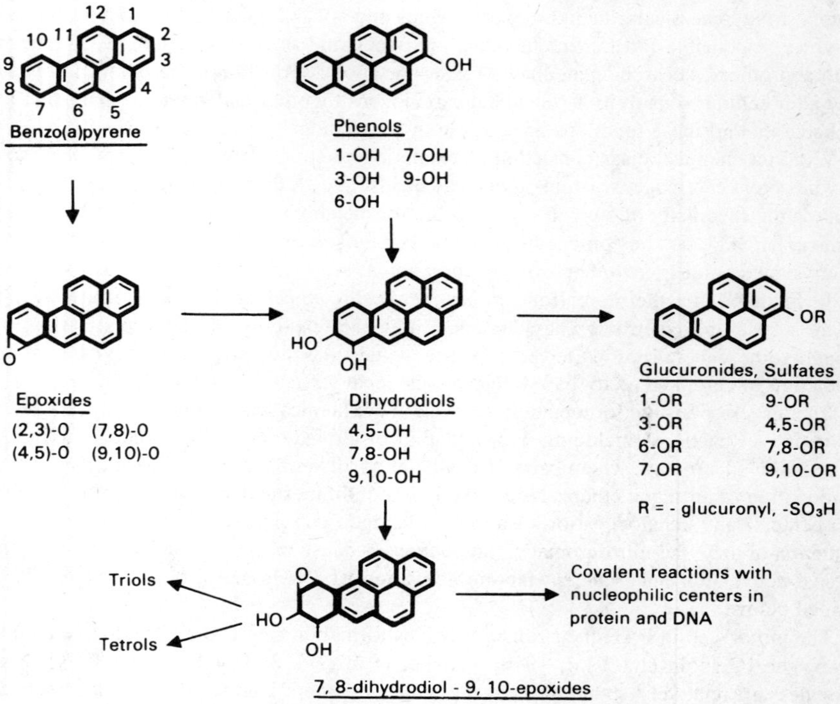

Figure 5.2 Metabolism of benzo[a]pyrene (adapted from Yang et al., 1978).

stituents, the substituents may increase the reactivity of epoxides and diol-epoxides, or the substituents may enhance carcinogenicity by favoring enzymatic formation of certain epoxides and diol-epoxides.

Structural requirements for tumorigenic activity of alkylated phenanthrenes illustrate this phenomenon. Figure 5.3, adapted from La Voie et al. (1981b), shows three phenanthrene isomers, each with a bay-region methylene unit thought to be requisite for carcinogenic activity. The two isomers on the left (1,4- and 4,10-dimethylphenanthrene) are mutagenic and tumorigenic, but the 4,8-dimethylphenanthrene isomer is essentially inactive. Of the monomethyl isomers tested, none were tumorigenic in dermal initiation/promotion assays with mice (La Voie et al., 1981c). Both 1,2,4-trimethylephenanthrene and 1,2,3,4-tetramethylphenanthrene were carcinogenic in chronic mouse skin-painting studies (Badger et al, 1942; Batenandt and Dannenberg, 1953). Carcinogenic azaarenes are also activated by formation of diol-epoxides, and the position of the heterocyclic nitrogen atom markedly influences carcinogenic and mutagenic activities (Levin et al., 1983; Wood et al., 1983).

Metabolic activation of amino-PAHs involves addition of oxygen to the amino nitrogen atom rather than at carbon atoms as for PAHs (Fig. 5.4). The initial metabolites, N-hydroxy aryl amines, are subject to several subsequent enzymatic and nonenzymatic reactions that produce electrophilic aryl nitronium ions, which form covalent adducts with DNA and other cellular constituents (Arcos and Argus, 1974; Krieck, 1974; Radomski, 1979; Searle, 1984). The carcinogenic activity of structural isomers of amino-PAHs often differ greatly. For example, 2-aminonaphthalene and 4-aminobiphenyl are potent bladder carcinogens, while

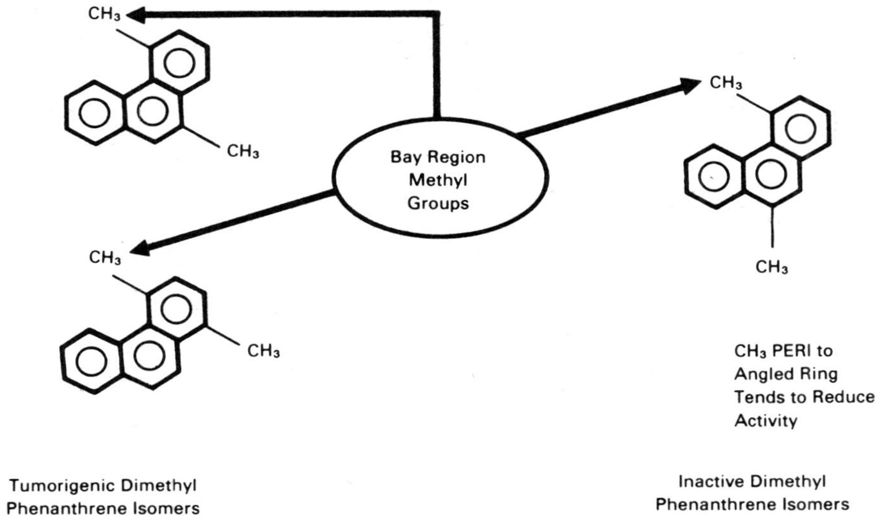

Figure 5.3 Structural requirements for activity in methylated-polycyclic aromatic hydrocarbons (adapted from La Voie et al., 1981b).

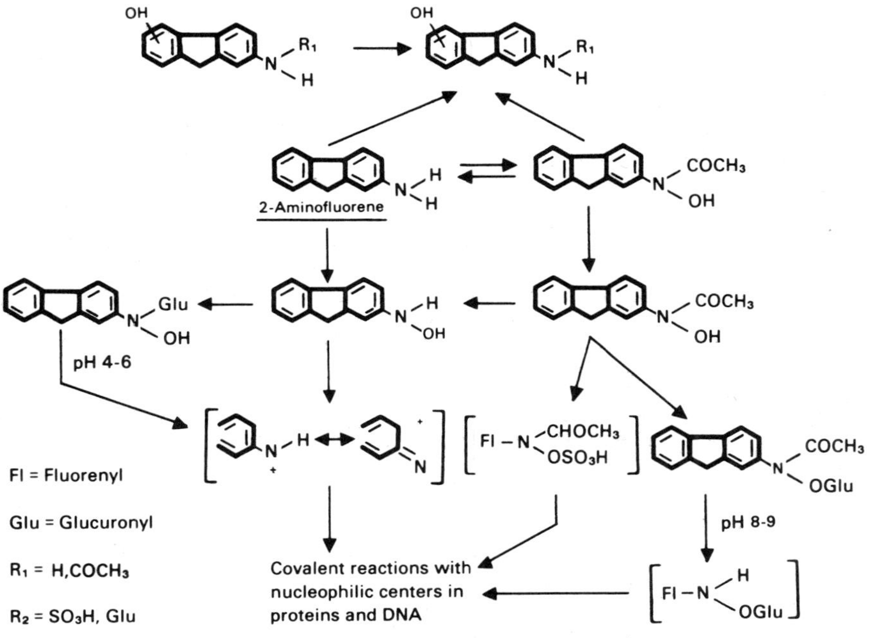

Figure 5.4 Metabolism of 2-aminofluorene. Reactions indicated as occurring at pH 4–6 and 8–9 are believed to be important in the urine of animals with acid or alkaline urine, respectively. Compounds in brackets are especially reactive with cellular nucleophils (adapted from Irving, 1979).

their isomers 1-aminonaphthalene and 2-aminobiphenyl have negligible activity. These and other amino-PAHs are present in coal-conversion materials. Differences in the carcinogenic potency of isomeric amines reflect both differential susceptibilty to enzymatic N-oxidation- and reactivity of oxidation products formed. Alkylation of the aromatic ring also affects carcinogenic activity. *Ortho*-methyl homologs of some amino-PAHs (e.g., 2-amino-naphthalene and 4-aminobiphenyl) are substantially more carcinogenic and mutagenic than are unsubstituted amines. Alkyl substituents at other positions may diminish carcinogenicity or alter organ specificity for tumorigenic risk.

5.4.1 Determination of Mutagenic and Carcinogenic PACs in Coal-Conversion Materials

Studies emphasizing short-term, in vitro measures of genotoxicity as an indicator of carcinogenic potential for coal-conversion materials yielded results that initially appeared inconsistent regarding the relative importance of PAHs. Reasons for the initial inconsistencies included the following: (1) Small amounts of amino-PAHs in PAH-containing fractions were highly mutagenic in the Ames *Salmonella*/mammalian microsome mutagenicity assay and initially deemphasized the importance of the more abundant ''neutral'' fractions that were only minimally or moderately mutagenic. (2) Toxicity (cell killing) of unfractionated or minimally fractionated materials hindered in vitro mutagenesis assays, and the use of exogenous metabolic activation in mammalian-cell mutagenesis assays was not well understood. (3) Results of in vitro assays with single compounds could not readily be interpreted for complex mixtures of related chemicals.

A growing body of evidence indicates that the presence of nonmutagenic components (e.g., two- to four-ring PAHs) in a complex mixture inhibits biological responses to mutagenic components (e.g., BaP). Thus, even though all the materials tested by in vitro methods contained PAHs and were expected to be mutagenic, the apparent contribution of PAH-containing neutral fractions to the overall mutagenicity of fractionated coal liquids ranged from negligible to about 50%.

In vivo skin-painting studies with various coal-conversion materials showed, however, that a major portion of their carcinogenic activity was associated with neutral components (Holland et al., 1981; Mahlum et al., 1984). Neutral fractions of coal-gasification by-product tar were isolated by liquid–liquid partitioning (Haugen et al., 1983), and the larger aromatic bases were removed using the techniques of Boparai et al. (1983). These neutral fractions were mutagenic in the Ames assay and the Chinese hamster ovary (CHO) cell mutation assay and also caused dermal carcinogenesis in mouse skin-painting assays. The neutral fractions were further resolved using a series of reverse- and normal-phase high-performance liquid chromatography (HPLC) systems. Mutagenesis assays with *Salmonella* showed that, for the first level of normal-phase chromatographic isolates, most activity was associated with the PAH fraction; less activity was associated with unidentified, more-polar constituents (Haugen et al., 1983; Haugen and Stamoudis, 1986).

PAH-enriched chromatographic fractions of coal liquids were isolated using alumina-column chromatography (Later et al., 1981a) without initial liquid–liquid partitioning. These fractions were potent initiators of dermal carcinogenicity with mice (Mahlum et al., 1984) and were also active as complete dermal carcinogens in chronic assays (Mahlum et al., 1983). Neutral-PAHs were further resolved by HPLC, and subfractions were assayed for mutagenic activity in *Salmonella* and as initiators of dermal carcinogenesis in mice. Most biological activity was associated with subfractions containing primarily five- ring PAHs such as BaP and their C_1 and C_2 homologs. Although the initial PAH-enriched fractions obtained by alumina-column chromatography were carcinogenic, they showed negligible mutagenic activity in *Salmonella* (Later et al., 1982a,b; Mahlum et al., 1983). However, they were mutagenic in the CHO cell mutation assay (Mahlum et al., 1983; Wilson et al., 1984) and caused transformation in culture of Syrian hamster embryo (SHE) cells (Mahlum et al., 1983). Lack of mutagenic response to fractions known to contain mutagenic PAHs is believed to reflect chemical and biochemical interactions (e.g. antagonism or competing metabolism) that are apparently more evident in short-term bacterial assays than in higher-level mammalian cell or animal assays.

Neutral fractions of coal liquids and related materials have also been prepared by removing basic and acidic components, using successive extractions in ether or methylene chloride with 1 *N* HCl and 1 *N* NaOH (Epler, 1980). PAH-enriched neutral subfractions, isolated by Fluorisil-column absorption chromatography, were carcinogenic (Holland et al., 1981). Although quantitative data indicated that the PAH fraction was responsible for most of the carcinogenic activity of the crude oil, this conclusion was not supported by carcinogenicity data for other fractions. Neutral fractions obtained by liquid–liquid partitioning were further resolved, using LH-20 column chromatography, to obtain fractions designated aliphatic, aromatic, polyaromatic, and polar (Guerin et al., 1981). Data for numerous coal-conversion materials indicated that when the basic fraction (sum of *ether-soluble* bases and *-insoluble* bases) had high mutagenicity, the *polyaromatic* neutral LH-20 subfraction also had high activity. Mutagenic components in the polyaromatic fractions were assumed to be neutral, four- to six-ring PAHs (e.g., benz[a]anthracene or BaP). Possible explanations are that (1) coal conversion and/or postconversion treatment led to formation of mutagenic aromatic bases and mutagenic neutral PAHs or (2) mutagens in neutral polyaromatic chromatographic fractions were highly active aromatic bases that were not removed from the sample by extraction with the mild 1 *N* HCl. The second explanation is supported by data showing that, with a highly mutagenic *basic* fraction (obtained by liquid partitioning), activity of the corresponding *neutral* fraction was associated with both nonpolar and polar subfractions obtained by Fluorisil-column chromatography.

As discussed above, many Cl_1, C_2, and, in certain cases, C_3 alkyl-PAHs are more potent carcinogens than the corresponding unsubstituted compounds. Despite the potential contribution of C_1–C_3 alkyl-PAHs to carcinogenesis of coal-conversion materials, their role has not been well defined and many alkylated PAH isomers having three or more rings have not been identified. Several factors are responsible: (1) Although judicious preparative liquid chromatography and capillary-column gas chromatography–mass spectrometry (GC-MS) allows generic identification of alkyl-PAHs, such as C_2-benz[a]anthracene, preparative techniques have not been used (and may be inadequate) to separate sufficient quantities for structural determination. (2) The large number of multialkylated PAH isomers prohibits their synthesis and use as chromatographic standards; for example, there are 66 isomers of dimethylbenz[a]anthracene. Among these, 7,12-dimethylbenz[a]anthracene is a potent carcinogen, although benz[a]anthracene is only weakly carcinogenic. (3) Currently available gas chromatography (GC) systems fail to resolve some isomers in complex mixtures, leaving them unidentified. (4) There is no established theoretical framework from which to predict the multialkylated isomers most likely to be present.

For PAH compounds, bioactivity has been related to structures having K or bay regions. There is an increase in activity of certain four-ring catacondensed structures with addition of two methyl groups that form bay-region-like areas. PAH compounds classed as peri/catacondensed or catacondensed (especially those alkylated in certain positions) are active both as pure compounds and when separated (but not purified) from complex mixtures. Fully pericondensed structures, such as pyrene, perylene, and coronene, are less active than peri/catacondensed five-ring compounds, C_1 and C_2 catacondensed four-ring compounds, and cyclopenta[cd]pyrene. Certain indeno compounds are also active. For example, cyclopenta[cd]pyrene is active as a pure compound and is found in certain active fractions of coal-gasification tars and soot. However, it is not an important contributor to the toxicological activity of coal liquids.

Because of the way PAHs are formed during coal conversion and the incomplete combustion of carbonaceous materials, only a limited number of the structurally possible four- to six-ring PAH isomers are present in coal-conversion materials. Of these, some are potent carcinogens while others have negligible activity (e.g., BaP and benzo[e]pyrene, respectively). The structures of the PAHS affect their carcinogenicity in several ways. A combination of shape and electron-density distributions governs the degree to which enzymatically catalyzed epoxidation occurs at critical molecular sites. When formed, the reactivity of diol-epoxides depends on the degree to which the electron density at the benzylic carbon is diminished, thus facilitating electrophilic reaction with cellular constituents. Reactivity and carcinogenicity also depend on the presence of alkyl substituents. When present at ring positions essential for metabolic activation, alkyl substituents render carcinogenic PAHs inactive. However, when present at some other sites, the reactivity of epoxides is increased dramatically. For example, chrysene has minimal carcinogenic activity, while 5-methylchrysene is a potent carcinogen. Although alkyl-PAH homologs are abundant in some coal-conversion materials, little information is available about the positions of the alkyl substituents in the many four- and six-ring PAHs. General relations between the degree of PAH alkylation in coal-conversion materials and their carcinogenicity have not been established.

In summary, all studies of coal-conversion products and by-products have been consistent with regard to the conclusion that four- to six-ring PAHs are largely responsible for dermal carcinogenic activity. Cellular measures of mutagenicity support this conclusion but not necessarily in a quantitative manner. Results of the cellular assays indicate association of mutagenicity with specific known carcinogens and closely related compounds.

5.4.2 Determination of Amino-PAHs as Principal Mutagens

Basic fractions isolated by liquid partitioning generally consitute less than 5% of coal-conversion materials by weight and contain, principally, azaarenes, aromatic amines, and their respective alkyl homologs (Epler, 1980; Guerin et al., 1981; Kimball and Munro, 1981; Stamoudis et al., 1983). N-PAC fractions isolated by alumina-column chromatography often account for 10–20% of the crude coal liquid by weight (Later et al., 1981a,b, 1982b). Together, azaarenes and alkyl-azaarenes usually account for more than 50 wt % of the basic fraction. Amino-PAHs and azaarenes are also present in N-PAC fractions obtained by alumina-column chromatography. Use of this technique avoids the incomplete recovery that sometimes occurs with liquid-partitioning procedures of hydrophobic compounds (because of their aromaticity or alkylation). Nitrogen-specific GC detectors have also been used to analyze amino-PAHs and other N-PACs in the presence of interfering amounts of hydrocarbons and other heteroatomic compounds. Sections 5.6.4.3 and 5.6.4.5 discuss amino-PAH and azaarene content, respectively, in various materials. Amino-PAHs typically constitute less than 1% of coal-conversion materials. For example, concentrations of individual aminonaphthalenes are less than 500 $\mu g/g$; those of higher molecular weight amino-PAHs are even lower, except in high-boiling-point (bp) distillates.

Thus, amino-PAHs are difficult to detect and identify in unfractionated materials even with capillary-column GC and GC-MS. Futhermore, the electron impact mass spectra of C_n-aromatic amines and isomeric C_{n+1}-azaarenes (e.g., aminonaphthalene and methylquinoline) are often indistinguishable. Without selective derivatization, many isomeric azaarenes and amino-PAHs cannot be distinguished by GC-MS from standards data unless chromatographic retention times are known or can be predicted (Wilson et al., 1980; Toste et al., 1981; Stamoudis et al., 1983). Thus, enrichment of aromatic amines by liquid partitioning or adsorption liquid chromatography has been essential to elucidate their toxicological role.

Bacterial mutagenicity of basic or N-PAC fractions often ranges from 10- to 50-fold greater than that of unfractionated coal-conversion materials (Epler et al., 1978a; Guerin et al., 1980, 1981; Kimball and Munro, 1981; Later et al., 1981b; Mahlum et al., 1983; Stamoudis et al., 1983). Basic and N-PAC fractions, which were also mutagenic in mammalian-cell systems (Hsie et al., 1980), have been shown to be dermal carcinogens in chronic mouse skin-painting and dermal initiation/promotion assays for tumorigenicity (Mahlum et al., 1984). The basic fraction obtained by liquid partitioning of solvent-refined coal II (SRC-II) heavy distillate (HD) was also carcinogenic in chronic skin-painting assays. However, because PAH content of the fraction was 20 wt %, activity could not be attributed specifically to basic components alone.

Because basic components were implicated in the mutagenic activity of coal-conversion materials, several independent studies were initiated to identify principal mutagens. Basic fractions from low-bp materials contained mono- and bicyclic amino-PAHs and 2°-N-PACs (Wilson and Pelroy, 1979; Guerin et al., 1980; Wilson et al., 1980, 1983a,c; Stamoudis et al., 1983). Although some bicyclic compounds are known mutagens and carcinogens, mutagenicity of pure standards indicated that bicyclic compounds in basic fractions (aminonaphthalenes and quinolines) of complex mixtures were not responsible for their high mutagenicity. Because members of both classes of aromatic bases might be responsible, it was essential to separate the amines and azaarenes in order to distinguish their relative toxicity.

Initial evidence for the role of amino-PAHs in the mutagenicity of coal-liquefaction materials was obtained by Pelroy and Gandolfi (1980), using a primary amine-specific activating enzyme in the *Salmonella* TA98 assay instead of the mixed-function aryl oxidases found in standard S9 rat liver preparations. These studies demonstrated that mutagenic responses to several thin-layer chromatography (TLC) subfractions of SRC-II HD basic fraction were similar to those using S9 and suggested that amino-PAHs were important mutagens. The presence of amino-PAHs in the active TLC fraction was confirmed by Wilson et al. (1980), who separated amino-PAHs from their azaarene isobars, using GC-MS.

These findings were further confirmed using high-resolution mass spectrometry (HRMS) to obtain elemental compositions on some 50 amino-PAHs that were selectively isolated from SRC-II HD and derivatized with pentafluoropropionic (PFP) anhydride to form the corresponding fluoramides (Later et al., 1982b). HRMS studies also showed the presence of aminocarbazoles and aminoazaarenes in mutagenically active subfractions. Accurate retention times were obtained for many amino-PAH isomers and their PFP derivatives, using capillary GC.

Relative contributions of amines and azaarenes to the mutagenicity of liquefaction (Pelroy and Stewart, 1981) and gasification (Haugen et al., 1981a) basic fractions were evaluated using nitrous acid. Nitrous acid reacts with amino-PAHs, forming the corresponding diazonium salts, which yield nonmutagenic phenols. Azaarenes do not react with nitrous acid. Addition of nitrous acid to basic fractions of coal-conversion materials decreased their mutagenicity by 50–90%. Although the effect of nitrous acid treatment confirms the role of amino-PAH compounds in mutagenicity, results should be interpreted cautiously for mixtures. Reaction conditions may lead to formation of nitro-PAH homologs (e.g., nitrobenzo[a]pyrene) and nitroso hydroxy-PAC homologs, some of which are potent mutagens in the Ames assay (Haugen et al., 1981a).

Alumina-column chromatography was used to obtain two basic subfractions, successively eluted with benzene and ethanol; the latter fraction was enriched in aromatic amines (Guerin et al., 1980; Buchanan et al., 1983). The aromatic amine fraction was further resolved by chromatography on Sephadex LH-20. Successive elution with isopropanol and acetone showed that most of the mutagenicity was in the acetone fraction, which contained less than 5% of basic fraction and primarily one- to four-ring alkylated aromatic amines. Subsequent analysis with Sephadex LH-20 under different conditions revealed that the mutagenicity was associated mostly with three-to five-ring aromatic amines.

Basic fractions obtained by liquid–liquid partitioning (Boparai et al., 1983; Stamoudis et al., 1983) of coal-gasification and -liquefaction materials were resolved by semipreparative cation-exchange HPLC according to base strength (Haugen et al., 1982a,b, 1986; Stamoudis et al., 1983). Mutagenicity assays of chromatographic fractions revealed that aromatic amines were responsible for mutagenicity; occasionally, a well-defined, mutagenic chromatographic fraction predominated. Further resolution of this fraction, according to carbon number, with reverse-phase HPLC, indicated that mutagenicity was associated with three-ring aminophenanthrenes (Stamoudis et al, 1983; Haugen et al., 1986). The presence of aminophenanthrenes and closely related amines correlated semiquantitatively with the mutagenicity of basic fractions.

Later et al. (1981a) developed and applied alumina-column chromatography to previously unfractionated coal-conversion materials. Fractions of increasing polarity were isolated by alumina-column chromatography: aliphatic hydrocarbons (AHs), neutral-PACs, N-PACs, and hydroxy-PACs. Mutagenic activity in *Salmonella* was detected almost exclusively in the N-PAC fraction (Later et al., 1981a,b, 1983a,b, 1984b; Wilson et al., 1983c). The N-PAC fraction was also mutagenic in mammalian-cell cultures (Mahlum et al., 1983) and carcinogenic in dermal initiation/promotion assays (Mahlum et al., 1984). The N-PAC fraction was further resolved by silica-column chromatography to obtain three subfractions of increasing polarity, enriched in secondary heterocyclic-PACs (e.g., carbazoles), amino-PAHs (e.g., aminoanthracene), and tertiary heterocyclic-PACs, respectively (Later et al., 1981a, 1982b). The amino-PAH fraction contained the most mutagenicity, with a smaller contribution from the 3°-N-PAC fraction (Wilson et al., 1982c; Mahlum et al., 1983). The two- and three-ring amino-PAHs appeared to be responsible for the mutagenicity of amino-PAH fractions based on (1) the predominance, in the amino-PAH fractions, of aminonaphthalenes, aminobiphenyls, and aminophenanthrenes; and (2) the relative mutagenic potencies of these compounds.

In mice, dermal carcinogenicity caused by basic (or N-PAC) fractions of coal-conversion materials may reflect the presence of azaarenes having more than three rings and their alkyl homologs. Dermal carcinogenicity of N-PAC fractions may be due to larger amino-PAHs (e.g., 2-aminoanthracene), although the best-characterized diaromatic amino-PAHs are not dermal carcinogens. Preparative separation of amino-PAHs and azaarenes to evaluate their importance in the dermal carcinogenesis of coal-conversion materials has not been achieved. Nitrous acid treatment of an N-PAC fraction of SRC-II distillate (bp > 850°F), prepared by alumina-column chromatography (Later et al., 1981a), diminished dermal carcinogenicity in the initiation/promotion assay (Mahlum et al., 1984). The N-PAC fraction was free of PAHs and contained predominantly carbazoles, azaarenes, and amino-PAHs; only the latter react with nitrous acid. Results suggest that amino-PAHs play a role in dermal carcinogenesis, assuming reaction products were separated from unreactive components.

Highly mutagenic, three- and four-ring amino-PAHs were isolated recently (Kosuge et al., 1982) from coal tar (presumably from the coking industry). Workers exposed to coal-carbonization products showed a higher than normal incidence of bladder cancer (Doll et al., 1965), as did those exposed to known amino-PAHs (Parkes and Evans, 1981; Garner et al., 1984).

5.5 COMPOSITIONAL CHARACTERISTICS OF COAL AND COAL-CONVERSION MATERIALS

5.5.1 Coal Chemistry

Coal is a sedimentary rock composed principally of macerals (phytogenic organic substances) and subordinately of minerals and trace elements. Water and gases are also contained within submicroscopic pores of the coal structure. Feed-coal characteristics are important in the design of coal-conversion processes. Properties such as reactivity, caloric value, mineral content, and caking and grinding characteristics must be considered when establishing process conditions or the suitability of various coals for use in a given conversion technology. Coal properties and characteristics are, among other things, a function of its chemical composition. In this section, both the macroscopic properties as well as the organic and inorganic constituencies of coal and products derived from coal are considered.

5.5.1.1 *Classification of Coals*

Numerous coal classification schemes have been devised because of the wide variety of coals available and the density of their end uses. Coals have been classified according to structure, rank, geological age, recognizable coal-forming plant materials, volatile matter and/or fixed carbon content, petrographic constituents, and so on. Baughman (1978) has reviewed many of the standard classification schemes as they specifically related to coal conversion.

Classification of coals, according to the American Society for Testing and Materials (ASTM, 1986a), specifies coal rank based on fixed carbon content and/or calorific value calculated on a mineral-matter-free basis. The four major coal ranks according to this classification system, in order of increasing heat value, are lignite < subbituminous < bituminous < anthracite. The coalification diagenesis pathway follows this same ranking sequence, with peat preceding lignite (Hayatsu et al., 1978). In general, increasing carbon and aromaticity (rank) and decreasing oxygen and hydrogen content correlate with increasing geological age and coal rank. There are also additional subgroups of these four coal ranks. For example, the three subgroups of anthracite are meta-anthracite, anthracite, and semi-anthracite. Secondary classification of higher-rank coals is on a dry basis according to fixed carbon content and volatile matter and, for lower-rank coals, according to heating value on a moist basis. This ASTM classification scheme and the parameters that determine coal rank are shown in Table 5.1. Ultimate analyses for carbon, hydrogen, oxygen, nitrogen, and sulfur of coals according to ranks as defined in Table 5.1 are given in Table 5.2.

Another method for classifying coals is based on their reflectance properties, which are directly determined by a coal's petrographic characteristics and features. Reflectance values can be related to caloric value or heat content (Baughman, 1978) and/or carbon content (Neavel, 1981) and are therefore useful in classifying and ranking coals with fixed carbon levels below those shown in Table 5.1 (~69%).

Lithological classification of coals specifies three major groups: banded, nonbanded, and impure (Baughman, 1978; ASTM 1986b). Banded coal contains visible bands of vitrain (shiny black vitreous material), which are single large fragments of ancient plants. Banded coals also contain layers of plant chips called fusain. In banded coal, fusain and vitrain are found in a matrix of attrital coal. Nonbanded coals display a uniform, lusterless, fine-grained texture and are composed of comminuted and compacted plant detritus with resinous components of spore and/or algal remains. Impure coal is heavily impregnated with mineral matter and contains 25–50 wt % ash content on a dry basis.

Table 5.1 Classification of Coals by Rank[a]

Class/Group	Fixed Carbon Limits (%), Dry, Mineral-Matter-Free Basis		Volatile Matter Limits (%), Dry, Mineral-Matter-Free Basis		Calorific Value Limits (Btu/lb), Moist,[b] Mineral-Matter-Free Basis	
	Equal or Greater than	Less than	Greater than	Equal or less than	Equal or Greater than	Less than
Anthracite						
Meta-anthracite	98	—	—	2	—	—
Anthracite	92	98	2	8	—	—
Semi-anthracite[c]	86	92	8	14	—	—
Bituminous						
Low-volatile bituminous coal	78	86	14	22	—	—
Medium-volatile bituminous coal	69	78	22	31	14,000[d]	—
High-volatile A bituminous coal	—	69	31	—	13,000[d]	14,000
High-volatile B bituminous coal	—	—	—	—	11,500	13,000
High-volatile C bituminous coal	—	—	—	—	10,500	11,500
Subbituminous						
Subbituminous A coal	—	—	—	—	10,500	11,500
Subbituminous B coal	—	—	—	—	9,500	10,500
Subbituminous C coal	—	—	—	—	8,300	9,500
Lignitic						
Lignite A	—	—	—	—	6,300	8,300
Lignite B	—	—	—	—	—	6,300

[a]This classification does not include a few coals, principally nonbanded varieties, which have unusual physical and chemical properties and which come within the limits of fixed carbon or calorific value of the high-volatile bituminous and subbituminous ranks. All these coals either contain less than 48% dry, mineral-matter-free fixed carbon or have more than 15,00 moist, mineral-matter-free British thermal units per pound.

[b]Moist refers to coal containing its natural inherent moisture but not including visible water on the surface of the coal.

[c]If agglomerating, classify in low-volatile group of the bituminous class.

[d]Coals having 69% or more fixed carbon on the dry, mineral-matter-free basis are classified according to fixed carbon, regardless of calorific value.

Source: Adapted from Baughman (1978).

Table 5.2 Ultimate Analyses of Coal Samples of Differing Ranks

Coal Type (Typical Sample)	Element (wt %), MAF[a]				
	Carbon	Hydrogen	Oxygen	Nitrogen	Sulfur
Meta-anthracite[b]	97.9	0.21	1.7	0.2	—
Anthracite[b]	95.9	0.89	1.8	0.3	1.8
Anthracite	92.8	2.7	2.9	1.0	0.6
Semi-anthracite	90.5	3.9	3.4	1.5	0.7
Low-volatile bituminous	90.8	4.6	3.3	0.7	0.6
Medium-volatile bituminous	89.1	5.0	3.6	1.7	0.6
High-volatile A bituminous	84.9	5.6	6.9	1.6	1.0
High-volatile B bituminous	81.9	5.1	10.5	1.9	0.6
High-volatile C bituminous	77.3	4.9	14.3	1.2	2.3
Subbituminous A	78.5	5.3	13.9	1.5	0.8
Subbituminous B	72.3	4.7	21.0	1.7	0.3
Subbituminous C	70.6	4.8	23.3	0.7	0.6
Lignite	70.6	4.7	23.4	0.7	0.6

[a] Moisture- and ash-free basis.

[b] Coals that yield three-dimensional (*hkl*) crystalline X-ray patterns of graphite to a significant extent.

Source: Neavel (1981).

The maceral classification system allows for differentiation among coals (Baughman, 1978) based on the source and composition of their organic constituents. Maceral properties help predict the suitability or behavior of coals in conversion processes. Three main classes of macerals are recognized: vitrinite, exinite (or liptinite), and inertinite (see Baughman, 1978; Neavel, 1981; Schweighardt, 1981). Maceral subclasses in the exinite/liptinite group include alginite, sporinite, resinite, and cutinite and are composed of organic materials derived from the implied plant origin. Exinites constitute the major source of extractable organics in coal and are the most chemically reactive and hydrogen rich of the maceral groups. Inertinite macerals, such as fusinite, are thought to be the remains of woody substances (lignin or cellulose) that may have been pyrolyzed prior to deposition and coalification and therefore are relatively depleted in hydrogen compared to other groups. Much more commonly though, the woody tissues and substances were accumulated under water, where microbial processes decomposed and converted them to humic substances of variable composition. These highly functionalized, aromatic, humic substances were altered by metamorphic processes (heat or pressure) into what are now classified as vitrinite macerals. Typical U.S. Coals are relatively vitrinite rich. The major maceral classes and subclasses, as well as their postulated source and characteristics, are shown in Fig. 5.5. Each of the various maceral types has characteristic reflectance ranges as shown in Table 5.3, which enable differentiation of the constituent macerals in coals and hence their classification.

Although the methods considered above for coal classification have been widely used, none is universally accepted and a need for new procedures for classifying coals has been expressed by some, particularly in conjunction with defining feed-coal properties for predicting responses of coal samples in conversion processes. Neavel (1981) has suggested that a new classification procedure should be multivariate and statistically treated with consideration of such factors as coal type (proportion of various macerals and minerals), coal rank (stage of metamorphic development), fundamental properties such as reflectance, and structure on both molecular and elemental levels.

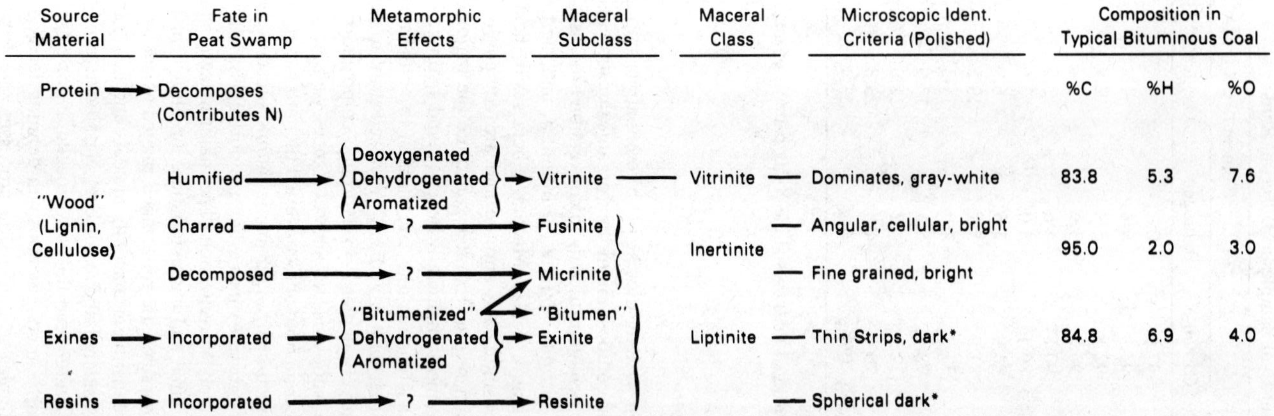

Figure 5.5 Principal features of the major classes of macerals (adapted from Neavel, 1981). (*In low-volume bituminous and anthracites, liptinite is indistinguishable from vitrinite.)

Table 5.3 Reflectance and Distinguishing Optical Characteristics of Coal Macerals

Maceral Suite	Maceral Group	Range of Maximum Reflectance % Under Oil	Other Distinguishing Characteristics[a]
Vitrinite			
	Anthrinoid	2.50–10.00	Opaque in transmitted light; grayish-white in reflected light
	Vitrinoid	0.40–2.40	Translucent in transmitted light; usually yellow, red, or brown. Gray in reflected light
	Xylinoid	0.10–0.39	Translucent in transmitted light; usually buff, whitish-yellow to yellowish-brown. Dark gray in reflected light
	Exinoid	0.05–1.50	Coalified spore, pollen, cuticular or endodermal materials. Translucent in transmitted light; whitish-yellow, yellow, golden yellow, or red. Black, dark gray to light gray in reflected light
Liptinite			
	Fusinoid	4.00–10.00	Characteristics essentially those of fusinite (Stopes–Heerlen system)
Inertinite			
	Semifusinoid	0.20–3.99	Characteristics essentially those of semifusinite (Stopes–Heerlen system)
	Micrinoid	0.20–8.00	Characteristics essentially those of micrinite (Stopes–Heerlen system)

[a] Descriptions for transmitted light refer to thin sections of standard thickness.

Source: Adapted from Baughman (1978).

5.5.1.2 *Inorganic Constituents in Coals*

The inorganic portion of coal can range from a few percent to more than 50% (by weight) and is composed of carbonates, sulfides, sulfates, and other mineral matter (King, 1978). Baughman (1978) reviewed six groups characteristic of mineral matter in U.S. coals, which are shown in Table 5.4. The analysis of mineral matter in coal is generally done by ashing (combusting) the coal and analyzing the chemical composition of ash in terms of the oxides of silicon, aluminum, iron, calcium, magnesium, titanium, sodium, sulfur, and potassium. The average ash analysis of U.S. coals for SiO, A1O, and FeO content, as determined by the U.S. Bureau of Mines, has been reported at 45.7, 26.0, and 18.1%, respectively. Generally, 90% of bituminous coal ash is composed of these three constituents. Table 5.5 shows the variation of coal ash composition as a function of coal rank.

Coal mineral matter also contains appreciable quantities of nearly all elements in the periodic table (Koppenaal and Manahan, 1976; King, 1978). Research, to date, has not been as extensive on trace elements as on the organic constituency of coal. As with mineral matter, the levels of inorganic trace elements in coal are determined in ashed coal and are

Table 5.4 Groups of Inorganic Minerals in U.S. Coals

Group	Mineral Species
Shale	Muscovite, montmorillonite, bravaisite, illite, hydromuscovite
Kaolin	Kaoliniate, levisite, methahalloysite
Sulfide	Pyrite, marcasite
Carbonate	Calcite, siderite, dolomite, ankerite
Chloride	Sylvite, halite
Accessory minerals	Quartz, gypsum magnetite, hematite, sphalerite, feldspar, garnet, hornblende, apatite, zircon, epidote, biotite, augite, prochlorite, chlorite, diaspore, lepidocrocite, barite, kyanite, staurolite, topax, tourmaline, pyrophyllite, penninite

Source: Adapted from Baughman (1978).

a function of coal type and rank. Typical elements found at ppm levels in coal and its ash include, but may not be limited to, arsenic, boron, beryllium, bromine, cadmium, cobalt, chromium, copper, gallium, germanium, mercury, manganese, molybdenum, nickel, phosphorous, lead, selenium, antimony, vanadium, zinc, and zirconium.

It is generally appreciated that mineral matter and trace elements are associated with coal impact conversion and utilization processes. Reactions involving minerals during combustion include oxidation of iron pyrite to iron oxide, loss of HO and CO from silicates and carbonates, respectively, and formation of sulfur oxides that may combine with calcium. The involvement of mineral matter in catalytic processes in coal liquefaction is well established (Bockrath and Schroder, 1981). The role of pyrite, for example, in enhancing catalytic processes is fairly well documented; however, the importance of clay and/or other mineral matter is not as well defined.

5.5.1.3 *Organic Constituents in Coal*

Several generic structures have been suggested for the organic constituents in coals. Organics are thought to be present as short-chain-length aliphatic or cyclic saturates with condensed aromatic structures consisting primarily of two to six (or more) rings. Figure 5.6 shows typical structural units assumed for organic coal constituents from four different ranks. Hydrocarbon structures are covalently cross-linked by oxygen or sulfur functionalities (Wiser, 1978; Herod et al., 1981) and methylene bridges. Oxygen content and aromaticity vary widely, depending on coal type and source, and can affect gasification and liquefaction behavior. Support for these general types of structure comes from nuclear magnetic resonance (NMR), mass spectral analyses of coal and coal-conversion materials (Bodzek et al., 1979; Deno et al., 1980; Herod et al., 1981).

An alternate view is that coal structure consists of condensed aromatic components but that tertiary coal structure is more dependent on hydrogen bonding and van der Waals forces than covalent linkages. This view is supported by results from fairly low-temperature solvation of coals, which yields liquids with components in the 200–800-amu range (Bodzek and Marzec, 1981). They maintain that since insufficient thermal energy was available dur-

Table 5.5 Variation in Coal Ash Composition with Rank

Rank	SiO_2 (%)	Al_2O_3 (%)	Fe_2O_3 (%)	TiO_2 (%)	CaO (%)	MgO (%)	Na_2O (%)	K_2O (%)	SO_3 (%)
Anthracite	48–68	25–44	2–10	1.0–2	0.2–4	0.2–1	—	—	0.1–1
Bituminous	7–68	4–39	2–44	0.5–4	0.7–36	0.1–4	0.2–3	0.2–4	0.1–32
Subbituminous	17–58	4–35	3–19	0.6–2	2.2–52	0.5–8	—	—	3.0–16
Lignite	6–40	4–26	1–34	0.0–0.8	12.4–52	2.8–14	0.2–28	0.1–1.3	8.3–32

Source: Adapted from Baughman (1978).

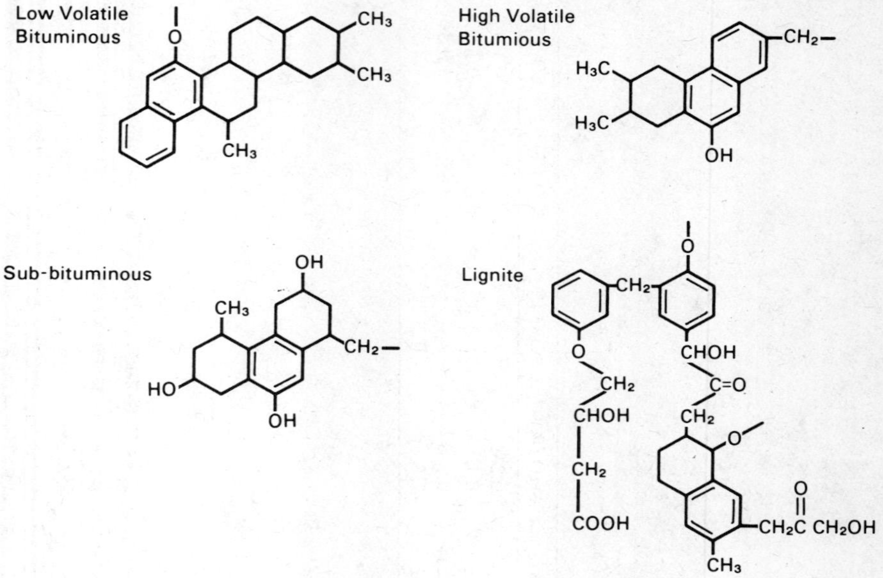

Figure 5.6 Typical organic structural units in various coal ranks (adapted from Wender, 1975).

ing reaction to allow covalent bond cleavage, such bonds are less important than previously thought in determining coal structure.

Several workers (White and Lee, 1980; Bartle et al., 1981; Hayatsu et al., 1981; Snape et al., 1981) have reported aromatic species trapped in coal. Results of mild solvation studies, in which organics were extracted from coal with benzene (Ouchi and Imuta, 1963a) or pH-adjusted aqueous solvents (Willey et al., 1981b) at room temperature, confirm this observation. This finding has been central to the development of low-severity liquefaction procedures.

Bartle et al. (1982) used electrochemical methods and postulated that the basic aromatic structure in coals primarily consists of two or three rings. This group and others reported on the presence of biological marker compounds in coal; these compounds exhibit long-term stability to diagenesis and possess skeletal features related to biosynthetic pathways. For example, some saturated isoprenoids may result from the diagenesis of the precusor phytol, which is widely associated with plant pigments (White et al., 1977). White and Lee (1980) reported on the possible geochemical origin of other coal-derived organic compounds, particularly S-PACS. They identified 80 PAH compounds or isomers by capillary-column GC-MS in benzene/hexane extracts of Kentucky coal. Some alkylated indanones identified were thought to originate from the intermolecular acetylation of phenyl alkanoic acids; these types of compounds have been identified in silicified conifers. Other PACs were identified in plant debris, pine resins and pigments, and from reactions of elemental sulfur or pyrite with hydrocarbons.

In light of these findings, uncertainty about the exact structure of coals persists because direct analysis of organics in coal is not yet well developed. Solid-state analytical techniques, including magic-angle NMR, alternate ionization mass spectroscopic methods such as field ionization (Bodzek and Marzec, 1981), and ^{252}Cf plasma (Lytle et al., 1982) desorption have been used. Physical isolation and direct analysis of macerals from the solid coal matrix have also been achieved by electron microscopy (Harris and Yust, 1981) and magic-angle NMR spectroscopy (Pugmire et al., 1981).

According to data obtained from field ionization mass spectrometry (FIMS) (Bodzek and Marzec, 1981), most organics in coal have molecular weights ranging from 70 to 800 amu. High-resolution field ionization mass spectrometry (HRFIMS) allows determination of heteroatomic species with elemental compositions such as C_xH_y, C_xH_yO, $C_xH_yO_2$, C_xH_yN, $C_xH_yN_2$, C_xH_yON, $C_xH_yON_2$, $C_xH_yO_2N$, C_xH_yS, C_xH_yOS, and C_xH_yNS. Such heteroatomic PACs are also found in coal liquids and are often important in determining certain biological effects. Analyses of coals by ^{252}Cf plasma desorption mass spectrometry (PDMS) yield data that were generally in agreement with those from FIMS regarding the molecular weight range for organics: Ions were detected in the mass range from about 200 to 860 amu. Bodzek and Marzec (1981) have provided a compendium of elemental compositions for organic species in coal-derived products. Data were derived from 1500 publications prior to 1963 by Anderson and Wu (1963) and were compared with elemental compositions obtained from organic extracts of coal at room temperature (i.e., temperatures that did not allow thermolytic bond cleavage). Results of Herod et al. (1981), who studied supercritical gas and solvent extracts of coal, are typical of those in the literature: PAHs were the most abundant compound type; C_{16} compounds with z numbers from 6 to 22 had the highest concentrations. These findings were related to coal structure by comparing NMR and mass spectral data.

5.5.2 Chemical Reactions in Coal Conversion

Conversion of organic coal constituents to liquid fuels is characterized by thermolytic and hydrolytic cleavage of carbon–carbon and carbon–heteroatom bonds, with release of CO_2, H_2S, NH_3, and H_2O. Hydrogen; available either from the reactor headspace or as donatable hydrogen from the recycle solvent, stabilizes or "caps" the resulting radical hydrocarbon units formed from thermolytic bond scission. These stabilized species, which are characterized by a lower molecular weight and fewer heteroatomic substituents than those of the original feed coal, make up the oil portion of the liquefaction product. Uncapped radicals can participate in retrograde reactions to reform preasphaltene or asphaltene components. These materials, together with unconverted components and mineral and ash matter, are chemically distinct from the product oils.

The sequence discussed above summarizes hydrogenation and concurrent denitrogenation, dehydration, and desulfurization during liquefaction. Typical solid-coal hydrogen : carbon ratios range from 0.6 to 1.0 : 1; those of coal-derived fuels range from 1.2 to 2.0 : 1. Because the reactions involved do not go to completion during liquefaction, there is some oxygen and residual amounts of organic-bound nitrogen and sulfur in coal liquids. The extent to which the oils are formed is roughly proportional to the quality of the coal liquids obtained. Thus, higher-quality materials obtained from hydrotreating distillates have lower heteroatom content and higher hydrogen : carbon ratios than do feedstocks.

Liquefaction reactions can be categorized as solvent extraction, noncatalytic (mainly thermolytic; e.g., SRC-I and SRC-II), and catalytic conversion (H-Coal). Two-stage processes, such as integrated two-stage liquefaction (ITSL), rely primarily on extraction/thermal conversion in the first stage and catalytic hydrogenative conversion in the second stage. A model for understanding coal extraction by solvents, proposed by Marzec et al. (1979), has macromolecular coal structures (i.e., extractable substances) linked by electron donor–acceptor bonds. Extraction is viewed as a substitution reaction, where extractable substances are replaced by solvent molecules in the electron donor–acceptor bonds.

Of particular interest here are reactions that result in biologically active compounds.

However, only a relatively small number of chemical reactions have been identified in this regard. Reactions of greatest biological significance involve formation of PAHs and N-PACs (particularly amino-PAHs) and those that specify the degree of saturation, condensation, and/or alkylation of the neutral PAHs.

Reaction sequences giving rise to PAHs and heterocyclic PACs during gasification have many pathways in common with those in liquefaction. Thermolytic bond cleavages typify gasification reactions that are sustained in an oxidizing and thermally severe environment rather than in the reducing environment encountered in liquefaction. Gasification reactions that form toxicologically significant PACs were discussed by Fillo et al. (1983). Haggin (1982) and others proposed the coal model depicted in Fig. 5.7. Primary fragments in the model are shown in Fig. 5.8, which depicts the initial breakup of the hypothetical coal model under pyrolytic conditions. Incomplete pyrolytic destruction of aromatic fragments from coal may yield many PACs found in condensates from gasification. However, relatively high concentrations of stable PAHs, such as phenanthrene and pyrene, in gasification tars and oils provide evidence that reactions involving condensation of radical species also contribute to the formation and composition of PAHs in these materials.

5.5.2.1 *Formation of PACs*

Formation of monocyclic and polycyclic aromatic compounds results from condensation of organic radical species, such as ethylene fragments, at high temperatures (Badger and Buttery, 1958; Badger and Kimber, 1958a, b; Badger et al., 1958; Crittenden and Long, 1976).

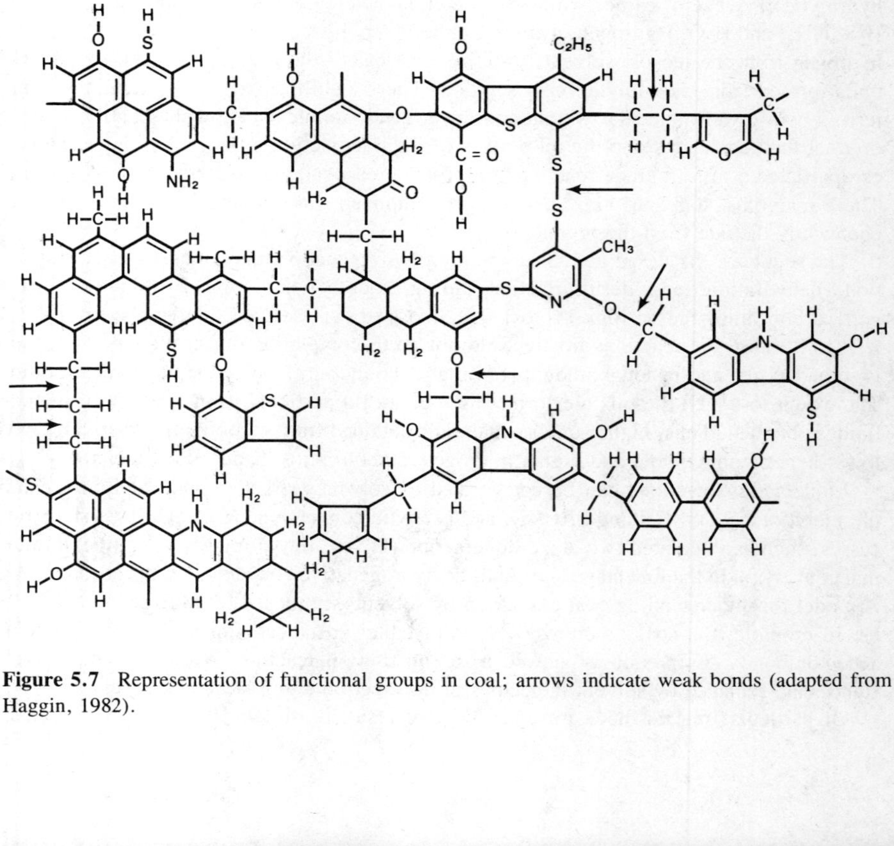

Figure 5.7 Representation of functional groups in coal; arrows indicate weak bonds (adapted from Haggin, 1982).

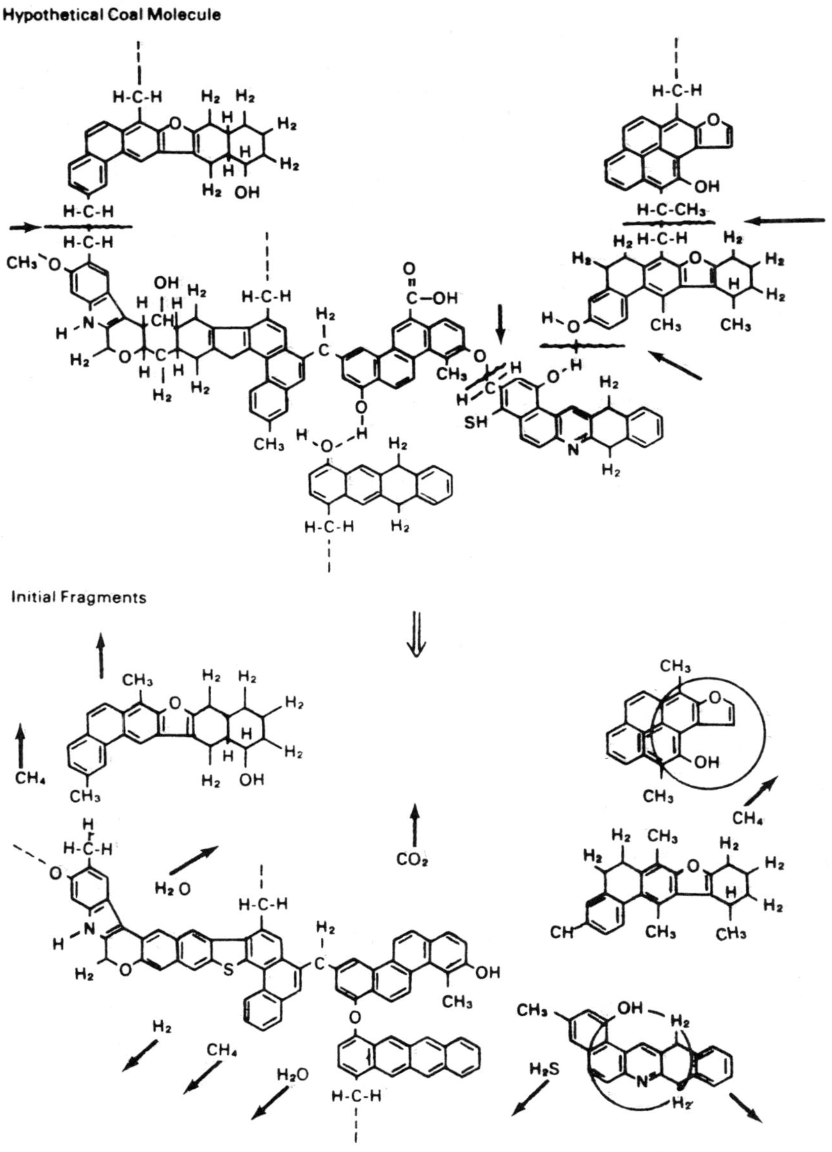

Figure 5.8 Primary reactions in thermal conversion of coal (adapted from Haggin, 1982).

Although this mechanism is better understood for combustion or pyrolysis than for coal conversion, the processes have several common features. Although the amounts of PAHs formed during efficient combustion of well-defined fuels (e.g., methane, butane, acetylene) may be low, a qualitatively similar array of PAHs is produced during combustion of complex fuels such as tobacco, wood, petroleum, and coal. [See M. L. Lee et al. (1980, 1981) and references contained therein.]

Formation of BaP by combustion of acetylene was described by Badger et al. (1960) (see Fig. 5.9). Studies of the pyrolysis of simple hydrocarbons led to an understanding of the mechanism and governing principles of PAH formation. Carbon–hydrogen and carbon–carbon fissions of aliphatic hydrocarbons (AHs) and simple aromatic hydrocarbons, which occur at high temperatures, apparently form primary radicals that recombine to yield less-hydrogenated, more-condensed products. Analytical characterization of lignite and subbituminous coal extracts shows that most aromatic carbon is contained in systems with three or less aromatic rings; roughly half is acyclic.

Liquefaction of bituminous coal in the low-severity short-contact-time (SCT) reactor of the Lummus ITSL process yields compounds with three or more aromatic rings in the product oils. Formation of more extensive ring systems in coal liquids is assumed to proceed by multiple condensation and cyclization of PAH nuclei with available acyclic C_3–C_5 compounds. Furthermore, alkyl- and heteroatomic-bridged aromatic ring systems are present in coal and may condense during liquefaction to form larger PAC compounds.

High levels of alkylated parent PACs are a distinguishing characteristic of coal-conversion materials as compared with other PAC-bearing matrices. Coal liquids produced by different processes vary widely in their degree of alkylation. One measure of the degree of alkylation for coal liquids is the maximum number of methylene units associated with a given parent PAC. Alkylated PACs with up to 18 or more methylene units have been reported in Synthoil materials (J. L. Schultz et al., 1979). Certain nonhydrotreated SRC-II heavy-end materials also contained alkylated PACs with alkyl homologs as high as C_{12}–C_{16}, as detected by low-voltage probe-inlet mass spectrometry (Wilson et al., 1982a). Griest et al. (1979) used GC and CG-MS to identify alkylated homologous series with up to six carbon units of parent PAHs including biphenyl, naphthalene, fluorene, phenanthrene, fluoranthene, and pyrene. Later et al. (1984b) reported alkylated parent PAHs up to C_{12} in EDS process materials using GC, reconstructed GC-MS ion chromatograms of homologous series, and low-voltage mass spectrometry (LVMS). Several ions with mass assignments

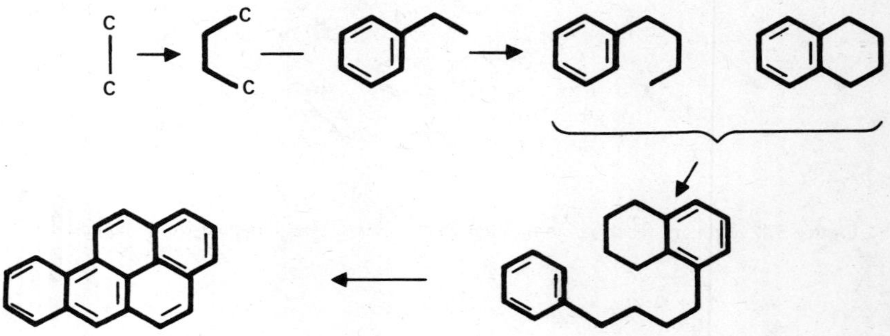

Figure 5.9 Proposed stepwise synthesis of benzo[a]pyrene during pyrolysis (adapted from Badger et al., 1960).

made by these techniques may arise partially or completely from hydroaromatic PAHs. For example, mass 206 could either be tetrahydropyrene or C_2-phenanthrene or some other equivalent isomer. It is most likely that a complex combination of hydroaromaticity and alkylation exists in coal-liquefaction materials. A second measure of alkylation is the relative concentration of the parent to the sum of its alkyl homologs. For example, EDS materials (Later et al., 1984b) show a high degree of alkylation (with some C_1 species in higher concentrations than their nonalkylated parent structures); Lummus ITSL materials (Wilson et al. 1983a) showed little alkylation. C_1 species in the latter materials are typically less than 30% of their nonalkylated parent structures. The extent to which PACs are alkylated appears to influence toxicological effects (Griest et al., 1979; Mahlum et al., 1984). As discussed in Section 5.4, certain methyl, dimethyl, and trimethyl homologs of three- to six-ring PAHs are more carcinogenic than their parent structures.

Sturm et al. (1981) discussed aromatic ring opening and breakage of alkyl linkages in the coal matrix as two possible mechanisms for the formation of alkylated PACs during coal conversion. Probably the best illustration of the second mechanism is the increase in alkylation associated with increased residence time in the liquefaction reactor or increased liquefaction severity. The role of $(CH_2)_n$ linkages in the coal matrix as a source of PAC alkylation may be estimated by comparing liquids produced from coals of different types or ranks that have differing degrees of alkyl linkage. For example, in the EDS process, coal liquids produced from lignite or subbituminous coals were more aliphatic than those produced using bituminous feed coals (Later et al., 1984b).

Because aliphatic–aromatic carbon bonds are weaker than aromatic–aromatic bonds, the degree of PAH alkylation depends on the temperature during formation: PAHs formed at higher temperatures tend to be less alkylated than those formed at lower temperatures. The optimum temperature for formation of PAHs is 650–750°C. At temperatures below this range, tar formed from non-oxidative pyrolysis of coal consists mainly of paraffins, cycloparaffins, olefins, and phenols. At temperatures above this range, PAHs begin to undergo thermal degradation. Because the aromatic–alkyl carbon bonds are more labile than aromatic–aromatic bonds, PAHs present in tars and soots formed at higher temperatures (e.g., coke oven tar, 800–1400°C; carbon black furnace, 1900°C) have little or no alkyl substitution, while tars formed during coal gasification at 600–800°C have an abundance of alkylated PAHs. The relative abundance of the PAHs formed from combustion of complex materials decreases with increasing molecular weights in a manner similar to that described above, regardless of carbon source.

Compared to the C_{10}–C_{16} PAHs, relatively few of the possible structural C_{16}–C_{24} PAH isomers are formed in significant quantities. The occurrence of various PAH species is qualitatively similar regardless of carbon source.

5.5.2.2 *Formation of Heteroatomic PACs*

Many mechanisms governing formation of PAHs also apply to the formation of nitrogen-, oxygen-, and sulfur-containing heterocyclic compounds. In feed coals, oxygen concentration is higher than that of sulfur. Within the coal matrix, oxygen and sulfur are thought to be important in forming covalent linkages between aromatic moieties. While these are often present as heterocyclic components (see Fig. 5.7), they may also be found as ether, thio, or disulfide linkages between cyclic or aromatic moieties. Hence, disruption of the coal structure via thermolytic or hydrogenative bond breakage can result in the formation of not only endocyclic heterocyclic PACs but also exocyclic hydroxy- and thiol-functional compounds. Nitrogen is thought to be present almost exclusively as heterocyclic functionalities.

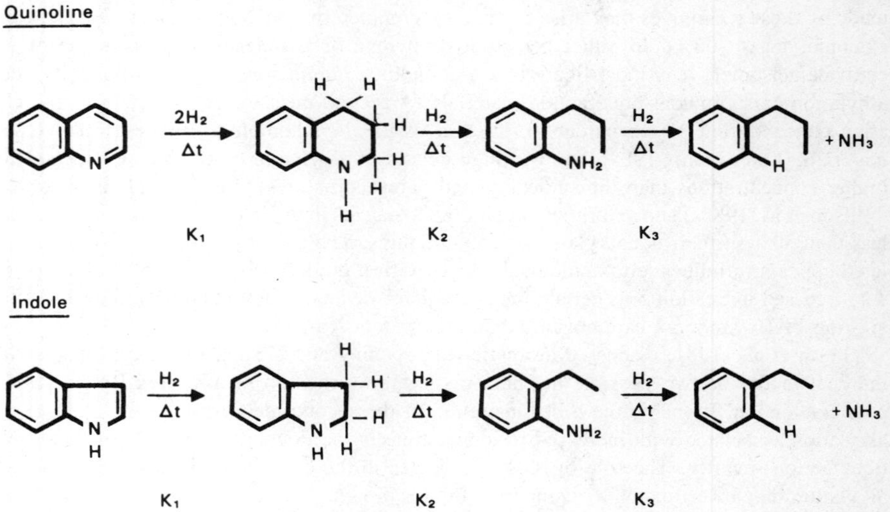

Figure 5.10 Denitrogenation of quiniline and indole (adapted from White et al., 1978).

Thus, formation of amino-PAHs requires carbon–nitrogen bond scission, which increases with increasing temperatures. Therefore, the amount of amino-PAHs in single-stage coal-conversion products often increases with increasing liquefaction severity.

Hydroliquefaction of coals carried out at sufficient severity, even in a noncatalytic, single-stage, thermal reactor, may be accompanied by substantial deoxygenation, desulfurization, and denitrogenation, as demonstrated by the release of carbon oxides, hydrogen sulfide, and ammonia gases from such reactions. Use of appropriate catalysts increases the rates of these reactions. Catalysts designed specifically for this purpose are generally used in the second stage of two-stage processes to remove heteroatoms, particularly nitrogen and sulfur, remaining after first-stage thermal processes.

Denitrogenation reactions are believed to proceed as shown in Fig. 5.10. Note that unless the k_3 rate constants are considerably greater than the k_2 rate constants, amino-PAH by-products may be present in the resultant coal liquid. Schmitter et al. (1984) investigated the fate of basic nitrogen compounds during catalytic hydrotreatment. They found that β-methylated azaarenes were particularly stable, that higher benzologs and unsubstituted parent compounds were the most susceptible to denitrogenation, and that alkyl anilines, the last intermediate before denitrogenation, were the most abundant bases in hydrotreated samples. The nonbasic or pyrrolic functional aza compounds are more resistant to denitrogenation during catalytic hydrogenation than the basic aza compounds (Aczel et al., 1977; Dorbon et al.,1984; Later and Wilson, 1985).

5.6 COMPOUND CLASSES IN COAL-CONVERSION MATERIALS

The remainder of this section is a brief discussion of the classes of compounds found in coal-conversion materials. Included are the concentration ranges in which such compounds are generally found, as well as some discussion of their toxicity in laboratory bioassays.

5.6.1 Aliphatic Hydrocarbons

The aliphatic hydrocabons (AHs) generally occur in coal liquids at concentrations between 5 and 60%. In solid products, such as SRC-I, AHs may constitute less than 1% by weight. Concentrations of AHs vary with distillation temperatures and are generally higher in lower-boiling distillates; relative concentration of AHs decreases monotonically with distillation temperature. Heavy-end (>800°F) materials seldom contain more than 5% AHs unless they have undergone catalytic hydrogenation. The AHs produced during noncatalytic, direct liquefaction are generally straight-chain (normal) with minimal branching or unsaturation. Conversely, AHs from two-stage liquefaction (TSL) products, with catalyst in one or both stages, show a high degree of cyclization; saturated cyclic compounds are components of such products. Although certain AHs are known promoters of carcinogenesis, purified AH fractions from coal liquids seldom show activity in mutational or tumorigenesis assays (Stamoudis et al., 1981; Wilson et al., 1983a, 1986; Mahlum et al., 1984).

5.6.2 PAHs and Their Subclasses

PAH concentrations in coal-conversion materials commonly range from 20 to 70% by weight (Koppenaal and Manahan, 1976; Boduszynski et al., 1983b, 1984). Donath (1908) recognized the existence of anthracene and chrysene derivatives in coal extracts in 1908. Pictet and Ramseyer (1911) isolated some partially hydrogenated alkyl benzenes and hexahydrofluorenes from benzene coal extracts. Ouchi and Imuta (1963a, b) isolated several individual PAHs from benzene extracts of a bituminous coal and provided absolute identification, using UV and infrared (IR) spectroscopy, of naphthalene, phenanthrene, chrysene, picene, fluorene, anthracene, pyrene, benzofluorenes, fluoranthene, benz[a]anthracene, a few oxygen heterocyclic PACs, and several alkylated homologs of these compounds. Recent reviews of the analytical chemistry of coal-derived PAHs include those by White (1983).

The importance of PAHs as carcinogens was apparent from the work by Kennaway, Cook, and co-workers (Cook et al., 1933; Kennaway, 1955). Tars produced during coking and coal gasification were carcinogenic in laboratory animals. Occupational exposure to these materials was associated with an increased incidence of dermal carcinogenesis. Isolation, synthesis, and bioassay of benzanthracenes, dibenzanthracenes, and other PAHs required years of work.

Representative structures of PAH classes found in coal liquids and gasifier condensates are shown in Fig. 5.11. Several structural properties allow parent PAHs with different molecular features to be described uniquely. These descriptors are used when referring to PAH structural features to avoid the ambiguities in the common usage of certain terms. Structures such as cyclopenta[cd]pyrene and fluoranthene, for example, which incorporate five-membered ring systems, are both commonly referred to as having cyclopenteno functionality (Vassilaros et al., 1982a). Condensed ring systems can be classified into four types: catacondensed (every aromatic ring in the structure has no more than two nonadjacent faces in common with other rings; e.g., phenanthrene), pericondensed (every aromatic ring in the structure has two or more adjacent faces in common with other rings; e.g., pyrene), peri/catacondensed (mixture of single and multiple faces in common, with one or more carbon atoms common to three rings; e.g., BaP), or no condensation (no faces in common; connected by carbon–carbon bonds or alkyl linkage; e.g., biphenyl). The base or secondary functional group for neutral PAHs describes the distinctive features of the ring system itself, particularly with respect to five-membered rings. Cyclopenteno, for example, refers strictly to the five-membered ring structure shown for cyclopenta[cd]pyrene: one having a double bond between the two additional carbons that form the five-membered ring, so that there

Name	Structure	Condensation	Base Component or Secondary Functional Component	Z Number
Benzene		None	Benzeno	-6
Indan		Catacondensed	Indeno	-10
Naphthalene	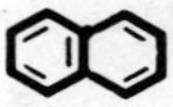	Catacondensed	Naphtho	-12
Biphenyl		Carbon-Carbon Bridged	Biphenyl	-14
Fluorene		Catacondensed	Indeno	-14
4,5-Methanophenanthrene	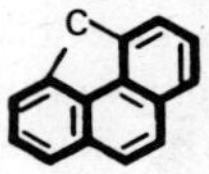	Pericondensed	Methylene Bridged	-20
Fluoranthene	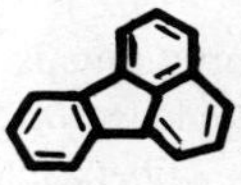	Peri/catacondensed	Indeno	-22
Benzo(ghi)fluoranthene	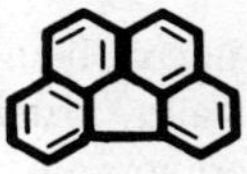	Pericondensed	Indeno	-26
Cyclopenta(cd)pyrene	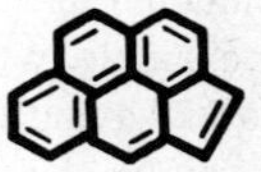	Pericondensed	Cyclopenteno	-26
Benzo(a)pyrene	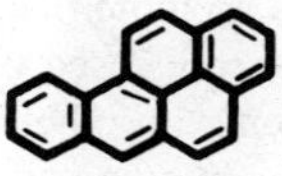	Peri/catacondensed	Pyreno	-28
Anthanthrene	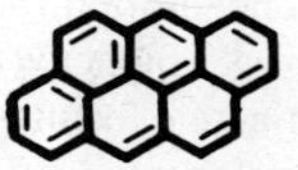	Pericondensed	Anthreno	-32

Figure 5.11 Structures and designations for the various polycyclic aromatic hydrocarbons (PAHs) in coal-conversion materials.

are no saturated carbon atoms. Five-membered systems having saturated carbon atoms, such as 4,5-methanophenanthrene, are considered alkyl bridged; their name is derived from the two positions on the ring system to which the alkyl bridge connects and the number of carbon atoms forming the bridge (e.g. one carbon, methano; two carbons, ethano; etc.). Interior five-membered ring systems are referred to as indeno structures; whether they have a saturated carbon depends on the type of condensation (e.g., compare fluorene and fluoranthene).

Tumorigenic activity is exhibited primarily by PAHs having four or five aromatic rings. Some alkyl homologs of certain catacondensed, four- and five-ring species are also active. (See Section 5.1 for further discussion of structure–activity relations of neutral PAHs.) Alkylated PAHs in coal-derived liquids have been investigated by Karr and Chang (1958), Chang and Karr (1961, 1962), Goeckner and Griest (1979), Griest et al. (1979), Tomkins et al. (1981, 1982), and White (1982). Highly alkylated PAH fractions isolated from carcinogenic, coal-derived materials generally exhibit less activity than fractions having PAH structures with little or no alkylation (Later et al., 1984b; Mahlum et al., 1984).

PAH fractions of coal liquids from several processes have been characterized according to relative concentrations of compounds with common parent ring structures, degree of alkylation, and hydroaromaticity. The effect of alkylation in decreasing the overall biological activity of a complex mixture is well documented (Mahlum et al., 1984). PAH fractions from SRC-II distillates were separated by HPLC to yield essentially nonalkylated, monoalkylated, and dialkylated subfractions of compounds such as benzopyrenes, perylenes, and benzanthracenes. The latter [together with some higher-molecular-weight (C_1 or C_2) species] were distinctly isolated from highly alkylated compounds having three or more methylene units, which constituted about 70 wt % of the PAH fraction. Analogous results were obtained in studies with EDS materials (Later et al., 1984b).

The initiation/promotion skin tumorigenicity and microbial mutagenicity assays of fractions showed that the highest specific activity resided in fractions containing five-ring compounds and C_1 five-ring compounds (Lucke et al., 1985). Thus, the majority of the biological activity resided in about 10% of the material by weight; some 70 wt % of the highly alkylated original material contributed little to the initiation activity of the fractionated complex mixture (Mahlum et al., 1984).

5.6.3 Hydroaromatics

Partially saturated PAHs or hydroaromatics are components of feed coals and coal-conversion product fuels (Ostermayer and Fittig, 1872; Haworth, 1932; Bodzek and Marzec, 1981), including materials from single-stage processes that have not undergone catalytic hydrotreatment (Weimer et al., 1980). Incorporation of gas-phase hydrogen into organic structures in single-stage reactors contributes to the formation of hydroaromatic compounds. Fully saturated ring structures are referred to as naphthenics.

Hydroaromatics are generally major components of coal liquids from two-stage processes incorporating catalytic hydrogenation or hydrotreatment. Catalytic hydrogenation, a refining operation common to the petroleum industry, has been evaluated for both in-process and postproduction upgrading of coal-based liquids. Because of the aromatic character of coal liquids, a consequence of hydrotreatment is the production of hydroaromatics.

Incorporation of additional hydrogen gives rise to conformational changes in many essentially planar PAHs (Wozniak and Hites, 1983). The large number of congeners possible for a single PAH compound or isomer greatly increases the chemical complexity of materials containing hydroaromatics. Conformational changes of hydroaromatic compounds are important in the separation of hydroaromatics from aromatics and may also affect the biological activity of hydroaromatics (Later et al., 1986).

Compared with the PAHs, relatively less work has been done on the biological activity of hydroaromatic compounds. The cytotoxicity of certain hydroaromatics is greater than that of their aromatic analogs (Griest et al., 1983; Hites, 1983). Mahlum et al. (1983), however, showed that mutagenic and carcinogenic activities of coal liquids are decreased by postproduction catalytic hydrotreatment. As pure compounds, certain hydroaromatics were carcinogenic in mouse skin-painting assays (Lijinsky et al., 1965a) but, in most cases, were less active than their fully aromatic analogs (Lijinsky et al., 1965b). Later et al. (1986) showed that mutagenic and tumorigenic activities of hydrogenated BaP mixtures were decreased compared to those of nonhydrogenated BaP.

5.6.4 Heteroatomic PACs

Coal liquids contain heteroatomic PACs that are formed from incorporation of either one nitrogen, one sulfur, or one oxygen atom into the aromatic ring system (Wilson et al., 1983c). PACs with more than one heteroatom have also been found in coal liquids and are considered in this section (Anderson and Wu, 1963; Paudler and Chaeplen, 1979; Burchill et al., 1982a, 1983b; Later et al., 1982b).

5.6.4.1 *Sulfur-Containing PACs*

S-PACs are found in relatively low concentrations in direct liquefaction products and gasifier condensates (Willey et al., 1981a; Burchill et al., 1982a,b). The polycyclic aromatic sulfur heterocycles (PASHs), which include thiophene and its benzologs, are generally the most prevalent S-PAC species in most coal-derived materials (Fig. 5.12) (Willey et al., 1981a). PASHs are also important by-products of incomplete combustion of hydrocarbon fuels containing sulfur (Lee and Hites, 1976). Heterocyclic sulfur may also exist in six-membered rings with oxygen, nitrogen, or other sulfur atoms in compounds such as phenothiazoles or thianthrenes. Evidence for the latter compounds in coal-conversion products has come mainly from HRMS analyses of coal tars (Pichler and Herlan, 1973) and from other data (McKay et al., 1976; Hunt and Shabanowitz, 1982) showing the presence of compounds whose elemental composition can be rationalized to such structures (Wilson et al., 1983c).

Several three-, four-, and five-ring PASHs have recently been synthesized (Iwao et al., 1980; Karcher et al., 1981; Pratap et al., 1981; Thompson et al., 1981; Tominaga et al., 1981a, b) and evaluated for mutagenicity and carcinogenicity (Lee and Hites, 1976; Karcher et al., 1981; Pelroy et al., 1983). Of the three- and four-ring PASHs synthesized, several showed mutagenic activity in the Ames assay using TA98 or TA100 tests strains. PASHs studied by Karcher et al. (1981) (Fig. 5.12) included one that was carcinogenic. While PASHs, as a class, do not appear to be particularly mutagenic, these compounds may bioaccumulate in aquatic vertebrates to a greater extent than neutral PAHs (Vassilaros et al., 1982b).

There is some analytical evidence for the presence of sulfur in the form of aromatic thiols or thiol ethers in SRC liquids. Because disulfide bridges exist in coal macromolecular structure, the presence of thiols or thiol ethers in coal liquids is not unexpected. However, their isolation from coal liquids has not yet been reported. Aromatic thiols have not been investigated, as a class, for their mutagenic activity.

Compound	Structure	Occurrence	Biological Activity
Phenanthro (4, 5-bcd) thiophene	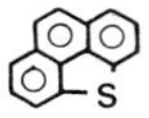	Coal Tars Carbon Black	Weak Mutagen
Benzo(b)naphtho(2, 1-d) thiophene	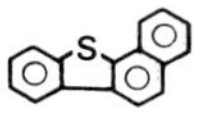	Coal Tar Derivatives	Weak Mutagen
Benzo(b)naphtho(2, 3-d) thiophene	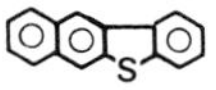	Coking Plant Effluents	Weak Mutagen
Benzo(b)naphtho(1, 2-d)thiophene	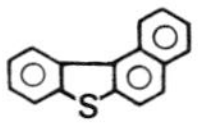	Coal Tar Derivatives	
Benzo(2, 3)phenanthro(4, 5-bcd) thiophene	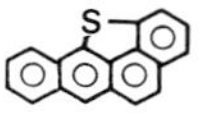	Cooking Plant Effluents	Potent Mutagen
Chryseno(4, 5-bcd)thiophene	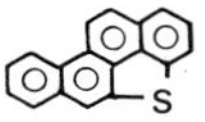	Coal Tar Derivatives	Not Determined
Benzo(b)phenanthro(3, 4-d)thiophene	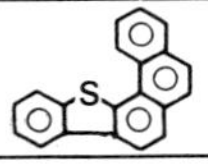		Potent Carcinogen

Figure 5.12 Sulfur heterocycles from coal-conversion processes (adapted from Karcher et al., 1981).

5.6.4.2 *Nitrogen-Containing PACs*

Relative to petroleum products, coal-conversion materials are distinguished by their higher concentrations of N-PACs. There are four subclasses of heteroatomic aromatic nitrogen compounds commonly found in coal-derived materials (Fig. 5.13). Azaarenes have moderate base strength, while aromatic amines are weaker bases; indoles and carbazoles are extremely weak acids; and cyano-functional compounds have essentially neutral characteristics. Acid–base strengths and polarities (the latter defined by interaction with adsorbents) allow separation of these subclasses of N-PACs from each other and from other PACs. Difunctional N-PACs with two nitrogen heteroatoms also occur in coal-derived materials but are typically found in much lower concentration than their mononitrogen counterparts.

Nitrogen-functional compounds range in concentration from less than 1% to greater than 25% of coal-conversion materials. The most abundant azaarenes and aromatic amines in lower-boiling coal-conversion materials are the monoaromatic pyridines and anilines, respectively, and their alkyl homologs. Because of volatility, losses during sample preparation (e.g., removal of solvents) contribute to uncertain quantitation of anilines and pyridines, and the amounts present depend on process conditions, particularly the temperature at which the materials were distilled.

5.6.4.3 *Amino-PAHs*

Karr and Chang (1958) first identified specific amino-PAHs in coal-derived materials using modern analytical instrumentation. N-benzylaniline, phenylaniline, and aminonaphthalene were identified spectrophotometrically in coal-tar base fractions. Anilines and alkylanilines were reported in gas oils from coal-liquefaction materials by White et al. (1978), using GC-

Structure	Number of Isomers	Name	Nitrogen Functionality
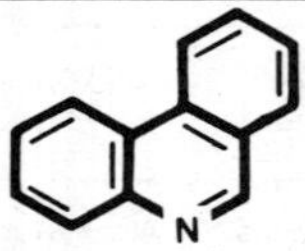	8	Phenanthridine or 9-Azaphenanthrene	Pyridine -N=
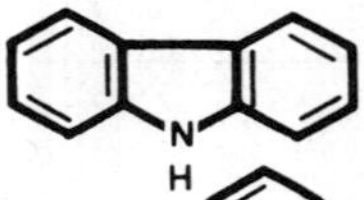	10	Carbazole or 9H-Azafluorene	Pyrrole -NH-
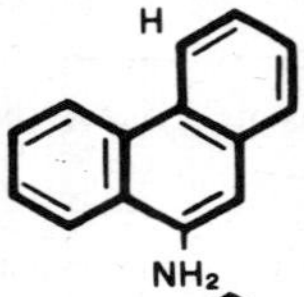	8	9-Aminophenanthrene or 9-Phenanthrylamine or 9-Phenanthrenamine	Amino $-NH_2$
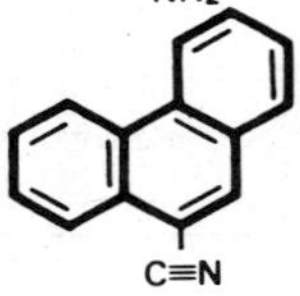	8	9-Cyanophenanthrene 9-Phenanthrylcyanide or 9-Phenanthrenecarbonitrile	Cyano -C≡N

Figure 5.13 Possible classes of nitrogen-containing polycyclic aromatic compounds (N-PACs).

MS. Other authors, prior to 1979 (Schiller and Mathiason, 1977; Paudler and Chaeplen, 1979), reported anilines, aminoindans, and aminonaphthalenes in various direct coal-liquefaction materials. Amino-PAHs and their alkyl homologs are relatively minor components of coal-conversion materials, generally less than 1% by weight, with isometric amino-PAHs having up to four aromatic rings commonly detected. The work of Tomkins et al. (1982) with HPLC and glow discharge GC showed that amines were present in shale and coal oils and two petroleum crudes. Using his HPLC method, he reported levels of 2-aminonaphthalene from 0.4 ppm in Wilmington crude petroleum to 82 ppm in a coal-based oil.

Wilson et al. (1980, 1982c) and Ho et al. (1981a, c) reported amino-PAHs with three and four rings in SRC-I, SRC-II, and H-Coal materials. Concentrations of amino-PAHs were correlated directly with mutagenic activity of TLC subfractions of an SRC-II HD basic fraction (Wilson et al., 1980). Purified amino-PAH fractions from several coal liquids were shown to be the most mutagenically active components (Later et al., 1981b, 1982b; Wilson et al., 1982c).

Experimentally, many two- and three-ring aromatic amines cause cancer in the bladder and other tissues of several species. Information is available on structure–activity relations for two- to five-ring amino-PAHs (Ho et al., 1981a; Later et al., 1984a). Later (1985a) compiled data (shown in Fig. 5.14) that relate mutagenic and carcinogenic activities of amino-PAHs to their PAH and N-PAC analogs as a function of aromatic ring number. The mutagenicity and carcinogenicity of the biologically active PAHs and N-PACs generally increase with increasing number of aromatic rings. However, the position of the amino group influences mutagenic and carcinogenic activities. Later et al. (1984a) synthesized two-, three-, and four-ring amino-PAHs and evaluated mutagenic activity in several tester strains with metabolic activation. Specific activities for several parent ring structures varied as a function of the amine substituent position. An example of this isomer specificity is

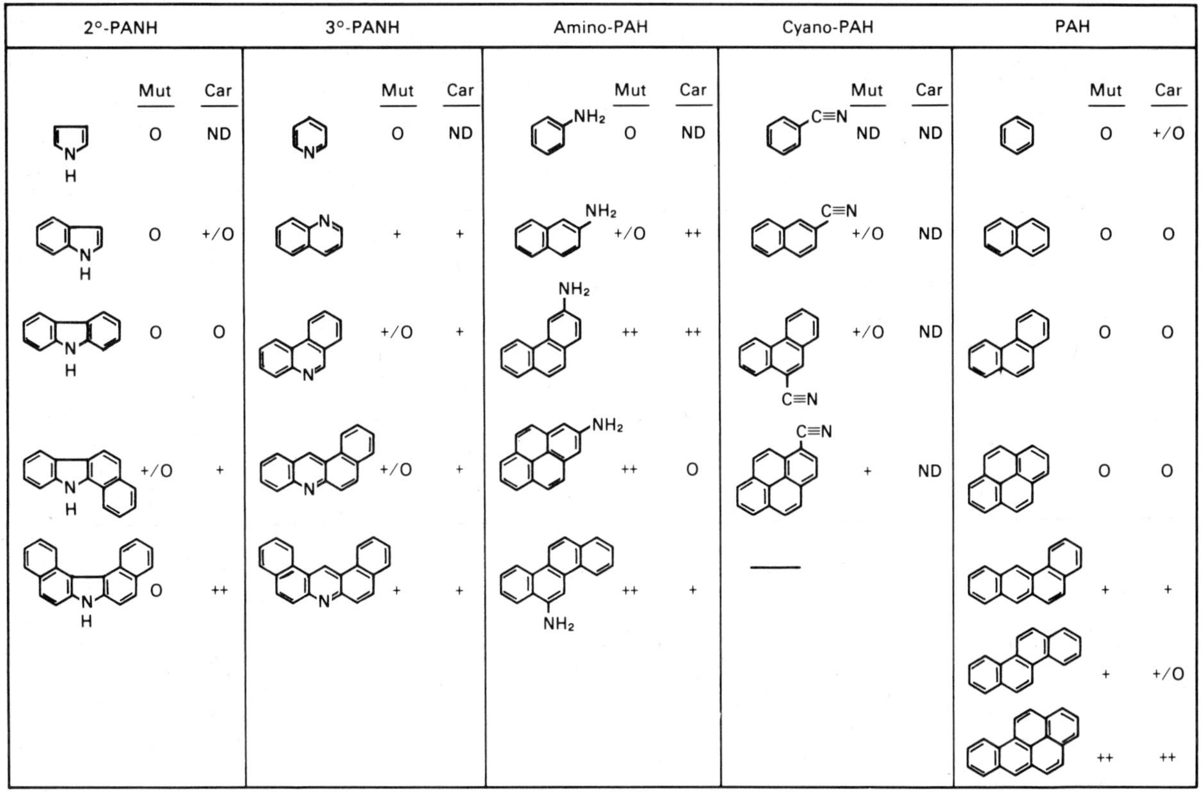

Figure 5.14 Mutagenic and carcinogenic activity in selected polycyclic aromatic hydrocarbons (PAH) (from Later, 1985a). Key to mutagenic (mut) or carcinogenic (car) potency: 0 = not active; +/0 = weakly active; + = moderately active; ++ = strongly active; ND = not determined. 2° PANHs = secondary polycyclic aromatic nitrogen heterocycles; 3° PANHs = tertiary PANHs.

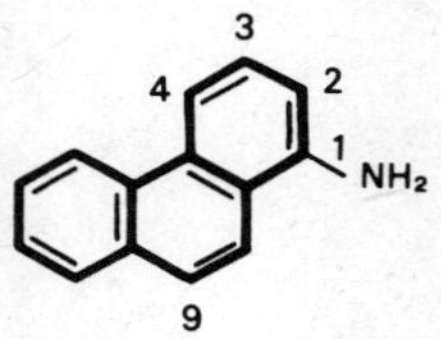

Isomer	Revertants/μg TA98	TA100	TA1535
1-Aminophenanthrene	1,232	317	1
2-Aminophenanthrene	1,504	2,818	108
3-Aminophenanthrene	30,300	2,387	76
4-Aminophenanthrene	0	4	0
9-Aminophenanthrene	4,920	4,862	63

Figure 5.15 Mutagenicities of aminophenanthrene isomers (from Later et al., 1984a).

shown in Fig. 5.15 for the five aminophenanthrene isomers. Isomers with amino substitution alpha to a point of ring fusion generally showed lower activity than did their beta-substituted counterparts. The most active isomers had the highest degree of linear electron displacement, as defined by the para principle, and provided for the greatest stearic accessibility of the amino substituent. Greater stearic exposure of the amino group probably enhances N-hydroxylation reactions thought to form proximal mutagens (Krieck and Westra, 1979).

N-alkylated amino-PAHs have not been studied extensively. Ho et al. (1981b) showed that N-methylation had little effect on the mutagenic activity of aminofluorene and aminopyrene isomers. However, N-phenylation and N-dimethylation eliminated mutagenic activity of 2-aminonaphthalene against *S. typhimurium* TA98. *Ortho*-methyl substituents markedly enhance carcinogenicity of amino-PAHs.

Amino-PAHs in N-PAC isolates from SRC-II materials are thought to be important tumor initiators (Mahlum et al., 1983) and are relatively active in chronic skin-painting assays (Cunningham et al., 1985). The Ames assay readily detects carcinogenic amino-PAHs either as pure compounds or when present in complex coal-derived mixtures. Certain three- and four-ring aminopyrenes, aminofluoranthenes, aminochrysenes, and aminophenanthrenes are extremely mutagenic.

5.6.4.4 *Pyrrolic-PAHs*

Indoles and carbazoles are the least basic of the major nitrogen-containing compound classes found in coal liquids; carbazole is often the single most abundant nitrogen-containing compound in higher-boiling coal-liquid distillates. Alkylated carbazoles, particularly the methylcarbazoles, are also found in these materials at relatively high levels. Concentrations of 4.5, 11.9, and 5.2% (by weight) were reported for the pyrrolic-PAH isolates of an SRC-II HD, an SRC-I process solvent (PS), and a gasification tar, respectively (Later and Lee, 1983).

Benzologs of carbazole, such as dibenzocarbazole, exhibit carcinogenic activity in rodent skin-painting tests, and active metabolites have been isolated. As pure compounds, benzocarbazole and dibenzocarbazole exhibited activity in the microbial mutagenicity assay (Ho et al., 1979). Ho et al. (1980b) reported that pyrrolic-PAH isolates of coal oils had twice the mutagenic activity of their analogous PAH isolates. However, carbazole fractions from SRC liquids (Later and Lee, 1983) exhibited much lower mutagenic activity in *S. typhimurium* TA98 than the corresponding amino-PAH-enriched fractions; certain fractions demonstrated no activity.

5.6.4.5 *Azaarenes*

Azaarenes are major components of nitrogen-rich fractions of most direct liquefaction and gasification liquids and tars. Weight concentrations for the same coal-derived materials cited in the pyrrolic-PAH section (SRC-II HD, SRC-I PS, and gasification tar) were 2.3, 3.0, and 6.9% respectively. Predominant azaarenes formed during coal conversion are the mono- and bicyclic compounds, pyridines and quinolines, and their alkyl homologs. Because of their volatility, amounts of lower-molecular-weight bases relative to each other and to higher-molecular-weight azaarenes depend on thermal history. Nitrogen-rich isolates of higher-boiling coal liquids are primarily composed of benzoquinolines and naphthoquinolines. As an N-PAC subclass, azaarenes are the most basic aromatic nitrogen compounds generally found in coal liquids.

Compared to neutral PAHs having the same ring system, azaarenes appear to show increased microbial mutagenicity (see Fig. 5.14) but less than that observed for amino-PAHs. As a class, azaarenes are weakly carcinogenic; many compounds show no activity in rodent skin-painting assays. Azaarene-rich fractions from SRC-II distillates boiling between 750 and 850°F exhibit about 20% of the skin-tumor-initiating activity of the PAH fractions from the same materials (Mahlum et al., 1984). However, in recent chronic mouse skin-painting assays with high-boiling SRC-II distillates, N-PAC fractions were as active as their PAH counterparts, but had a longer latency period prior to expression (Renne et al., 1986). Azaarenes showing slight carcinogenic activity as pure compounds include phenanthridine, benz[a]acridine, benz[c]acridine, and several dimethyl benzacridines and methyl quinolines. Certain methyl benzacridines and methyl dibenzacridines also show moderate to strong carcinogenic activity.

The azaarene-rich HPLC fraction from SRC-II HD was also active in the CHO assay (Mahlum et al., 1983). The HPLC fraction enriched in amino-PAHs was less active than the azaarene-rich fraction. Response to azaarenes in the CHO system was roughly comparable to that from a PAH fraction containing about 1300 ppm BaP. Dimethylbenzacridine, which is moderately mutagenic in the Ames assay, is highly potent in CHO assays.

5.6.4.6 *Cyano-PAHs*

Certain coal-conversion liquids contain minor concentrations of cyano-PAHs (Later et al., 1985a). For example, cyano-PAHs have been found in carbon blacks and gasifier condensate from Lurgi processes. In general, the cyano-PAHs are more abundant in combustion products or in materials from processes with more severe oxidative conditions than direct

Table 5.6 Comparative Chemical Composition of Cyano-Polycyclic Aromatic Hydrocarbons (PAHs) and other Nitrogen-Containing Polycyclic Aromatic Compounds (N-PACs) in Gasification-Derived Materials

	Weight % of Crude	
Chemical Class	Heavy Creosote	Medium Oil
PAHs	39.8	40.1
N-PACs	14.9	14.7
Pyrrolic-PAHs	2.2	2.7
Amino-PAHs	0.5	0.4
Aza-PAHs	9.8	9.6
Cyano-PAHs	1.9	1.2

Source: Later et al. (1983b).

liquefaction. Table 5.6 compares the relative weight composition of a gasifier-derived heavy creosote and medium oil. The cyano-PAHs were less concentrated than either the aza-PAHs or pyrrolic-PAHs but more concentrated than the amino-PAHs. Typical structures of cyano-PAHs identified in coal-derived materials include cyanonaphthalenes, cyanobiphenyls, cyanoanthracenes, cyanophenanthrenes, cyanofluorenes, cyanopyrenes, and cyanofluoranthenes, as well as alkylated species of these compounds.

Relatively little is known about the biological activity of the cyano-PAHs (see Fig. 5.14) as compared with the other N-PAC subclasses. Pure 1- and 2-cyanonaphthalene and 9-cyanophenanthrene showed low-level mutagenicity in the fluctuation assay against TA98, a variation of the microbial Ames assay (Later et al., 1984b). Krishan et al. (1979) also reported low-level mutagenicity for cyanoacenaphthenes and cyanoacenaphthylenes in the 8-azaguanine forward-mutation assay with *S. typhimurium*. Cyano-PAH standard compounds exhibited acute toxicity with comparable or lower LD_{50} values than the analogous parent PAHs (Later et al., 1984b).

5.6.4.7 *PACs with More than One Nitrogen Atom*

Aromatic compounds containing multiple nitrogen atoms occur in low concentrations in coal liquids produced by hydrogenation or pyrolysis (Pichler and Herlan, 1973; Bodzek et al., 1979; Kosuge et al., 1982; Later et al., 1982b). For compounds with two, four, or any even number of nitrogen atoms, analytical determination by MS is complicated because the odd-integer, nominal-mass molecular ions are not present in their spectra. Even-integer molecular weights of these compounds make them difficult to distinguish from non-N-PACs in complex mixtures without the aid of HRMS. Dinitrogen heterocycles such as benzocinnolines and phenanthrolines are particularly refractory to low-resolution mass spectral analysis because their mass spectra are similar to those of the oxygen-containing fluorenones. Both compound types have even-numbered molecular ions and M-28 as a major fragment ion. However, aminoazaarenes and aminocarbazoles may be distinguished by specific derivatization of the primary amino functionality (Later et al., 1982b). Compounds such as phenazines have been distinguished in complex mixtures by their characteristic metastable ion energy spectra.

Ho et al. (1979) evaluated the mutagenicity of several aromatic compounds containing two nitrogens. Unlike anthracene, the addition of a primary amino functionality in the two position on acridine did not alter mutagenic activity. Several multinitrogen PAHs exhibit mutagenic or comutagenic activity (Nagoa et al., 1977); N-nitrosamines and norharman are two examples. Later et al. (1982b) identified aminocarbazoles and aminoazaarenes in SRC liquids. The presence of amino-carbazoles and alkylaminocarbazoles in coal liquids is of interest because of their structural similarity to compounds such as norharman, which is a demonstrated comutagen (Nagao et al., 1977). La Voie et al. (1981c) showed that several aminocarbazole isomers were mutagenically active.

5.6.4.8 *Oxygen-Containing PACs*

The major subclasses of O-PACs found in coal liquids are shown in Fig. 5.16. Hydroxy-PAHs and their alkyl homologs are generally the principal oxygenated compounds in coal liquids (Uden et al., 1978; Schabron et al., 1979; Hurtubise et al., 1981; Later et al., 1981a; White et al., 1983; Lucke et al., 1986). As with the N-PACs, the single-ring compounds such as phenol and its alkylated derivatives are found at higher concentrations in undistilled coal-derived liquids than the multiple-ring hydroxy-PAHs. Some hydroxy-PAHs in shale-oil fractions exhibit promotional activity in mouse skin-painting assays (Bogowski and Mirme, 1979). However, complex mixtures of these compounds, as derived from coal liquids, are not particularly mutagenic or carcinogenic.

Hydroxylated compounds in several coal liquids, including EDS (Later et al., 1984b) and Char-Oil Energy Development (COED) derived (Benson et al., 1979), are hydroxy analogs of the same structures found in the PAH fractions and reflect the same distribution as their parent structures. In distillate materials, decreased volatility conferred by addition of a hydroxyl group lowers the molecular weight distribution of the hydroxy-PAH parent structures, as compared to the parent structures of the unsubstituted PAHs. Hydroxy-PAH components of coal liquids appear to be formed by addition of from one to four hydroxyl groups on the PACs. Monohydroxy compounds are generally an order of magnitude more concentrated than are the dihydroxy-PAHs, which, in turn, are more concentrated than the trihydroxy- and tetrahydroxy-PAHs.

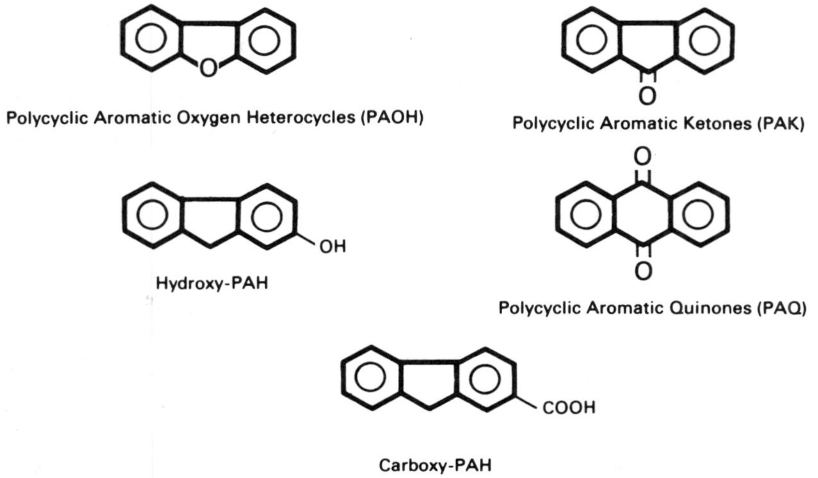

Figure 5.16 Oxygen-containing polycyclic aromatic compounds found in coal-derived materials.

Hydroxy-PAH compounds are highly polar, particularly at lower molecular weights, or when multiple hydroxyl groups are present on a given ring system. Polarity facilitates transport in aqueous systems, making hydroxy-PAHs of environmental concern. Dauble et al. (1983) and others (Buryan et al., 1978; Guenther et al., 1981) showed that phenols were major components in water-soluble fractions of coal liquids. Phenols are highly toxic to aquatic organisms such as the water flea, *Daphnia magna*.

Other oxygenated species in coal-conversion products include carboxyl compounds (see Fig. 5.16), such as polycyclic aromatic quinones and polycyclic aromatic ketones (Later et al., 1981b; Ho et al., 1983). Carbonyl functionalities, if present, occur in low concentrations in coal-hydrogenation products (Later et al., 1981b; Ho et al., 1983) and are often present in higher concentrations in pyrolysis and gasification condensate materials. Compounds such as fluorenone and phenalenone are mutagenic and may account, in part, for the activity expressed in neutral polar fractions of H-coal materials (Ho et al., 1979) and fossil-fuel combustion products (Leary et al., 1983). Furans and aryl ethers are often present, particularly in H-Coal and SRC materials as well as in gasifier condensates. Aromatic carboxylic acids occur in some coal liquids from pyrolysis processes. Organic acids have also been reported (Duty et al., 1980). Carboxylic acids are apparently absent from hydrogenation products; if present, their concentrations are below detection limits of the IR methods used (Aczel et al., 1977).

5.6.4.9 *Mixed-Element Heteroatomic PAHs*

Mixed-element heteroatomic PAHs with more than one type of heteroatom are also found in coal liquids. Because standards are generally unavailable, few of these have been identified specifically, although mass spectrometric and other data allow their characterization. Figure 5.17 shows structures for some mixed-element heteroatomic PAHs in coal liquids.

For example, thiazoles and phenothiazines have been identified in raw coal liquids from the British National Coal Board liquid solvent extraction direct liquefaction process (Burchill et al., 1982 a, b). Mixed nitrogen–sulfur heterocycles were also reported in coal tars by Pichler and Herlan (1973). Thiophenic sulfur was also detected in conjunction with both pyrrolic and pyridinic nitrogen structures.

The presence of mixed sulfur or nitrogen heterocycles in coal liquids has been confirmed by HRMS, with matches for such elemental compositions as $C_{11}H_7NS$ (azadibenzothiophene) and $C_{12}H_9NS$ (phenothiazine) in the SRC-II basic fraction. These elemental compositions can be rationalized to the same types of nitrogen–sulfur-containing-PAHs reported in British National Coal Board process materials (Burchill et al., 1982a, b). Nishioka and co-workers (1985a–c) recently isolated and identified aminothiophenic-PAHs such as the aminodibenzothiophenes by adsorption and GC, using synthesized standard compounds.

Mass spectral evidence is also available regarding occurrence of nitrogen–oxygen and sulfur–oxygen multiheteroatomic compounds. HPLC fractions from SRC-II HD contained both highly polar compounds such as hydroxy-PAHs and multiheteroatomic constituents. Using HRMS, these compounds yielded elemental compositions suggesting that nitrogen and sulfur occurred together with oxygen in some of them. Phenoxazines were identified in SRC-I materials by Paudler and Chaeplen (1979). Sulfur in hydroxy-PAH fractions is likely to occur as hydroxylated thiophenes; nitrogen species occur as hydroxylated aza-compounds or hydroxycarbazoles. Hydroxybenzothiophenes (Jewell and Haotung, 1964) occur in coal tar, as do hydroxycarbazoles and hydroxy aza compounds such as hydroxyazapyrene. Nishioka et al. (1985a–c) identified hydroxythiophenic-PAHs such as hydroxyphenylthiophene, hydroxybenzothiophene, and hydroxydibenzothiophene, hydroxybenzothiophene, and hydroxydibenzothiophene in SRC-II HD. Dioxygenated, as well as mixed

Type	Structure	Determined By	Biological Activity
Phenoxazine	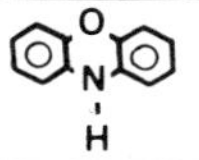	GCMS	Not Known
Phenothiazene	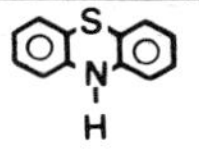	HRMS	Not Known
Thiazole	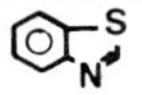	S-Specific GC HRMS GCMS	Not Known
Amino Phenol	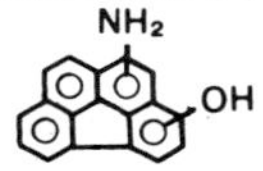	Metastable Ion Analysis, HRMS	Appear to be Active in Complex Mixtures
Hydroxy Azaarene/ Carbazole	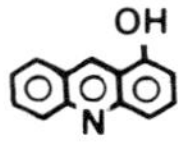	LC Off-Line	Not Known
Hydroxy Thiophenes	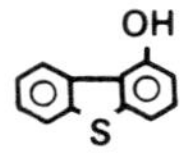	HRMS	Not Known
Amino Thiophenes	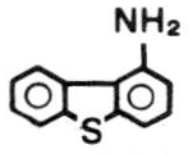	S/N-Specific GC, GC/MS	Potent Microbial Mutagens

Figure 5.17 Examples of mixed heterocyclic polycyclic aromatic compounds in coal-conversion materials (adapted from Wilson et al., 1983c). GC-MS = gas chromatography–mass spectrometry; HRMS = high-resolution mass spectrometry; LC = liquid chromatography.

nitrogen–oxygen and sulfur–oxygen heteroatomic PACs were identified in isolates from SRC-II HD by Lucke et al. (1986); possible structures include dihydroxy-PAHs and hydroxyfurans, hydroxythiophenes, and hydroxyazaarenes.

Oxygen in N-PAC fractions is likely to occur as phenoxazenes (Paudler and Chaeplen, 1979) and pyridofurans. Compounds such as the methoxycarbazoles reported in hydrogenated coal tars would, if present, also appear in N-PAC fractions.

5.6.5 Free Radicals

Free radicals are uncharged molecular species with unpaired electrons. Such electrons represent unpaired spins, which may be detected and quantitated by electron spin resonance spectroscopy. Free radicals result from homolytic covalent bond cleavage with lifetimes ranging from milliseconds to years or longer, and may be induced by thermal energy (thermolytic bond cleavage), impinging radiation (radiolysis), or chemical reactions. Free radicals are recognized as important intermediates in chemical reactions and are generally more reactive than their precursor molecular structures. Free radicals are of interest relative to health effects of coal liquids because of their ability to enter into reactions leading to covalent binding with DNA and formation of DNA adducts which lead to mutation.

Free radical concentrations in coal vary with rank: inertinite (or fusinite) macerals have relatively high free radical concentrations, liptinites have low concentrations, and vitrinites show intermediate values. Coal liquids may also contain substantial concentrations of free radicals (Mrochek et al., 1983; Kalkwarf et al., 1984), which appear to be long-lived and are stable during alumina adsorption chromatographic separation.

Mrochek et al. (1983) measured relative radical concentrations in ITSL, H-Coal and SRC-I PSs using a high-temperature, high-pressure flow cell on the electron spin resonance spectrometer. Free radical concentrations increased with increasing temperatures up to between 200 and 350°C, followed by a decrease at higher temperatures. H-Coal solvents from Illinois No. 6 coal showed 14-fold increases in relative concentrations through this temperature range, while Wilsonville SRC-I and Lummus ITSL samples showed threefold and fivefold increases, respectively. Petrakis and Grandy (1980) showed that coals are liquified more efficiently using solvents with higher residual concentrations of free radicals, such as tetralin, than by solvents with low free radical concentrations, such as naphthalene.

Chemical class fractions of SRC-II coal liquids showed variations up to three orders of magnitude in free radical concentrations. Kalkwarf et al. (1984) showed that mutagenic activity of chemical class fractions from an SRC-II fuel-oil blend (FOB) varied with free radical concentrations. The AH fraction had unpaired spin concentrations of $< 10^{13}$ spins/g and was not mutagenic. The amino-PAH fraction was the most mutagenic and contained the highest free radical concentrations (4.5×10^{17} unpaired spins/g).

5.6.6 Trace Elements

Although Fruchter et al. (1978) described an interlaboratory effort to measure trace elements in coal liquids, little work has been reported in this area. Table 5.7 shows concentrations of selected trace elements in SRC-II crude shale oil, residual fuel oil, and coal. Mercury, arsenic, and cadmium are among the most hazardous trace elements. Mercury has been of

Table 5.7 Trace Elements (ppm) in Solvent-Refined Coal II (SRC-II) Fuel-Oil Materials, Crude Shale Oil, and Residual Fuel Oil

Element	Fuel-Oil Blend[a]	SRC-II	Illinois Coal	Crude Shale Oil	Residual Fuel Oil
As	0.015	0.015	16.3	15.6	0.055
Sb	0.00099	0.0026	1.10	0.008	0.004
Se	0.0179	0.0284	2.52	0.86	0.09
Hg	0.0024	0.0034	0.114	0.200	0.004
Br	0.055	0.092	4.14	0.079	0.22
Cl	42.4	6.9	275	—	40
V	0.21	0.179	29.7	—	87
Ni	0.060[b]	0.099	13.7	0.88	12.5
Co	0.0047	0.0030	5.54	0.37	0.32
Cr	1.21	2.57	14.3	0.04	0.070
Fe	4.14	4.14	2.25	0.003	0.0005
Ti	2.99	<10	595	<10	<10
Na	0.18	0.512	155	19.4	33
Rb	<0.1	<0.014	7.69	—	—
Cs	0.0012	0.0003	0.83	<0.0002	—
K	<0.1	0.403	1790	—	—

[a] Fuel oil blend, 2.9 : 1 heavy distillate : middle distillate for Powhatan-derived oil; 5.75 : 1 for Pittsburgh seam coal.

[b] Mean of two determinations.

Source: Adapted from Filby et al. (1982).

concern relative to coal liquefaction because of its potential toxicity in organic complexes and its relatively high volatility. However, mercury concentrations in coal are low, and although mass balance for this element in coal liquefaction facilities has been difficult to achieve, analysis of products and waste streams showed no evidence of toxicity from mercury or its organic complexes. Trace elements considered hazardous (mercury, arsenic, cadmium, antimony, lead) are found in low concentrations or are absent in coal liquids. After distillation, trace element concentrations are generally highest in higher-boiling-range materials.

Semiquantitative spark source mass spectrometry of SRC-II naptha, middle distillate (MD), HD, and vacuum bottoms obtained from Ft. Lewis showed large differences in trace metal content of grab samples from the same stream taken 13 months apart. The variations were attributed to different coals being used during the two sampling periods and to variability in grab-sampling techniques and analysis. However, concentrations of nonvolatile trace metals increased with boiling point in both sampling periods, with highest concentrations in vacuum bottoms. Mercury was below limits of detection in all cases.

Trace minerals were found in the most polar fractions of HPLC-separated SRC coal liquids. (A. P. Toste, Pacific Northwest Laboratory, unpublished data; Wilson et al., 1983c), suggesting they may be complexed with acidic organic functionalites in coal liquids. Koppenaal and Manahan (1976) suggested that metal carbonyls, organic hydroxides, metallocenes, metal alkyls, and metal chelates may be present in coal liquids. Organic complexation of certain trace elements may increase toxicity and genetic activity, as shown for mercury.

5.7 EFFECTS OF ENVIRONMENTAL WEATHERING AND AGING ON THE COMPOSITION OF COAL-DERIVED MATERIALS

Chemical stability is an important consideration for coal-derived fuels, which may tend to be reactive because of their relatively high heteroatom content. Degradation reactions, including precipitate or sludge formation, can adversely affect combustion characteristics and/or severely limit the usefulness of fuels. Changes in chemical composition resulting from degradation may also affect the toxicological characteristics of these materials. The storage stability and aging characteristics of several materials are discussed in Chapter 2, Section 2.6.2.

5.8 EFFECTS OF PROCESS CONDITIONS ON CHEMICAL COMPOSITION OF COAL-CONVERSION MATERIALS

Differences in process conditions for a given technology may also affect the biological activity of resultant materials. An example is the comparison of ITSL and nonintegrated two-stage liquefaction (NTSL) materials [runs 241, (Pannell and Sood, 1982) and 242], which were produced at the Wilsonville facility using different configurations of the same three major process units: a short-contact-time (SCT) liquefaction reactor, the Kerr–McGee critical solvent deasher, and an H-Oil hydrotreater. In the NTSL mode, the product stream from the SCT reactor goes to the hydrotreater via the critical solvent deashing unit; the reactor and hydrotreater each provide their own recycle streams. Thus, the reactor does not receive the highly hydrogenated product from the hydrotreating units as a component of the feed coal slurry solvent. In the ITSL mode, a portion of the hydrotreater product is recycled to the reactor to act as recycle solvent. Chemical analysis and biological assays of these materials showed significant differences in composition and toxicological activity (Wilson et al., 1983a).

5.8.1 Process Parameters and Catalysts

Chemical composition of coal-conversion materials generally reflects the process involved and the conditions under which it is operated. Schiller (1977a) studied various SRC coal liquids and concluded the following: (1) Increased liquefaction severity increased the concentration of oxygen-containing compounds and decreased concentrations of nitrogen compounds in distillable products; (2) catacondensed compounds, such as naphthalenes and phenanthrenes, were the major aromatic compounds resulting from mild or moderate hydrogenation, while compounds such as acenaphthene and biphenyl increased in concentrations with severity; (3) as liquefaction severity increased, furan concentrations decreased, and hydroxy-PAH concentrations increased; and (4) average molecular weight of SRC materials decreased with increasing liquefaction severity.

These conclusions are consistent with those from studies on materials from the ITSL process operated under low-severity liquefaction conditions (Later, 1985a). Low severity was achieved by reducing reaction contact time and reactor temperature. Average molecular weight of ITSL materials was higher than that of SRC materials and distilled mainly in the 650–850°F range.

The use of catalysts is an important consideration in both single- and multiple-stage processes. Iron catalysts used in the H-Coal process, for example, appear to increase alkylation of PACs relative to noncatalytic single-stage processes. Higher-quality catalysts, such as those used in the EDS ITSL and catalytic two-stage liquefaction (CTSL) processes, also facilitate denitrogenation and desulfurization and yield products that are higher in aliphatic and hydroaromatic content. In the CTSL process, a cobalt–molybdenum desulfurization catalyst is used in the first stage, while a nickel–molybdenum catalyst designed primarily for cracking is used in the second stage. This combination yields a product that is high in aliphatic and saturate compounds relative to other coal liquids.

Although operating temperatures may also affect chemical composition, sufficient data are not available to generalize in this regard. For example, the chemical characteristics of materials from the Wilsonville plant were studied during ITSL operations. Several process streams were compared at differing liquefaction (first-stage) and hydrotreater (second-stage) temperatures. Operating temperatures that yielded the highest rates of suitable product also yielded the most ammonia gas and resulted in materials with the lowest mutagenic activity (Later, 1985a).

Although specific relations among process variables and biological activity have not been determined, certain general aspects of the process appear important in determining mutagenic activity; these include coal slurry solvent quality, hydrotreater catalyst quality (Wilson et al., 1984), and liquefaction reactor severity (Wilson et al., 1983b; Later, 1985a). Neutral PAHs, which are the most important carcinogens, do not appear to be as sensitive to changes in process conditions as do the amino-PAHs and other moderately polar mutagens. Carcinogenicity of coal liquids is therefore largely a function of upper distillation cut point temperature and H:C ratio and not of specific process conditions.

5.8.2 Feed Coal Composition

Variations in feed coal rank may also result in compositional differences in coal-liquid products. Sturm et al. (1981) conducted a comparative study of liquids from six feeds rang-

Table 5.8 Summary of Ring Distributions of the Saturate Concentrates from Eight Coal Liquids, Weight % Total Coal Liquid Basis

Mass Z Series	Ring Number	Pittsburgh	Illinois No. 6		Western Kentucky	Colstrip	Wyodak		Lignite
			Run 1	Run 2			Run 1	Run 2	
+2	0	1.79	2.07	1.86	4.79	4.42	5.69	6.22	3.59
0	1	10.02	8.74	9.60	12.32	9.79	14.06	19.62	11.92
−2	2	7.63	5.02	7.28	7.87	7.24	9.91	15.95	7.80
−4	3	4.67	2.22	3.77	4.22	4.13	4.94	10.04	3.42
−6	4	2.17	1.38	1.89	2.17	2.21	2.33	5.25	1.72
−8	5	1.01	0.06	0.70	0.84	0.82	1.30	2.13	0.66
−10	6	0.25	0.06	0.09	0.30	0.25	0.28	0.67	0.17
MA	1	0.10	0.23	0.09	0.20	0.06	0.06	0.06	0.07

Source: Sturm et al. (1981).

ing in rank from lignite to highly volatile bituminous A coal. Extensive analytical data are contained in this report. The authors concluded among other things that "coal rank alone could not account for the observed variations in complexity of the coal liquids." Table 5.8 is a summary of ring distribution from saturate concentrates from eight coal liquids.

In studies on coal liquids from the EDS process, which compared materials produced from operation with lignite, bituminous, and subbituminous coals, compositional changes consistent with the relative aromaticity of the three feedstocks were observed. Table 5.9 (Later et al. 1984b) gives chemical class fractionation data during bottoms recycle operation for EDS recycle solvent derived from the feeds mentioned above. The relative mutagenic activity of comparable distillate materials produced from these three feeds were in a rank order consistent with the relative aromaticity of the organic material in the feed coals (B. W. Wilson et al., Pacific Northwest Laboratory, unpublished data).

5.8.3 Unit Operations

Molecular weight, which can be controlled largely by distillation, is an important parameter in determining both the mutagenic and carcinogenic activities of coal liquids (see Fig. 5.18).

Table 5.9 Chemical Class Composition, During Bottoms Recycle Operation, of Recycle Solvents as a Function of Coal Type in the EDS Process

Compound[a]	Chemical Class Composition Versus Coal Type (wt %/Fraction) Illinois bituminous	Wyoming Subbituminous	Texas Lignite
AHs	46	57	59
PAHs	37	25	25
N-PACs	7	3	3
H-PAHs	13	11	9
Recovery	103	97	95

[a] AHs = aromatic hydrocarbons; H-PAHs = hydroxy-polycyclic aromatic hydrocarbons; N-PACs = nitrogen-containing polycyclic aromatic compounds; PAHs = polycyclic aromatic hydrocarbons.

Source: Later et al., 1984b.

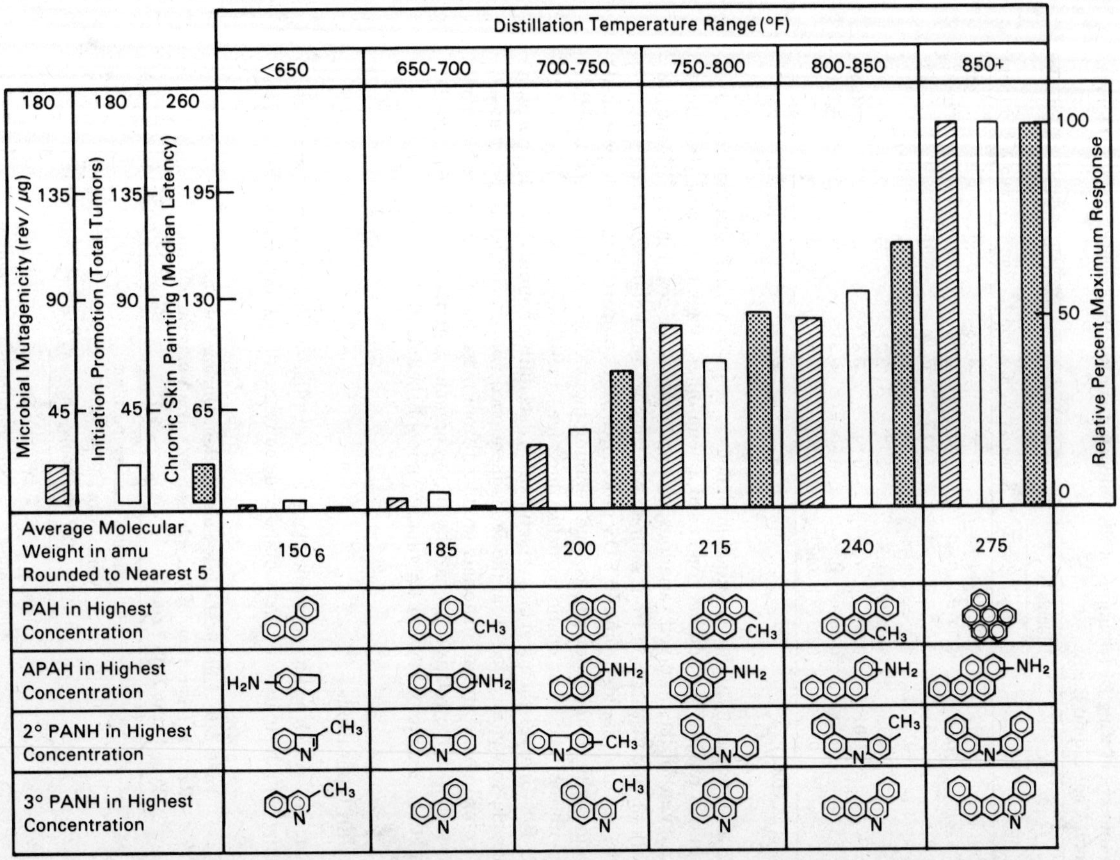

Figure 5.18 Compounds found in highest concentrations in polycyclic aromatic hydrocarbon (PAH) and nitrogen-containing polycyclic aromatic compound (N-PAC) fractions of solvent-refined coal II (SRC-II) 50°F boiling-point cuts shown below a histogram depicting response of those fractions in three different bioassays. A-PAHs = amino-PAHs; 2° PANHs = secondary polycyclic aromatic nitrogen heterocycles; 3° PANHs = tertiary PANHs. Average molecular weight of the crude distillate fractions determined by low-voltage probe-inlet mass spectrometric analysis (from Wilson et al., 1986).

Below each bp range in Fig. 5.18 are compound structures from each major chemical class that appears in highest concentration for that temperature range. Note that PAHs, with more than 16 aromatic carbon atoms (including those known to be carcinogenic), are essentially confined to cuts boiling above 650°F.

There was little difference between the mutagenic and tumorigenic activities of 600–700°F bp materials among the various coal liquids studied. One exception is the mutagenic activity found in the 650–700°F PAH fraction of EDS bottoms recycle solvent (Later et al., 1984b). Chemical analyses show that <650°F ITSL materials essentially lack amino-PAHs such as 2-aminonaphthalene and 4-aminobiphenyl (less than 1 ppm under normal operation with catalysts in good condition); nonhydrotreated materials, such as SRC-II, have higher concentrations of these compounds below the upper cut point 700°F distillation temperature.

Postproduction catalytic hydrogenation has been performed on materials from several processes, including SRC-I (Potts et al., 1981), H-Coal syncrude (O'Rear et al., 1981), SRC-II (Wilson et al., 1981a), SRC-II syncrude (Frumkin et al., 1981), and EDS (Epperly, 1982). Biodirected analyses were eventually carried out on SRC-II, H-Coal, and EDS materials. In each case there was a decrease in average molecular weight with substantial decrease in heteroatomic species concentrations. Aliphatic and hydroaromatic contents tended to increase. Sturm et al. (1981) summarized results from these types of investigation by concluding that given sufficient upgrading, coal-liquefaction materials from any source could be made to look like coal-liquefaction materials from any other source.

Postproduction hydrotreatment reduces PAH and N-PAC concentrations up to 80% (Wilson et al., 1981a; Epperly, 1982), while amino-PAHs can virtually be eliminated (Wilson et al., 1981a). Several investigators have shown that the reduction in amino-PAH content and the dearomatization that occurs during postproduction hydrotreatment of distillate feeds can significantly decrease mutagenic (Weimer et al., 1980; Wilson et al., 1981a) and carcinogenic (Mahlum et al., 1984) activities of coal-liquefaction materials.

On-line hydrotreatment, as in the ITSL process, has less effect on chemical composition and thus on biological activity. This approach is used mainly to improve the liquefaction process by increasing recycle solvent quality, coal-conversion rates, and final product quality. Like postproduction hydrotreatment, it sharply reduces N-PAC concentrations, particularly in <700°F bp materials. Reductions in N-PAC content accompany decreased mutagenicity in *S. typhimurium* and CHO cell assays (Weimer et al., 1980; Wilson et al., 1981a; Mahlum et al., 1983; Later, 1985a). However, neutral PAH content is reduced less by on-line hydrotreatment than by postproduction hydrotreatment. Skin tumorigenicity, which primarily reflects PAH content, is also less affected by on-line hydrotreatment (Wilson et al., 1983c; Later, 1985a).

On-line hydrotreatment is probably less effective in decreasing concentrations of mutagens and carcinogens for two reasons: (1) Feeds to on-line hydrotreaters often contain heavy-end, insoluble organic, and unconverted material components. (2) On-line feeds have not been returned to ambient temperature following thermal conversion; thus, the possibility of retrograde reactions involving thermally activated species is higher than for those materials that have been cooled and thus stabilized (H. Schindler, personal communication).

Data on materials from the SRC, EDS, and ITSL processes have been used to evaluate relations between toxicity and chemical composition for coal-liquefaction materials (Wilson et al., 1986; Wright et al., 1985). From review of the chemical composition of coal liquids as it relates to their biological activity, seven chemical properties were identified that were related to mutagenic, carcinogenic (tumorigenic), and, to some extent, teratogenic activities. These are (1) molecular weight, (2) aromaticity, (3) PAH content, (4) N-PAC content,

(5) amino-PAH content, (6) degree of alkylation, and (7) hydroaromatic-PAH content (Weil and Condra, 1960; Guerin et al., 1980; Calkins et al., 1981; Pelroy and Wilson, 1981; Wilson et al., 1984; Wright et al., 1985). The first five properties usually correlate directly with one or more of these toxicological endpoints. In contrast, hydroaromatic content and increased alkylation of PACs are generally associated with reduced effect.

5.9 SEPARATION METHODS FOR COAL-CONVERSION MATERIALS

Coal-derived materials are complex mixtures of several compound classes, including saturated and unsaturated hydrocarbons, PAHs, and their N-PAC, O-PAC, and S-PAC heteroatomic derivatives. Relative concentrations of these compound classes and individual components differ among materials and reflect process and distillation history; rank, age, and type of feed coal; conversion technology; and other factors. Separation of constituents of coal-derived products is essential to fundamental and applied investigations to elucidate chemical composition and biological properties.

Two major separation approaches have been applied to coal-liquid mixtures: (1) *class-specific* fractionations and (2) *biodirected* chemical separations. In the first approach, method development has been directed by the goal of achieving separations of homogeneous, class-specific isolates for chemical analysis and biological characterization. For example, methods could be developed to separate monoaromatic and diaromatic compounds or PAHs and N-PACs. In the second approach, a biological endpoint such as acute toxicity, mutagenicity, or tumorigenicity has been selected and successive separations are performed until a simplified mixture responsible for the observed biological activity endpoint is obtained. These approaches are not necessarily distinct or mutually exclusive; methods have been reported that satisfy both requirements.

Because of their chemical complexity, some preseparation is usually required before detailed analyses of synfuels are feasible. Preferably, the initial separation should reproducibly yield discrete components according to chemical functionality or biological endpoint. Other desired aspects of preseparation include high recovery rates, separation of adequate samples for biological testing and chemical characterization, and preservation of potential biological activity throughout the fractionation process. Successful methods are generally simple and amenable to the routine analysis of numerous samples.

Both classical and contemporary separation techniques have been applied to coal-derived materials. Separation methods applied to coal tars 50 years ago are still used, although improvements in resolution, speed, and sensitivity have been achieved. Separation methods used to analyze other complex matrices, such as petroleum-derived materials and cigarette smoke condensates, have also been adapted for separation of coal product constituents. Methods used to isolate chemical fractions of coal products include distillation, liquid–liquid chromatography, liquid–solid chromatography, or a combination of these techniques. Separation methods applied to investigate the chemical and toxicological characteristics of coal-derived materials are reviewed below. Those methods that have been used to correlate chemical composition with biological effects are emphasized.

5.9.1 Distillation

Distillation, which separates chemical constituents according to bp or volatility, was one of the first techniques used to isolate and identify components of coal tar. Reviews, such as those of Kennaway (1955), describe the isolation of carcinogenic PAHs from coal tar by distillation.

Relations between bp and chemical structure are generally predictable for major compound classes found in coal products; namely, the AHs, PAHs, and monoheteroatomic PACs. The bp of components within a class of PACs typically increase with increasing molecular weight and/or ring number. Table 5.10 shows this relation for the PAHs. Additional factors that affect bp include degree of branching in AHs or alkyl substitution on aromatic compounds, degree and position of alkylated substituents on aromatic compounds, structural arrangement of carbon atoms in polycyclic compounds, and substitution and position of nitrogen, oxygen, and sulfur heteroatoms. For example, addition of methyl and/or ethyl units to the parent PAH naphthalene (Table 5.10) increases the bp of the resulting compound. Nitrogen- and sulfur- heteroatom-substituted fluorenes have markedly increased bp compared to the nonfunctional fluorene; incorporation of oxygen into the fluorene structure has little effect on bp temperature. Because PACs with more than four condensed aromatic rings boil only at reduced pressure, heavy-end coal liquids require vacuum distillation.

Coal-derived materials may contain hundreds of thousands of components from many chemical classes. Overlap in bp ranges of PACs is to be expected; furthermore, a specific component may be distributed over two or three adjacent distillate fractions. Both preparative and analytical distillations of these coal-derived materials yield multi-class mixtures of compounds having comparable bp. In general, however, the molecular weight of distillates increases with increasing distillation temperature and/or decreasing pressure.

Fractional distillation was applied to direct coal-liquefaction products to effect a molecular weight separation and isolate biologically active components in higher-boiling-range

Table 5.10 Boiling Points of Selected Polycyclic Aromatic Compounds

Compound	Molecular Weight (amu)	Boiling Point[a] °C	Boiling Point[a] °F
Naphthalene	128	218	424
1-Methylnaphthalene	142	244	471
2-Methylnaphthalene	142	241	466
1-Ethylnaphthalene	156	259	498
2-Ethylnaphthalene	156	258	496
1,4-Dimethylnaphthalene	156	268	514
2,3-Dimethylnaphthalene	156	268	514
2,3,6-Trimethylnaphthalene	170	286	507
Biphenyl	154	256	493
Fluorene	166	294	561
Dibenzofuran	168	287	549
Dibenzothiophene	184	333	631
Carbazole	167	355	671
Anthracene	178	342	648
Phenanthrene	178	340	644
Pyrene	202	393	739
Fluoranthene	202	375	707
Naphthacene	228	450	675
Chrysene	228	448	838
Triphenylene	228	435	815
Benz[a]anthracene	228	425	797

[a] Boiling points at ~1 atm or 760 mm Hg.

materials (Pelroy and Wilson, 1981; Wilson et al., 1982a,b). Figure 5.19 shows the weight distributions of the various chemical classes present in the 50°F distillate cuts of reblended end products from the SRC-II and EDS coal-liquefaction processes, respectively. For each process, more than 80% of the material is distilled below 700°F. AH content decreases, and levels of N-PACs and hydroxy-PAHs increase with increasing bp temperature. Neutral PAH concentrations do not change markedly with increased temperature above about 700°F.

The effect of distillation temperature on molecular weight is shown in Fig. 5.20 for the PAH fraction of SRC-II bp cuts. Compounds of increasing molecular weight and thus of decreased volatility are found in distillate fractions of increased temperature. Since the retention of a compound in GC is a function of volatility, compounds with higher bp will have longer GC retention times than lower-boiling compounds. For example, pyrene, which is noncarcinogenic, is found in maximum concentration in the 700–750°F cut; the potent carcinogenic compounds, such as BaP, are distilled in the >800°F cuts. This trend is also observed for the mutagenic N-PAC and amino-PAH components of distillate cuts.

Studies with coal-liquefaction materials indicate that components exhibiting genotoxic activity generally distill at temperatures above 700–750°F (see Chapter 4). Analogous findings have been obtained with other hydrocarbon mixtures such as petroleum distillates and cigarette smoke condensate.

5.9.2 Solvent-Partitioning Chromatography

Separations by solvent-partitioning chromatography are based on the differential solubility of sample components in liquid solvents of differing polarity or chemical properties. Solvent extraction techniques have been used to classify coal-derived products according to their solubility in organic solvents. Solvents of increasing polarity or specific chemical property are used to sequentially extract solutes of similar polarity or chemical properties while leaving the insoluble or precipitated constituents. Several solvent extraction procedures for coal or coal-derived oils have been reported (Mima et al., 1978; Schwager and Yen, 1978; Schweighardt and Thames, 1978; Burke et al., 1979; Ruberto and Cronaur, 1979; Steffen et al., 1979; Schultz and Mima, 1980; Boduszynski et al., 1982a,b). The most widely used liquid extraction method is shown in Fig. 5.21 and classifies constituents as follows: *n*-pentane- or *n*-hexane-soluble residues (oils); benzene- or toluene-soluble and *n*-pentane- or *n*-hexane-insoluble residues (asphaltenes); pyridine-soluble and benzene- or toluene-insoluble residues (preasphaltenes); and pyridine-insoluble residues (Schweighardt and Thames, 1978; Ruberto and Cronaur, 1979; Boduszynski et al., 1982a,b). SRC process-derived distillable liquids and nondistillable vacuum bottoms were separated into solvent fractions by this method, then class fractionated by adsorption chromatography (Boduszynski et al., 1982a,b). Results indicated that the oils and asphaltenes of both materials were composed of the same chemical classes: AHs, PAHs, N-PACs, and hydroxy-PAHs. The main differences in solvent fractions of the SRC materials were in the relative concentrations of these classes. Oils were rich in PAHs, while asphaltenes had higher levels of N-PACs and hydroxy-PAHs. Generally, solvent fractionations of this type do not produce homogeneous, class-specific isolates.

A variation on solvent extraction involves the use of supercritical fluids to volatilize high-boiling, high-molecular-weight materials from coal and coal-derived products (Bartle et al., 1975; Smith and Udseth, 1983a,b; Snape, 1983). Bartle et al. (1975), who extracted coal with supercritical toluene, separated the extract by adsorption chromatography and, after chemical analyses of extract subfractions by GC, IR, and NMR, concluded that the supercritical extract was composed of unaltered materials from the coal matrix, which were products of coalification of plant chemicals.

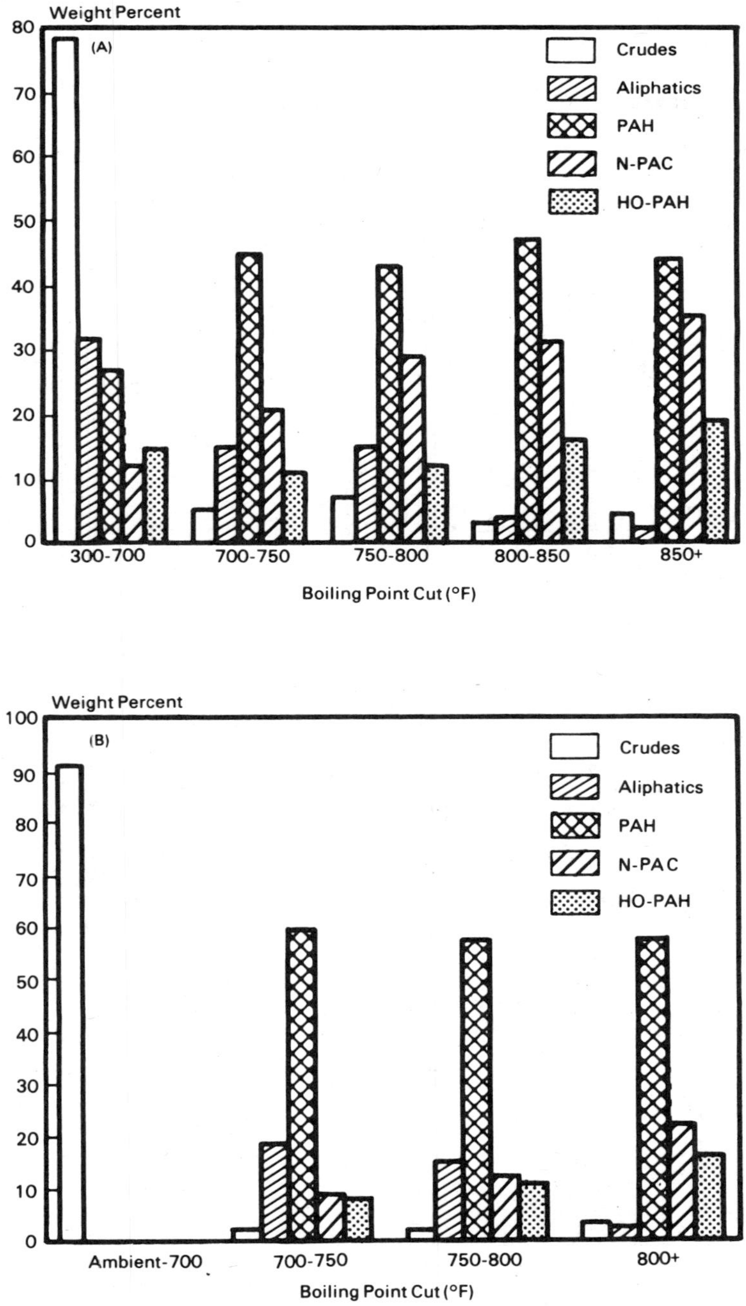

Figure 5.19 Weight distribution of chemical classes in the (A) solvent-refined coal II and (B) EDS distillate fractions (Later et al., 1983b).

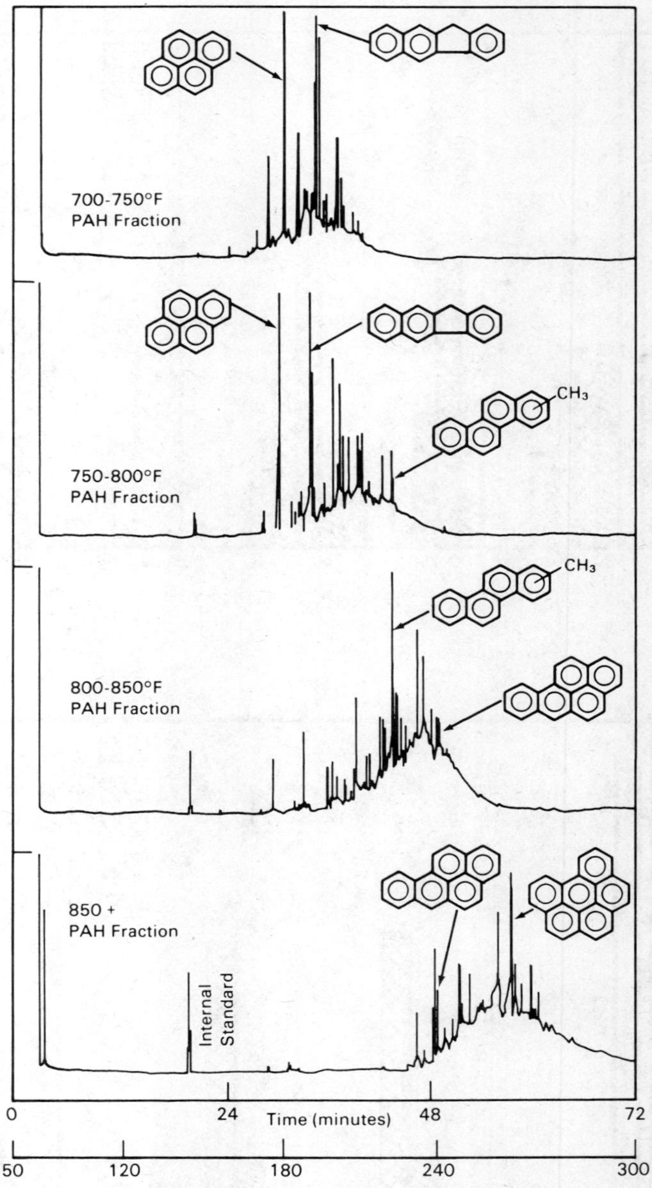

Figure 5.20 Capillary-column gas chromatograms of polycyclic aromatic hydrocarbon (PAH) fractions from the solvent-refined coal II (SRC-II) 50°F distillate cuts (from Later et al., 1983b).

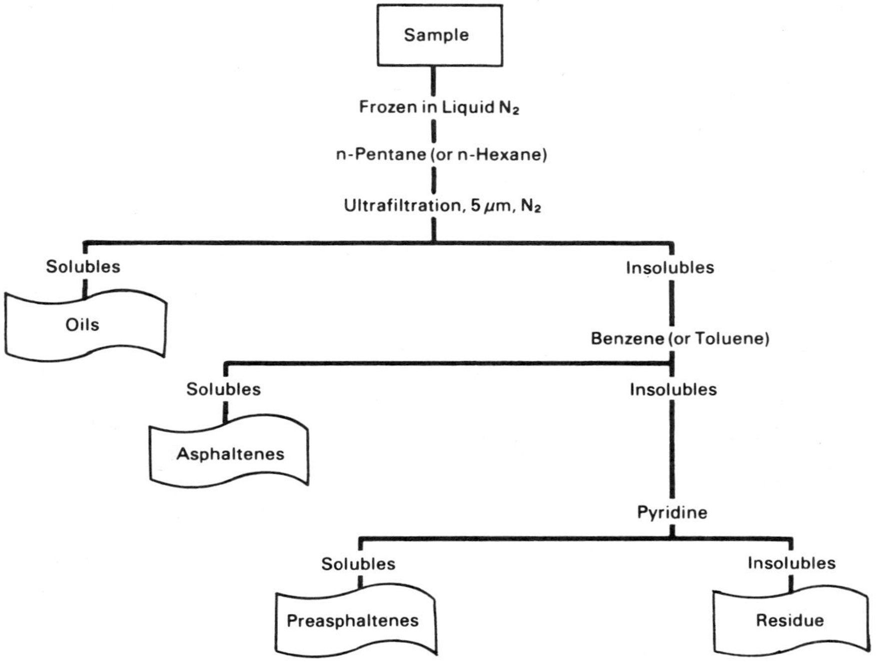

Figure 5.21 Solvent separation scheme for classification of coal-derived materials (adapted from Schweighardt and Thames, 1978).

Smith and Udseth (1983a) extracted coal with a mixture of supercritical pentane and 2-propanol. The extraction products were analyzed continuously by direct fluid injection into a mass spectrometer. Supercritical fluid extraction chromatography was also used to characterize the high-molecular-weight, nondistillable portion of coal-liquefaction products according to volatility in a manner similar to that in which vacuum distillation is used for low-molecular-weight, distillable materials (Warzinski and Ruether, 1983). Plots of extracted mass versus fluid conditions were obtained that related extractability of crude material to pressure and temperature parameters of the supercritical fluid. The mass distribution information provided by these plots was analogous to that provided by the more conventional distillation curves.

Solvent-partition chromatography has been used frequently to isolate chemical classes of coal-derived products. In this technique, sample solutes are partitioned between two immiscible liquid phases. Usually, one phase is an organic solvent and the other phase is aqueous. The pH of the aqueous phase can be adjusted sequentially to partition compounds of different acidity, basicity, and/or chemical functionality into either the organic or aqueous phases, where they are recovered in unique fractions. Partition chromatography, as applied to coal liquids, has been referred to as acid–base extraction chromatography. Extracted solutes are generically classified as acids, bases, or neutrals.

A generalized acid–base extraction scheme is shown in Fig. 5.22. Typical organic solvents in which the acid–base extractions are performed include methylene chloride, diethyl ether, isooctane, and benzene. All are immiscible in aqueous HCl and NaOH phases typi-

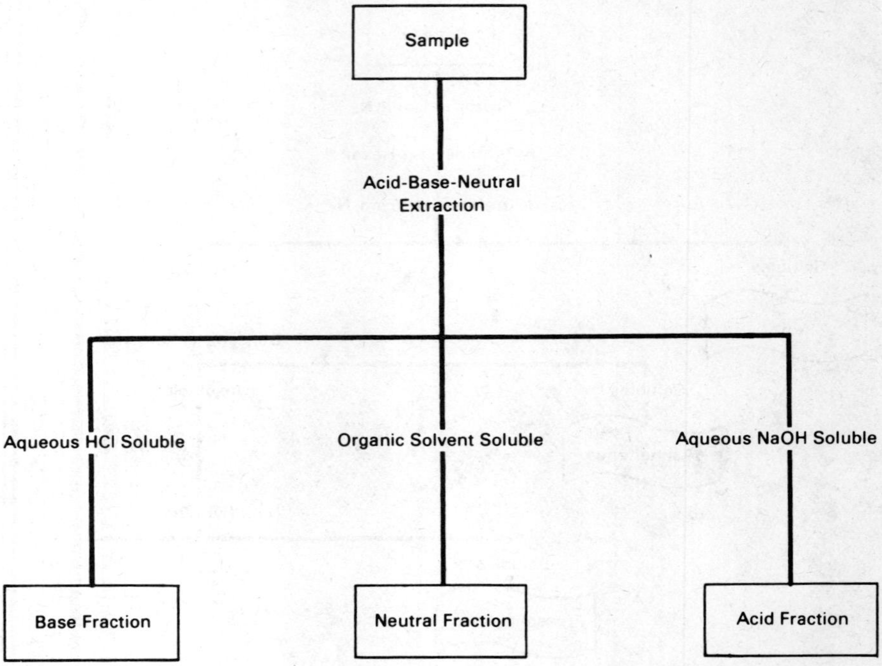

Figure 5.22 Generalized acid–base neutral fractionation isolation scheme for coal-derived materials (from Later, 1985a).

cally used to extract acidic and basic components from the crude sample material. Phase separations are carried out in separatory funnels, vials, or test tubes. After basic components are extracted with aqueous HCl, the pH of the acid extract is readjusted to >10 with NaOH to deprotonate the basic constituents. The basic constituents are then back-extracted, with an organic solvent, as uncharged neutral compounds. Conversely, acid compounds are extracted into the aqueous phase with basic solutions, acidified, and back-extracted, with organic solvent, as neutral species. Acid–base extractions are usually repetitive to ensure efficient recoveries. Once acids, bases, and neutrals are in organic solvents, the sample volume can easily be reduced to concentrate the solutes prior to analyses.

Several workers (Rubin et al., 1976; Guerin et al., 1980; Ho et al., 1980b) have described acid–base extraction as the initial isolation step in multiple-step chromatographic methods. Acid, base, and neutral fractions of coal liquids were produced by dissolving crude materials in diethyl ether, extracting acids with 1 *M* NaOH, and extracting bases with 1 *M* HCl. This procedure was adapted by Rubin et al. (1976) from a method originally developed by Swain et al. (1969) to fractionate cigarette smoke condensate. More-basic N-PACs, such as 3°-polycyclic aromatic nitrogen heterocycles (PANHs) and amino-PAHs, are extracted in the basic fraction, while PAHs and neutral aza-PAHs (like indole and carbazole) remain in the neutral fraction (Ho et al., 1980a,b). Hydroxy-functional compounds, such as the phenols, are isolated in the acid fraction.

Another solvent fractionation method for separating crude coal-liquefaction products was described by Petersen et al. (1976) and applied by Wilson and co-workers (Wilson and Pelroy, 1979; Wilson et al., 1980, 1981a) to study hydrotreated SRC materials. The method involves dissolving the coal-derived liquid in isooctane, then sequentially partitioning with 1 *N* HCl, 1 *N* NaOH, and dimethylsulfoxide. Resulting fractions were designated as isooctane-insoluble tar; basic, acid-induced tar; acidic, base-induced tar; neutral; and PAHs. As indicated by the fraction names, formation of emulsions and tars are generally unavoidable with the isooctane–water system, making other solvent-partitioning methods more attractive.

Haugen et al. (1982a,b) isolated PAC constituents of gasification condensates by acid–base partition chromatography prior to subsequent class fractionation of amino-PAHs and 3°-PANHs by ion-exchange HPLC. A solution of nonvolatile condensate in methylene chloride was extracted three times with 1 *N* HCl. Aqueous phases were combined and washed with diethyl ether. The pH was adjusted to 11–12 with NaOH, and bases were extracted from the aqueous phase with methylene chloride. The back-extracted organic phase was dried over anhydrous Na_2SO_4 to remove residual water, and the solvent was evaporated to obtain a purified base fraction.

Novotny and co-workers (Novotny et al., 1974, 1981; Schultz et al., 1979) used acid–base partition chromatography to isolate PAC classes from coal tar and SRC materials (Fig. 5.23). Acids or hydroxy-PAHs were isolated by aqueous NaOH extraction. N-PACs or

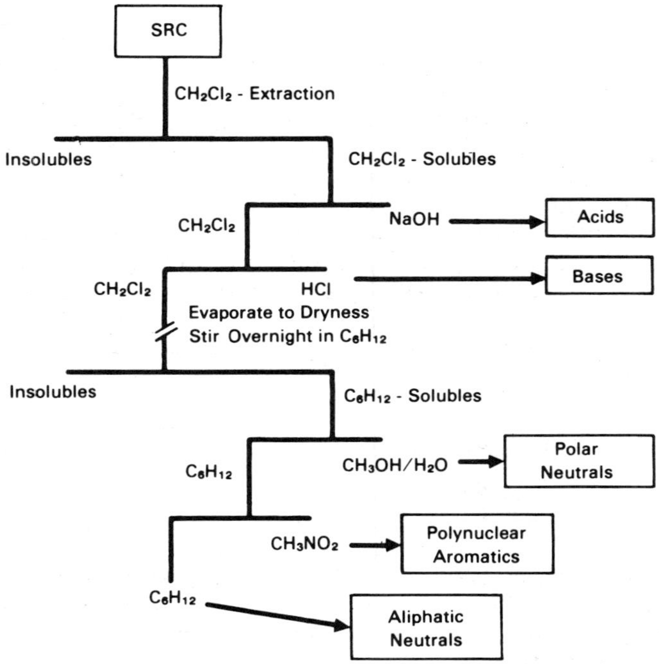

Figure 5.23 Solvent-partition scheme for the isolation of polycyclic aromatic compound classes from coal tar and solvent-refined coal (SRC) materials (adapted from Schultz et al., 1979).

bases were isolated from neutral PAC classes by extracting the dichloromethane-soluble portion of the crude material with aqueous HCl. The resulting benzene-soluble fraction was separated into polar neutrals, PAHs, and aliphatic neutrals by sequential elution with methanol/water and nitromethane.

Problems associated with acid–base extractions of coal-derived materials include formation of emulsions and tars at the liquid–liquid interface, and low recoveries of hydrophobic three- and four-ring PACs. Refinements in the acid–base extraction technique that enable higher recoveries of PACs, particularly the nitrogen bases, were suggested by Boparai et al. (1983). The effect of HCl concentration on extraction efficiency was investigated for representative N-PACs of 3°-PANH and amino-PAH classes. Only 1-aminonaphthalene was extracted in high yield using 1 *N* HCl; recovery of other N-PAC bases was lower. The portion of the N-PAC standards not recovered in the base fraction was found in the neutral fraction. Increasing the normality of HCl did not significantly enhance recoveries. However, when the N-PACs were extracted with a binary 3.4 *N* HCl/43% methanol phase, solubility of hydrophobic bases in the aqueous phase also markedly increased; recovery of three-, four-, and five-ring 3°-PANHs and amino-PAHs in the basic fraction was improved. Boparai et al. (1983) pointed out that with methanol as a cosolvent, polar compounds (e.g., phenols) may be transferred to the aqueous phase by solvent partitioning. Therefore, it was recommended that samples be extracted with NaOH to remove acidic compounds prior to isolating the N-PAC base fraction. Neutral PAHs, like methylnaphthalene, biphenyl, pyrene, and diphenylanthracene, were not extracted by the methanol/HCl phase and remained in the neutral fraction. Results of this study emphasize the need to verify recoveries of representative compounds in proposed extraction procedures and to use cosolvents to enhance yields of PAC bases.

A hybrid solvent extraction and acid–base partition chromatographic method has been developed in which N-PAC components are precipitated, with HCl, from other PAC in coal-conversion products (Sternberg et al., 1975; White et al., 1977, 1978; Schweighardt et al., 1978; J. L. Schultz et al., 1979). First, HCl gas is bubbled through a benzene solution of crude material. N-PAC bases form insoluble HCl salts that precipitate out of benzene solvent; the other PAC classes remain in solution. The precipitate is centrifuged and separated from the mother liquor, slurried in an organic solvent such as benzene or ether, and gaseous ammonia or aqueous NaOH is used to regenerate the original N-PAC bases, which are soluble in the organic solvent. Resulting isolates can then be analyzed to identify specific compounds. Finseth et al. (1982) reported higher levels of residual acidic or phenolic compounds in N-PAC fractions isolated by this HCl precipitation method than by a cation-exchange chromatography method.

5.9.3 Liquid–Solid Chromatography

Separations based on interaction of a solute between liquid and solid phases have widely been used to separate PAC classes in coal-derived products. Examples include adsorption, gel-permeation, sieving, and ion-exchange chromatography. Separations are commonly performed in open columns at ambient or low pressure, although HPLC has also been used. Various solid materials are used as packing: for example, alumina, silicic acid, Sephadex LH-20, Bio-Beads, Fluorisil, XAD and Amberlite resins, silica gel, cellulose acetate, and other gels and adsorbents. Solutes are separated based on (1) physicochemical interactions between the compound and the packing and (2) the solute's solubility in liquid solvents used as mobile phases. Inorganic metals and salts, and chemically bonded functional groups such as amino and cyano groups, can be used as surface modifiers to enhance the selectivity of liquid–solid separations.

Major advantages of liquid–solid chromatography include (1) reproducible and precise class fractionations, (2) elimination of emulsions and tars encountered in liquid–liquid chromatography, (3) ease of performance of preparative-scale separations, and (4) quantitative recoveries of PACs. Major liquid–solid chromatographic methods used to isolate PAC classes from coal-derived products are reviewed below.

5.9.4 Gel-Permeation Chromatography

Liquid–solid chromatography on Sephadex LH-20 and Bio-Bead SX series has been used as a fractionation technique for preparing coal-derived materials for toxicity testing. Sephadex LH-20, a lipophilic, dextran-based material, and Bio-Beads, a styrene-divinylbenzene polymer, can be used in various chromatographic modes, depending on the solvent used (Streuli, 1971; Jones et al., 1977). For example, retention mechanisms that prevail with tetrahydrofuran (THF) and LH-20 are molecular sieving (nonpolar compounds are separated by size, the largest compounds eluting first) and hydrogen bonding (polar compounds such as the N-PACs are eluted in reverse order of hydrogen bonding strengths). With an alcohol eluent like isopropanol, the primary retention mechanism is π bonding or adsorption (the larger the aromatic π system, the greater the retention; compounds generally elute in order of increasing number of condensed aromatic rings). In the adsorption chromatographic mode, a heteroatom in an aromatic compound increases its retention compared to that of a related, nonfunctional neutral compound. Major advantages of this technique are the inertness of gel materials, the quantitative elution of even polar functional PACs, and the relatively large loading capacities of gel columns (Klimisch and Stader, 1972).

Snook (1978) studied the behavior of PACs on Bio-Beads. Bartle and co-workers (1983, 1984) demonstrated the use of an evaporative analyzer as a mass detector in size-exclusion gel-permeation chromatography for analysis of high-molecular-weight coal extracts. Dutcher et al. (1983a,b) and Royer et al. (1983) devised a separation scheme for coal-gasification condensates, using Sephadex LH-20. Five fractions were collected with a tetrahydrofuran eluent (Royer et al., 1983); although chemical classes overlapped, some separation of polar PACs and neutral PAHs was achieved by their method. Jones et al. (1977) used an isopropanol solvent to perform a chemical class type separation of nonfunctional hydrocarbons of crude coal liquids on LH-20. Their scheme is shown in Fig. 5.24. Other investigators have used Sephadex LH-20 column chromatography in multiple-step separation schemes to isolate PAC constituents in coal-derived products (Ho et al., 1980a,b; Rao et al., 1981; Toste et al., 1982, 1983).

Caution must be exercised when using gel-permeation chromatography to isolate PACs from complex mixtures. Polar moieties of certain functional compounds may be subject to adsorption mechanisms regardless of eluent used. Thus, elution patterns of polar compounds, such as hydroxy-PAHs and N-PACs, can be interdispersed with those of neutral compounds that elute strictly according to molecular size.

5.9.5 Alumina/Silica Adsorption Chromatography

Open-column, ambient-pressure, liquid–solid chromatography on alumina and/or silica adsorbents has also been used for separating coal-derived PACs. Application of adsorption chromatography to fossil fuels originated in the petroleum industry (Rosen and Middleton, 1955; Snyder and Buell, 1968; Snyder et al., 1968) and has been adapted successfully to separate chemical classes of coal-derived materials. Alumina and silica are available in various grades, mesh sizes, pH, and activities (i.e., water content). Class separations are

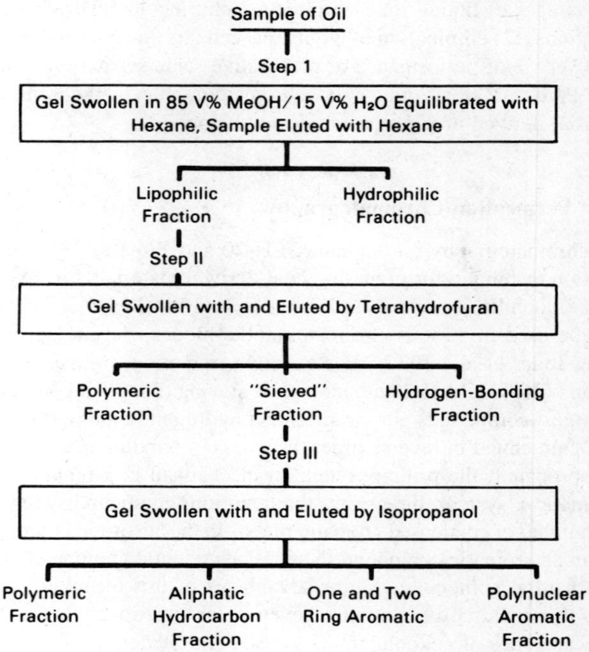

Figure 5.24 Sephadex LH-20-isopropanol class separation scheme for synthetic crude oils (adapted from Jones et al., 1977).

achieved by sequential elution of the adsorbed sample with organic solvents of increasing polarity.

A disadvantage with alumina and/or silica is that polar compounds such as N-PACs, hydroxy-PAHs, and mixed-functional PACs may be irreversibly adsorbed or undergo catalytic degradation on the adsorbent surface. However, Later and Lee (1983) reported high recoveries for two-, three-, and four-ring neutral and polar PACs on both silicic acid and neutral alumina (Table 5.11). Excellent recoveries were obtained for AH, PAH, most N-PAC, and hydroxy-PAH chemical classes. Recoveries for the amino-PAHs were generally poorer: 60–80%. Thus, when this method is used, potential adsorption or degradation must be considered carefully, particularly for more reactive or polar compound classes, when quantitative assessments are required for PACs isolated from coal-derived products by adsorption chromatography.

Farcasiu (1977) reported a separation scheme for SRC based on chromatographic fractionation by sequential elution with specific solvents on silica gel columns. Neutral PAHs were concentrated in the initial two fractions; nonbasic and basic nitrogen heterocycles were concentrated in the third and fifth fractions, respectively, of a nine-fraction scheme. A similar method, using alumina adsorbent, was described by Callen et al. (1976, 1978) to separate coal-liquefaction materials into 12–14 fractions that were analyzed by spectrometric methods to determine the nature of PACs in each.

An alumina adsorption chromatographic method initially developed by Schiller and Mathiason (Schiller, 1977b; Schiller and Mathiason, 1977) and modified by others (Boduszynski et al., 1982a,b; Later et al., 1981a; Leary et al., 1983) has been widely used to isolate

Table 5.11 Recoveries of Selected Polycyclic Aromatic Compounds from Neutral Alumina and Silicic Acid

		Recoveries (%)	
Compound	Chemical Class[a]	Al_2O_3	SiO_2
n-C_{18}	AHs	~100	—
Phenanthrene	PAHs	97.6[b]	—
Triphenylene	PAHs	94.0[c]	—
Benzo[a]pyrene	PAHs	96.8	—
Carbazole	2°-PANHs	87.0[b]	87.8[c]
Benzo[a]carbazole	2°-PANHs	84.2[c]	62.5
Quinoline	3°-PANHs	99.8	81.1[b]
Acridine	3°-PANHs	99.2[b]	90.4[b]
Phenanthridine	3°-PANHs	94.4[c]	97.7[b]
1-Azapyrene	3°-PANHs	92.8[c]	94.6[b]
Benz[c]acridine	3°-PANHs	97.8[c]	93.5[b]
1-Aminonaphthalene	Amino-PAHs	83.7[b]	73.8[b]
9-Aminophenanthrene	Amino-PAHs	56.5[c]	38.9[b]
3-Aminofluoranthene	Amino-PAHs	64.1[b]	64.0[b]
6-Aminochrysene	Amino-PAHs	73.7[b]	57.5[c]
Acenaphthenol	Hydroxy-PAHs	85.0	—
8-Hydroxybenzo[a]pyrene	Hydroxy-PAHs	93.6	—

[a] AHs = aliphatic hydrocarbons; PAHs = polycyclic aromatic hydrocarbons; PANHs = polycyclic aromatic nitrogen heterocycles.

[b] Average of three or more determinations.

[c] Average of two determinations.

Source: Later and Lee (1983).

homogeneous PAC fractions rapidly from various coal-conversion materials in a single column chromatographic step (Fig. 5.25). Either basic or neutral alumina can be used; neutral PAHs are eluted prior to the N-PAC and hydroxy-PAH chemical classes. Popl et al. (1972) discussed the mechanisms and advantages of using pH-controlled alumina to separate aromatic hydrocarbons and heterocycles. Wozniak and Hites (1983) further modified the alumina method shown in Fig. 5.25 by doping the adsorbent with 10–15% picric acid, a charge-transfer complexing agent that enabled a class-specific separation of AHs, hydroaromatic compounds, and PAHs.

Figures 5.26 and 5.27 demonstrate the selectivity of the alumina adsorption chromatographic method for separating PAC components of complex mixtures. Most O-PACs in coal-derived materials are hydroxy-PAHs with relatively low levels of keto-functional PACs (group E; Fig. 5.26). Thus, discrete, class-specific fractions are generally obtainable under optimized elution conditions. Excellent reproducibility can be achieved when activity (i.e., water content) of the alumina adsorbent is optimized for a specified separation and closely monitored to ensure that the adsorbent activity remains within limits required (Later et al., 1985a).

Alumina and silica adsorption chromatography have also been used to separate subclasses of N-PACs, in particular the 3°-PANHs and amino-PAHs (Ho et al., 1980a, 1981a–c, 1983; Later et al., 1981a,b, 1982a,b; Later and Lee, 1983). Ho et al. (1981a,c) separated unsubstituted amino-PAHs from N-substituted amino-PAHs and 3°-PANHs on basic alumina. Later and co-workers (Later et al., 1981a,b; Later and Lee, 1983) demonstrated the feasibility of separating 2°-PANHs, amino-PAHs, and 3°-PANHs by gradient-elution chro-

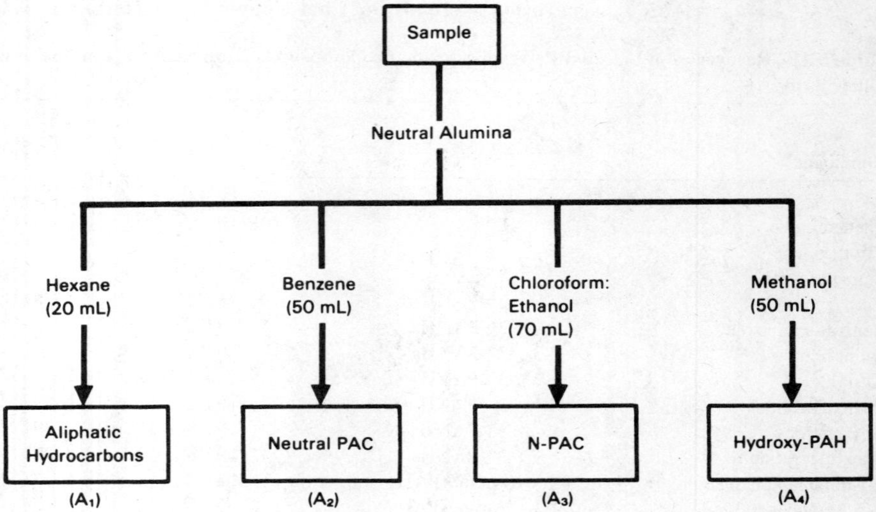

Figure 5.25 Chemical class separation of polycyclic aromatic compounds (PACs) on neutral alumina (from Later and Lee, 1983). N-PACs = nitrogen-containing PACs; PAHs = polycyclic aromatic hydrocarbons.

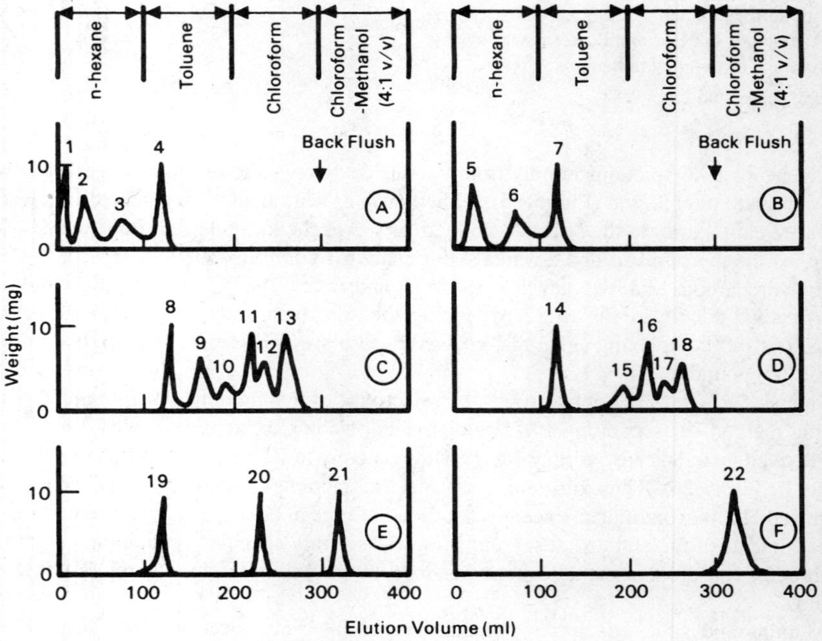

Figure 5.26 Elution of model compounds on basic alumina: (A) saturated and hydroaromatic hydrocarbons; (B) polycyclic aromatic hydrocarbons; (C) basic nitrogen heterocycles; (D) nonbasic nitrogen heterocycles; (E) oxygen-containing compounds; and (F) hydroxyl aromatics. Peak numbers and corresponding model compound structures are given in Fig. 5.27 (adapted from Boduszynski et al., 1982a,b).

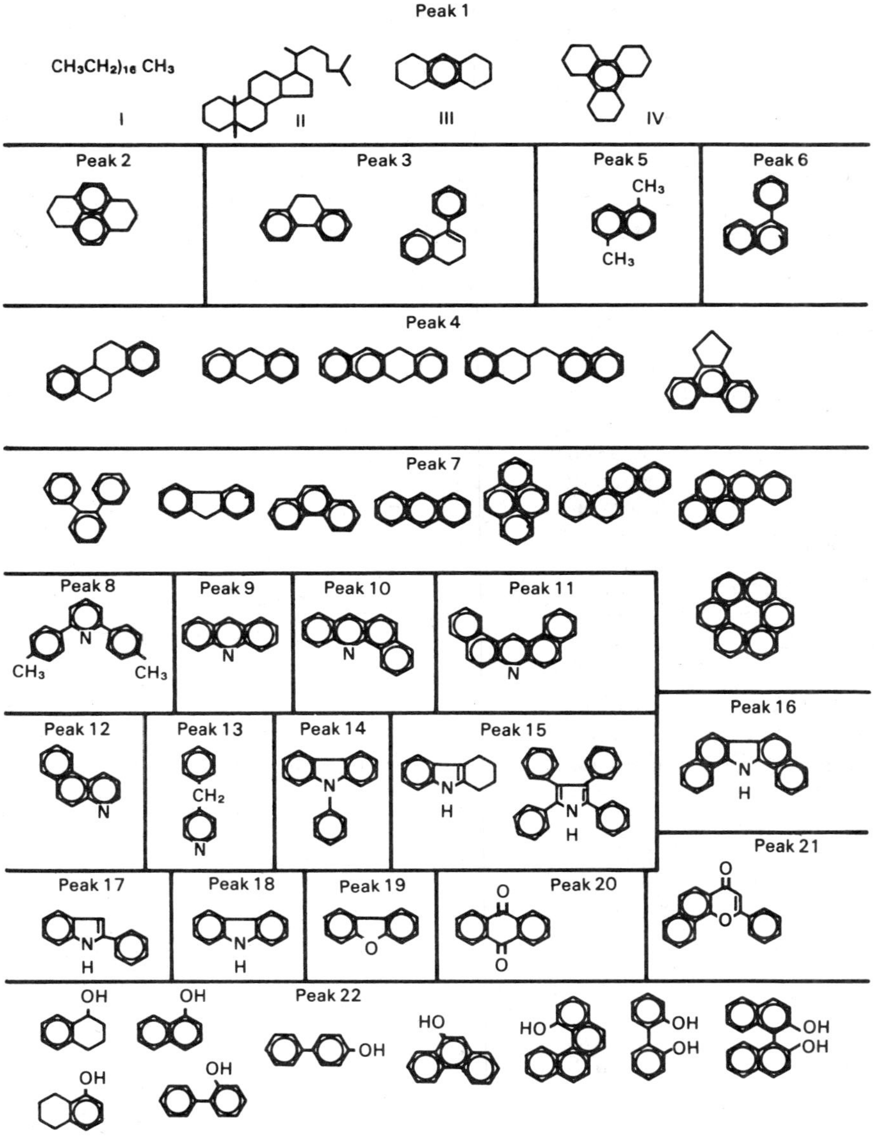

Figure 5.27 Standard compounds eluted from a basic alumina column. Peaks refer to Fig. 5.26 (adapted from Boduszynski et al., 1982a).

matography on silicic acid. Activity-regulated and picric-acid-doped alumina adsorption chromatography have also been used to isolate cyano-PAHs from other N-PAC classes (Dubay and Hites, 1978; Later et al., 1985a).

The thiophenic-PAHs (PASHs; see Table 2.8) elute with other neutral PACs on alumina. Subsequent chemical modifications, followed by iterative adsorption chromatographic steps using silica gel, have been used to isolate the PASHs from the more predominant PAHs in the neutral PAC fraction (Lee et al., 1980; Willey et al., 1981a,b). More recently, ligand-exchange chromatography, using silica gel impregnated with palladium chloride, has been used to isolate the PASHs from other neutral species in the PAH fraction (Nishioka et al., 1985a–c). This procedure is shown schematically in Fig. 5.28. This method eliminates the lengthy sulfonation and reduction steps and provides a homogeneous chemical class fraction in two separation steps. The PASHs isolated from coal-derived materials by either of these methods can be characterized chemically and assayed biologically without interference from other neutral PACs: that is, the furanic-PAHs and nonfunctional PAHs.

Other methods have been used to isolate the PASHs from complex fossil fuels. Joyce and Uden (1983) described a liquid chromatographic method for isolating thiophenic compounds by argentation chromatography (silver nitrate-coated silica column). Virtually complete separation was achieved for the one-ring thiophenes; benzothiophenes were enriched 96% as compared to the naphthalenes. These workers also compared the efficiency of their method with the sulfoxide/sulfone-oxidation-based isolation technique. Burchill et al. (1982a,b) avoided lengthy separation procedures by directly identifying the PASHs in coal tars and pitches by selective flame photometric detection GC.

5.9.6 Ion-Exchange Chromatography

Application of ion-exchange chromatography to separate neutral, acid, and basic PACs from fossil-fuel materials was initiated in the petroleum industry (Jewell and Snyder, 1968; Jewell et al., 1972) and has been adapted for use with coal-derived products (Schweighardt

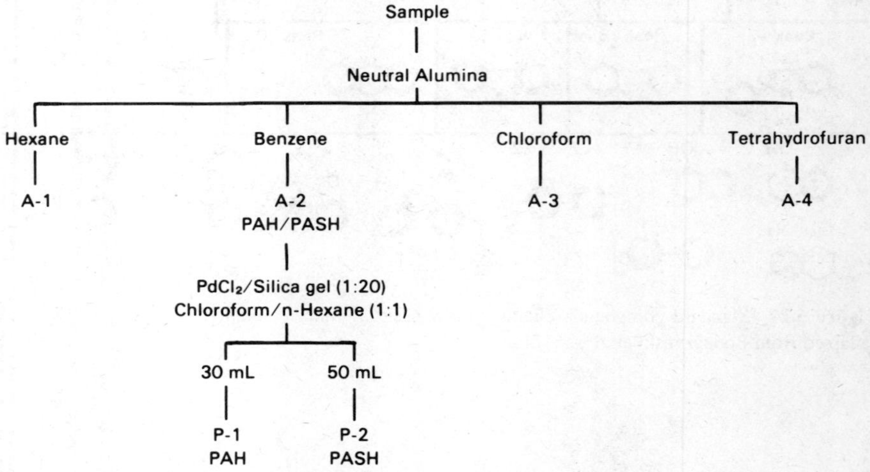

Figure 5.28 Schematic of polycyclic aromatic sulfur heterocycle (PASH) isolation procedure (adapted from Nishioka et al., 1985a–c).

et al., 1978; J. L. Schultz et al., 1979; Green et al., 1984). Coordination or complexation using anion/cation-exchange resins or metal salts has been used to separate O-PACs and N-PACs from other neutral PACs. Formation of complexes between solutes and functional sites on the surface of the solid phase by ionic and/or coordination mechanisms results in selective retention of polar functional compounds. After eluting the nonfunctional compounds with an appropriate nonpolar solvent, a polar mobile phase (such as dichloroethane, dichloromethane, methanol, or chloroform) disrupts the complex and elutes the acids and bases. Amberlyst and XAD resins are the ion-exchange solids most often used for basic and acidic compounds; ferric chloride adsorbed on a kaolin or clay substrate complexes the neutral or nonbasic nitrogen compounds (Jewell and Snyder, 1968; Jewell et al., 1972; J. L. Schultz et al., 1979; Schweighardt et al., 1978).

Schweighardt and co-workers (Schweighardt et al., 1978; J. L. Schultz et al., 1979) separated several pentane-soluble (oils) SRC liquids on a column of cation/anion-exchange resins, ferric-chloride-impregnated clay, and silica gel adsorbent. This scheme, known as the saturates–aromatics–resins–asphaltenes (SARA) method (Jewell and Snyder, 1968), was used to obtain saturates, aromatics, acids, bases, and neutral nitrogen compound classes. Amberlyst-A29 anion-exchange resin was used to remove acidic components; basic compounds were removed by the Amberlyst-15 cation-exchange resin. Neutral nitrogen compounds were removed by complexation with the $FeCl_3$/Attapulgus clay phase. Finally, neutral saturates and aromatics were separated on silica gel.

5.9.7 High-Performance Liquid Chromatography

The liquid–solid chromatographic techniques discussed above mainly involve separations at ambient or low pressure. Gross separations of crude materials into chemical classes, using the same techniques but at higher pressures, have also been accomplished with high-performance liquid chromatography (HPLC). In addition to the adsorption-type HPLC packings, chemically bonded phases (e.g., amino-, cyano-, and alkyl silanes) have also been used in liquid–liquid HPLC separations of coal liquids.

Of the types of packing material available, bonded reverse-phase packings provide superior reproducibility and are most generally applicable to analysis of the aromatic chemicals present in coal-conversion materials. These packings are composed of hydrophobic moieties (most commonly octadecyl, C_{18}) covalently bonded to silica particles. In these reverse-phase systems, aromatic compounds bound to the hydrophobic surface are eluted with mixtures of water and organic solvents (typically methanol or acetonitrile) of decreasing polarity, that is, with solvents of decreasing water content.

To a first approximation, the members of a class of aromatic compounds (e.g., PAHs or amino-PAHs) are eluted in order of increasing molecular weight, with the strength of hydrophobic interactions with the bonded phase depending on the number of both aromatic and aliphatic carbons. For compounds of identical or similar molecular weight, the molecular shape plays a secondary role in the order of elution in a manner dependent on the geometry and configuration of the bonded phase resulting from the manufacturing process. For PAHs, these principles and their application have been reviewed by M. L. Lee et al. (1981) and Wise (1983); they are also applicable to separation of azaarenes (Dong et al., 1977; Colin et al., 1981) and amino-PAHs (Haugen et al., 1986).

In general, PAHs can be resolved from their analogous heteroatomic PACs, for example, amino-PAHs, hydroxy-PAHs, and azaarenes, having a comparable range of carbon content. However, resolution of groups (e.g., two- to four-ring) of these polar PACs from each other is not readily achieved. Furthermore, smaller PAHs (e.g., naphthalene) and larger polar PACs (e.g., aminochrysene) are eluted at comparable solvent strengths. Thus, preseparation of classes of PACs is a desirable prerequisite for analytical reverse-phase HPLC.

Burke et al. (1981) used reverse-phase HPLC with an octadecylsilane column to separate SRC and H-coal liquefaction PSs into several fractions. Polar PACs, such as hydroxy-PAHs and N-PACs, eluted first, followed by neutral PAHs and finally AHs. Retention times of various functional PACs were correlated with structural parameters to assess the solvent quality of different process materials. Ogan and Katz (1982) devised a multiple-stage HPLC system using reverse-phase and size-exclusion chromatographic techniques to separate coal liquids and oils. The first stage separated components according to chemical similarity; the second stage separated them according to molecular size; and the final stage used a high-resolution, reverse-phase C_{18} bonded-phase column to resolve individual compounds. Dark et al. (1977) and Dark and McFadden (1978) also reported a multiple-step HPLC method for preparative-scale class fractionation and compound-specific analytical determinations of constituents in liquefied coal products. The first of two HPLC systems had a silica gel column; the second had a reverse-phase C_{18} column. Structural characterization of the second column's effluents was achieved by coupled HPLC-MS.

Chromatographic separation of compound classes on a plain HPLC silica adsorption column with aprotic, dipolar solvents was investigated by Chmielowiec (1983). These solvents (like dimethylsulfoxide) strongly moderated the adsorbent solid surface and selectivity complex with various PAC functional groups to determine their retention. Evaluation of moderators was based on their ability to discriminate among chemically distinct compound classes with specific functional groups. Capacity factors, ln k', of several PAC classes are given in Fig. 5.29, which shows the class-grouping tendencies, system-resolving ability, and possible overlaps between compound classes. Resolution of hydrocarbons, PAHs, aza-compounds, pyrrolic-PAHs, and polar difunctional compounds was generally achieved. The method was used to separate coal-derived naphtha into chemically distinct classes (Chmielowiec, 1983).

An HPLC method that is widely used for the class-specific separation of PACs in alternate fuels employs a normal-phase NH_2 column with a multiple-solvent mobile-phase system (Wise et al., 1977; Hertz et al., 1980; Tomkins et al., 1982; Toste et al., 1982, 1983). Wise et al. (1977) and Hertz et al. (1980) used a preparative-scale aminosilane column and modified the mobile-phase composition from 100% hexane to 100% methylene chloride to separate nonpolar and polar PACs. Similarly, Toste et al. (1982, 1983) used an aminosilane-column HPLC separation scheme to isolate chemically distinct PAC fractions from coal-liquefaction materials. A ternary solvent system of hexane/methylene chloride/isopropanol was used to elute the samples. Three elution regions were defined: nonpolar, moderately polar, and polar. Neutral PAHs eluted in the nonpolar region; N-PACs and hydroxy-PAHs eluted in the moderately polar and polar regions, respectively. Retentions of several standard PACs on this system are shown in Fig. 5.30. Separation of an SRC-II distillate cut with this method is shown in Fig. 5.31.

Like low-pressure techniques, HPLC has been used to separate heterocyclic-PAC classes. Haugen et al. (1982a, b) described a cation-exchange HPLC method to separate two classes of N-PACs: 3°-PANHs and amino-PAHs. Both were readily resolved except in unusual cases where structural properties resulted in similar basicities. The HPLC system consisted of a silica-based cation-exchange column with a 40% acetonitrile/10 mM aqueous sodium citrate mobile phase. Elution of N-PACs appeared to be primarily a function of pK_a values. Figure 5.32 shows an HPLC separation of several standard amino-PAHs and 3°-PANHs by this method. This HPLC scheme was applied to the separation of the basic fractions of

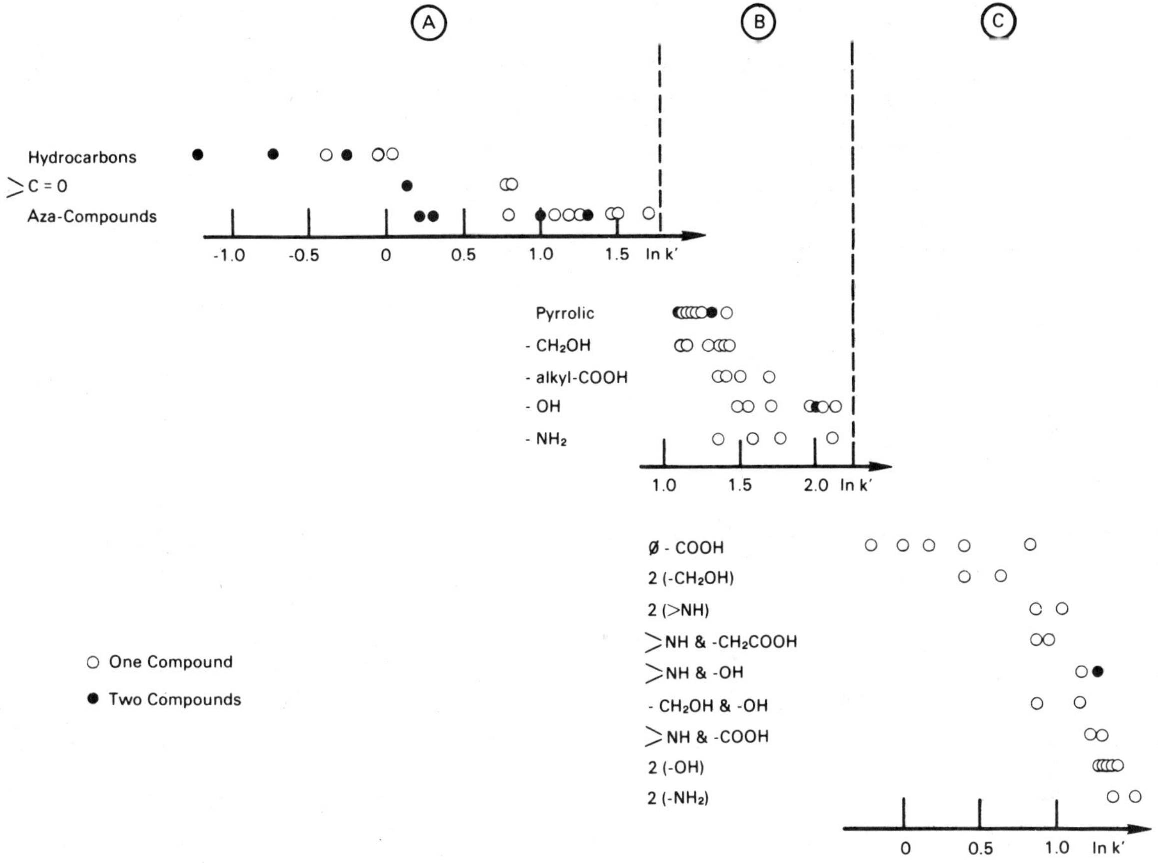

Figure 5.29 Retentions of model compounds on a silica high-performance liquid chromatography column (adapted from Chmielowiec, 1983). Eluents were (A) 0.05% dimethylsulfoxide (DMSO) in CCl_4; (B) 1% DMSO in CCl_4; (C) 10% DMSO in CCl_4.

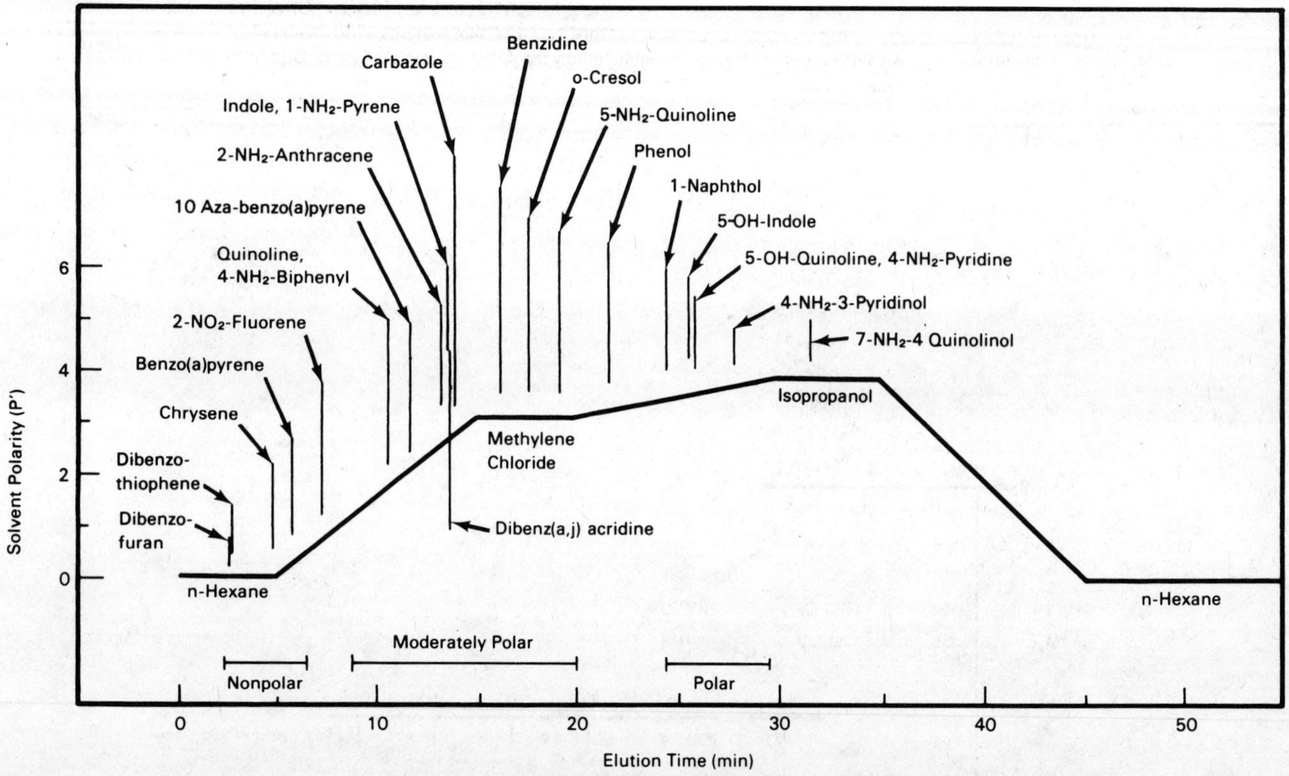

Figure 5.30 High-performance liquid chromatography of standards on a μBond Pack NH_2 column. Each standard (1 mg/mL in methanol or chloroform) was chromatographed, using the three-solvent, mobile-phase system depicted above, at a flow rate of 5 mL/min. Organic species typically eluted in three distinct regions of polarity: nonpolar, moderately polar, and polar (adapted from Toste et al., 1982).

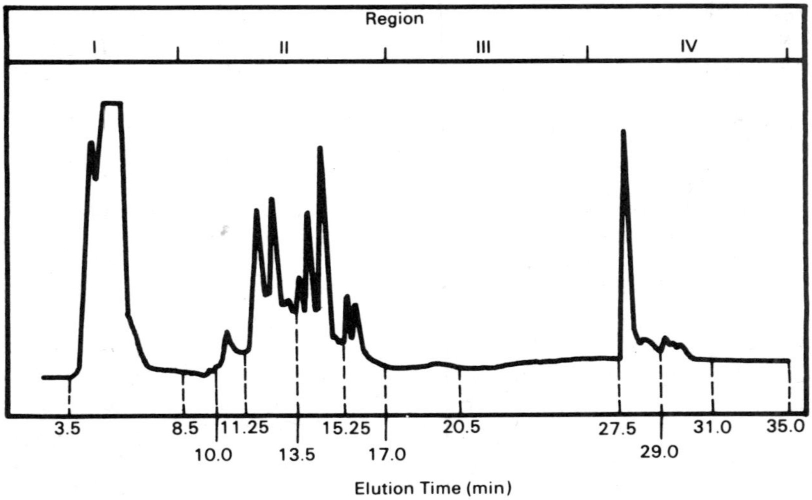

Figure 5.31 High-performance liquid chromatography chromatogram of a solvent-refined coal II distillate cut (639–689°F). Elution times for various regions are indicated for nonpolar, I; moderately polar, II and III; and polar, IV. Conditions are as in Fig. 5.30. Detection was by ultraviolet absorption at 250 nm (from Pelroy and Wilson, 1981).

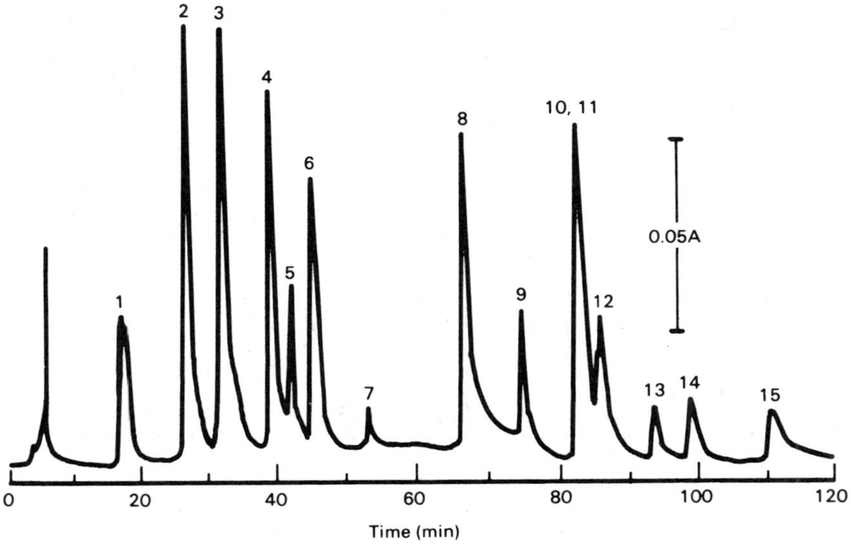

Figure 5.32 Separation of a standard mixture of primary aromatic amines and aromatic amines by high-performance liquid chromatography. Standards were (1) 1-aminopyrene, (2) 1-aminoanthracene, (3) 2-aminoanthracene, (4) 4-aminobiphenyl, (5) 2-aminonaphthalene, (6) 2-aminofluorene, (7) aniline, (8) benzo[f] quinoline, (9) quinoline, (10) acridine, (11) isoquinoline, (12) pyridine, (13) 3-methylpyridine, (14) 2-methylpyridine, and (15) 2, 4-dimethylpyridine (from Haugen et al., 1982a).

coal-gasification and -liquefaction materials. Several cuts collected during HPLC separations were assayed for mutagenicity; HRGC techniques were applied for the detailed characterization of each subfraction, allowing determination of mutagenic fractions and compounds.

HPLC techniques have also been valuable in multiple-step separations to analyze neutral PAH fractions isolated by either acid–base partition or adsorption chromatography. These methods have been used primarily to isolate and determine the mutagenic and carcinogenic constituents of neutral PAH isolates of coal-liquefaction and gasification materials. Haugen et al. (1983) and Haugen and Stamoudis (1986) described a three-step HPLC procedure for fractionating mutagenic constituents of the neutral extract of a fixed-bed gasifier condensate tar. The chromatogram and mutagenic activity from the second HPLC step (a reverse-phase octadecylsilane-column system) are shown in Fig. 5.33. Mutagenic fractions were subsequently separated by a third HPLC step (normal-phase Partisil 10 PAC-column system), and subfractions were collected for identification of constituents and for additional biological testing. Figure 5.34 is an HPLC trace of the neutral PAHs isolated by alumina adsorption chromatography from an SRC-II distillate cut. Each of eight subfractions from this sample were characterized chemically (Lucke et al., 1985) and tested for tumorigenicity in the mouse skin initiation/promotion assay (Mahlum et al., 1984). Highest dermal tumorigenicity was typically observed for fractions (generally fractions 1–4) containing the four- and five-ring parent and methylated PAHs, such as benzopyrenes and methylchrysenes.

In summary, HPLC separations have widely been used for chemical class separation of compounds in coal-derived materials. The HPLC techniques described are amenable to precise, reproducible analysis, and preparative-scale separations provide the volumes of fractionated samples required for biological assays.

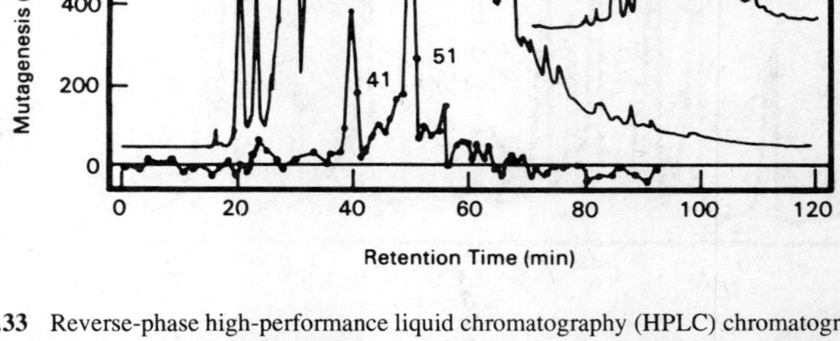

Figure 5.33 Reverse-phase high-performance liquid chromatography (HPLC) chromatogram of silica HPLC fraction 6. The principal chromatogram is for 21-mg material; the inset is for 0.5-mg material on an analytical column. Lighter line, A_{280}. Mutagenesis (heavy line) is shown as net revertants per plate (rev/plate) for 0.1-mL portions of the 4-mL fractions. Numbers of the most mutagenic fractions are indicated (adapted from Boduszynski et al., 1983a).

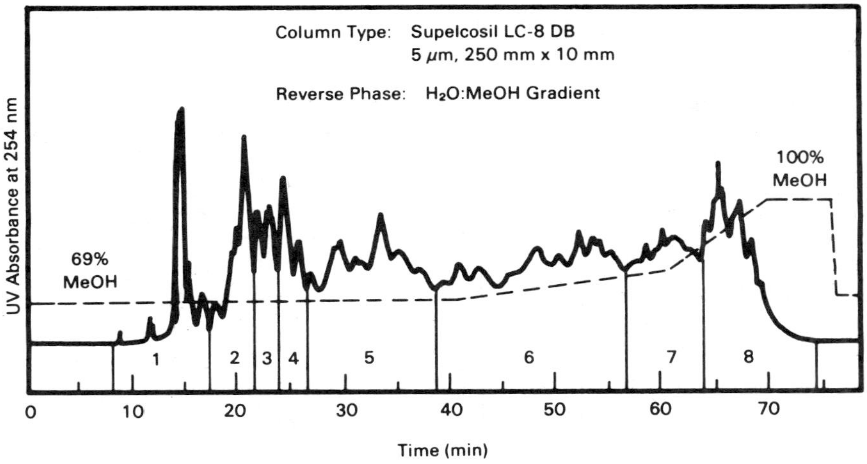

Figure 5.34 High-performance liquid chromatography separation of neutral polycyclic aromatic hydrocarbon isolate of a solvent-refined coal II 800–850°F Distillate Cut (adapted from Lucke et al., 1985).

5.9.8 Multiple-Step Chromatographic Methods

Complete isolation of PAC classes from complex matrices, such as coal-derived materials, in one chromatographic step, is not generally feasible. Therefore, multiple-step procedures have been developed that combine the techniques discussed above to isolate homogeneous class-specific fractions that can subsequently be characterized by high-performance analytical techniques.

PAC separation according to chemical functionality has several advantages. First, different PAC classes in coal-derived products are typically present in different concentrations; thus, detection interferences and obscuring of certain classes can occur. Biological activities of various PAC classes also differ, depending on chemical functionality, molecular weight, or structure. Separating PAC classes enhances understanding of the role and effect of these compounds on coal product stability, processing technologies, end-product usefulness, coal-conversion chemistry, and coal genesis and origin. Thus, multiple-step chromatographic methods have been developed to isolate PAC classes, as discussed below.

Figures 5.35 and 5.36 show a combined acid–base extraction, Sephaex LH-20 gel chromatography, and alumina and silica adsorption chromatography, used to isolate PAC classes. Initial separation by acid–base extraction (Fig. 5.35) yields three fractions containing PACs. Basic compounds (3°-PANHs and amino-PAHs) are in the ether-soluble base fraction; neutral or nonbasic PACs, such as neutral N-PACs (pyrrolic-PAHs), PASHs, furanic-PAHs, and PAHs, are in the neutral fraction (Rubin et al., 1976; Guerin et al., 1980; Ho et al., 1980a,b); and acidic PAHs (phenols and hydroxy-PAHs) are in the acid fraction. Subsequent separation of the neutral fraction, first on Sephadex LH-20, then by silicic acid adsorption chromatography, isolates a PAH fraction, I (see Figure 5.35); a neutral N-PAH fraction, II; and a polar fraction, III (Ho et al., 1980a,b). Fraction I generally contains nonfunctional PAHs, furanic and thiophenic compounds. Primary components of fraction II from synthetic crude oils were pyrrolic-PAHs, for example, indoles, phenylpyrroles,

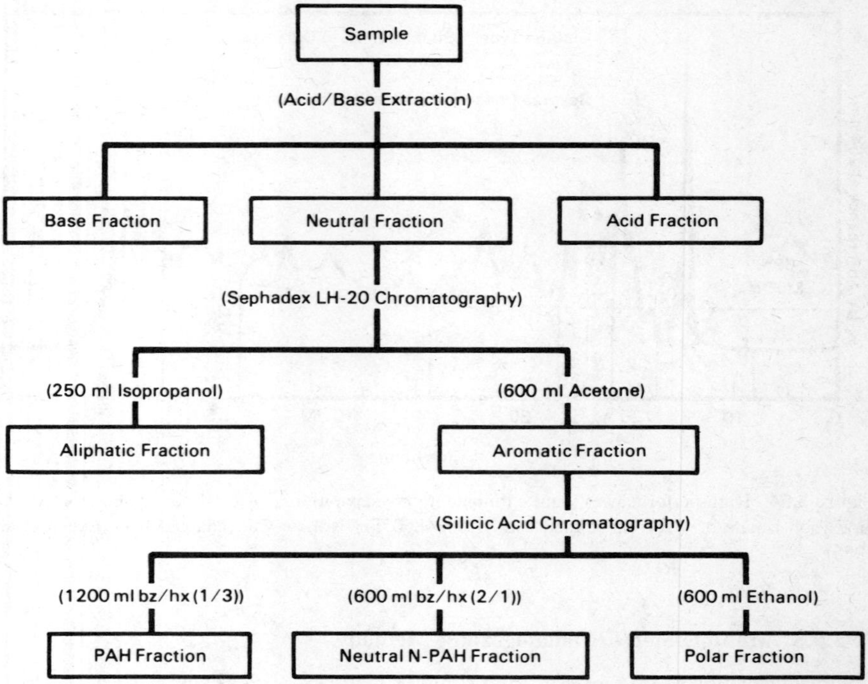

Figure 5.35 Isolation of polycyclic aromatic compounds from coal-derived crude oils (adapted from Ho et al., 1980a). PAHs = polycyclic aromatic hydrocarbons; N-PAHs = nitrogen-containing PAHs.

carbazoles, phenylindoles, benzocarbazoles, and naphthylindoles (Ho et al., 1979, 1981b). Fraction III contained mostly hydroxy-PAHs, although trace levels of aza-PAHs and amino-PAHs (e.g., aminophenylnaphthalenes) were also detected (Ho et al., 1979, 1981b, 1983). The PACs in the neutral fraction presumably reflected incomplete partitioning of these compounds in the initial acid–base extraction. Figure 5.36 shows secondary separation of the ether-soluble base fraction, containing mostly 3°-PANHs and amino-PAHs. Purification of the N-PACs involves two chromatographic steps: the first on basic alumina, followed by Sephadex LH-20 (Ho et al., 1980a,b, 1981b,c, 1983). The benzene fraction contained mainly nitrogen-substitute amino-PAHs and strongly basic aza-PAHs, such as pyridines, quinolines, acridines, benzacridines, and their alkylated homologs (Ho et al., 1979, 1981b, 1983). The alumina–ethanol fraction was purified on Sephadex LH-20 (see Fig. 5.36) to give an isopropanol subfraction, containing partially hydrogenated one- and two-ring aza-PAHs, and an acetone subfraction, containing primarily multiring aza-PAHs and amino-PAHs ranging from one to four aromatic rings (Ho et al., 1979, 1983; Buchanan et al., 1981). Ho et al. (1983) added a third step to separate aza- and amino-PAHs by basic alumina adsorption chromatography. Detailed, compound-specific analyses were performed on each subfraction by GC, MS, and spectrophotometric (UV and IR) techniques. Bioassays have also been conducted with fractions from this separation scheme for various coal-, petroleum-, and shale-derived materials.

Haugen et al. (1982a,b) used an initial acid–base extraction in conjunction with normal-

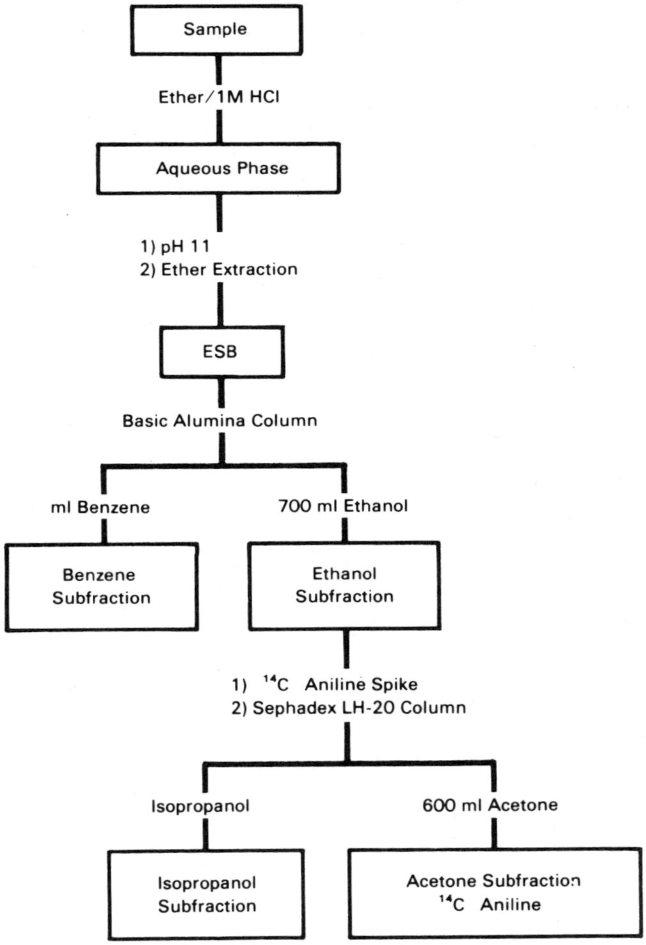

Figure 5.36 Isolation of ether-soluble bases or nitrogen-containing polycyclic aromatic compounds (adapted from Ho et al., 1983). ESB = ether soluble base.

phase, reverse-phase, and ion-exchange HPLC of neutral and basic fractions to characterize PAHs and N-PACs in coal-gasification and -liquefaction process materials. The method incorporated modifications in the acid–base extraction step suggested by Boparai et al. (1983) to ensure quantitative recoveries of N-PACs in the basic fraction.

Multiple-step separation methods that initially involve Sephadex LH-20 gel chromatography have also been reported (Toste et al., 1982, 1983; Dutcher et al., 1983a,b; Royer et al., 1983). Toste et al. (1982, 1983) fractionated SRC liquids by LH-20 partition chromatography, then subfractionated each of three eluates from the gel column, using an amino-silane column and a ternary-solvent mobile-phase HPLC system. PAHs and N-PACs eluted in the nonpolar and moderately polar regions, respectively; hydroxy-PAHs eluted in the polar region (see Figs. 5.29 and 5.30). Another scheme initially involving Sephadex LH-20 chromatography is shown in Fig. 5.37 (Dutcher et al., 1983a,b; Royer et al., 1983).

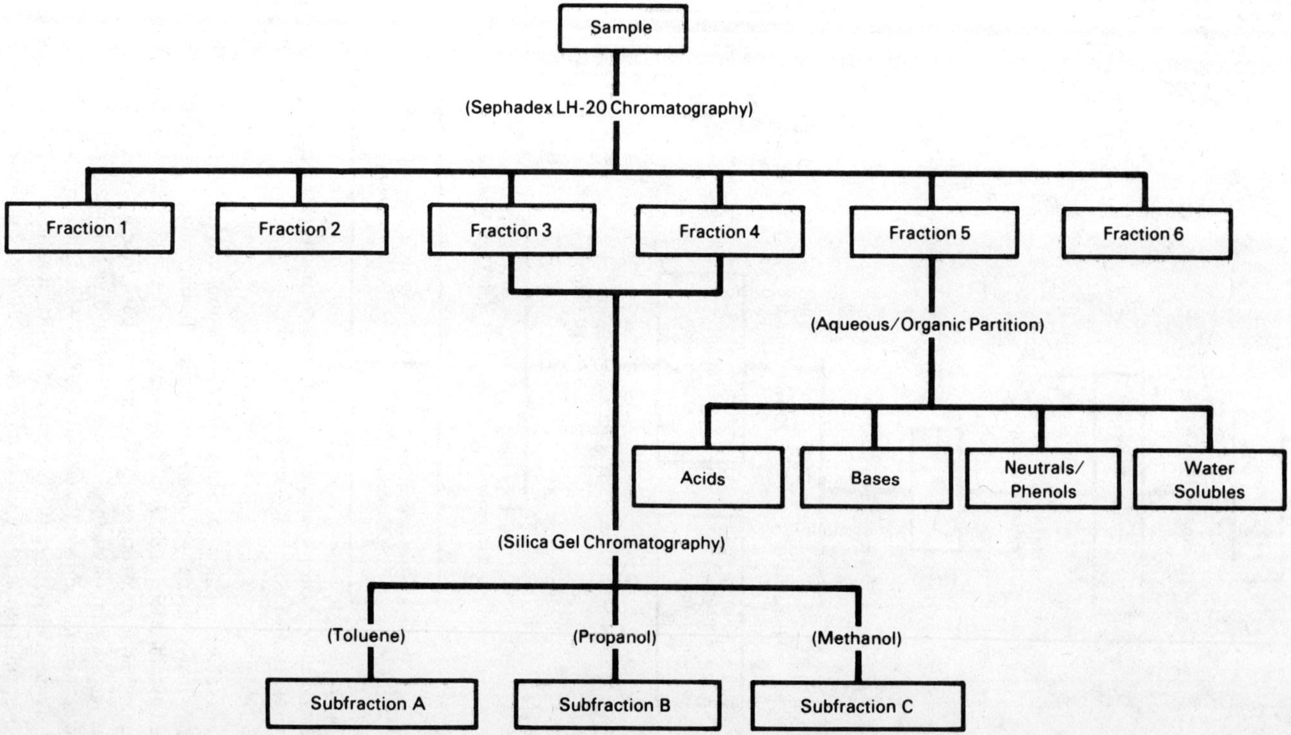

Figure 5.37 Fractionation of coal-gasification condensate tars (adapted from Royer et al., 1983).

Neutral PAHs were isolated in subfraction A. Quinoline/benzoquinoline-type aza-PAHs eluted in the basic isolate of the fifth LH-20 gel fraction. Pyrrolic-PAHs were fractionated into a neutral/phenol isolate by acid–base extraction chromatography. Both mutagenicity and cytotoxicity assays were conducted on these isolates (Hill et al., 1983; Royer et al., 1983).

Multiple-step column chromatographic schemes have also been reported that separate not only PAC classes but also subclasses, such as N-PACs and PAHs. In one example (Figs. 5.38), neutral PACs were isolated initially from N-PACs and hydroxy-PACs by adsorption chromatography on neutral alumina. Next, pyrrolic-PAHs were isolated from N-PACs, and partial separation of amino-PAHs and 3°-PANHs was achieved by silicic acid adsorption chromatography. Subsequently, the amino-PAH-rich isolate was derivatized with pentafluoropropionic acid anhydride, quantitatively forming fluoroalkylamide derivatives of only amino-PAHs. The latter were separated by gel-permeation chromatography. The derivatized amino-PAHs and remaining 3°-PANHs were separated completely (Later and Lee, 1983; Later et al., 1983a). The derivatized amino-PAH isolate was then hydrolyzed to regenerate free amino compounds (Later et al., 1983a). Neutral PACs were subfractionated by reverse-phase HPLC (Fig. 5.40). This separation procedure isolates neutral PAH and N-PAC subclasses into homogeneous fractions that can be tested biologically and characterized chemically, with minimal interference from other closely related PAC classes. Furthermore, chemical class distribution of various PACs in coal products can then be obtained by gravimetric measurements. Detailed analysis of PAC class-specific fractions is achieved by capillary column GC and MS (Later et al., 1981a, 1983a; Later and Lee, 1983; Lucke et al., 1985).

Table 5.12 lists the principal organic compound classes found in coal-conversion materials and shows into which fractions they elute, or would be likely to elute, for the various separation methods discussed here. This table is meant as an overview only. The molecular weight, degree of alkylation, and isomerism of PACs can have an effect on their separation characteristics. In the solvent–solvent type fractionations, insoluble or tar fractions, made up mainly of high-molecular-weight and/or very polar compounds, including multiheteroatomics, often contain lower-molecular-weight, less polar compounds that have been intercalated or "trapped" in the tars.

In summary, isolation of hydrocarbon compounds in fossil fuels has received considerable attention. Currently available methods allow class-specific separations of PAC classes from organically complex coal-liquefaction, -gasification, and -carbonization materials. Analytical techniques that have been used for compound-specific characterizations of isolates from these separation methods are described below.

5.10 CHEMICAL DERIVATIZATION AND MODIFICATION

5.10.1 Oxidation/Reduction Reactions

Oxidation reactions with chemical modifying agents such as sodium dichromate or potassium permanganate have been used to elucidate the structure and composition of coal and its derived products. Friedman et al. (1965) developed a method involving aqueous $Na_2Cr_2O_7$ oxidation of alkyl-substituted PACs to corresponding acids with a minimum of ring destruction (i.e., 2,6-dimethylnaphthalene is oxidized to corresponding 2,6-naphthalenedicarboxylic acid, with >99% yield). Duty et al. (1980) and Hayatsu et al. (1978) used the method of oxidize alkyl-PACs in lignite and its solvent-refined liquefaction product. Reaction prod-

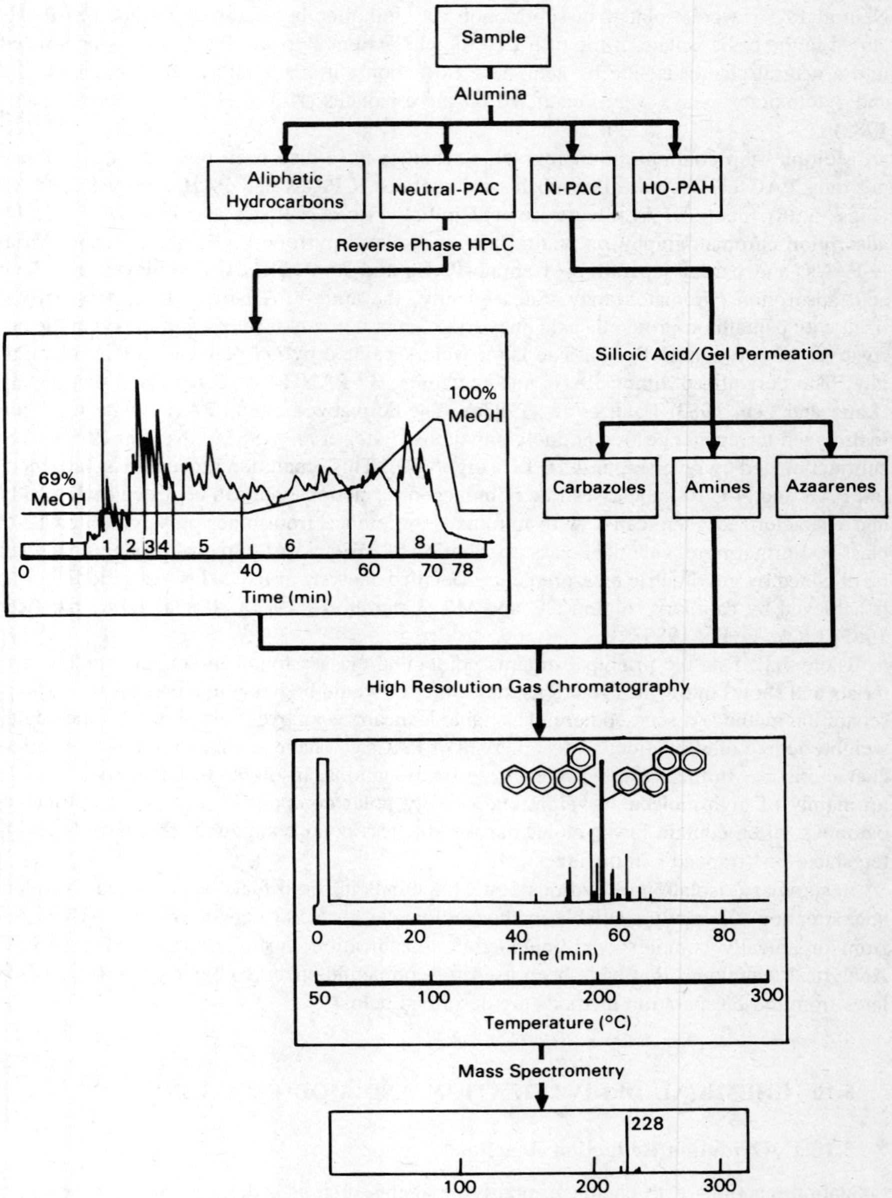

Figure 5.38 Overview of a comprehensive separation scheme (adapted from Later et al., 1983b). PACs = polycyclic aromatic compounds; N-PACs = nitrogen-containing PACs; HO-PAHs = hydroxy polycyclic aromatic hydrocarbons; HPLC = high-performance liquid chromatography.

Table 5.12 Comparison of Chemical Separation Procedures Used in Coal-Conversion Materials Separations

	Separation Method				
Compound Class	Isooctane	Ether	LH-20	Alumina	HPLC
AHs	Neutral	Neutral	Toluene/hexane	AHs	Hexane
Mono- and diaromatics	Neutral/PAHs	Neutral/aromatic	Toluene/hexane	AHs/PAHs	Hexane
Hydroaromatics	Neutral			AHs/PAHs	Hexane
Neutral thiophenes	Neutral/NT	Neutral/aromatic	—	PAHs	—
Carbazoles	Basic/BT	ESB/neutral/aromatic	—	N-PACs/S1	MeCl
Azaarenes	Basic/BT	ESB	—	N-PACs/S3	MeCl
Amino-PAHs	Basic/BT	ESB	—	N-PACs/S2	MeCl
Cyano-PAHs	Basic	ESB/WBS	Methanol	N-PACs/S1	MeCl
Phenols	Acid	WSA	MeOH	H-PACs	Isopropanol
Hydroxy-PACs	Acid/tar	WSA/ESA	MeOH	H-PACs	Isopropanol
Multiheteroatomic (high MW)	NT/BT/AT	Insol A&B	MeOH	H-PACs	Isopropanol

Key: AH = aliphatic hydrocarbons; AT = acid tar; BT = basic tar; ESA = ether-soluble acids; ESB = ether-soluble bases; HPLC = high-performance liquid chromatography; Insol A&B = insoluble acids and bases; MW = molecular weight; N-PACs = nitrogen-containing polycyclic aromatic compounds (PACs); H-PACs = hydrogen-containing PACs; NT = neutral tar; PAHs = polycyclic aromatic hydrocarbons; WSA = water-soluble acids; WSB = water-soluble bases.

ucts were analyzed by GC-MS, and the following conclusions regarding structure of each material were made: (1) Solvent-refined lignite had greater aromatic condensation than feed coal; thus, polynuclear aromatic rings were produced during solvent refining. (2) O-PACs, such as dibenzofuran and anthraquinone, were components of both feed coal and liquefaction product. (3) Dibenzothiophenes were abundantly produced during solvent refining. (4) Pyridine was detected in feed coal and product. (5) Carbazole and benzoquinolines were absent in feed coal but were detected in solvent-refined lignite, indicating that they were products of conversion.

Hayatsu et al. (1981) used a similar oxidative technique with $KMnO_4$ to study structural properties of different coal types. Lignite contained two types of macromolecular material: a kerogenlike material (high hydrogen : carbon ratio) and an aromatic material (low hydrogen : carbon ratio). By comparison, bituminous coal contained only limited long-chain, hydrocarbon or kerogenlike material.

Another oxidative reagent used to study bituminous coal structure was a mixture of hydrogen peroxide and trifluoroacetic (TFA) acid [or trifluoroperoxyacetic (TFPA) acid] (Deno et al., 1978, 1980). This reagent was favored over oxidations with O_2, HNO_3, $KMnO_4$, and $NaCr_2O_7$ because benzylic hydrogen was attacked by the latter, leading to greater destruction of the aliphatic structure. The H_2O_2/TFPA acid method preserves the aliphatic structure despite destruction of aromatic rings. A more correct assessment of the AH constituents that occur in the undisturbed coal lattice can be gained by chemical analyses of the reaction products.

Reductive alkylation of coal has also been used to elucidate composition and structure of coal and coal-derived materials (Sternberg et al., 1971). Coal is believed to contain clusters of condensed aromatic rings (PACs); therefore, introducing alkyl groups into the coal molecule by forming aromatic hydrocarbon anions and then reacting the anions with alkyl halides should theoretically disrupt the coal structure. This method should produce extractable, analyzable fundamental components that can be related to the original coal structure. However, Franz and Skiens (1978) points out difficulties in extracting meaningful information from reductive alkylation experiments with lignites, subbituminous, and other low-rank coals. These problems may reflect side reactions, such as condensation of two radical aromatic hydrocarbons to form a reaction by-product, which may lead to erroneous conclusions about the original coal structure.

Oxidation of thiopenic-PACs in coal-liquefaction and -gasification products to their corresponding sulfonyl derivatives has enabled their complete isolation and biological characterization. This method was discussed above.

5.10.2 Acetylation

Derivatization of reactive functional PACs, such as hydroxy and amino groups, is an alternative for characterizing constituents in coal-derived materials. Derivatization by acetylation has widely been used for polar functional compounds and has also been used to characterize coal products. The technique has helped elucidate the composition of mutagenic amino-PAHs found in coal-conversion products.

Derivatization with acid anhydrides, such as acetic anhydride, TFA anhydride, or PFP anhydride, quantitatively converts amino-PAHs to their corresponding amide derivatives, as shown below.

The reaction is selective for amino-PAHs in the presence of pyridine-type aza-PAHs and has been used to differentiate these N-PAC classes. Reaction yields with complex mixtures are quantitative and typically exceed 90% conversion, depending on amino-PAH structure (Masuda and Hoffman, 1969; Later et al., 1982b).

Felice (1982) derivatized aniline and its alkyl homologs with acetic anhydride to distinguish them from alkyl pyridines in N-PAC isolates of an SRC-II blend and its aqueous extracts. Later et al. (1981a, 1982b, 1983a) derivatized amino-PAH-enriched isolates of SRC liquefaction and Lurgi gasification materials (Later and Lee, 1983) with TFA anhydride and PFP anhydride. The method enabled selective detection and identification of derivatized amino compounds by capillary-column GC in conjunction with electron-capture detection and HRMS, in the presence of more concentrated 3°-PANHs like benzoquinolines (Later et al., 1982b). Complete separation of derivatized amino-PAHs from other N-PAC classes was achieved by gel-permeation chromatography (Later et al., 1982b, 1983a). Catalytic hydrolysis of fluoralkylamides allowed regeneration of "free" amino compounds. Thus, bioassays of complex amino-PAH mixtures isolated from coal-derived materials were possible. Similarly, TFA anhydride was used as the derivatizing agent to determine qualitatively and quantitatively the amino-PAHs in basic fractions of gasifier tars, shale- and coal-derived oils, and petroleum crudes (Tomkins and Feldman, 1980; Tomkins and Ho, 1982).

5.10.3 Nitrous Acid Treatment

Nitrous acid treatment substantially decreases amino-PAH-induced mutagenicity in complex mixtures by converting amino-PAHs to their corresponding phenols, which are mutagenically inactive (Yoshida and Matsumoto, 1978). Such treatment has been useful in attributing the mutagenic activity of certain complex mixtures to amino-PAHs (Pelroy and Wilson, 1981).

Haugen et al. (1981a) suggested the possibility of side reactions forming nitro-PAHs in test mixtures. Completeness and specificity of the reaction were addressed by B. W. Wilson and co-workers (unpublished data; see Fig. 5.39). They compared selected ion gas chromatograms of nitrous-acid-treated and -untreated amino-PAH standards and SRC-II chemical fractions. Evidence from nitrous acid reaction with secondary and tertiary nitrogen functionalities was also considered. Nitrous acid treatment essentially eliminated by C_1-aminophenanthrene ion at m/z 207 in the basic tar fraction (Fig. 5.39); there was little, if any, effect on earlier-eluting carbazole and aza compounds. Mutagenicity assays generally indicated loss of indirect activity for most *S. typhimurium* strains investigated. However, for some compounds, direct-acting mutagens were formed by nitrous acid treatment. Regardless of whether nitrosation of PAHs occurred, substantial reduction or elimination of indirect-acting mutagenic response after nitrous acid treatment appears to be diagnostic for mutagenic amino-PAHs and demonstrates their role as chemical mutagens.

5.11 ORGANIC ANALYSIS

5.11.1 High-Performance Liquid Chromatography

High-efficiency reverse-phase HPLC separations of PAHs have been achieved (Fig. 5.40). Reliable application of such procedures to actual samples often requires (1) preseparation procedures to isolate the desired class of aromatics and (2) use of optical detection systems that provide selectivity. An abundance of alkylated PACs within a given class markedly

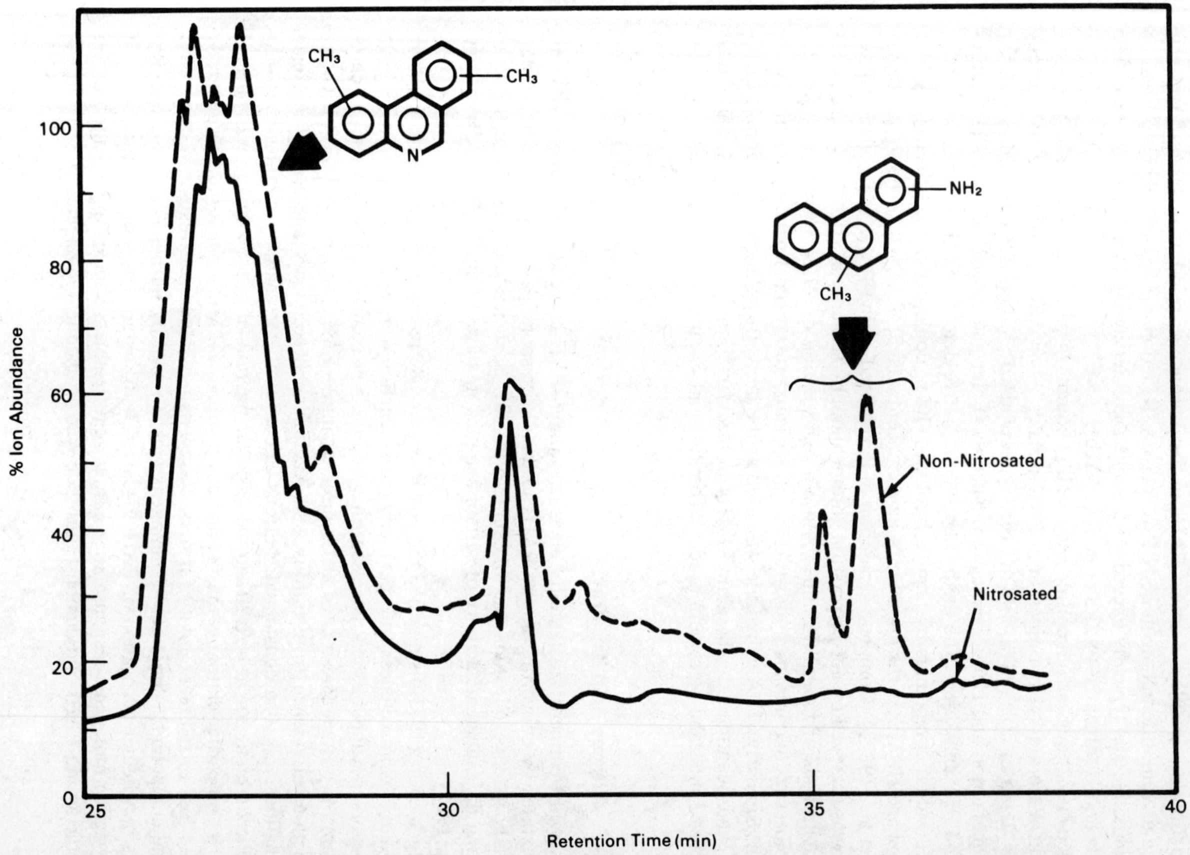

Figure 5.39 Selected ion chromatogram of the m/z 207 for nitrosated (solid line) and nonnitrosated (dashed line) basic fraction superimposed to illustrate the absence of methylaminophenanthrene in the nitrosated material (B.W. Wilson and R.A. Pelroy, unpublished data).

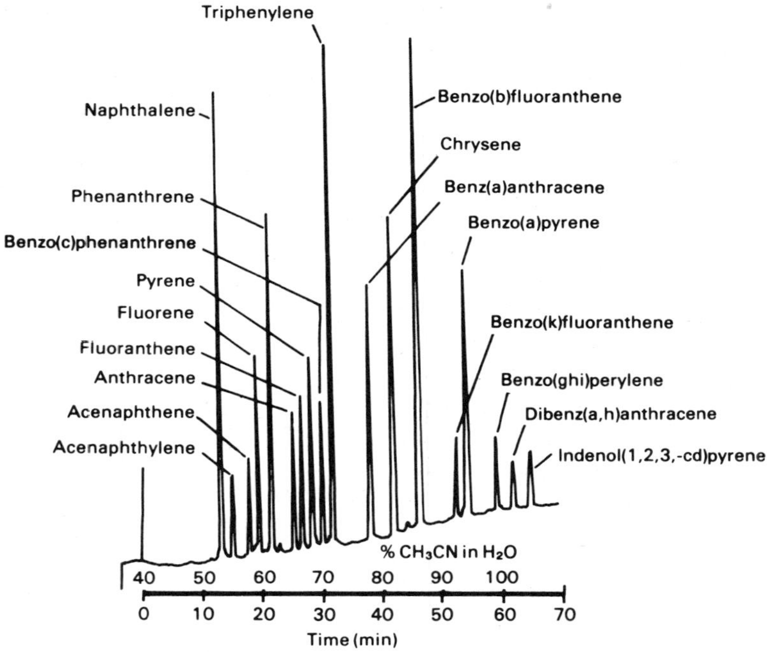

Figure 5.40 Reverse-phase liquid chromatographic separation of polycyclic aromatic hydrocarbons. Column: Vydac 201TP reverse phase; detection: ultraviolet absorbance at 254 nm; conditions: linear gradient from 40 to 100% acetonitrile in water at 1%/min at 1 mL/min (adapted from Wise et al., 1980).

limits resolution. For example, although phenanthrene and chrysene can readily be resolved (C_{14} and C_{18}, respectively), the retention of C_4-phenanthrenes is comparable to that for chrysene. Under these circumstances, definitive qualitative analysis requires subsequent complementary analyses such as GC and GC-MS.

Routine application of HPLC for chemical analysis was greatly accelerated by the parallel development not only of packing materials but also solvent delivery systems and detection devices. Theoretical and practical aspects of HPLC analysis are described in a comprehensive text by Snyder and Kirkland (1979). Commercially available instruments allow automated gradient elution of chemicals with day-to-day variation in retention times of $<2\%$ and variation in relative retention times of $<0.5\%$. This precise delivery solvent, coupled with the use of commercially packed columns (void volume 2–3 mL) containing particles of uniform size (e.g., 5 or 10 μm, $\pm 10\%$), shape, and porosity, allows routine chromatographic peak widths to be less than 10 s, with typical flow rates of 1 mL/min. Optical detection systems necessarily have flow cells with internal volumes of ≤ 10 μL and appropriate response times and stability.

HPLC analyses are facilitated by the use of selective detection systems using (1) dual-wavelength absorbance detection to record or compute characteristic absorbance ratios and (2) fluorometric detectors with excitation and emission wavelengths selected to maximize response to the desired analyte and minimize response to chromatographically similar compounds. Although not yet routinely applied, mass spectral as well as rapid scanning absorbance and fluorescence detectors can be used for qualitative and quantitative analysis of numerous compounds during a single separation.

The sensitivities of absorbance detection systems (typically at 254 nm) are less than 50 ng. Depending on their UV spectra, however, the response factors for single-wavelength absorption detection vary widely among the PAHs. Refractive index monitors provide a more consistent measure of mass, but their relative insensitivity rules out their use on an analytical scale. The sensitivities for selective fluorometric detection of PAHs are 10–100 pg; absorbance and fluorescence are often used in series to take advantage of their respective universal and selective capabilities. Electrochemical detectors, while modifying the analyte, provide both selectivity and sensitivity. They have generally been useful for analysis of amino-PAHs and hydroxy-PAHs, but principal applications have not yet included coal-conversion materials.

Heterocyclic PASH fractions of an SRC coal liquid were characterized by HPLC techniques (Wise et al., 1982). S-PACs were initially isolated by the sulfone oxidation reduction method of Willey et al. (1981a). The S-PAC fraction was then separated into four subfractions according to number of aromatic rings, using normal-phase HPLC on a semipreparative aminosilane column. Finally, compound-specific identifications were achieved by comparing retention data of sample components and standard compounds with an HPLC reverse-phase C_{18} column. Several three-, four-, and five-ring parent S-PACs, and their alkylated analogs, were identified.

5.11.2 Gas Chromatography

Perhaps the most detailed characterization of lower-molecular-weight species (<300 amu) in coal-conversion materials has been achieved by capillary-column gas chromatography (GC). Lee and Wright (1980) reviewed operational variables of capillary GC (column materials and dimensions, stationary phases, carrier-gas types, and velocities) for analysis of PACs in coal tar. Generally, nonpolar to slightly polar stationary phases, such as OV-1, OV-101, SP-2100, SE-52, and SE-54, coated on glass and fused silica capillary columns, have been used for coal tar analyses. Polarity is determined by increasing percentages of phenyl groups in methylsiloxane polymer. A phase of cross-linked 70% phenyl, 4% vinyl, and methylphenylpolysiloxane provides excellent thermal stability and has been used to separate, at temperatures of 400°C, low-volatility PACs with molecular weights up to 400 amu (Peaden et al., 1982). Schomburg et al. (1978) and Borwitzky and Schomburg (1979) used polar phenyl phases of OV-7 (20% phenyl and methylsilicone), OV-61 (33% phenyl and methylsilicone), and Poly S 179 (polyphenylether sulfone) on various glass capillary columns to separate coal tar PACs. The latter phase demonstrated high temperature stability (390°C) and eluted several dibenzofluoranthene/pyrene (molecular weight = 302) and coronene (molecular weight = 300) isomers.

Kong et al. (1982a,b) applied a novel mesogenic polysiloxane phase to separate difficult-to-resolve coal tar PAC isomers such as triphenylene/chrysene and benzofluoranthenes. Novotny et al. (1981) fractionated high-temperature coal tar with a liquid–liquid partitioning scheme and determined the presence of AHs, PAHs, PANHs, PASHs, and hydroxylated-PAHs by GC on capillary columns.

Schultz et al. (1979) separated polynuclear aromatic and aliphatic compounds in an SRC recycle oil with a solvent partition procedure and identified over 146 constituents by capillary GM-MS. Likewise, GC-MS with a methylsilicone-coated, fused-silica capillary column allowed characterization of PACs, AHs, and phenols in fractions of an SRC-I PS obtained by preseparation using HPLC (Burke et al., 1981).

Oxygen is important in the cross-linking of aromatic moieties in coal macerals. Compounds with various oxygen functionalities are prevalent in coal-derived materials. Snape et al. (1981) used supercritical gas extraction to isolate aliphatic carboxylic acids from lignite and subbituminous coals. After cold diethylether precipitation and methylation of extracts, capillary GC on an OV-1 WCOT column allowed the determination of C_{20}–C_{34} acids. These workers postulated that some related, straight-chain alkanes were derived from the straight-chain alipathic carboxylic acids during maturation of coal. High concentrations of phenolic compounds found in coal-derived materials are of concern because of their toxicity. SE-52-coated capillary columns have been used to analyze hydroxylated aromatic compounds (Later et al., 1981a) and are adequate for the rapid screening of such fractions. However, resolution of all isomeric alkyl phenols has not been achieved with slightly polar phases such as SE-52; thus, polar stationary phases are required.

Because oxygen-, nitrogen-, and sulfur-functional aromatic compounds are often found, element-specific or -selective detectors, in conjunction with capillary GC, have been of value in characterizing coal-derived materials. Kapila and Vogt (1981) used tandem flame ionization detection (FID) and photoionization detection to distinguish aromatic from aliphatic by-products of gasifier tars and water residues. Classification of aromatic, alkane, or alkene hydrocarbons was based on the normalized relative molar response to the eluted peak in both detectors. Brady (1975) described a tandem capillary GC vapor-phase IR spectrometry system used to differentiate cresol isomers in a coal tar distillate. Components isolated by capillary-column GC were trapped separately in a vapor cell, and their IR spectra were recorded. Similarly, matrix isolation of capillary effluents on cryostated disks and Fourier transform IR (FTIR) spectrometric analysis of PACs in SRC materials were reported by Hembree et al. (1981). Perhaps the most versatile selective detectors for coal-derived materials are the nitrogen/phosphorus detector (NPD) and the flame photodetection (FPD, sulfur and phosphorus element-selective). Isolated sulfur heterocycles from coal liquids were analyzed by dual-trace FID-FPD (Lee et al., 1980; Willey et al., 1981a,b). Yang and Cram (1979) used FPD in both sulfur- and phosphorus-selective modes to analyze unfractionated coal-hydrogenation products.

NPD was used to analyze anilines and pyridines in SRC distillates and aqueous extracts (Felice, 1982). N-PACs in SRC-II HD were analyzed by FID-NPD dual-trace GC, using a fused-silica capillary effluent splitter (Later et al., 1981c). Likewise, Gangwal (1981) used dual detectors, both FID-FPD and FID-NPD, for the capillary analysis of bituminous coal and lignite gasification tars that had been prefractionated by solvent partitioning.

Use of element- or functional-group-selective detection methods enhances the data obtained from high-resolution capillary separations of coal-derived materials by providing compound composition information. For example, selective detection of amino-PAHs in coal liquids can be achieved by converting these compounds to fluoropropylamides by derivatization with pentalfluoropropionic anhydride. Subsequent separation is achieved by capillary chromatography and detection and identification are achieved with an ECD and mass spectrometer (Later et al., 1982b, 1983a).

Capillary GC can be used to determine physical and chemical properties and "fingerprint" coal and coal-derived materials. Solash et al. (1978) used an Apiezon L capillary column to study the relation between fuel properties and chemical composition of jet fuels refined from coal oils. Fingerprinting analysis of coal extracts and profiling of AHs and aromatic hydrocarbons on capillary columns coated with OV-1 were used to relate chemical composition to coal rank (Alexander and Hazai, 1981). Capillary-column GC has been used to study process treatments and modification reactions of coal and coal-derived materials. Examples include (1) oxidative reactions to correlate reaction products to coal structure

(Hayatsu et al., 1978, 1981), (2) reductive alkylation for the same purpose (Franz and Skiens, 1978), (3) hydrogenation of coal oil to improve fuel quality (Philip et al., 1981; Wilson et al., 1981a), and (4) selective distillation to reduce concentrations of mutagenic components of coal oils (Wilson et al., 1982b).

5.11.3 Supercritical Fluid Chromatography

The use of capillary GC for the analysis of coal-derived compounds provides excellent separation for AHs and PACs that have sufficient volatility to be amenable to this technique. In general, the upper molecular weight range for PACs that can be determined effectively by GC is approximately 300 daltons; equivalent to six to seven condensed aromatic rings. Peaden et al. (1980) reported the separation and identification of coal-tar-derived PACs of molecular weight greater than 400 daltons by reverse-phase HPLC. With supercritical fluid chromatography (SFC), however, even higher-molecular-weight PACs can be separated and analyzed. In SFC, the chromatographic mobile phase, which can be either an inorganic or organic medium, is subjected to temperatures and pressures near or above its critical parameters. A supercritical fluid has chemical, chromatographic, and physical properties intermediate between those of a gas and a liquid and can be used to perform separation of nonvolatile and high-molecular-weight PACs.

Smith and Udseth (1983b) interfaced on supercritical fluid extraction cell directly to the mass spectrometer for on-line introduction of extracted coal. Pentane modified with 5% isopropanol was used for direct fluid injection of the extract into the mass spectrometer; analyses of compounds with molecular weights in excess of 600 daltons were observed in the mass spectra. Smith and co-workers (1984) also reported on the rapid analysis of coal-derived PACs by capillary SFC-MS; analyses of compounds having from two to six condensed aromatic rings were achieved in less than 6 min.

Wright and Smith (1984) used capillary SFC for the analysis of the chemical class fractions of MD fuels. A supercritical pentane system equipped with fluorescence detection has been used for the analysis of a PAH standard with up to nine condensed aromatic rings. The system was also used to separate high-molecular-weight carbon blacks, coal tars, and coal-derived vacuum still bottoms (Fjeldsted et al., 1983). The vacuum still bottoms had a molecular weight range of 200–800 daltons, as determined by field desorption MS.

5.11.4 Mass Spectrometry

Mass spectrometry (MS) is probably the most important analytical tool for elucidating the chemical basis for toxicological responses to coal-derived materials (Ho et al., 1979, 1980b, 1981a,b; Wilson and Pelroy, 1979; Wilson et al., 1980, 1981a,b, 1982c; Willey et al., 1981a,b; Buchanan, 1982). MS has also contributed to our knowledge of the organic constituents in coal itself (but it is not discussed here). Many generic coal structures presented earlier were deduced from mass spectral data. MS techniques are varied, providing different information and allowing several approaches to confirm results. Of principal interest here are techniques that yield chemical data relative to biological activity of coal liquids.

HRMS analysis uses magnetic sector instruments to yield elemental compositions for organic compounds. High-resolution instrument designs generally incorporate an electrostatic sector that reduces energy dispersion of the ion beam. Because of double focusing, these instruments operate with decreased momentum dispersion, allowing more precise measurement of ion momentum and therefore of ion mass. Modern high-resolution, double-focusing instruments resolve in excess of $\Delta m/m = 100{,}000$. For example, a mass resolution of 56,000 is required to distinguish a compound such as C_4-benzothiophene ($C_{12}H_{14}S$), with a molecular weight of 190.0816 amu, from 4,5-methanophenanthrene, which has the same nominal molecular weight but an exact mass of 190.0782 amu.

Techniques are available to introduce samples into the MS ion source in the solid, liquid, and gas phases. In GC-MS, the sample is introduced into the MS ion source directly from the capillary column of a gas chromatograph. High-resolution, small-bore GC columns generally have carrier gas effluent rates that are compatible with current available MS high-vacuum systems and generally do not require an interface to remove the bulk of the carrier gas, as is necessary with higher flow rates associated with packed columns. However, MS scan rates in this mode must be sufficient to acquire several spectra during the elution time of a single GC peak, which for HRGC columns may be only a few seconds in duration.

Solids and liquids may be introduced via a heated inlet probe or heated batch inlet system for EI, chemical ionization (CI), or FI operation. Condensed-phase samples may also be loaded directly on an emitter surface for field desorption mass spectrometry (FDMS) or onto a flat surface probe for fast-atom bombardment (FAB) (Barber et al., 1982). Ionization modes of interest for coal-derived products are conventional EI, at both low ionization potential (10–20 eV) and at 70 eV, as well as FI, CI, and, to a lesser extent, particle bombardment techniques, such as ^{252}Cf plasma desorption (PD) and FAB.

Conventional mass spectral analyses use a collimated electron beam at 70 eV to impact sample molecules in the vapor phase to strip away an electron, leaving a positively charged molecular ion, which remains intact or fragments to yield one or more product ions and their associated neutral species. Product ions are generally specific, to some extent, by molecular ion structure and therefore often convey structural information. Because of the aromatic nature of coal-derived materials, mass spectra obtained from coal liquids or their chemical fractions may be relatively simple compared with those from petroleum or oil shale materials. For example, with conventional EI ionization, most nonalkylated-PAHs yield intense molecular ions with little fragmentation.

Conventional EIMS, at both 70 eV and low ionization potential, have been widely used to characterize petroleum liquids for engineering and process evaluation. Numerous studies (Scheppele et al., 1976; Aczel et al., 1977; Anbar, 1977; Schiller, 1977; J. L. Schultz et al., 1979; R. V. Schultz et al., 1979, Scheppele, 1978a,b; Benson et al., 1979; Bodzek et al., 1979; Bodzek and Marzec, 1981) have extended these methods to analysis of coal-hydrogenation products. Development and use of 70-eV and low-voltage ionization techniques to study coal-derived materials was reviewed by Scheppele (1978b).

Swansiger et al. (1982) used low-resolution MS group-type analysis to quantitate transferable hydrogen in SRC liquids. Transferable hydrogen corresponds roughly to hydroaromatic content. Thus, values obtained for transferable hydrogen by MS methods can be compared roughly to those obtained by NMR. An advantage of MS in this type of analysis is that it allows identification of specific compounds bearing transferable hydrogen.

Reducing the ionization potential in the source from 70 to less than 20 eV substantially reduces the fragmentation of many compounds found in fossil fuels. This approach, termed LVMS analysis, has been extensively in the petroleum industry and has been applied to coal-derived materials (Aczel et al., 1977; Kekin et al., 1977; Later et al., 1984b). Later et al. (1984b) used LVMS to characterize EDS materials and determine more accurately the degree of alkylation associated with certain carcinogenically active PAH constituents. Scheppele et al. (1976) compared the relative sensitivities of LVMS and FIMS for the analysis of coal-derived materials.

FIMS, in which molecules in the gas phase are ionized with local electrical potentials on the order of 16^6 V/m (Beckey, 1971), has been used to characterize high-boiling and nondistillable coal liquids. Here, ionization takes place at energies equal to or less than the ionization potential of the sample compound. Therefore, there is little or no fragmentation of molecular ions compared to that obtained by EI, particularly at 70 eV.

Use of FIMS with coal or coal liquids was first reported by Anbar (1977). Since then, the method has been used extensively, especially to analyze HPLC fractions from coal liquids that were separated according to the number of double bonds. Assuming the separation procedure was effective, FIMS analyses of the fractions identified several polycyclic structures having more than six rings. High-resolution analysis is difficult with FIMS because of inherent uncertainties in the point of ionization and the large energy spread. Therefore, identifications based on FIMS alone are not generally confirmed by elemental composition data. Use of FIMS to determine and characterize compounds in coal-derived materials has been reported by Scheppele et al. (1977), Whitehurst et al. (1982), and Boduszynski et al. (1983a,b).

FDMS is closely related to FIMS and uses essentially the same ion source configuration. However, with FDMS, the sample is loaded directly on the emitter and desorbed directly from the condensed phase in an electrostatic field. The method does not require heating the sample on an inlet probe, although heating the emitter is generally necessary for less-volatile samples (St. John et al., 1977).

FAB has been evaluated as an ionization method to identify N-PACs in coal liquids (Grigsby et al., 1982). $(M + H)^+$ and $(M - H)^+$ species were formed for standard N-PAC compounds, along with the M^+ molecular ion. However, except for losses from alkyl or hydroxyl substituents, little fragmentation occurred.

Several approaches have been investigated for analyzing complex mixtures using CI. For example, Hunt et al. (1972) showed that primary, secondary, and tertiary amino functionalities can be distinguished using isotopically labeled reagent gases. Figure 5.41 shows

Primary Amines				
	$RNH_2 + NH_4^+$ M	$\Rightarrow$	RNH_3^+ M + 1	
	$+ ND_4^+$	$\Rightarrow$	RND_3^+ M + 4	>3
Secondary Amines				
	$R_2NH + NH_4^+$ M	$\Rightarrow$	$R_2NH_2^+$ M + 1	
	$+ ND_4^+$	$\Rightarrow$	$R_2ND_2^+$ M + 3	>2
Tertiary Amines				
	$R_3N + NH_4^+$ M	$\Rightarrow$	R_3NH^+ M + 1	
				>1
	$+ ND_4^+$	$\Rightarrow$	R_3ND^+ M + 2	

Figure 5.41 Chemical ionization of nitrogen-containing aromatics using ammonia and perdeuteroammonia (adapted from Buchanan et al., 1983).

reactions that distinguish three types of N-PAC. Buchanan et al. (1983) applied CI to analyze N-PACs in several coal liquids and expanded the use of isotopically labeled reagent gases to analyze oxygenated species. Figure 5.42 shows an analogous scheme for oxygenated species, using methanol. Hydroxyl aromatics exhibit a difference of 2 amu between the molecular ion formed with deuterated and nondeuterated methanol. However, aromatic ethers have only 1 amu difference between molecular ions formed with deuterated and nondeuterated methanol. Because of their ability to eliminate H_2O (or O_2), substituted alcohols show no difference between base peak ions formed with deuterated and nondeuterated methanol.

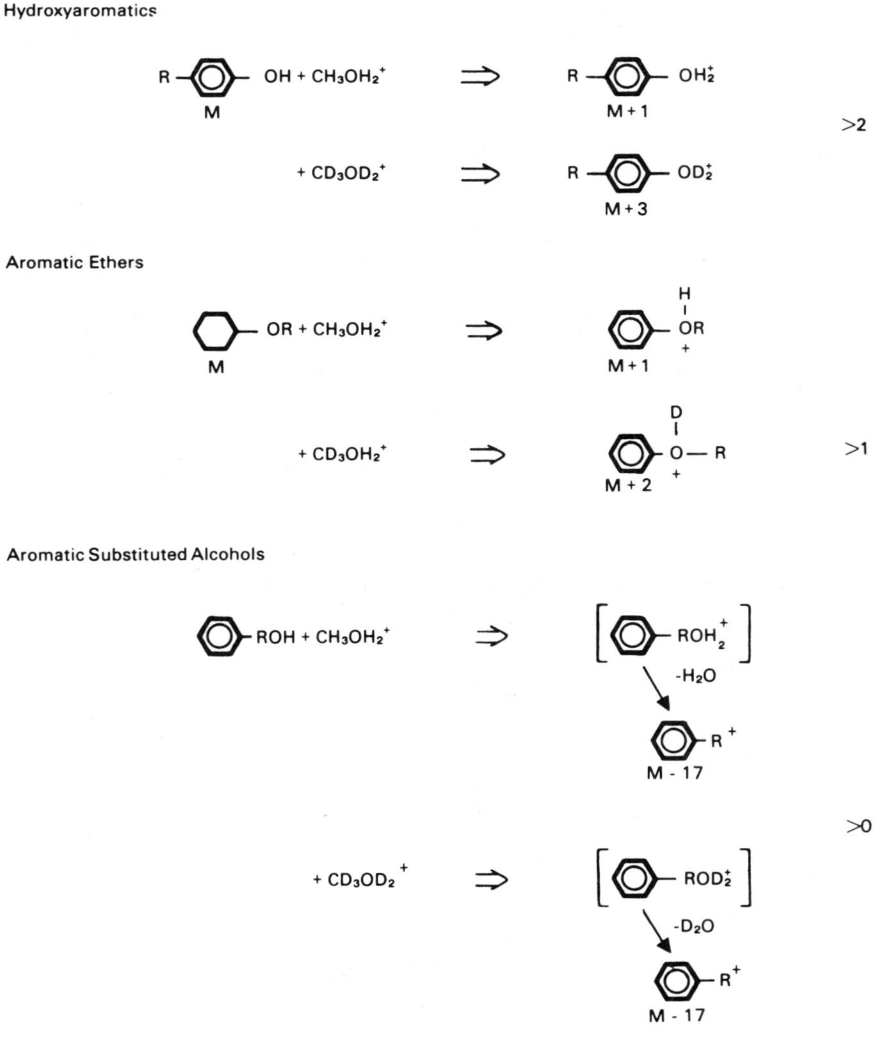

Figure 5.42 Chemical ionization of oxygen-containing aromatics using methanol and perdeuteromethanol (adapted from Buchanan et al., 1983).

Buchanan et al. (1983) recently compiled mass spectral data from N-PAC compounds in coal-conversion materials. Analysis of metastable ions arising from unimolecular or collisionally induced decomposition has been used to determine N-PACs (Hunt and Shabanowitz, 1982), PASHs, and multiheteroatomic species (Wilson et al., 1981b, 1983a) in several hydrocarbon matrices, including coal liquids.

Figure 5.43 shows how certain primary, secondary, and tertiary nitrogen compounds, in which normal mass spectra are essentially indistinguishable, may be determined by metastable-ion analysis. Zackett et al. (1979) and Zackett and Cooks (1981) used metastable-ion energy analysis to distinguish quinoline isomers in SRC liquids. Ciupek et al. (1982a,b) compared low- and high-energy collision-induced spectra of some aza and amine compounds. Wilson et al. (1981b) shows that amino-PAHs may be distinguished from their aza and pyrrolic isobars in complex mixtures by energy analysis of unimolecular ion decompositions.

5.11.5 Infrared Spectrophotometry

Modern infrared (IR) spectroscopy has not been widely used to analyze coal-derived materials with regard to health effect issues. Because of numerous rotational and vibrational degrees of freedom available even to relatively simple organic molecules, IR spectra may be quite complex. The inherent complexity, coupled with the complexity of coal-derived materials and their chemical fractions, limits the application of IR in identifying specific compounds.

Kennaway (1955) identified certain absorption bands that he related to relative carcinogenic potency of coal pitch distillates. These were termed *carcinogen bands* and were later shown to be typical for methylated, catacondensed, four-ring compounds such as benzanthracenes.

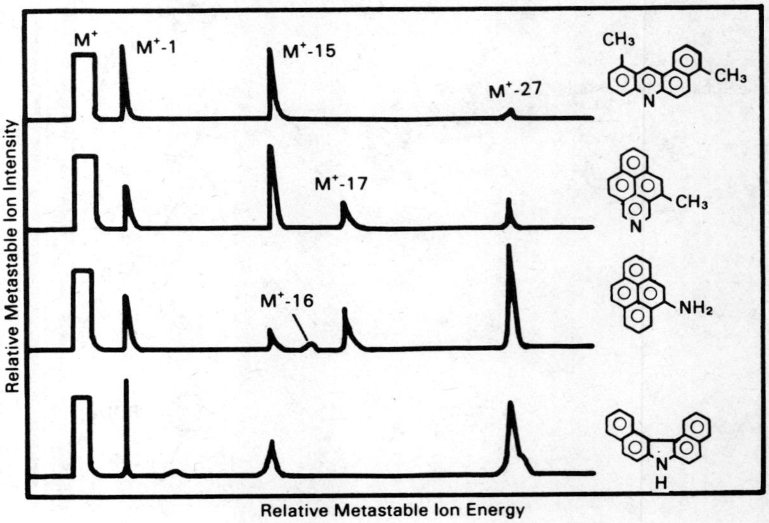

Figure 5.43 Essential features from metastable ion kinetic energy spectra for dimethylbenzacridine, methylazapyrene, aminopyrene, and dibenzacridine. Relative intensities of M-15, M-17, and M-27 ions are characteristic of the various structures illustrated (from Wilson et al., 1981b).

Recent improvements, such as FTIR, increase resolution and the speed of analysis. FTIR has been used as a detector for GC and size-exclusion chromatographic (Brown et al., 1981) separations of coal-liquid fractions. GC-FTIR has also been used to investigate gasifier-derived semivolatiles (Erickson et al., 1976), coal gases (Erickson et al., 1981), and polar compounds (Brown and Taylor, 1983).

IR has also been used to screen certain specific functional groups within coal liquids (Aczel et al., 1977; Ho et al., 1979). Robbins et al. (1984) used an IR method for determining phenolics in the THF-soluble portion of coal liquids. Their method was demonstrated to be rapid (3–4 determinations/h, including sample preparation), reproducible (± 0.1 meg/g, 2σ) on coal-derived distillates and residuum, and accurate (± 0.1 meq/g, 2σ) on a wide range of phenol types. Reports of IR absorption characteristics for the primary amino functionality in the biological active, ether-soluble, base fraction from SRC preceded actual identification of higher-molecular-weight amino-PAHs in these coal liquids by GC-MS (Ho et al., 1979).

5.11.6 Nuclear Magnetic Resonance Spectroscopy

Nuclear magnetic resonance (NMR) techniques have been used to study coal and coal-solvent characteristics in support of process development and engineering. Comparison of data from proton magnetic resonance with those from carbon-13 magnetic resonance (Pugmire et al., 1977; Retcofsky et al., 1977) spectroscopy, using FT techniques, allows accurate estimates of hydrogen : carbon ratios and characterization of proton environments in coal-derived materials.

Carbon-13 magnetic resonance spectra exhibit 40–50-fold increases in chemical shift values compared with proton spectra and allow precise determination of certain structural details. The natural abundance of the ^{13}C isotope is only about 1% that of ^{12}C. Thus, the change of adjacent ^{13}C nuclei is about 1 in 10,000, and interactions among ^{13}C nuclei are not detectable. (^{13}C–^{12}C interactions do not occur because the ^{12}C nucleus is not magnetic.) The low abundance of ^{13}C is also a major disadvantage with this technique: ^{13}C has relatively small magnetic moment, which makes signals difficult to detect. Also, relaxation times for ^{13}C are relatively long. Analyses such as FTIR are generally required to obtain useful spectra. Fischer et al. (1978) and Sklenar et al. (1980) used ^{13}C NMR to study the structure of coal-derived asphaltenes. Hydrogen transfer during distillation of coal-derived materials has also been studied using ^{13}C NMR (Stadelhofer et al., 1980), as have hydroaromatic species in coal liquids (Seshadri et al., 1978a).

NMR has been of particular value in assessing solvent quality and hydrogen transfer during liquefaction or upgrading (Stadelhofer et al., 1980) and in studying hydroaromatics (Seshadri et al., 1978a,b). NMR has also been used to study high-molecular-weight asphaltenes (Fischer et al., 1978; Sklenar et al., 1980), solids (Coleman et al., 1977), and pitches (Drake et al., 1978).

Coleman and Boyd (1982) applied NMR to study hydroxy compounds in coal-derived materials. NMR has also confirmed the relative absence of carbonyl from coal-liquefaction materials.

M. R. Guerin and co-workers (unpublished data) showed a positive correlation between aromaticity of coal liquids, as determined by NMR, and mutagenic activity. Increased aromatic proton values (7.25 ppm downfield from tetramethyloctane as internal standard) correlated with increased mutagenic activity; low activity was found for samples with aromatic proton content below 10%, moderate activity for samples with 10–20% aromatic proton content, and high activity for samples with aromatic proton content above 20%.

5.11.7 Electron Spin Resonance Spectroscopy

Unpaired electron spins, such as those associated with free radicals, can be detected and their environment characterized by their microwave absorbance behavior in an external magnetic field. The effect, termed electron spin resonance (ESR) or electron paramagnetic resonance (EPR), arises from unequal population densities of states available to electrons at ambient, or near-ambient, temperatures. As is the case for other types of electronic spectroscopy, ESR spectra arise when energy is absorbed by electrons—in this case, during transition to a different spin state.

ESR spectrometers are generally capable of detecting free radicals in concentrations of 1.0×10^{11} spins/g or greater. Of primary interest here is the use of ESR to detect and characterize organic free radicals present in coal-derived materials. Free radicals are generally more chemically reactive than their precursor molecules and are thought to have a major role in sediment and sludge formation in fossil fuels, as well as in the process of hydroliquefaction itself. More important is the involvement of free radicals in chemical binding reactions with DNA and the possibility that coal-derived free radical species may have a role in the genotoxicity exhibited by certain coal-derived materials.

ESR techniques are described in detail by Alger (1968). Petrakis and Grandy (1980, 1981), Kalkwarf et al. (1984), and others have measured free radical concentrations in coals and/or various coal liquids, using ESR. Quantitative determination of free radicals in complex mixtures can be achieved using external calibration standards containing known quantities of free radicals.

5.11.8 Shpol'skii and Low-Temperature Luminescence Spectroscopy

Broadened peaks that are characteristic of PAC fluorescence spectra obtained in liquid solutions at room temperature often hinder analysis of these types of compound, particularly when complex mixtures are analyzed by this technique. Although synchronous room-temperature luminescence has been used to narrow spectral lines of PACs in coal liquids (Vo-Dinh and Martinez, 1981), a more attractive alternate approach is low-temperature luminescence or Shpol'skii spectroscopy. With this technique, increased resolution is obtained by line-narrowing of characteristic vibrational features in fluorescence spectra of analyte solutes that are frozen at low temperatures in solutions of *n*-alkanes, alcohols, ethers, or organic gases. A generally accepted advantage of this technique is that qualitative and quantitative determinations can be performed without extensive sample preparation (Yang et al., 1980). For example, the technique is particularly sensitive for unambiguous isomeric determinations of such carcinogenic compounds as BaP, benzo[ghi]perylene, and 7,12-dimethylbenz[a]anthracene (D'Silva et al., 1976, 1983; Drake et al., 1978). This technique has particular advantages for distinguishing isomeric parent and alkylated PACs (Brown et al., 1980; Woo et al., 1980; Garrigues and Ewald, 1983; Garrigues et al., 1983).

Shpol'skii spectroscopy has been used to identify PAHs in unfractionated coal extracts (Drake et al., 1978; Woo et al., 1978) and SRC (Yang et al., 1980; D'Silva et al., 1983). Sulfur and nitrogen heterocyclic PACs can also be identified effectively by low-temperature Shpol'skii luminescence in complex fossil-fuel matrices (Woo et al., 1980; Garrigues et al., 1983).

Both the intensities and half-widths of individual "quasilines" in Shpol'skii spectra of PAC solutes can depend on the freezing rates of the sample, the final temperture of the frozen solution, and the solvent system used for the analysis. The utility of other forms of low-temperature spectroscopy, such as matrix isolation, have therefore been investigated (Tokoubalides et al., 1977). In matrix isolation, gas-phase samples of PACs are mixed with a large excess of inert-matrix gas, such as nitrogen, xenon, or argon, and deposited onto a cold finger or surface. The deposited frozen solid is then subjected to spectroscopic analysis, usually at temperatures lower than 20°K (Stroupe et al., 1977). Low-temperature fluorescence spectroscopy has been reported for coal-derived materials such as coking plant effluents (Maple et al., 1980) and Synthoil coal liquids (Stroupe et al., 1977). Finally, matrix-isolation techniques are particularly effective for the off-line and on-line analyses of effluents of chromatographic systems such as GC and HPLC.

5.12 INORGANIC ANALYSIS

Although methods such as spark source mass spectrometry (Koppenaal et al., 1980), X-ray fluorescence (Sanders et al., 1983), atomic absorption, and atomic emission spectroscopy, as well as neutron and gamma-ray activation, have been used to determine trace metal species in either coal or coal-derived materials, these analyses have not played an important part in understanding their toxicity.

Although there is little information on the toxicity of residual metals or organometallic complexes in coal-derived liquids, there is no direct evidence that metal species influence toxicity of coal-derived liquids in laboratory assays for mutagenicity and carcinogenicity. Because inorganic determinations appear to be of secondary importance, they are not discussed. Comparative trace-element concentration data for several synfuel materials were presented earlier.

5.13 COMPOSITION OF PROCESS MATERIALS

This section describes the chemical composition of coal-conversion materials, with emphasis on compounds or compound types implicated in mutagenic and carcinogenic responses. Many of the samples discussed in Chapter 3 were not analyzed chemically, hence no data on composition are available. In addition, much of the information reviewed was acquired for purposes other than to elucidate the chemical basis of toxicological response. Those data on chemical composition obtained during toxicological investigations often involved methods designed to address specific questions, which varied among laboratories.

Specifications for process and product streams in coal-liquefaction facilities vary. Thus, direct comparisons of generically similar streams from different processes may be misleading. Direct liquefaction streams are designated according to general categories based on their function within a process, with special attention to their upper distillation temperature. Elemental composition data are presented for representative materials. Quantitative data on concentrations of specific compounds from various separation schemes and high-resolution gas chromatographic analysis are also given. Coal-conversion processes and materials from which these data were derived are reviewed in Chapters 2, 3, and 4, and in the appendixes to Chapters 3 and 4.

Separations by acid–base–neutral and Swain–Steadman LH-20 methods have been used extensively to prepare samples for biological assay. Information from these separations constitutes an important contribution to the database for coal liquids, although there are few quantitative data on specific compound types in various fractions. Other separation methods used specifically during health-effect studies were designed to provide quantitative determinations on specific compounds and a compound-specific overall characterization of coal-derived materials.

The initial set of tables and figures in this section are intended to give an overall comparison of the materials tested. The processes are then considered individually. The National Bureau of Standards (NBS) has conducted a program to compare analytical methods applied to synfuels. Table 5.13 (May et al., 1983) shows results of efforts to quantitate selected compounds in SRC-II HD; interlaboratory variations among quantitative determinations are shown. NBS has also made available certified reference materials (CRM) for certain coal liquids, such as SRC-II FOB, for use as quantitative standards for measurement of specific compounds in these complex mixtures. CRM samples are designated as such in the tables contained in this chapter.

Tables 5.14–5.17 provide comparative data from the same laboratory for several processes. Table 5.14 compares the composition of Synthoil, H-Coal, EDS, Char-Oil Energy Development (COED), and SRC materials, using acid–base separation, followed by LH-20 or Fluorisil chromatography. Table 5.15 shows concentrations of specific PAH carcinogens in materials from several conversion technologies. Table 5.16 shows quantitative data for specific PAHs in the PAH fraction of selected materials. Table 5.17 shows analogous data for N-PAC fractions of the same material.

Figures 5.44–5.46 are gas chromatograms of PAH fractions from several coal-derived fuels, a shale oil, and a petroleum crude (Griest et al., 1987). These data show the types and relative concentrations of neutral PAHs found in these fossil fuels. The relative lack of definition and increased unresolved baseline associated with the full-boiling-range petroleum- and shale-derived materials reflect increased concentrations and varieties of alkylated and partially saturated species compared to coal-derived materials. Additional data comparing upgraded coal-derived materials with analogous petroleum products are contained in Guerin et al. (1986).

5.13.1 Comparative Analysis of Coal-Liquefaction Materials

5.13.1.1 *SRC-II Materials*

Chemical composition relative to toxicological activity of SRC-II materials has been extensively studied (Fruchter et al., 1978; Wilson and Pelroy, 1979; Felix et al., 1980; Mahlum et al., 1980; Wilson et al., 1980; Guerin et al., 1981; Kim et al., 1981; Kimball and Munro, 1981; Willey et al., 1981a, b; Later et al., 1982b, 1983a). The relatively large data base reflects the early availability of large quantities of SRC-II material. SRC-II HD was used, for example, in NBS ''round-robin'' studies to compare analytical techniques for coal liquids (May et al., 1983).

SRC-II materials discussed below were obtained from the Ft. Lewis, Washington pilot plant, the Harmarville, Pennsylvania process development unit (PDU; P-99), or the bench-scale operation at Gulf Research & Development Company, Merriam, Kansas. Data reported by Filby et al. (1982), Pacific Northwest Laboratory (PNL; 1979), Rhodes (1981), and Weimer et al. (1980) are for materials from the Ft. Lewis plant, including HD, MD, and naphtha. Materials used in distillation studies were from the Harmarville PDU. The HD recycle materials (Wilson et al., 1983b) were from the Merriam, Kansas facility.

PAHs in SRC-II HD are mainly pericondensed, peri/catacondensed, and catacondensed species. Fluorenes, fluoranthrenes, and other compounds containing five-member rings comprise most of the remaining material in the PAH fractions. Alkylated species having more than 12 methylene units (C_{12}) have been detected in SRC-II materials. Generally, hydroaromatic content of SRC-II materials is low ($\sim 5\%$), compared to that of analogous materials from other processes.

N-PACs found in SRC-II materials include azaarenes, carbazoles, and amino-PAHs,

Table 5.13 Interlaboratory Comparison of Quantitative Analyses of Specific Organic Compounds in Solvent-Refined Coal II (SRC-II) Material[a]

	Laboratory Number							
Compound	National Bureau of Standards	2	3	4	5	6	7	(% RSD)[b]
Phenol	23.8 ± 1.2[c]	—	24.5	19.8	28.9 ± 0.3	21.0	16.8 ± 2.5	22.4 ± 4.2 (19)
O-Cresol	12.5 ± 0.5	—	13.9	11.2	46.2[d] ± 0.5	8.0	12.4 ± 2.0	11.6 ± 2.0 (17)
Acridine	0.14 ± 0.02	—	0.43	0.118	0.145 ± 0.02	0.7	0.13 ± 0.02	0.28 ± 0.24 (86)
Carbazole	1.96 ± 0.30	5.0	2.23	2.44	1.87 ± 0.22	0.9	—	2.4 ± 1.4 (57)
Dibenzothiophene	1.02 ± 0.07	1.0	6.4[d]	1.33	1.02 ± 0.01	1.8	2.1 ± 0.2	1.4 ± 0.5 (34)
Anthracene	0.811 ± 0.013	1.15 ± 0.15	—	—	—	—	—	—
2-Methylanthracene	0.504 ± 0.030	0.55 ± 0.15	—	—	—	—	—	—
Fluoranthene	3.30 ± 0.16	5 ± 1.5	—	—	—	—	—	—
Pyrene	6.0 ± 2	6.0 ± 2	10.6	4.6	6.49 ± 0.06	6.17	4.9 ± 0.5	6.4 ± 2.0 (31)
Benzo[a]pyrene	0.134 ± 0.007	0.129 ± 0.03	0.126	0.145	0.173 ± 0.02	0.094	0.18 ± 0.02	0.14 ± 0.03 (19)
Benzo[e]pyrene	0.143 ± 0.005	—	0.207	0.126	0.162	0.128	0.23 ± 0.02	0.17 ± 0.04 (26)

[a] Concentrations are in mg/g.
[b] RSD = relative standard deviation (%).
[c] Uncertainty equals 1 standard deviation.
[d] Values not used in computations.

Source: Adapted from May et al. (1983).

Table 5.14 Summary of Approximate Weight Distributions of Selected Petroleum- and Coal-Derived Oils by Chemical Class

		Weight Distribution				
		Percentage of Total				
Sample	Fractionation Scheme	Acids	Bases	Neutrals	Other	Total Recovered (%)
Petroleum Crude Oils						
Composite No. 5107	Acid–base	1	2	85	1	89
	Acid–base	1	2	84	1	89
	Acid–base/LH-20	2	3	72	15	92
Wilmington No. 5301	Acid–base/LH-20	2	1	90	7	99
Recluse No. 5305	Acid–base/LH-20	<1	<1	65	20	85
South Swan Hills No. 5106	Acid–base/LH-20	1	2	66	19	88
Gach Saran No. 4104	Acid–base/LH-20	1	4	77	11	93
Prudhoe Bay No. 5105	Acid–base/LH-20	2	2	89	8	101
ABL No. 5102	Acid–base/LH-20	2	3	84	13	102
Shale oils						
Paraho No. 4601	Acid–base/LH-20	1	4	95	0	100
Paraho HDT No. 4602	Acid–base/LH-20	<1	2	88	1	92
LETC No. 4101	Acid–base	2	7	85	1	95
Coal-Derived Oils						
Synthoil No. 1202 (L6121)[a]	Acid–base	7	8	74	13	102
	Acid–base	8	10	65	20	103
	LH-20	10	13	56	26	105

SRC II FO No. 1701 (L1211)	Acid–base/LH-20	4	5	80	3	92
H-Coal ASB (Syn) No. 1309 (L3142)	Acid–base/LH-20	1	3	95	1	99
H-Coal VSOH (Syn) No. 1310 (L3143)	Acid–base/LH-20	1	3	95	1	100
H-Coal VSB (Syn) No. 1311 (L3144)	Acid–base/LH-20	2	5	30	65	102
H-Coal ASOH (FO) No. 1312 (L3151)	Acid–base/LH-20	1	2	55	35	93
H-Coal ASB (FO) No. 1313 (L3152)	Acid–base/LH-20	1	2	95	1	99
H-Coal VSOH (FO) No. 1314 (L3153)	Acid–base/LH-20	1	2	95	1	99
H-Coal VSB (FO) No. 1315 (L3154)	Acid–base/LH-20	1	10	30	60	101
H-Coal "Dist" No. 1601 (L3112)	Acid–base/LH-20	1	1	85	4	91
H-Coal "Dist" HDT-L No. 1602 (H0321/24)	Acid–base/LH-20	1	2	85	10	98
H-Coal "Dist" HDT-M No. 1603 (H0320/23)	Acid–base/LH-20	1	2	80	10	93
H-Coal "Dist" HDT-H No. 1603 (H0319/22)	Acid–base/LH-20	1	1	80	15	97
EDS A_2 VGO (L4213)	DMSO-extract	—	—	—	—	—
EDS A_3 spent solvent (L4212)	Acid–base	5.41	1.64	78.11	1.01	—
EDS A_6 recycle solvent	Acid–base	1.08	0.66	82.66	1.05	—
EDS VGO No. 97210	Acid–base/LH-20	7.09	13.13	82.43	—	—
EDS recycle solvent (L4215)	Acid–base/LH-20	0.18	0.74	98.65	—	—

ASB = atmospheric still bottoms; ASOH = atmospheric still overheads; COED = Char-Oil Energy Development, DMSO = dimethylsulfoxide, FO = fuel oil; HDT = hydrotreated; LETC = Livermore Energy Technology Center; Syn = syncrude; VGO = vacuum gas oil; VSB = vacuum still bottoms; VSOH = vacuum still overheads.

[a]Our sample number given in parentheses.

Source: Data from Guerin et al. (1981).

Table 5.15 Four- to Six-Ring Polycyclic Aromatic Hydrocarbon Dermal Tumorigens

			Dermal Tumorigens (ppm)							
Name	Molecular Weight	Carcino-genicity[a]	SRC-II FOB CRM1[b]	SCR-II MD:HD 915[c]	H-Coal HFO 1327[d]	H-Coal HFO 1331[d]	EDS Bottoms 1654[e]	EDS Spent Solvent 1657[e]	EDS Recycle Solvent 1658[e]	UMD Gasifier ORM4
Benz[a]anthracene	228	+	340	260	200	270	3	590	12	580
Chrysene	228	0/+	490	310	250	570	75	650	150	580
2-Methylchrysene[f]	242	+	95	110	81	82	23	240	78	300
Benzo[b, j]fluoranthenes	252	++	190	97	120	360	54	350	230	220
Benzo[k]fluoranthene	252	++	+	+	+	89	—	180	—	—
Benzo[a]fluoranthene	252	NA	—	—	4	—	—	24	—	—
Benzo[e]pyrene	252	0/+	280	110	160	250	43	460	340	180
Benzo[a]pyrene	252	++	170	200	140	160	8	640	58	370
3-Methylcholanthrene[f]	268	++	—	—	50	43	12	290	88	—
Dibenz[a, j]anthracene[f]	278	++	—	—	17	18	—	45	100	—
Indenol[1,2,3-cd]pyrene	276	+	95	43	47	78	8	77	140	—
Dibenzo[a,h-a,c]anthracenes[f]	278	+	33	29	16	23	—	37	52	—
Benzo[g,h,l]perylene	276	+	120	51	80	100	—	90	130	—
Total ($\mu g/g$)		+	1813	1210	1165	2043	214	3673	1290	1680

Key: SRC = solvent refined coal; FOB = fuel-oil blend; CRM = certified research material; HD = heavy distillate; HFO = heavy fuel oil; MD = middle distillate; NA = not available; UMD = University of Minnesota–Duluth.

[a]References: A. Dipple, in C.E. Searle (ed.), *Chemical Carcinogens*, ACS Monograph 173, American Chemical Society, Washington, DC, 1976, p. 245. M.L. Lee et al., in *Analytical Chemistry of Polycyclic Aromatic Compounds*, Academic Press, New York, 1981, p. 441. National Research Council, in *Polycyclic Aromatic Hydrocarbons: Evaluation of Sources and Effects*, National Academy Press, Washington, DC, 1983.

[b]SRC-II FOB was derived from Sample L1211.

[c]SRC-II 915 was derived from Sample L1261.

[d]H-Coal samples were derived from Sample L3152 and Sample L3154.

[e]EDS Samples were derived from Sample L98-101.

[f]Tentative identification.

Source: Adapted from Griest et al., 1987.

Table 5.16 Quantitative Determinations (ppt or mg/g) of Selected Polycyclic Aromatic Hydrocarbons (PAHs) in Crude Coal Liquids[a]

Compound	SRC-I Process Solvent (L2111)[b]	TSL MD (L12)	SRC-II HD (L1111)	SRC-II FOB (L1251)	High H_2 FOB (L1251)	SRC-II H-Coal Blend (L70A)
Tetraline	—[c]	—	—	0.65	0.38	—
Naphthalene	0.18	0.23	—	11	0.027	3.7
2-Methylnaphthalene	7.8	1.2	0.19	26	0.023	8.2
1-Methylnaphthalene	1.5	0.27	—	5.6	—	1.6
Biphenyl	5.7	1.7	0.14	8.6	5.3	0.80
2-Ethylnaphthalene	—	0.53	—	9.2	—	2.4
2,6- or 2,7-Dimethylnaphthalene	—	0.93	—	15	—	2.3
1,7-Dimethylnaphthalene	1.9	0.28	0.10	3.2	—	0.77
1,6 or 1,3-Dimethylnaphthalene	—	0.33	—	2.1	—	0.56
Acenaphthene	5.4	0.29	0.52	4.5	0.24	1.1
Dibenzofuran	12	0.90	1.35	5.25	2.6	0.72
Fluorene	13	2.2	10	4.75	1.4	2.85
Dihydrofluorene	—	—	—	—	—	7.5
1-Methylfluorene[d]	2.7	1.4	4.6	3.0	0.52	4.7
Dibenzothiophene	5.0	1.4	8.9	2.3	—	—
Phenanthrene	48	3.4	62	13	1.4	3.4
3-Methylphenanthrene	6.8	1.2	18	2.1	—	1.6
2-Methylphenanthrene	7.3	1.4	21	2.65	—	1.85
-H-Cyclopenta[def]phenanthrene	1.3	—	1.1	0.65	—	—
1-Methylphenanthrene	2.5	0.66	4.6	0.79	—	0.71
Dihydrofluoranthene	—	—	—	0.50	0.20	0.83
Fluoranthene	6.9	0.071	7.3	3.0	—	0.50
Pyrene	7.6	—	34	5.45	0.605	3.25
Benzo[a]fluorene	1.7	—	5.9	0.74	—	0.24
Benzo[b]fluorene	0.96	—	22	2.6	—	1.5
1-Methylpyrene	0.74	—	2.3	0.31	—	0.13
Benzo[b]naphtho[2,1-d]thiophene	0.46	—	1.0	0.28	—	—
Benz[a]anthracene	0.71	—	2.2	0.44	—	0.071
Chrysene	0.96	—	4.3	0.63	—	0.25
3-Methylchrysene	0.39	—	1.4	0.10	—	—
2-Methylchrysene	0.21	—	0.84	—	—	—
6- or 4-Methylchrysene	0.32	—	1.6	0.41	—	0.24
1-Methylchrysene	0.16	—	0.76	0.11	—	—
Benzo[j or b]fluoranthene	0.091	—	1.2	0.17	—	0.079
Benzo[k]fluoranthene	0.020	—	0.74	0.043	—	—
Benzo[e]pyrene	0.037	—	1.1	0.195	—	—
Benzo[a]pyrene	0.12	—	0.65	0.185	—	—
Indeno[1,2,3-cd]pyrene	0.032	—	0.33	—	—	—
Dibenz[a,h or a,c]anthracene	—	—	0.12	—	—	—
Benzo[ghi]perylene	0.12	—	0.63	0.10	—	—

[a] Concentrations were determined by gas chromatographic analysis of the PAH fractions, calculated to crude concentrations based on the weight % that eluted in the PAH fractions. Relative standard deviation ranged from approximately 5% to 20%. Column heading abbreviations are as follows: SRC = solvent-refined coal; TSL = two-stage liquefaction; MD = middle distillate; HD = heavy distillate; FOB = fuel-oil blend.

[b] Our sample number given in parentheses.

[c] Dash means no data available.

[d] Methyldihydrofluorene (m/z 182) for H-Coal sample.

Source: Compiled from Later et al. (1984b), Later (1985b), and Wilson et al. (1981a, 1982b, 1983a, b, 1985).

Table 5.17 Quantitative Determinations of Selected Nitrogen-Containing Polycyclic Aromatic Compounds (N-PACs) in Crude Coal Liquids (ppm or $\mu g/g$)[a]

Compounds	SRC-I Process Solvent (L2111)[b]	SRC-I TSL MD (L10)	SRC-II HD (L1243)	SRC-II FOB (L1251)	SRC-II High H_2 FOB (H0104)
Aniline	—[c]	—	—	400	—
Methylaniline	—	—	—	1,800	—
Tetrahydroquinoline	—	4.9	—	910	—
Quinoline	340	4.7	11	1,500	—
Isoquinoline	58.5	—	—	200	—
Indole	580	76	14	2,300	—
2-Methylquinoline	860	—	44	2,100	—
3-Methylquinoline	600	—	35	1,000	—
3-Methylindole	—	—	—	1,300	—
2-Methylindole	—	—	—	680	—
2,6-Dimethylquinoline	890	—	150	960	—
2,5-Dimethylindole	580	—	25	590	—
2,3-Dimethylindole	180	—	31	500	—
4-Azafluorene	130	—	130	—	—
Benzo[h]quinoline	770	290	1,700	310	—
Acridine	550	240	960	125	—
Benzo[f]quinoline[d]	1,100	680	3,500	460	—
Carbazole	4,500	810	15,000	2,200	31
2- or 3-Methylbenzo[f]quinoline	390	360	1,200	190	—
1-Methylcarbazole	540	390	2,200	310	16
3-Methylcarbazole	1,300	340	5,400	720	12
2-Methylcarbazole	—	230	—	500	8.6
4-Methylcarbazole	—	120	—	170	6.8
1- or 2-Azafluoranthene	175	—	300	245	—
7-Azafluoranthene	105	—	350	290	—
4-Azapyrene	250	—	2,400	670	—
Benz[c]acridine	180	—	280	—	—
Benz[a]acridine	210	—	145	—	—
Benzo[a]carbazole	330	—	1,300	480	—
1-Azachrysene	22	—	90	—	—
Benzo[b]carbazole	230	—	660	270	—
Benzo[c]carbazole	250	—	1,100	430	—
10-Azabenzo[a]pyrene	16	—	105	—	—
Dibenzo[a,i]carbazole[d]	—	—	60	—	—
Dibenzo[a,g or c,g]carbazole	—	—	95	—	—

[a] Relative standard deviation ranged from 5% to 25%; quantitation was determined by gas chromatography analysis of N-PAC fractions calculated to crude concentrations by the weight that eluted in the N-PAC fraction. Column heading abbreviations are as follows: SRC = solvent-refined coal; TSL = two-stage liquefaction; MD = middle distillate; HD = heavy distillate; FOB = fuel-oil blend.

[b] Sample number given in parentheses.

[c] Dash means no data available.

[d] Coelutes with phenanthridine.

Source: Compiled from Later et al. (1984b), Later (1985b), and Wilson et al. (1981a, 1983b, 1984).

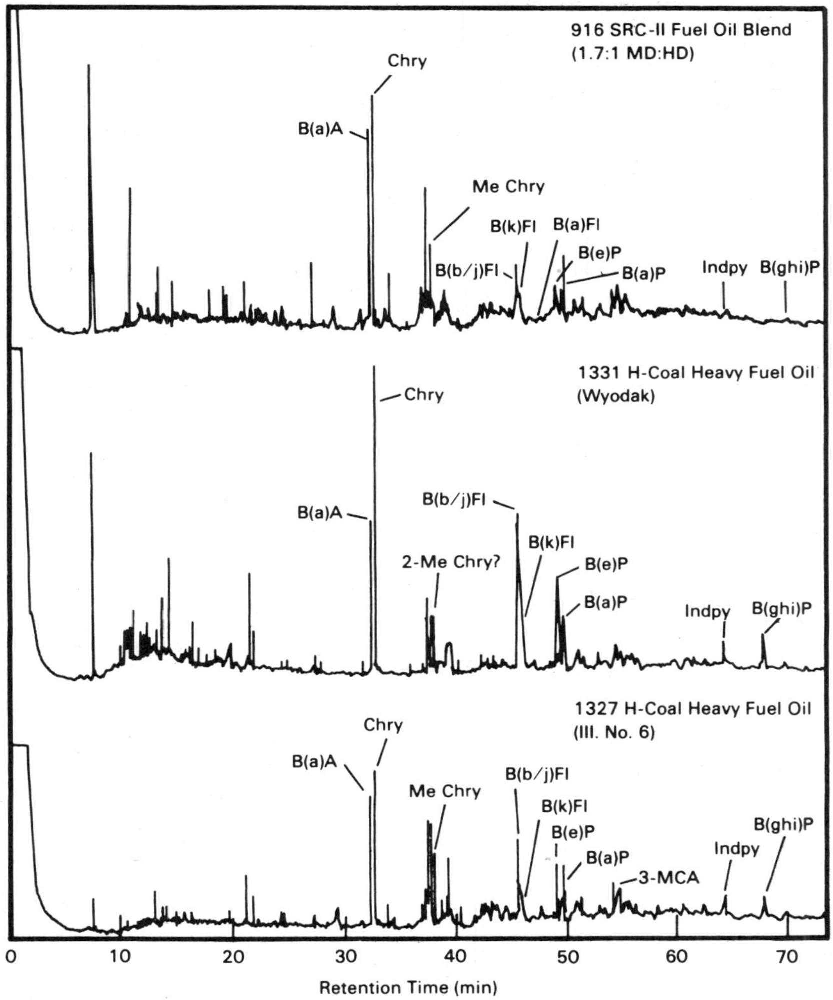

Figure 5.44 Gas chromatograms of polycyclic aromatic hydrocarbon (PAH) fractions from solvent-refined coal II (SRC-II) fuel-oil blend (FOB; Sample L1262) and H-Coal heavy oils derived from Wyodak (subbituminous) and Illinois No. 6 (bituminous) coals. Numerical designations are repository numbers (B.A. Tomkins et al., unpublished data). B(a)A = benz[a]anthracene; Chry = chrysene; MeChry = methyl chrysene; B(b/j)Fl = benzo[b/j]fluoranthene; B(a)P = benzo[a]pyrene; B(e)P = benzo[e]pyrene; Indpy = indenopyrene; B(ghi)P = benzo[ghi]perylene (adapted from Griest et al., 1987).

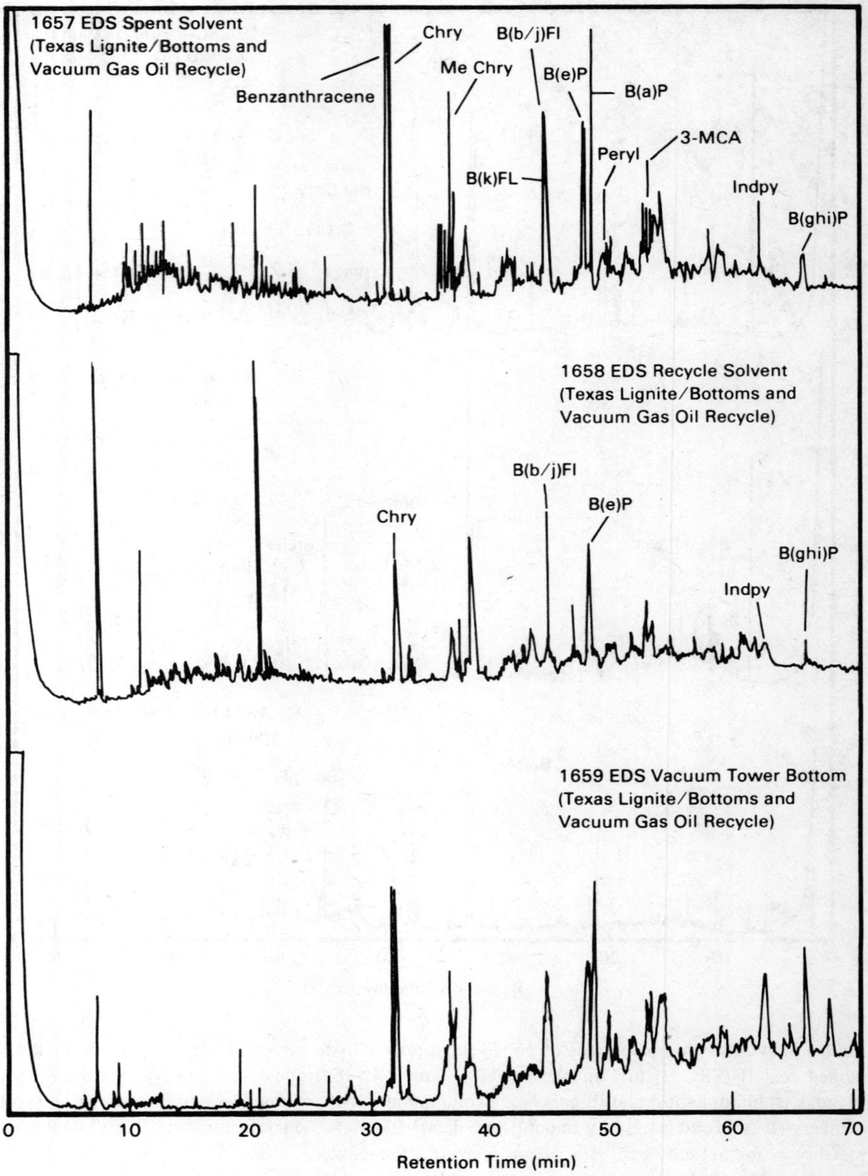

Figure 5.45 Gas chromatograms of polycyclic aromatic hydrocarbon (PAH) fractions from three EDS coal-liquefaction plant materials obtained during operation with Texas lignite. The top chromatogram is from spent solvent (Sample L4141); the middle chromatogram is from hydrogenated recycle solvent (Sample L4142); the bottom chromatogram is from vacuum tower bottoms (Sample L4143). B(a)A = benz[a]anthracene; Chry = chrysene; B(b/j)Fl = benzo[b/j]fluoranthene; B(a)P = benzo[a]pyrene; B(e)P = benzo[e]pyrene; Indpy = indenopyrene; B(ghi)P = benzo[ghi]perylene (adapted from Griest et al., 1987).

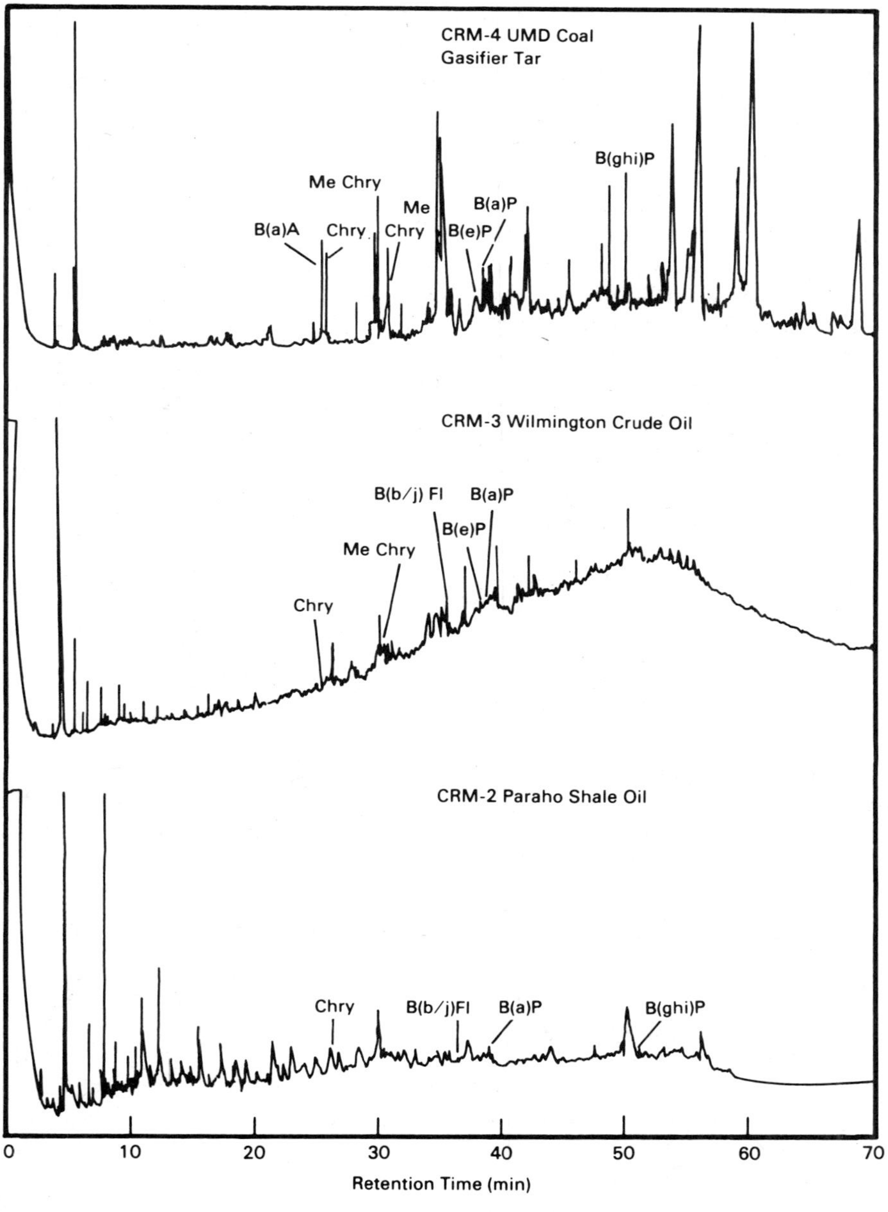

Figure 5.46 Gas chromatograms of polycyclic aromatic hydrocarbon fractions from University of Minnesota-Duluth (UMD) coal gasifier tar, Wilmington crude oil, and Paraho shale oil. B(a)A = benz[a]anthracene; Chry = chrysene; B(b/j)Fl = benzo[b/j]fluoranthene; B(a)P = benzo[a]pyrene; B(e)P = benzo[e]pyrene; Indpy = indenopyrene; B(ghi)P = benzo[ghi]perylene; CRM = certified reference material (adapted from Griest et al., 1987).

together with their alkylated homologs, some greater than C_{12}. N-alkylation has not been determined for SRC materials to the same extent as it has for H-Coal liquids. Nonetheless, no reports were found of N-alkylated compounds present in SRC materials.

Blends of SRC-II MD and HD were prepared to represent eventual fuel oils and were characterized chemically and toxicologically (See Kimball and Munro 1981; Weimer et al. 1980). A blend of 2.9 : 1 MD : HD, designated fuel oil blend (FOB), was used as feedstock for catalytic hydrogenation in fuel upgrading studies. MS data for the feedstock material are shown in Table 5.18. The 6 : 1 and 1.7 : 1 blends of MD : HD (see Fig. 5.44) were also characterized and hydrotreated. (Chemical data are shown in Table 5.15.) Toxicological activity (mutagenicity or carcinogenicity) of the blends was less than that of SRC-II HD, primarily reflecting dilution of carcinogenic compounds by lower-boiling MD.

Selected HPLC data and specific mutagenic activity of SRC-II HD subfractions are shown in Fig. 5.47. Mutagenicity of each fraction is indicated below the chromatogram. There was some separation of PAHs based on ring size and substantial separation of carbazoles and amino-PAHs in the chromatographic region where bases eluted. Microbial mutagenic activity with *S. typhimurium* strain TA98 resided mainly in the amino-PAH-rich subfraction 6. Subfraction 2 contained high concentrations of BaP.

Fractionation of SRC-II HD according to chemical class was reported by Later et al. (1981a). The fractions (AHs, PAHs, N-PACs, and hydroxy-PAHs, as well carbazole, amino-PAH, and azaarene subfractions of N-PACs) were shown by element-specific gas chromotographic detection to be highly purified.

For the chemical class fractions isolated from SRC-II HD by the various fractionation schemes, PAH isolates were generally most active in initiation/promotion and chronic skin-painting studies and showed some activity (depending on separation scheme) in microbial mutagenicity assays. Depending on separation type, acid- or hydroxy-PAH fractions also showed low-level activity in microbial mutagenicity assays. The AH fractions in all cases were inactive.

In summary, SRC-II materials boiling above 700°F are highly aromatic relative to product streams from other processes. SRC-II HD from Ft. Lewis (bp range, 550 to >850°F) was active in both mutational and carcinogenicity assays. Biological activity largely reflected the inclusion of >850°F bottoms materials.

5.13.1.2 *SRC-I Materials from the Ft. Lewis Pilot Plant*

SRC-I solid product, PS (analogous to HD in the SRC-II process; Sample L2213), wash solvent (WS) (an MD cut; Sample L2212), and LO (naphtha cut; Sample L2211) from Ft. Lewis were also evaluated for potential toxicological effects (PNL, 1979). Samples were obtained from the pilot plant in 1976 during processing of Kentucky coal. Table 5.19 shows GC data on the composition of unfractionated WS (bp, 350–500°F) and PS (bp, 500–850°F). SRC-I liquids were generally less aromatic than their SRC-II counterparts from the same plant. This reflected, in part, the fact that most heavy ends produced in the SRC-I mode went into the solid product, whereas the end product in the SRC-II mode was liquid. Toste et al. (1982), who performed HPLC separations on LH-20 fractions of SRC-I PS and solid products (Fig. 5.17), reported mutagenic activities of the subfractions.

No comprehensive quantitative analyses were performed regarding composition of the SRC-I solid product. In the separation scheme used, neutral PAHs eluted first, followed by compounds of increasing polarity. Last to elute were higher-molecular-weight hydroxy- and dihydroxy-PAHs and mixed heteroatomic species having hydroxyl functionality. Note the substantial increase in spectrophotometric detector response in the highly polar regions of

Table 5.18 High-Resolution Mass Spectrometric Characterization of Solvent Refined Coal II (SRC-II) Fuel-Oil Blend, Sample L1251

Z Number	Compound Type Analysis	Weight %	Average Molecular Weight
	Hydrocarbons (C_nH_{2n-z})[a]		
−2	Paraffins/Naphthenes	3.6	226
0	1-ring	1.3	224
2	2-rings	1.8	208
4	3-rings	1.3	206
6	4-rings	0.4	190
		8.4	216
	Aromatic Hydrocarbons (C_nH_{2n-z})[a]		
6	Benzenes	0.9	158
8	Indans/tetralins	10.3	165
10	Dinaphthene benzenes	2.7	185
12	Naphthalenes	12.1	159
14	Acenaphthenes/biphenyls	7.5	185
16	Fluorenes	4.4	205
18	Phenanthrenes	6.7	197
20	Acenanthrenes	2.6	227
22	Pyrenes	4.7	223
24	Chrysenes	0.9	246
26	Benzofluoranthenes	0.3	276
28	Benzopyrenes	0.6	280
30	Dibenzanthracenes	<0.1	292
32	Benzoperylenes	T[b]	
34	Dibenzopyrenes	T	
36	Coronenes	T	
		53.8	186
	Aromatic Oxygens ($C_nH_{2n=z}O$)[a]		
6	Phenols	14.4	131
8	Naphthenophenols	5.7	154
10	Benzofurans	0.6	183
12	Naphthols	<0.1	180
14	Naphthenonaphthols	3.2	180
16	Dibenzofurans	2.6	202
18	Hydroxyanthracenes	0.3	230
20	Hydroxynaphthenoanthracenes	0.2	233
22	Hydroxypyrenes	0.5	232
24	Hydroxychrysenes	T	
		27.6	153
	Aromatic Dioxygenates ($C_nH_{2n-z}O_2$)[a]		
6	Hydroxyphenols	0.3[c]	120
14	Hydroxynaphthenonaphthols	0.7[c]	218
16	Hydroxydibenzofurans	0.5[c]	215
18	Dihydroxyanthracenes	<0.1[c]	247
20	Dihydroxynaphthenanthracenes	<0.1[c]	278
		1.7[c]	203

Table 5.18 (*Continued*)

Z Number	Compound Type Analysis	Weight %	Average Molecular Weight
	Aromatic Trioxygenates ($C_nH_{2n-z}O_3$)[a]		
6	Dihydroxyphenols	<0.1[c]	204
12	Dihydroxynaphthols	0.2[c]	211
14	Dihydroxynaphthenonaphthols	T	
		0.3[c]	212
	Aromatic Nitrogens ($C_nH_{2n=z}N$)[a]		
5	Pyridines	1.5	130
7	Naphthenopyridines	1.1	153
9	Indoles	0.6	152
11	Quinolines	1.3	156
13	Naphthenoquinolines	0.6	198
15	Carbazoles	0.6	191
17	Acridines	0.6	202
19	Naphthenobenzoquinolines	<0.1	222
21	Benzocarbazoles	0.5	228
23	Benzacridines	<0.1	254
25	Naphthenobenzacridines	T	
27	Dibenzocarbazoles	T	
		7.0	167
	Aromatic Sulfurs ($C_nH_{2n-z}S$)[a]		
6	Benzenethiols	T	
8	Dihydrothionaphthenes	0.2	199
10	Benzothiophenes	<0.1	172
12	Naphthenobenzothiophenes	T	
14	Indenothiophenes	T	
16	Dibenzothiophenes	0.6	205
18	Acenaphthenothiophenes	<0.1	239
20	Fluorenothiophenes	T	
22	Tribenzothiophenes	<0.1	250
24	Acephenanthrylenethiophenes	T	
26	Pyrenothiophenes	T	
		1.2	209

Summary of Table 5.18

Compound Type	Weight %	Average *C* Number	Average Molecular Weight	Average *Z* Number
		Sample Analysis Summary (C_nH_{2n-z})[a]		
Saturated hydrocarbons (paraffins/naphthenes)	8.4	15.5	216	0.7
Aromatic hydrocarbons	53.8	14.3	186	14.1
Aromatic nitrogens	7.0	11.7	167	10.8
Aromatic oxygens	27.6	10.4	153	8.9
Aromatic sulfurs	1.2	13.7	209	14.6
Aromatic dioxygenates	1.7	13.1	203	13.5
Aromatic trioxygenates	0.3	12.5	212	10.8

Summary of Table 5.18 *(Continued)*

	Elemental Analysis by Mass Spectrometry
Atomic carbon	87.72
Atomic hydrogen	8.29
Atomic nitrogen	0.59
Atomic oxygen	3.22
Atomic sulfur	0.18
	100.00

[a] Suggested compound types. Multiheteroaromatics normalized out of the results.
[b] T = trace (≤0.05 wt %).
[c] Estimates, owing to lack of sensitivity data.
Source: Adapted from Riedle and deRosset (1979).

the solid product, indicating an increase in polarity and molecular weight of solid product constituents, relative to those of the PS.

Subfractions showing the highest specific mutagenic activity were enriched in amino-PAHs. However, highly polar subfractions from the SRC-I solid product, consisting mainly of hydroxylated-PAHs, were also mutagenic. Mutagenic activity could not be related positively to hydroxy-PAHs because HRMS analyses showed diheteroatomic and multiheteroatomic PACs, some containing nitrogen and/or sulfur heteroatoms, in the same subfraction (Wilson et al., 1983c). Pelroy (1981) performed mutagenicity assays on SRC-I materials from Ft. Lewis; additional data on these materials may be found in Wilson et al. (1982b). SRC-I PS samples (L2121) were also subjected to fractional distillation, then chemically analyzed prior to assay for toxicological activity.

5.13.1.3 *Materials from the Advanced Coal Liquefaction Research Facility*

The Advanced Coal Liquefaction Research Facility in Wilsonville, Alabama has been used to investigate several different processes or process modifications. Originally an SRC-I pilot plant, the facility produced many samples used in early toxicological evaluations. Modifications to the plant, including addition of critical solvent deashing and H-Oil hydrotreater units, provided the capability for TSL operation. During testing of additional process units prior to 1984, the plant was operated in both NTSL and ITSL modes, and samples were collected representing both configurations. Samples collected during process runs made for the International Coal Refining Company (ICRC) were designated SRC and TSL in reports describing toxicological studies on these materials (Drozdowicz and Kelly, 1983). Blends obtained during SRC-I, ITSL, SRC, and TSL operations were distilled into 50°F bp cuts. Data from analyses of these materials are described in the section on fractionally distilled materials.

Wilson et al. (1984) studied effects of catalyst age and operating conditions on chemical composition as well as the tumorigenic and mutagenic activity of ITSL materials. These studies showed greater mutagenic activity during upset conditions than under normal operation. Materials from ITSL were also compared to those from the single-stage SRC-II process with regard to chemical composition and toxicological activity.

5.13.1.4 *Materials from Early Single-Stage Operation at Wilsonville*

Initially, operation of the Wilsonville pilot plant focused on developing a process to produce a low-sulfur, ashless, solid fuel for use in electric power generation. Although PS and MD

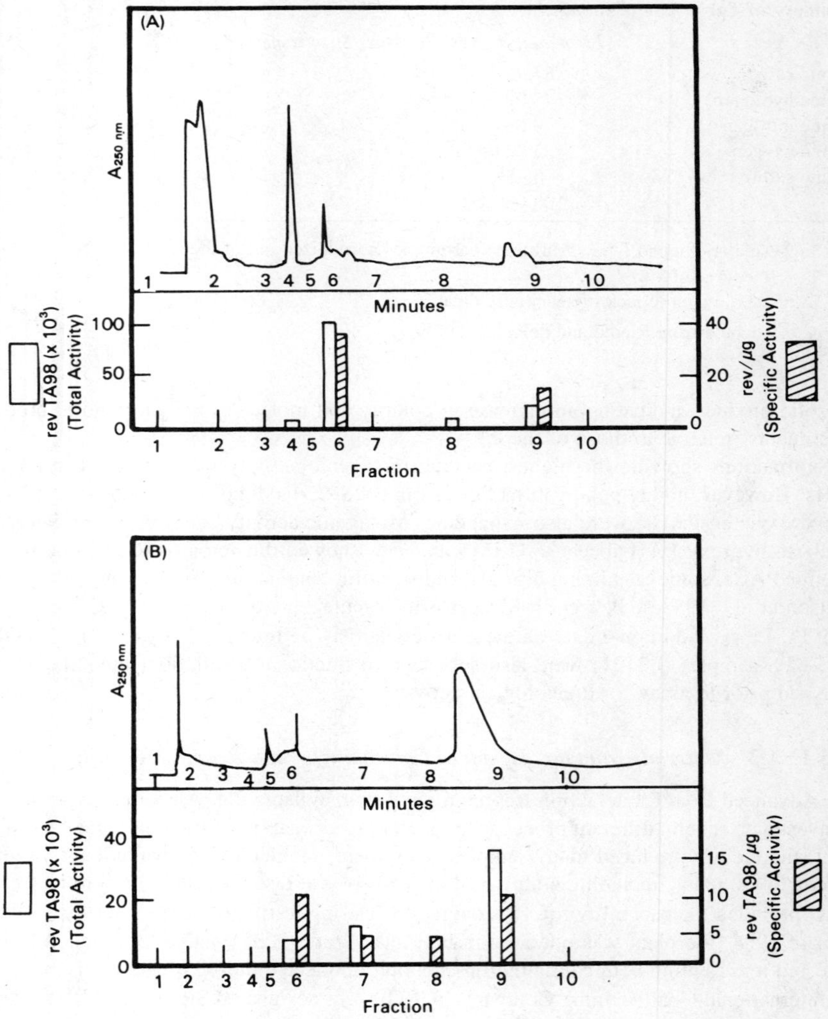

Figure 5.47 High-performance liquid chromatography (HPLC) analysis and mutagenicity of solvent-refined coal I materials: (A) process solvent (Sample L2213) and (B) solid product (Sample S1). Mass recovery of each sample after HPLC (% dry weight recovered) was 100% (adapted from Toste et al., 1982).

were also produced, they were generally considered as process streams as opposed to product.

Paudler and Chaeplen (1979) determined N-PACs in SRC-I distillates from Wilsonville by GC and GC-MS. They reported the presence of a novel phenoxazine compound and concluded that the most probable route for its formation was by condensation of aniline with phenol. SRC-I distillates were also characterized by the Demonstration Plant Analyt-

Table 5.19 Major Components of Solvent-Refined Coal I Wash Solvent and Process Solvent, Based on the Response of Internal Standards[a]

Compound	Wash Solvent[b] (L2212)[d]	Process Solvent[c] (L2213)
C_2-Benzene	5.90 ± 0.50	—
	4.91 ± 0.43	—
C_3-Benzene	6.58 ± 0.34	—
Phenol	34.2 ± 1.8	—
C_3-Benzene	7.67 ± 0.77	—
Indan	16.1 ± 1.0	—
o-Cresol	26.8 ± 1.10	—
C_4-Benzene	4.78 ± 0.44	—
m- and *p*-Cresol, C_1-indan	57.6 ± 4.1	—
n-Undecane	9.19 ± 0.71	—
C_2-Phenol	5.76 ± 0.27	—
C_1-Indan	10.5 ± 0.5	—
	17.4 ± 0.8	—
C_2-Phenol	—	—
Tetralin	46.9 ± 2.7	—
C_2-Phenol	—	—
Naphthalene	84.1 ± 7.3	7.77 ± 0.20
C_2-Phenol	11.7 ± 2.6	—
C_2-Phenol	5.80 ± 0.35	—
C_1-Dihydronaphthalene	7.50 ± 0.25	—
C_1-Tetralin	10.3 ± 2.5	—
	12.1 ± 0.5	—
C_1 Tetralin, C_4-phenol	1.77 ± 1.35	—
C_1-Dihydronaphthalene	4.79 ± 0.19	—
C_3-Phenol	4.44 ± 0.16	—
	4.78 ± 0.18	—
C_2-Indan	5.66 ± 0.36	—
	12.1 ± 0.5	6.33 ± 0.20
	6.81 ± 0.24	12.3 ± 0.20
2-Methylnaphthalene	9.21 ± 0.25	38.3 ± 0.6
1-Methylnaphthalene	8.40 ± 0.38	7.77 ± 0.13
C_3-Indan	—	—
Indanol	—	4.55 ± 0.14
C_2-Tetralin, C_4-phenol	—	6.80 ± 0.20
C_4-Phenol, C_3-indan	—	5.65 ± 0.21
Biphenyl	—	12.8 ± 0.4
2-Ethylnaphthalene	—	10.9 ± 0.3
2,6- or 2,7-Dimethylnaphthalene	—	11.4 ± 0.3
1,7-Dimethylnaphthalene	—	3.11 ± 0.14
C_2-Naphthalene, diphenylether	—	13.0 ± 0.4
Acenaphthene	—	8.78 ± 1.03
C_3-Naphthalene	—	7.33 ± 0.34
	—	4.79 ± 0.21
Dibenzofuran	—	12.8 ± 0.5
Fluorene	—	11.9 ± 0.5
C_1-Acenaphthene, C_2-biphenyl	—	6.58 ± 0.23
C_4-Naphthalene	—	5.09 ± 0.15
	—	14.9 ± 0.5
Xanthene	—	5.17 ± 0.23

Table 5.19 (*Continued*)

Compound	Wash Solvent[b] (L2212)[d]	Process Solvent[c] (L2213)
C_1-Fluorene	—	4.98 ± 0.46
	—	—
	—	5.09 ± 0.15
	—	—
Dibenzothiophene	—	—
Tetrahydrophenanthrene	—	—
Phenanthrene	—	53.0 ± 2.2
C_2-Fluorene	—	—
	—	—
	—	—
Carbazole	—	—
3-Methylphenanthrene	—	5.13 ± 0.26
2-Methylphenanthrene	—	6.35 ± 0.24
1-Methylphenanthrene	—	—
C_2-Xanthene	—	—
Phenylnaphthalene	—	—
C_2-Phenanthrene	—	—
Fluoranthene	—	6.85 ± 0.31
Pyrene	—	7.77 ± 0.40
Dihydropyrene	—	—
Benzo[b]fluorene	—	8.09 ± 1.29
C_2-Pyrene, C_1-benzo[b]fluorene	—	—

[a] Concentrations are in parts per thousand.
[b] Distillation boiling-point range: 350–500°F.
[c] Distillation boiling-point range: 500–850°F.
[d] Sample number.

Source: Rhodes (1981).

ical Laboratory in Allentown, Pennsylvania. Data from these investigations, including capillary-column GC analyses, are contained in quarterly technical reports, one of which (Wood et al., 1981) tentatively identified several thiophene compounds by GC-FPD in the 350–850°F distillate. Thiophenes present in materials from the single-stage process were in direct contrast to the near-absence of these compounds in two-stage and hybrid catalytic processes (particularly EDS). Tables 5.17 and 5.19 list concentrations of major compounds in SRC-I PS PAH and N-PAC fractions.

Coleman et al. (1977) conducted NMR studies on the composition of the solid SRC-I materials. The SRC-I solid product was size fractionated with Bio-Beads, using THF. Estimates showed that the average molecular weight of about 6% of the material was 259 amu; 48% was 388 amu; about 24% was 487 amu; and about 19% was 711 amu. Hydrogen : carbon ratios and the number of heteroatoms per carbon atom were relatively constant across the molecular weight range represented by these fractions. NMR studies did not detect carbonyl functionalities in these materials, a finding consistent with GC, GC-MS, and MS data. Excess hydrogen occurred largely in monocyclic aliphatic ring structures, possibly associated with aromatic systems.

5.13.1.5 *Materials from the ITSL and NTSL Processes*

TSL has been operated in both integrated or coupled (ITSL) and noncoupled (NTSL) modes at the Wilsonville, Alabama, Advanced Coal Liquefaction Research Facility. During operation in the ITSL mode, the high-pressure bottoms stream (550–950°F plus ~40% resids) from the flash separator following the hydrotreater was recycled and used as solvent to slurry feed coal entering the first-stage liquefaction reactor. During NTSL operation, a 350–650°F distillate of the first stage and a 450–1000°F distillate of the second stage were recycled separately and blended with feed material entering the reactor from which they were produced. Although not a candidate for commercialization, NTSL provides a base from which to evaluate the effects of hydrogenating coal slurry recycle solvent. Both operational modes yielded products that displayed significantly reduced genotoxicity compared to products from single-stage processes, such as SRC. Deamination, which is only partial in the low-severity liquefaction step, is virtually completed in the hydrotreatment stage, thus reducing overall concentrations of genotoxic N-PACs or amino-PAHs in process materials and products. ITSL materials generally exhibit lower microbial mutagenicity than their NTSL counterparts. Optimum conditions for reduced mutagenic activity of ITSL materials were achieved during ''lined-out'' operation at 860°F in the liquefaction reactor and 720°F in the hydrotreater (Later, 1985a).

ITSL (Wilsonville run No. 242) and NTSL (Wilsonville run No. 241) samples of various process streams were analyzed to determine the effect of hydrotreated recycle pasting solvent on gross chemical class composition. The greatest differences between these operational modes involved the composition of deashed hydrotreater feed (see Table 5.20). The concentration of hydroxy-PACs in the NTSL hydrotreater feed was higher than that in ITSL hydrotreater feed. The net result of improved conversion in the first liquefaction stage is reflected in higher aliphatic concentration and lower hydroxy-PAC concentration in ITSL distillation bottoms compared to those from NTSL (Table 5.20).

Differences in specific components of NTSL and ITSL materials were determined using qualitative and quantitative high-resolution capillary-column GC (see Table 5.21). Although deashed hydrotreater feed materials from NTSL and ITSL processes were comparable regarding distillation parameters, a marked difference was observed regarding individual components of these materials. Neutral PAHs from the ITSL material contained primarily nonalkylated compounds; the same fraction from the NTSL material had a higher alkylated versus parent PAH content. In summary, comparison of NTSL and ITSL materials showed that coupled operation yielded products with lower heteroatom content and lower mutagenic activity.

Table 5.20 Chemical Class Composition (wt %) of Selected Integrated Two-Stage Liquefaction (ITSL) and Nonintegrated Two-Stage Liquefaction (NTSL) Process Streams as Determined by Alumina-Column Separation

	Deashed Hydrotreater Feed		Blended Hydrotreater Feed		Distillation Bottoms	
Compound Class[a]	NTSL (L5214)[b]	ITSL (L5344)	NTSL (L5216)	ITSL (L5346)	NTSL (H0513)	ITSL (H0583)
AHs	<1	11	22	14	45	63
PAHs	10	36	24	38	34	26
N-PACs	33	32	19	27	7	5
Hydroxy-PAHs	36	22	35	23	15	9
Total Recovery (%)	80	101	100	102	101	103

[a] AHs = aliphatic hydrocarbons; N-PACs = nitrogen-containing polycyclic aromatic compounds; PAHs = polycyclic aromatic hydrocarbons.

[b] Sample number given in parentheses.

Table 5.21 Concentrations of Selected Polycyclic Aromatic Hydrocarbons (PAHs) in Nonintegrated Two-Stage Liquefaction (NTSL) and Integrated Two-Stage Liquefaction (ITSL) Deashed Hydrotreater Feed (DHF)

	Concentrations[a] ($\mu g/g$)	
Compound	NTSL DHF (L5214)[b]	ITSL DHF (L5344)
Phenanthrene	78 ± 10	ND[c]
2-Methylphenanthrene	66 ± 9	ND
Fluoranthene	79 ± 3	ND
Pyrene	170 ± 10	280 ± 10
1-Methylpyrene	53 ± 13	ND
Benzo[b]fluorene	260 ± 20	ND
Chrysene	300 ± 20	<24
3-Methylchrysene	260 ± 40	ND
Benzo[a]pyrene	300 ± 60	540 ± 40

[a] Based on gas chromatography analysis of PAH fractions calculated to crude concentration using the weight % that eluted in PAH fractions.

[b] Sample number given in parentheses.

[c] Not determined.

Source: Adapted from Later (1985b).

5.13.1.6 *Comparative Analysis of Wilsonville SRC-I and TSL Distillates*

Chemical analyses were conducted on SRC materials collected from Wilsonville for toxicological evaluations. Materials included SRC solids, hydrotreated SRC, SRC-I MD, TSL MD, and 50°F distillate cuts from these materials. Distillate cuts spanned the range from 400 to 750°F for SRC materials and 350 to 750°F for TSL materials. Raw GS and UV data for each sample were reported by Drozdowicz and Kelly (1983). More detailed chemical analyses of these materials as well as studies evaluating the stability of these samples after storage, are reviewed below.

The TSL MD appeared to have a higher aliphatic content owing to the presence of certain hydroaromatic species that eluted with hexane and thus were recorded in the AH fraction. TSL MD had a corresponding lower weight % in the other three fractions obtained. Both SRC-I and TSL MD had a higher PAH weight % content than an SRC-II MD (380–550°F nominal boiling range; 11% in the PAH fraction). SRC-I and TSL MD also had a lower phenolic- or hydroxylated-PAH weight % content than the SRC-II MD (35% in the hydroxylated-PAH fraction). The TSL MD had a lower N-PAC fraction weight than both the SRC-I and SRC-II MD. The solid TSL product also showed a higher weight content in the aliphatic fraction compared to that of the SRC-I solid.

Table 5.22 shows chemical class fraction data for PAH fractions of the SRC-I and TSL MD and solid products, as determined by HRGC. The TSL sample appeared to have a slightly higher nominal boiling range than the MD. Concentrations of hydrogenated materials, relative to their respective parent compounds in PAH fractions, were higher for TSL than for SRC-I MD. The degree of alkylation appears similar for both materials when comparing methylphenanthrenes and phenanthrene concentrations. Higher PAH concentrations were recovered from SRC-I MD. Compounds in highest concentration in both samples were 2-methylnaphthalene (MD, SRC-I) and phenanthrene (MD, TSL).

Quantitative determinations of selected nitrogen compounds in N-PAC fractions of SRC-I solid product and TSL MD samples showed that the most abundant species were pyrrole-

Table 5.22 Chemical Class Fractionation Weight Distribution from Alumina-Column Chromatography

	Middle Distillate (wt %)			Solid Product Fuel (wt %)	
Fraction[a]	SRC-I[b]	TSL[c]	SRC-II	SRC-I	TSL
AHs	31	61	36	0.8	3
PAHs	39	29	11	28	18
N-PACs	13	5	11	42	41
Hydroxy-PAHs	17	5	35	23	23
Total recovery (%)	100	100	93	93.8	85

[a] AHs = aliphatic hydrocarbons; N-PACs = nitrogen-containing polycyclic aromatic compounds; PAHs = polycyclic aromatic hydrocarbons.

[b] SRC = solvent-refined coal.

[c] TSL = two-stage liquefaction.

Source: Compiled from Later (1985b).

type compounds; carbazole was the most abundant in both samples. Comparing methylcarbazoles to carbazoles, the concentration ratios of alkylated to nonalkylated compounds appeared similar to both samples. However, the ratio of quinoline-type compounds was higher than that of pyrrole-type in TSL versus that from SRC-I MD. These determinations further indicated that TSL materials had lower overall N-PAC content than did SRC-I materials.

The solid TSL product yielded more chromatographable material than the SRC-I solid. Alkylation of both solid products was similar, as determined by comparing methylchrysenes with chrysenes. TSL solid product had lower concentrations of lower-molecular-weight species, such as alkylated naphthalenes, fluorenes, and phenanthrenes, than did SRC-I solid product. A methylchrysene isomer was the most concentrated species in the SRC-I solid product. Benzo[a]pyrene content was similar in both products: 400 ppm in SRC-I and 670 ppm in TSL.

5.13.1.7 *H-Coal Process Materials*

H-Coal materials have been separated by several methods, including HPLC, and analyzed by GC and GC-MS (Wong and Gladstone, 1983). Table 5.14 shows the weight distribution for H-Coal materials by chemical class, determined by the Swain–Steadman LH-20 fractionation scheme (Kimball and Munro, 1981).

Scheppele (1978a, b) compared H-Coal materials derived from Illinois No. 6 coal, during operation in the syncrude mode, with those produced during fuel-oil mode operations. Endpoint temperatures of syncrude- and fuel-oil-mode still overheads were 197 and 203°C, respectively; endpoints of high-boiling distillates were 493 and 413°C, respectively. For higher-boiling distillates, saturates, monoaromatics, and polyaromatic polar compounds accounted for about 8, 20, and 19%, respectively, of fuel-oil materials and 5, 11, and 33%, respectively, of syncrude materials.

Fuel-oil-mode materials had higher concentrations of condensed polyaromatics and lower concentrations of lower-boiling saturates than syncrude-mode material. Syncrude overheads were more hydrogenated and contained more polyaromatic polar and heteroatomic compounds than fuel-oil overheads. Table 5.23 shows data for several PAHs in H-Coal process streams. Table 5.24 shows concentrations of organonitrogen compounds in several H-Coal materials.

Table 5.23 Concentrations ($\mu g/g$) of Selected Compounds in the Polycyclic Aromatic Hydrocarbon (PAH) Isolates of H-Coal Process Development Unit Materials[a]

			ASOH		ASB		VSOH		VSB	
Compound	Molecular Weight	Carcino-genicity[b]	Syn 1308[c] (L3141)[d]	FO 1312 (L3151)	Syn 1309 (L3142)	FO 1313 (L3151)	Syn 1310 (L3143	FO 1314 (L3153)	Syn 1311 (L3144)	FO 1315 (L3154)
Benz[ghi]fluoranthene	226	0	—	7	690	60	1300	—		—
Benz[c]phenanthrene	226	+	0.1	1	520	200	1200	110		30
9-Phenylanthracene	254	0	1	6	—	—	—	1000		300
Benz[a]anthracene	228	+	1	4	380	70	1500	500		220
Chrysene	228	0/+	—	1	—	—	450	250		110
2-Methylchrysene	242	+	0.4	1	40	20	70	—		—
Benzo[b]fluoranthene	252	++	1	0.5	—	20	280	270		330
Benzo[a]pyrene	252	++	1	1	—	40	—	—		700
Perylene	252	0	0.4	1	50	20	500	330		400
9,10-Diphenylanthracene	330	0	—	0.5	70	2	—	—		200
3-Methylcholanthrene	268	++	—	1	520	110	—	410		850
Dibenz[ah/ac]anthracenes	278	+	0.3	0.1		3	280	80		300
Picene	278	0	0.1	—	50	1	—	14		—
Benz[ghi]perylene	276	+	—	—	40	10	—	50		140
Anthanthrene	276	0	—	0.2	—	0.7	14	19		16

[a]Column heading abbreviations are as follows: ASB = atmospheric still bottoms; ASOH = atmospheric still overheads; FO = fuel-oil mode; Syn = syncrude mode; VSB = vacuum still bottoms; VSOH = vacuum still overhead.

[b]From the appendix in M. L. Lee et al. (1981).

[c]Oak Ridge National Laboratory material repository designations for each sample.

[d]Sample number given in parentheses.

Source: B. A. Tomkins and W. H. Griest, Oak Ridge National Laboratories (unpublished data).

Table 5.24 Organonitrogen Gas Chromatography/Nitrogen/Phosphorus Detector Screening of H-Coal Materials ($\mu g/g$ oil)

	Naphtha		Light Fuel Oil		Heavy Fuel Oil		
Compound	587[a] (L3211)[b]	886 (L3241)	589 (L3212)	887 (L3242)	591 (L3213)	888 (L3243)	Blend 930
Pyridine	20	137	—	—	—	—	—
3-MePyridine; 2,6-DiMePyridine	140	220	—	—	—	—	—
2,5- +2,3-DiMePyr, 3-EtPyr	300	670	129	101	—	—	7
Aniline	—	80	—	14	—	—	2
3,4-DiMePyr	—	90	—	11	—	—	7
o-MeAniline	2000	1700	—	2000	—	—	300
m- +*p*-MeAniline	—	—	68	—	—	—	—
2,4- +2,5-DiMeAniline	3200	730	90	—	—	—	—
2,4,6-TriMeAniline	50	25	64	1000	—	—	400
2,3-DiMeAniline	—	—	—	—	—	—	360
Quinoline	—	—	—	1000	-	-	160
Isoquin, Indole, 1-MeIndole	—	13	350	1700	—	160	6400
8-MeQuinoline	4	—	140	204	—	110	140
Tetrahydroquinoline	12	—	80	220	—	90	50
1-Meisoquinoline	—	7	70	19	—	70	80
3-Me +6-MeQuinoline, 7-MeIndole	20	40	—	—	—	—	180
2-Meindole	—	—	530	270	—	280	240
1,2-DiMeIndol; 2,7-DiMeQuinoline	70	—	500	70	—	580	410
2,3-DiMeIndole	100	—	511	—	13	420	—
1-Aminonaphthalene	70	130	23	—	—	63	—
2-Aminonaphthalene	—	—	800	46	—	920	—
2,3,5-TriMeIndole	—	—	50	3	—	13	10
Acridine	—	—	81	170	110	290	74
Carbazole	—	—	120	10	700	1200	140
n'-Ethcarbazole	—	6	4	—	—	92	—
2-Aminofluorene	—	—	—	—	—	—	200
2-Aminoanthracene	—	—	—	—	—	340	200
2,3-Bencarbazole	—	—	—	—	80	140	70
2-Aminofluoranthene	—	—	—	—	80	96	—
1-Aminopyrene	—	—	—	—	130	320	—

[a] Three-digit numbers at the head of each column are Oak Ridge National Laboratory material repository designations.
[b] Sample number given in parentheses.

Source: B. A. Tomkins and W. H. Griest, Oak Ridge National Laboratory (unpublished data).

In summary, H-Coal materials are less aromatic overall than SRC-II materials, and their PAC alkylation is greater than that of SRC-II and ITSL materials but less than that of EDS materials. H-Coal N-PAC concentrations are less than those of SRC-II materials but greater than those in EDS and ITSL materials of comparable bp range.

5.13.1.8 *EDS Process Materials*

Physical properties of EDS materials have been analyzed extensively. Results can be found in a series of progress reports, and an excellent summary is available (Epperly, 1982). However, most data discussed here are taken from Later et al. (1984b) because they were obtained in support of toxicological evaluations. EDS recycle solvents analyzed included those from once-through (OT), bottoms recycle (BR), and bottoms/vacuum gas oil (VGO) recycle (BVR) operations; they were obtained using bituminous and subbituminous coals and lignite.

Relative chemical compositions, determined by alumina-column chemical class fractionations of five EDS process materials, are shown in Table 5.25. EDS naphtha from BR operations was mainly aliphatic, with more than 65% of the recoverable materials eluted in the AH fraction. Neutral PAHs were low in concentration in raw naphtha, with few N-PAC components. High concentrations of the hydroxy-PAH fraction (presumably phenol, alkyl phenols, and related low-molecular-weight hydroxy-PAHs) were also found in raw naphtha.

The OT recycle solvent had more AHs than did the BVR-derived materials. Hydroaromatics, determined according to methods modified from Wozniak and Hites (1983), constituted 40% for OT recycle solvent compared to 26% for BVR recycle sovlent. The hydroaromatic PAH portion of each recycle solvent was 36% of the crude. AHs and hydroaromatic PAHs made up 76% of OT recycle solvent and 63% of BVR material.

Figure 5.48 shows gas chromatograms comparing PAH fractions of BVR and OT recycle sovlent; qualitative and quantitative peak information is shown in Table 5.26. Generally, EDS materials and their 50°F distillates contained higher levels of alkylated PAHs than previously observed in SRC materials. Alkylation up to C_{10} was typically observed with GC, HPLC, GC-MS, and probe-inlet MS for parent PAHs such as biphenyls, fluorenes, phenanthrenes, pyrenes, chrysenes, and benzopyrenes/fluoranthenes. The high alkylation observed with capillary-column GC for neutral PAH alumina fractions of recycle solvent was confirmed by low-voltage, low-resolution probe-inlet MS.

Table 5.25 Chemical Composition of Selected EDS Process Materials

Compound Class	Weight Distribution (%)[a]				
	Raw Naphtha (L4131)[d]	Recycle Solvent[b] (L4132)	Vacuum Still Bottoms[b] (L4133)	Recycle Solvent (L4111)	Heavy VGO[c] (L4112)
		Alumina Separations			
AHs	28	46*	2	69*	5
Neutral PAHs	4	37	7	21	31
N-PACs	<1	7	24	2	22
Hydroxy-PAHs	11	13	27	8	41
Total recovery (%)	43	103	60	100	99
		Picric Acid–Alumina Separation			
AHs		26		40	
Hydroaromatic-PAHs		37		36	
Dihydro-PAHs + PAHs		15		8	
>4-ring-PAHs		9		2	
Total recovery (%)		87		86	

AHs = aliphatic hydrocarbons; N-PACs = nitrogen-containing polycyclic aromatic compounds; PAHs = polycyclic aromatic hydrocarbons.

[a] Values are the average of two individual determinations except for the two values marked with asterisks, which are single values.

[b] Samples collected during bottoms and vacuum gas oil (VGO) recycle operations with Illinois No. 6 coal.

[c] Samples collected during once-through operation with Illinois No. 6 coal.

[d] Sample number given in parentheses.

Source: Adapted from Later et al. (1984b).

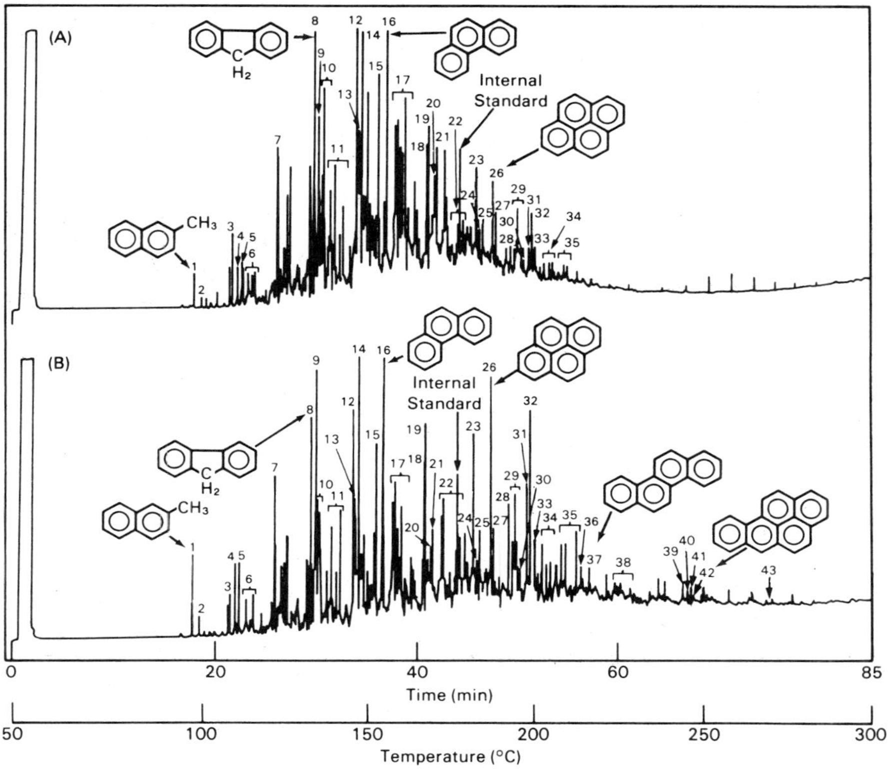

Figure 5.48 Capillary-column gas chromatograms of the polycyclic aromatic hydrocarbon fractions from the (A) once-through recycle solvent and (B) bottoms and vacuum gas oil recycle solvent from the EDS coal-liquefaction process. See Table 5.26 for quantitative information (adapted from Later et al., 1984b).

Heteroatom content of both recycle solvents follows trends similar to those discussed for neutral PAHs. N-PAC (BVR, 7%; OT, 2%) and hydroxy-PAH (BVR, 13%; OT, 8%) concentrations are lower in OT recycle solvents than in BVR recycle solvents. Heterocyclic PACs tend to be excised from organic compounds during hydrogenation of recycle solvent. Addition of vacuum still bottoms having high heterocyclic PAC concentrations increases overall concentration of these compounds in recycle solvent. Major N-PAC species in recycle solvents were indole/carbazole-type compounds, with lower concentrations of quinoline/benzoquinoline azaarenes. The N-PACs were highly alkylated. Concentrations of alkylated PACs were higher than for the nonalkylated parent structures. These observations were confirmed by probe-inlet MS, with alkylation detected up to C_8 for carbazoles.

The chromatogram in Fig. 5.49 shows that major species of the hydroxy-PAH fraction of recycle solvents are alkylated phenols and hydroxybiphenyls (Table 5.27). Corresponding qualitative information is available in Later et al. (1984b). Differences in hydroxy-PAH concentrations of these recycle solvents were detailed by Dauble et al. (1983). Briefly, phenol and alkylated phenols comprised a higher percentage of the total hydroxy-PAH fraction of OT recycle solvent than that of BVR solvent.

Table 5.26 Peak Identifications and Concentrations of Polycyclic Aromatic Hydrocarbons (PAHs) from the Bottoms and Vacuum Gas Oil (VGO) Recycle and Once-Through Recycle Solvents

Peak Number[b]	Molecular Weight	Compound[c]	Concentration[a] ($\mu g/g$) Bottoms and VGO Recycle (L4132)[d]	Once-Through (L4111)
1	142	2-Methylnaphthalene	2000 ± 100	420 ± 20
2	142	1-Methylnaphthalene	480 ± 20	130 ± 10
3	154	Biphenyl	1900 ± 700	570 ± 20
4	156	2-Ethylnaphthalene	ND[e]	ND
5	156	2,3 or 2,6-Dimethylnaphthalene	2300 ± 100	670 ± 40
6	156	C_2-Naphthalenes	ND	ND
7	170	C_3-Naphthalene	1850 ± 480	1300 ± 100
8	166	Fluorene	4700 ± 600	3600 ± 200
9	168	Dihydrofluorene	6200 ± 1100	1400 ± 100
10	168	C_1-Biphenyls	ND	ND
11	182	C_2-Biphenyls	ND	ND
12	180	2-Methylfluorene	4100 ± 500	2800 ± 400
13	180	1-Methylfluorene	2300 ± 600	3100 ± 600
14	180	Methylfluorene	5700 ± 400	3600 ± 500
15	196	C_3-Biphenyls	ND	ND
16	178	Phenanthrene	8000 ± 100	2900 ± 500
17	194	C_2-Fluorenes	8500 ± 900[f]	5500 ± 400[f]
18	192	3-Methylphenanthrene	3100 ± 100	1300 ± 200
19	192	2-Methylphenanthrene	2800 ± 300	1200 ± 100
20	192	9- or 4-Methylphenanthrene	1400 ± 100	950 ± 20
21	192	1-Methylphenanthrene	1850 ± 100	1100 ± 100
22	206	C_2-Phenanthrenes	ND	ND
23	204	Dihydrofluoranthene	2600 ± 100	970 ± 40
24	202	Fluoranthene	520 ± 70	250 ± 40
25	220	C_3-Phenanthrene	ND	ND
26	202	Pyrene	3500 ± 300	690 ± 210
27	204	Dihydropyrene	1100 ± 100	320 ± 20
28	218	C_1-Dihydrofluoranthene/pyrenes	ND	ND
29	234	C_4-Phenanthrene	ND	ND
30	216	Benzo[a]fluorene	370 ± 300	53 ± 10
31	216	Benzo[b]fluorene, 2- or 4-methylpyrene	2100 ± 300	270 ± 30
32	218	Methylphenylnaphthalene	ND	ND
33	216	1-Methylpyrene	440 ± 70	360 ± 80
34	234	Benzonaphthothiophenes	ND	ND
35	232	C_2-Dihydrofluoranthene/pyrenes	ND	ND
36	228	Benz[a]anthracene	330 ± 70	21[g]
37	228	Chrysene	630 ± 40	80[g]
38	242	C_1-Chrysenes	1600 ± 90	140[f,g]
39	252	Benzo[j or b]fluoranthene	110 ± 40	12[g]
40	252	Benzo[k]fluoranthene	ND	23[g]
41	252	Benzo[e]pyrene	210 ± 40	ND
42	252	Benzo[a]pyrene	63 ± 22	ND
43	276	Benzo[ghi]perylene	120 ± 10	ND

[a]Based on GC analysis of PAH fractions, calculated to crude concentrations using the weight % that eluted in the PAH fraction.

[b]Refers to peaks in Fig. 5.48, gas chromatography (GC) analysis of PAH fractions.

[c]Tentative identification by gas chromatography–mass spectrometry (GC-MS) and retention times.

[d]Sample number given in parentheses.

[e]No quantitative data.

[f]A composite of more than one peak.

[g]Value derived from one determination only.

Source: Later et al. (1984b).

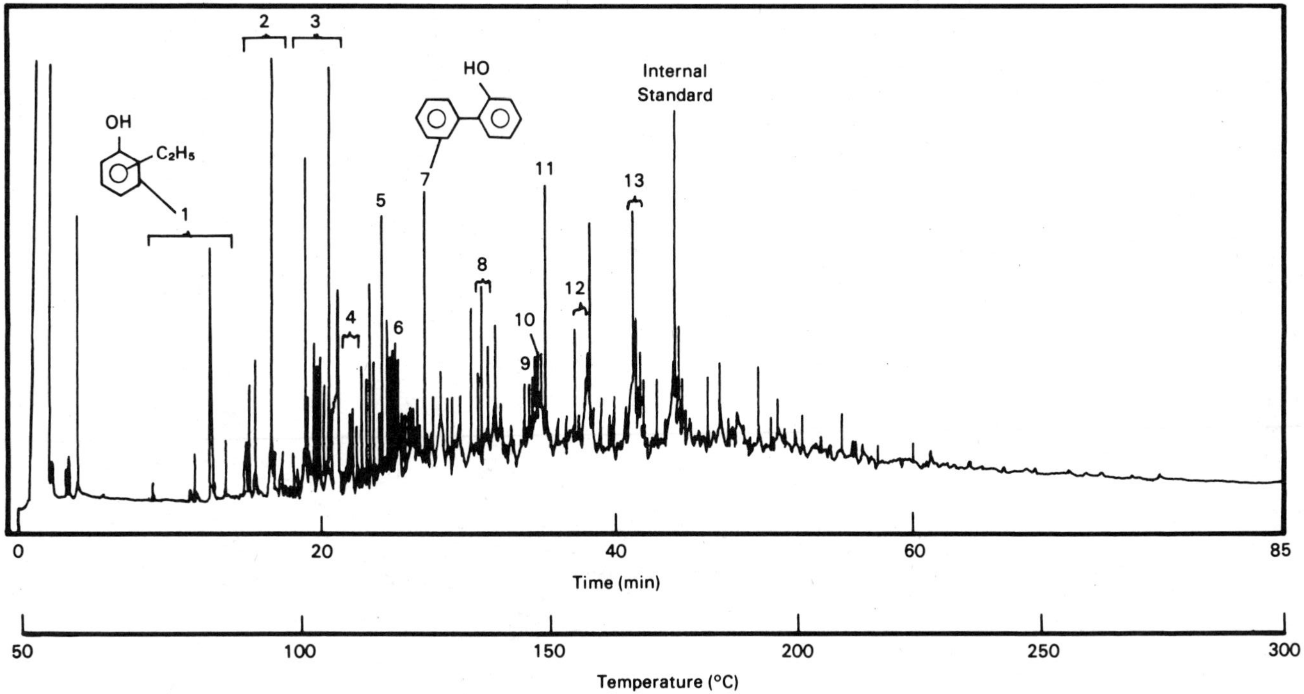

Figure 5.49 High-resolution gas chromatogram of the A_4 fraction from EDS bottoms and vacuum gas oil recycle solvent. Peak identifications are shown in Table 5.27 (adapted from Later et al., 1984b).

Table 5.27 Peak Identifications of Hydroxy-Polycyclic Aromatic Hydrocarbons from Bottoms and Vacuum Gas Oil Recycle Solvent

Peak Number[a]	Molecular Weight	Compound[b]
1	122	C_2-Phenols
2	136	C_3-Phenols
3	150	C_4-Phenols
4	164	C_5-Phenols
5	164	C_5-Phenol
6	162	C_2-Indanol
7	170	2-Hydroxybiphenyl
8	184	C_1-Hydroxybiphenyl
9	198	C_2-Hydroxybiphenyl
10	170	Hydroxybiphenyl
11	190	C_4-Indanol
12	184	C_1-Hydroxybiphenyl
13	198	C_2-Hydroxybiphenyl

[a] Refers to peaks in chromatogram, Fig. 5.49.

[b] Tentatively identified by gas chromatography–mass spectrometry.

Source: Later et al. (1984b).

Heavy VGO and vacuum still bottoms, from Illinois No. 6 coal, were also evaluated. The latter was a high-molecular-weight material with relatively high heterocyclic PAC content. Nearly 40% of the material appeared to be composed of the average molecular weights for neutral PAH and N-PAC fractions: 350 and 380 amu, respectively. Low-voltage mass spectra indicate that major constituents were highly alkylated four-, five-, and six-ring hydroxy-PAHs; compounds with more than six aromatic rings were also present.

Heavy VGO was also mostly polar and aromatic; greater than 60% of the gross chemical composition was attributable to the N-PACs and hydroxy-PAHs. Vacuum still bottoms were higher in molecular weight, with the distribution starting around chrysene (m/z 228). Both materials were highly alkylated. Major N-PAC constituents in vacuum still bottoms and heavy VGO were carbazoles, alkylated carbazoles, benzocarbazoles, and alkylated benzocarbazoles. Concentrations of specific neutral PAHs are shown in Table 5.28 (see also Fig. 5.50). Hydroxy-PAHs contained highly alkylated three- to six-ring hydroxy-PAHs.

To quantitate mutagenic compounds, quantitative and qualitative analyses were also conducted on amino-PAH isolates of four EDS process materials. Table 5.29 lists concentrations of 19 amino-PAHs, determined by comparative retention and response-factor data. As predicted from relative concentrations and molecular weights of amino-PAHs determined, the BVR solvent N-PAC fraction was more mutagenic (i.e., dose response and maximum response) than OT materials.

Amino-PAH data for vacuum still bottoms and heavy VGO are consistent with the mutagenic activity observed for these materials. Vacuum still bottoms contained mostly high-molecular-weight amino-PAHs; VGO contained a wider range of amino-PAHs with more higher-molecular-weight compounds. Both materials showed strong activity as determined by dose response and maximum mutagenic response in *S. typhimurium* TA98 assays.

Acid–base–neutral separations data in Table 5.14 compare hydrogenated recycle solvent

Table 5.28 Peak Identifications and Concentrations for Selected Neutral Polycyclic Aromatic Hydrocarbons (PAHs) in EDS Vacuum Still Bottoms (VBs) and Heavy Vacuum Gas Oil (VGO)

			Concentration ($\mu g/g$)[a]	
Peak Number[b]	Molecular Weight (m/z)	Compound[c]	Heavy VGO (L4112)[d]	VB (L4133)
1	166	Fluorene	47[e]	
2	180	Methylfluorene	230 ± 20	28 ± 7
3	178	Phenanthrene	1300 ± 100	
4	178	Anthracene	190 ± 20	
5	191	3-Methylphenanthrene	840 ± 120	
6	192	2-Methylphenanthrene	1000 ± 100	15 ± 3
7	192	9- or 4-Methylphenanthrene	560 ± 120	
8	192	1-Methylphenanthrene	650 ± 90	
9	206	C_2-Phenanthrenes	ND[f]	
10	202	Fluoranthene	1600 ± 400	25 ± 2
11	202	Pyrene	4000 ± 900	150 ± 10
12	218	C_1-Phenylnaphthalene	ND	ND
13	216	Benzo[a]fluorene	1800 ± 200	35 ± 7
14	216	Benzo[b]fluorene, 2- or 4-methylpyrene	2500 ± 500	150 ± 10
15	218	C_1-Phenylnaphthalene	ND	ND
16	216	1-Methylpyrene	3000 ± 700	70 ± 7
17	230	C_1-Benzofluorenes, C_2-pyrenes	ND	ND
18	228	Benzo[a]anthracene	1400 ± 200	84 ± 4
19	228	Chrysene	1400 ± 200	84 ± 28
20	242	C_1-Chrysenes[g]	3700 ± 500	390 ± 20
21	256	C_2-Chrysenes	ND	ND
22	252	Benzo[j or b]fluoranthene	310 ± 30	130 ± 10
23	252	Benzo[k]fluoranthene	120 ± 30	110 ± 10
24	256	7,12-Dimethylbenzanthracene	29	ND
25	252	Benzo[e]pyrene	310 ± 190	130 ± 20
26	252	Benzo[a]pyrene	590 ± 110	230 ± 10
27	266	C_1-Benzopyrenes, C_1-benzofluoranthenes	ND	
28	276	Indeno[1,2,3-cd]pyrene	ND	170 ± 20
29	278	Dibenz[a,c or a,h]anthracene	150 ± 40	110 ± 40
30	276	Benzo[ghi]perylene	56 ± 20	280 ± 30

[a] Based on GC analysis of PAH fractions calculated to crude concentrations using the weight % that eluted in the PAH fractions.

[b] Refers to peaks in the isolated PAH fraction shown in Fig. 5.50.

[c] Tentatively identified by gas chromatography (GC)–mass spectrometry and GC retention times.

[d] Sample number given in parentheses.

[e] Value derived from one determination.

[f] No quantitative data.

[g] Composite of four peaks.

Source: Later et al. (1984b).

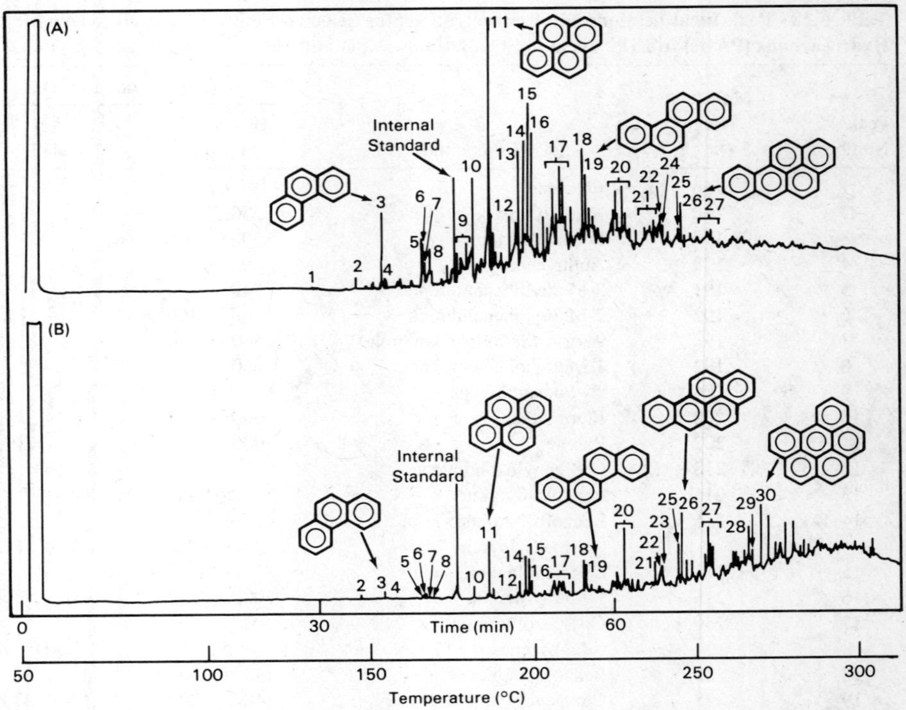

Figure 5.50 High-resolution gas chromatogram of polycyclic aromatic hydrocarbon fractions from (A) EDS heavy vacuum gas oil and (B) EDS vacuum still bottoms (from Later et al., 1984b).

with the hydrogen-depleted or ''spent'' solvent. The spent solvent was lower in mutagenic activity than its hydrotreated counterpart.

Studies to evaluate chemical class composition of EDS recycle solvents and VGO, as a function of feed coal type, have been limited. Recycle solvents and VGO from BVR operation with Illinois No. 6 (bituminous) coal were more aromatic (see Table 5.30), according to chemical class fractionation, than those obtained from Wyodak (subbituminous) or Texas lignite feeds. Recycle solvents from BVR operation with bituminous coal were also more mutagenic than those from BVR operation with Wyodak and Texas lignite. Illinois No. 6 VGO was more mutagenic than that from Wyodak. VGO from Texas lignite feed was not available (Later et al., 1984b).

5.13.1.9 *Materials from the C. E. Lummus Integrated Two-Stage Liquefaction Process*

The term *integrated two-stage liquefaction* (ITSL) has been used to refer to at least two different processes. In the United States, ITSL was initially developed to the pilot stage by the C. E. Lummus Company in New Brunswick, New Jersey. The term *integrated* was applied to indicate coupling of liquefaction and hydrogenation unit operations; coal pasting solvent is recycled directly from the hydrotreater after separation. Hydrogen-rich solvent is essential to optimize short-contact-time reactor operations.

Table 5.29 Concentrations of Selected Amino-Polycyclic Aromatic Hydrocarbons in EDS Process Materials[a]

	Concentration (μ/g crude)[a]			
Compound[b]	Recycle Solvent BVR (L4141)[c]	VB (L4133)	Recycle Solvent OT (L4111)	Heavy VGO (L4112)
1-Aminonaphthalene	1.6	ND	ND	ND
2-Aminobiphenyl	1.1	ND	ND	0.6
2-Aminonaphthalene	1.7	ND	0.2	1.1
3-Methyl-2-aminonaphthalene	0.8	ND	0.1	1.1
4-Aminobiphenyl	2.6	ND	0.4	1.1
4-Aminofluorene	2.0	ND	0.4	0.8
2-Aminofluorene	3.7	ND	1.4	ND
9-Aminophenanthrene, and/or 1-aminophenanthrene, and/or 1-aminoanthracene	9.6	3.3	1.3	2.9
2-Aminophenanthrene, and/or 3-aminophenanthrene	3.1	1.5	1.0	5.4
2-Aminoanthracene	2.7	0.6	2.1	3.1
3-Aminofluoranthrene	4.5	ND	1.2	3.8
1-Aminopyrene	2.9	ND	2.5	48.8
2-Aminopyrene	2.5	4.4	0.3	2.8
8-Aminofluoranthene	3.9	9.4	2.1	9.0
3-Aminochrysene	3.8	13.0	0.5	29.8
6-Aminochrysene	1.9	13.0	0.7	5.1

BVR = bottoms and vacuum gas oil recycle; ND = not detected; OT = once through; VB = vacuum still bottoms; VGO = vacuum gas oil.

[a]Values are the averages of three independent measurements at three concentration levels.

[b]Tentatively identified by gas chromatography (GC)–mass spectrometry and GC retention times.

[c]Sample number given in parentheses.

Source: Later et al. (1984b).

Table 5.30 Chemical Class Composition of Recycle Solvent as a Function of Coal Type

	Percent by Weight in Each Fraction			
	Illinois #6 Bituminous		Wyoming Subbituminous	Texas Lignite
Compound	BVR (L4132)[a]	OT (L4111)	BVR (L4121)	BVR (L4141)
AHs	46	69	57	59
PAHs	37	21	25	25
N-PACs	7	2	3	3
Hydroxy-PAHs	13	8	11	9
Recovery (%)	103	100	97	95

AHs = aliphatic hydrocarbons; BVR = bottoms and vacuum gas oil recycle, N-PACs = nitrogen-containing polycyclic aromatic compounds; OT = once-through operation, PAHs = polycyclic aromatic hydrocarbons.

[a]Sample numbers given in parentheses.

Source: Wright and Dauble (1986).

ITSL operations at Wilsonville incorporate essentially the same components as the Lummus process. At the Wilsonville plant, 40–60% of the ITSL distillate produced is derived thermolytically in the first-stage liquefaction reactor, whereas almost all of the Lummus ITSL distillate is made in the hydrotreater. The difference reflects lower liquefaction reactor severity resulting from the short-contact-time involved in the Lummus process.

Elemental composition, distillation ranges, heteroatom content, viscosity, and other properties of ITSL materials from the Lummus PDU are contained in Schindler and Chen (1982). Catalyst studies, coal-type comparisons, and other research data relative to ITSL operation are also available from the U.S. Department of Energy (DOE).

Lummus PDU samples were obtained between 1980 and 1983 for toxicological evaluation and chemical analysis during run conditions designated as *normal* or *off-normal* by the process developer. Data reported here, mainly from Wilson et al. (1983a, 1984), provide some insight concerning the influence of operating conditions on chemical and toxicological properties of resultant materials.

Generally, PAH fractions from the PDU process stream entering the hydrotreater second stage are less alkylated than SRC-II and EDS recycle solvents. Fully aromatic, nonalkylated peri-, peri/cata-, and catacondensed PAHs comprise about 40% by weight of the PAH fraction; partially hydrogenated indeno and alkylated species make up most remaining material.

N-PAC concentrations in ITSL materials are generally lower than in materials from hybrid processes and substantially lower than those from single-stage processes. Hydrotreater effectiveness in removing nitrogen is shown in two chromatograms (Fig. 5.51). N-PAC species in ITSL streams were primarily azaarenes and carbazoles. The parent structures appear to be alkylated to about the same extent as their PAH analogs. Hydroaromatic N-PACs have also been detected in ITSL materials.

Concentrations of selected N-PACs and hydroaromatic species in materials collected after the first stage are designated LC Finer feed (LCFF); those collected after the second stage are designated LC Finer total liquid product (LCF TLP) (Table 5.31). Note the increase in hydroaromatics and concurrent decrease in N-PAC concentrations after second-stage operation.

Table 5.32 shows alumina-column chromatographic separation data for the >850°F component of ITSL LCFF and LCF TLP collected during normal and off-normal process run conditions. Results for the >850°F component of SRC-II are included for comparison. The data indicate that LC Fining hydrogenates the >850°F feed component and lower-boiling components and that hydrogenation of heavy ends (>850°F materials) suffers more from reduction in catalyst activity than does hydrogenation of <850°F materials (compare LCF TLP from 3LCF7-90 to LCFF from 3LCF7-37). Quantitative differences in specific compounds are included in Wilson et al. (1984).

Table 5.33 shows concentrations (ppm) of PAH compounds monitored by HRGC in PAH fractions of selected ITSL samples. Estimated concentrations for several hydroaromatic species, based on HRGC and GC-MS analysis, are shown in Table 5.34. Relative concentrations for fully aromatic five-ring compounds such as BaP and benzo[e]pyrene were substantially lower in materials produced under normal conditions compared to those from off-normal conditions. BaP content was about 2200 ppm in LCF TLP from 3LCF7-37 (normal operation), compared to about 5360 ppm in LCF TLP 3LCF7-90 (off-normal operation). Although the trend was similar, differences were not as great for four-ring aromatic compounds (i.e., normal operation for a given coal type resulted in lower concentrations of aromatics than did off-normal conditions).

Table 5.31 shows concentrations of nitrogen-containing compounds in N-PAC fractions of several ITSL materials. Concentrations of amino-PAHs were not detectable without fur-

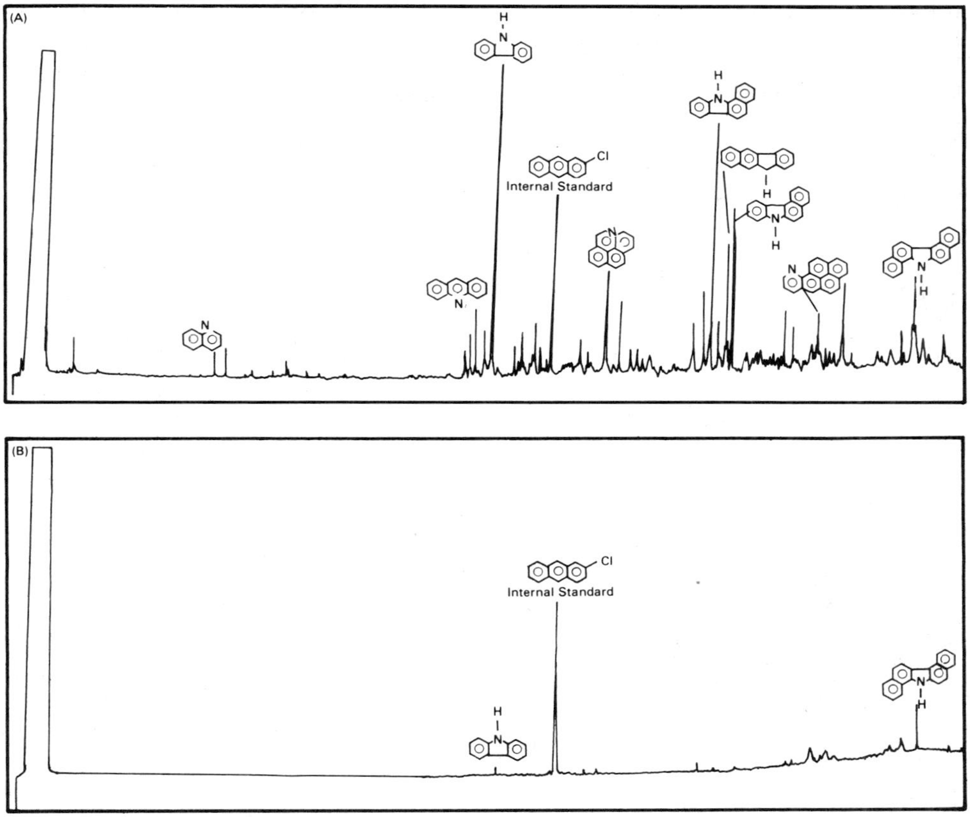

Figure 5.51 Chromatogram of the nitrogen-containing polycyclic aromatic compound fraction of integrated two-stage liquefaction material collected from run No. 3LCF7-90 (Lummus process development unit): (A) LC Finer feed (Sample L5121) and (B) total liquid product (Sample H0621) (from Wilson et al., 1983b).

ther separation and derivatization. Amino-PAH concentrations detected by this method in ITSL LCFF and recycle solvent are compared to those in SRC-II HD in Table 5.35. Higher concentrations of organic nitrogen were found in higher-boiling ITSL cuts than in similar-boiling-range SRC liquids (Fig. 5.52). Besides increased molecular weights of N-PACs in ITSL materials, there appeared to be differences in nitrogen functionalities relative to SRC.

Organic nitrogen in ITSL materials produced during normal operating conditions was found mainly as carbazoles and azaarenes, with some diaza compounds and trace levels of amino-PAHs. Partially hydrogenated analogs of these compounds (including amino-PAHs) were also present, particularly in LCF TLP. For comparison, nitrogen occurred in SRC materials as aminoazaarenes, aminocarbazoles (Later et al., 1982b) thiazoles, cyano-PAHs, phenoxazines, phenothiazines, and phenazines at low concentrations. Pyrrolic, azaarene, diazaarene, and amino-PAH compounds were present in SRC materials at relatively higher levels. These differences may reflect feed coal differences. However, it is more likely that decreased reaction severity in the ITSL short-contact-time reactor may also minimize or

Table 5.31 Concentrations of Selected Nitrogen-Containing Polycyclic Aromatic Compounds (N-PACs) from the C. E. Lummus Integrated Two-Stage Liquefaction Process[a] (ppm or $\mu g/g$)

Compound	3LCF-37 LCFF (L5111)[b]	3LCF-37 LCFTLP (H0611)	3LCF-90 LCFF (L5121)	3LCF-90 LCFTLP (H0621)	3LCF8-52 Recycle Solvent (H0632)	3LCF8-52 Heavy Oil (L5132)
Quinoline	36	—[c]	—	—	—	5.9
Isoquinoline	37	—	—	—	—	—
Indole	13	—	—	—	—	—
2-Methylquinoline	22	—	—	—	—	—
3-Methylquinoline	8.0	—	—	—	—	—
4-Methylquinoline	22	—	—	—	—	9.8
2,6-Dimethylquinoline	22	—	—	—	7.2	12
2,5-Dimethylindole	8.9	—	—	—	2.9	4.9
2,3-Dimethylindole	14	—	—	—	4.8	11
4-Azafluorene	8.6	—	17	70	9.7	2.7
Benzo[h]quinoline	130	—	140	—	83	86
Acridine	220	—	190	46	110	120
Benzo[f]quinoline[d]	170	—	190	32	24	92
Carbazole	1200	13	1100	270	780	830
2- or 3-Methylbenzo[f]-quinoline	68	—	44	38	13	38
1-Methylcarbazole	49	2.2	51	51	43	34
3-Methylcarbazole	11	1.1	120	57	83	90
1- or 2-Azafluoranthene	130	9.1	150	42	53	52
7-Azafluoranthene	79	10	160	44	50	33
4-Azapyrene	320	5.0	440	230	160	260
Benz[c]acridine	280	11	430	490	300	312
Benz[a]acridine	380	4.1	990	210	180	180
Benzo[a]carbazole	1200	8.5	2900	510	650	715
Benzo[b]carbazole	700	—	1500	290	380	440
Benzo[c]carbazole	860	9.1	2000	590	540	600
10-Azabenzo[a]pyrene	300	2.0	600	150	180	520
Dibenzo[a, i]carbazole[e]	67	—	110	29	36	6.9
Dibenzo[a,g or c,g] carbazole	470	130	600	250	380	455

[a]Relative standard deviation ranged from 5% to 50%; quantitation was determined from gas chromatography analysis of N-PAC fractions and calculated to crude concentrations using the weight % that eluted in the N-PAC fraction. Column heading abbreviations are as follows: LCFF = LC Finer feed; LCFTLP = LC Finer total liquid product.

[b]Sample number given in parentheses.

[c]Dashes indicate no data.

[d]Coelutes with phenanthridine.

[e]Coelutes with dibenz[a,j]acridine.

Source: Wilson et al. (1983a).

Table 5.32 Percentage of Integrated Two-Stage Liquefaction Materials Partitioning into Various Chemical Class Fractions During High-Performance Liquid Chromatography Separation

		Chemical Class Fraction				
Run Number	Material	AHs	PAHs	N-PACs	Hydroxy-PAHs	Recovery (%)
3LCF7-37	LCFF >850°F	1	15	18	45	79
(normal operation)	LCFTLP >850°	1	37	32	24	98
3LCF7-90						
(off-normal operation)	LCFTLP >850°F	1	20	36	45	101
SRC-II HD	>850°F	1	46	36	25	107

AHs = aliphatic hydrocarbons; PAHs = polycyclic aromatic hydrocarbons; N-PACs = nitrogen-containing polycyclic aromatic compounds, LCFF = LC Finer feed, LCFTLP = LC Finer total liquid product, SRC = solvent refined coal, HD = heavy distillate.

Source: Wilson et al. (1983a).

preclude formation of these compounds. For example, phenoxazines, phenazines, and phenothiazines are thought to be condensation products of a phenol and an aniline, two anilines, and an aryl thiol with aniline, respectively. Overall reaction severity and, in particular, residence time in the short-contact-time reactor are probably insufficient to produce the precursors of these condensation products at rates required for the condensation reactions to occur. Similarly, the short-contact-time reactor environment is probably sufficiently oxidative to produce cyano-PAHs.

Compared to SRC-I PS and SRC-II HD, amino-PAH concentrations in ITSL materials were low. There was evidence that primary amino groups may be attached to partially hydrogenated and/or alkylated rings. Although total amino-PAH content in ITSL samples was an order of magnitude less than that in the SRC-II sample, microbial mutagenicity was not reduced accordingly.

Although heavier pericondensed PAHs such as pyrenes, benzopyrenes, anthanthrenes, and benzo[ghi]perylene appeared as partially hydrogenated species in many of the process streams, these parent structures were resistant to carbon–carbon bond scission during catalytic hydrogenation. Because of interest in hydrogenated pericondensed species as hydrogen donors, these compounds and their hydroaromatic analogs were analyzed in various process streams and under several operating conditions. Results are discussed below.

Table 5.36 shows concentration ratios of hydroaromatic to aromatic species in materials from run No. 3LCF7-90 before and after LC Fining. Benzofluoranthenes were more hydrogenated in LCFF than were the more highly condensed benzopyrene structures. During LC Fining, fluoranthenes and benzofluoranthenes apparently undergo ring opening to a greater extent than do pyrenes and benzopyrenes; concentrations of the former compounds (particularly fluoranthenes) decreased significantly during LC Fining. Susceptibility of catacondensed compounds such as benzanthracenes to hydrogenation in complex mixtures and under ITSL conditions appears to be intermediate between that of pericondensed pyrenes/benzopyrenes and that of fluoranthenes/benzofluoranthenes.

5.13.1.10 *Materials from the Veba Oel/Ruhrkohle AG Process*

Heavy oil and middle oil from the coal-liquefaction facility operated by Veba Oel/Ruhrkohle AG in Bottrop, Federal Republic of Germany, were assayed for chemical composi-

Table 5.33 Concentrations of Selected Polycyclic Aromatic Hydrocarbon (PAH) Compounds from the Integrated Two-Stage Liquefaction Process (parts/thousand, mg/g)[a]

Compound	3LCF-37 LCFF (L5111)[b]	3LCF-37 TLP (H0611)	3LCF-90 LCFF (L5121)	3LCF-90 TLP (H0621)	3LCF8-52 Recycle Solvent (H0632)	3LCF8-52 Heavy Oil (L5132)
Naphthalene	1.1	0.24	0.063	0.58	0.16	0.090
2-Methylnaphthalene	1.3	1.0	0.035	1.8	0.42	0.38
1-Methylnaphthalene	0.69	0.46	—[c]	0.67	0.23	0.21
Biphenyl	0.30	0.34	—	0.28	0.10	0.10
1,7-Dimethylnaphthalene	0.53	0.24	0.035	0.37	0.21	0.21
Acenaphthene	1.7	0.17	0.31	0.42	0.81	0.73
Dibenzofuran	1.4	0.45	0.22	0.15	1.2	0.80
Fluorene	2.7	0.64	0.70	1.0	3.3	2.1
1-Methylfluorene	0.49	0.40	0.063	0.71	0.49	0.41
Dibenzothiophene	3.1	0.58	0.83	1.2	1.6	1.6
Phenanthrene	15	1.2	8.5	3.8	14.5	10
3-Methylphenanthrene	2.3	0.37	1.2	0.85	2.4	1.9
2-Methylphenanthrene	2.6	0.98	1.5	1.1	2.6	2.1
4-H-Cyclopenta[def]-phenanthrene	1.9	0.24	0.90	0.87	1.6	1.2
1-Methylphenanthrene	1.3	0.15	0.73	0.55	1.1	0.93
Fluoranthene	16	0.36	14	3.3	11	9.1
Pyrene	29	11	20	18	30	24
Benzo[a]fluorene	3.0	0.45	3.7	1.1	2.5	2.2
Benzo[b]fluorene[d]	5.4	3.0	4.1	5.4	6.8	6.3
1-Methylpyrene	2.4	1.1	1.9	1.6	2.3	1.9
Benzo[b]naphtho[2,1-d]-thiophene	1.4	—	2.3	0.096	0.62	0.37
Benz[a]anthracene	6.7	0.21	12	1.4	4.2	3.8
Chrysene[e]	7.6	0.19	14	4.3	6.3	5.45
3-Methylchrysene	2.0	0.41	4.2	1.8	2.6	1.9
2-Methylchrysene	1.3	0.049	2.4	0.78	1.5	1.3
6- or 4-Methylchrysene	2.0	2.2	1.0	3.1	4.8	2.7
1-Methylchrysene	1.9	0.32	2.5	0.79	1.9	1.8
Benzo[j or b]fluoranthene	6.4	0.66	16	5.7	6.6	6.0
Benzo[k]fluoranthene	2.9	0.26	5.7	1.4	3.7	3.6
Benzo[e]pyrene	4.6	0.66	8.0	4.5	6.1	5.1
Benzo[a]pyrene	5.3	0.59	11	2.0	4.3	4.55
Perylene	1.9	0.085	3.1	0.71	1.9	2.5
Indeno[1,2,3-cd]pyrene	1.8	0.60	3.6	1.8	2.0	1.6
Dibenz[a,h or a,c]anthracene	0.56	0.15	1.1	0.47	0.61	0.50
Benzo[ghi]perylene	2.2	1.6	3.5	3.2	4.3	2.9

[a] Relative standard deviation ranged from approximately 5% to 20%; concentrations were calculated from gas chromatography analysis of PAH fractions calculated to concentrations in crude materials using the weight % that eluted in the PAH fraction. Column heading abbreviations are as follows: LCFF = LC Finer feed; TLP = total liquid product.

[b] Sample numbers given in parentheses.

[c] Dash indicates no data.

[d] Coelutes with 2- or 4-methylpyrene.

[e] Coelutes with triphenylene.

Source: Wilson et al. (1983a).

Table 5.34 Concentrations (ppm) of Hydroaromatic Analogs of Major Polycyclic Aromatic Hydrocarbons (PAHs) in Integrated Two-Stage Liquefaction LC Finer Feed (LCFF) and LC Finer Total Liquid Product (LCFTLP) from C. E. Lummus Process Development Unit[a]

Hydroaromatic Analog/Parent Compound Molecular Weight	Molecular Weight (m/z)	LCFF (L5121)[c]	LCFTLP[b] (H0621)
Octahydro 178	186	21	1400 ± 100
Hexahydro 178	184	240 ± 40	1350 ± 70
Octa plus dihydro 178	1, 180	70 ± 21	930 ± 70
Dihydro 178	180	91 ± 14	720 ± 30
Dihydro 178	180	63 ± 28	330 ± 20
Octahydro 178	186	150 ± 40	1300 ± 100
Tetrahydro 178	182	850 ± 90	2500 ± 100
Decahydro 202	212	190 ± 60	2100 ± 200
Decahydro 202	210	120 ± 40	410 ± 20
Decahydro 202	212	160 ± 100	1200 ± 100
Octahydro 202	210	190 ± 30	980 ± 70
Octahydro 202	210	200 ± 30	610 ± 40
Octahydro 202	210	190 ± 40	440 ± 40
Hexahydro 202	208	420 ± 270	1600 ± 100
Tetrahydro	206	56	2300 ± 100
Hexahydro 202	208	130 ± 40	2500 ± 100
Hexahydro 202	208	130 ± 90	2800 ± 100
Dihydro 202	204	550 ± 50	560 ± 60
Octahydro 228	236	1800 ± 350	1250 ± 200
Tetrahydro 228	232	2500 ± 400	5100 ± 600
Dihydro 228	230	620 ± 110	890 ± 120
Tetrahydro 228	232	1900 ± 700	3400 ± 400
Dodecahydro 252[d]	264	670 ± 20	470 ± 120
Dodecahydro 252[d]	264	1500 ± 300	840 ± 120
Decahydro 252[d]	262	410 ± 80	290 ± 50
Hexahydro 252	258	5500 ± 1100	7800 ± 1200
Tetrahydro 252	256	1650 ± 470	2300 ± 400
Hexahydro 252	258	—	1700 ± 200
Tetrahydro 252	256	1100 ± 400	3000 ± 700
Tetrahydro 252	256	600 ± 220	1100 ± 300

[a] Concentrations are based on gas chromatography analysis of PAH fractions calculated to crude concentrations using the weight % that eluted in the PAH fractions.

[b] Samples from run No. SLCF7-90.

[c] Sample number given in parentheses.

[d] These compounds have been identified as benzofluoranthene isomers by their relative retention times.

Source: Wilson et al. (1983a).

Table 5.35 Estimated Concentrations[a] of Selected Amino-Polycyclic Aromatic Hydrocarbons (PAHs) in the Amino-PAH Subfractions and Crude Materials from Selected Integrated Two-Stage Liquefaction (ITSL) and Solvent-Refined Coal (SRC) Samples

	Concentrations in Amino-PAH Fraction (ppm)			Concentrations in Crude Sample (ppm)		
Compound	3LCF7-37 LCFF[b] (L5111)[e]	3LCF8-52 Recycle Solvent (H0632)	SRC-II HD[c] (L1111)	LCFTLP[d] (L5111)	Recycle Solvent (H0632)	SRC-II HD (L1111)
5-Aminoindan	150	220	150	1.5	2.3	2
1-Aminonaphthalene	160	270	3,900	1.7	2.9	70
2-Aminobiphenyl	70	130	1,480	0.7	1.9	20
2-Aminonaphthalene	60	230	17,100	0.6	2.4	290
4-Aminobiphenyl	80	370	3,160	0.8	3.9	510
2-Aminofluorene	160	430	58,000	1.6	4.5	980
1-Aminoanthracene	130	700	17,350	1.3	7.4	290
9-Aminophenanthrene	110			1.1		
2-Aminophenanthrene	120	150	19,750	1.2	151.0	330
3-Aminofluoranthene	140	430	2,500	1.4	4.5	40
1-Aminopyrene	90	950	4,720	0.9	10.0	80
6-Aminochrysene	570	1770	0	6.0	32.0	

[a]Rounded to two significant figures.
[b]LCFF = LC Finer feed from run No. 3LCF-37 (on normal conditions); ITSL recycle solvent from run No. 3LCF-52 (on normal conditions). All ITSL materials from the C. E. Lummus pilot plant.
[c]HD = heavy distillate.
[d]LCFTLP = LC Finer total liquid product.
[e]Sample number given in parentheses.
Source: Wilson et al. (1983a).

tion, microbial mutagenicity, and prophage induction. The German process uses an iron-rich, red mud catalyst in the liquefaction reactor and operates at lower pressure than does the Bergius–Pier process from which it was derived. Wilson and Pelroy (1983) found that Veba Oel heavy oil (bp, >320°C) and middle oil (bp, 200–320°C) had higher concentrations of hydroaromatics and lower concentrations of amino-PAHs, although the overall weight distributions by chemical class were similar, compared to those of analogous SRC-II materials. The higher hydroaromatic content found by GC and GC-MS in Veba Oel materials was confirmed by GC FTIR. For example, tetralin concentrations in the middle oils were estimated at 35,000 ppm, compared with naphthalene concentrations of about 77,000 ppm. Amino-PAHs in Veba Oel heavy oil detected by GC-MS of the PFP derivatives included aminophenanthrenes, aminopyrenes, and aminochrysenes, and their alkyl homologs up to C_2 or C_3. Overall alkylation of PACs in the Veba Oel samples was not determined. Mutagenic activity of Veba Oel heavy oil was concentrated in the amino-PAH fractions; middle oil was not mutagenically active.

5.13.1.11 *Materials from the Synthoil Process*

Chemical composition of Synthoil materials has been characterized extensively (Aczel et al., 1977). PAH percentages in Synthoil materials (from several analytical methods) varied from 24 to 53%, depending on feed coal and batch number. Synthoil coal liquids also had

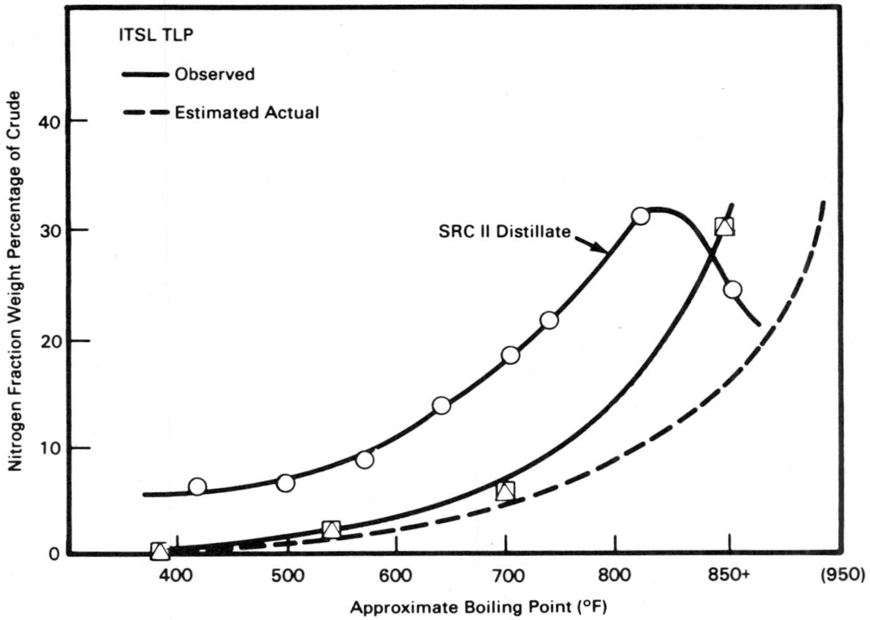

Figure 5.52 Nitrogen compounds (wt %) as a function of boiling point: comparison of integrated two-stage liquefaction (ITSL) and solvent-refined coal II (SRC-II) (Sample L1111) full-boiling-range distillate materials. TLP = total liquid product (from Wilson et al., 1983b).

Table 5.36 Concentration Ratios of Selected Polycyclic Aromatic Hydrocarbon (PAH) Groups Versus Their Hydroaromatic Derivatives for Lummus Integrated Two-Stage Liquefaction Materials Before and After LC Fining

	Concentration Ratios	
PAH Group	LC Finer Feed (L5121)[a]	LC Finer Total Liquid Product (H0621)
Σ Phenanthrene/anthracene / Σ Hydrophenanthrenes/hydroanthracenes	6.9	0.5
Σ Pyrene/fluoranthrene / Σ Hydropyrenes/hydrofluoranthenes	14.9	1.4
Σ Chrysene/benzanthracene / Σ Hydrochrysenes/hydrobenzanthracenes	3.8	0.5
Σ Benzofluoranthenes/benzopyrenes/perylenes / Σ Hydrobenzofluoranthenes/hydrobenzopyrenes/hydroperylenes	3.8	0.8

[a] Sample number given in parentheses.

Source: Wilson et al. (1983a).

substantial hydroaromatic content. For example, dihydropyrenes averaged about 2500 ppm in five samples derived from Kentucky Homestead coal, while pyrene itself had an average concentration of about 3200 ppm. Fluoranthene concentrations were about 17% those of pyrene; thus, the ratio of pericondensed pyreno-PAHs to indeno-PAHs was relatively high. BaP concentrations ranged from 50 to 260 ppm, averaging 150 ppm for material from Kentucky Homestead feed coal and 270 ppm for three batches from Ireland mine coal.

Heterocyclic N-PACs in Synthoil materials were highly alkylated compared to ITSL materials. The C_3 carbazoles, for example, had the highest concentration of any $C_nH_{2n-15}N$ series compound at an average 860 ppm; carbazole averaged 430 ppm in five samples derived from Kentucky Homestead coal. Total carbazoles and their alkyl homologs averaged about 7000 ppm; total acridines and their alkyl homologs averaged about 5600 ppm. Other mononitrogen-containing compounds and their alkyl homologs averaged 31,000 ppm. Nitrogen–oxygen mixed heterocycles averaged 2600 ppm in these samples. Minor series nitrogen–oxygen heterocycles and nitrogen–sulfur heterocycles averaged 1600 ppm.

Schiller (1977a) compared the composition of several coal-liquefaction materials, including SRC-I, COED, and Synthoil. Results from mass spectral analyses of Synthoil materials are also contained in Scheppele (1978a, b). J. L. Schultz et al. (1979) characterized Synthoil materials and described 195 molecular formulas for N-PACs, most of which represented more than one isomer. Compounds reported ranged from pyridines to azadibenzopyrenes. Both aza and pyrrolic nitrogen species were identified; no amino-PAHs were reported. It was noted that the use of a cobalt–molybdenum catalyst in the 1/2-ton/day Synthoil reactor, under 2000 psi hydrogen at 450°F, yielded a product having nitrogen fractions with gas chromatographic and mass spectral properties essentially indistinguishable from those of products produced without catalyst.

5.13.1.12 *Catalytic Two-Stage Liquefaction (CTSL) Process*

Materials from an early demonstration run (Lawrenceville, New Jersey PDU) of the Hydrocarbon Research, Inc. CTSL process were chemically analyzed by Later and Wilson (1985). A catalyst is used in both stages of the process, and separation of insoluble organic matter and ash is accomplished by distillation and filtration after the second stage rather than between stages. Samples studied included THF extracts of first- and second-stage products and a second-stage pressure-filter liquid.

The CTSL materials were relatively high in AH content and had high concentrations of hydroaromatics that contributed, in part, to the relatively high AH fraction weights. N-PAC fraction weights were low, varying between 5 and 15%, depending on the run period.

Table 5.37 shows data from chemical class fractionation of selected CTSL materials. Separations of the AH and PAH neutral fractions were carried out on picric acid columns to determine better the hydroaromatics present in a set of the CTSL samples. In general, the CTSL samples appeared to be more "petroleumlike" than other two-stage materials tested in terms of their saturate and partial saturate content relative to their aromatic content. In marked contrast to other two-stage processes, gas chromatograms of PAH fractions of materials from the two CTSL stages looked remarkably similar. Thus, while the relative concentration of PAHs and hydroaromatic PAHs changed from stage to stage, the chemical constituency of the PAH fraction remained fairly constant. The materials themselves, however, could be distinguished by the presence of a benzene-insoluble precipitate in the first-stage material that was absent in the second-stage extract.

Table 5.37 Data from Chemical Class Fractionation of Selected Catalytic Two-Stage Liquefaction Materials

Sample Description	Sample Number	Fraction Weight Percentage				Recovery (%)
		AHs	PAHs	N-PACs	H-PAHs	
First Stage Condition 1 Period 3	L5411	53	29	6	13	101
Second Stage Condition 1 Period 3	L5412	60	35	7	11	113
First Stage Condition 3 Period 9	L5431	54	22	8	13	97
Second Stage Condition 3 Period 9	L5432	69	24	7	11	111

AHs = aliphatic hydrocarbons; H-PACs = hydrogen-containing polycyclic aromatic hydrocarbons; N-PACs = nitrogen-containing polycyclic aromatic compounds; PAHs = polycyclic aromatic hydrocarbons.

Source: Later and Wilson (1984).

5.13.2 Comparative Analysis of Coal-Gasification Materials

The organic condensates arising from coal-gasification processes are by-products that, by process design, are produced in relatively small quantity. In a commercial gasifier, they would be used primarily to supplement fuels for steam generators and other plant operations. Although analytical data for gasification condensates are not as extensive as for coal-liquefaction products, it is possible to evaluate the similarities and differences between the two.

In this section, we present analytical data for eight or nine gasifiers representing a range of process designs and scales. The data show that qualitative similarities predominate. Although a variety of process streams and by-products were examined for each gasifier, we consider here only information about organic condensates less volatile than benzene and xylenes.

5.13.2.1 *Materials from the COED Process*

Samples from the COED process were chemically fractionated in association with stability studies (Brinkman et al., 1979), biological testing (Rubin et al., 1976; Rao et al., 1981), process characterization, and development of analytical methods (Vogh and Thomson, 1981). Mass spectral and IR data for COED materials are available in Benson et al. (1979).

Epler et al. (1978a, b) fractionated COED and Synthoil samples according to the method of Swain et al. (1969), as modified by Bell et al. (1969), and found substantial differences in partitioning. Hexane solubles were substantially higher for COED (74%) than for Synthoil syncrudes (28%). The benzene ether fraction constituted 22% of the Synthoil material and only 4.7% of the COED material. Data from Swain–Steadman separation of several COED materials are also contained in Table 5.14.

5.13.2.2 *Materials from the Hygas Process*

Two analytical approaches were used for chemical characterization of condensed by-products from the Hygas process. Initially, process-oriented analyses were performed under standardized conditions for numerous samples obtained from a range of operating conditions. The extensive data were reported in a five-volume series of reports. The final report (Anastasia et al., 1982) includes information most relevant to toxicological characterization. The process design included a recycling system whereby LO condensates were used to slurry the coal for delivery to the gasifier. Although the recycle oil was a major process stream, analytical data for the oil represented the integrated composition of the gasifier by-products. To obtain analyses representative of individual gasifier runs, samples were obtained from the low-temperature reactor freeboard. Analytical data for low-temperature reactor samples, appropriately corrected for LO content, represented gasification products that had minimal interference from the recycle and product gas quenching operations.

The low-temperature reactor samples were obtained from an internal gasifier stream. Although they are not environmentally important, the compositions of these samples are presented because (1) they are most relevant to comparison of conversion processes, (2) they represent gasifier by-products that would eventually enter environmentally important streams, and (3) they are the most mutagenic type of Hygas sample (Reilly et al., 1981). The data in Table 5.38 indicate that toluene, benzene, other one-ring aromatic hydrocarbons, and oxygen-containing compounds (predominantly phenols) together constituted 80–90% of the product oils. These compounds, although toxic, are not mutagenic or carcinogenic. Thus, subsequent characterization was focused on the ''nonvolatile'' portion [bp, >200°C (390°F)] of the low-temperature reactor condensates.

Table 5.38 Composition of Oil Produced in the Hygas Pilot Plant[a]

	Estimated Composition of Oils (wt %)				
Compounds	82[b]	84	85	86	87
Toluene	5.9	10.1	11.6	11.3	10.1
Benzene	22.0	35.4	47.7	43.4	76.2
Other compounds					
1-ring	13.1	9.7	5.8	7.8	2.1
2-ring	6.0	6.5	7.6	3.9	2.8
3-ring	1.5	1.8	2.6	1.5	1.3
>4-ring	0.4	0.7	0.3	0.6	0.3
Oxygen-containing	42.6	27.8	17.2	23.3	5.1
Nitrogen-containing	3.8	2.9	2.7	2.1	0.8
Sulfur-containing	4.1	3.9	3.2	3.1	0.8
Aliphatic	0.1	0.1	0.2	0.2	0.1
Unidentified compounds	0.5	1.1	1.2	2.8	0.3
Total for all components	100	100	100	100	100
Total for major components	84	83	82	86	94
Duration of run (days)	0.2	3	4	2	0.2
Number of samples	5	5	9	5	5

[a] Samples (10–20 min) were collected periodically throughout the indicated duration of each run. Each value is the mean for the indicated number of samples. Illinois No. 6 coal was used for test 82; Kentucky No. 9 and No. 11 coal for tests 84–87. Compositional data were obtained for samples that were probably different from those assayed for toxicity.

[b] Test number.

Source: Adapted from Anastasia et al. (1982).

The distribution of components in the nonvolatile organics of the low-temperature reactor condensate from test 84 (Sample G2119, Table 5.39) was determined by liquid partitioning and subsequent GC and GC-MS analysis. These results and those in Table 5.38, obtained by two independent analytical approaches, are in reasonable agreement after accounting for removal of the volatile one-ring components. The major discrepancy is in the content of PAHs; the greater abundance indicated in Table 5.39 for the nonvolatile organics appears more typical of other gasifier by-products.

As shown in Table 5.40, the PACs in the basic and neutral fractions ranged from one to five rings and were predominantly alkyl-substituted components. As is typical of products of pyrolysis and combustion, the greatest degree of alkyl substitution occurred for the smaller PACs. Results of more detailed analyses of the neutral and basic fractions are presented in Tables 5.41 and 5.42, respectively.

For the liquid-partitioning fractions described above (as well as those from other pilot-plant tests), the mutagenic activity was recovered in approximately equal amounts in the basic and neutral fractions, with negligible activity appearing in the acidic fraction (Stamoudis et al., 1981). The mutagenicity associated with the basic fraction appeared to be due primarily to amino-PAHs, as indicated by substantial decrease in mutagenicity after treatment with nitrous acid (Haugen et al., 1981a, b). Resolution of amino-PAHs and azaarenes, using ion-exchange HPLC, more definitively showed that a narrowly defined subclass of amino-PAHs was responsible for most of the mutagenic activity associated with the basic fraction (Haugen et al., 1982a, b). The reported presence of amino (phenanthrenes/anthracenes) in the most mutagenic subfractions isolated is consistent with subsequent data from

Table 5.39 Distribution of Components in Liquid-Partitioning Fractions of Oil (Sample G2119) Produced in the Hygas Pilot Plant[a]

	Distribution in Fraction (wt %)		
Compound Class	Neutral	Acidic	Basic
(Recovery from nonvolatile organics in low-temperature reactor condensate)	(66)	(28)	(6)
Aliphatics	>1	—	—
Aromatics	88	>95	100
PAHs	73	—	—
Amino-PAHs	—	—	21
Azaarenes	—	—	79
Hydroxy-PAHs	—	>95	—
O-PACs	7	—	—
N-PACs	<1	—	—
S-PACs	7	—	—

N-PACs = nitrogen-containing polycyclic aromatic compounds; O-PACs = oxygen-containing polycyclic aromatic compounds; PAHs = polycyclic aromatic hydrocarbons; S-PACs = sulfur-containing polycyclic aromatic compounds.

[a]The values in parentheses indicate recovery in the fractions obtained by liquid partitioning (using 1 *N* HCl to extract organic bases). Subsequent data in each column represent the percent distribution within each fraction as determined by gas chromatography (GC) and GC–mass spectrometry analyses.

Source: Adapted from Stamoudis et al. (1981).

Table 5.40 Distribution of Components According to Number of Aromatic Rings and Degree of Alkyl Substitution[a]

Compound Type	Neutral (wt %)	Basic (wt %)
Number of Aromatic Rings		
1	6	28
2	60	61
3	23	10
4	4	0.6
5	0.4	—
Number of Alkyl Carbons		
0	30	35
1	31	32
2	19	16
3	12	7
4	4	3
5	—	4

[a] Qualitatively reported information also indicated that one- to three-ring hydroxy-polycyclic aromatic hydrocarbons were present with up to four alkyl carbons.

Source: Adapted from Stamoudis et al. (1981).

several sources, which show the role of these compounds in the unusually high mutagenicity of basic components of coal-conversion materials.

5.13.2.3 *Materials from the U.S. Environmental Protection Agency's Research Triangle Institute Gasifier*

Yields and compositions were compared for by-product tars produced from high-sulfur caking Illinois bituminous coal, low-sulfur, noncaking western lignites and subbituminous coals, a wood–coal mixture, and peat in a laboratory fixed-bed, continuous flow, low-Btu gasifier constructed from a 4-ft long, 3-in. diameter steel pipe (Gangwal and Nichols, 1981). The relation of the coal composition and its physical properties to the composition of the tars was examined. The distribution of the components in liquid-partitioning fractions varied with coal type (Table 5.43). For example, substantially greater amounts of organic acids, and correspondingly smaller amounts of neutral components, were produced with Illinois No. 6 coal. Based on preliminary analyses of the liquid-partitioning fractions, the crude tars were analyzed by capillary-column GC—MS and by GC with FID, NPD, and FPD. As determined by a simulated GC distillation procedure, the bp range for the tars was 150–650°C. Vapor-phase osmometry indicated a mean molecular weight of 180–210.

As shown in Table 5.44, the distribution of the components differed in the various tars. Only a few firm relations with coal composition were identified. Compared with the other coals, the tar produced from gasification of the high-sulfur Illinois coal had more sulfur compounds, more azaarenes, a greater abundance of alkyl-PAHs, and more components

Table 5.41 Composition of Neutral Fraction of Hygas Low-Temperature Reactor Condensate[a] (Sample G2119)

Compounds	Concentration (mg/g)
Aliphatics	
Alkanes (C_{25})	5
PAHs	
(C_2–C_3)-Indans	17
(C_1–C_3)-Indenes	6
Naphthalene	23
1-Methylnaphthalene	30
2-Methylnaphthalene	63
(C_2–C_4)-Naphthalenes	130
Biphenyl	23
(C_1–C_4)-Biphenyl	50
Acenaphthene	9
(C_1–C_3)-Acenaphthene	23
(C_1–C_2)-Fluorene	68
Phenanthrene	51
Anthracene	23
(C_1–C_2)-(Phenanthrene/anthracene)	41
Fluoranthene	11
Pyrene	10
(C_1–C_2)-(Fluoranthene/pyrene)	18
Phenylnaphthalene	14
Methylphenylnaphthalene	19
Chrysene	10
Benzopyrenes	3
Heterocyclic PACs	
(C_0–C_4)-Benzothiophene	36
Dibenzothiophene	8
Indole	2
Dibenzofuran	38
Methyldibenzofuran	21
Total	752

PACs = polycyclic aromatic compounds. PAH = polycyclic aromatic hydrocarbons.

[a] Data for low-temperature reactor condensate, test 84. For this presentation, the sum of the concentrations for individually reported alkylated compounds is indicated.

Source: Adapted from Stamoudis et al. (1981).

that were not amenable to GC. For example, the total amount of C_1- and C_2-naphthalenes in this tar was 75% that of naphthalene; the value was less than 5% for the other tars. The authors reported that inspection of the gas chromatograms revealed these differences in level of alkylation among the tars.

The tars contained two-to six-ring PAHs, the amounts decreasing with increasing molecular weight in a pattern characteristic of that for gasification tars and "coal tar." The

Table 5.42 Composition of Basic Fraction of Hygas Low-Temperature Reactor Condensate[a]

Compounds	Concentration (mg/g)
Amino-PAHs	
Aniline	5
(C_2–C_3)-Anilines	113
1-Aminonaphthalene	44
2-Aminonaphthalene	8
Methylaminonaphthalene	37
Aminoanthracenes and aminophenanthrenes	7
N-PACs	
(C_1–C_5)-Pyridines	171
Quinoline and isoquinoline (azanaphthalenes)	150
(C_1–C_3)-Azanaphthalenes	274
Azabiphenyls	29
(C_1–C_2)-Azabiphenyls	39
Azafluorenes	20
(C_1–C_2)-Azafluorenes	4
Azaphenanthrenes or azaanthracenes	77
Methylazaphenanthrenes	29
Azapyrene	6
Total	1010

N-PACs = nitrogen-containing polycyclic aromatic compounds. PAH = polycyclic aromatic hydrocarbons.

[a]The sum of the concentrations for individually reported alkylated compounds is indicated.

Source: Adapted from Stamoudis et al. (1981).

Table 5.43 Distribution of Components in Gasifier Condensate

	Coal Type (amount, wt %)					
Fraction	North Dakota Lignite	Wyoming Subbituminous	Montana Rosebud	Illinois No. 6 and $CaCO_3$[a]	Illinois No. 6 and Wood[b]	Peat Pellets
Organic acids[c]	1.2	0.6	0.3	22.0	9.7	4.1
Organic bases	3.2	2.2	1.6	2.5	2.7	1.82
Total neutrals	95.6	97.0	98.1	75.5	87.6	94.1
Nonpolar neutrals	16.7	12.1	13.4	16.1	18.8	3.9
PAHs[d]	62	46.8	30.5	20.7	36.6	50.4
Polar neutrals	5	5.5	2.3	3.9	18.9	24.3
Cyclohexane insolubles	11.2	32.6	51.9	34.8	13.3	15.4

[a]Illinois No. 6 coal, 70%, limestone, 30%.

[b]Illinois No. 6 coal, 79%, wood (Woodex), 21%.

[c]The measurements for organic acids in the tar do not reflect total production of phenols because a substantial portion of these compounds remained in the aqueous phase of the product stream, which was only 1–3% tar.

[d]PAHs = polycyclic aromatic hydrocarbons.

Source: Adapted from Gangwal and Nichols (1981).

Table 5.44 Composition of Tars Produced During Fixed-Bed Gasification in a Laboratory Reactor[a]

Compound Number[b]	Compound	North Dakota Lignite	Wyoming Subbituminous	Montana Rosebud	Illinois No. 6 Plus CaCO$_3$	Illinois No. 6 Plus Wood	Peat Pellets
	Tar (mg/g coal)	4.7	4.4	4.4	21	3.5	9.4
PAHs							
	Naphthalene	320	200	320	42	68	270
	1-Methylnaphthalene	3.0	1.7	1.3	9.5	5.4	4.5
	2-Methylnaphthalene	3.0	2.7	2.0	8.1	9.4	4.7
	C_2-Naphthalenes	1.5	1.1	1.1	9.0	6.5	2.8
1	Biphenyl	5.3	4.5	2.7	5.2	4.0	5.6
2	Acenaphthylene	68	41	52	13	15	36
3	Acenaphthene	7.0	9.3	6.4	1.8	1.8	6.4
	Dibenzofuran	19	12	4.1	9.0	8.9	9.8
4	Fluorene	30	19	10	5.7	8.2	11
5	Phenanthrene	94	89	84	20	27	57
6	Anthracene	17	15	11	10	9.4	11
7	Fluroanthene	64	73	68	13	10	41
8	Pyrene	47	57	91	11	8.9	43
9	Benzo[a]fluorene	4.3	3.9	1.0	3.1	1.2	1.2
10	Benzo[b]fluorene	4.3	3.0	1.2	1.9	0.9	0.9
11	Benz[a]anthracene	7.7	11	11	4.7	1.8	7.1
12	Chrysene	4.0	7.3	6.8	3.3	1.1	3.6
13	Triphenylene	1.9	3.2	3.6	1.4	0.5	2.3
14	Benzo[b]fluoranthene	4.9	7.3	12	2.0	0.9	3.8
15	Benzo[k]fluoranthene	1.5	1.8	2.5	1.1	0.5	1.9
16	Benzo[e]pyrene	1.6	3.2	8.4	0.9	0.9	2.6
17	Benzo[a]pyrene	3.6	5.2	12	1.8	1.1	3.7
18	Perylene	6.2	2.5	6.1	1.8	0.4	3.2
19	Indeno[1,2,3-cd]pyrene	1.0	0.5	1.0	0.2	<0.2	0.1
20	Dibenz[a,h]anthracene	1.1	2.5	7.5	0.8	<0.2	1.2
21	Benzo[ghi]perylene	0.6	1.1	7.7	0.5	<0.2	0.8
Amino-PAHs							
22	Aniline and/or dimethylpyridine	0.01	0.1	0.9	2.5	1.3	0.6
23	1-Aminonaphthalene	0.02	0.03	<0.01	0.3	0.06	0.03
24	2-Aminonaphthalene	<0.01	<0.02	<0.01	0.08	0.04	0.02
25	C_2-Aminonaphthalene and/or phenoxyaniline	0.2	0.1	0.1	0.3	0.3	0.07
Azaarenes							
	Quinoline	1.5	1.4	0.5	4.2	1.9	0.8
	Isoquinoline	0.3	0.2	0.2	1.8	0.6	—
	2-Methylquinoline	0.07	0.06	0.04	0.7	0.4	0.3
	Dimethylisoquinolines and/or dimethylquinolines	0.7	0.8	1.7	0.8	0.5	0.8
	Azabiphenyl	0.1	0.1	0.1	0.3	0.3	0.08
	Azabenzene and/or bipyridyl	0.1	0.2	0.3	0.5	0.3	0.1
	7,8-Benzoquinoline	0.2	0.2	0.1	1.1	0.6	0.06

Table 5.44 ***(Continued)***

Compound Number[b]	Compound	North Dakota Lignite	Wyoming Subbituminous	Montana Rosebud	Illinois No. 6 Plus $CaCO_3$	Illinois No. 6 Plus Wood	Peat Pellets
	Acridine	0.8	0.6	1.1	1.1	0.7	0.2
	Phenanthridine and/or 5,6-benzoquinoline	0.6	0.4	0.3	1.2	0.5	0.3
	Azafluoranthene and/or naphthylphenylamine	1.1	1.5	2.5	0.7	0.5	0.6
	Azapyrene	0.3	0.2	0.9	0.7	0.6	0.2
	Benzacridine isomers	0.3	0.7	1.4	0.7	0.9	0.2
3°-N-PACs							
	Indole	1.1	1.0	0.1	2.8	0.8	0.6
	Methylindoles	0.02	0.2	0.3	0.3	0.1	0.2
	Carbazole	2.8	3.4	4.3	3.1	2.8	1.0
S-PACs							
	Benzenethiol	<0.4	<0.4	<0.4	<0.1	<0.6	<0.2
	Toluene thiol and/or benzylmercaptan	<0.4	<0.4	<0.4	<0.1	<0.6	<0.2
	Benzothiophene	7.0	5.0	5.2	9.0	7.1	1.1
	Dihydrobenzothiophene	<0.4	0.5	<0.4	<0.1	<0.6	<0.2
	Methylbenzothiophenes	<0.4	0.6	0.3	1.3	1.0	0.3
	C_2-Benzothiophenes	<0.4	<0.4	<0.4	0.9	1.3	<0.2
	C_3-Benzothiophenes	<0.4	<0.4	<0.4	0.7	0.6	<0.2
	Dibenzothiophene	2.6	2.5	3.6	6.7	4.6	0.8
	Naphthothiophenes	0.8	0.9	1.8	0.2	1.9	0.5
	Methyldibenzothiophenes	<0.4	<0.4	0.5	1.6	<0.6	<0.2
	Phenanthro[4,5-bcd] thiophenes	1.5	0.6	3.6	3.5	2.0	0.8
	Naphthobenzothiophenes	<0.4	<0.4	<0.4	3.0	1.1	<0.2
	Dinaphthothiophenes	<0.4	<0.4	<0.4	0.7	<0.6	<0.2

PAH = polycyclic aromatic hydrocarbons; N-PACs = nitrogen-containing polycyclic aromatic compounds; S-PACs = sulfur-containing polycyclic aromatic compounds.

[a] The published concentrations (Gangwal and Nichols, 1981) were expressed as micrograms of compound per gram of coal. Based on the reported yields of tar, the published concentrations were converted to milligrams of compound per gram of tar.

[b] The compound numbers (1–25) were assigned for the graphic comparison shown in Fig. 5.55

major PAHs were naphthalene, acenaphthylene, fluorene, phenanthrene, fluoranthrene, and pyrene.

As determined by analysis of the tars using GC and element-specific detectors, the heteroatomic PACs included: azaarenes and thiophenes having up to five rings, benzenethiols, indoles, and carbazoles. Predominant azaarenes included quinoline and benzoquinoline isomers. Based on the data reported, the abundance of amino-PAHs ranged from 5 to 20% of that for azaarenes. 1- and 2-aminonaphthalene were also detected.

The comprehensive set of data for gasification of a variety of materials allows comparisons of composition in a manner suggested by Munro et al. (1983). The compositions of four of the six tars described in Table 5.44 are compared graphically in Fig. 5.53. Four tars were selected for presentation; those not illustrated have similar profiles. It is apparent

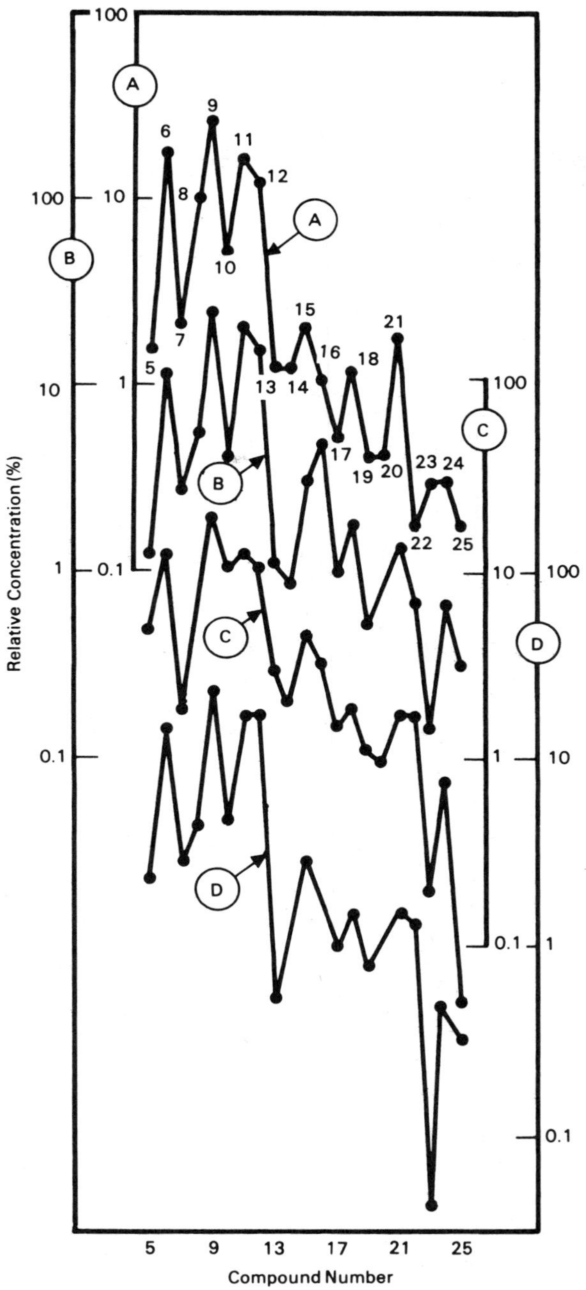

Figure 5.53 Relative concentrations of components in tars produced during gasification of (A) North Dakota lignite coal, (B) Wyoming subbituminous coal, (C) Illinois No. 6 coal plus $CaCO_3$, and (D) peat pellets. For each tar the concentration of individual compounds is expressed relative to the sum of concentrations of the indicated compounds. The compound numbers and data used for computation of relative concentrations are presented in Table 5.45 (compiled from Gangwal and Nichols, 1981).

that the tars produced during gasification of three coal types, and peat, had distinct qualitative and quantitative similarities despite the diverse nature of the fuels. The composition of one of these tars (from Wyoming subbituminous coal, sample B in Fig. 5.53) is compared to that of tar produced in a small commercial gasifier, and in a coke oven, Fig. 5.54. The obvious similarities in the relative concentrations of major components illustrate that the composition is governed by chemical principles that are largely independent of the specific process.

This conclusion was supported by the authors' subsequent comparison of the compositions of tars from the RTI gasifier and several other gasifiers of larger scale (Table 5.45).

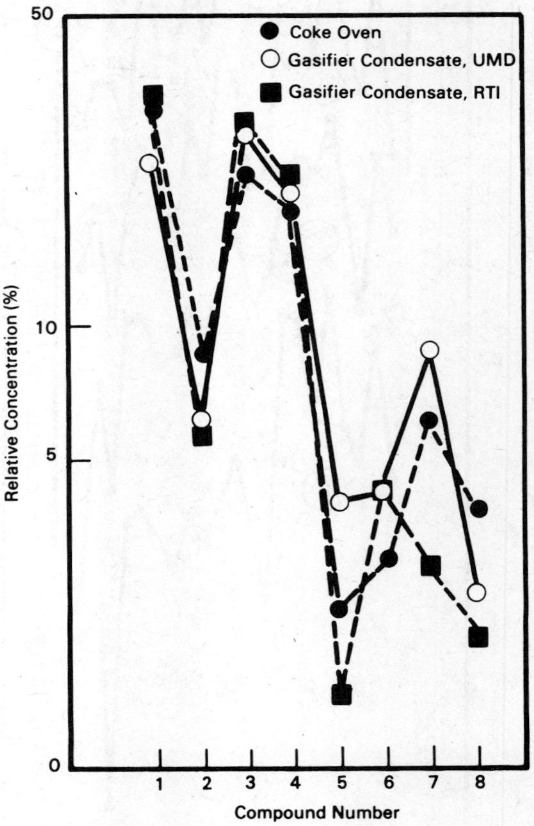

Figure 5.54 Relative concentrations of components in tars produced in a coking oven, a laboratory fixed-bed gasifier, and a low-Btu gasifier. For each material, the amounts of individual compounds are expressed relative to the sum of the concentrations for the indicated compounds. Data for tars from a coke oven and the University of Minnesota–Duluth (UMD) gasifier are from Munro et al. (1983); the second tar was produced during gasification of Wyoming subbituminous coal in a laboratory reactor at Research Triangle Institute (RTI; Gangwal and Nichols, 1981). The compounds are (1) phenanthrene, (2) anthracene, (3) fluoranthene, (4) pyrene, (5) benzo[a]fluorene, (6) benz[a]anthracene, (7) chrysene, and (8) benzo[a]pyrene.

Table 5.45 Concentration of Components in Gasifier Tars, Log (μg/g tar)

Component	RTI (95% confidence range)	Synthane	METC	Chapman Wilputte	GFETC	Coal Tar	
						1	2
Total tar (mg/g coal)	15–23	37	26–34	100	12–36	—	—
Naphthalene	4.6–5.0	4.8	4.4	3.3	4.8	5.0	
Acenaphthene	3.5–3.8	5.1	—	—	—	3.5	—
Acenaphthylene	4.2–4.5	—	—	—	—	4.3	—
Biphenyl	3.4–3.6	—	—	—	4.2	3.6	—
Phenanthrene	4.3–4.6	5.1[a]	—	—	4.6	4.7	4.3
Pyrene	3.9–4.4	4.6	—	—	4.4	4.3	4.0
Benz[a]anthracene	3.4–3.8	—	—	—	—	—	3.8
Chrysene	3.4–3.8	4.9[b]	—	3.5	4.2	4.3	4.5
Benzo[a]pyrene	3.2–3.4	—	3.6	2.9	3.8[c]	—	3.3
Perylene	3.0–3.4	—	—	2.9	—	—	3.9
Benzo[ghi]perylene	2.6–3.2	—	—	—	—	—	3.3
Dibenz[a,h]anthracene	2.9–3.3	—	—	—	—	—	2.4
Phenol	3.6–4.0	4.1	2.6	3.3	5.0	3.6	—
Cresols	4.0–4.4	—	4.3	—	—	3.9	—
Xylenols	3.8–4.2	—	—	—	—	3.3	—

RTI = Research Triangle Institute; METC = Morgantown Energy Technology Center; GFETC = Grand Forks Energy Technology Center.

[a] Includes anthracene.

[b] Includes benz[a]anthracene.

[c] Includes benzo[e]pyrene.

Source: Mixon et al. (1984).

5.13.2.4 *Industrial Gasifier at the University of Minnesota–Duluth*

Data are available on the chemical and toxicological characteristics of materials from a low-Btu coal gasifier operated at the University of Minnesota–Duluth. Information is presented here concerning the tar collected on electrostatic precipitators in the gasification plant (Samples G1301–G1323). These tars were also the subject of toxicological characterization.

Numerous samples of the tar were subjected to a liquid-partitioning procedure, in which 1 *N* HCl was used to extract the aromatic bases (Epler et al., 1978a, b). The relative distributions of the acidic, basic, and neutral components of 23 samples placed them in two groups, as defined by the yield of aromatic bases (Table 5.46). Although some samples with high base content also had high mutagenicity, this was not a consistent characteristic of the data. Results of related studies (Ho et al., 1983) give insight as to the probable reason for this inconsistency. These studies showed that, for the fractionation scheme employed, mutagenic activity ascribed to ''polycyclic aromatic hydrocarbons'' may have been, in fact, due to polycyclic aromatics with polar functional groups (e.g., amino-PAHs).

Five of the 23 samples were also fractionated using a procedure whereby the aromatic bases were more effectively extracted with 6 *N* HCl, and the amino-PAHs were enriched by subsequent chromatography on alumina and Sephadex LH-20 (Tomkins and Ho, 1982). The results reported for three of the samples show that the tars contained two- and three-ring amino-PAHs and their alkyl homologs (Table 5.47). Except for negligible concentrations of four-ring amino-PAHs in the gasification tars, the composition of the tars and a coking-derived oil (NBS certified reference material) were qualitatively and quantitatively

Table 5.46 Composition of Tar Produced in the Gasifier at the University of Minnesota–Duluth[a]

Electrostatic Precipitator Tar	Composition (% of total)		
	Acids	Bases	Neutrals
Group A (n = 17)	14 (3–19)	5 (2–11)	80 (72–95)
Group B (n = 6)	7 (3–15)	28 (19–36)	56 (44–72)

[a] Data were reported for 23 samples (G1301–G1323) of tar collected on an electrostatic precipitator. Samples assigned to groups A and B had relatively low (2–11%) and high (19–36%) contents of bases, respectively. For each group, the mean value is presented (range in parentheses).

Source: Adapted from Epler et al. (1983).

similar. The most abundant amines for the five tars examined were aminonaphthalenes and methylaminonaphthalenes; the concentrations of the aminonaphthalenes were indicative of the total amino-PAH concentrations (Cowser, 1981).

Because the mutagenicity of amino-PAHs in the Ames assay attracted considerable attention for studies of coal gasification and liquefaction, it is of interest to compare the occurrence of amino-PAHs in gasification by-products. Their presence in condensates (tars) from five gasifiers and two coke oven tars is compared in Table 5.48. Each gasification condensate contained aminonaphthalenes, the amino-PAHs most likely to be unambiguously identified and quantitated.

Until recently, analyses of basic N-PACs in coke oven tar indicated only the presence of azaarenes (Burchill et al., 1983a, b). However, Kosuge et al. (1982) isolated mutagenic aminophenanthrenes, aminopyrene, and aminobenzanthracene from an unspecified coal tar, and a family of two- to four-ring amino-PAHs was detected in a reference coke oven tar (Table 5.48). Thus, besides the qualitative and quantitative similarities among the distri-

Table 5.47 Concentration of Amino-Polycyclic Aromatic Hydrocarbons in Tar from Low-Btu Gasifier (University of Minnesota–Duluth)[a]

Compound	Concentration (μg/g)			
	Tar Sample Number			
	17 (G1304)[c]	18 (G1305)	83 (G1319)	CRM-1[b] (L1211)
1-Aminonaphthalene	14	18	14	52
2-Aminonaphthalene	41	55	27	69
Methylaminonaphthalenes	87	98	80	96
C_2-Aminonaphthalenes	—	46	32	—
Aminobiphenyls	—	—	—	31
(C_1–C_4)-Aminobiphenyls	21	—	12	71
Aminofluorenes	50	—	—	6
(C_1–C_2)-Aminofluorenes	26	—	—	7
Amino(anthracenes/phenanthrenes)	—	11	6	37
Methylamino(anthracenes/phenanthrenes)	—	7	—	10
Aminopyrenes	—	—	—	14

[a] For unspecified isomers, the sum of their individual concentrations is shown (Tomkins and Ho, 1982).

[b] For comparison, data for a certified reference material (CRM) coal oil are shown.

[c] Sample number given in parentheses.

Table 5.48 Occurrence of Amino-Polycyclic Aromatic Hydrocarbons (PAHs) in Coal-Gasification Condensates and Coke Oven Tar

	Occurrence ($\mu g/g$)[a]							
							Coal Tar	
Compounds[b]	UMD	HYGAS	UNDERC	METC	RTI	MI	A	B
1-Aminonaphthalene	9–18	870	x_1	y_1	<10–300	z_1		+
2-Aminonaphthalene	23–55	170	x_2	y_2	<10–80	z_2		+
Aminoacenaphthenes or aminoacenaphthylene		+						
Aminobiphenyls	+	+	+			+		+
Aminofluorenes	+	+				+		+
Amino (phenanthrenes/ anthracenes)	+	+	+			+	+	+
Aminopyrenes							+	+

[a]The data sources are as follows: University of Minnesota–Duluth (UMD), low-Btu gasifier (Cowser, 1981; Tomkins and Ho, 1982). HYGAS, nonvolatile fraction of low-temperature reactor in HYGAS pilot plant, occurrence (Haugen et al. 1982b); quantitation (Stamoudis et al., 1981). University of North Dakota Energy Research Center (UNDERC) process tar from pilot plant, occurrence (Haugen et al., 1986); quantitation (Stamoudis et al., 1981). Morgantown Energy Technology Center (METC), low-Btu gasifier (Stamoudis et al., 1986); Research Triangle Institute (RTI), laboratory-scale pyrolysis unit (Fillo et al., 1983). Coal tar A, coal tar of unspecified orgin from which mutagenic amino-PAHs were isolated; the isolated mutagens are indicated (Kosuge et al., 1982). Coal tar B, National Bureau of Standards coal tar, analyzed as described by Later et al. (1981a, 1982b); unpublished data provided by C. W. Wright, Pacific Northwest Laboratory.

[b]For 1- and 2-aminonaphthalene, quantitative results are cited; for all other compounds, the + symbol indicates that it was identified in the material. Absence of + symbol in the table does not necessarily indicate that the compound was not present.

butions of PAHs in gasification condensates (including those from the UMD gasifier) and coke oven tar (Fig. 5.54), these materials are also similar with respect to the presence of amino-PAHs. The family of azaarenes present in coke oven tar (Burchill et al., 1983a, b) is also similar to that found in the gasification condensates described in this section.

5.13.2.5 *A Slagging, Fixed-Bed Gasification Pilot Plant at the University of North Dakota Energy Research Center*

By-product tars from the pilot plant in Grand Forks, North Dakota have been the subject of two independent research efforts. Investigators at UNDERC (formerly Grand Forks Energy Technology Center, GFETC) examined chemical characteristics of tars as a function of operating conditions, while others focused their efforts on more detailed chemical and toxicological characterization of representative materials. Miller et al. (1981) described the dependence of the composition of the nonvolatile portion of tars produced during gasification of subbituminous and various lignite coals on coal type, the operating pressure, and the oxygen feed rate. The tars were fractionated by alumina-column chromatography and analyzed by MS and by packed-column GC-MS. The composition of tar produced during a prolonged test of gasification of lignite is shown in Table 5.49. Characteristic two-to five-ring PAHs and heteroatomic PACs were present, with alkylated homologs being more abundant than unsubstituted compounds.

Table 5.49 Composition of Tar Produced in the University of North Dakota Energy Research Center Gasifier[a]

Probable Structural Types	z Number	Concentration (mg/mL)																
		6	7	8	9	10	11	12	13	14	15	16	17	18	19	20	21	Total
Benzenes	−6	0.4	1.7	2.2	4.0	0.0												8.3
Indans/tetralins	−8				0.7	7.1	13.	10	6.9	4.0								42
Indenes	−10				0.0	3.2	0.0											3.2
Naphthalenes	−12					30	22	13	9.8	8.7	6.6	4.7	3.1	2.4				100
Acenaphthenes/biphenyls	−14					8.3	6.2	4.8	3.7	2.4	1.6	1.1						28
Fluorene/acenaphthylenes	−16							20	29	6.9	4.3	3.9	2.5	2.1	1.8			70
Phenanthracene/anthracene	−18									23	9.1	6.1	3.3	3.1	2.1	1.3		48
Pyrene/fluoranthenes	−22											18	5.8	2.5	1.6	0.8		29
Chrysenes	−24													5.9	2.3	1.4		9.6
Benzopyrenes	−28															9.3	2.6	12
Pyridines	−5 (nitrogen)	2.6	43	21	5.6	2.2	0.8											76
Quinolines	−11 nitrogen				7.1	6.1	7.2	5.5	1.9	0.5								28
Phenols	−6 (oxygen)	16	40	38	28	6.5	0.8											130
Indanols	−8 (oxygen)				6.7	7.6	7.2	8.9	3.5	4.8								34
Naphthols	−12 (oxygen)					28	16	11	5.1	2.6	1.8							64
Dibenzofurans	−16 (oxygen)							6.2	4.8	3.7	2.4	1.6	1.1					20
Total																		720

[a] Results were obtained by low-voltage mass spectrographic analysis of tar after water and light oil had been removed. The z number is defined from the molecular expressed as C_nH_{n+z}; values for n are at the top of each column.

Source: Miller et al. (1981).

Results of statistical analyses showed no correlation between the gasification pressure and the composition of the tar (three tests with lignite: 860, 1380, and 2760 kPa). For these three tests, there was a positive correlation between the hydrogen content of the product gas and the yields of individual PACs in the tar. The yields of individual compounds increased 20–120% as the hydrogen content of the product gas increased from 25 to 31%.

The data in Table 5.50 illustrate the dependence of tar composition on the feedstock for four coals gasified under identical pressures and oxygen feed rates. Data are shown for the 20 compounds for which the abundance was most dramatically dependent on the type of coal.

There was little or no correlation between coal type and the average molecular weight of the heteroatomic PACs. The degree of alkylation remained constant except for naphthalene, fluorene, and chrysene isomers, for which the average number of alkyl carbon atoms

Table 5.50 Effect of Coal Type on Composition of Tar from the University of North Dakota Energy Research Center Slagging, Fixed-Bed Gasifier

	Ultimate Analysis (wt %, moisture- and-ash-free)			
Element	Indianhead Lignite	Gascoyne Lignite	Baukol-Noonan Lignite	Rosebud Subbituminous
H	4.8	2.4	2.7	2.9
C	70.8	69.9	73.7	73.3
N	0.9	1.0	0.9	1.3
S	1.8	1.4	0.6	2.3
O (by difference)	21.7	25.3	22.1	20.3

Compound[a]		Concentration of Component in Tar (mg/g)			
z Number	Molecular Weight	Indianhead Lignite	Gascoyne Lignite	Baukol-Noonan Lignite	Rosebud Subbituminous
−12	128	30	5.8	0.0	0.0
	142	22	12	3.3	6.9
	170	9.7	11	16	23
	184	8.7	12	17	22
−16	152	20	18	15	9.3
	166	28	30	50	37
	180	6.9	9.1	15	12
−18	178	23	32	51	50
	192	9.1	12	20	18
	206	6.1	8.9	13	13
−22	202	33	28	18	24
−5 (nitrogen)	107	43	23	3.7	3.4
	121	21	19	6.1	6.4
−6 (oxygen)	94	16	2.7	2.0	0.7
	108	40	8.4	1.2	0.8
	136	28	22	5.7	7.7
	150	6.5	7.0	3.8	4.4
−12 (oxygen)	134	28	25	13	2.6
	148	16	17	11	2.0
	162	11	14	12	1.6

[a] Refer to Table 5.49 for compounds corresponding to the indicated values for *z* and molecular weight.

Source: Adapted from Miller et al. (1981).

was directly correlated to the moisture-and-ash-free (MAF) hydrogen:carbon ratio. The total concentration for naphthalenes and three-ring PACs as inversely correlated with the MAF hydrogen content of the coal. The ratio of total nitrogen to total oxygen compounds was inversely correlated with the MAF nitrogen content of the coal. Thus, some characteristics of the tars were dependent on the coal type, while others were mostly invariant, determined more by the gasification conditions than by the coal type.

Boparai et al. (1983) measured toxicological effects of tars and other by-products from the gasifier (Chapter 3) and selected a single tar (test 87, Sample G1121) for detailed, biodirected, chemical characterization. Acidic, basic, and neutral components were isolated using a liquid-partitioning procedure in which the recovery of higher azaarenes and amino-PAHs in the basic fraction was enhanced by extraction with 3 *N* HCl in 40% aqueous methanol (Haugen et al., 1986). On a moisture-free basis, the yields of neutral, acidic, basic and insoluble (CH_2Cl_2) components were 55, 26, 2.6, and 17%, respectively. The yield of acidic components (primarily hydroxy-PAHs) is consistent with the content of hydroxy-PAHs determined in an independent analysis of tar produced during test 52 (Table 5.49). The yield of neutral components is also consistent, but the concentrations of pyridines and quinolines indicated in Table 5.50 suggest that the yield of basic components in the liquid-partitioning fraction should be greater than 2–3%. This difference may be attributed to actual differences in the azaarene content of the tars, but it may also partially be due to differences in fractionation techniques that cause varying degrees of volatilization of pyridines.

The distribution of PAHs in tar from test 52 (Table 5.49) was nearly identical to the distribution determined in another laboratory by GC and GC-MS analyses of the neutral fraction from test 87 (Fig. 5.55). Thus, it is reasonable to assume that more detailed chemical and toxicological properties of the tar from test 87 are shared by other tars from the gasifier.

Of the components in the neutral fraction, the three- to six-ring PAHs were determined by silica-column chromatography to be the predominant mutagens as measured in the Ames assay. Further fractionation of PAHs by complementary reverse-phase and bonded normal-phase HPLC, along with mutagenicity assays of fractions, allowed resolution of numerous mutagens, as well as detection by GC and GC-MS of components obscured by the complexity of the less-purified fractions (Haugen et al., 1983; Haugen and Stamoudis, 1986). The most prominent mutagens isolated by this approach are indicated in Table 5.51. Among these are methylene-bridged PAHs, which were identified by detailed analyses of coal tar (Borwitzky and Schomburg, 1979). These compounds were part of a series of components having molecular weights (MWs) consistent with their identity as methylene-bridged compounds, including MW 254 (methylenebenzopyrene) and MW 290 (methylenedibenzanthracenes).

In addition to the compounds indicated in Table 5.51, other unidentified, poorly resolved, and less-active mutagens were present. However, the data show that the short-term mutagenesis assay indicated carcinogenic risk associated with the same family of compounds to which the dermal carcinogenicities of tobacco smoke condensate and coke oven tar have been attributed.

The components of the basic fraction were further resolved by sequential cation-exchange and reverse-phase HPLC (Stamoudis et al., 1983; Haugen et al., 1986). The results showed that the unusually high mutagenic activity of the basic fraction was due to the presence of extremely potent aminophenanthrene isomers. Although the relative contribution of these compounds to the mutagenicity of different materials may vary, results obtained by similar fractionation of low-temperature reactor condensate from the HYGAS and

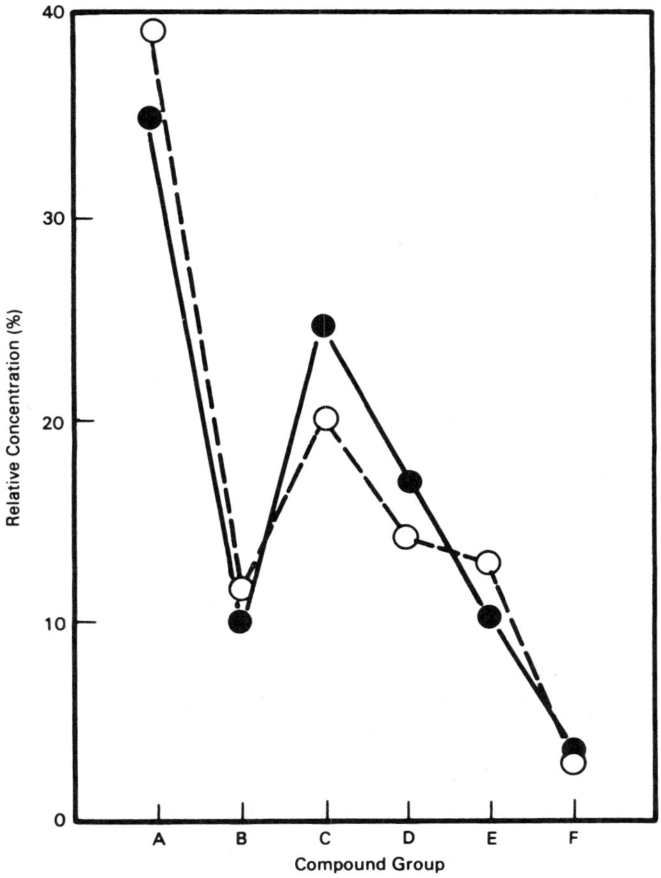

Figure 5.55 Comparison of relative concentrations of polycyclic aromatic hydrocarbons as determined by independent analyses of two tars from the University of North Dakota Energy Research Center pilot plant. The compound groups (including all alkyl homologs) are A, naphthalenes; B, acenapthene/biphenyls; C, fluorene/acenaphthylenes; D, phenanthrene/anthracenes; E, fluoranthene/pyrenes; and F, chrysene isomers. The concentration of each group is expressed relative to the total concentration for all groups, A to F. Data points indicated as closed circles are for gasifier test 52 (performed between 1976 and 1978) with lignite coal (Miller et al., 1981). Data points indicated as open circles are for gasifier test 87 performed with Indianhead lignite coal (V.C. Stamoudis et al., unpublished data). Gasification conditions are described in Wilzbach et al. (1982).

a reference coal oil (CRM-1) showed that aminophenanthrenes were also major contributors to the mutagenicity in these materials (Haugen et al., 1982a, b; Stamoudis et al., 1983). Azaarenes were also active as mutagens, but the activity was not attributed to specific compounds. The results were consistent with other studies of coal-conversion materials showing that three- and four-ring amino-PAHs are primarily responsible for the substantial mutagenic activity of basic components. The much more abundant neutral PAHs, however, are the components principally responsible for the dermal carcinogenicity of the coal-conversion materials.

Table 5.51 Principal Mutagenic Polycyclic Aromatic Hydrocarbons Chromatographically Resolved from the University of North Dakota Energy Research Center Gasifier Tar

Compound	Molecular Weight
Fluoranthene[a]	202
Cyclopenta[cd]pyrene	226
Benz[a]anthracene	228
1,12-Methylenebenz[a]anthracene[b]	240
4,5-Methylenechrysene[b]	240
Methyl-(chrysenes/benz[a]anthracenes)	242
Benzo[a]pyrene	252
Anthanthrene and isomers	276
Dibenzopyrene isomers	302

[a] Although the specific mutagenicity of fluoranthene was low, its chemical abundance was relatively high.
[b] Tentative identification.

Source: Haugen et al. (1986).

5.13.2.6 *Research-Scale Low-Btu Gasifier Operated at the Morgantown Energy Technology Center*

Materials were obtained from a low-Btu gasifier, which is a pressurized version of the McDowell–Wellman (Wellman–Galusha) atmospheric pressure stirred-bed device, with an inner diameter of 1.1 m. A major reason for operating the gasifier has been to evaluate methods for cleanup of the product gas. Although a variety of samples have been characterized chemically and toxicologically, this presentation focuses on the organic condensates. Two independent approaches were used to characterize the tars chemically.

Tar (Sample G1211) from a process stream tar trap was fractionated, using Sephadex LH-20 column chromatography, as originally described by Jones et al. (1977). Polymeric material was eluted initially, followed by aliphatics, PAHs, polar PACs, and organics capable of strong hydrogen bonding. The PAH-enriched fractions were further resolved by silica-column chromatography, and the fraction enriched in polar PACs was subfractionated into acidic, basic, and neutral components. The basic components of the polar LH-20 fraction were obtained by extracting dichloromethane solutions with 1 *N* H_3PO_4 (mild conditions), and the acidic components (not including phenols) were extracted with 5% Na_2CO_3. Some properties of these fractions are shown in Table 5.52.

Consistent with their high average molecular weight, fractions 1 and 2, as well as silica subfractions B and C did not contain components that could be analyzed by capillary-column GC-MS [maximum temperature, 270°C (518°F)]. As shown in Table 5.53, GC-MS analysis of the silica subfraction A indicated the presence of a series of two- to five-ring PAHs and their alkyl homologs; two- to four-ring azaarenes, but not amino-PAHs, were detected in the basic fraction. O-PACs and S-PACs were also present in the tar.

In another laboratory, organic condensates from the METC gasifier were characterized by GC and GC-MS analyses of acidic, basic, and neutral liquid-partitioning fractions. One of these was a condensate (designated here as "heavy tar," Sample G1221), collected by a process-stream tar trap. The second was an aqueous/organic condensate (here designated as "light tar," Sample G1225), collected in a side-stream sampler located upstream from

Table 5.52 Characteristics of Sephadex LH-20 Chromatographic Fractions of Morgantown Energy Technology Center Gasifier Tar

Material	Compound Type	Recovery (wt %)	Specific Mutagenicity revertants/μg	Average Molecular Weight[a]
Whole tar	—	(100)	8	437
Fraction				
1	Polymeric	9	3	833
2	Aliphatic	21	8	437
3	Polycyclic aromatic	27	10	299
4	Polycyclic aromatic	19	11	299
Silica[b]				
A		20	13	291
B		22	42	373
C		1	85	502
5	Polar polycyclic aromatic	19	13	320
Acids[c]		1.5	0	417
Bases[c]		0.8	53	307
Neutrals/phenols[c]		9.1	7	312
H_2O solubles[c]		7.6	0	347
6	Hydrogen-bonding components	5	12	444

[a] Determined by vapor-phase osmometry.
[b] Subfractions of Fraction 3 and 4 pooled.
[c] Subfractions of Fraction 5.

Source: Adapted from Royer et al. (1983).

devices designed to remove condensable organics. Material was collected at this site to determine more accurately the composition of the raw off-gas, unaltered by cleanup devices. Characterization of samples from the low-temperature reactor in the Hygas pilot plant served a similar purpose.

Bases were extracted from methylene chloride solutions of the whole tar with methanol/aqueous HCl to enhance recovery of the compounds with more than three rings. The distribution of material and mutagenic activity in the liquid-partitioning fractions is shown in Table 5.54. Nearly half the heavy tar was insoluble in methylene chloride, probably reflecting the high molecular weight or polymeric nature of some of its major components (see Table 5.52).

The major components in the acidic fraction from the light tar were phenols, hydroxyindans, -indenes, and -naphthalenes, as well as their alkyl homologs. Phenols and naphthols were also among the major components of the acidic fraction from heavy tar, but other major components were three- and four-ring phenolic compounds and their alkyl homologs. The larger abundance of acidic components in the light tar (Table 5.54) can be attributed to the greater abundance among the gasification products of the more volatile one- and two-ring hydroxy-PAHs, which were not as prevalent in the heavy tar.

The mutagenic activities of the heavy tar and its fractions were distinctly greater than for the corresponding fractions of light tar (Table 5.54). This difference presumably reflects the greater relative abundance of the PACs having more than three-rings in the heavy tar, as shown by comparative GC and GC-MS analyses of the unfractionated tars and their basic fractions (Tables 5.55 and 5.56). Consistent with results obtained by Royer et al. (1983),

Table 5.53 Compounds Identified in Subfractions of Morgantown Energy Technology Center Gasifier Tar[a]

Compound	Number of Isomers
PAHs[b]	
(C_2–C_3)-Naphthalenes	7
Acenaphthene	—
Fluorene	—
Methylfluorenes	3
Phenanthrene	—
Anthracene	—
(C_1–C_3)-(Phenanthrene/anthracene)	10
Fluoranthene	—
Pyrene	—
Methyl(fluoranthene/pyrene)	1
Chrysene	—
Benzo[a]fluorene	—
Perylene	—
Benzo[a]pyrene	—
Azaarenes[c]	
(C_1–C_3)-Quinolines	15
Benzo[h]quinoline	—
Acridine	—
Phenanthridine	—
Benzo[f]quinoline	—
(C_1–C_2)Benzoquinolines	13
Aza(pyrene/fluoranthene)	—
Dibenzoquinoline	—
S-PAHs[d]	
Dibenzothiophene	—
Methyldibenzothiophene	—
Hydroxy-PAHs[e]	
(C_0–C_2)-Hydrofluorenes	—
(C_1–C_3)-Hydroxybiphenyls	—
(C_0–C_1)-Hydroxyphenanthrenes	—
O-PACs[f]	
(C_0–C_1)-Carbazole	—
Phenanthridone	—

[a] For compounds not specifically identified, the number of isomers is indicated, where reported.

[b] PAHs = polycyclic aromatic hydrocarbons, the major components of silica fraction A.

[c] Major components of basic fraction.

[d] S-PAHs = sulfur-containing polycyclic aromatic hydrocarbons, the major components of neutral/phenol fraction.

[e] Major component of neutral/phenol fraction.

[f] O-PACs = oxygen-containing polycyclic aromatic hydrocarbons, the major components of neutral/phenol fraction.

Source: Adapted from Hanson et al. (1979), Hill et al. (1983), and Royer et al. (1983).

Table 5.54 Distribution of Material and Mutagenic Activity in Light and Heavy Tars from the Morgantown Energy Technology Center Gasifier

	Recovery (wt %)		Mutagenic Activity (revertants/μg)	
Fraction	Light Tar[a]	Heavy Tar[b]	Light Tar[a]	Heavy Tar[b]
Tar	100[c]	100	2	71
Acids	34	6.2	<0.1	6
Bases	2.2	2.4	170	700
Neutrals	30	47	2	20
Insolubles	—[d]	44	—	NT

[a]Sample G1225.

[b]Sample G1221.

[c]For the light tar, 4% of the material used for partitioning was the nonvolatile organic fraction of the aqueous phase that accompanied the tar.

[d]Not reported.

Source: Stamoudis (1986).

Table 5.55 Composition of Light and Heavy Tars from the Morgantown Energy Technology Center (METC) Gasifier[a]

	Concentration (mg/g)	
Compound[b]	Light Tar	Heavy Tar
PAHs		
(C_3–C_4)-Benzenes	2.9	—
Indene*	3.3	—
(C_2–C_2)-Indenes	10	—
(C_1–C_5)-Indans	2.8	—
Naphthalene*	17	0.6
1-Methylnaphthalene	12	0.2
2-Methylnaphthalene	11	0.4
(C_2–C_4)-Naphthalenes*	30	<1
Acenaphthylene*	6.0	0.2
Acenaphthene*	2.7	0.3
Methylacenaphthene	3.6	—
Biphenyl	3.7	<0.1
(C_1–C_3)-Biphenyls	3.5	0.8
Fluorene*	3.8	1.2
(C_1–C_2)-Fluorenes*	6.8	0.8
Phenanthrene*	4.4	3.0
Anthracene*	2.9	1.1
(C_1–C_2)-Phenanthrene/anthracene*	4.1	3.4
Acephenanthrylene	0.2	<0.3
Aceanthrylene	<0.1	0.5
4,5-Methylenephenanthrene	—	0.6
Methyl-4,5-methylenephenanthrene and/or acephenanthrene/aceanthracene	0.2	0.4
2-Phenylnaphthalene*	0.3	0.6
Fluoranthene*	1.0	3.4
Pyrene*	0.9	2.5

Table 5.55 (*Continued*)

Compound[b]	Concentration (mg/g) Light Tar	Heavy Tar
Methyl(dihydropyrene/acephenanthrene/ aceanthracene)	0.9	1.6
(C_1–C_2)-Fluoranthene/pyrene*	0.8	0.8
Benzofluorenes*	1.1	4.1
Methylbenzofluorenes	0.2	0.8
Cyclopenta[cd]pyrene	<0.1	<0.1
Benzo[ghi]fluoranthene	—	0.6
Benzo[c]phenanthrene	—	0.6
Benzo[a]anthracene*	0.2	2.5
Chrysene*	0.2	2.2
Benz(phenanthrene/anthracene)*	0.4	1.3
Methylbenz(phenanthrene/anthracene)*	—	1.2
Methylenechrysenes	—	2.5
Benzofluoranthenes*	0.2	3.6
Benzo[e]pyrene*	—	1.5
Benzo[a]pyrene*	—	1.5
Perylene*	—	0.3
Methylbenzo(fluoranthene/pyrene)	—	2.5
Methylenebenzo(fluoranthene/pyrene)	—	0.2
Indeno[1,2,3-cd]pyrene	—	0.7
Benzochrysene	—	0.3
Benzo[ghi]perylene	—	0.4
Dibenzo(phenanthrene/anthracene)	—	0.6
(C_{11}–C_{27})-Alkanes	33	0
*Hydroxy-PAHs**		
Phenol	32	0.6
Cresols	59	1.0
(C_2–C_3)-Phenols	32	0.9
O-PACs		
Benzofurans	1.0	—
(C_1–C_2)-Benzofurans	11.0	—
Dibenzofuran*	1.8	0.5
Methyldibenzofuran*	1.6	0.2
S-PACs		
Benzothiophene	3.1	—
Dibenzothiophenes*	—	0.8
Methyldibenzothiophenes*	—	0.1
Carbazole*	—	0.8
Total	311	56
Total analyzed by gas chromatography	345	65

PAHs = polycyclic aromatic hydrocarbons; O-PACs = oxygen-containing polycyclic aromatic compounds; S-PACs = sulfur-containing polycyclic aromatic compounds.

[a]The sum of concentrations for unspecified isomers is indicated in the table.

[b]Compounds indicated by * were detected in by-products of the METC gasifier by independent analyses at another laboratory (Hanson et al., 1979; Hill et al., 1983; Royer et al., 1983).

Source: Stamoudis (1986).

Table 5.56 Composition of Basic Fractions of Tar from the Morgantown Energy Technology Center (METC) Gasifier[a]

	Concentration (mg/g)	
Compound[b]	Light Tar	Heavy Tar
(C_1–C_4)-Pyridines	21	ND[c]
(C_2–C_4)-Diazoles, (C_2–C_3)-pyridines, and methylanilines	120	ND
(C_0–C_2)-Azaindan and C_2-aniline	16	ND
Quinoline	34	ND
Isoquinoline	15	ND
(C_1–C_4)-Quinolines*, (C_0–C_2)-azabiphenyls, (C_0–C_1)-aminonaphthalenes, and (C_0–C_1)-azafluorenes	150	4.9
Aza (phenanthrene/anthracene)*	10	12
(C_1–C_2)-Aza(phenanthrene/anthracene)*	6.6	39
Aza(fluoranthene/pyrene)*	0.6	19
(C_1–C_2)-Aza(fluoranthene/pyrene)	<0.1	24
Azabenz(phenanthrene/anthracene)*	<0.1	25
Methylazabenz(phenanthrene/anthracene)	<0.1	19
Azabenz(fluoranthene/pyrene)	<0.1	7.9
Naphthalenes, phenanthrenes	<10	ND
Total	380	150

[a] The sum of concentrations for unspecified isomers is indicated in the table.

[b] Compounds indicated by * were detected in by-products of the METC gasifier by independent analyses at another laboratory (Hanson et al., 1979; Hill et al., 1983; Royer et al., 1983).

[c] ND = not detected.

Source: Stamoudis (1986).

more than 90% of the material in the heavy tar could not be analyzed under the conditions employed (Table 5.55). Tables 5.55 and 5.56 also show that most of the major components of the gasifier by-product condensates that could be analyzed by GC were detected by independent analyses.

Although all the GC-detectable major components of the basic fractions were azaarenes, single-ion monitoring of the GC-MS responses clearly revealed the presence of aminonaphthalenes and other amino-PAHs. Although they were not detected, the unusually great mutagenicity of the basic fraction from the heavy tar was presumably due to the enrichment in this material of amino-PAHs having more than three rings by virtue of the absence of more volatile one- and two-ring aromatic bases. This is supported by the observation that about 90% of the mutagenic activity of the basic fraction was lost after mild treatment with nitrous acid, which readily converts amino-PAHs to phenols and other products (via formation of diazonium salts) but does not react with azaarenes.

The series of azaarenes shown in Table 5.56 is similar to that for other coal-conversion materials described in this chapter. Although the relative contribution of the lighter bases (especially anilines and pyridines) depends on the sampling site and the degree to which volatiles are removed in fractionation procedures, only minor differences appear between coal-conversion devices with respect to the series of gas chromatographable aromatic bases and their relative abundance in the products and by-products.

5.13.2.7 *Industrial Chapman Low-Btu Gasifier Operated at the Holston Army Base*

A source test and evaluation program was conducted for an industrial, fixed-bed, atmospheric pressure, single-state Chapman low-Btu gasifier operated at the Holston Army base, Kingsport, Tennessee (Page and Rhodes, 1978). Organic and inorganic chemical analyses were performed on gases, dusts, process and waste streams, and by-product tar. Parallel toxicological studies were conducted. Results are presented here for the by-product tar (Sample G1811) that was the most potentially hazardous material tested.

The tar was analyzed by EPA Level 1 methods, whereby fractions obtained by silica-column chromatography were analyzed by IR and probe-inlet low-resolution MS. Individual components were not quantitated, and qualitative analysis was based on molecular weight, IR spectra (which were complex), and the chromatographic behavior on silica.

Qualitatively, the profile of major components in the tar from the commercial gasifier has no distinct difference from profiles for pilot plants and laboratory-scale reactors described in this section. Although the data do not allow accurate quantitative comparisons, it is apparent that the relative amounts of PAHs, hydroxy-PAHs, basic nitrogen compounds, and other heteroatomic PACs did not differ remarkably from those for other gasification tars.

5.13.2.8 *Commercial Medium-Btu Gasifier Operated in the Kosovo Region of Yugoslavia*

Materials from the Kosovo gasifier were the subject of an EPA source test and evaluation (K. W. Lee et al., 1981). Heavy tar (Sample G1412) was among the discharge media with the highest priority for pollution control. Light tar (Sample G1414) and medium oil (Sample G1413) were present in materials discharged to the atmosphere, for example, via the coal lock vent discharge, which contained particulates at a concentration of 8100 mg/m^3, of which 7800 were tars and oils. Among the components of the tar and oil were four- and five-ring PAHs. The concentration of BaP in the light tar and medium oil was 0.2 and 0.07 mg/g, respectively. The total concentration of BaP isomers was about four to five times greater.

5.13.2.9 *Laboratory-Scale Pyrolysis Unit Operated at the Mellon Institute*

The effects of temperature [600, 800, or 1000°C (1112, 1472, or 1832°F)], pressure (3 or 10 atm), gas phase (He, CO_2, H_2, steam), residence time (0.2 or 1.3 s), and coal type (lignite or subbituminous) on the composition and toxicological characteristics of oil and tar produced during devolatilization of coal in a laboratory-scale, 2-lb/h, entrained-flow reactor were investigated (Fillo et al., 1983). Data for a few of the 49 reactor tests are presented here, and general trends are summarized.

The chemical compositions of the tars and oils were determined by capillary-column GC, supported by results of GC-MS analyses of representative samples and by a computer-based system for recognition of components by their relative retention indices. Selected samples were fractionated by liquid-partitioning to obtain acidic, basic, and neutral fractions prior to toxicological assays and chemical characterization.

As the reactor temperature increased from 600 to 1000°C (1112 to 1832°F), the following events occurred: (1) The yield of oil/tar diminished; (2) the major aromatic components of the oil/tar changed from more than four rings to one to three rings; and (3) the relative abundance of hydroxy-PAHs and alkylated PACs decreased drastically. The mutagenic activities varied widely among the oil/tar products produced under a range of reactor conditions. For reactions at 600°C (1112°F) the mutagenic activities were low or negligible; they were greatest at 800°C (1472°F). The mutagenic activities of the oil/tar were related (but not necessarily causally related) to temperature of formation, bulk molecular weight (as evidenced by atomic hydrogen : carbon ratios), degree of aromatic compound alkylation, ring structure, and the content of *N*-containing aromatic bases (azaarenes plus amino-PAHs).

The observations were consistent with a model for which primary fragments from coal devolatilization form secondary products at elevated temperatures, phenols released from coal are decomposed, heterocyclic O-PACs are converted to phenols, and heterocyclic N-PACs are converted to phenols, and heterocyclic N-PACs are decomposed to form amino-PAHs and subsequent products.

The lengthy, comprehensive report includes a comparative presentation, beyond the scope of this section, of the relations between reaction conditions and the composition of oil/tar by-products for the Hygas, METC, and UNDERC gasifiers, as well as for the MI reactor. Observations made for these units are discussed in relation to the model for coal gasification summarized above.

The chromatograms in Fig. 5.56 illustrate the temperature dependence on the content of phenols/cresols, and the relative abundance of one- to three-ring PACs versus four- to five-ring PACs. The compositions of the combined oils for the 800°C test are shown in Table 5.57.

Liquid-partitioning fractions were prepared for condensates from a similar test, differing only in reactor pressure. As for other gasification condensates, the mutagenic activity of the acidic fraction was negligible, while the total activities of the basic and neutral fractions were comparable. The yields of basic and neutral fractions were 0.5–4% and 40–60%, respectively. However, the specific mutagenicities of the basic fractions were 15–30 times greater than for the neutral fractions. The composition of the neutral fraction was similar to that indicated in Table 5.57, except that four- and five-ring PAHs were more abundant. The major constituents of the basic fraction were (1) azaarene analogs of naphthalene, biphenyl, acenaphthalene, phenanthrene/anthracene, pyrene, and dihydropyrene; (2) amino-PAH analogs of naphthalene, biphenyl, fluorene, and phenanthracene/anthracene; and (3) their respective alkyl homologs. Their relative abundance characteristically diminished from the two- to four-ring components, with the degree of alkylation the greatest for the smaller N-PACs. The major components in the acidic fraction were phenols and other hydroxy-PAH analogs of the two- to four-ring PAHs (including hydroxy-anthracene/phenanthrene and hydroxy-fluoranthene/pyrene).

5.13.3 Chemical Analyses of Fractionally Distilled Materials

Fractional distillation of complex hydrocarbon mixtures has been discussed. Chemical composition and toxicological activity of bp cuts are of interest because distillation is used in coal-liquefaction processes to effect material separation.

Weimer et al. (1980) studied crude bp cuts of an SRC-II material and showed that coal liquids boiling below a certain temperature range were inactive, while higher-boiling materials were active in laboratory mutagenicity assays. Furthermore, the boundary temperature separating inactive from active liquids could be fairly well defined. This observation gave rise to a set of experiments in which coal liquids representing several processes were selectively distilled into 50°F bp cuts and the cuts subjected to chemical analysis and bioassay (Wilson et al., 1982b; Mahlum, 1983).

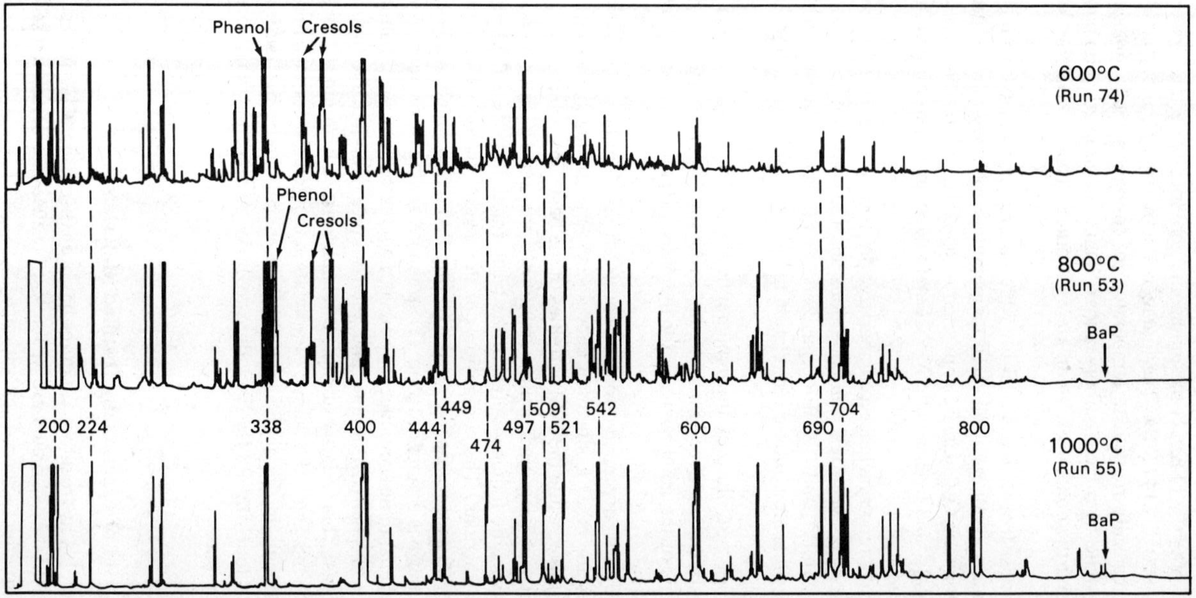

Figure 5.56 Typical gas chromatograms for combined oils produced at 600, 800, and 1000°C (1112, 1472, and 1832°F). For each test, condensates collected in each of a series of traps were proportionately combined. The pressure (atm), residence time (s), and gas phase for the 600 (Sample G3174), 800 (Sample G3153), and 1000°C (Sample G3155) tests were, respectively, (3, 1.3, He), (10, 0.2, H), and (10, 0.2, He). The components, labeled with relative retention indices, are as follows: 200, benzene; 224, toluene; 338, indene; 400, naphthalene; 444, 2-methylnaphthalene; 449, 1-methylnaphthalene; 474, biphenyl; 497, acenaphthylene; 509, acenaphthene; 521, dibenzofuran; 542, fluorene; 600, phenanthrene; 690, fluoranthene; 704, pyrene; and 800, chrysene (from Fillo et al., 1983).

Table 5.57 Components in By-Product Tar (Sample G1811) from an Industrial Chapman Low-Btu Gasifier

Silica Fraction Number	Fraction[a] (mg/g tar)	Compound	(Relative Amount in Fraction)[b]
1	100	(C_{10}–C_{18})-AHs	(100)
2	39	AHs	(10)
		Aromatic hydrocarbons, benzene naphthalene, alkyl series	(100)
3	280	PAHs, (C_0–C_5)-naphthalenes, (C_0–C_2)-acenaphthalene, (C_0–C_2)-phenanthracene, (C_0–C_2)-fluor(anthene/pyrene), chrysenes and alkyl series, benzopyrenes and alkyl series	(100)
4	100	PAHs, coronene and alkyl series	(100)
		Heterocyclic nitrogen compounds, carbazole and alkyl series	(100)
		Heterocyclic oxygen compounds	(100)
5	83	Heterocyclic sulfur compounds, dibenzothiophene and alkyl series	(100)
		Heterocyclic nitrogen compounds, carbazole and alkyl series	(10)
		Heterocyclic nitrogen compounds, benzonaphthofuran and alkyl series	(10)
6	310	Phenols, naphthols, and alkyl series	(100)
		Heterocyclic nitrogen compounds	(100)
		Heterocyclic sulfur compounds	(100)
		Heterocyclic oxygen compounds	(100)
		Amines	(10)
7	95	Amines, aniline and alkyl series	(100)
		Heterocyclic nitrogen compounds, quinoline and alkyl series, zanaphthalene and alkyl series	(100)
8	89	Alkyl halides	(10)
		Carboxylic acids and esters	(10)

AHs = aliphatic hydrocarbons. PAHs = polycyclic aromatic hydrocarbons.

[a]Gravimetric.

[b]For each compound class or subclass, an estimate of relative amounts was made to the nearest order of magnitude, based on the mass spectral intensities.

Source: Adapted from Page and Rhodes (1978).

5.13.3.1 *SRC-I*

Chemical analyses of SRC-I liquid bp cuts from the Wilsonville, Alabama pilot plant were reported by Wilson et al. (1982b, c). The analyses focused on determining amino-PAHs in the cuts. Table 5.58 shows the distribution of amino-PAHs among distillate cuts obtained from a blend of two SRC-I process streams.

Distributions of certain PAHs and their alkylated homologs in the 50°F cuts were also determined. Concentrations of selected PAHs in the SRC-I bp cuts are shown in Table 5.59. Between 600 and 800°F, each 50°F temperature increment corresponded roughly to one methylene unit, or 16 amu, average increase in molecular weight for the PAHs present.

Table 5.58 Listing of the Molecular Ions and Possible Compound Structures Detected from the Pentafluoropropionic (PFP) Derivatives of the Amino-Polycyclic Aromatic Hydrocarbons by Gas Chromatography–Mass Spectrometry (MS) and Probe-Inlet MS, in the Various Distillation Fractions from the Solvent-Refined Coal I Process Solvent

Compound	Molecular Weight		Boiling Range (°C) [°F]					Representative Structure
			315-345 [89-94]	345-370 [94-99]	370-400 [99-104]	400-425 [104-109]	425->455 [109->114]	
			Cumulative Weight Percent Distilled					
	Underivatized	PFP	50.2	74.8	81.8	87.8	100	
C_3 Aniline	135	281	X					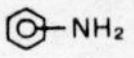
C_4 Aniline	149	295	X					
C_1 Aminoindan	145	291	X					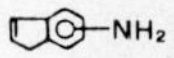
C_2 Aminoindan	159	305	X					
Aminonaphthalene	143	289	X					
C_1 Aminonaphthalene	157	303	X	X				
C_2 Aminonaphthalene	171	317	X	X				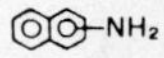
C_3 Aminonaphthalene	185	331	X	X				
C_4 Aminonaphthalene	199	345		X				
Aminobiphenyl	169	315		X				
C_1 Aminobiphenyl	183	329		X				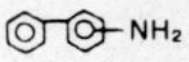
C_2 Aminobiphenyl	197	343		X				
C_3 Aminobiphenyl	211	357		X	X			
C_4 Aminobiphenyl	225	371		X	X			
Aminofluorene	181	327		X				
C_1 Aminofluorene	195	341			X			
C_2 Aminofluorene	209	355			X	X		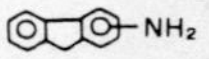
C_3 Aminofluorene	223	369			X	X		
C_4 Aminofluorene	237	383				X		
Aminophenanthrene	193	339			X	X		
C_1 Aminophenanthrene	207	353			X	X		
C_2 Aminophenanthrene	221	367			X	X		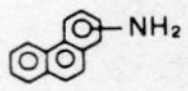
C_3 Aminophenanthrene	235	381			X	X		
C_4 Aminophenanthrene	249	395				X	X	
Aminophenylnaphthalene	219	365			X	X		
C_1 Aminophenylnaphthalene	233	379				X		
C_2 Aminophenylnaphthalene	247	393				X		
C_3 Aminophenylnaphthalene	261	407						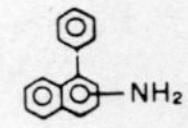
C_4 Aminophenylnaphthalene	275	421					X	
Aminopyrene	217	363				X		
C_1 Aminopyrene	231	377				X		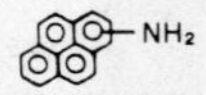
C_2 Aminopyrene	245	391				X		
C_3 Aminopyrene	259	405				X	X	
C_4 Aminopyrene	273	419					X	
Aminocyclopentapyrene	241	387				X	X	
C_1 Aminocyclopentapyrene	255	401				X	X	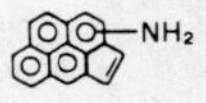
C_2 Aminocyclopentapyrene	269	415					X	
Aminobenzanthracene	243	389				X	X	
C_1 Aminobenzanthracene	257	403				X	X	
C_2 Aminobenzanthracene	271	417				X	X	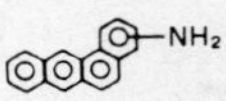
C_3 Aminobenzanthracene	285	431					X	
C_4 Aminobenzanthracene	299	445					X	
C_5 Aminobenzanthracene	313	459					X	
C_6 Aminobenzanthracene	327	473					X	
Aminobinaphthyl	269	415					X	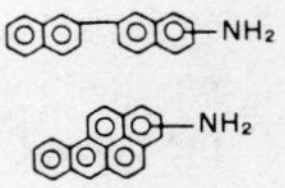
C_1 Aminobinaphthyl	283	429					X	
C_2 Aminobinaphthyl	297	443					X	
C_3 Aminobinaphthyl	311	457					X	
C_4 Aminobinaphthyl	325	471					X	
Aminobenzopyrene	267	413					X	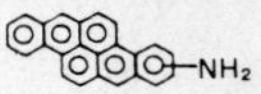
C_1 Aminobenzopyrene	281	427					X	
Aminodibenzopyrene	317	463					X	

Source: Wilson et al. (1982c).

Table 5.59 Major Polycyclic Aromatic Hydrocarbon (PAH) Components in Solvent-Refined Coal I (Sample L2121) Boiling-Point Cuts

		Concentration[a] (mg/g)		
Compound	Molecular Weight	700–750°F (N2107)[b]	750–800°F (N2108)	800 to >850°F (N2109)
Naphthalene	128	2	2	—
Acenaphthylene	152	4.9	3.4	—
Fluorene	154	36 ± 5	43 ± 8	—
Fluorene	166	38 ± 6	44 ± 6	—
Phenanthrene	178	58 ± 27	76 ± 8	12 ± 1
Anthracene	178	—	—	—
Fluoranthene	202	2,400 ± 300	26 ± 2	—
Pyrene	202	15,000 ± 2000	77 ± 9	6
9,10-Dimethylanthracene	206	1,500 ± 200	8.7	—
Benzo[b]fluorene	216	39,000 ± 3000	2,000 ± 100	—
1-Methylpyrene	216	40,000 ± 6000	1,000 ± 200	—
Benz[a]anthracene	228	1,700 ± 100	1,450 ± 140	—
Chrysene	228	400 ± 280	3,700 ± 200	110 ± 20
Methylchrysene	242	—	1,200 ± 100	650 ± 60
Benzo[b]fluoranthene	252	—	—	460 ± 90
Benzo[k]fluoranthene	252	—	—	250
7,12-Dimethylbenzanthracene	256	—	—	19
Benzo[e]pyrene	252	—	—	460 ± 30
Benzo[a]pyrene	252	—	—	340 ± 50
Indeno[1,2,3-cd]pyrene	276	—	—	86
Dibenz[a,c]anthracene	278	—	—	—
Benzo[g,h,i]perylene	276	—	—	73
Coronene	300	—	—	—

[a]Concentration based on gas chromatography analysis of PAH fractions calculated to crude concentration, using the weight % that eluted in the PAH fraction.

[b]Sample number given in parentheses.

Thus, C_2-pyrene concentrations peaked in the 50°F cut above that where C_1-pyrene concentrations were maximum.

5.13.3.2 *SRC-II*

Chemical composition (Pelroy and Wilson, 1981) and toxicological activity (Mahlum et al., 1983) of SRC-II bp cuts were also studied. Materials studied were blended from process streams produced at the Harmarville, Pennsylvania PDU during operation with Powhatan No. 5 mine coal. Blends used for biological testing were distilled into 50°F cuts. Other selectively distilled materials, with various and occasionally overlapping boiling ranges between about 600 and 1060°F, were also studied (Pelroy et al., 1985). Data from these studies showed that N-PAC and hydroxy-PAH fraction weights generally increased with increasing distillation temperature, while AH fraction weight decreased.

Figure 5.57 shows the structures of compounds in highest concentrations in PAH fractions and the carbazole, azaarene, and N-PAC and amino-PAH subfractions from 50°F SRC-II bp cuts from the Harmarville PDU. Table 5.60 shows concentrations of selected

Distillation Temperature Range °F	<650	650-700	700-750	750-800	800-850	850+
Average Molecular Weight in AMU Rounded to Nearest 5	160	185	200	215	240	275
PAH in Highest Concentration	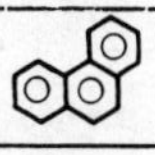	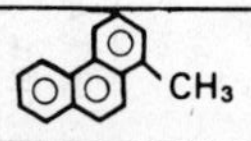	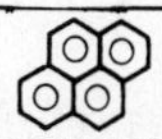	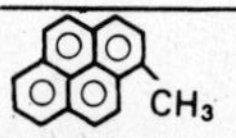	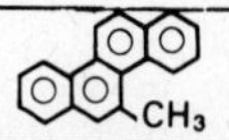	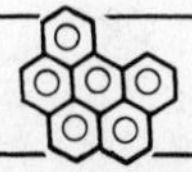
APAH in Highest Concentration	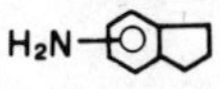	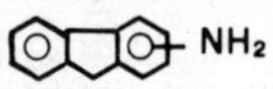	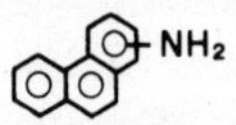	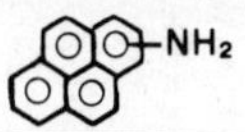	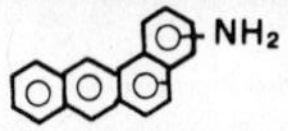	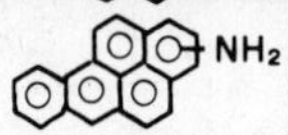
2° PANH in Highest Concentration	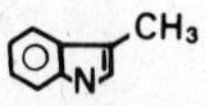		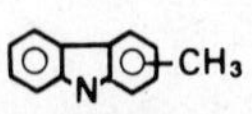	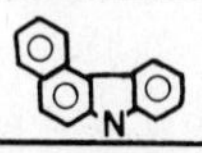	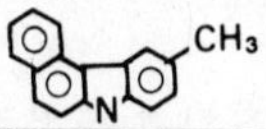	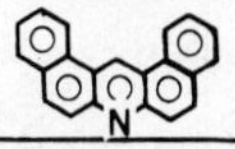
3° PANH in Highest Concentration	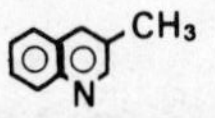	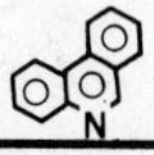	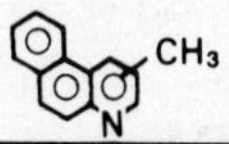	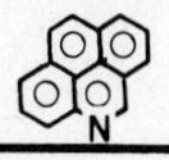	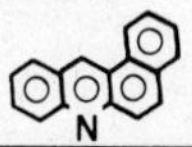	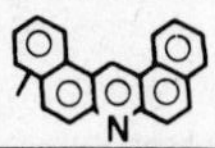

Figure 5.57 Structures in highest concentration in polycyclic aromatic hydrocarbon (PAH), amino-PAH carbazole (2°-polycyclic aromatic nitrogen heterocycle, PANH), and azaarene (3°-PANH) fractions of solvent-refined coal II 50°F boiling-point cuts of a blend from the Harmarville, Pennsylvania process development unit (from Wilson et al., 1986).

Table 5.60 Concentrations of Representative Polycyclic Aromatic Hydrocarbons (PAHs) in Solvent-Refined Coal II (Sample L1251) Boiling-Point Cuts

Compound	Weight	Concentration[a] ($\mu g/g$) 700–750°F (N2107)[b]	750–800°F (N2108)	800–850°F (N2109)	>850°F
Naphthalene	128	—	—	—	—
Acenaphthylene	152	—	—	4.9 ± 2.9	4.9 ± 1.6
Fluorene	154	27 ± 1	16 ± 3	20 ± 5	—
Phenanthrene	178	640 ± 140	230 ± 20	100 ± 10	23 ± 8
Anthracene	178	—	—	—	—
Fluoranthene	202	18,000 ± 1,000	1,300 ± 100	110 ± 10	12
Pyrene	202	160,000 ± 20,000	18,000 ± 1,000	1,200 ± 100	95 ± 22
9,10-Dimethylanthracene	206	65,000	420 ± 60	11	—
2- or 4-Methylpyrene or benzo[b]fluorene	216	150,000 ± 40,000	37,000 ± 4,000	2,800 ± 100	—
1-Methylpyrene	216	61,000 ± 16,000	12,000 ± 2,000	640 ± 90	—
Benz[a]anthracene	228	2,400 ± 550	3,100 ± 200	1,500 ± 100	—
Chrysene	228	1,200 ± 300	3,800 ± 100	3,600 ± 300	13
Methylchrysene	242	360	4,700 ± 350	17,000 ± 2,500	87 ± 17
Benzo[b]fluoranthene	252	—	—	4,700 ± 1,300	960 ± 90
Benzo[k]fluoranthene	252	160 ± 30	46 ± 28	1,950 ± 250	120 ± 10
7,12-Dimethylbenzanthracene	256	—	120	2,200 ± 650	ND
Benzo[e]pyrene	252	—	—	3,000 ± 350	2,300 ± 200
Benzo[a]pyrene	252	—	54 ± 10	1,700 ± 200	1,500 ± 150
Indeno[1,2,3-cd]pyrene	276	—	—	—	2,900 ± 900
Dibenz[a,c]anthracene or dibenz[a,c]anthracene	278	—	—	—	800 ± 470
Benzo[g,h,i]perylene	276	—	—	—	6,200
Coronene	300	—	—	—	430 ± 60

[a]Concentration based on gas chromatography analysis of PAH fractions calculated to crude concentration using the weight % that eluted in the PAH fraction.

[b]Sample number given in parentheses.

Source: Pelroy and Wilson (1981).

Table 5.61 Chemical Class Composition of Selected EDS Boiling-Point Cuts

Compound Class	Distribution (wt %) 650–700°F (N4102)[a]	700–750°F (N4103)	750–800°F (N4104)	>800°F (N4105)
AHs	23	19	16	4
Neutral PAHs	52	60	58	58
N-PACs	10	9	13	23
Hydroxy-PAHs	8	9	12	17
Total recovery	93	97	99	102

AHs = aliphatic hydrocarbons; N-PACs = nitrogen-containing polycyclic aromatic compounds; PAHs = polycyclic aromatic hydrocarbons.

[a] Sample number given in parentheses.

Source: Later et al. (1984b).

PAHs in bp cuts, determined by HRGC. Weight percentages of the bp cuts that partitioned into various chemical class fractions are shown in Table 5.61.

Table 5.62 shows N-PACs that were in high concentration or were of toxicological importance in the 50°F bp cuts from SRC-II HD. Carcinogenic and mutagenic compounds were essentially confined to cuts boiling above 700° F. Table 5.63 shows concentrations of amino-PAHs in SRC-II bp cuts determined by HRGC (C. W. Wright and D. W. Later, unpublished data).

Anderson and Freel (1983) reported experiments where materials that distilled above a specified temperature were recycled to extinction in the SRC-II process; this was termed HD recycle operation and was evaluated at Harmarville. Products formed had upper temperature cut points of 270, 290, and 310°C (528, 554, and 590°F). The <270°C materials were produced from bituminous and subbituminous coals. Weight distribution of the four major chemical class fractions for <270°C and 310°C materials are shown in Table 5.64.

N-PAC concentrations in fractions from the −270 and −310°C runs are compared to those in an SRC-II FOB in Table 5.64. Table 5.65 shows data for compounds in PAH fractions of HD recycle distillates (Wilson et al., 1983b). No product distillates derived during HD recycle were mutagenic in *S. typhimurium* TA98.

Table 5.62 Major Nitrogen-Containing Polycyclic Aromatic Compound (N-PAC) Constituents in Solvent-Refined Coal II Boiling-Point Cuts

Compound	Concentration[a] (μg/g) 700–750°F (N1102)[b]	750–800°F (N1103)	800–850°F (N1104)	>850°F (N1105)
Carbazole	100	6.0	—	—
Benzo[a]carbazole	140	3300	5400	—
Phenanthridine	33	12	—	—
1-Azapyrene	1200	290	130	—
2-Azafluoranthene	5300	5200	990	—
Benz[c]acridine	100	220	620	—

[a] Concentrations based on gas chromatography analysis of N-PAC fractions calculated to crude concentrations using the weight % that eluted in the N-PAC fractions.

[b] Sample number given in parentheses.

Source: Adapted from Pelroy and Wilson (1981).

Table 5.63 Composition (ppm) of Amino-Polycyclic Aromatic Hydrocarbons in Solvent-Refined Coal II Boiling Point Cuts

Compound	Molecular Weight	Distillation Temperature Range (°F) 700–750	750–800	800–850	>850
1-Aminonaphthalene	143	0.6	1.5	2.9	0.6
2-Aminobiphenyl	169	0.3	0.3	0.9	0.3
2-Aminonaphthalene	143	1.0	3.3	5.8	1.1
3-Methyl-2-aminonaphthalene	157	0.5	2.1	3.1	0.3
4-Aminobiphenyl	169	1.2	5.2	7.3	0.8
4-Aminofluorene	181	8.8	19.0	16.8	1.6
2-Aminofluorene	181	124.5	41.6	21.0	3.0
9-Aminophenanthrene	193	40.4	48.8	12.8	2.6
3-Aminophenanthrene	193	13.1	253.0	37.4	4.1
2-Aminoanthracene	217	120.3	107.6	23.7	1.5
3-Aminofluoranthene	217	10.6	412.5	161.8	3.2
1-Aminopyrene	217	7.2	155.0	239.4	3.1
2-Aminopyrene	217	5.4	137.8	239.1	5.5
8-Aminofluoranthene	217	2.0	316.7	661.0	18.9
6-Aminochrysene	243	4.5	17.8	203.1	11.9
3-Aminochrysene	243	0.6	34.5	319.8	41.8
7-Aminobenzo[a]pyrene	267	—	—	22.8	144.2

Source: Pelroy and Wilson (1981).

Table 5.64 Concentrations (ppm) of Major Components in Nitrogen-Containing Polycyclic Aromatic Compound (N-PAC) Fractions of Two Distillates Produced by Solvent-Refined Coal (SRC-II) Heavy Distillate (HD) Recycle Operation Compared to Those from SRC-II Fuel-Oil Blend (FOB)

	HD Recycle Runs		
	<270°C (518°F)	<310°C (590°F)	FOB
Compound	(467RA, L1331)[a]	(467RC, L1351)	(L1251)
Tetrahydroquinoline	1100 ± 100	440 ± 40	910 ± 300
Quinoline	910 ± 110	360 ± 80	1500 ± 300
Indole	2700 ± 200	1800 ± 100	2300 ± 400
2-Methylquinoline	1700 ± 200	600 ± 100	2100 ± 400
8-Methylquinoline	320 ± 70	200 ± 40	280 ± 30
Methylindole	1700 ± 200	1600 ± 50	1300 ± 300
2,5-Dimethylindole	490 ± 120	980 ± 100	590 ± 120
2,3-Dimethylindole	510 ± 90	920 ± 90	500 ± 90
Benzo[h]quinoline	69 ± 7	220 ± 30	310 ± 10
Benzo[f]quinoline	66 ± 5	230 ± 30	460 ± 20
Carbazole	280 ± 10	790 ± 40	2200 ± 100
1-Methylcarbazole	84 ± 2	750 ± 10	720 ± 30
Methylcarbazole	56 ± 2	130 ± 10	500 ± 20
Benzo[a]carbazole[b]	—	—	390 ± 150
Benzo[b]carbazole[b]	—	—	220 ± 80
Benzo[c]carbazole[b]	—	—	370 ± 180

[a] Sample number given in parentheses.

[b] N-PACs with known mutagenic, carcinogenic, or tumorigenic activity.

Source: Wilson et al. (1983b).

Table 5.65 Concentrations (ppm) of Major Components in Polycyclic Aromatic Hydrocarbon Fractions of Various Solvent-Refined Coal II (SRC-II) Heavy Distillate (HD) Recycle Distillates Compared to Those from SRC-II Fuel-Oil Blend (FOB)

	HD Recycle Runs		FOB
Compound	<270°C (518°F)	<310°C (590°F)	2.9:1 (MD[a]:HD)
Naphthalene	23 ± 2	24	12,080 ± 300
2-Methylnaphthalene	3,300 ± 500	2,000 ± 300	28,000 ± 400
1-Methylnaphthalene	760 ± 110	490 ± 70	5,900 ± 100
Biphenyl	—	—	9,000 ± 200
2-Ethylnaphthalene	4,600 ± 600	2,900 ± 450	9,600 ± 250
2,6- or 2,7-Dimethylnaphthalene	6,000 ± 900	3,400 ± 500	15,000 ± 1,400
1,7-Dimethylnaphthalene	1,700 ± 200	1,100 ± 150	3,400 ± 300
1,6- or 1,3-Dimethylnaphthalene	1,600 ± 200	1,250 ± 240	2,200 ± 100
Acenaphthene	2,600 ± 300	2,200 ± 450	4,750 ± 60
C_3-Naphthalene	5,900 ± 900	4,600 ± 800	6,900 ± 700
C_3-Naphthalene	5,200 ± 600	4,500 ± 800	2,900 ± 1,200
Dibenzofuran	2,100 ± 200	1,650 ± 160	5,500 ± 100
Fluorene	3,200 ± 300	3,900 ± 500	5,000 ± 100
9-Methylfluorene	2,100 ± 600	2,500 ± 500	2,750 ± 200
2-Methylfluorene	1,600 ± 100	3,200 ± 400	3,100 ± 100
1-Methylfluorene	1,600 ± 100	2,700 ± 300	235 ± 30
Dibenzothiophene	1,500 ± 200	2,950 ± 400	2,400 ± 100
Phenanthrene	5,200 ± 400	7,900 ± 900	14,000 ± 650
3-Methylphenanthrene	1,100 ± 100	1,700 ± 100	2,200 ± 40
2-Methylphenanthrene	1,300 ± 50	2,000 ± 50	2,800 ± 100
1-Methylphenanthrene	410 ± 20	710 ± 80	830 ± 20
Dihydrofluoranthene	360 ± 10	820 ± 10	520 ± 10
Fluoranthene	190 ± 10	270 ± 20	3,100 ± 100
Pyrene	930 ± 30	1,600 ± 30	5,700 ± 10
Benzo[a]fluorene	74 ± 15	90 ± 27	270 ± 10
Benzo[b]fluorene, 2- or 4-methylpyrene	300 ± 10	290 ± 220	2,700 ± 100
1-Methylpyrene[b]	81 ± 10	96 ± 22	650 ± 90
Benz[a]anthracene[b]	—	—	460 ± 30
Chrysene	29	37	660 ± 40
Benzo[j or b]fluoranthene[b]	—	—	170 ± 10
Benzo[k]fluoranthene[b]	—	—	45 ± 7
Benzo[e]pyrene	—	—	200 ±10
Benzo[a]pyrene[b]	—	—	190 ± 10

[a] MD = middle distillate.

[b] Nitrogen-containing polycyclic aromatic compounds with known mutagenic, carcinogenic, or tumorigenic activity.

Source: Wilson et al. (1983b).

HPLC subfractions of SRC-II 50°F bp cuts were studied to determine the effect of chemical composition on expression of PAH carcinogens. Figure 5.58 shows an HPLC chromatogram of the 850°F cut of Harmarville PS. Most mutagenic and tumorigenic activity elutes in fractions 1 and 2. Compounds in these fractions were mainly five-ring PAHs and their C_1 or C_2 isomers. The bp cut with highest initiation/promotion and mutagenic activity also had the highest concentration of BaP.

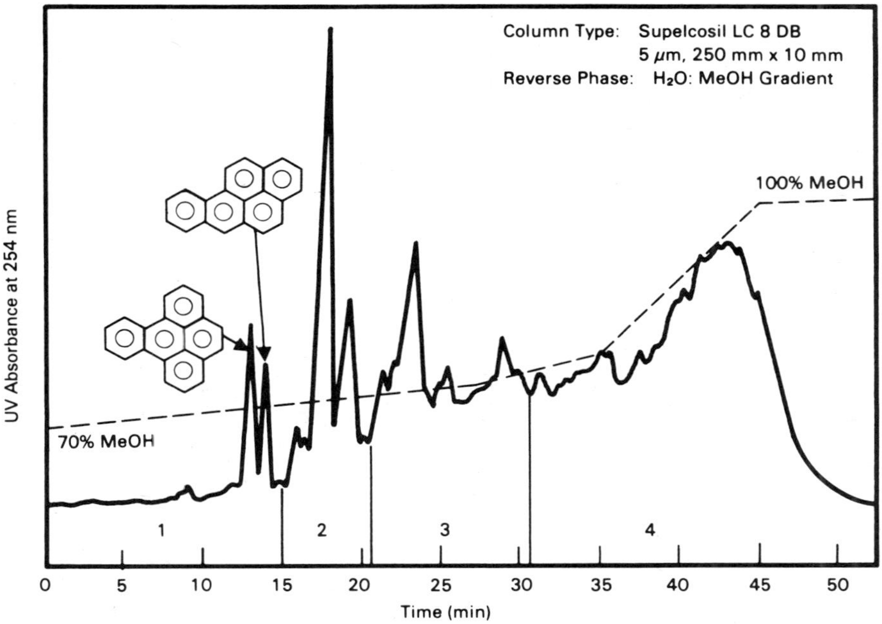

Figure 5.58 High-performance liquid chromatography chromatogram of the 850°F cut (Sample N1105) from fractionally distilled solvent-refined coal II product from the Harmarville, Pennsylvania process development unit (from Mahlum et al., 1983).

5.13.3.3 *EDS*

The chemical compositions of several EDS bp cuts were also investigated (Later et al., 1984b). A full-boiling-range, process-derived EDS liquid was simulated by blending raw EDS naphtha and recycle solvent. This blend, containing compounds in the 150 to >800°F boiling range, was then distilled into 50°F cuts. Weight distributions of the four major chemical classes are shown in Table 5.66 for the four highest-boiling cuts. About 88% of the reblended material was distilled below 700°F. Although the four bp cuts shown in Table 5.61 represent only 12% of the total weight of feed blend, they contain most of the biological activity of the full-boiling-range material. Chemically, AH content decreased, with a concomitant increase in N-PACs and polar hydroxy-PAHs with increasing distillation temperature. Neutral PAH content in each bp cut generally remained constant (52–60%) with increasing temperature.

GC, GC-MS, and probe-inlet MS studies showed the effect of distillation temperature on molecular weight. Increasing molecular weight compounds were distributed in higher-temperature bp cuts. For example, pyrene was about the same maximum concentration in 650–700°F and 700–750°F cuts but decreased in the two higher distillates. Carcinogenic compounds with high molecular weights, such as BaP, were distilled entirely in the >800°F cut (see Table 5.66).

A similar trend was observed for N-PACs, amino-PAHs, and hydroxy-PAHs. Table 5.67 shows concentrations of major amino-PAHs in N-PAC fractions of the EDS bp cuts. Other species in the N-PAC fraction included pyrrolic-PAHs, such as indoles, carbazoles,

Table 5.66 Peak Assignments for Components in Polycyclic Aromatic Hydrocarbon (PAH) Fraction and Concentrations for PAHs for Four EDS Boiling-Point Cuts

Peak Number	Molecular Weight	Compound[b]	Concentration ($\mu g/g$)[a] 650–700°F (N4102)[c]	700–750°F (N4103)	750–800°F (N4104)	>800°F (N4105)
1	178	Phenanthrene	420 ± 40	—	—	—
2	192	3-Methylphenanthrene	2,000 ± 200	—	—	—
3	192	2-Methylphenanthrene	3,400 ± 400	—	—	—
4	192	1-Methylphenanthrene	1,500 ± 100	—	—	—
5	204	2-Phenylnaphthalene	3,100 ± 300	—	—	—
6	206	C_2-Phenanthrene	12,500 ± 500	—	—	—
7	208	C_3-Fluorene	2,800 ± 1,300	—	—	—
8	202	Fluoranthene	5,100 ± 500	660 ± 10	41[d]	—
9	220	C_3-Phenanthrene	11,000 ± 900	ND[e]	ND	—
10	218	C_1-Phenylnaphthalene	4,600 ± 500	300 ± 40	ND	—
11	202	Pyrene	16,000 ± 1,000	18,000 ± 600	700 ± 10	99 ± 3
12	234	C_4-Phenanthrene	3,000 ± 300	ND	ND	—
13	216	Benzo[b]fluorene	—	28,000 ± 7,000	1,500 ± 100	—
14	218	C_1-Phenylnaphthalene	—	33,000 ± 7,000	1,300 ± 200	—
15	216	1-Methylpyrene	—	16,000 ± 4,000	810 ± 60	—
16	232	C_2-Phenylnaphthalene	—	ND	ND	—
17	230	C_1-Benzofluorene, C_2-pyrene	—	23,000 ± 1,000	9,300 ± 350	—
18	232	C_2-Phenylnaphthalene	—	ND	ND	—
19	244	C_2-Benzofluorene, C_3-pyrene	—	ND	5,000 ± 300	—
20	228	Benz[a]anthracene	—	180 ± 10	3,200 ± 1,200	—
21	228	Chrysene	—	960 ± 200	14,000 ± 1,500	—
22	258	C_3-Benzofluorene, C_4-pyrene[d]	—	—	12,000 ± 4,000	—

23	242	3-Methylchrysene	—	—	5,500 ± 2,500	—
24	242	2-Methylchrysene	—	—	3,300 ± 2,900	170
25	242	6- or 5-Methylchrysene	—	—	12,000 ± 7,500	2,400 ± 200
26	256	C_2-Chrysene	—	—	—	2,700 ± 300
27	252	Benzo[j or b]fluoranthene	—	—	—	3,700 ± 2,100
28	252	Benzo[k]fluoranthene	—	—	—	530 ± 200
29	256	C_2-Chrysene	—	—	—	2,600 ± 200
30	252	Benzo[e]pyrene	—	—	—	2,100 ± 500
31	252	Benzo[a]pyrene	—	—	—	1,300 ± 700
32	252	Perylene	—	—	—	93 ± 40
33	262	C_6-Phenanthrene	—	—	—	ND
34	276	Indeno[1,2,3-cd]pyrene	—	—	—	390 ± 90
35	278	Dibenz[a,h, or a,c]anthracene	—	—	—	21
36	276	Benzo[g,h,i]perylene	—	—	—	1,000 ± 200

[a]Concentration based on gas chromatography (GC) analysis of PAH fractions calculated in μg/g of crude using weight % that eluted in the PAH fraction.

[b]Tentatively identified by gas chromatography–mass spectrometry and retention times.

[c]Sample number given in parentheses.

[d]Value derived from one determination only.

[e]No quantitative data.

Source: Later et al. (1984b).

Table 5.67 Concentrations of Selected Amino-Polycyclic Aromatic Hydrocarbons in EDS Boiling-Point Cuts

Compound[b]	Molecular Weight[c]	Concentration (μg/g N-PAC fraction)[a] 650–700°F (N4102)[d]	700–750°F (N4103)	750–800°F (N4104)	>800°F (N4105)
1-Aminonaphthalene	289	—	2.1	2.8	—
2-Aminobiphenyl	315	0.3	3.1	3.3	—
2-Aminonaphthalene	289	0.3	2.5	2.9	0.7
3-Methyl-2-aminonaphthalene	293	2.2	1.5	1.8	—
4-Aminobiphenyl	315	10.3	4.2	3.3	—
4-Aminofluorene	327	25.0	15.5	5.0	—
2-Aminofluorene	327	—	80.7	7.2	1.1
C_2-Aminofluorene	355	—	901.5	—	—
9-Aminophenanthrene and/or 1-aminophenanthrene and/or 1-aminoanthracene	339	22.9	194.0	16.2	13.8
C_3-Aminofluorene	369	—	1680.2	—	—
3-Aminophenanthrene and/or 2-aminophenanthrene	339	10.6	130.0	73.2	1.8
C_3-Aminobiphenyl	357	—	756.8	—	—
2-Aminoanthracene	339	17.2	305.1	73.1	—
C_2-Aminophenanthrene	367	—	648.1	1174.1	—
C_3-Aminophenanthrene	381	—	—	601.2	—
C_3-Aminophenanthrene	381	—	—	894.5	—
3-Aminofluoranthene	363	8.4	41.2	388.6	3.0
1-Aminopyrene	363	6.9	14.8	172.6	5.1
2-Aminopyrene	363	—	44.6	192.9	5.1
8-Aminofluoranthene	363	—	28.0	157.4	12.6
6-Aminochrysene	389	—	—	33.1	46.9
3-Aminochrysene	389	—	—	12.8	20.3
7-Aminobenzo[a]pyrene	413	—	—	—	8.2

[a] N-PAC = nitrogen-containing polycyclic aromatic compound; values reported are the average of three separate determinations.
[b] Tentatively identified by gas chromatography–mass spectrometry (GC–MS) and GC retention times.
[c] Pentafluoropropylamide-derived mass as determined by GC-MS.
[d] Sample number given in parentheses.
Source: Later et al. (1984b).

benzocarbazoles, dibenzocarbazoles, and their alkylated homologs. Tertiary amines or azaarenes, such as the quinolines, acridines, phenanthridines, and benzoquinolines, were detected in much lower concentrations than the pyrrolic-PAH components. (See Table 5.68.)

Figure 5.59 shows a probe-inlet mass spectrum of an HPLC subfraction from an EDS >800°F bp cut. The spectrum is simpler than that from crude EDS materials, shown in Fig. 5.60. The gas chromatogram of the HPLC subfraction (Fig. 5.59) shows that the ion peak at m/z 252 arises from at least four different isomers, three of which are identified. Bioassay results of such simplified subfractions can be correlated to toxicological activity of specific chemical structures in coal liquids.

Figure 5.61 shows probe-inlet mass spectra of hydroxy-PAH fractions from three EDS

Table 5.68 Concentrations of Nitrogen-Containing Polycyclic Aromatic Compounds (N-PACs) for Four EDS Boiling-Point (bp) Cuts

		Concentration[a] (μg/g; bp Range, °F)			
Molecular Weight	Compound[b]	650–700 (N4102)[c]	700–750 (N4103)	750–800 (N4104)	>800 (N4105)
129	Quinoline	1.5[d]	—	—	—
179	Benzo[h]quinoline	28 ± 4	13	—	—
179	Acridine	17 ± 2	—	—	—
179	Phenanthridine and/or benzo[f]quinoline	85 ± 17	—	—	—
167	Carbazole	410 ± 30	24 ± 2	—	—
181	1-Methylcarbazole	51 ± 12	—	—	—
181	3-Methylcarbazole	1800 ± 200	60 ± 1	—	—
181	2-Methylcarbazole	1400 ± 600	59 ± 2	—	—
181	4-Methylcarbazole	970 ± 70	29 ± 3	—	—
195	C_2-Carbazole[e]	2800 ± 300	240 ± 10	—	—
209	C_3-Carbazole[e]	1700 ± 100	3100 ± 500	160 ± 10	—
223	C_4-Carbazole[e]	—	4100 ± 900	260 ± 60	—
237	C_5-Carbazole	—	370 ± 40	220 ± 10	—
221	C_3-Benzoquinolines[e]	—	440 ± 110	6000 ± 300	—
235	C_4-Benzoquinolines[f]	—	—	2300 ± 100	—
217	Benzo[c]carbazole	—	—	4000 ± 1000	92 ± 15
249	C_5-Benzoquinolines[f]	—	—	1400 ± 200	—
231	C_1-Benzocarbazoles[e]	—	—	300 ± 40	550 ± 80
245	C_2-Benzocarbazoles[f]	—	—	—	300 ± 70
253	Azabenzopyrene	—	—	—	170 ± 110
259	C_3-Benzocarbazoles	—	—	—	100 ± 70
273	C_4-Benzocarbazoles[e]	—	—	—	460 ± 70
287	C_5-Benzocarbazoles	—	—	—	92 ± 120
267	Dibenz[c,g or a,g] carbazole	—	—	—	41

[a] Concentration based on gas chromatography (GC) analysis of N-PAC fractions calculated as μg/g of crude using weight % that eluted in the N-PAC fractions.
[b] Cation identification tentative; based on mass spectrometry and retention times.
[c] Sample number given in parentheses.
[d] Value derived from one determination only.
[e] Composite of three peaks.
[f] Composite of two peaks.

Source: Later et al. (1984b).

bp cuts and illustrates the variety of hydroxy-PAHs present. Hydroxy-PAH fractions were not active in the bioassays used on these materials.

No mutagenic response was obtained with PAH fractions from SRC-I, SRC-II, ITSL, or H-Coal distillates that boiled below 700°F. However, EDS bottoms recycle solvent did show mutagenic activity, which prompted HPLC separations. Figure 5.62 shows an HPLC chromatogram of the 650–700°F bp cut from EDS bottoms recycle solvent; HPLC subfractions are indicated by number. Mutagenic activity for each fraction is determined by slope and maximum response.

Isolation of mutagenic activity in HPLC fraction 5 indicated that a relatively well-de-

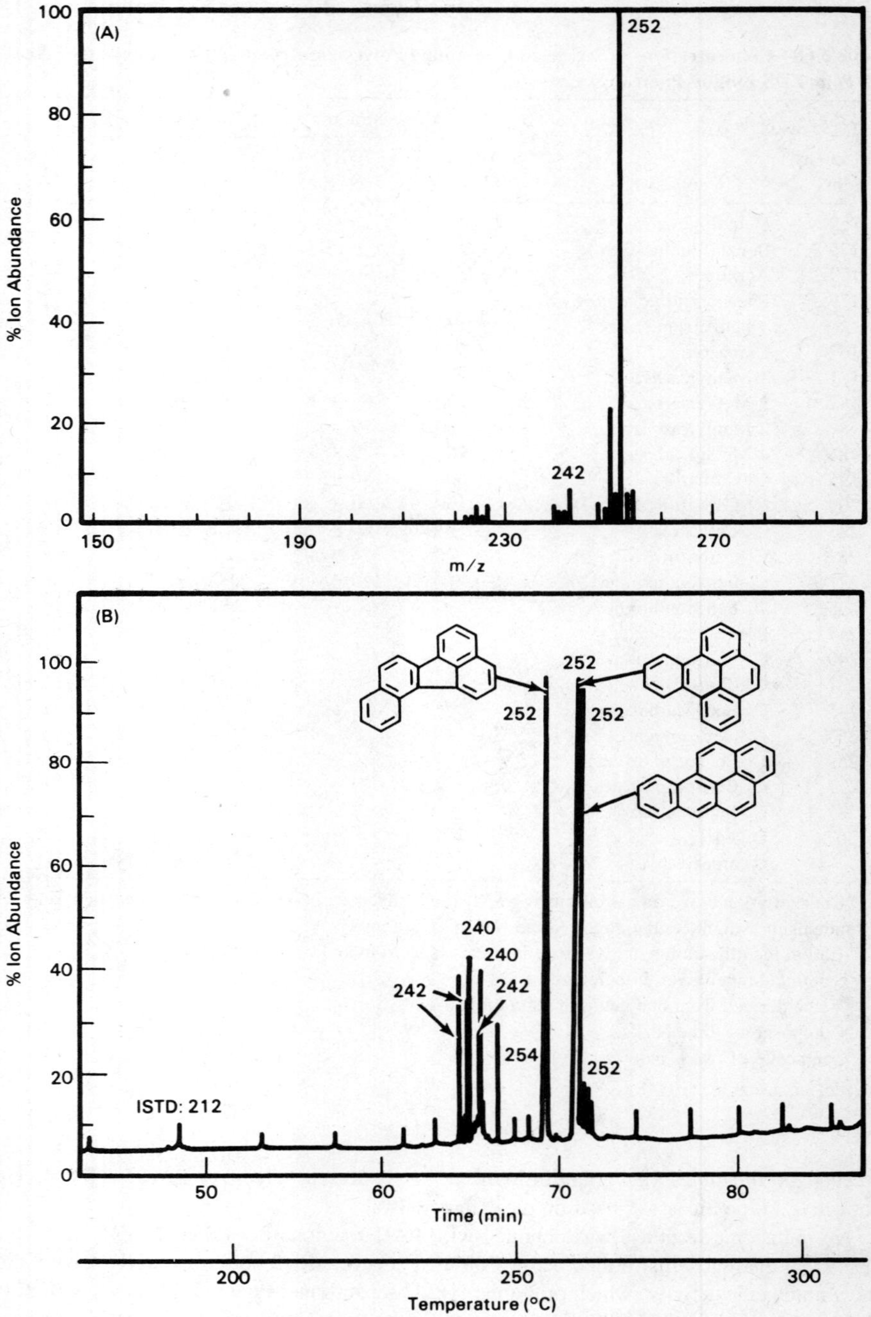

Figure 5.59 (A) Gas chromatogram from high-performance liquid chromatography (HPLC) subfraction of polycyclic aromatic hydrocarbon fraction from >800°F boiling-point cut (Sample N4105) from EDS recycle solvent. (B) Mass spectrum of cuts showing that most of the materials yield molecular ions at m/z 252 (from Later et al., 1984b).

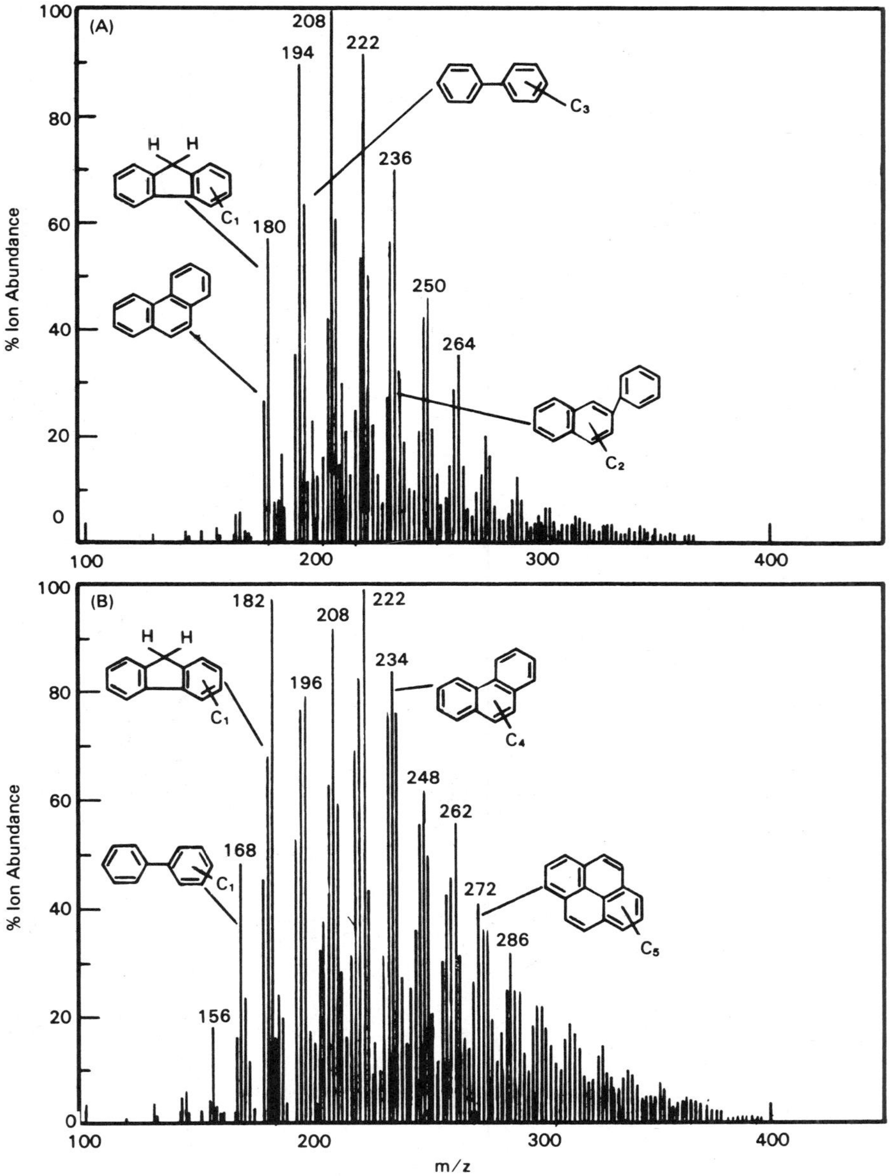

Figure 5.60 Mass spectra of EDS recycle solvents (A) once-through (Sample L4111) and (B) bottoms recycle (Sample L4132) (from Later et al., 1984b).

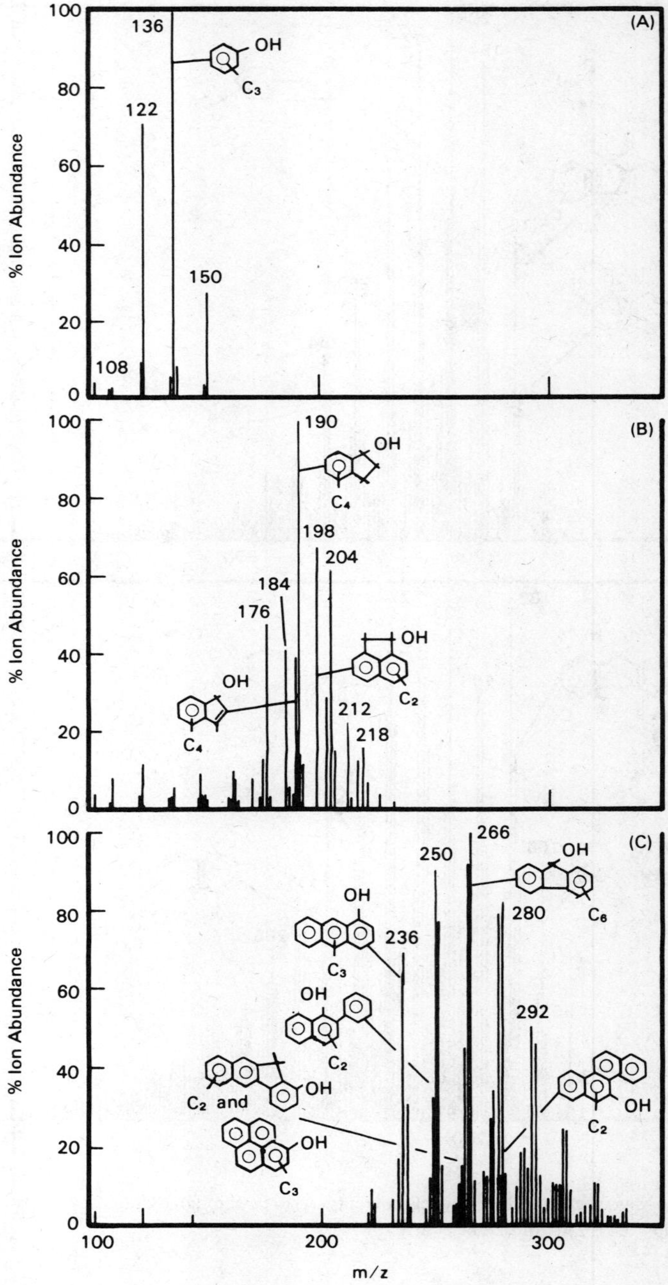

Figure 5.61 The 12-eV mass spectra of selected hydroxy polycyclic aromatic hydrocarbon (PAH) fractions A_4 from the EDS distillate cuts for boiling ranges (A) 400–450°F and (B) 550–600°F, and (C) 750–800°F (from Later et al., 1984b).

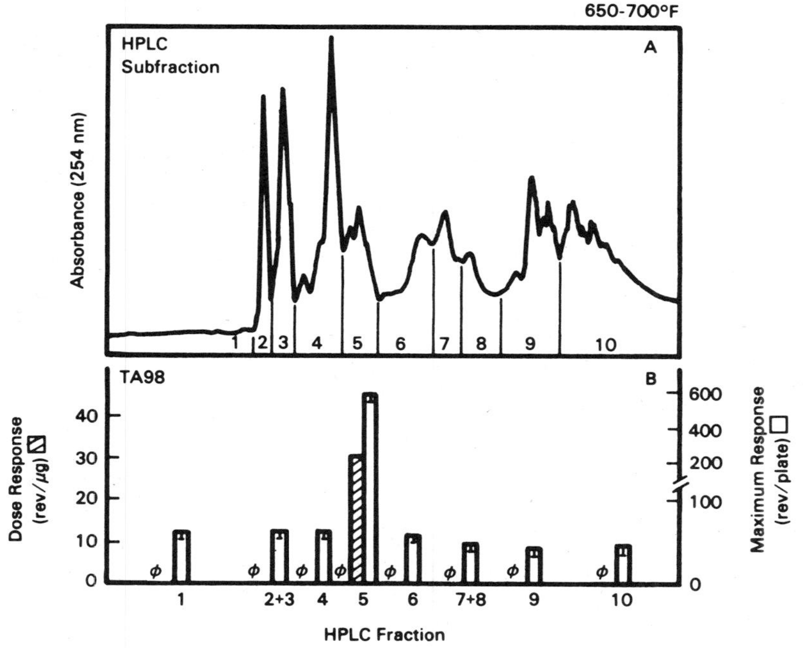

Figure 5.62 (A) High-performance liquid chromatography (HPLC) chromatogram of EDS 650–700°F boiling-point cut. (B) Histogram depicting microbial mutagenic activity of fractions indicated (from Later et al., 1984b).

fined class of compounds was probably responsible. GC-MS analysis showed the fractions contained C_2-phenanthrenes (Lucke et al., 1985), some isomers of which are mutagenic and tumorigenic in initiation/promotion assays (La Voie et al., 1981a, b).

5.13.3.4 *H-Coal Blend*

A 1.4 : 1.0 blend of an H-Coal light oil (N-103) and heavy oil (N-202) from the Oak Ridge repository was fractionally distilled by Chevron Research to yield cuts with 50°F bp ranges from 650 to >850°F. Physical characterization and chemical analysis data on the blend and distillate products were reported by Wilson et al. (1985). Results from chemical class fractionation of the distillate cuts are shown in Table 5.69. Table 5.70 shows the concentrations of selected PAHs in the distillate cuts, as determined from the GC. Analogous data for the N-PAC fractions are shown in Table 5.71. Figure 5.63 shows mass spectra of derivatized amino-PAHs extracted from the distillate cuts. The groups of peaks on the right of each figure are the PFP derivatives; selected structures consistent with the mass spectra are also shown. Concentrations for selected amino-PAHs are given in Table 5.72. Figure 5.64 shows the distribution of heteroatoms in the H-Coal bp cuts as a function of distillation temperature.

The types of compound found in the various distillate cuts from the H-Coal materials were analogous to those found in studies on distillate cuts from other single-stage processes. H-Coal materials were more highly alkylated than the other single-stage materials tested.

Table 5.69 Relative Chemical Class Composition of the H-Coal Distillate Fractions

		Weight Distribution (%)[a]					
Fraction	Compound Type	HO : LO Blend[b] (L3243 : L3242)[c]	650–700°F (N3201)	700–750°F (N3202)	750–800°F (N3203)	800–850°F (N3204)	>850°F (N3205)
A_1	AHs	38.9 ± 1.6	15.8 ± 0.5	11.2 ± 0.6	16.4 ± 1.6	16.6 ± 1.2	3.1 ± 0.6
A	Neutral PAHs	36.0 ± 1.6	54.5 ± 0.3	59.1 ± 0.6	50.4 ± 1.9	44.5 ± 0.2	40.1 ± 0.1
A	N-PACs	8.7 ± 0.2	14.1 ± 0.7	13.9 ± 0.6	15.9 ± 1.2	21.4 ± 4.0	31.4 ± 0.6
A_4	Hydroxy-PAHs	15.6 ± 0.3	15.3 ± 0.5	15.6 ± 0.2	18.0 ± 0.1	20.0 ± 0.9	25.7 ± 1.2

AHs = aliphatic hydrocarbons; HO : LO = heavy oil : light oil, N-PAC = nitrogen-containing polycyclic aromatic compounds; PAHs = polycyclic aromatic hydrocarbons.

[a] Values reported are the average of two or three separate analyses and are normalized for 100% recovery.

[b] HO/LO blend at 1.4 : 1.0; full-boiling-range material (Sample L3233).

[c] Sample number given in parentheses.

Source: Wilson et al. (1985).

Table 5.70 Quantitative Determinations of Selected Components in the H-Coal (Sample L3233) 50°F Boiling-Point-Cut Materials

Peak No.	Mol. Wt.	Compound[a]	Concentration (parts per thousand) 650°–700°F	700°–750°F	750°–800°F	800°–850°F	>860°F
1	180	1-Methylfluorene	0.20[b]	—	—	—	—
2	178	Phenanthrene	0.64	—	—	—	—
3	192	3-Methylphenanthrene	3.59	—	—	—	—
4	192	2-Methylphenanthrene	7.3	0.07	—	—	—
5	192	9- or 4-Methylphenanthrene	2.76	ND	—	—	—
6	192	1-Methylphenanthrene	2.75	ND	—	—	—
7	208	C_2-Fluorenes	ND[c]	ND	—	—	—
8	204	2-Phenylnaphthalene	12.21	ND	—	—	—
9	206	C_2-Phenanthrenes	ND	ND	—	—	—
10	222	C_3-Fluorene	ND	ND	—	—	—
11	204	Dihydrofluoranthene	18.32	0.32	—	—	—
12	202	Fluoranthene	5.24	1.07	—	—	—
13	218	C_1-Phenylnaphthalene	6.05	ND	—	—	—
14	202	Pyrene	16.5	58.7	3.76	—	—
15	204	Dihydropyrene	2.50	4.53	0.20	—	—
16	220	C_3-Phenanthrenes	ND	ND	—	—	—
17	216	C_1-Pyrene/fluoranthene	1.24	8.32	—	—	—
18	216	Benzo[a]fluorene	—	5.77	4.02	—	—
19	216	Benzo[b]fluorene[d]	—	48.97	27.25	—	—
20	218	Dihydro 216	—	24.49	10.2	—	—
21	216	C_1-Pyrene/fluoranthene	—	ND	—	—	—
22	216	1-Methylpyrene	—	3.41	2.52	—	—
23	232	C_2-Phenylnaphthalene	—	ND	—	—	—
24	234	Benzonaphthothiophenes	—	ND	—	—	—
25	230	C_1-Benzofluorenes, C_1-pyrenes[e]	—	14.40	20.65	0.11	—
26	232	C_1-Dihydro 216	—	ND	—	—	—
27	232	C_1-Dihydro 216	—	2.1	5.6	0.11	—
28	244	C_2-Benzofluorene, C_3-pyrene	—	0.95	3.95	0.47	—

Table 5.70 (*Continued*)

Peak No.	Mol. Wt.	Compound[a]	Concentration (parts per thousand)				
			650°–700°F	700°–750°F	750°–800°F	800°–850°F	>860°F
29	228	Benz[a]apthracene	—	0.49	1.31	1.03	0.24
30	228, 224	Chrysene[f]	—	1.11	5.7	2.57	0.53
31	242	C_1-Benz[a]anthraccnc	—	ND	—	—	—
32	258	C_3-Benzofluorene, C_4-pyrene	—	ND	3.59	7.26	0.64
33	242	3-Methylchrysene	—	ND	1.39	3.44	0.75
34	242	2-Methylchrysene	—	ND	0.69	1.38	0.36
35	242	6- or 4-Methylchrysene	—	0.32	3.02	14.70	2.51
36	242	1-Methylchrysene	—	—	—	1.08	0.30
37	256	C_2-Chrysenes/benzanthracenes[e]	—	—	—	0.46	1.32
38	272	C_4-Benzofluorene, C_5-pyrene	—	—	—	1.09	0.35
39	256	C_2-Chrysene/benzanthracene	—	—	—	3.27	1.17
40	252	Benzo[j or b]fluoranthene	—	—	—	1.91	1.63
41	252	Benzo[k]fluoranthene	—	—	—	1.12	0.93
42	252	Benzo[e]pyrene	—	—	—	1.46	1.45
43	252	Benzo[a]pyrene	—	—	—	1.02	1.56
44	276	Indeno[1,2,3-c,d]pyrene	—	—	—	0.17	1.42
45	278	Dibenz[a,c or a,h]anthracene	—	—	—	—	0.53
46	276	Benzo[g,h,i]perylene	—	—	—	0.19	2.16

[a] Tentative identifications based on retention time and gas chromatography–mass spectrometry data.
[b] Based on three determinations.
[c] No quantitative data.
[d] Coelutes with 2- or 4-methylpyrene.
[e] Quantification is the composite of two peaks.
[f] Coelutes with C_2-benzofluorene or C_3-pyrene.

Source: Wilson et al. (1985).

Table 5.71 Quantitative Determinations of Selected Components in the Nitrogen-Containing Polycyclic Aromatic Compound Fractions of the H-Coal (Sample L3233) 50°F Boiling-Point-Cut Materials

Peak No.	Mol. Wt.	Compound[a]	Concentration (Parts per Thousand) 650°–700°F	700°–750°F	750°–800°F	800°–850°F
1	179	Acridine	0.385 ± 0.159[b]	—	—	—
2	179	Benzo[f]quinoline, phenanthridine	2.07 ± 0.68	—	—	—
3	167	Carbazole	6.07 ± 1.36	—	—	—
4	193	2- or 3-Methylbenzo[f]quinoline	14.0 ± 3.8	—	—	—
5	193	C_1-Benzoquinolines	—[c]	—	—	—
6	181	1-Methylcarbazole	10.7 ± 0.5	—	—	—
7	193	C_1-Benzoquinoline	10.5 ± 4.2	—	—	—
8	181	3-Methylcarbazole	47.5 ± 1.1	0.294 ± 0.035	—	—
9	181	2-Methylcarbazole	35.1 ± 2.3	0.099	—	—
10	181	4-Methylcarbazole	18.1 ± 1.9	—	—	—
11	211	C_4-Azabiphenyls	—	—	—	—
12	207	C_2-Benzoquinoline	6.36 ± 2.72	0.312 ± 0.034	—	—
13	207, 195	C_2-Benzoquinolines, C_2-carbazoles	—	—	—	—
14	195	C_2-Carbazole	6.25 ± 1.82	—	—	—
15	195	C_2-Carbazole	11.6 ± 1.7	—	—	—
16	195	C_2-Carbazole	9.95 ± 1.42	—	—	—
17	203	1- or 4-Azapyrene	—	25.7 ± 0.7	22.2 ± 0.3	—
18	209	C_2-Carbazole	—	7.75 ± 0.25	—	—
19	221, 209	C_3-Benzoquinolines, C_3-carbazoles	—	—	—	—
20	223	C_4-Carbazoles	—	—	—	—
21	217	C_1-Azapyrene/fluoranthene	—	5.94 ± 0.53	9.67 ±1.13	—
22	217	C_1-Azapyrene/fluoranthene	—	5.53 ± 0.46	6.55 ± 1.90	—
23	223	C_4-Carbazole	—	5.56 ± 0.28	—	—
24	235	C_4-Benzoquinoline	—	—	—	—
25	223	C_4-Carbazole	—	—	—	—
26	217	C_1-Azapyrene/fluoranthene	—	4.22 ± 0.24	13.8 ± 1.7	1.17 ± 0.14
27	223	C_4-Carbazole	—	4.50	—	—
28	217	C_1-Azapyrene/fluoranthene	—	—	18.3 ±2.8	1.65 ± 0.17

Table 5.71 (*Continued*)

Peak No.	Mol. Wt.	Compound[a]	Concentration (Parts per Thousand)			
			650°–700°F	700°–750°F	750°–800°F	800°–850°F
29	237	C_5-Carbazole	—	2.81 ± 0.06	6.12 ± 0.41	—
30	231	C_2-Azapyrenes[d]	—	—	7.07 ± 1.96	0.368 ± 0.003
31	249	C_5-Benzoquinoline	—	—	—	—
32	231	C_2-Azapyrenes	—	—	—	—
33	263	C_6-Benzoquinolines	—	—	—	—
34	217	Benzo[a]carbazole	—	—	5.65 ± 1.91	9.17 ± 1.51
35	245	C_3-Azapyrene	—	—	3.20 ± 0.15	6.51 ± 0.85
36	217	Benzo[b]carbazole	—	—	1.37 ± 0.08	9.88 ± 1.81
37	217	Benzo[c]carbazole	—	—	2.53 ± 0.46	12.1 ± 0.9
38	231	C_1-Benzocarbazole	—	—	—	5.55 ± 0.79
39	231	C_1-Benzocarbazoles	—	—	—	—

[a] Tentative identifications based on retention time and gas chromatography-mass spectrometry data.
[b] Based on three determinations.
[c] No quantitative data.
[d] Quantification is the composite of three peaks.

Source: Wilson et al. 1985.

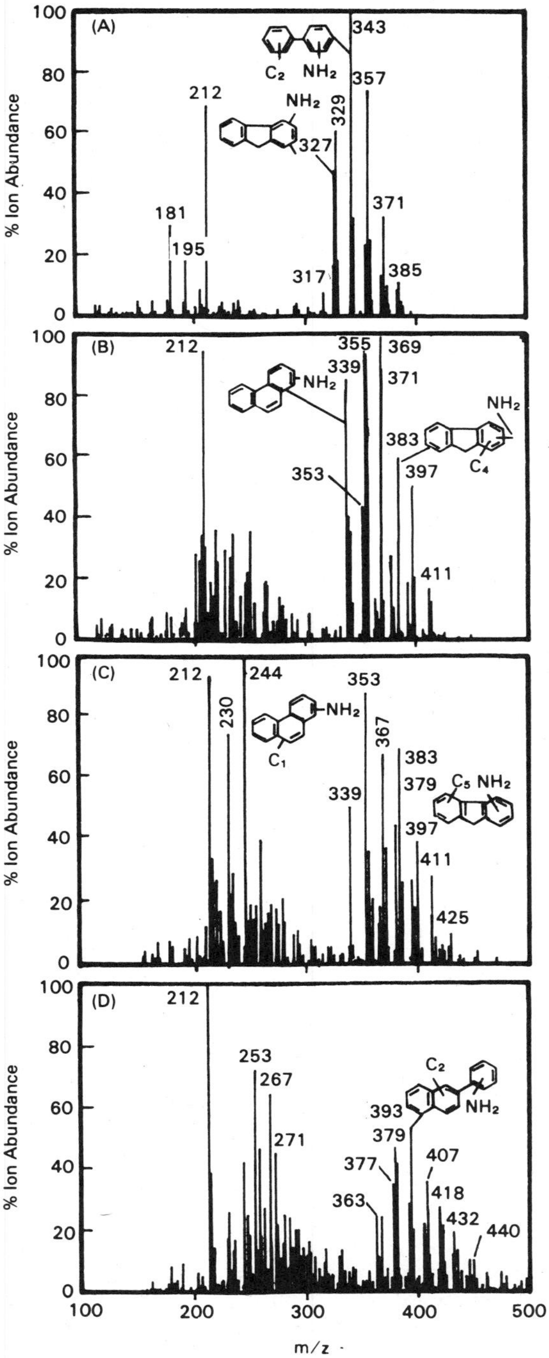

Figure 5.63 The 12-eV direct inlet probe mass spectra of selected pentafluoropropionic-derivatized amino polycyclic aromatic hydrocarbon isolates of the H-Coal (Sample L3233) 50°F distillate cuts: (A) 650–700°F; (B) 700–750°F; (C) 750–800°F; and (D) 800–850°F (from Wilson et al., 1985).

Table 5.72 Concentrations for Selected Amino-Polycyclic Aromatic Hydrocarbons (PAHs) in Nitrogen-Containing Polycyclic Aromatic Compound (N-PAC) Fractions (A_3) of the H-Coal (Sample L3233) Boiling-Point Cuts

	Amino-PAH Concentrations (μg/g N-PAC fraction)				
Compound[a]	650–700°F	700–750°F	750–800°F	800–850°F	>850°F
3-Methyl-2-aminonaphthalene	28	—	—	—	—
4-Aminobiphenyl	44	—	—	—	—
4-Aminofluorene	570	—	—	—	—
2-Aminofluorene	764	174	—	—	—
9-Aminophenanthrene	—	1458	42	—	—
3-Aminophenanthrene	—	2915	693	—	—
2-Aminoanthracene	—	2954	1031	—	—
3-Aminofluoranthene	—	291	899	69	—
2-Aminopyrene	927	—	365	—	—
1-Aminopyrene	—	736	413	—	—
8-Aminofluoranthene	—	—	863	3239	—
6-Aminochrysene	—	—	62	311	56
3-Aminochrysene	—	—	—	404	177
7-Aminobenzo[a]pyrene	—	—	—	62	362

[a] Peak identification assigned by gas chromatography/electron-capture detection retention data and confirmed by negative-ion chemical ionization gas chromatography–mass spectrometry.

Source: Wilson et al. (1985).

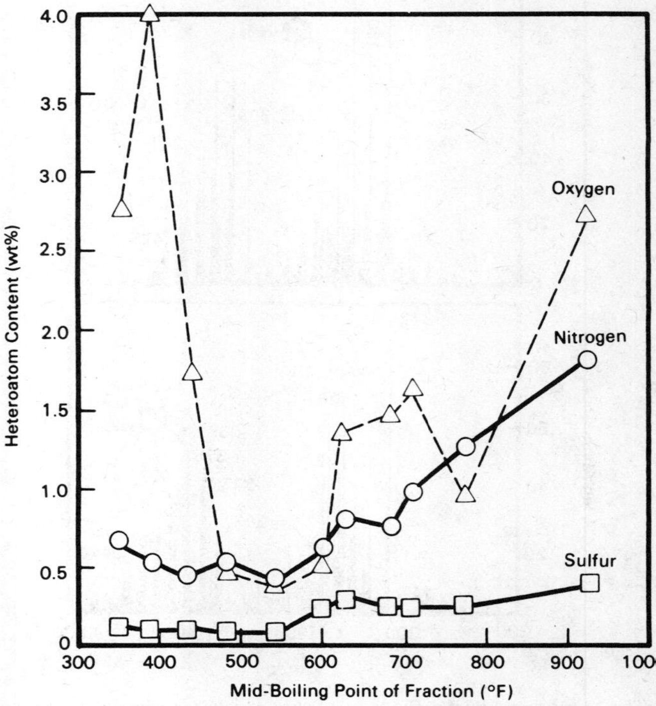

Figure 5.64 Distribution of heteroatom content in H-Coal (Sample L3233) distillation fractions (from Wilson et al., 1985).

As predicted, based on these data, mutagenic activity increased with bp temperature from about 650 up to 850°F, then decreased slightly in the >850°F material.

5.13.3.5 *Lummus ITSL*

Fractionally distilled coal liquids from the Lummus ITSL process were assayed for microbial mutagenicity (Wilson et al., 1983a, 1984). The <650°F material and its 150°F bp cuts were essentially inactive; the N-PAC fraction from the 500–650°F cut showed slight activity. Therefore, chemical analyses were performed only on <650°F, 650–850°F, and >850°F bp cuts.

Amino-PAH concentration in >650°F ITSL materials was low or undetectable. AH and PAH fractions contained certain hydroaromatic species not detected in single-stage and hybrid process distillate cuts.

5.14 EFFECTS OF POSTPRODUCTION HYDROGENATION ON CHEMISTRY OF COAL LIQUIDS

Postproduction catalytic hydrogenation of various direct-liquefaction materials was evaluated to improve their value as fuels or feedstocks and reduce their toxicity. Catalytic hydrotreatment of coal liquids generally lowers their aromaticity, heteroatom content, and molecular weight. As a result of catalytic hydrotreatment, alkylation of PACs and relative concentrations of highly stable PAHs such as pericondensed pyrenes often increase. The hydroaromatic content of all PAH structures also increases (Wilson et al., 1981b, 1983a; Buchanan et al., 1983; Griest et al., 1983). Although partially hydrogenated PAHs comprise most of the hydroaromatics found in coal liquids, partially saturated heteroatomic species have been determined (Later et al., 1984b). Reactions leading to these effects are shown in Fig. 5.65.

Although there may be substantial differences in catalyst type and reaction severity between in-stream and postproduction hydrotreatment, major determinants of chemical composition of products from both hydrotreatment techniques are probably the characteristics of the feed stream itself. A primary function of in-process hydrotreaters is coal conversion. Thus, feed streams to these units generally contain substantial amounts of unconverted heavy-end (>850°F) materials. For example, in the ITSL process at Wilsonville, Alabama, operators estimate that about half the coal-to-distillate conversion occurs in the H-Oil hydrotreater. In contrast, postproduction hydrotreatment is used exclusively for product upgrading; thus, feedstocks are often distillates ranging from 400°F to 600°F and contain little or no >850°F material and essentially no unconverted materials.

Wilson et al. (1981b) showed that mutagenicity of a catalytically hydrotreated SRC-II blend (2.9:1, MD:HD), designated FOB, was substantially reduced compared to that of the nonhydrotreated feedstock. Chemical analyses showed that severe hydrotreatment (1870 SCF H_2/bbl feed) lowered phenolic (from about 13 to 1.7 wt %), PAH (from about 11 to 0.75 wt %), and N-PAC content (from about 2.8 to 0.1 wt %). Reductions in N-PACs generally result in decreased mutagenic activity and reflect lowering of amino-PAH concentrations, with concurrent formation of ammonia gas and alkylated aryl compound (Fig. 5.65). Likewise, ring opening during hydrotreatment lowers concentrations of PAHs and reduces carcinogenic activity.

Catalytic hydrogenation of several H-Coal and EDS materials and the SRC-II FOB was performed by Universal Oil Products. Results, including some MS data for raw and upgraded materials, are contained in monthly reports (Tan and deRosset, 1978).

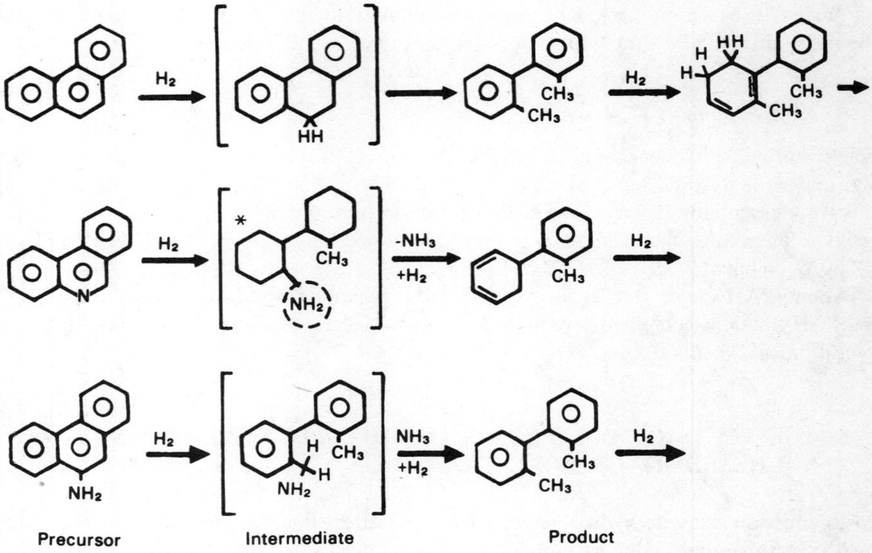

Figure 5.65 Reaction sequences in hydrogenation of polycyclic aromatic hydrocarbons and nitrogen-containing polycyclic aromatic compounds. Bracketed intermediates for amino-PAHs are thought to be short-lived, particularly under catalytic conditions (from Weimer et al., 1980).

In toxicological studies of hydrotreated SRC-II and H-Coal, materials were subjected to low-, medium-, and high-severity hydrotreatment, and their toxicological properties were compared to those of starting materials and petroleum reference materials. Concentrations of selected N-PACs in feedstock and hydrotreated SRC-II blends are shown in Table 5.73.

Effects of postproduction hydrotreatment on ecotoxicity of EDS, SRC-II, and H-Coal materials were evaluated by Griest et al. (1983). SRC-II samples included 1.7 : 1 and 6.0 : 1 blends of MD : HD designated CRM-1 and Sample 916. Process data and chemical and toxicological characteristics of hydrotreated and source materials (including effects of water-washing) were reported.

In another set of studies (Griest et al., 1983), hydrotreatment was conducted at three different severities determined by the amount of nitrogen (ppm) remaining. For the SRC-II liquid, samples were hydrotreated to 5300 ppm nitrogen (low severity), 1710 ppm nitrogen (medium severity), and 505 ppm nitrogen (high severity). Corresponding values for H-Coal were 2650, 535, and 0.2 ppm nitrogen. Table 5.74 shows the reduction in concentrations of PAH compounds from hydrotreatment. Substantial reductions in acidic, basic, and neutral polar compounds also occurred, with relative increases in volatile and neutral saturates after hydrotreating SRC-II and H-Coal materials. More recent data from studies on these materials are contained in Guerin et al. (1986).

Exxon Research & Engineering Company blended several EDS fuels to evaluate hydrogenation and combustion. Chemical data are available for these materials, including their polynuclear aromatic content and polynuclear aromatic emissions from laboratory combustion tests (Epperly, 1982). Table 5.75 shows the PAH contents of prehydrotreated and hydrotreated EDS materials derived from Wyodak coals.

Table 5.73 Gas Chromatography/Nitrogen/Phosphorus Detector Screening for Organonitrogen (μg/g oil)

	EDS		SRC-II MD:HD				
					HDT		
Compound	Blend 923	HDT 924	Blend 915[a] (L1261)[b]	Blend AWW 916 (L1262)	Low 917 (H0201)	Mid 918 (H0202)	High 919 (H0203)
Pyridine	—	—	—	—	20	3	4
3-Methylpyridine; 2,6-dimethylpyridine	—	—	200	190	72	—	—
2,5- + 2,3-Dimethylpyridine; 3-ethylpridine	—	—	300	360	150	20	47
3,5-Dimethylpyridine; 2,4,6-trimethylpyridine	2	2	—	3100	600	140	5
Aniline	0.6	1	60	—	—	—	—
3,4-Dimethylpyridine	—	4	—	—	—	4	50
o-Methylaniline	—	1	3000	3200	2100	1050	400
m- + *p*-Methylaniline	4	—	—	2600	800	700	—
2,4- + 2,5-Dimethylaniline	—	—	4200	4400	2300	1400	300
2,4,6-Trimethylaniline	—	0.3	520	410	200	30	13
2,3-Dimethylaniline	1	—	—	—	—	—	—
Quinoline	1	0.3	1300	1300	200	—	—
Isoquin, indole, 1-methylindole	—	2	3100	300	550	100	—
8-Methylquinoline	—	0.4	800	900	9	14	1
Tetrahydroquinoline	—	—	400	430	220	130	60
1-Methylisoquinoline	1	—	200	230	240	13	—
3-Methyl + 6-methylquinoline, 7-methylindole	0.1	0.3	900	800	40	10	—
2-Methylindole	—	—	1500	1500	520	210	50
1,2-Dimethylindole; 2,7-dimethylquinoline	0.4	0.1	2000	1900	—	180	50
2,3-Dimethylindole	1	0.1	—	—	—	—	—
1-Aminonaphthalene	—	0.2	160	180	50	60	40
2-Aminonaphthalene	0.3	—	2200	2100	180	160	70
2,3,5-Trimethylindole	0.5	0.2	—	—	10	—	—
Acridine	10	0.8	154	140	400	111	20
Carbazole	2	1	1600	1500	400	300	10
n-Ethylcarbazole	61	0.1	21	30	33	—	4
2-Aminofluorene	—	0.2	—	—	8	5	—
2-Aminoanthracene	8	0.5	200	180	6	—	—
2,3-Benzcarbazole	40	—	120	93	41	20	—
2-Aminofluoranthene	11	—	90	90	15	20	—
1-Aminopyrene	43	—	—	130	—	40	—
2-Aminochrysene	—	—	—	—	—	—	—

AWW = after water wash, HD = heavy distillate, HDT = hydrotreated, MD = middle distillate, SRC = solvent refined coal.

[a] Oak Ridge National Laboratory repository number.

[b] Sample number given in parentheses.

Source: B.A. Tomkins and W.H. Griest (unpublished data).

Table 5.74 Polycyclic Aromatic Hydrocarbon Isolates in Gas Chromatography Analyses ($\mu g/g$)

			Mobile/EPRI H-Coal Distillate			
Compound	Mol. W.	Carcino-genicity	Raw 1601[a] (L3232)[b]	HDT/L 1602 (H0204)	HDT/M 1603 (H0205)	HDT/H 1604 (H0206)
Benz[g,h,l]fluoranthene	226	0	—	1400	—	—
Benz[c]phenanthrene	226	+	620	—	26	17
9-Phenylanthracene	228	0	330	—	180	83
Benz[a]anthracene	228	+	100	170	130	120
Chrysene	228	0/+	100	—	—	30
2-Methylchrysene	242	+	—	20	17	—
Benzo[b]fluoranthene	252	++	100	—	17	—
Benzo[a]pyrene	252	++	—	—	—	—
Perylene	252	0	40	15	9	2
9,10-Diphenylanthracene	330	0	20	27	—	—
3-Methylcholanthrene	268	++	250	24	57	60
Dibenz[a,h/a,c]anthracenes	278	+	12	—	13	9
Picene	278	0	4	3	—	—
Benz[g,h,l]perylene	276	+	15	1	9	40
Anthanthrene	276	0	2	3	13	—

EPRI = Electric Power Research Institute HDT/L = low severity hydrotreatment, HDT/M = moderate severity hydrotreatment, HDT/H = high severity hydrotreatment.

[a] Oak Ridge National Laboratory repository number.

[b] Sample number given in parentheses.

Source: B.A. Tomkins and W. H. Griest (unpublished data).

5.15 COMPARISON OF COAL-, SHALE-, AND PETROLEUM-DERIVED MATERIALS

Compared to petroleum-derived fuels, coal liquids are more aromatic and have higher nitrogen content. While petroleums are composed primarily of saturates, including aliphatics and naphthenics, coal liquids are composed primarily of PACs (Table 5.66). While the nitrogen content of petroleum oils is generally low, sulfur content may vary over an order of magnitude or more and is present mainly in the form of thiopenes. The PAH content in petroleum distillates is lower than for nonhydrotreated coal liquids and is likely to be lower than that for hydrotreated coal-derived materials.

The toxicology of petroleum distillates relative to that of coal liquids is consistent overall with the relative chemical composition of these two materials. Thus, petroleum crudes are generally found to be carcinogenic in chronic skin-painting assays; this is presumably due to their neutral PAH content and possibly the contributing promotional effects of long-chain hydrocarbons. In the microbial assays, petroleum crudes are not particularly active, a characteristic attributable to their lack of significant N-PACs, particularly their amino-PAH content. Available data indicate that crude petroleums are less potent than coal liquids of comparable distillation temperature range in the chronic skin-painting assays.

Shale-derived oils are by-products of kerogen pyrolysis and, as such, are generally more highly oxygenated than coal-derived oils. Shale oils contain many more individual compounds and isomers than do coal-derived materials of comparable distillation temperature range. PACs in shales are highly alkylated, compared to both coal- and petroleum-derived

Table 5.75 Polycyclic Aromatic Hydrocarbon Content of EDS Fuel

Compound	Illinois 350/1000°F	Illinois >400°F	Wyodak 350/1000°F	Wyodak >350°F	Illinois 350/1000°F Plus Wyodak Coker Liquids	Hydrotreated Wyodak 350/1000°F	Hydortreated Wyodak >350°F
Three-Ring							
Fluorene	3,900	2,640	9,120	1,710	1,530	3,800	976
Phenanthrene	11,050	9,410	40,670	12,650	5,340	2,010	557
Anthracene	800	666	474	474	442	8.5	<11
Four-Ring							
Fluoranthene	2,570	1,320	4,300	3,620	1,820	138	95.5
Pyrene	4,120	1,200	6,950	6,810	2,920	2,590	1,600
Triphenylene	112	181	358	241	82	11.4	4.2
Chrysene	448	270	774	924	382	10.4	40.5
Benz[a]anthracene	398	411	1,165	1,530	438	5.6	4.5
Methyl	229	285	579	862	200	<2.8	<2.2
Dimethyl/ethyl	139	2.5	174	329	70	<2.8	<2.2
Five-Ring							
Benzo[g,h,i]fluoranthene	<100	13.3	119	214	<32	<15.2	<12
Benzo[b]fluoranthene	125	111	298	147	174	<6.3	<5
Benzo[g]fluoranthene	<119	58.9	142	91	64	<12.1	<9
Benzo[h]fluoranthene	115	65.6	210	41	136	<9	<7
Perylene	92	128	244	554	132	<6	<5
Benzo[a]pyrene	329	406	834	1,320	282	7.2	13.6
Benzo[e]pyrene	191	216	802	629	208	84.5	14.8
Methylbenzo[a]pyrene	239	495	337	802	236	<10	<8
Methylbenzo[e]pyrene	85	192	176	223	88	102	182
Bibenzo[a,h]anthracene	22	1.7	<19	20	<24	<2	<1.5
Six-Ring							
1,2,3-Indeno[c,d]pyrene	9.1	10.0	10.8	11.1	8	<1	<1
Benzo[g,h,i]perylene	71	119	217	154	100	235	390
Coronene	<4	0.9	<10.4	10.4	<16	23	29
Total	25,000	18,200	67,900	33,400	14,700	9,960	3,740

Source: Adapted from Epperly (1982).

Table 5.76 Comparison of Composition for 550–700°F Distillate Cuts from a Coal Liquid and Two Petroleum Crudes

Compound	Coal Liquid	Petroleum A	Petroleum B
Hydrocarbons			
Saturates	14	56	64
Aromatics	77	42	33
Other	9	2	3
Nonhydrocarbon Elements			
Nitrogen	0.59	0.003	0.007
Oxygen	0.6	0.1	<0.05
Sulfur	0.19	—	2.2

materials, and lower-temperature distillates are characterized by extended series of alkyl homologs for low-molecular-weight heteroatomic compounds, including pyrroles, phenols, anilines, and thiophenes. Shale oils also contain more multiheteroatomic and multifunctional compounds than either coal or petroleum fuels.

The carcinogenic activity of shale oils tested has generally been greater than that of commonly used crude petroleum crudes but less than that of nonhydrotreated coal liquids of comparable distillation temperature range (Table 5.76). The table is a compilation from selected published literature tables, illustrating the range of compositions found for coal, shale, and petroleum materials. As is the case for coal liquids, hydrotreatment significantly reduces the carcinogenic activity of shale oils.

The mutagenic activity of raw shale oils is greater than that of petroleum crudes, but generally less than that of nonhydrotreated coal liquids. This difference appears to arise from the lower amino-PAH content of the shales compared to coal liquids. While restricting the upper distillation temperature cut point for coal liquids has a significant effect on the mutagenic and carcinogenic activity of these materials, analogous distillations of Paraho shale oils have shown no such effect on mutagenicity.

As is the case for petroleum crudes, coal liquids from different processes may vary significantly relative to the main compound classes of which they are composed. An objective of this chapter has been to document these differences with regard to organic constituency (for more detail, consult the cited literature). However, compositional differences found in nonhydrotreated coal liquids do not preclude them from being distinguished chemically from both petroleum- and shale-derived materials using instrumental analytical techniques.

5.16 ANALYTICAL CHEMISTRY OF COAL-CONVERSION MATERIALS RELATED TO TOXICOLOGICAL, ENVIRONMENTAL, AND ENGINEERING ISSUES

5.16.1 Determination of Toxicologically Important Chemical Classes in Coal-Conversion Materials

Biodirected analytical chemistry investigations reviewed in this chapter have centered mainly on two classes of compounds: the PAH and the amino-PAH. This arises from the activity of these two classes of compounds in the toxicological assays (see Chapter 3) used to evaluate coal-conversion materials. This section provides some detail on the work leading to identification of these compound classes as important toxicants.

5.16.1.1 *Identification of PAHs as Mutagens and Carcinogens in Coal-Conversion Materials*

The conclusion that PAHs are the major contributors to the mouse skin tumorigenicity of coal-conversion materials has been supported by dermal carcinogenesis and tumor-initiating assays and by short-term mutagenesis assays using both bacterial and mammalian cells in vitro. For experiments in which PAH subfractions were used, the specific importance of the four- and five-ring PAHs is apparent. This conclusion is consistent with earlier studies of both coal tars and cigarette-smoke condensate; PAHs are principally responsible for the dermal carcinogenicity of both.

Three independent approaches have provided complementary and confirming evidence for the importance of four- to six-ring PAHs for the dermal carcinogenicity of coal-conversion materials. Each approach included fractionation, chemical analysis, assays of dermal carcinogenesis or dermal tumor-initiating activity, and cellular assay for mutagenic activity.

Holland et al. (1981) showed that the dermal carcinogenicities of the respective PAH-enriched fractions of crude Synthoil and shale oil were substantially greater than for the crudes. More-defined fractionation and chemical characterization of carcinogenic H-Coal and SRC-II oils provided fractions containing predominantly two- to five-ring PAHs and their alkyl homologs, along wih less than 10% (w/w) of 3°-N-PACs (principally dihydroindoles, indoles, carbazoles, and benzocarbazoles) and traces of aminoindenes, aminofluorenes, quinolines, and acridines. The PAH-enriched fraction was obtained by subsequent aqueous/organic liquid-partitioning (neutral fraction), Sephadex LH-20 column chromatography (polycyclic aromatic plus polar), and silica-column chromatography (PAHs plus neutral nitrogen compounds). For both oils, the tumorigenic activity was clearly concentrated in the PAH fraction, while the neutral polar and aliphatic fractions had relatively little or no activity, respectively.

In other studies, SRC-I and SRC-II distillates were fractionated by alumina-column chromatography to obtain four fractions highly enriched in AHs, PAHs, N-PACs, and O-PACs, respectively. The AH and O-PAC fractions of three distillates (SRC-I, 800 to >850°F; SRC-II, 800 to >850°F and >850°F) had little dermal tumor-initiating activity in mice. Although the N-PAC fractions had some initiating activity, the activity was clearly concentrated in the respective PAH fractions (Mahlum et al., 1984). For each of these fractions, detailed chemical analysis showed that the predominant components were four- to six-ring PAHs and their alkyl homologs, including compounds known to be carcinogens when examined individually. Because of the distillation process, the relative concentration of the generally noncarcinogenic two- and three-ring PAHs was negligible to minor. Further resolution of the PAH fractions by HPLC (Lucke et al., 1985) showed that fractions containing benzo[a]pyrene were most active as initiators of dermal tumorigenesis.

As determined by chronic treatment of mice, the neutral, PAH-containing fraction of a coal-gasification condensate was substantially more tumorigenic than the corresponding basic or acidic fractions obtained by aqueous/organic liquid partitioning (Cunningham et al., 1985). The neutral fraction was resolved by silica-column chromatography to obtain two fractions containing, respectively, (1) aliphatic and aromatic hydrocarbons and (2) more-polar heteroatomic aromatic compounds. The aliphatic/aromatic fraction was substantially more active as a dermal tumor initiator than the polar neutral fraction (Haugen et al., 1986).

The aliphatic/aromatic fraction contained two- to five-ring PAHs and their alkyl homologs, as well as alkanes and alkenes with up to 35 carbon atoms.

It is apparent that results of independent approaches to studying different coal-conversion materials consistently demonstrate that four- to six-ring PAHs are primarily responsible for dermal tumorigenesis. Because of the effort and time required for tumorigenesis assays, short-term in vitro mutagenicity assays have also been used to identify toxicologically important components of coal-conversion materials. The two in vitro assays most widely used are measures of mutagenicity in bacteria (*S. typhimurium*) and in mammalian cells (CHO cells). For both of these systems, rat liver enzymes are added for metabolic activation. In general, the results of these assays are consistent with the identification of PAHs as the principal dermal carcinogens. However, these assays also revealed the presence of amino-PAHs, which are highly mutagenic in the Ames assay.

When evaluated individually, most carcinogenic PAHs have readily measurable mutagenic activity in the Ames assay. However, the activity of mutagenic PAHs is suppressed in the presence of complex mixtures of PACs (Pelroy and Petersen, 1979; Haugen et al., 1981a). Thus, some investigators have observed little or no bacterial mutagenicity in highly enriched PAH fractions of coal liquids (Later et al., 1983a; Wilson et al., 1984). However, measurement of mutagenicity in CHO cells showed that the PAH (and to a lesser degree) the N-PAC alumina-column fractions of SRC-II HD were the most active (Wilson et al., 1984). Further subfractionation by HPLC and bioassay of the PAH fraction (which has negligible bacterial mutagenicity) allowed five-ring PAHs to be detected as the most active bacterial mutagens.

Silica-column chromatography of the neutral fractions of coal-gasification condensate and a coal liquid (Synthoil) showed that bacterial mutagenicity was associated with fractions containing three- to five-ring PAHs (Ho et al., 1980a, b; Haugen et al., 1983; Haugen and Stamoudis, in press). Further chromatographic subfractionation, isolation, and bioassay showed that four- to five-ring PAHs were the predominant mutagens as measured by both the bacterial and mammalian cell assays (Haugen et al., 1983; Cunningham et al., 1985; Haugen and Stamoudis, 1986).

In summary, PAHs are the principal dermal tumorigens in coal-conversion materials. The importance of PAHs is also revealed by their predominance as mutagens for CHO cells. When PAHs are sufficiently purified to remove interfering components, their bacterial mutagenicity is also revealed.

5.16.1.2 *Determination of Amino-PAHs as Principal Mutagens*

The specific mutagenicities of basic fractions, obtained by liquid partitioning, and of N-PAC chromatographic fractions, as measured in the Ames assay, were typically 10- to 50-fold greater than those of unfractionated coal-conversion materials (Epler et al., 1978a, b; Guerin et al., 1981; Kimball and Munro, 1981; Later et al., 1982b; Mahlum et al., 1983; Stamoudis et al., 1983). The mutagenicity in the fractions accounted for 30–100% of the activity detected in the unfractionated materials. Basic and chromatographic N-PAC fractions of coal-conversion materials are also (1) mutagenic for CHO cells (Hsie et al., 1980; Wilson et al., 1984; Cunningham et al., 1985), (2) initiators of dermal tumorigenesis (Mahlum et al., 1984), and (3) dermal carcinogens as demonstrated by long-term chronic treatment (Renne et al., 1981; Cunningham et al., 1985). The tumorigenic activity of the PAH-containing fractions, however, was considerably greater.

Basic fractions obtained by liquid partitioning of coal-conversion materials are composed principally of amino-PAHs, azaarenes, and their respective alkyl homologs, with the azaarenes predominating. Nitrous acid, which converts amino-PAHs to the corresponding phenols, markedly diminished the mutagenicity of basic fractions of coal liquids and gasifier condensates (Haugen et al., 1981b; Pelroy and Stewart, 1981) as well as the dermal tumor-initiating activity of chromatographic N-PAC fractions (Mahlum et al., 1983). A purified enzyme, amine oxidase, was used for metabolic activation of basic fractions as measured in the *Salmonella* assay (Pelroy and Gandolfi, 1980). These two observations provided indirect evidence that the amino-PAHs, not the azaarenes, were principally responsible for the bacterial mutagenicity.

Fractionations of SRC-II coal liquids, other synthetic coal liquids, and coal-gasification condensates, using thin-layer chromatography (Wilson et al., 1980) and various column chromatographic procedures, each demonstrated that three- to five-ring amino-PAHs were principally responsible for the high specific mutagenicity of the basic components. The diversity of the methods, summarized below, provides strong complementary evidence for the contribution of the amino-PAHs.

Alumina-column chromatography of a basic fraction of a synthetic coal liquid yielded two subfractions. The most mutagenic fraction was enriched in aromatic amines (Guerin et al., 1980; Buchanan et al., 1983) and was further resolved by chromatography on Sephadex LH-20. Most of the mutagenicity was isolated in a fraction containing less than 5% of the basic components and was composed primarily of one- to four-ring alkylated aromatic amines. Additional fractionation and bioassay showed that the mutagenicity was associated mostly with three- to five-ring aromatic amines.

Basic fractions obtained by liquid–liquid partitioning (Boparai et al., 1983; Stamoudis et al., 1983) of coal-gasification and -liquefaction materials were resolved by semipreparative cation-exchange HPLC according to base strength (Haugen et al., 1982a, b; Stamoudis et al., 1983). Chromatographic fractions enriched in amino-PAHs, but not in azaarenes, were principally responsible for the mutagenicity. For some materials, a characteristic mutagenic chromatographic fraction predominated. Further resolution of this fraction by reverse-phase HPLC showed that highly mutagenic three-ring aminophenanthrenes were responsible for the activity (Stamoudis et al., 1983; Haugen et al., 1986).

Alumina-column chromatography of previously unfractionated coal-liquid distillates showed that bacterial mutagenicity was detected almost exclusively in the N-PAC fraction containing amino-PAHs, azaarenes, and 2°-PANHs (e.g., carbazoles) (Later and Lee, 1983; Later et al., 1983a; Wilson et al., 1984). The N-PAC fractions were also mutagenic in mammalian-cell cultures (Mahlum et al., 1983; Wilson et al., 1983a, 1984) and were initiators of dermal tumorigenesis (Mahlum et al., 1984). The N-PAC fraction was further resolved by silica-column chromatography to obtain three subfractions of increasing polarity, which were enriched in secondary heterocyclic-PACs (e.g., carbazoles), amino-PAHs (e.g., aminoanthracene), and tertiary heterocyclic-PACs, respectively (Later et al., 1981a, 1982b, 1983a). The amino-PAH fraction contained the most mutagenicity, with a smaller contribution from the 3°-N-PAC fraction (Later et al., 1981b; Later and Lee, 1983; Mahlum et al., 1983).

These observations of coal-conversion materials are consistent with those of Kosuge et al. (1982), who found that three- and four-ring amino-PAHs were responsible for the bacterial mutagenicity of ''coal tar'' (presumably from a commercial process).

In summary, the evidence demonstrated that three- to five-ring amino-PAHs are responsible for the relatively high bacterial mutagenicity of the basic, nitrogen-containing components of coal-conversion materials. The mutagenicity of some materials is almost exclusively attributed to amino-PAHs. Comparison of the relative mutagenicities of two- to five-ring amino-PAHs showed that some isomers of three- and four-ring amino-PAHs are more than 1000-fold more active in the Ames assay than are the generally more abundant two-ring aminonaphthalenes and aminobiphenyls (Ho et al., 1981c; Later et al., 1984a), which are also present in coal-conversion materials. Thus, 2-aminonaphthalene and 4-aminobiphenyl, known human bladder carcinogens present in coal-conversion materials, were not detected either by the assays of bacterial mutagenicity or assays of dermal tumorigenicity.

The dermal tumorigenicity of basic and N-PAC fractions may be due to azaarenes with more than three rings (Dipple et al., 1984) or to amino-PAHs. The role of amino-PAHs is supported by the observation that 2-aminoanthracene is a dermal tumorigen (Lennox, 1958) and that the tumorigenic activity of the N-PAC fraction of an SRC-II material was diminished by treatment with nitrous acid (Mahlum et al., 1983; Pelroy et al., 1984).

5.17 CONCLUSIONS

The chapter has been concerned primarily with the chemical properties of various coal-derived materials as they relate to toxicological activity. To a lesser extent, certain process strategies have been related to chemical composition. With minor exceptions, the same classes of chemicals are present in the coal-liquefaction products and -gasification by-products studied. Gasification by-products tended to show the presence of oxidative reaction products, such as PACs with carbonyl and cyano functionalities, in addition to those heteroatomic functional groups found in PACs from the more reductive coal-liquefaction processes. Particular similarities exist in the PAH compounds found in these materials prior to upgrading. Quantitatively, however, there may be substantial differences in the relative concentrations of the compounds and chemical classes present. Hydrotreated and nonhydrotreated materials, for example, vary substantially from one another in both heteroatom content, hydroaromaticity, and aliphatic hydrocarbon content.

Distillation can be used to separate carcinogenic and mutagenic components of coal-conversion materials from those components that show no toxicological activity. For a wide variety of materials studied, distillates with boiling ranges above 700°F were toxicologically active, while those with boiling ranges below 650°F were essentially inactive.

Catacondensed, cata/pericondensed, and certain C_1–C_3 catacondensed PAHs having from four to six aromatic rings are responsible for most of the carcinogenic activity found in coal-derived materials in laboratory assay systems. In general, PAH isolates of coal-derived materials containing primarily four- to six-ring, alkylated PAHs having three or more methylene units show decreased mutagenic and tumorigenic potency compared with PAH fractions whose primary constituents are four- to six-ring C_0–C_3 PAHs. Expression of specific PAH mutagens and carcinogens can be greatly affected by the matrix in which they are found.

PAH fractions of coal-derived materials with significant levels of hydroaromatic PAH analogs resulting from process hydrotreatment generally display reduced potency in both microbial mutagenicity and mouse skin tumor-initiating assays. Hence, hydrogenation has been used to reduce the adverse toxicological effects of coal-derived materials. Mild to moderate hydrotreatment decreases the mutagenic activity of coal-conversion materials by reduction of primary amino functionalities and, to a lesser extent, by dearomatization of PAH structures through saturation and/or carbon–carbon bond cleavage. More-severe hydrotreatment significantly decreases skin carcinogenicity of these materials, primarily from dearomatization of four- to six-ring PAH structures. Process parameters aside from those that determine upper distillation cut-point temperature and H : C ratios do not appear to have a significant effect on the skin carcinogenicity of coal-conversion materials.

Review of the literature indicates that the presence of amino-PAH in a coal liquid is readily detected by a positive response in the microbial mutagenicity assay, using *S. typhimurium* TA98 or other strains; in most cases, regardless of the matrix composition. Hence, amino-PAH are expressed from a variety of coal-conversion materials, including single- and two-stage liquefaction products and gasification condensates. In microbial mutagenicity assays, three- and four-ring amino-PAHs are more active than those having less than three or more than four aromatic rings, with potency being a function of amino substituent position. Amino-PAHs are not as potent as PAH in dermal carcinogenicity assays. Several specific amino-PAH, such as 2-aminonaphthalene and 4-aminobiphenyl, which are not particularly active in the microbial assays, are known to be carcinogenic in humans. However, the presence of these compounds in middle distillate range materials may not be detected by either the in vitro or in vivo tests.

Alterations in process conditions (within a specific process), such as catalyst quality, reaction severity and residence time, can substantially affect the chemical composition of coal-derived liquids, particularly the toxicologically active constituents such as amino-PAH and PAH. However, the effects of process conditions are expressed predominantly through their influence on boiling point and aromaticity. Coal type may also affect the chemical composition of coal liquids. However, because coal conversion process conditions are optimized for each coal used, direct relationships between coal type and toxicological activity cannot be determined.

Other N-PAC classes also include compounds that are mutagenic and/or carcinogenic. Data suggest that these compounds, such as benzacridines, dibenzcarbazoles, dimethylbenzacridines, and dibenzacridines, also contribute to the mutagenic and carcinogenic activities of coal-conversion materials. N-PAC isolates produce observable responses in tumor-initiating and chronic skin-painting assays, but at a reduced potency compared to PAH. Differentiation of the N-PAC subclass(es) producing these responses in these assays have not been reported to date. Effects of the complex matrix upon carcinogenic expression of these compounds have not been studied.

REFERENCES

Aczel, T. A., R. B. Williams, R. J. Pancirov, and J. H. Karchmer, *Chemical Properties of Synthoil Products and Feeds, Summary of Final Report*, MERC-8007-1, National Technical Information Service, Springfield, VA, 1977.

Alexander, G., and I. Hazai, Chromatographic Fingerprinting of Coal Extracts. *J. Chromatogr.*, **217**, 19–38 (1981).

Alger, R. S., *Electron Paramagnetic Resonance: Techniques and Applications.* Wiley, New York, 1968.

Anastasia, L. J., R. J. Evans, R. H. Karst, and C. S. Juan, Environmental Assessment of the HYGAS Process, in *Final Report on HYGAS Environmental Characterization: HYGAS Tests 77 through 87*, Vol. V, FE-2433-28, National Technical Information Service, Springfield, VA, 1982.

Anbar, M., *Molecular Profile Analysis of Coal Products*, EPRI-AF-390, Electric Power Research Institute, Palo Alto, CA, 1977.

Anderson, R. P., and J. Freel, Improving Coal Liquid Quality by Heavy Distillate Recycle, in H. E. Lebowitz and R. C. Rovesti (eds.), *Proceedings of Eighth Annual EPRI Contractor's Conference on Coal Liquefaction*, EPRI AP-5366-SR, Electric Power Research Institute, Palo Alto, CA, 1983.

Anderson, H. C., and W. R. Wu, *Properties of Compounds in Coal Carbonization Products*, Bulletin 606, U.S. Bureau of Mines, Washington, DC, 1963.

Arcos, J. C., and M. F. Argus, *Chemical Induction of Cancer*, vols. IIA and IIB, Academic Press, New York, 1974.

ASTM, *Standard Classification of Coals by Rank*, ASTM D-388-77, American Society for Testing and Materials, Philadelphia, PA, 1986a.

ASTM, *Standard Definitions of Terms Relating to Megascopic Descriptions of Coals and Coal Seams and Microscopic Descriptions and Analysis of Coals*, ASTM D-2796-77, American Society for Testing and Materials, Philadelphia, PA, 1986b.

Badger, G. M., and R. G. Buttery, The Formation of Aromatic Hydrocarbons at High Temperatures. Part 4. The Pyrolysis of Styrene. *J. Chem. Soc.*, 2458–2463 (1958).

Badger, G. M., and R. W. L. Kimber, The Formation of Aromatic Hydrocarbons at High Temperatures. Part 2. An Examination of the "Schroeter" Tar. *J. Chem. Soc.*, 2453–2554 (1958a).

Badger, G. M., and R. W. L. Kimber, The Formation of Aromatic Hydrocarbons at High Temperatures. Part 3. The Pyrolysis of 1,4-Phenylbutylnaphthalene. *J. Chem. Soc.*, 2455–2458 (1958b).

Badger, G. M., J. W. Cook, C. L. Hewett, E. L. Kennaway, N. M. Kennaway, and R. H. Martin, The Production of Cancer by Pure Carcinogens. *Proc. R. Soc. Lond. Ser. B. Biol. Sci.*, **131,** 170–182 (1942).

Badger, G. M., G. R. Buttery, R. W. L. Kimber, G. E. Lewis, A. G. Moritz, and I. M. Napier, The Formation of Aromatic Hydrocarbons at High Temperatures. Part 1. Introduction. *J. Chem. Soc.*, 2449–2452 (1958).

Badger, G. M., R. W. L. Kimber, and T. N. Spotswood, Mode of Formation of 3,4-Benzopyrene in Human Environments. *Nature*, **187,** 663–665 (1960).

Barber, M., R. S. Bordoli, G. J. Elliot, R. D. Sedgwick, and A. N. Taylor, Fast Atom Bombardment Mass Spectrometry. *Anal. Chem.*, **54,** 645A (1982).

Bartle, K. D., T. G. Martin, and D. F. Williams, Chemical Nature of a Supercritical-Gas Extract of Coal at 350°C. *Fuel*, **54,** 226–235 (1975).

Bartle, K. D., M. L. Lee, and S. A. Wise, Modern Analytical Methods for Environmental Polycyclic Aromatic Compounds. *Chem. Soc. Rev.*, **10,** 113–158 (1981).

Bartle, K. D., C. Gibson, D. Mills, M. J. Mulligan, N. Taylor, T. G. Martin, and C. E. Snape, Differential-Pulse Voltammetry at the Hanging-Mercury-Drop Electrode for Identification of Aromatic Structures in Coal Extracts. *Anal. Chem.*, **54,** 1730–1733 (1982).

Bartle, K. D., N. Taylor, M. J. Mulligan, D. G. Mills, and C. Gibson, Evaporative Analyser as a Mass Detector in the Size-Exclusion Chromatography of Coal Extracts. *Fuel*, **62,** 1181–1185 (1983).

Bartle, K. D., M. J. Mulligan, N. Taylor, T. G. Martin, and C. E. Snape, Molecular Mass Calibration in Size-Exclusion Chromatography of Coal Derivatives. *Fuel*, **63,** 1556–1560 (1984).

Batenandt, A., and H. Dannenberg, *Arch. Geschwulstforsch.*, **6,** 1 (1953).

Baughman, G. L., *Synthetic Fuels Data Handbook*, 2nd ed. Cameron Engineers, Denver, CO, 1978.

Beckey, H. D., *Field Ionization Mass Spectrometry*, Pergamon Press, New York, 1971.

Bell, J. H., S. Ireland, and A. W. Spear, Identification of Aromatic Ketones in Cigarette Smoke Condensate. *Anal. Chem.*, **41,** 310–313 (1969).

Benson, P. A., S. E. Scheppele, G. J. Greenwood, T. Aczel, Q. G. Grindstaff, and B. F. Beier, *Consequences of the Mass Spectrometric and Infrared Analysis of Oils and Asphaltenes for the Chemistry of Coal Liquefaction: Characterization of Coal-Derived Liquids and Other Fossil-Fuel-Related Materials Employing Mass Spectrometry*, FE-2537-10, U.S. Department of Energy, Washington, DC, 1979.

Bloch, B., and W. Dreifuss, *Schweiz. Med. Wochenschr.*, **51,** 1033 (1921).

Bockrath, B. C., and K. T. Schroder, Coal Mineral Matter Catalysis in Liquefaction, in B. D. Blaustein, B. C. Bockrath, and S. Friedman (eds.), *New Approaches to Coal Chemistry*, American Chemical Society, Washington, DC, 1981, pp. 191–200.

Boduszynski, M. M., R. J. Hurtubise, and H. F. Silver, Separation of Solvent-Refined Coal into Solvent-Derived Fractions. *Anal. Chem.*, **54,** 372–375 (1982a).

Boduszynski, M. M., R. J. Hurtubise, and H. F. Silver, Separation of Solvent-Refined Coal into Compound-Class Fractions. *Anal. Chem.*, **54,** 375–381 (1982b).

Boduszynski, M. M., R. J. Hurtubise, T. W. Allen, and H. F. Silver, Liquid Chromatography/Field Ionization Mass Spectrometry in the Analysis of High-Boiling and Nondistillable Coal Liquids for Hydrocarbons. *Anal. Chem.*, **55,** 225–231 (1983a).

Boduszynski, M. M., R. J. Hurtubise, T. W. Allen, and H. F. Silver, Determination of Hydrocarbon Composition in High-Boiling and Nondistillable Coal Liquids by Liquid Chromatography/Field Ionization Mass Spectrometry. *Anal. Chem.*, **55,** 232–241 (1983b).

Boduszynski, M. M., R. J. Hurtubise, and H. F. Silver, Relationship Between Solvent-Derived and Compound-Class Fractions in Coal-Derived Distillates and Vacuum Still Bottoms. *Fuel*, **63,** 93–100 (1984).

Bodzek, D., T. Krzuzanowska, and A. Marzec, Heterocompounds Present in Asphaltenes from Various Products of Coal Hydrogenation. *Fuel*, **58,** 196–202 (1979).

Bodzek, D., and A. Marzec, Molecular Components of Coal and Coal Structure. *Fuel*, **60,** 47–51 (1981).

Bogowski, P., and M. Mirme, Cocarcinogenicity of Phenols from Estonian Shale Tars (Oils). *Environ. Health Perspect.*, **30,** 177–178 (1979).

Boparai, A. S., D. A. Haugen, K. M. Suhrbier, and J. F. Schneider, An Improved Procedure for Extraction of Aromatic Bases from Synfuel Materials, in C. W. Wright, W. C. Weimer, and W. D. Felix (eds.), *Advanced Techniques in Synthetic Fuels Analysis*, CONF-811160, National Technical Information Service, Springfield, VA, 1983.

Borwitzky, H., and G. Schomburg, Separation and Identification of Polynuclear Aromatic Compounds in Coal Tar by Using Glass Capillary Chromatography Including Combined Gas Chromatography–Mass Spectrometry. *J. Chromatogr.*, **170,** 99–124 (1979).

Brady, R. F., Analysis of Microgram Quantities of Organic Vapors by Combined Capillary-Column Gas Chromatography and Vapor Phase Infrared Spectrometry. *Anal. Chem.*, **47,** 1425–1428 (1975).

Brinkman, D. W., M. L. Whisman, and J. N. Bowden, *Stability Characteristics of Hydrocarbon Fuels from Alternative Sources*, BETC/RI-78/28, U.S. Department of Energy, National Technical Information Service, Springfield, VA, 1979.

Brown, J. C., J. A. Duncanson, and G. J. Small, Fluorescence Line Narrowing Spectrometry in Glasses for Direct Determination of Polycyclic Aromatic Hydrocarbons in Solvent-Refined Coal. *Anal. Chem.*, **52,** 1711–1715 (1980).

Brown, R. S., and L. Taylor, Speciation of Intermediate Polar Material in Coal Derived Process Solvents by Liquid Chromatography/Fourier Transform Infrared Spectrometry. *Anal. Chem.*, **55,** 723–773 (1983).

Brown, R. S., D. W. Hausler, L. T. Taylor, and R. C. Carter, Fourier Transform Infrared Spectrometric Detection in Size-Exclusion Chromatographic Separation of Polar Synfuel Material. *Anal. Chem.*, **53,** 197–201 (1981).

Buchanan, M. V., Mass Spectral Characterization of Nitrogen-Containing Compounds with Ammonia Chemical Ionization. *Anal. Chem.*, **54,** 570–574 (1982).

Buchanan, M. V., C.-H. Ho, M. R. Guerin, and B. R. Clark, Chemical Characterization of Mutagenic Nitrogen-Containing Polycyclic Aromatic Hydrocarbons in Fossil Fuel, in M. Cook and A. J. Dennis (eds.), *Polynuclear Aromatic Hydrocarbons: Chemical Analysis and Biological Fate*, Battelle Press, Columbus, OH, 1981, pp. 133–144.

Buchanan, M. V., M. R. Guerin, G. L. Kao, L. B. Rubin, and J. E. Caton, Biodirectional Chemical Characterization of Synfuel Materials, in C. W. Wright, W. C. Weimer, and W. D. Felix (eds.), *Advanced Techniques in Synthetic Fuels Analysis*, CONF-811160, National Technical Information Service, Springfield, VA, 1983.

Burchill, P., A. A. Herod, and E. Pritchard, Identification of Sulphur Heterocycles in Coal Tar and Pitch. *J. Chromatogr.*, **242,** 51–64 (1982a).

Burchill, P., A. A. Herod, and E. Pritchard, Determination of Nitrogen–Sulphur Mixed Heteroatomic Compounds and Sulphur Heterocycles in an Anthracene Oil. *J. Chromatogr.*, **242,** 65–76 (1982b).

Burchill, P., A. A. Herod, and E. Pritchard, Investigation of Nitrogen Compounds in Coal Tar Products. 1. Unfractionated Materials. *Fuel*, **62,** 11–19 (1983a).

Burchill, P., A. A. Herod, and E. Pritchard, Investigation of Nitrogen Compounds in Coal Tar Products. 2. Basic Fractions. *Fuel*, **62,** 20–29 (1983b).

Burke, F. P., R. A. Winschel, and D. L. Wooton, Liquid Column Fractionation: A Method of Solvent Fractionation of Coal Liquefaction and Petroleum Products. *Fuel*, **58,** 539–541 (1979).

Burke, F. P., R. A. Winschel, and T. C. Pochapsky, Composition and Performance of Distillate Recycle Solvents from the SRC-I Process. *Fuel*, **60,** 562–572 (1981).

Buryan, P., J. Macak, and V. M. Nabivach, Investigation of the Composition of Coal-Tar Phenols and Xylenols by Capillary Chromatography. *J. Chromatogr.*, **148,** 203–210 (1978).

Calkins, W. H., J. F. Deye, R. W. Hartgrove, C. F. King, and D. F. Krahn, Synthetic Crude Oils from Coal: Mutagenicity and Tumor-Initiation Screening Tests, in D. D. Mahlum, R. H. Gray, and W. D. Felix (eds.), *Coal Conversion and the Environment*, CONF-801039, National Technical Information Service, Springfield, VA, 1981, pp. 462–470.

Callen, R. B., C. A. Simpson, J. G. Bendoraitis, and S. E. Votz, Upgrading Coal Liquids to Gas Turbine Fuels. 1. Analytical Characterization of Coal Liquids. *Ind. Eng. Chem. Prod. Res. Dev.*, **15,** 223–233 (1976).

Callen, R. B., C. A. Simpson, and J. G. Bendoraitis, Analytical Characterization of Solvent-Refined Coal Comparison with Petroleum Residua, in P. C. Uden, S. Siggia, and H. B. Jensen (eds.), *Analytical Chemistry of Liquid Fuel Sources, Advances in Chemistry Series 170*, American Chemical Society, Washington, DC, 1978, pp. 307–322.

Case, R. A. M., and M. E. Hosker, Tumor of the Urinary Bladder as an Occupational Disease in the Rubber Industry in England. *Br. J. Prev. Soc. Med.*, **8,** 39–50 (1954).

Case, R. A. M., M. E. Hosker, D. B. McDonald, and J. T. Person, Tumors of the Urinary Bladder in Workmen Engaged in the Manufacture and Use of Certain Dye Intermediates in the British Chemical Industry. I. The Role of Aniline, Benzidine, 1-Naphthylamine and 2-Naphthylamine. *Br. J. Ind. Med.*, **11,** 75–104 (1954).

Chang, T.-C. L., and C. Karr, Gas–Liquid Chromatographic Analysis of Aromatic Hydrocarbons Boiling Between 202 and 280 in a Low-Temperature Coal Tar. *Anal. Chim. Acta,* **24,** 343–356 (1961).

Chang, T.-C. L., and C. Karr, Gas–Liquid Chromatographic Analysis of C_{10}–C_{16} N-Paraffins, Isoparaffins and Olefins in a Low-Temperature Coal Tar. *Anal. Chim. Acta*, **26,** 410–418 (1962).

Chmielowiec, J., Separation of Polar Compound Classes in Liquid Fossil Fuels by Liquid Chromatography with Aprotic Dipolar Solvents. *Anal. Chem.*, **55,** 2367–2374 (1983).

Ciupek, J. D., D. Zackett, R. G. Cooks, and K. V. Wood, High and Low Energy Collision Mass Spectrometry of Aza and Amino-Polynuclear Aromatic Compounds in Coal-Derived Liquids. *Anal. Chem.*, **54,** 2215–2219 (1982a).

Ciupek, J. D., D. Zackett, R. G. Cooks, and K. V. Wood, Determination of Transferable Hydrogen in Coal Liquids by Mass Spectrometry. *Anal. Chem.*, **54,** 2576–2582 (1982b).

Clar, E., *Polycyclic Hydrocarbons*, vols. 1 and 2, Academic Press, New York, 1964.

Coleman, W. M., and A. R. Boyd, Determination of Hydroxyl-Containing Compounds in Synthetic Fuels by Silicon-29 Nuclear Magnetic Resonance Spectrometry. *Anal. Chem.*, **54,** 133–134 (1982).

Coleman, W. M., D. L. Wooton, H. C. Dorn, and L. T. Taylor, Preparative Gel Permeation Chromatographic Separation of Solvent Refined Coal. *Anal. Chem.*, **49,** 533–537 (1977).

Colin, H., J.-M. Schmitter, and G. Gulochon, Liquid Chromatography of Azaarenes. *Anal. Chem.*, **53,** 625–631 (1981).

Cook, J. W., C. L. Hewett, and I. Heiger, The Isolation of a Cancer-Producing Hydrocarbon from Coal Tar, Parts I, II, and III. *Chem. Soc. J.*, **106,** 395–405 (1933).

Cowser, K. E., *University of Minnesota–Duluth Coal Gasification Project Semiannual Progress Report for the Period Ending December 31, 1980*, ORNL/TM-7730, Oak Ridge National Laboratory, Oak Ridge, TN, 1981.

Critenden, B. D., and R. Long, The Mechanism of Formation of Polynuclear Aromatic Compounds in Combustion Systems, in R. I. Frendenthal and P. W. Jones (eds.), *Carcinogenesis*, Vol. 1, Raven Press, New York, 1976, pp. 209–223.

Cunningham, M. L., D. A. Haugen, F. R. Kirchner, and C. A. Reilly, Jr., Toxicologic Responses to a Complex Coal Conversion By-Product: Mammalian Cell Mutagenicity and Dermal Carcinogenicity, in M. D. Waters, S. S. Sandhu, J. Lewtas, L. Claxton, G. Strauss, and S. Newnow (eds.), *Short-Term Bioassays in the Analysis of Complex Environmental Mixtures IV*, Plenum Press, New York, 1985, pp. 113–123.

Cunningham, M. L., M. S. Swanson, C. A. Reilly, Jr., and D. A. Haugen, Improvements in the CHO/HGPRT Assay and Sample Preparation for Mutagenesis Testing of Complex Mixtures, in R. H. Gray, E. K. Chess, P. J. Mellinger, R. G. Riley, and D. L. Springer (eds.), *Health and Environmental Research on Complex Organic Mixtures*, National Technical Information Service, Springfield, VA, in press.

Dark, W. A., and W. H. McFadden, The Role of HPLC and LC-MS in the Separation and Characterization of Coal Liquefaction Products. *J. Chromatogr. Sci.*, **16,** 289–293 (1978).

Dark, W. A., W. H. McFadden, and D. L. Bradford, Fractionation of Coal Liquids by HPLC with Structural Characterization by LC-MS. *J. Chromatogr. Sci.*, **15,** 454–460 (1977).

Dauble, D. D., A. J. Scott, E. W. Lusty, B. L. Thomas, and R. W. Hanf, Jr., *Ecotoxicity of Materials from Integrated Two Stage Liquefaction and Exxon Donor Solvent Processes*, PNL-4675, Pacific Northwest Laboratory. National Technical Information Service, Springfield, VA, 1983.

Deno, N. C., B. A. Griegger, and S. G. Stroud, New Method for Elucidating the Structures of Coal. *Fuel*, **57,** 455–459 (1978).

Deno, N. C., K. W. Curry, B. A. Griegger, A. D. Jones, W. G. Rakitsky, K. A. Smith, K. Wagner, and R. D. Minard, Dihydroaromatic Structure of Illinois No. 6 Monterey Coal. *Fuel*, **59,** 694–698 (1980).

Dipple, A., R. C. Moschel, and C. A. H. Bigger, Polynuclear Aromatic Carcinogens, in C. E. Searle (ed.), *Chemical Carcinogens*, vol. 1, 2nd ed., American Chemical Society, Washington, DC, 1984, pp. 41–163.

Doll, R., R. E. Fisher, E. J. Gammon, W. Gunn, G. O. Hughes, F. H. Tyrer, and W. Wilson, Mortality of Gas Workers with Special Reference to Cancers of Lung and Bladder, Chronic Bronchitis, and Pneumoconiosis. *Br. J. Ind. Med.*, **22,** 1–12 (1965).

Donath, E., Contribution to the Knowledge of Coals. *Chem. Ztg.*, **32,** 1271, (1908).

Dong, M., D. C. Locke, and D. Hoffman, Separation of Aza-arenes by High Pressure Liquid Chromatography. *J. Chromatogr. Sci.*, **15,** 32–35 (1977).

Dorbon, M., I. Ignatiadis, J. M. Schmitter, P. Arpino, G. Guiochon, H. Toulhoat, and A. Huc, Identification of Carbazoles and Benzocarbazoles in a Coker Gas Oil and Influence of Catalytic Hydrotreatment on Their Distribution. *Fuel*, **63,** 565–570 (1984).

Drake, J. A. G., D. W. Jones, B. S. Causey, and G. F. Kirkbright, Shpol'skii Luminescence Spectroscopy of Extracts of Coal and Coal-Tar Pitch. *Fuel*, **57,** 663–666 (1978).

Drozdowicz, B. Z., and C. M. Kelly, *Interim Report on the Genetic and Animal Toxicology Testing of SRC-I Products, Intermediates and Waste Materials*, DOE/OR-03054-50, International Coal Refining Company, Allentown, PA, 1983.

D'Silva, A. P., G. J. Oestreich, and V. A. Fassel, X-Ray Excited Optical Luminescence of Polynuclear Aromatic Hydrocarbons. *Anal. Chem.*, **48,** 915–917 (1976).

D'Silva, A. P., V. A. Fassel, Y. Yang, and M. Iles, Laser Excited Shpol'skii Spectrometry (LESS) for the Direct Determination of Polynuclear Aromatic Hydrocarbons in Liquid Fuels, in C. W. Wright, W. C. Weimer, and W. D. Felix (eds.), *Advanced Techniques for Synthetic Fuel Analysis*, Chap. 20, CONF-81160, National Technical Information Service, Springfield, VA, 1983, pp. 317–333.

Dubay, G. R., and R. A. Hites, Cyano-arenes Produced by Combustion of Nitrogen Containing Fuels. *Environ. Sci. Technol.*, **12,** 965–968 (1978).

Dutcher, J. S., R. E. Royer, J. D. Hill, C. E. Mitchell, and R. L. Hanson, Identification of Components of Biologically Active Fractions of Low Btu Gasifier Coal Tars, in C. W. Wright, W. C. Weimer, and W. D. Felix (eds.), *Advanced Techniques in Synthetic Fuels Analysis*, Chap. 8, CONF-81160, National Technical Information Service, Springfield, VA, 1983a, pp. 133–144.

Dutcher, J. S., R. E. Royer, C. E. Mitchell, and A. R. Dahl, Fractionation of Coal Tar from Low Btu Gasification, in C. W. Wright, W. C. Weimer, and W. D. Felix (eds.), *Advanced Techniques in Synthetic Fuels Analysis*, Chap. 2, CONF-81160, National Technical Information Service, Springfield, VA, 1983b, pp. 12–23.

Duty, R. C., R. Hayatsu, R. G. Scott, L. P. Moore, R. E. Winans, and M. H. Studier, Characterization of a Complex Mixture of Aromatic and Heterocyclic Acids Obtained from Oxidation of Lignite Coal and Its Solvent-Refined Products. *Fuel*, **59,** 97–101 (1980).

Epler, J. L., The Use of Short-Term Tests in the Isolation and Identification of Chemical Mutagens in Complex Mixtures, in F. J. de Serres and A. Hollaender (eds.), *Chemical Mutagens: Principles and Methods for their Detection*, Vol. 6, Plenum Press, New York, 1980, pp. 239–270.

Epler, J. L., J. A. Young, A. A. Hardigree, T. K. Rao, M. R. Guerin, I. B. Rabin, C.-H. Ho, and B. R. Clark, Analytical and Biological Analyses of Test Materials from the Synthetic Fuel Technologies. *Mutat. Res.*, **57,** 265–276 (1978a).

Epler, J. L., F. W. Larimer, T. K. Rao, C. E. Nix, and T. Ho, Energy-Related Pollutants in the Environment: Use of Short-Term Tests for Mutagenicity in the Isolation and Identification of Biohazards. *Environ. Health Perspect.*, **27,** 11–20 (1978b).

Epler, J. L., R. J. M. Fry, T. K. Rao, F. W. Larimer, J. N. Dumont, T. W. Schultz, L. B. Russell, W. M. Generoso, H. Witschi, L. H. Smith, W. H. Haschek, and J. M. Holland, *Biomedical Response to Products and Effluents from the University of Minnesota–Duluth Gasifier*, ORNL/TM-8821, Oak Ridge National Laboratory. National Technical Information Service, Springfield, VA, 1983.

Epperly, W. R., *EDS Coal Liquefaction Process Development Phase V, EDS Product Quality Final Report*, FE-2893-97, Exxon Research & Engineering Company. National Technical Information Service, Springfield, VA, 1982.

Erickson, M. D., S. D. Cooper, C. M. Sparacino, and R. A. Zweidinger, Gas Chromatography/Fourier Transform Infrared Spectrometry Analysis of Semivolatile Coal Gasification Pollutants. *Appl. Spectrosc.*, **33,** 575–577 (1976).

Erickson, M. D., S. E. Frazier, and C. M. Sparacino, High-Resolution FTIR Analysis of Coal Gases. *Fuel*, **60,** 263–266 (1981).

Farcasiu, M., Fractionation and Structural Characterization of Coal Liquids. *Fuel*, **56,** 9–14 (1977).

Felice, L. J., Determination of Alkyl Anilines and Alkyl Pyridines in Solvent Refined Coal Distillates and Aqueous Extracts by Gas Chromatography/Mass Spectrometry. *Anal. Chem.*, **54,** 869–872 (1982).

Felix, W. D., D. D. Mahlum, W. C. Weimer, R. A. Pelroy, and B. W. Wilson, Chemical and Biological Characteristics of SRC-II Products and By-Products, in *Environmental Aspects of Fuel Conversion Technology*, EPA-600/9-81-006, U.S. Environmental Protection Agency, Washington, DC, 1980, pp. 134–158.

Filby, R. H., V. Ekambaram, F. S. Jacobs, B. Blomquist, K. L. Ho, F. Iskander, and C. A. Grimm, Use of Nuclear Techniques to Study Metal Speciation in Coal, Coal Conversion Products and Petroleum. *Trans. Am. Nucl. Sci.*, **41,** 22–233 (1982).

Fillo, J. P., V. C. Stamoudis, J. R. Stetter, and S. W. Vance, *Influence of Coal Devolatilization Conditions on the Yield, Chemistry and Toxicology of By-Product Oils and Tars*, DOE/ET/14746-11, ANL/SER-2, National Technical Information Service, Springfield, VA, 1983.

Finseth, D. H., Z. T. Przybylski, and C. E. Schmidt, Comparison of Nitrogen Bases Generated Via HCl Precipitation and Ion Exchange Chromatography. *Fuel*, **61,** 1174–1176 (1982).

Fischer, P., J. W. Stadelhofer, and M. Zander, Structural Investigation of Coal-Tar Pitches and Coal Extracts by C-13 NMR Spectroscopy. *Fuel*, **57,** 345–352 (1978).

Fjeldsted, J. C., W. P. Jackson, P. A. Peaden, and M. L. Lee, Density Programming in Capillary Supercritical Fluid Chromatography. *J. Chromatogr. Sci.*, **279,** 222–225 (1983).

Franz, J. A., and W. E. Skiens, Side Reactions in the Reductive Alkylation of Low-Rank Coal. *Fuel*, **57,** 502–504 (1978).

Friedman, L., D. L. Fishel, and H. Shecter, Oxidation of Alkylarenes with Aqueous Sodium Dichromate. A Useful Method for Preparing Mono- and Polyaromatic Carboxylic Acids. *J. Org. Chem.*, **30,** 1453–1465 (1965).

Fruchter, J. S., J. C. Laul, M. R. Petersen, P. W. Ryan, and M. E. Turner, High-Precision Trace Element and Organic Constituent Analysis of Oil Shale and Solvent-Refined Coal Materials, in P. C. Uden, S. Siggia, and H. B. Lensen (eds.), *Analytical Chemistry of Liquid Fuel Sources*, American Chemical Society, Washington, DC, 1978, pp. 255–281.

Frumkin, H. A., R. A. Sullivan, and B. E. Stangland, Catalytic Upgrading of SRC-II Syncrude, in R. F. Sullivan (ed.), *Upgrading of Coal Liquids*, American Chemical Society, Washington, DC, 1981.

Gangwal, S. K., Gas Chromatographic Investigation of Raw Wastewater from Coal Gasification. *J. Chromatogr.*, **204,** 439–444 (1981).

Gangwal, S. K., and D. G. Nichols, Chemical Characterization of Tars from Fixed-Bed Gasification of Eastern and Western Coals, in D. D. Mahlum, R. H. Gray, and W. D. Felix (eds.), *Coal Conversion and the Environment*, CONF-801039, National Technical Information Service, Springfield, VA, 1981, pp. 16–33.

Garner, R. C., C. N. Martin, and D. B. Clayton, Carcinogenic Aromatic Amines and Related Compounds, in C. E. Searle (ed.), *Chemical Carcinogens*, vol. 1, 2nd ed., American Chemical Society, Washington, DC, 1984, pp. 175–276.

Garrigues, P., and M. Ewald, Identification of Monomethylated Polycyclic Aromatic Hydrocarbons in Crude Oils by Liquid Chromatography and High-Resolution Shpol'skii Effect Fluorescence Spectrometry. *Anal. Chem.*, **55,** 2155–2159 (1983).

Garrigues, P., R. DeVazelhes, M. Ewald, J. Joussot-Dubien, J.-M. Schmitter, and G. Gutochon, Identification of Triaromatic Azaarenes in Crude Oils by High-Resolution Spectrofluorimetry in Shpol'skii Matrices. *Anal. Chem.*, **55,** 138–140 (1983).

Gelboin, H. V., and P. O. P. Ts'o (eds.), *Polycyclic Hydrocarbons and Cancer*, Vol. 1, Academic Press, New York, 1978.

Gelboin, H. V., and P. O. P. Ts'o (eds.), *Polycyclic Hydrocarbons and Cancer*, Vol. 2, Academic Press, New York, 1979.

Gelboin, H. V., and P. O. P. Ts'o (eds.), *Polycyclic Hydrocarbons and Cancer*, Vol. 3, Academic Press, New York, 1981.

Goeckner, N. A., and W. H. Griest, Determination of Methyl Chrysenes in a Liquefaction Product. *Sci. Tot. Environ.*, **8,** 187–193 (1979).

Green, J. B., R. J. Hoff, P. W. Woodward, and L. L. Stevens, Separation of Liquid Fossil Fuels into Acid, Base and Neutral Concentrates. *Fuel*, **63,** 1209–1301 (1984).

Griest, W. H., B. A. Tomkins, J. L. Epler, and T. K. Rao, Characterization of Multialkylated Polycyclic Aromatic Hydrocarbons in Energy-Related Materials, in P. W. Jones and P. Leber (eds.), *Polynuclear Aromatic Hydrocarbons*, Ann Arbor Science Publishers, Ann Arbor, MI, 1979, pp. 395–409.

Griest, W. H., J. M. Giddings, and J. A. Klein, *Effects of Hydrotreatment on the Properties of Coal Derived Liquid Products: A Status Report*, ORNL-TM-8836, Oak Ridge National Laboratory. National Technical Information Service, Springfield, VA, 1983.

Griest, W. H., B. A. Tomkins, R. R. Regan, and J. W. Russell, Comparison of Selected 4- to 6-ring Polycyclic Aromatic Hydrocarbon Dermal Tumorigens in Coal Derived Liquids and Solids. *Fuel*, **66,** 408–411 (1987).

Grigsby, R. D., S. E. Scheppele, Q. G. Gridstaff, G. P. Sturm, L. C. E. Taylor, and H. W. Tudge, Evaluation of Fast Atom Bombardment Mass Spectrometry for Identification of Nitrogen-Containing Compounds in Fossil Fuels. *Anal. Chem.*, **54,** 1108–1113 (1982).

Guengerich, F. P. and D. C. Liebler, Enzymatic Activation of Chemicals to Toxic Metabolites. *CRC Crit. Rev. Toxicol.*, **14,** 259–307 (1985).

Guenther, F. R., R. M. Parris, S. N. Chesler, and L. R. Hilpert, Determination of Phenolic Compounds in Alternate Fuel Matrices. *J. Chromatogr.*, **207,** 256–261 (1981).

Guerin, M. R., C.-H. Ho, T. K. Rao, B. R. Clark, and J. L. Epler, Separation and Identification of Mutagenic Constituents of Petroleum Substitutes. *Int. J. Environ. Anal. Chem.*, **8,** 217–225 (1980).

Guerin, M. R., I. B. Rubin, T. K. Rao, B. R. Clark, and J. L. Epler, Distribution of Mutagenic Activity in Petroleum and Petroleum Substitutes. *Fuel*, **60,** 282–288 (1981).

Guerin, M. R., W. H. Griest, C.-H. Ho, L. H. Smith, and H. P. Witschi, *Integrated Report on the Toxicological Mitigation of Coal Liquids by Hydrotreatment and Other Processes*, ORNL-ITM-10070, Oak Ridge National Laboratory, National Technical Information Service, Springfield, VA, 1986.

Haggin, J., Interest in Coal Chemistry Intensifies. *Chem. Eng. News.*, **60,** 17–26 (1982).

Hanson, R. L., R. E. Royer, R. L. Carpenter, and G. J. Newton, Characterization of Potential Organic Emissions from Low Btu Gasifier for Coal Conversion, in P. W. Jones and P. Leber (eds.), *Polynuclear Aromatic Hydrocarbons*, Ann Arbor Science Publishers, Ann Arbor, MI, 1979, pp. 3–19.

Harris, L. A., and C. S. Yust, The Ultrafine Structure of Coal Determined by Electron Microscopy, in M. L. Gorbaty and K. Ouchi (eds.), *Coal Structure*, Chap. 21, Advances in Chemistry Series 192, American Chemical Society, Washington DC, 1981, p. 321.

Haugen, D. A., M. J. Peak, and C. A. Reilly, Use of Nitrous Acid Dependent Decrease in Mutagenicity as an Indication of the Presence of Mutagenic Primary Aromatic-Amines. Non-Specific Reactions with Phenols and Benzo[a]pyrene. *Mutat. Res.*, **82,** 59–67 (1981a).

Haugen, D. A., M. J. Peak, and C. A. Reilly, Chemical and Biological Characterization of High-Btu Coal Gasification (the Hygas Process) II. Nitrous Acid Treatment for Detection of Mutagenic Primary Aromatic Amines: Nonspecific Reactions, in D. D. Mahlum, R. H. Gray, and W. D. Felix (eds.), *Coal Conversion and the Environment*, CONF-801039, National Technical Information Service, Springfield, VA, 1981b, pp. 115–127.

Haugen, D. A., M. J. Peak, K. M. Suhrbier, and V. C. Stamoudis, Isolation of Mutagenic Aromatic Amines from a Coal Conversion Oil by Cation Exchange Chromatography. *Anal. Chem.*, **54,** 32–37 (1982a).

Haugen, D. A., V. C. Stamoudis, M. J. Peak, and A. S. Boparai, Isolation and Identification of Mutagenic Primary Aromatic Amines from Synthetic Fuel Materials, in M. Cooke, A. J. Dennis, and G. L. Fisher (eds.), *Polynuclear Aromatic Hydrocarbons: Physical and Biological Chemistry*, Battelle Press, Columbus, OH, 1982b, pp. 347–356.

Haugen, D. A., V. C. Stamoudis, M. J. Peak, and A. S. Boparai, Isolation of Mutagenic PAH, in M. Cooke and A. J. Dennis (eds.), *Polynuclear Aromatic Hydrocarbons: Formation, Metabolism, and Measurement*, Battelle Press, Columbus, OH, 1982c, pp. 607–620.

Haugen, D. A., M. J. Peak, and C. A. Reilly, Mixtures of Polycyclic Aromatic Compounds Inhibit Mutagenesis in the *Salmonella*/Microsome Assay by Inhibition of Metabolic Activation. *Mutat. Res.*, **116,** 257–269 (1983).

Haugen, D. A., V. C. Stamoudis, M. J. Peak, and A. S. Boparai, Isolation of a Highly Mutagenic Aminophenanthrene from a Coal Gasification Process Tar. *Environ. Res.*, **39,** 60–73 (1986).

Haugen, D. A., and V. C. Stamoudis, Isolation and Identification of Mutagenic Polycyclic Aromatic Hydrocarbons from a Coal Gasifier Condensate. *Environ. Res.*, **41,** 400–419 (1986).

Haugen, D. A., M. S. Swanson, F. R. Kirchner, V. C. Stamoudis, and C. A. Reilly, Toxicological Interactions Between Carcinogenic and Weakly Carcinogenic Mixtures in the Mouse-Skin Initiation–Promotion System, in R. H. Gray, E. K. Chess, P. J. Mellinger, R. G. Riley, and D. L. Springer (eds.), *Health and Environmental Research on Complex Organic Mixtures*, National Technical Information Service, Springfield, VA (in press).

Haworth, R. D., Synthesis of Alkylphenanthrenes. Parts 1, 2, 3, and 4—Methylphenanthrenes. *J. Chem. Soc.*, 1125–1133 (1932).

Hayatsu, R., R. E. Winans, R. G. Scott, L. P. Moore, and M. H. Studier, Trapped Organic Compounds and Aromatic Units in Coal. *Fuel*, **57,** 541–548 (1978).

Hayatsu, R., R. E. Winans, R. G. Scott, and R. L. McBeth, Is Kerogen-like Material Present in Coal? 1. Buffer-Controlled Permanganate Oxidation of Coal. *Fuel*, **60,** 158–161 (1981).

Heidelberger, C., Chemical Carcinogenesis. *Ann. Rev. Biochem.*, **44,** 79–121 (1975).

Heiger, I., The Spectra of Cancer Producing Tars and Oils and of Related Substances. *Biochem. J.*, **24,** 505–511 (1930).

Hembree, D. M.l, A. A. Garrison, R. A. Crocombe, R. A. Yokley, E. L. Wehry, and G. Amantov, Matrix Isolation Fourier Transform Infrared Spectrometric Detection in the Open Tubular Column Gas Chromatography of Polycyclic Aromatic Hydrocarbons. *Anal. Chem.*, **53,** 1783–1788 (1981).

Herod, A. A., W. R. Ladner, and C. E. Snape, Chemical and Physical Structure of Coal. *Philos. Trans. R. Soc. London Ser. A*, **300,** 3–14 (1981).

Hertz, H. S., J. M. Brown, S. N. Chesler, F. R. Guenther, L. R. Hilpert, W. E. May, R. M. Parris, and S. A. Wise, Determination of Individual Organic Compounds in Shale Oil. *Anal. Chem.*, **52,** 1650–1657 (1980).

Hill, J. O., R. E. Royer, C. E. Mitchell, and J. S. Dutcher, In Vitro Cytotoxicity to Alveolar Macrophages of Tar from a Low-Btu Coal Gasifier. *Environ. Res.*, **31,** 484–492 (1983).

Ho, C.-H., B. R. Clark, M. R. Guerin, C. Ma, and T. K. Rao, Aromatic Nitrogen Compounds in Fossil Fuels—A Potential Hazard. *Am. Chem. Soc. Div. Fuel Chem. Prepr.*, **24,** 59 (1979).

Ho, C.-H., B. R. Clark, and M. R. Guerin, Direct Analysis of Organic Compounds in Aqueous By-Products from Fossil Fuel Conversion Processes: Oil Shale Retorting, Synthane Coal Gasification and COED Coal Liquefaction. *J. Environ. Sci. Health*, **A11,** 481–489 (1980a).

Ho, C.-H., C. Y. Ma, B. R. Clark, M. R. Guerin, T. K. Rao, and J. L. Epler, Separation of Neutral Nitrogen Compounds from Synthetic Crude Oils for Biological Testing. *Environ. Res.*, **22,** 412–422 (1980b).

Ho, C.-H., B. R. Clark, M. R. Guerin, B. D. Barkenbus, T. K. Rao, and J. L. Epler, Analytical and Biological Analyses of Test Materials from the Synthetic Fuel Technologies. IV. Studies of Chemical Structure–Mutagenic Activity Relationships of Aromatic Nitrogen Compounds Relevant to Synfuels. *Mutat. Res.*, 85, 335–345 (1981a).

Ho, C.-H., M. V. Buchanan, B. R. Clark, and M. R. Guerin, Chemical Characterization of Synfuel Neutral Fractions, in D. D. Mahlum, R. H. Gray, and W. D. Felix (eds.), *Coal Conversion and the Environment, Chemical, Biomedical and Ecological Considerations*, CONF-801039, National Technical Information Service, Springfield, VA, 1981b.

Ho, C.-H., M. R. Guerin, B. R. Clark, T. K. Rao, and J. L. Epler, Isolation of Alkaline Mutagens from Complex Mixtures. *J. Anal. Toxicol.*, **5,** 143–147 (1981c).

Ho, C.-H., M. R. Guerin, and M. J. Buchanan, Isolation and Identification of Bioactive Constituents in Synfuels, in C. W. Wright, W. C. Weimer, and W. D. Felix (eds.), *Advanced Techniques in Synfuel Analysis*, CONF-811160, National Technical Information Service, Springfield, VA, 1983, pp. 24–43.

Holland, J. M., D. A. Wolf, and B. R. Clark, Relative Potency Estimation for Synthetic Petroleum Skin Carcinogens. *Environ. Health Perspect.*, **38,** 149–155 (1981).

Hsie, A. W., P. A. Brimer, J. P. O'Neill, J. L. Epler, M. R. Guerin, and M. H. Hsie, Mutagenicity of Alkaline Constituents of a Coal-Liquified Crude Oil in Mammalian Cells. *Mutat. Res.*, **78,** 79–84 (1980).

Hunt, D. F., and J. Shabanowitz, Determination of Organosulfur Compounds in Hydrocarbon Matrices by Collision Activated Dissociation Mass Spectrometry. *Anal. Chem.*, **54,** 574–578 (1982).

Hunt, D. F., C. N. McEwen, and R. A. Upham, Determination of Active Hydrogen in Organic Compounds by Chemical Ionization Mass Spectrometry. *Anal. Chem.*, **44,** 1292–1294 (1972).

Hurtubise, R. J., T. W. Allen, J. F. Schabron, and H. F. Silver, Estimation of Alkylphenol Content in Coal Recycle Solvents. *Fuel*, **60,** 385–389 (1981).

Hyden, H. J., C. A. Horr, A. T. Myers, J. S. Ball, and W. J. Wenger, Metal Content of Twenty-Four Petroleums. *J. Chem. Eng. Data*, **5,** 553–557 (1960).

Irving, C. C., Species and Tissue Variations in the Metabolic Activation of Aromatic Amines, in A. C. Griffin and C. R. Shaw (eds.), *Carcinogens: Identification and Mechanisms of Action*, Raven Press, New York, 1979, pp. 211–227.

Iwao, M., M. L. Lee, and R. N. Castle, Synthesis of Phenanthro[b]thiophenes. *J. Heterocycl. Chem.*, **17,** 1259–1264 (1980).

Jakoby, W. B. (ed.), *Enzymatic Basis of Detoxication*, Vols. I and II, Academic Press, New York, 1980.

Jakoby, W. B., J. R. Bend, and J. Caldwell (eds.), *Metabolic Basis of Detoxication: Metabolism of Functional Groups*, Academic Press, New York, 1982.

Jewell, D. M., and G. K. Haotung, Identification of Nitrogen Bases in Heavy Gas Oil; Chromatographic Methods of Separation. *J. Chem. Eng. Data*, **9,** 297–304 (1964).

Jewell, D. M., and R. E. Snyder, Selective Separation of "Nonbasic" Nitrogen Compounds from Petroleum by Anion Exchange of Ferric Chloride Complexes. *J. Chromatogr.*, **38,** 351–354 (1968).

Jewell, D. M., J. H. Weber, J. W. Bunger, H. Plancher, and D. R. Latham, Ion-Exchange, Coordination, and Adsorption Chromatographic Separation of Heavy-end Petroleum Distillates. *Anal. Chem.*, **44,** 1391–1395 (1972).

Jones, A. R., M. R. Guerin, and B. R. Clark, Preparative-Scale Liquid Chromatographic Fractionation of Crude Oils Derived from Coal and Shale. *Anal. Chem.*, **49,** 1766–1771 (1977).

Joyce, W. F., and P. C. Uden, Isolation of Thiophenic Compounds by Argentation Liquid Chromatography. *Anal. Chem.*, **55,** 540–543 (1983).

Kalkwarf, D. R., D. L. Stewart, R. A. Pelroy, and W. C. Weimer, *Photodegeneration of Mutagens in SRC Liquids*, PNL-4982, Pacific Northwest Laboratory, Richland, WA. National Technical Information Service, Springfield, VA, 1984.

Kapila, M. S., and C. R. Vogt, A Gas-Tight Low Volume Photoionization Detector for Capillary Gas Chromatography. *HRC&C*, **4,** 233–235 (1981).

Karcher, W., A. Nelen, R. Dupaus, J. VonEijk, P. Glaude, and J. Jacob, New Results in the Detection, Identification, and Mutagenic Testing of Heterocyclic Polycyclic Aromatic Hydrocarbons, in M. Cooke and A. J. Dennis (eds.), *Polynuclear Aromatic Hydrocarbons*, Battelle Press, Columbus, OH, 1981, pp. 312–327.

Karr, C., and T. L. Chang, Spectrophotometric Analysis of the Distillable Low-Temperature Tar Bases. *J. Inst. Fuel*, **31,** 522–527 (1958).

Kekin, N. A., V. M. Shmal'ko, and V. V. Vodolazhchenko, Use of Low Voltage Mass Spectrometry in Analysis of Coal Tar Processing Products. *Ind. Lab.*, **43,** 639–641 (1977).

Kennaway, E., The Identification of a Carcinogenic Compound in Coal-Tar. *Br. Med. J.*, **2,** 749–752 (1955).

Kim, J. I., M. D. Notich, H. C. Higman, and K. M. Wagner, *Environmental Assessment Source Test and Evaluation Report, Fort Lewis SRC Coal Liquefaction Pilot Plant*, H-C1016/002-81-996, Hittman Associates. U.S. Environmental Protection Agency, Research Triangle Park, NC. National Technical Information Service, Springfield, VA, 1981.

Kimball, R. F., and N. B. Munro, *A Critical Review of the Mutagenic and Other Genotoxic Effects of Direct Coal Liquefaction*, ORNL-5721, Oak Ridge National Laboratory, Oak Ridge, TN. National Technical Information Service, Springfield, VA, 1981.

King, F. T. C., Petrographic Techniques in Coal Analysis, in C. Karr Jr. (ed.), *Analytical Methods for Coal and Coal Products*, Vol. 1, Academic Press, New York, 1978, pp. 3–26.

Klimisch, H.-J., and L. Stader, Trennung Komplexer Stoffgemische Lipophiler und Hydrophiler Substanzen Durch Gel-Verteilungschromatographie an Sephadex LH-20. *J. Chromatogr.*, **67,** 291–297 (1972).

Kong, R. C., M. L. Lee, Y. Tominaga, and R. Pratrap, Capillary Column Gas Chromatographic Resolution of Isomeric Polycyclic Aromatic Sulfur Heterocyclic in a Coal Liquid. *J. Chromatogr. Sci.*, **20,** 502–510 (1982a).

Kong, R. C., M. L. Lee, Y. Tominaga, R. Pratrap, M. Iwao, and R. N. Castle, Mesogenic Polysiloxane Stationary Phase for High-Resolution Gas-Chromatography of Isomeric Polycyclic Aromatic Compounds. *Anal. Chem.*, **54,** 1802–1806 (1982b).

Koppenaal, D. W., and S. E. Manahan, Hazardous Chemicals from Coal Conversion Processes? *Environ. Sci. Technol.*, **10,** 1104–1107 (1976).

Koppenaal, D., R. G. Lett, F. R. Brown, and S. E. Manahan, Determination of Nickel, Copper, Selenium, Cadmium, Thallium, and Lead in Coal Gasification Products by Isotope Dilution Spark Source Mass Spectrometry. *Anal. Chem.*, **52,** 44–49 (1980).

Kosuge, T., H. Zenda, H. Nukaya, A. Terada, T. Okamoto, K. Shudo, K. Yamaguchi, Y. Iitaka, T. Sugimura, M. Nagao, and K. Wakabayashi, Isolation and Structural Determination of Mutagenic Substances in Coal-Tar. *Chem. Pharmacol. Bull.*, **30,** 1535–1538 (1982).

Krieck, E., Carcinogenesis by Aromatic-Amines. *Biochem. Biophys. Acta*, **355,** 177–203 (1974).

Krieck, E., and J. G. Westra, Metabolic Activation of Aromatic Amines and Amides and Interactions with Nucleic Acids, in P. L. Grover (ed.), *Chemical Carcinogens and DNA*, Vol. 2, CRC Press, Boca Raton, FL, 1979, pp. 1–28.

Krishan, S. K., D. A. Kaden, W. G. Thilly, and R. A. Hites, Cyanoarenes in Soot. Synthesis and Mutagenicity of Cyanoacenaphthylenes. *Environ. Sci. Technol.*, **13,** 1532–1534 (1979).

Later, D. W., Nitrogen Containing Polycyclic Aromatic Compounds in Coal Derived Materials, in A. Borseth and T. Ramdahl (eds.), *Handbook of Polycyclic Aromatic-Hydrocarbons*, Vol. II, Chap. 9, Marcel Dekker, New York, 1985a.

Later, D. W., *Two-Stage Coal Liquefaction Materials from the Wilsonville Facility Operated in the Non-Integrated and Integrated Modes: Chemical Analyses and Biological Testing*, PNL-5215, Pacific Northwest Laboratory, Richland, WA. National Technical Information Service, Springfield, VA, 1985b.

Later, D. W., and M. L. Lee, Chromatographic Methods for the Chemical and Biological Characterization of Polycyclic Aromatic Compounds in Synfuel Materials, in C. W. Wright, W. C. Weimer, and W. D. Felix (eds.), *Advanced Techniques in Synthetic Fuels Analysis*, Chap. 4, CONF-81160, National Technical Information Service, Springfield, VA, 1983, pp. 44–73.

Later, D. W., and B. W. Wilson, *Preliminary Chemical Analysis and Biological Testing of Materials from the HRI Catalytic Two-Stage Liquefaction (CTSL) Process*, PNL-5274, Pacific Northwest Laboratory, Richland, WA. National Technical Information Service, Springfield, VA, 1985.

Later, D. W., M. L. Lee, K. D. Bartle, R. C. Kong, and D. L. Vassilaros, Chemical Class Separation and Characterization of Organic Compounds in Synthetic Fuels. *Anal. Chem.*, **53,** 1612–1620 (1981a).

Later, D. W., M. Lee, R. A. Pelroy, and B. W. Wilson, Identification and Mutagenicity of Nitrogen-Containing Polycyclic Aromatic Compounds in Synthetic Fuels, in M. W. Cooke, A. J. Dennis, and G. L. Fisher (eds.), *Polynuclear Aromatic Hydrocarbons: Physical and Biological Chemistry*, Battelle Press, Columbus, OH, 1981b, pp. 427–438.

Later, D. W., B. W. Wright, and M. L. Lee, Construction of an Efficient Fused Silica Capillary Column Effluent Splitter for Gas Chromatography. *HRC&C*, **4,** 406–408 (1981c).

Later, D. W., M. L. Lee, R. A. Pelroy, D. D. Mahlum, C. W. Wright, M. L. Lee, W. C. Weimer, and B. W. Wilson, Identification and Comparative Genotoxicity of Polycyclic Aromatic Hydrogen and Related Nitrogen-Containing Heteroatomic Species in Products from Coal Liquefaction Processes, in M. W. Cooke and A. J. Dennis (eds.), *Polynuclear Aromatic Hydrocarbons: Formation, Metabolism and Measurement*, Battelle Press, Columbus, OH, 1982a, pp. 771–783.

Later, D. W., M. L. Lee, and B. W. Wilson, Selective Detection of Amino Polycyclic Aromatic Compounds in Solvent Refined Coal. *Anal. Chem.*, **54,** 117–123 (1982b).

Later, D. W., T. C. Andros, and M. L. Lee, Isolation and Identification of Amino Polycyclic Aromatic Hydrocarbons from Coal-Derived Products. *Anal. Chem.*, **55,** 2126–2132 (1983a).

Later, D. W., C. W. Wright, and B. W. Wilson, The Analytical Chemistry of Products from Process Strategies Designed to Reduce the Biological Activity of Direct Coal Liquefaction Materials. *Am. Chem. Soc. Div. Fuel Chem. Prepr.*, **28,** 273–284 (1983b).

Later, D. W., R. A. Pelroy, D. L. Stewart, T. McFall, G. M. Booth, M. L. Lee, M. Tedjamulia, and R. N. Castle, Microbial Mutagenicity of Isomeric 2-Ring, 3-Ring, and 4-Ring Amino Polycyclic Aromatic Hydrocarbons. *Environ. Mutagen.*, **6,** 497–515 (1984a).

Later, D. W., R. A. Pelroy, and B. W. Wilson, *Chemical Analysis and Biological Testing of Materials from the EDS Coal Liquefaction Process: A Status Report*, PNL-4690, Pacific Northwest Laboratory, Richland, WA. National Technical Information Service, Springfield, VA, 1984b.

Later, D. W., S. A. Barraclough, R. B. Lucke, R. A. Pelroy, P. Stoker, and G. M. Booth, Analytical and Chromatographic Methods for Determination of Cyano Polycyclic Aromatic Hydrocarbons in Coal-Derived Materials, in M. W. Cooke and A. J. Dennis (eds.), *Polynuclear Aromatic Hydrocarbons, Mechanisms, Methods and Metabolism*, Battelle Press, Columbus, OH, 1985a, pp. 827–841.

Later, D. W., W. Wilson, and M. L. Lee, Standardization of Alumina and Silica Adsorbents Used for Chemical Class Separations of Polycyclic Aromatic Compounds. *Anal. Chem.*, **57,** 2979–2984 (1985b).

Later, D. W., T. J. Wozniak, R. B. Lucke, R. A. Pelroy, D. D. Mahlum, and R. A. Hites, Hydrogenated Polycyclic Aromatic Hydrocarbons in Coal Liquefaction Materials: Analytical Chemistry and Biological Activity, in M. Cooke and A. J. Dennis (eds.), *Polynuclear Aromatic Hydrocarbons: Chemistry, Characterization and Carcinogenesis*, Battelle Press, Columbus, OH, 1986, pp. 473–488.

La Voie, E. J., A. Govil, G. Briggs, and D. Hoffman, Mutagenicity of Amino Carbazoles and Nitrocarbazoles. *Mutat. Res.*, **90,** 337–344 (1981a).

La Voie, E. J., L. Tully-Freiler, V. Bedenko, and D. Hoffman, Mutagenicity, Tumor-Initiating Activity, and Metabolism of Methylphenanthrenes. *Cancer Res.*, **41,** 3441–3447 (1981b).

La Voie, E. J., L. Thilley, V. Bendenko, and D. Hoffman, Mutagenicity of Methylated Fluorenes and Benzofluorenes. *Mutat. Res.*, **91,** 167–176 (1981c).

Leary, J. A., A. L. Lafleur, H. L. Liber, and K. Bieman, Chemical and Toxicologic Characterization of Fossil Fuel Combustion Product Phenalen-1-one. *Anal. Chem.*, **55,** 758–761 (1983).

Lee, K. W., W. S. Seames, R. V. Collins, K. J. Bombaugh, and G. C. Page, *EPA Environmental Assessment Source Test and Evaluation Report—Lurgi Kosovo Medium Btu Gasification*, EPA-600/7-81-142, National Technical Information Service, Springfield, VA, 1981.

Lee, M. L., and R. Hites, Characterization of Sulfur-Containing Polycyclic Aromatic Compounds in Carbon Blacks. *Anal. Chem.*, **48,** 1890–1893 (1976).

Lee, M. L., and B. W. Wright, Capillary Column Gas Chromatography of Polycyclic Aromatic Compounds: A Review. *J. Chromatogr. Sci.*, **18,** 345–358 (1980).

Lee, M. L., C. Willey, R. N. Castle, and C. M. White, Separation and Identification of Sulphur Heterocycles in Coal Derived Products, in A. Bjorseth and A. J. Dennis (eds.), *Polynuclear Aromatic Hydrocarbons: Chemistry and Biological Effects*, Battelle Press, Columbus, OH, 1980, pp. 59–73.

Lee, M. L., M. V. Novotny, and K. D. Bartle, *Analytical Chemistry of Polycyclic Aromatic Compounds*, Academic Press, New York, 1981.

Lennox, B., The Production of a Variety of Skin Tumours in Rats with 2-Anthramine, and a Comparison with the Effects in Mice. *Br. J. Cancer*, **9,** 631–639 (1958).

Levin, W., A. W. Wood, R. L. Chang, S. Kumar, H. Yagi, D. M. Jerina, R. E. Lehr, and A. H. Conney, Tumor-Initiating Activity of Benz[c]acridine and 12 of Its Derivatives on Mouse Skin. *Cancer Res.*, **43,** 4625–4628 (1983).

Lijinsky, W., F. Cefis, H. Garcia, and U. Saffiotti, Skin Tumorigenesis by Dibenz[a,h]anthracene Modified by Its Hydrogenated Distillates. *J. Natl. Cancer Inst.*, **34,** 7–12 (1965a).

Lijinsky, W., H. Garcia, B. Terracini, and U. Saffiotti, Tumorigenic Activity of Hydrogenated Derivatives of Dibenz[a]anthracene. *J. Natl. Cancer Inst.*, **34,** 1–6 (1965b).

Lucke, R. B., D. W. Later, C. W. Wright, E. K. Chess, and W. C. Weimer, Integrated, Multiple-Stage Chromatographic Method for the Separation and Identification of Polycyclic Aromatic Hydrocarbons in Complex Coal Liquids. *Anal. Chem.*, **57,** 633–639 (1985).

Lucke, R. B., D. W. Later, and E. K. Chess, Separation and Determination of Hydroxy Functional Polycyclic Aromatic Hydrocarbons in Coal Liquefaction Materials, in M. Cooke and A. J. Dennis (eds.), *Polycyclic Aromatic Hydrocarbons: Chemistry, Characterization and Carcinogenesis*, Battelle Press, Columbus, OH, 1986, pp. 505–516.

Lytle, J. M., G. L. Tingey, and R. D. MacFarlane, Californium-252 Plasma Desorption Mass Spectrometry, PDMS, of Solid Coal. *Anal. Chem.*, **54,** 1881–1883 (1982).

Mahlum, D. D., Skin Tumor Initiation Activity of Coal Liquids with Different Boiling Point Ranges. *J. Appl. Toxicol.*, **3,** 254–258 (1983).

Mahlum, D. D., R. A. Pelroy, B. W. Wilson, and H. Drucker, Toxicologic Studies of SRC Materials, in *Proceedings of the Second U.S. Department of Energy Environmental Control Symposium*, March 17–19, 1980, Vol. 1, *Fossil Energy*, CONF-800334/1, National Technical Information Service, Springfield, VA, 1980, pp. 634–646.

Mahlum, D. D., M. E. Frazier, R. A. Pelroy, and R. A. Renne, *Comparison of in Vitro and in Vivo Studies with Coal Liquids from the Solvent Refined Coal-II Process*, PNL-4836, Pacific Northwest Laboratory, Richland, WA. National Technical Information Service, Springfield, VA, 1983.

Mahlum, D. D., C. W. Wright, E. K. Chess, and B. W. Wilson, Fractionation of Skin Tumor Initiating Activity in Coal Liquids. *Cancer Res.*, **44,** 5176–5181 (1984).

Maple, J. R., E. L. Wehry, and G. Mamantov, Laser-Induced Fluorescence Spectrometry of Polycyclic Aromatic Hydrocarbons Isolated in Vapor Deposited Normal Alkane Matrices. *Anal. Chem.*, **52,** 920–924 (1980).

Marzec, A., M. Juzwa, K. Betlej, and M. Sobkowiak, Bituminous Coal Extraction in Terms of Electron-Donor and Electron-Acceptor Interactions in the Solvent Refined Coal System. *Fuel Process Technol.*, **2,** 35–44 (1979).

Masuda, Y., and D. Hoffman, A Method for the Determination of Primary Amines of Polynuclear Aromatic Hydrocarbons. *J. Chromatogr. Sci.*, **7,** 694–697 (1969).

May, W. E., J. M. Brown-Thomas, S. N. Chesler, F. R. Guenther, L. R. Hilpert, R. M. Paris, K. L. Richie, S. A. Wise, and H. S. Hertz, Interlaboratory Comparisons of Quantitative Analysis of Individual Compounds in Simple and Complex Mixtures, in C. W. Wright, W. C. Weimer, and W. D. Felix (eds.), *Advanced Techniques in Synthetic Fuels Analysis*, CONF-811160, National Technical Information Service, Springfield, VA, 1983, pp. 381–394.

McKay, J. F., J. H. Weber, and D. R. Latham, Characterization of Nitrogen Bases in High-Boiling Petroleum Distillates. *Anal. Chem.*, **48,** 891–898 (1976).

Miller, D. J., J. K. Olson, and H. H. Schobert, Mass Spectroscopic Characterization of Tars from the Gasification of Low-Rank Coals. *Fuel*, **60,** 370–374 (1981).

Miller, E. C., and J. A. Miller, Mechanisms of Chemical Carcinogenesis. *Cancer*, **47,** 1055–1064 (1981).

Miller, J. A., and E. C. Miller, The Metabolic Activation of Chemical Carcinogens to Reactive Electrophiles, in J. M. Yuhas, R. W. Tennant, and J. D. Regan (eds.), *Biology of Radiation Carcinogenesis*, Raven Press, New York, 1976, pp. 147–164.

Mima, M. J., H. Schultz, and W. E. McKinstry, Methods for the Determination of Benzene Insolubles, Asphaltenes and Oils for the Coal-Derived Liquids, in C. Karr (ed.), *Analytical Methods for Coal and Coal Products*, Vol. 1, Chap. 19, Academic Press, New York, 1978.

Mixon, F. O., J. G. Cleland, T. J. Hughes, D. A. Green, and S. K. Gangwal, Laboratory-Scale Simulation of Coal Gasification for Biological Studies, in K. E. Cowser (ed.), *Synthetic Fossil Fuel Technologies, Results of Health and Environmental Studies*, Butterworth, Stoneham, MA, 1984, pp. 199–227.

Mrochek, J. E., R. Livingston, and H. Zeldes, Measurement of Solvent Quality by Electron Spin Resonance, in *Proceedings of the Department of Energy Direct Coal Liquefaction Contractor's Review Meeting*, November 16–17, 1983, Pittsburgh Energy Technology Center, Pittsburgh, PA, 1983, pp. 10-1–10-35.

Munro, N. B., R. J. M. Fry, R. B. Gammage, W. M. Haschek, E. E. Colle, J. A. Klein, and T. W. Schultz, *Indirect Coal Liquefaction: A Review of Potential Health Effects and Worker Exposure During Gasification and Synthesis*, ORNL-5938, Oak Ridge National Laboratory, Oak Ridge, TN. National Technical Information Service, Springfield, VA, 1983.

Nagoa, M., T. Yahagi, T. Kawachi, T. Sugimura, T. Kosuge, K. Tsuji, K. Wakabayashi, S. Mizusaki, and T. Matsumoto, Comutagenic Action of Norharman and Harman. *Proc. Jpn. Acad. Sci.*, **53,** 95–98 (1977).

Neavel, R. C., Coal Structure and Coal Science: Overview and Recommendations, in M. L. Gorbaty and K. Ouchi (eds.), *Coal Structure*, Chap. 1, Advances in Chemistry Series 192, American Chemical Society, Washington, DC, 1981.

Nelson, S. D., Metabolic Activation and Drug Toxicity. *J. Med. Chem.*, **25,** 753–765 (1982).

Nishioka, M., M. L. Lee, H. Kudo, D. R. Muchiri, L. J. Baldwin, S. Pakray, J. G. Stuart, and R. N. Castle, Determination of Hydroxylated Thiophenic Compounds in a Coal Liquid. *Anal. Chem.*, **57,** 1327–1330 (1985a).

Nishioka, M., R. M. Campbell, W. R. West, P. A. Smith, G. M. Booth, M. L. Lee, H. Kudo, and R. N. Castle, Determination of Aminobenzothiophenes in a Coal Liquid. *Anal. Chem.*, **57,** 1868–1871 (1985b).

Nishioka, M., R. M. Campbell, M. L. Lee, D. R. Muchiri, J. G. Stuart, and R. N. Castle, Isolation and Determination of Hydroxylated Nitrogen Heterocycles in a Coal Liquid. *Anal. Chem.*, **57,** 2211–2215 (1985c).

Novotny, M., M. L. Lee, and K. D. Bartle, The Methods for Fractionation, Analysis, and Identification of Polynuclear Aromatic Hydrocarbons in Complex Mixtures. *J. Chromatogr. Sci.*, **12,** 606–612 (1974).

Novotny, M., J. W. Strand, S. L. Smith, D. Wiesler, and F. J. Schwende, Compositional Studies of Coal Tar by Capillary Gas Chromatography/Mass Spectrometry. *Fuel*, **60,** 213–220 (1981).

Ogan, K., and E. Katz, Analysis of Complex Samples by Coupled-Column Chromatography. *Anal. Chem.*, **54,** 163–173 (1982).

O'Rear, D. J., R. F. Sullivan, and B. E. Stangland, Catalytic Upgrading of H-Coal Syncrudes, in R. F. Sullivan (ed.), *Upgrading Coal Liquids*, American Chemical Society Monograph Series 156, American Chemical Society, Washingon, DC, 1981.

Ostermayer, E., and R. Fittig, Uber Einen Neuen Kohlerwasserstoff Aus Dem Steinkohlevtheer. *Ber. Dtsch. Chem. Ges.*, **5,** 933–937 (1872).

Ouchi, K., and K. Imuta, The Analysis of Benzene Extracts of Yūbari Coal I—Chromatographic Fractionation of Petroleum Ether–Methanol Soluble Portion and Ultraviolet and Infrared Spectra of Fractions. *Fuel*, **42,** 25–35 (1963a).

Ouchi, K., and K. Imuta, The Analysis of Benzene Extracts of Yūbari Coal II—Analysis by Gas Chromatography. *Fuel*, **42,** 445–446 (1963b).

Page, G. C., and W. J. Rhodes, *Environmental Assessment: Source Test and Evaluation Report—Chapman Low-Btu Gasification*, EPA-600/7-78-202, U.S. Enivronmental Protection Agency, Cincinnati, OH, 1978.

Pannell, R. B., and A. Sood, Simulated Distillation of Coal Liquids. *J. Chromatogr. Sci.*, **20,** 443–445 (1982).

Parkes, H. G., and A. E. J. Evans, Epidemiology of the Aromatic Amine Cancers, in C. E. Searles (ed.), *Chemical Carcinogens*, Vol. 1, American Chemical Society Monograph 182, American Chemical Society, Washington, DC, 1981, pp. 277–301.

Paudler, W. W., and M. Chaeplen, Nitrogen Bases in Solvent-Refined Coal. *Fuel*, **58,** 775–778 (1979).

Peaden, P. A., M. L. Lee, Y. Hicvata, and M. Novotny, High-Performance Liquid Chromatographic Separation of High-Molecular Weight Polycyclic Aromatic Compounds in Carbon Black. *Anal. Chem.*, **52,** 2268–2271 (1980).

Peaden, P. A., B. W. Wright, and M. L. Lee, The Preparation of Non-Extractable Methylphenylpolysiloxane Stationary Phase for Capillary Column Gas Chromatography. *Chromatographia*, **15,** 335–340 (1982).

Pelroy, R. A., *The Mutagenic and Chemical Properties of SRC-I Materials: A Status Report*, PNL-3604, Pacific Northwest Laboratory, Richland, WA. National Technical Information Service, Springfield, VA, 1981.

Pelroy, R. A., and A. Gandolfi, Use of Mixed Function Amine Oxidase for Metabolic Activation in the Ames/*Salmonella* Assay System. *Mutat. Res.*, **72,** 329–334 (1980).

Pelroy, R. A., and M. R. Petersen, Use of Ames Test in Evaluation of Shale Oil Fractions. *Environ. Health Perspect.*, **30,** 191–203 (1979).

Pelroy, R. A., and D. L. Stewart, The Effects of Nitrous Acid on the Mutagenicity of Two Coal Liquids and Their Genetically Active Chemical Fractions. *Mutat. Res.*, **90,** 297–308 (1981).

Pelroy, R. A., and B. W. Wilson, *Fractional Distillation as a Strategy for Reducing the Genotoxic Potential of SRC-II Coal Liquids: A Status Report*, PNL-3787, Pacific Northwest Laboratory, Richland, WA. National Technical Information Service, Springfield, VA, 1981.

Pelroy, R. A., D. L. Stewart, Y. Tominaga, M. Iwao, R. W. Castle, and M. L. Lee, Microbial Mutagenicity of 3 Ring and 4 Ring Polycyclic Aromatic Sulfur Heterocycles. *Mutat. Res.*, **117,** 31–40 (1983).

Pelroy, R. A., D. D. Mahlum, M. E. Frazier, R. A. Renne, D. L. Stewart, and E. K. Chess, Comparison of In Vivo Carcinogenesis and In Vitro Genotoxicity of Complex Hydrocarbon Mixtures, in K. E. Cowser (ed.), *Synthetic Fossil Fuel Technologies, Results of Health and Environmental Studies*, Butterworth, Stoneham, MA, 1984, pp. 267-302.

Pelroy, R. A., M. E. Frazier, D. W. Later, C. W. Wright, and B. W. Wilson, Effect of Process Distillation on Mutagenicity and Cell-Transformation Activity of Solvent-Refined, Coal-Derived Liquids. *Fuel*, **64,** 639–643 (1985).

Petersen, M. R., J. Fruchter, and J. C. Laul, *Characterization of Substances in Products, Effluents, and Wastes from Synthetic Fuel Production Tests*, BNWL-2131, Pacific Northwest Laboratory, Richland, WA, 1976.

Petrakis, L., and D. W. Grandy, Free Radicals on Coal and Coal Conversion. 2. Effect of Liquefaction Processing Conditions on the Formation and Quenching of Coal Free Radicals. *Fuel*, **59,** 227–232 (1980).

Petrakis, L., and D. W. Grandy, Free Radicals in Coal and Coal Conversion. 4. Investigation of Free Radicals in Selected Macerals Upon Liquefaction. *Fuel*, **60,** 120–124 (1981).

Philip, R. P., N. J. Russell, and T. D. Gilbert, Hydrogenation of Tasmanian Alginate—Analysis of Hexane-Soluble Products by Thermal Distillation–Gas Chromatography–Mass Spectrometry. *Fuel*, **60,** 937–943 (1981).

Pichler, V. H., and A. Herlan, Untersuchung Eine Hochtemperaturteers Mit Hochauf Massenspektrometrie. *Erdoel Kohle, Erdgas, Petrochem.*, **26,** 401–407 (1973).

Pictet, A., and L. Ramseyer, Uber Einen Bestandteil der Steinkohle. *Ber. Dtsch. Chem. Ges.*, **44,** 2486 (1911).

PNL, *Biomedical Studies on Solvent Refined Coal (SRC-11) Liquefaction Materials: A Status Report and Appendix*, PNL-3189, Pacific Northwest Laboratory, Richland, WA. National Technical Information Service, Springfield, VA, 1979.

Popl, M., V. Dolansky, and J. Mostecky, Absorption Elution Chromatography of Nitrogen-Containing Heterocycles on Alumina. *J. Chromatogr.*, **74,** 51–57 (1972).

Pott, P., Cancer Scroti, in Hawes, Clark, and Collins (eds.), *Chirurgical Observations*, London, 1775, pp. 63–67. [Reprinted in *Natl. Cancer Inst.*, **10,** 7 (1963).]

Potts, J. D., K. D. Hastings, R. S. Chillingworth, and H. Unger, LC Fining of Solvent Refined Coal, in F. R. Sullivan (ed.), *Upgrading Coal Liquids*, American Chemical Society Symposium Series 156, American Chemical Society, Washington, DC, 1981.

Pratap, R., Y. Tominaga, M. L. Lee, and R. N. Castle, Synthesis of Pyreno[b]thiophenes. *J. Heterocycl. Chem.*, **18,** 973–975 (1981).

Pugmire, R. J., D. M. Grant, K. W. Zilm, L. L. Anderson, A. G. Oblad, and R. E. Wood, Carbon-13 Magnetic Resonance of Coal-Derived Liquids. *Fuel*, **56,** 295–301 (1977).

Pugmire, R. J., K. W. Zilm, D. M. Grant, S. R. Larter, J. Allen, J. T. Senftle, A. Davis, and W. Spackman, Carbon-13 CP/MS Study of Coal Macerals of Varying Rank, in B. D. Blaustein, B. C. Bockrath, and S. Friedman (eds.), *New Approaches to Coal Chemistry*, Chap. 2, American Chemical Society Symposium Series 169, American Chemical Society, Washington, DC, 1981.

Radomski, J. L., The Primary Aromatic Amines: Their Biological Properties and Structure–Activity Relationships. *Ann. Rev. Pharmacol. Toxicol.*, **19,** 129–157 (1979).

Rao, T. K., B. E. Allen, S. W. Ramey, J. L. Epler, I. B. Rubin, M. R. Guerin, and B. R. Clark, Analytical and Biological Analyses of Test Materials from the Synthetic Fuel Technologies. III. Use of Sephadex LH-20 Gel Chromatography Technique for the Bioassay of Crude Synthetic Fuels. *Mutat. Res.*, **85,** 29–39 (1981).

Rehn, L., *Arch. Klin. Chir.*, **50,** 588 (1895).

Reilly, Jr., C. A., M. J. Peak, T. Matsushita, F. R. Kirchner, and D. A. Haugen, Chemical and Biological Characterization of High-Btu Coal Gasification (the HYGAS Process). IV. Biological

Activity, in D. D. Mahlum, R. H. Gray, and W. D. Felix (eds.), *Coal Conversion and the Environment, Chemical, Biological and Ecological Considerations*, CONF-801039, National Technical Information Service, Springfield, VA, 1981, pp. 310–324.

Renne, R. A., L. G. Smith, and D. D. Mahlum, Epidermal Carcinogenicity of Some Crude Fossil Fuels in Mice, in D. D. Mahlum, R. H. Gray, and W. D. Felix (eds.), *Coal Conversion and the Environment, Chemical, Biological and Ecological Considerations*, CONF-801039, National Technical Information Service, Springfield, VA, 1981, pp. 471–481.

Renne, R. A., C. W. Wright, L. G. Smith, and R. L. Buschbom, Epidermal Carcinogenesis Studies of Synthetic Fossil Fuel Materials in Mice. *Toxicology*, **40,** 311–325 (1986).

Retcofsky, H. L., F. K. Schweighardt, and M. Hough, Determination of Aromaticities of Coal Derivatives by Nuclear Magnetic Resonance Spectrometry and the Brown–Ladner Equation. *Anal. Chem.*, **49,** 585–588 (1977).

Rhodes, W. J., *Environmental Assessment Source Test and Evaluation Report: Fort Lewis Coal Liquefaction Pilot Plant*, H-C/016/002-81-996, Hittmann Associates, U.S. Environmental Protection Agency, Industrial Environmental Research Laboratory. National Technical Information Service, Springfield, VA, 1981.

Riedle F. J., and A. J. deRosset, *Upgrading of Coal Liquids, Interim Report, Hydrotreating and Fluid Catalysis Cracking of EDS Process Derived Gas Oil*, FE-2566-30, National Technical Information Service, Springfield, VA, 1979.

Robbins, G. A., R. A. Winschel, and F. P. Burke, *Proceedings of the DOE Direct Coal Liquefaction Contractors' Review Meeting*, October 17–18, 1984, Pittsburgh Energy Technology Center. National Technical Information Service, Springfield, VA, 1984.

Rosen, A. A., and F. M. Middleton, Identification of Petroleum Refinery Wastes in Surface Waters. *Anal. Chem.*, **27,** 790–794 (1955).

Royer, R. E., C. E. Mitchell, R. L. Hanson, J. S. Dutcher, and W. E. Bechtold, Fractionation, Chemical Analysis, and Mutagenicity Testing of Low-Btu Coal Gasifier Tar. *Environ. Res.*, **31,** 460–471 (1983).

Ruberto, R. G., and D. C. Cronaur, Effect of Oxygen Function Groups in Coal and Liquefied Coal Solvent Extraction Sequence on the Apparent Distribution of Coal Liquid Products. *Fuel Process. Technol.*, **3,** 215–219 (1979).

Rubin, I. B., M. R. Guerin, A. A. Hardigree, and J. L. Epler, Fractionation of Synthetic Crude Oils from Coal for Biological Testing. *Environ. Res.*, **12,** 358–365 (1976).

Sanders, R. W., K. B. Olsen, W. S. Weimer, and K. K. Neilson, Multielement Analysis of Unweighed Oil Samples by X-Ray Fluorescence Spectrometry with Two Excitation Sources. *Anal. Chem.*, **55,** 1911–1914 (1983).

Schabron, J. F., R. J. Hurtubise, and H. F. Silver, Chromatographic and Spectrometric Methods for the Separation, Characterization, and Identification of Alkylphenols in Coal-Derived Solvents. *Anal. Chem.*, **51,** 1426–1433 (1979).

Scheppele, S. E., *Characterization of Coal-Derived Liquids and Other Fossil Fuel Related Materials Employing Mass Spectrometry*, FE-2537-8, National Technical Information Service, Springfield, VA, 1978a.

Scheppele, S. E., *Characterization of Coal-Derived Liquids and Other Fossil Fuel Related Materials Employing Mass Spectrometry; Mass Spectrometry and Fossil-Energy Conversion Technology: A Review*, FE-2537-7, National Technical Information Service, Springfield, VA, 1978b.

Scheppele, S. E., P. L. Grizzle, S. J. Greenwood, T. D. Marriott, and N. B. Perreira, Determination of Field-Ionization Relative Sensitivities for the Analysis of Coal-Derived Liquids and the IR Correlation with Low-Voltage Electron-Impact Relative Sensitivities. *Anal. Chem.*, **48,** 2105–2113 (1976).

Scheppele, S. E., G. J. Greenwood, and P. A. Benson, Field-Ionization Relative Sensitivities for Analysis of Coal-Derived Liquids Determined as a Function of Ion-Source Temperature and Binary Mixture Composition. *Anal. Chem.*, **49,** 1847–1850 (1977).

Schiller, J. E., Composition of Coal Liquefaction Products. *Hydrocarbon Process.*, **56,** 147–152 (1977a).

Schiller, J. E., Nitrogen Compounds in Coal-Derived Liquids. *Anal. Chem.*, **49,** 2292–2294 (1977b).

Schiller, J. E., and D. R. Mathiason, Separation Method for Coal-Derived Solids and Heavy Liquids. *Anal. Chem.*, **49,** 1225–1228 (1977).

Schindler, H. D., and J. M. Chen, *Integrated Two Stage Liquefaction Topical Technical Progress Report*, Lummus 14804-Q7 and 14804-QB, The Lummus Company Engineering and Development Center, Bloomfield, NJ, 1982.

Schmitter, J. M., I. Ignatiadis, M. Dorbon, P. Arpino, G. Guiochon, H. Toulhoat, and A. Hoc, Identification of Nitrogen Bases in a Coker Gas Oil and Influence of Catalytic Hydrotreatment on Their Composition. *Fuel*, **63,** 557–564 (1984).

Schomburg, G., R. Dielman, H. Borwitzky, and J. Husmann, Capillary Column Gas Chromatography of Compounds of Low Volatility Temperature Stability of Stationary Liquids on Various Glass Surfaces. *J. Chromatogr.*, **167,** 337–354 (1978).

Schultz, J. L., C. M. White, F. K. Schweighardt, and A. G. Sharkey, *Characterization of the Heterocyclic Compounds in Coal Liquefaction Products, Part I. Nitrogen Compounds*, PERC/RI-77/7, National Technical Information Service, Springfield, VA, 1979.

Schultz, R. V., J. W. Jorgenson, M. P. Mackarinec, M. Novotny, and L. J. Todd, Characterization of Polynuclear Aromatic and Aliphatic Hydrocarbon Fractions of Solvent-Refined Coal by Glass Capillary Gas Chromatography/Mass Spectrometry. *Fuel*, **58,** 783–789 (1979).

Schultz, H., and M. J. Mima, Comparison of Methods for the Determination of Asphaltenes, Oils and Insolubles—Part II. *Am. Chem. Soc. Div. Fuel Chem. Prepr.*, **25,** 18–24 (1980).

Schwager, I., and T. F. Yen, Coal-Liquefaction Products from Major Demonstration Processes. 1. Separation and Analysis. *Fuel*, **57,** 100–104 (1978).

Schweighardt, F. K., Coal Liquefaction: Process Description and Effluent Characterization, in D. D. Mahlum, R. H. Gray, and W. D. Felix (eds.), *Coal Conversion and the Environment, Chemical, Biomedical and Ecological Consideration*, CONF-801039, National Technical Information Service, Springfield, VA, 1981, pp. 1–15.

Schweighardt, F. K., and B. M. Thames, Solvent Extraction of Coal-Derived Products. *Anal. Chem.*, **50,** 1381–1382 (1978).

Schweighardt, F. K., C. M. White, S. Friedman, and J. L. Schultz, Heteroatom Species in Coal Liquefaction Products, in J. W. Larsen (ed.), *Organic Chemistry of Coals*, American Chemical Society Symposium Series 71, American Chemical Society, Washington, DC, 1978, pp. 240–257.

Searle, C. E. (ed.), *Chemical Carcinogens*, Vols. 1 and 2, American Chemical Society, Washington, DC, 1984.

Seshadri, K. S., R. G. Ruberto, D. M. Jewell, and H. P. Malone, Carbon-13 Nuclear Magnetic Resonance Spectra of Model Hydroaromatic Hydrocarbons and Solvents. *Fuel*, **57,** 111–116 (1978a).

Seshadri, K. S., R. G. Ruberto, D. M. Jewell, and H. P. Malone, Application of Carbon-13 Nuclear Magnetic Resonance Spectrometry to Coal Chemistry: Calculation of Transferable Hydrogen. *Fuel*, **57,** 549–553 (1978b).

Sklenar, V., M. Hajek, G. Sebor, I. Lang, M. Suchanek, and Z. Starcuk, Analysis of Coal Asphaltenes by Carbon-13 Fourier Transform Nuclear Magnetic Resonance Spectrometry. *Anal. Chem.*, **52,** 1794–1797 (1980).

Slaga, T. J. (ed.), *Tumor Promotion in Internal Organs*, CRC Press, Boca Raton, FL, 1983.

Slaga, T. J. (ed.), *Cellular Responses to Tumor Promoters*, CRC Press, Boca Raton, FL, 1984a.

Slaga, T. J. (ed.), *Tumor Promotion and Carcinogenesis in Vitro*, CRC Press, Boca Raton, FL, 1984b.

Slaga, T. J. (ed.), *Tumor Promotion and Skin Carcinogenesis*, CRC Press, Boca Raton, FL, 1984c.

Slaga, T. J., A. Sivak, and R. K. Boutwell (eds.), *Mechanisms of Tumor Promotion and Cocarcinogenesis*, Raven Press, New York, 1978.

Smith, R. D., and H. R. Udseth, New Method for the Direct Analysis of Supercritical Fluid Coal Extraction and Liquefaction. *Fuel*, **62,** 466–468 (1983a).

Smith, R. D., and H. R. Udseth, Direct Mass Spectrometric Analysis of Supercritical Fluid Extract Products. *Sep. Sci. Technol.*, **18,** 245–252 (1983b).

Smith, R. D., H. T. Kalinoski, H. R. Udseth, and B. W. Wright, Capillary Supercritical Fluid Chromatography—Mass Spectrometry with Electron Impact Ionization. *Anal. Chem.*, **56,** 2476–2480 (1984).

Snape, C. E., Estimation of Aliphatic H/C Ratios for Coal Liquefaction Products by Spin-Echo C.N.M.R. *Fuel*, **62,** 621–624 (1983).

Snape, C. E., B. J. Stokes, and K. D. Bartle, Identification of Straight-Chain Fatty Acids in Coal Extracts and Their Geochemical Relation with Straight Chain Alkanes. *Fuel*, **60,** 903–911 (1981).

Snook, M. E., Gel Elution of Heterocyclic Analogues of Polynuclear Aromatic Hydrocarbons from Bio-Beads. *Anal. Chim. Acta*, **99,** 299–304 (1978).

Snyder, L. R., and B. E. Buell, Nitrogen and Oxygen Compound Types in Petroleum Total Analysis of a 700–850°F Distillate from a California Crude Oil. *Anal. Chem.*, **40,** 1295–1303 (1968).

Snyder, L. R., and J. J. Kirkland, *Introduction to Modern Liquid Chromatography*, Wiley, New York, 1979.

Snyder, L. R., B. E. Buell, and A. E. Howard, Nitrogen and Oxygen Compound Types in Petroleum: A General Separation Scheme. *Anal. Chem.*, **40,** 1303–1317 (1968).

Solash, J., R. N. Hazlett, J. M. Hall, and C. J. Nowack, Relation Between Fuel Properties and Chemical Composition. 1. Jet Fuels from Coal, Oil Shale and Tar Sands. *Fuel*, **57,** 521–528 (1978).

St. John, G. A., S. E. Buttrill, and M. Anbar, Field Ionization and Field Desorption Mass Spectrometry Applied to Coal Research, in *Proceedings*, *American Chemical Society Meeting*, August 29–September 2, 1977, Stanford Research Institute, Palo Alto, CA. American Chemical Society, Washington, DC, 1977, pp. 141–155.

Stadelhofer, J. W., M. Zander, and R. Gerhards, C-13 N.M.R. Study on the Hydrogen Transfer During the Distillations of Crude Coal Tar. *Fuel*, **59,** 604–605 (1980).

Stamoudis, V. C., Chemical and Toxicological Characterization of Morgantown Energy Technology Center Fixed-Bed Gasifier Condensates, DOE/MC/49533-1809 (ANL/SER-6), Argonne National Laboratory, Argonne, IL. National Technical Information Service, Springfield, VA, 1986.

Stamoudis, V. C., S. Bourne, D. A. Haugen, M. J. Peak, C. A. Reilly, J. R. Stetter, and K. Wilzbach, Chemical and Biological Characterization of High-Btu Coal Gasification (the HYGAS Process). I. Chemical Characterization of Mutagenic Fractions, in D. D. Mahlum, R. H. Gray, and W. D. Felix (eds.), *Coal Conversion and the Environment, Chemical, Biomedical and Ecological Considerations*, CONF-801039, National Technical Information Service, Springfield, VA, 1981, pp. 67–95.

Stamoudis, V. C., D. A. Haugen, M. J. Peak, and K. Wilzbach, Biodirected Chemical Characterization of Synfuel Materials, in C. W. Wright, W. C. Weimer, and W. D. Felix (eds.), *Advanced Techniques in Synthetic Fuels Analysis*, CONF-811160, National Technical Information Service, Springfield, VA, 1983, pp. 201–214.

Steffen, F. W., K. T. Schroeder, and B. C. Bockrath, Determination of Asphaltenes in Fuels from Coal Hydrogenation. *Anal. Chem.*, **51,** 1164–1167 (1979).

Sternberg, H. W., C. L. Delle Donne, P. Pantages, E. C. Moroni, and R. E. Markby, Solubilization of a LUB Coal by Reductive Alkylation. *Fuel*, **50,** 432–442 (1971).

Sternberg, H. W., R. Raymond, and F. K. Schweighardt, Acid–Base Structure of Coal-Derived Asphaltenes. *Science*, **188,** 49–51 (1975).

Streuli, C. A., The Effect of Solvent Change on the Separation Process of Sephadex LH-20 Modified Dextran. *J. Chromatogr.*, **56,** 225–229 (1971).

Stroupe, R. C., P. Tokoubalides, R. B. Dickinson, E. L. Wehry, and G. Mamantov, Low-Temperature Fluorescence Spectrometric Determination of Polycyclic Hydrocarbons by Matrix Isolation. *Anal. Chem.*, **49,** 701–705 (1977).

Sturm, G. P., J. E. Dooley, J. S. Thompson, P. W. Woodward, and J. W. Vogh, The Composition of Liquids from Coals of Different Rank, in R. F. Sullivan (ed.), *Upgrading of Coal Liquids*, American Chemical Society Symposium Series 156, American Chemical Society, Washington, DC, 1981.

Swain, A. P., J. E. Cooper, and R. L. Stedman, Large-Scale Fractionation of Cigarette Smoke Condensate for Chemical and Biological Investigations. *Cancer Res.*, **29,** 579–583 (1969).

Swansiger, J. T., H. T. Best, D. A. Danner, and T. L. Youngless, Determination of Transferable Hydrogen in Coal Liquids by Mass Spectrometry. *Anal. Chem.*, **54,** 2576–2582 (1982).

Tan, T., and A. J. deRosset, *Hydrotreating and Reforming H-Coal Process-Derived Naphthas*, Interim Report, UOP Inc. Corporate Research Center, Des Plaines, IL, 1978.

Thakker, D. R., H. Yagi, M. Nordquist, R. E. Lehr, W. Levin, A. W. Wood, R. L. Change, A. H. Cooney, and D. M. Jerina, Polycyclic Aromatic Hydrocarbons and Carcinogenesis: The Bay-Region Theory, in J. C. Arcos, Y.-T. Woo, and M. F. Argus (eds.), *Chemical Induction of Cancer*, Vol. IIIA, Academic Press, New York, 1982, pp. 727–747.

Thompson, R. D., M. Iwao, M. L. Lee, and R. N. Castle, Synthesis of Benzo[3,2]phenanthro[4,5-bed]thiophene and Chryseno[4,5-bed]thiophene. *J. Heterocycl. Chem.*, **18,** 981–984 (1981).

Tokoubalides, P., E. L. Wehry, and G. Mamtov, Observation of Quasilinear Fluorescence Spectra (the Shpol'skii Effect) in Matrix-Isolated Polycyclic Aromatic Hydrocarbons. *J. Phys. Chem.*, **81,** 1769–1772 (1977).

Tominaga, Y., M. L. Lee, and R. N. Castle, Synthesis of *Anthra[b]thiophenes* and *Benzo[b]naphtho[d]thiophenes. J. Heterocycl. Chem.*, **18,** 967–972 (1981a).

Tominaga, Y., M. L. Lee, and R. N. Castle, Synthesis of Phenaleno[1,9-bc]thiophene. *J. Heterocycl. Chem.*, **18,** 977–979 (1981b).

Tomkins, B. A., and C. Feldman, Detection of Primary and Secondary Amines in Energy-Related Materials Using an Element-Selective Glow-Discharge Detector. *Anal. Chim. Acta*, **119,** 283–290 (1980).

Tomkins, B. A., and C.-H. Ho, Determination of Polycyclic Aromatic Amines in Natural and Synthetic Crudes. *Anal. Chem.*, **54,** 91–96 (1982).

Tomkins, B. A., R. R. Reagan, J. E. Caton, and W. H. Griest, Liquid Chromatographic Determination of Benzo[a]pyrene in Natural, Synthetic, and Refined Crudes. *Anal. Chem.*, **53,** 1213–1217 (1981).

Tomkins, B. A., W. H. Griest, J. E. Caton, and R. R. Reagan, Multicomponent Isolation and Analysis of Polynuclear Aromatics, in M. Cooke, A. J. Dennis, and G. L. Fisher (eds.), *Polynuclear Aromatic Hydrocarbons: Physical & Biological Chemistry*, Battelle Press, Columbus, OH, 1982, pp. 813–824.

Toste, A. P., D. S. Sklarew, and R. A. Pelroy, Comparison of Chemical and Mutagenic Properties of a Coal Liquid and Shale Oil, in D. D. Mahlum, R. H. Gray, and W. D. Felix (eds.), *Coal Conversion and the Environment, Chemical, Biomedical and Ecological Considerations*, CONF-801039, National Technical Information Service, Springfield, VA, 1981, pp. 96–114.

Toste, A. P., D. S. Sklarew, and R. A. Pelroy, Partition Chromatography–High-Performance Liquid Chromatography Facilitates the Organic Analysis and Biotesting of Synfuels. *J. Chromatogr.*, **249,** 267–282 (1982).

Toste, A. P., D. S. Sklarew, and R. A. Pelroy, The Role of Partition Chromatography/HPLC in the Organic and Mutagenic Characterization of Synfuels, in C. W. Wright, W. C. Weimer, and W. D. Felix (eds.), *Advanced Techniques in Synthetic Fuels Analysis*, CONF-81160, National Technical Information Service, Springfield, VA, 1983, pp. 74–94.

Uden, P. C., A. Carpenter, Jr., F. P. DiSanzo, H. F. Hackett, S. Siggia, and D. E. Henderson, Aspects of Chromatographic Analysis of Oil Shale and Shale Oil, in P. C. Uden, S. Siggia, and H. B. Jensen (eds.), *Analytical Chemistry of Liquid Fuel Sources: Tar, Sands, Oil Shale, Coal and Petroleum*, American Chemical Society, Washington, DC, 1978, pp. 213–231.

Vassilaros, D. L., R. C. Kong, D. W. Later, and M. L. Lee, Linear Retention Index System for Polycyclic Aromatic Compounds—Critical Evaluation and Additional Indices. *J. Chromatogr.*, **252,** 1–20 (1982a).

Vassilaros, D. L., D. W. Stoker, G. B. Booth, and M. L. Lee, Capillary Gas Chromatographic Determination of Polycyclic Aromatic Compounds in Vertebrate Fish Tissue. *Anal. Chem.*, **54,** 106–112 (1982b).

Vo-Dinh, T., and P. R. Martinez. Direct Determination of Selected Polynuclear Aromatic Hydrocarbons in a Coal Liquefaction Product by Synchronous Luminescence Techniques. *Anal. Chim. Acta*, **125,** 13–19 (1981).

Vogh, J. W., and J. S. Thomson, Preparative Liquid Chromatography for Fractionation of Petroleum and Synthetic Crude Oils. *Anal. Chem.*, **53,** 1345–1350 (1981).

Warzinski, R. P., and J. A. Ruether, Fractional Destraction of Coal Liquefaction Products, in *Proceedings of the Department of Energy Direct Coal Liquefaction Contractors Review Meeting*, November 16–17, 1983, Pittsburgh Energy Technology Center, Pittsburgh, PA, 1983, pp. 11-1–11-32.

Weil, C. S., and N. I. Condra, The Hazards to Health in the Hydrogenation of Coal: II. Carcinogenic Effects of Materials on the Skin of Mice. *Arch. Environ. Health*, **1,** 187–193 (1960).

Weimer, W. C., and C. W. Wright, *Decommissioning Samples from the Ft. Lewis, WA Solvent Refined Coal Pilot Plant: Chemical Analysis and Biological Testing*, PNL-5556, Pacific Northwest Laboratory, Richland, WA. National Technical Information Service, Springfield, VA, 1985.

Weimer, W. C., B. W. Wilson, R. A. Pelroy, and J. C. Craun, *Initial Chemical and Biological Characterization of Hydrotreated SRC-II Liquids; A Status Report*, PNL-3464, Pacific Northwest Laboratory, Richland, WA. National Technical Information Service, Springfield, VA, 1980.

Wender, I., Catalytic Synthesis of Chemicals from Coal. *Am. Chem. Soc. Div. Fuel Chem. Prepr.*, **170,** 24(A) (1975).

White, C. M., Instrumental Methods for the Determination of PAH in Coal and Coal Derived Materials. *J. Am. Chem. Soc.*, **27,** 185–194 (1982).

White, C. M., Determination of Polynuclear Aromatic Hydrocarbons in Coal-Derived Materials, in A. Bjørseth (ed.), *Handbook of Polycyclic Aromatic Hydrocarbons*, Vol. 1, Marcel Dekker, New York, 1983, pp. 525–616.

White, C. M., and M. L. Lee, Identification and Geochemical Significance of Some Aromatic Components of Coal. *Geochim. Cosmochim. Acta*, **44,** 1825–1832 (1980).

White, C. M., J. L. Schultz, and A. G. Sharkey, Biological Markers in Coal and Coal Liquefaction Products. *Nature*, **268,** 620–622 (1977).

White, C. M., F. K. Schweighardt, and J. L. Schultz, Combined Gas Chromatographic–Mass Spectrometric Analyses of Nitrogen Bases in Light Oil from a Coal Liquefaction Product. *Fuel Process. Technol.*, **1,** 209–215 (1978).

White, C. M., L. Jones, N. C. Li, and K. C. Tewari, Aging of SRC-II Middle Distillate from Illinois No. 6 Coal II. Further Evidence of Phenolic Oxidative Coupling in Coal-Derived Liquids. *Fuel*, **62,** 1397–1403 (1983).

Whitehurst, D. D., S. E. Butrill, F. J. Derbyshire, M. Farcasiu, G. A. Odoerfer, and R. Rudnick, New Characterization Techniques for Coal-Derived Liquids. *Fuel*, **61,** 994–1006 (1982).

Willey, C., M. Iwao, R. N. Castle, and M. L. Lee, Determination of Sulfur Heterocycles in Coal Liquids and Shale Oils. *Anal. Chem.*, **53,** 400–407 (1981a).

Willey, C., R. A. Pelroy, D. S. Sklarew, and B. W. Wilson, Chemical Analysis and in Vitro Mutagenicity Testing of Water Soluble Constituents from an SRC Solid Product. *Am. Chem. Soc. Div. Environ. Chem. Prepr.*, **182,** 60–62 (1981b).

Wilson, B. W., and R. A. Pelroy, Mass Spectral Analysis of Mutagenically Active Subfractions from Coal Liquids, in *Proceedings of the 24th Annual Conference on Mass Spectrometry and Allied Topics*, American Society for Mass Spectrometry, East Lansing, MI, 1979, pp. 23–25.

Wilson, B. W., and R. A. Pelroy, *Biological and Biodirected Chemical Analysis of Veba Oel Coal Liquids*, prepared for Veba Oel A. G. Gelsenkirchen Beur, Federal Republic of Germany. Battelle, Pacific Northwest Laboratories, Richland, WA, 1983.

Wilson, B. W., R. A. Pelroy, and J. T. Cresto, Identification of Primary Aromatic Amines in Mutagenically Active Sub-Fractions from Coal Liquefaction Materials. *Mutat. Res.*, **79,** 193–202 (1980).

Wilson, B. W., M. R. Petersen, R. A. Pelroy, and J. T. Cresto, In-Vitro Assay for Mutagenic Activity and Gas Chromatographic–Mass Spectral Analysis of Coal Liquefaction Material and the Products Resulting from its Hydrogenation. *Fuel*, **60,** 289–294 (1981a).

Wilson, B. W., A. P. Toste, R. A. Pelroy, B. Vieux, and D. Wood, Accurate Mass/Metastable Ion Analysis of Higher Molecular Weight Nitrogen Compounds in Coal Liquids, in D. D. Mahlum, R. H. Gray, and W. D. Felix (eds.), *Coal Conversion and the Environment, Chemical, Biomedical and Ecological Considerations*, CONF-801039, National Technical Information Service, Springfield, VA, 1981b, pp. 148–158.

Wilson, B. W., E. K. Chess, and M. L. Gross, High Performance MS in the Optimization of Coal Liquefaction Processes, in *Abstracts, 30th Annual Conference on Mass Spectrometry and Allied Topics*, American Society for Mass Spectrometry, East Lansing, MI, 1982a, pp. 810–811.

Wilson, B. W., R. A. Pelroy, and D. D. Mahlum, *Chemical Characterization and Genotoxic Potential Related to Boiling Point for Fractionally Distilled SRC-I Coal Liquids*, PNL-4277, Pacific Northwest Laboratory, Richland, WA. National Technical Information Service, Springfield, VA, 1982b.

Wilson, B. W., C. Willey, D. W. Later, and M. L. Lee, Effect of Process Distillation on the Distribution of Amino Polycyclic Derived Liquids. *Fuel*, **61,** 473–477 (1982c).

Wilson, B. W., P. Buhl, and E. C. Moroni, *Biological Testing and Chemical Analysis of Materials from an Integrated Two-Stage Coal Liquefaction Process: A Status Report*, PNL-4553, Pacific Northwest Laboratory, Richland, WA. National Technical Information Service, Springfield, VA, 1983a.

Wilson, B. W., R. A. Pelroy, R. P. Anderson, and J. Freel, *Chemical and Biologicial Effects of Heavy Distillate Recycle in the SRC-II Process*, PNL-4815, Pacific Northwest Laboratory, Richland, WA. National Technical Information Service, Springfield, VA, 1983b.

Wilson, B. W., R. A. Pelroy, D. D. Mahlum, M. E. Frazier, and C. W. Wright, Heteroatomic Polycyclic Aromatic Hydrocarbons, in C. W. Wright, W. C. Weimer, and W. D. Felix (eds.),

Advanced Techniques in Synthetic Fuels Analysis, CONF-11160, National Technical Information Service, Springfield, VA, 1983c, pp. 231–250.

Wilson, B. W., R. A. Pelroy, D. D. Mahlum, M. E. Frazier, D. W. Later, and C. W. Wright, Comparative Chemical Composition and Biological Activity of Single- and Two-Stage Coal Liquefaction Process Streams. *Fuel*, **63,** 46–55 (1984).

Wilson, B. W., D. W. Later, C. W. Wright, and D. L. Stewart, *Chemical Analysis and Mutational Assay of Fractionally Distilled Oils from the H-Coal Direct Liquefaction Process*, PNL-5288, Pacific Northwest Laboratory, Richland, WA. National Technical Information Service, Springfield, VA, 1985.

Wilson, B. W., D. D. Mahlum, and R. A. Pelroy, Biomedical Implications of Altered Product Composition in Advanced Coal Liquefaction Processes. *Fuel Process. Technol.*, **13,** 1–16 (1986).

Wilzbach, K. E., J. R. Stetter, C. A. Reilly, Jr., and W. G. Wilson, *Environmental Research Program for Slagging Fixed-Bed Coal Gasification, Status Report*, ANL/SER-1, Argonne National Laboratory, Argonne, IL, 1982.

Wise, S. A., High Performance Liquid Chromatography for the Determination of Polycyclic Aromatic Hydrocarbons, in A. Bjørseth (ed.), *Handbook of Polycyclic Aromatic Hydrocarbons*, Marcel Dekker, New York, 1983, pp. 183–256.

Wise, S. A., S. N. Chesler, H. S. Hertz, L. R. Hilpert, and W. E. May, Chemically-Bonded Aminosilane Stationary Phase for the High-Performance Liquid Chromatographic Separation of Polynuclear Aromatic Compounds. *Anal. Chem.*, **49,** 2306–2310 (1977).

Wise, S. A., W. J. Bonnett, and W. E. May, Normal and Reverse Phase Liquid Chromatographic Separations of Polycyclic Aromatic Hydrocarbons, in A. Bjørseth and A. J. Dennis (eds.), *Polycyclic Aromatic Hydrocarbons: Chemistry and Biological Effects*, Battelle Press, Columbus, OH, 1980, pp. 791–806.

Wise, S. A., R. M. Campbell, W. E. May, M. L. Lee, and R. N. Castle, Normal and Reversed-Phase Liquid Chromatographic Separations of Polycyclic Aromatic Sulfur Heterocycles, in M. W. Cooke and A. J. Dennis (eds.), *Polynuclear Aromatic Hydrocarbons: Formation Metabolism, and Measurement*, Battelle Press, Columbus, OH, 1982, pp. 1247–1266.

Wiser, W. H., *Applied Research and Evaluation of Process Concepts for Liquefaction and Gasification of Western Coals,* Progress Report, U.S. Department of Energy, Fossil Energy, Department of Mining and Fuels Engineering, Salt Lake City, UT, 1978.

Wong, J. L., and F. M. Gladstone, Compositional Analysis of an H-Coal Liquid by Gas Chromatography–Mass Spectrometry–Computer Technique. *J. Chromatogr.*, **267,** 303–314 (1983).

Woo, C. S., A. P. DiSilva, and V. A. Fassel, Polynuclear Aromatic Hydrocarbon in Coal: Identification by Their X-Ray Excited Optical Luminescence. *Environ. Sci. Technol.*, **12,** 173–174 (1978).

Woo, C. S., A. P. DiSilva, and V. A. Fassel, Characterization of Environmental Samples for Polynuclear Aromatic Hydrocarbons by an X-Ray Excited Optical Luminescence Technique. *Anal. Chem.*, **52,** 159–164 (1980).

Wood, A. W., R. L. Chang, W. Levin, D. E. Ryan, P. E. Thomas, R. E. Lehr, S. Kumar, M. Shaefer-Ridder, U. Englehardt, H. Yagi, D. M. Jerina, and A. H. Conney, Mutagenicity of Diol-Epoxides and Tetrahydroepoxides of Benz [a] acridine and Benz [c] acridine in Bacteria and in Mammalian Cells. *Cancer Res.*, **43,** 1656–1662 (1983).

Wood, K. A., T. L. Slager, M. L. Klusariz, D. R. Elliott, K. C. Tewari, D. M. Parees, J. D. Moyer, A. Z. Kamelski, S. F. Palopoli, and R. C. Cusick, *SRC-I Demonstration Plant Analytical Laboratory*, DOE/OR/O3054-6, National Technical Information Service, Springfield, VA, 1981.

Wozniak, T. J., and R. A. Hites, Chemical and Biological Characterization of Phenanthrene Hydroaromatics, in C. W. Wright, W. C. Weimer, and W. D. Felix (eds.), *Advanced Techniques in Synthetic Fuels Analysis*, CONF-811160, National Technical Information Service, Springfield, VA, 1983, pp. 251–262.

Wright, B. W., and R. D. Smith, Application of Capillary Supercritical Fluid Chromatography to the Analysis of Middle Distillate Fuel. *Chromatographia,* **18,** 542–545 (1984).

Wright, C. W., and D. D. Dauble, *Effects of Coal Rank on the Chemical Composition and Toxicological Activity of Coal Liquefaction Materials*, PNL-5805, Pacific Northwest Laboratory, Richland, WA. National Technical Information Service, Springfield, VA, 1986.

Wright, C. W., D. W. Later, D. D. Dauble, and B. W. Wilson, *Fractionally Distilled SRC-I, SRC-II, EDS, H-Coal and ITSL Direct Coal Liquefaction Materials: A Comparative Summary of Chem-*

ical Analysis and Biological Testing, PNL-5528, Pacific Northwest Laboratory, Richland, WA. National Technical Information Service, Springfield, VA, 1985.

Yamagiwa, K., and K. Ichikawa, *Mitt. Med. Fak. Tokio*, **15,** 295 (1915).

Yang, F. J., and S. P. Cram, Characteristics and Performance of Gas Chromatographic Detectors with Glass Capillary Columns. *HRC&C*, **2,** 487–496 (1979).

Yang, S. K., J. Deutsch, and H. V. Gelboin, Benzo[a]pyrene Metabolism: Activation and Detoxification, in H. V. Gelboin and P. O. P. Ts'o (eds.), *Polycyclic Hydrocarbons and Cancer*, Vol. I, Academic Press, New York, 1978, pp. 205–231.

Yang, Y., A. P. D'Silva, V. A. Fassel, and M. Iles, Direct Determination of Polynuclear Aromatic Hydrocarbons in Coal Liquids and Shale Oil by Laser Excited Shpol'skii Spectrometry. *Anal. Chem.*, **52,** 1350–1351 (1980).

Yoshida, D., and T. Matsumoto, Changes in Mutagenicity of Protein Pyrolyzates by Reaction with Nitrate. *Mutat. Res.*, **58,** 35–40 (1978).

Zackett, D., and R. G. Cooks, Analysis of Synfuels by MS/MS, in B. D. Blaustein, B. C. Bockrath, and S. Friedman (eds.), *New Approaches in Coal Chemistry*, American Chemical Society Symposium Series 169, American Chemical Society, Washington, DC, 1981, pp. 267–288.

Zackett, D., V. M. Shaddock, and R. G. Cooks, Analysis of Coal Liquids by Mass-Analyzed Ion Kinetic Energy Spectrometry. *Anal. Chem.*, **51,** 1849–1852 (1979).

CHAPTER 6

Health Effects in Coal-Conversion Processes: Worker Health Experience

W. Hubis
Denver, Colorado

6.1 COAL COKING AND COMBUSTION

Historically, the two principal uses of coal have been its conversion to coke for use in the steel industry and its combustion for the production of heat and electric power generation. By-products from coking include coal tar, coal-tar pitch, and creosote; the latter two are derived from the distillation of coal tar. Coal-tar pitch is used as a binder for electrodes in the aluminum reduction process, while creosote is used primarily as a wood preservative. Crude coal tar is widely used in the clinical treatment of acute and chronic dermatosis, occupational and contact dermatitis, and psoriasis. Only in the last several decades has toxicological and epidemiological evidence been sufficient to conclude that coal tar and coal-tar pitch can cause lung, skin, and perhaps other cancers, and that creosote can cause skin and lung cancer (NIOSH, 1977; EPA, undated). Coke-oven emissions have been the subject of many toxicological and epidemiological studies that eventually culminated in an Occupational Safety and Health Administration (OSHA) standard on exposure to coke-oven emissions (OSHA, 1976; EPA, 1982).

Combustion of fossil fuels can produce polycyclic organic matter (NAS, 1972); the amount formed depends on combustion efficiency. Efficient, controlled combustion favors low emissions of polycyclic organic matter, whereas inefficient burning favors high emissions. Other studies (Tebbens et al., 1956; 1958) have shown that the incomplete combustion of any carbonaceous material forms condensed, multinuclear, aromatic hydrocarbons, many of which are toxic. For example, Hangebrauck et al. (1967) concluded that the most important source of benzo[a]pyrene (BaP), a known carcinogen, was inefficient coal combustion in hand-fired residential furnaces. In power plants that burn crushed or pulverized coal, BaP emissions are generally low, regardless of the fuel used.

6.1.1 Coal Coking

6.1.1.1 *Industrial Hygiene*

Prior to the mid-1950s, there is essentially no information in the literature relative to industrial hygiene practices during coal coking operations. One can only assume that minimal protection was available, that is, respirators and ventilation systems. In 1954 (O'Connor, 1971), 17 companies initiated a group research program to develop information as to whether inhalation of coal tar pitch volatiles (CTPV) might produce lung cancer, the concern being the health of coke-oven workers. Prior research had shown that tumors developed when coal tar pitch was repeatedly painted on the skin of mice. In 1967, the American Conference of Governmental Industrial Hygienists adopted a threshold limit value (TLV) of 0.20 mg/m^3 CTPV, described as a benzene-soluble fraction (BSF) of the particulates, and listed certain carcinogenic components of CTPV. The TLV was established to minimize exposures to the listed substances believed to be carcinogens: anthracene, BaP, phenanthrene, acridine, chrysene, and pyrene. The TLV was later promulgated as a Federal Standard (29 CFR 1910.1000) under the Occupational Safety and Health Act of 1970.

On June 8, 1971, the American Iron and Steel Institute petitioned the Secretary of Labor to develop a standard specific to coke-oven emissions and revoke the existing standard of 0.20 mg/m^3 CTPV on the grounds that it was an invalid measure for coke-oven worker exposure. On September 9, 1971, the Labor Department denied the petition and set out certain protective measures (use of respirators, protective skin creams, and medical examinations) pending installation of feasible engineering controls. In 1973, the National Institute for Occupational Safety and Health (NIOSH) published the *Criteria for a Recommended Standard—Occupational Exposure to Coke Oven Emissions* (NIOSH, 1973), recommending work practices to minimize the harmful effects from exposure to coke-oven emissions and inhalation of CTPV. In 1976 (OSHA, 1976), OSHA promulgated a Federal Standard on coke-oven emissions to reduce employee exposure to carcinogenic chemicals. The standard was based on data from epidemiological and animal studies indicating that chemicals in coke-oven emissions could produce skin and lung cancers in humans and animals.

The standard requires that the employer assure that no employee is exposed to coke-oven emissions at concentrations above 150 μg/m^3 of air, averaged over any 8-h period. It also requires that the employer control employee exposure to coke-oven emissions through engineering controls, work practices, and respiratory protection. Among the requirements are the following:

- Monitor exposure to coke-oven emissions of all employees working in regulated areas over an 8-h period at least every 3 months. This includes at least one full-shift personal sample for each period and each job classification.
- Notify employees of the monitoring results and, where necessary, take corrective actions to reduce exposure to or below the permissible exposure limit.
- Provide and assure the use of respirators when required.
- Provide and assure the use of appropriate protective clothing and equipment.
- Provide protective clothing in a clean and dry condition at least weekly, and clean, launder, or dispose of the clothing as required.
- Provide clean change rooms equipped with storage facilities for street clothes and separate storage facilities for protective clothing and equipment.
- Provide shower facilities and assure that employees working in regulated areas shower at the end of a work shift.

- Provide a medical examination at the time of initial assignment to a regulated area, and annually thereafter, with at least the following elements: work history, medical history, posterior–anterior chest X-ray, pulmonary function test, skin examination, urinalysis, sputum cytology, and urinary cytology.
- Provide a training program at least annually for all employees in the regulated area and assure their participation.
- Maintain accurate records of all measurements taken to monitor employee exposure to coke-oven emissions and of medical histories for each employee subject to medical surveillance.

6.1.1.2 *Epidemiology*

Many excellent and extensive reviews have been published (NIOSH, 1973, 1977; OSHA, 1976; Freudenthal et al., 1977; EPA, 1982, undated) on the numerous epidemiological studies conducted on workers exposed to coke-oven emissions, coal tar, coal-tar pitch, and creosote. A sampling of these studies is reviewed here to demonstrate the basis for one of the principal concerns in later coal-conversion pilot plants for potential adverse health effects to workers exposed to coal-conversion materials. These studies indicate that workers routinely exposed to coal carbonization, coal-tar fumes, and coke-oven emissions are at excess risk for lung cancer. Thus, since other coal-conversion processes would be expected to produce materials having similar components, considerable concern was evident that similar effects would be observed unless appropriate precautions were taken to protect the workers.

Doll et al. (1965) conducted a prospective mortality study of employees and pensioners who had worked at least 5 years for four British associations of gas companies (gas boards). The subjects were grouped into three classes: Class A—workers with high exposures to coal-tar products; Class B—maintenance workers with intermittent exposures to products in the gasworks; and Class C—workers exposed only to by-products or with minimal or no exposure. A total of 11,499 men were studied for 8 years; only 50 of these were not traced through the end of the study. Class A workers had the highest total death rate and the highest death rate from lung cancer, the latter being 69% higher than that for Class C or the national rate. The Class A death rate from bronchitis was 126% higher than that for Class C and 112% higher than the national rate. Death rates from the various causes in each of the four gas boards were also compared with regional death rates. Lung cancer and bronchitis mortalities in Class A workers were both higher than the regional rates in workers from all four gas boards; the increase in mortality ranged from 9 to 74% for lung cancer and from 7 to 144% for bronchitis. A study of Class A workers according to the type of retort house showed that workers in horizontal retort houses had higher lung cancer mortalities, while workers in vertical retort houses had mortalities from bronchitis higher than the national rates. Because ventilation in horizontal retort houses was inadequate, workers there probably had higher exposures. Doll et al. (1965) concluded that exposure to products of coal carbonization increased death rates from lung cancer and bronchitis, and that these rates increased with exposure intensity in the three occupational classes.

In a subsequent study, Doll et al. (1972) followed 3028 workers from the same four gas boards for an additional 4 years. The workers were divided into two groups: 2449 coal carbonization workers and 579 by-product and chemical plant workers. Another 4687 workers were added from four additional gas boards not previously studied. The workers were 40–65-year-old men who had been employed for at least 5 years. The cumulative data of 12 years, including those already reported, were also analyzed. The study confirmed earlier findings that exposure to products of coal carbonization can lead to cancer of the lungs. Work in the retort house also increased the risk of death from bladder and, to some extent, scrotal cancer. The data, contrary to earlier findings, provided some indication that bronchitis is a specific occupational hazard for gas workers. These findings were related by Doll et al. (1972) to changes in production methods and/or to long latency periods following poor working conditions during the war years. However, data from the four additional gas boards provided no evidence that by-product workers experienced increased risk of death related to their occupation.

Kawai et al. (1967) reported on an epidemiological study involving lung cancers among males with long exposures to tar fumes while working near the gas generator of a steel plant. Excess mortalities from lung cancers were noted only for subjects aged 45–54 years. Although subjects in this age group with 10–19 years experience in gas generator work showed a significantly increased risk of lung cancer, those under 45 years with the same work experience did not. The authors suggested that lung cancers of this origin become manifest only after workers grow old and reach a certain predisposed age. Kawai et al. (1967) concluded that workers exposed to gas generator operations showed an increased risk of lung cancer even after they stopped working. Tumors occurred only after exposures of 10 or more years to the carcinogenic agents and when workers reached 45–54 years of age. The longer the exposure, the greater the risk of dying from lung cancer.

Lloyd (1971) compared mortalities of coke-oven workers with those expected in a control population of steelworkers in seven plants in Allegheny County, Pennsylvania. The study compared the mortalities from 1953 to 1961, of coke-oven and non-coke-oven workers with expected deaths, by age-, race-, and nativity-specific rates for the total steelworker population. Results indicated that mortalities of coke-oven workers from respiratory cancer were 2.5 times that expected. Further breakdown of the study population by job location (topside, partial topside, side oven) showed a fivefold greater risk of lung cancer for topside workers. The risk after 5 or more years of full-time topside employment was 10 times the lung cancer rate experienced by other steelworkers and three times the rate for those with less than 5 years topside experience. It was concluded that mortalities for side-oven or partial-topside employees were not significantly different from what had been expected. In 1972, Redmond et al. (1972) expanded the study and considered mortality patterns among coke-oven workers from various geographical areas. The authors concluded that mortalities among coke-oven workers from respiratory cancers were over 2.5 times those predicted. The relative risks of lung cancer were 6.87 for men with 5 or more years full-topside employment, 3.22 for men with 5 or more years mixed-topside and side-oven experience, 2.10 for men with 5 or more years side-oven experience only, and 1.70 for all men with less than 5 years experience at start of follow-up. A new finding in this study was a risk 7.5-fold greater than normal of dying from kidney cancer among coke-oven workers.

Redmond et al. (1976) conducted an epidemiological study to determine the mortality risk from several organ cancers in steelworkers, including coke-plant workers. The 2543 coke-plant workers studied had a relative risk of 1.93 for all respiratory cancers; 2.01 for cancers of the lungs, bronchi, and trachea; 1.82 for genitourinary system cancer; and 5.00 for kidney cancer. The authors concluded that coke-oven workers with 5 years or more experience had a high risk of lung and kidney cancers, while non-coke-oven workers had a high risk of colon and pancreas cancer. Cancer of the buccal cavity and pharynx also appeared at a higher rate in non-coke-oven workers.

Hammond et al. (1976) described a cohort cancer mortality study of coal-tar pitch workers. Workers at a roofing site inhaled an average 53 μg BaP in 7 h, with concentrations ranging from undetectable to 135 μg BaP. The authors examined 5939 records for pitch workers who were union members between January 1, 1960 and December 31, 1971. The workers, aged 39–80 years, had 9–40 or more years work experience. The authors traced 5788 men for 12 years, while the remaining 151 were traced an averge of 5.5 years. The ratio of lung cancer deaths observed in pitch workers to the number of deaths expected from U.S. mortality data was 0.92 for workers exposed less than 20 years, 1.5 for those exposed 20–39 years, and 2.47 for those exposed 40 years or longer. Hammond et al. (1976) concluded that work exposure to BaP was associated with increased mortality from lung cancer. However, other agents to which the pitch workers were exposed, either singly or in combination, could have contributed to the increased lung-cancer mortality. Although smoking histories were not determined, many workers smoked cigarettes, and cigarette smoke and pitch fumes may have been concomitant or synergistic causes of lung cancer.

6.1.2 Coal Combustion

In 1775, Pott first observed that products from coal combustion could have potentially adverse human health effects (Pott, 1963). He attributed scrotal cancer in chimney sweeps to their occupational exposure to soot. Chimney sweeps in England did not bathe regularly and had long exposures to soot. Many sweeps began working at the early age of 4 or 5. Small children were needed to crawl through a chimney, sometimes while it was still hot. More recently, with the increase in gas and oil prices, wood and coal are again being used more routinely (Bagchi and Zimmerman, 1980). However, while coal is still used in small residential furnaces and fireplaces, the predominant coal combustion usage is in the utility industry to generate steam and electric power.

6.1.2.1 *Industrial Hygiene*

Efficiency of coal combustion is the primary factor in forming potentially hazardous by-products, mainly particulate organic material, including such undesirable compounds as polycyclic aromatic hydrocarbons (PAHs), some with known toxic properties (Tebbens et al., 1956, 1958; NAS, 1972). Production of these compounds decreases with increasing combustion efficiency. Because combustion efficiency is also a critical economic factor in energy production, the utility industry in general, has not been concerned with potentially adverse occupational health exposures other than to such things as dust, silica, and possibly trace elements. Generally, these emissions have been controlled effectively using dust-suppression materials and good ventilation. Thus, in the past, only standard industrial hygiene practices were adopted by utilities. Recently, however, several larger utilities have established more comprehensive industrial hygiene programs (EPRI, 1982). This may reflect a general awareness by industry of the need to decrease occupational health exposures to the extent possible. Extensive industrial hygiene programs were instituted at two utilities in preparation for and during test burns of two coal-liquefaction fuel oils that were being evaluated as potential replacements for petroleum fuel oils (EPRI, 1979, 1981).

6.1.2.2 *Solvent-Refined Coal II (SRC-II) Fuel-Oil Test Burn*

In September 1978 (EPRI, 1979), a combustion demonstration using distillate fuel oil from the SRC-II process was conducted in a utility boiler at the 74th Street Generating Station of the Consolidated Edison Company, New York. About 4500 bbl of a blended product, representing liquid fuels produced from several different coals at a coal-liquefacton pilot plant at Ft. Lewis, Washington, were burned. An extensive personal and ambient air monitoring program and an employee surveillance and industrial hygiene program were in effect during the burn. Plant operating personnel actively engaged in all phases of the burn were monitored using activated charcoal adsorption tubes and particulate samplers. Samples were also taken at numerous locations throughout the facility for airborne organic vapors, benzene, benzene solubles, phenols, and organic particulates. No measurable or identifiable compounds were detected during the test burn. Samples were also taken during fuel transfers involving company personnel and/or company-owned equipment (on or off company property), during maintenance operations on SRC-II fuel-oil wetted equipment and during any spill incident or where there was potential for contact with the fuel oil. Specific areas were designated as controlled access and limited only to personnel who had been medically examined and cleared to work on the project. An employee health education program included general information, first-aid procedures for eye and skin contamination, personal hygiene practices (general practices, showers, skin protection), personal protective equipment, physical examinations, industrial hygiene monitoring, and spill cleanup and control procedures. Medical and dermatological examinations were done on all employees involved in the test burn.

6.1.2.3 *EDS Fuel-Oil Test Burn*

Another test burn was performed with an EDS distillate fuel oil in a 44.5-MW utility boiler (EPRI, 1981). Industrial hygiene was emphasized during the test program by plant management and the Electric Power Research Institute (EPRI). Lectures on safety and how to handle the fuel were a major part of the pretest activities. Each program participant was required to change into work clothes in the morning and shower before changing back into street clothes at the end of each shift. One pair of work boots and three sets of coveralls and underclothes were provided to each man. Heavy sweat shirts were available for cold mornings, and an array of other protective clothing was also available. A local laundry picked up and washed soiled clothing daily. The carcinogenic potential of the test fuel led to considerable management and employee concern during the test program. Thus, an industrial hygienist familiar with the fuel was always at the station during testing. No fuel spills occurred, and no one was contaminated with the fuel. It was concluded that the attitudes of plant management and employees are critical to converting a large central station for use with coal-based fuels. Early and repetitive training is needed to provide information and overcome irrational fears prior to working with these fuels. The number of personnel needing protective clothing also needs to be carefully evaluated and efforts should be made to minimize the number of personnel having access to areas of potential hazardous exposures.

6.1.2.4 *EPRI*

In 1981, EPRI funded a study of health, personnel protection, and regulatory considerations for use of coal-derived fuels as a direct replacement for petroleum-derived fuels in electric power production (EPRI, 1982). The study examined occupational exposures to conventional petroleum-derived fuel oils in utility, transportation workers, and the general public; reviewed existing rules and regulations applicable to transporting and handling coal-derived fuels; estimated effects of exposure to coal-derived fuels; provided draft guidelines and procedures for transporting and handling coal-derived fuels; enumerated areas requiring additional research; and defined areas requiring regulatory and legislative clarification or modification. Briefly, the study concluded that, in general, there appear to be no serious impediments to the use of coal-derived liquids as direct replacements for petroleum fuels in electric utility generation plants. However, additional industrial hygiene and occupational health practices and possibly some protective engineering measures will be required for safe handling of these fuels.

6.1.3 Summary and Conclusions

Coking of coal has been shown epidemiologically to be a serious occupational hazard. Health concerns in other coal-conversion processes are based to some extent on the hazards related to coking operations. Health hazards in coal coking are associated with direct exposures to the coking operation, particularly topside, and to the by-product coal tar, coal-tar pitch, and creosote. The by-products have been the subjects of many epidemiological studies (Tebbens et al., 1958; Doll et al. 1965, 1972; Hangebrauck et al., 1967; Kawai et al., 1967; Lloyd, 1971; O'Connor, 1971; Redmond et al., 1972, 1976; NIOSH, 1973; Freudenthal et al., 1977). Standards and industrial hygiene and medical surveillance requirements have been mandated by OSHA to protect the coke-oven workers (NIOSH, 1973, 1977; OSHA, 1976).

Coal combustion is used primarily in the public utility industry to generate steam and electrical power. Because of high combustion efficiency, few if any PAHs are generated and these compounds are therefore not of principal concern. Dust, silica, and possibly trace elements, all of which can be effectively controlled through use of appropriate engineering controls and protective equipment, are the major concerns during coal combustion. While only standard industrial hygiene practices have generally been used in the past, there are recent indications of more extensive occupational health programs being employed (EPRI, 1982). This may partially reflect a general awareness of the need to decrease occupational exposures to the extent possible.

6.2 COAL LIQUEFACTION

Prior to start-up of the solvent-refined coal (SRC) pilot plant at Ft. Lewis, Washington, late in 1974, there was considerable concern for the potential health effects from exposure to organic compounds, primarily PAHs. Much of the concern was for the possibility of skin cancers in workers in the plant. The concern stemmed, to some degree, from documented exposure effects among coke-oven workers and, to a greater extent, on experiences at the Union Carbide Corporation's coal-hydrogenation plant at Institute, West Virginia, during 1952–1956 (Sexton et al., 1960). This concern became even greater after initial toxicological studies on coal-liquefaction material were published (Gray and Cowser, 1982).

Sexton et al. (1960) reported that the medical surveillance program involving 359 workers at the Institute plant revealed one worker with two cutaneous cancers, nine workers with one cutaneous cancer each, and 40 workers with one cutaneous precancerous lesion each. The 50 affected workers were followed in a mortality study by Palmer (1979). After 18–20 years, five cardiac-related deaths occurred in the group. Although the study involved a small sample size and only workers with skin lesions, Palmer concluded that the data demonstrated an increase in chronic disease and death, particularly from systemic cancers caused by exposure to high-boiling oils containing polycyclic hydrocarbons, coal tar, and pitch.

The medical surveillance program at the plant included a physical exam for all personnel working at the plant in the latter half of 1951, prior to the start-up, with annual physical exams thereafter (Sexton et al., 1960). However, skin examinations were not initiated until 1955. In 1953, after receiving information that materials in some process streams were capable of producing cancers in experimental animals, some protective measures were introduced to protect exposed workers. However, convincing supervisors and employees that there might be unusual health hazards associated with the coal-hydrogenation process was difficult. It was not until 1955, after the first suspected skin cancer was observed, that a complete precautionary program was strictly enforced by supervisors and rigidly followed by workers. A health education program was instituted, workmen were provided with protective equipment, and the particular work area was placed under specific precautionary restrictions. Beginning in 1954, selective placement of new workers in the process area was attempted but was difficult to accomplish. An attempt was also made to restrict the number of employees working in the coal-hydrogenation area, but this was also unsuccessful. It can be concluded from this experience that if worker exposure to these potentially hazardous materials is not properly managed, adverse health effects can occur.

Thus, based on available information at start-up of the SRC pilot plant, it was recognized that repeated exposure of plant personnel to products and by-products from coal-liquefaction processing could cause adverse health effects. Therefore, the only prudent approach was to implement an extensive employee protection plan involving good industrial hygiene practices, personal hygiene, education, medical surveillance, and engineering controls.

In 1981, recommendations were published (NIOSH, 1981a) for worker protection in coal-liquefaction facilities through the combined implementation of engineering controls, work practices, medical surveillance, exposure monitoring, education and training, and use of personal protective equipment. The recommendations were the following:

Medical Surveillance. Preplacement and interim medical histories, supplemented with preplacement and periodic exams emphasizing the lungs, upper respiratory tract, and skin; annual physical exams and complete physical exam at employment termination; annual pulmonary function tests, chest X-rays, and audiometric exams; periodic skin examinations.

Engineering Control. Various design considerations and engineering controls.

Work Practices. Emergency fire and medical services, prevention of inadvertent entry of inert or toxic materials into the work area, minimizing accumulation of hazardous materials, restricting access to process areas, and so on.

Informing Employees. All employees should be informed of known occupational exposure hazards associated with coal liquefaction; training should be repeated periodically; employees of contractors should be provided with similar information; warning signs and labels should be posted.

Personal Hygiene. Good personal hygiene practices should be encouraged; change rooms equipped with clean and dirty storage facilities should be provided; food should be discouraged in process areas; ultraviolet (UV) radiation should be used to check for skin contamination, but only under medical supervision.

Protective Equipment and Clothing. Clean work clothing, respiratory protection, hearing protection, workshoes or shoe coverings, and gloves should be provided.

Monitoring and Recordkeeping. A quarterly exposure monitoring program should be implemented, based on the results of an initial survey of potential exposures. The cyclohexane-soluble fraction of the sampled airborne particulates should be used for monitoring. Medical records and pertinent support documents, including environmental exposure records, should be established and maintained for all employees for the duration of employment plus 30 years.

Some precautionary comments are warranted concerning extrapolation of occupational exposure data from pilot plants to larger demonstration and commercial-scale facilities (Evans and Pallay, 1979). There are often considerable differences between pilot plants and larger-scale installations. Pilot plants are designed to obtain engineering data to estimate operating conditions or to provide information on specific process feasibility and practicality, whereas commercial plants are designed for economic operation. Pilot-plant equipment and operation are not optimized and are usually selected to allow varied test conditions. Equipment often does not function adequately under conditions found to exist during testing. Because pilot plants frequently go on-stream or off-stream to change programs, change layouts, or to repair equipment, potential occupational exposures at pilot plants are more frequent than at commercial facilities. The pilot-plant layout is more compact and does not necessarily utilize all elements that would normally be installed commercially. Pilot plants are not small-scale replicas of commercial facilities but are built to test certain defined parts of the process, using available equipment.

6.2.1 SRC Pilot Plant, Ft. Lewis, Washington

6.2.1.1 *Employee Health Protection Program*

Specific details of the employee health-protection program implemented at the SRC pilot plant at Ft. Lewis, Washington, can be found in Interim Reports (P&M Co., 1978, 1979, 1980) and a final report (P&M Co., 1982). A brief description and summary of program results are presented below.

6.2.1.1.1 Industrial Hygiene Monitoring. A comprehensive industrial hygiene program was initiated in 1975 to evaluate chemical and physical hazards and to prevent and control job-related diseases (P&M Co., 1978). Major program activities included in-plant air sampling, in-plant physical agents survey, and skin contamination evaluation.

In-plant air sampling included surveying the workplace and collecting personal samples to evaluate occupational exposures to existing or suspected airborne contaminants. Samples were analyzed for organic vapors, suspended particulates, PAHs, N-PACs, welding fumes, hydrogen sulfide, sulfur dioxide, phenols, carbon dioxide, carbon monoxide, dustfall, asbestos, and free silica. Concentrations of most contaminants were at acceptably low levels. Concentrations of PAHs in samples from the solidification area were higher than in samples from other process areas. Welding on parts contaminated with coal liquids resulted in increased exposure to benzene solubles. Hydrogen sulfide concentrations were determined to be a problem when transporting, transferring, or handling raw sour naphtha. Because of relatively high phenol concentrations in products, hazardous occupational exposure to phenols could occur through skin contact but not through inhalation. A problem was encountered initially when high carbon monoxide concentrations resulted from escape of gas through a worn rotary airlock, but this was effectively corrected.

In 1978, the health and safety programs at existing coal-conversion sites were reviewed for the Department of Energy (Husting, 1978). The Ft. Lewis SRC pilot-plant programs were included and were found to encompass all major elements required to provide worker protection, determine risk, conduct epidemiological studies, or possibly at some future date, establish cause–effect relations.

An industrial hygiene survey of the SRC pilot plant was performed for NIOSH in 1979 (NIOSH, 1979, 1981b, 1982). Air monitoring showed that aromatic compounds were the primary contaminants. Except for PAHs and nitrogen-containing polycylic aromatic compounds (N-PACs), the study concluded that other airborne contaminants would not constitute a health hazard at the measured levels. Also, it was observed that skin contact might be more critical than inhalation in determining worker exposure to PAHs, especially in view of the low levels of airborne two- and three-ring PAHs. Results of this survey confirmed data obtained under the pilot-plant's own monitoring program.

6.2.1.1.2 In-Plant Physical Agents. Plant noise was surveyed with a calibrated sound level meter (P&M Co., 1978). Overall, the pilot plant had minimal occupational noise levels. General space illumination and on-task lighting intensity were also surveyed (P&M Co., 1978). Some lighting levels for night-shift workers were less than adequate, indicating a need to establish a routine lighting-fixture maintenance and replacement program.

6.2.1.1.3 Skin Contamination Evaluation. Industrial hygiene techniques mainly assess problems associated with inhalation hazards. However, with coal liquids, skin contact is probably more critical than inhalation. Thus, there is a need to assess skin contamination quantitatively. Intensive research and development efforts to evaluate skin contamination began in 1976 with a skin-wipe sampling program (P&M Co., 1978, 1979) and continued in 1980 as a skin-wash sampling progam (P&M Co., 1980). Supplemental investigations resulted in developing a UV fluorescence photographic technique and a method of skin examination for black specks (P&M Co., 1982). Black specks, or deposits of SRC particulates, mixed and integrated with sebaceous secretions inside the skin pores, had appeared on many process area employees. Examinations for black specks showed that unit operators had higher skin contamination than did maintenance workers. Furthermore, workers were able to remove only 85% of the coal-tar contamination by showering (P&M Co., 1982). Analysis of the skin-wash and air-sampling data and workers' skin examination results showed that dermal exposure to coal-liquid contaminants correlated with respiratory exposure to airborne benzene-soluble particulates (P&M Co., 1982).

6.2.1.1.4 Personal Hygiene. Separate change houses, divided into clean and soiled areas, were provided for all male and female employees (P&M Co., 1978, 1979). The clean area was used for changing into company-provided clothing before going on-shift. The soiled area was used for removing and storing soiled clothing at the end of a shift and before showering. Employees were urged to shower at the end of the work shift, with excellent compliance (P&M Co., 1978). Soap, hand cleaners, and barrier creams were also provided.

6.2.1.1.5 Employee Education and Training. Health education provided accurate and up-to-date information on the hazards and their potentiating factors, possible effects of overexposure, preventive measures taken by the company to protect the employee, and the responsibility of the employee to protect himself. The program included an audiovisual slide presentation, a health protection manual, new employee orientations, and health protection continuing-education classes (P&M Co., 1978).

6.2.1.1.6 Clinical Program. The clinical program began in 1974 before the plant went into operation (P&M Co., 1978). Medical examinations were performed before the applicant was hired and annually thereafter and included a medical history and a complete physical examination. The skin was examined carefully for lesions, and employees were requested to report any unusual skin conditions. A quarterly skin exam was performed thereafter by the plant nurse. Any dermatological problems requiring a physician's care were referred by the plant nurse to a board-certified dermatologist. Some mild, transient cases of photodermatitis from exposure to coal-derived products were observed during the lifetime of the plant (P&M Co., 1982). Although two squamous cell skin carcinomas were identified on employees at the plant, it was concluded that they were not work-related based on their work histories and exposure (P&M Co., 1982).

6.2.1.2 *Industrial Hygiene Research*

6.2.1.2.1 Skin Contamination. In 1976 (P&M Co., 1978), a program was initiated with several objectives: quantitate occupational skin exposures; classify jobs and work locations according to frequency and severity of skin contamination; determine if there was a relation between the hazards of skin exposure and those of inhalation; educate employees on the importance of cleanliness; and add information to employees' medical and industrial hygiene records to provide a database for future epidemiological studies. Initially, a skin-wipe technique was evaluated, using a solvent-wetted pad, but this was abandoned because of inherent difficulties and uncontrollable factors, such as interference from natural skin oil, difficulty in controlling wetness of pad, inability to standardize wipe force, motion, area, and so on. An apparatus was then developed (P&M Co., 1980; Keenan and Cole, 1982), using a cyclohexane solvent spray over a confined area of skin, to collect quantitatively the solvent containing dissolved contaminants for laboratory analysis. The skin-wash sampling program was successfully implemented at the pilot plant (P&M Co., 1980). Skin contamination with coal-derived materials was quantified, and jobs were classified according to frequency and severity of contaminations encountered. Results indicated that unit operators received the equivalent contamination of 0.37 μg of SRC-II heavy distillate per cm^2 of exposed skin area, while that of maintenance workers was 0.28 $\mu g/cm^2$ of exposed skin area (P&M Co., 1982). The skin-wash technique also showed that only 85% of coal tar contamination was removed by showering.

6.2.1.2.2 Clothing. In 1980 (P&M Co., 1982), a program was initiated to evaluate various clothing materials and to develop a single-layered garment that could be worn as work trousers to prevent the occurrence of black specks in the skin, especially along the anterior thigh. To aid in evaluating the efficacy of these experimental garments, quarterly skin examinations, using white and UV light, were performed. While no single garment was found to be effective, an experimental trouser appeared to reduce visible contamination when worn over pajama pants. This confirmed Sexton's observations (Sexton et al., 1960) in an experiment where two men wore clean plants daily as underclothing. The experimental trousers in the SRC studies were made from Kimberly Clark Evolution II material and were designated Rubi-X. Although the garment appeared to reduce the number of black specks, the plant ceased operation before any firm conclusions could be reached.

6.2.1.2.3 Ultraviolet Fluorescence Photography. A combination of photographic equipment, UV lighting, and photographic techniques was developed to obtain pictures of the skin (P&M Co., 1982). Tiny specks of coal-liquid residue on the skin, invisible under normal lighting, can be detected by the UV photograph. Photographic evidence of skin contamination appeared to have a much greater impact than verbal descriptions and served to encourage employees to avoid excessive skin contamination and to wash thoroughly.

6.2.1.3 *Epidemiology*

In 1980 (P&M Co., 1982), an epidemiology program was initiated to collect and computerize all the medical examination data, medical histories, work histories, and personal data on worker exposures to coal-liquefaction materials. The program involved workers at the Ft. Lewis SRC pilot plant, the Harmarville, Pennsylvania P-99 process development unit, and the Merriam, Kansas research facility and included clinical, prospective, retrospective, and reproductive epidemiology. Clinical studies used data from available preplacement medical examinations, subsequent annual physical examinations and historical medical records. Prospective studies continually collected and periodically analyzed records of all cancer, morbidity (absenteeism), and mortality data among employees. For the retrospective studies, a cohort of Ft. Lewis SRC employees was established and information was gathered for the mortality and case-control studies. All workers ever employed at the pilot plant were identified, and demographic data were collected and added to the database. Reproductive studies involved collecting data on fertility, fetal wastage, and birth malformations and establishing a birth registry among employees. The epidemiology studies emphasized data collection to assess the short- and long-term health risk and expression of the magnitude of risk (cause-specific relative risk) and its impact on public and overall health (survival and quality of life). Although the industrial hygiene and medical surveillance programs protected the employees from potentially hazardous exposures to the coal-liquefaction materials, no information was available on long-term effects since the pilot plant operated for only 7 years. The prospective and retrospective studies are continuing to assess the long-term effects of exposure to coal-liquefaction materials on worker health.

6.2.2 EDS Coal-Liquefaction Plant (ECLP), Baytown, Texas

6.2.2.1 *Occupational Health Program*

The occupational health program for the ECLP at Baytown, Texas, included seven parts: engineering controls, industrial hygiene, operations and laboratory work practices, personal hygiene, medical surveillance, and health education (Montgomery, 1980; Shepard, 1981; Exxon, 1981, 1982, 1983). Objectives of the program were to assure a healthy environment at the pilot plant and to provide an expanded database to establish occupational health and safety criteria for future production facilities.

6.2.2.1.1 Engineering Controls. Engineering controls included a closed drain system, noise suppression, dust suppression, fume collection, and flare system, using appropriate equipment to minimize exposures (Montgomery, 1980; Shepard, 1981; Exxon, 1981, 1983).

6.2.2.1.2 Industrial Hygiene. Industrial hygiene emphasized monitoring exposures to coal dust, organic vapors, particulates, BSF of particulates, hydrogen sulfide, sulfur dioxide, and PAHs (Exxon, 1982, 1983). The program included baseline and monthly surveys, using personal and area sampling procedures. Operations monitoring was designed to evaluate employee exposure to toxic substances and physical agents. Monitoring, in conjunction with development and application of improved control measures, was expected to assure early detection and correction of potential health problems. Exposure to coal liquids and organic vapors of process personnel during pumpout and cleaning and of mechanical personnel during exchanger cleaning, pump and valve dismantling, equipment repairs, and welding was of primary concern.

An industrial hygiene sampling program was conducted from February 1980 through June 1981, during prestart-up, start-up/shakedown, and sustained operations of the ECLP with Illinois (Monterey mine) bituminous coal (Exxon, 1982). During prestart-up, ambient levels of most variables measured were low according to occupational health standards. Measurements for light hydrocarbons (aromatics and C_6–C_8 alkanes), PAHs, particulates, trace metals, and noise indicated that ambient conditions outside the plant, except under upset conditions at neighboring facilities, would not affect in-plant monitoring. Levels of BaP and anthracene, measured with a high-volume sampling system, were slightly higher than expected. However, levels of these PAHs were below detectable limits when sampled with the low-volume sampling system typical of industrial hygiene use. Therefore, the influence of these background levels on in-plant measurements was considered.

Measurements taken during shakedown and sustained pilot-plant operations showed the major sources of worker exposure to be coal-liquid fumes or aerosols generated by coal-liquid contamination of hot pump and hot valve surfaces, saturated insulation, and vacuum bottoms cooling belt emissions. Other sources of atmospheric emissions were from the atmospheric tower bottoms pump and packing leakoff collection barrels, and the vacuum tower bottoms pump. Coal-liquid fumes from these sources contained small but significant concentrations of four-ring-and-higher PAHs, including some of proven carcinogenic potential. Atmospheric emissions from reciprocating and from some centrifugal pumps were primarily hydrocarbon aerosols with low concentrations of hydrocarbon vapors. The average exposure of a process technician to coal-liquid fumes during the sustained run was 0.044 mg/m^3 (geometric mean, assuming regular rotation through each outside job post). The guideline for maximum allowable exposure to this material is 0.20 mg/m^3, 8-h, time-weighted, average exposure.

Personal sampling during this period indicated inhalation exposures were directly proportional to the emitting equipment in the job areas. The liquefaction distillation unit job posts had the highest average exposures (0.099 and 0.095 mg/m^3). Exposures for these posts decreased significantly between January, at the start of sustained run, and May, at the end of sustained run. Several measurements were above allowable guidelines, indicating that process operations and work practices significantly influence personal exposures. Analysis of personal samples showed no four-ring or higher PAHs except at high coal-liquid fume concentrations. All samples indicated no personal exposure to light aromatic compounds of benezene, toluene, xylene, or amino-PAHs. Exposures during start-up and sustained operations can be attributed to emissions from equipment operating at temperatures above 350–400°F. However, equipment at the pilot plant was not necessarily representative of a commercial facility.

A similar industrial hygiene sampling program was conducted during the second year of ECLP operations, from July 1981 to the shutdown of the pilot plant in August 1982 (Exxon, 1983). Data were gathered during operations on Wyodak subbituminous coal, Illinois No. 6 bituminous coal, and Texas lignite, all in the bottoms recycle mode.

Process operator exposures to coal-liquid fumes, as measured by the BSF of total particulates, were found to decrease steadily from an average 8-h geometric mean exposure of 0.023 mg/m^3 during the Wyodak and 0.008 mg/m^3 for the Illinois operations to 0.015 mg/m^3 for the lignite operation. Process operators were not exposed to measurable quantities of four-or-more ring PAHs. Naphthalene (0.003–0.028 mg/m^3), phenanthrene (0.001–0.009 mg/m^3), anthracene (0.001–0.026 mg/m^3), and fluorene (0.001–0.006 mg/m^3) were measured on some personal samples. Naphthalene has an 8-h exposure guideline of 50 mg/m^3, but no guidelines have been established for the other compounds. No measured exposures to light aromatics (benzene, toluene, and xylene) exceeded the detectable limit of 0.1 ppm. Neither phenol, cresol, nor N-PACs were detected in personal or area samples. No unusual exposures to coal-liquid fumes were measured on process operators or mechanical employees during shutdown and cleaning of process equipment for turnaround.

The mass median aerodynamic diameter (MMAD) of the BSF of the total aerosol emitted from the vacuum bottoms cooling belt was 1.30 μm; approximately 65% of the BSF aerosol was found to be respirable (less than 2.5 μm) and 35% inhalable (2.5–15 μm). PAH compounds appeared to be evenly distributed within the aerosol. The MMAD of the BSF aerosol emitted from the atmospheric bottoms pump was 2.2 μm, with approximately 42% in the respirable range and 47% inhalable.

Based on the results of the 2-year monitoring program, it was concluded that the ECLP industrial hygiene program demonstrated that, with proper engineering controls, equipment design, and proper operating and maintenance work practices, personnel exposure to potentially harmful and toxic materials can be avoided or maintained below desired levels.

NIOSH (1981c), in cooperation with project personnel, developed a comprehensive industrial hygiene test program for ECLP. In-plant testing was conducted in May 1981 to gather data on which to base an assessment of industrial hygiene criteria for commercial liquefaction plants. The assessment would be used by NIOSH as a basis for establishing technical recommendations to ensure that appropriate corrective and protective measures were built into coal-liquefaction technologies prior to commercialization. Results of this program were published in 1982 (NIOSH, 1982). PAH concentrations were measured throughout the plant with area and personal samplers. The PAHs detected were predominantly two- and three-ring aromatics, with only traces of four-ring compounds. The combined in-plant area sample geometric mean of 133.8 $\mu g/m^3$ was significantly higher than the mean of the personal samples for operators and maintenance workers. The mean area PAH level was 40% higher than that for operators and 240% higher than that for maintenance staff. The lower levels for PAH exposures of the two worker groups indicates that they were not exposed to sources of PAHs other than those found in the general work environment and were not in the process area long enough to receive the full burden of PAHs available. Operators spent about 60% of their time in the process area. Maintenance workers did not work on on-stream equipment during the survey. Operators had significantly higher exposure than those of the maintenance group. All major unit operations had significantly higher PAH concentrations than those at the plant perimeter. Although detectable amounts of N-PACs were sometimes found at the atmospheric stripper, vacuum stripper, and solvent fractionator, concentrations were all 0.6 mg/m^3 or less. Employee exposure to simple N-PACs was not a common occurrence, and measured concentrations were all below current health standards. Samples were also collected for benzene, toluene, xylene, phenols, cresols, and xylenols. None of the samples had concentrations of these compounds exceeding 1 ppm. Results indicate that exposure to the chemicals monitored does not present a health hazard to maintenance or operator personnel performing routine duties during normal plant operations.

6.2.2.1.3. Operations and Laboratory Work Practices. All work practices were evaluated to minimize occupational exposures, with emphasis on preplanning, adequate consideration of materials and equipment involved, and so on. (Montgomery, 1980; Exxon, 1981; Shepard, 1981). Where applicable, worker exposure was reduced through engineering, administrative, and/or personal protection measures. Exposures of laboratory personnel to coal liquids, organic vapors, and coal dust were minimized with proper laboratory procedures, extensive laboratory ventilation systems, and protective equipment.

6.2.2.1.4 Personal Hygiene. Protective gear was furnished to all process area workers (Montgomery, 1980; Exxon, 1981; Shepard 1981). Each employee received gloves, hard hats, safety glasses, and shoes. Face shields, dust masks, and respirators were available when required. Each morning, employees were supplied with clean coveralls and clean underwear. A dry-cleaner, washing machine and dryer were also available. The efficacy of the solvent in the dry cleaner was periodically tested by examining pieces of cleaned clothes for PAHs under UV light. At shift end, the worker left his soiled clothes on the "dirty" side of the locker room, shampooed and showered (first with waterless cleaner, then with soap), then entered the clean room to dress in street clothes. Disposable polyvinyl chloride shoe covers and long-sleeved jackets were provided visitors.

6.2.2.1.5 Medical Surveillance. The medical surveillance program was designed to monitor and detect changes in employee health during and following work at the pilot plant and to determine if any changes were the result of exposure to coal-derived materials (Montgomery, 1980; Shepard, 1981). Target organs included skin, lungs, and urinary tract. A preplacement medical examination was given to all employees and they were examined annually thereafter. The exam included comprehensive medical and work histories, with emphasis on skin, respiratory tract, liver, and kidney problems; a complete physical, with emphasis on the skin, mouth and respiratory and nervous systems, and evaluation of the worker's ability to use positive-pressure respirators. The employee was advised of his examination results, specifying any condition or abnormality found that could increase health risk from being exposed to materials in coal-liquefaction plants and any limitations that should be placed on his exposure. Medical records will be maintained for all employees for at least 30 years after the last occupational exposure.

6.2.2.1.6 Employee Education and Training. All employees and contractor-maintenance personnel were informed about the occupational health program through a slide/tape presentation in two languages, booklets outlining the program, and written tests (Montgomery, 1980; Exxon, 1981). Employees were also informed of industrial hygiene results through their supervisors or the resident industrial hygienist.

6.2.2.1.7 Epidemiology. A prospective epidemiology study that follows two matched groups of workers was included in the program (Montgomery, 1980; Shepard, 1981). Each coal-liquefaction employee was matched by age, race, sex, and smoking habits with a control process employee from an adjacent refinery. In addition to preplacement and periodic medical exams, both coal-liquefaction employees and the control group completed a work and exposure history questionnaire and had color photographs taken of potentially exposed skin surfaces: back of hands, each side of the face, and back of the neck. All skin surfaces were also examined visually, and any significant skin lesions were noted. Sputum and urine samples were collected for cytological evaluations on study employees and controls. Both groups received annual health exams, with special attention to charting skin lesions and cytological analysis of sputum and urine. Because coal-liquefaction employees have returned to the refinery, follow-up evaluation will continue as a part of the ongoing medical program for company employees.

6.2.3 H-Coal Pilot Plant, Catlettsburg, Kentucky

6.2.3.1 *Industrial Health and Safety Program*

A health and safety program was developed and implemented at the H-Coal pilot plant in Catlettsburg, Kentucky, to ensure protection of workers. The program included personal hygiene, including clothing, showering, and laundry provisions; medical surveillance; industrial hygiene monitoring and research; worker health-protection training; plant safety; and fire safety. Specific details of the program are described in topical reports (Ashland, 1982a–c, 1983a–e). The program is briefly summarized below.

6.2.3.1.1 Industrial Hygiene. Both area and personal samples were collected (Ashland, 1982a, 1983d). Workers were selected for monitoring by job classification and job activity, the latter based on high probability of maximum exposure. Area samples were collected with devices placed at locations selected for maximum exposure potential (Ashland, 1983d). Routine area monitoring for total and respirable fractions of particulates,

CTPV, BaP, and hydrogen sulfide were performed monthly. Continuous hydrogen sulfide monitors were located in certain areas. Monitoring for N-PACs, phenols, cresols, and cyanide was performed bimonthly. Personal monitoring involved the compounds listed above plus light aromatics, hexane, carbon disulfide, nitrogen oxides, carbon monoxide, and ammonia (Ashland, 1982a).

Generally, results indicated that worker exposures were below the TLV and that job activities where exposures exceeded the TLV could be controlled through engineering or personal-protection methods to reduce exposures. However, average exposure to CTPV slightly exceeded the TLV. Process sample technicians, pipefitters, and boilermakers consistently showed higher mean and peak exposure than other workers. Job categories with consistently low relative exposures included tank farm operators, wastewater operators, and laboratory technicians. A vent condenser substantially reduced emissions of benzene, toluene, xylene, and hexane from a hot oil storage tank vent.

An industrial hygiene survey (NIOSH, 1982) showed that mean PAH concentrations (133.8 $\mu g/m^3$ geometric mean) in combined area samples (mainly process areas but excluding plant perimeter) were three times the mean measured on maintenance workers (38.8 $\mu g/m^3$ geometric mean) and two times that of operators (96.9 $\mu g/m^3$ geometric mean). Workers in these groups apparently spent less time in the process area during the survey and received a lower time-weighted average PAH exposure. Results also indicated that the dissolver/coal preparation area (161.2 $\mu g/m^3$ geometric mean) and the fractionation area (155.7 $\mu g/m^3$ geometric mean) were major sources of PAH emissions; ambient PAH concentrations decreased with increased distance from these areas. N-PACs were not detected in area or personal samples. Benzene, toluene, and xylene concentrations were generally at the 0.01-ppm level and were always below 0.06 ppm.

6.2.3.1.2 Noise. A hearing conservation program was established that included noise surveys, audiometric testing, posting of high-noise areas, personal hearing protectors, and hearing conservation training for workers (Ashland, 1983c). Noise levels for certain activities indicated the need for a hearing conservation program. Hearing protectors were provided all workers where noise exposures exceeded the accepted level.

6.2.3.1.3 Personal Hygiene. The personal hygiene program included work practices, work clothing, and appropriate physical facilities (Ashland, 1983e). Work clothing was supplied to all process area employees likely to be exposed to process materials. Cleaning and laundering of soiled clothes was done by a commercial laundry. Workers were required to use clean and dirty locker rooms in the change house. Showers were required at the end of the work shift. Workers were allowed to eat, drink, and smoke only in designated areas. Gloves were required, whenever practical, to minimize skin contamination. Barrier creams and face shields were also provided to minimize contamination.

6.2.3.1.4 Medical Surveillance. Those who worked in plant process areas more than 90 days/year participated in the medical surveillance program (Ashland, 1983e). A preplacement medical exam was given, with annual exams thereafter. The exam included a comprehensive medical history questionnaire, skin examinations, audiometry, vision testing, posterior–anterior chest X-rays, electrocardiogram, pulmonary function, urinalysis, hematology, blood chemistry, and sputum and urine cytology. Exam results generally revealed no occupational health problems during the operation of the pilot plant.

6.2.3.1.5 Employee Education and Training. A slide/tape presentation and brochure were developed to convey health-protection information to process area workers

(Ashland, 1982c). Regular plant workers viewed the slide/tape presentation at least annually. Temporary workers reviewed the health information brochure before beginning work in process areas.

6.2.3.2 *Industrial Hygiene Research*

6.2.3.2.1 Cytological Monitoring. Use of the Ames *Salmonella*/mammalian microsome mutagenicity assay to monitor worker exposures to process materials in synfuels facilities was evaluated (Hannon et al., 1981; Ashland, 1983e; Hill and Enoch, 1983) as a means of detecting urinary tract disease because high-molecular-weight N-PACs were potentially present in coal-liquefaction process materials. Urine samples were collected from process area volunteers (experimental group) and office workers (control group), both of which were subdivided into smokers and nonsmokers. Air samples were collected concurrently during biological monitoring. Test subjects were monitored for 7–8 h during the work shift for airborne contaminant exposures, using appropriate personal samplers. Ames assay data showed a significant difference in mutagenic activity of urine collected from smokers and nonsmokers. Urine from smokers showed greater mutagenic activity, but no work-related increases in urine mutagenicity were observed. Results also showed low levels of mutagenicity in nonsmokers. There were no indications of increased mutagenic activity with increased exposure to airborne contaminants from process units.

Sputum cytological testing was also included during the semiannual medical examinations (Ashland, 1983e). Early-morning expectorant samples were collected at home, but this frequently failed to produce satisfactory sputum samples. The results were inconclusive and cast doubt on the value of this test in predicting the existence of pulmonary disease. It was finally concluded that the inclusion of cytological tests should carefully be considered as to their utility in general occupational medical surveillance programs.

6.2.3.2.2 Chemical Permeation Through Protective Clothing. Permeation of H-Coal heavy distillate through gloves of different materials was evaluated (Ashland, 1983a; Bennett et al., 1983). Results with nitrile, vinyl, and natural rubber showed that nitrile gloves had the longest breakthrough time for continuous exposure. Experiments showed that breakthrough occurred sooner with a soiled vinyl glove that was subjected to a cleaning process normally used for pilot-plant gloves than with a new glove. However, little permeation occurred with the cleaned glove until breakthrough was noticed with the new glove. Follow-up testing at the pilot plant indicated no difference in skin contamination of hands after wearing cleaned gloves or new gloves.

6.2.3.2.3 Skin Contamination. Fluorescence of coal-liquefaction materials caused by excitation with UV light has been used to monitor skin contamination (Vo-Dinh and Gammage, 1981; Schurosko et al., 1983). The degree of skin contamination occurring on process workers was monitored, using qualitative (black-light) and quantitative (Spill Spotter) methods to determine the efficacy of showering in removing contamination and to determine the quantitative relation between Spill Spotter response and skin-contamination concentration (Ashland, 1983b). Field measurements at the pilot plant showed a range of background fluorescence caused by differing skin types due to bodily location and pigmentation. Dirt on the skin partially quenched the fluorescence responses of the Spill Spotter. Nonetheless, using appropriate controls and discretion in field measurements to minimize the quenching factor, skin contamination was identified and quantified on selected process workers. One group of plant operators showed a 2- to 3-fold increase in fluorescence during the work shift. Showering generally reduced skin fluorescence to preshift levels, indicating that surface contamination could be removed by washing; however, the extent of removal was not readily quantifiable. Also, residual contamination after showering was greater for those workers with the greatest degree of original contamination.

6.2.3.2.4 Laundry Efficiency. The efficiency of a commercial laundry at cleaning work clothing from the pilot plant was evaluated (Ashland, 1983a). Laboratory testing and long-wave UV light were used to monitor skin contamination from recycled work clothing. Results showed that a dual-phase commercial-laundry dry cleaner was 98% effective in removing coal-liquefaction materials from outer clothing. Skin-fluorescence monitoring with UV light showed no skin contamination from recycled (cleaned by commercial laundry) work shirts and vinyl work gloves. Although quantifiable levels of process material were found in laundry wastewater, levels were low compared to the hydrocarbon content of wastewater streams.

6.2.4 Cresap Test Facility, West Virginia

The Cresap facility, located in Cresap, West Virginia, approximately 20 miles south of Wheeling, used the Consol synthetic fuel process to produce low-sulfur liquids of various boiling ranges.

6.2.4.1 *Industrial Hygiene and Occupational Health*

The industrial hygiene and occupational health program at Cresap includes personal hygiene, education, work practices and permit system, industrial hygiene monitoring, and medical surveillance (NIOSH, 1982). Workers are provided coveralls, shoes/boots, safety glasses, gloves, hard hats, and respirators. Clean coveralls are provided daily. Showers are required at the end of each shift. The shower facility has a single locker with no provisions for separation of clean and dirty clothes. Preemployment training and appropriate brochures and manuals are provided to each employee. Training includes the use and maintenance of respirators.

6.2.4.2 *Industrial Hygiene Monitoring*

Industrial hygiene monitoring is performed on an informal schedule, primarily for carbon monoxide, hydrogen sulfide, and noise. A comprehensive industrial hygiene survey was conducted by NIOSH in 1979 (NIOSH, 1982). Measurements were made for air concentrations of PAHs, light aromatic hydrocarbons, N-PACs, phenols, and gases. The combined area samples showed higher mean concentrations of PAHs than did the three worker groups monitored. Most workers were not in the process area long enough to accumulate PAH exposure equivalent to ambient levels. Uniformly low PAH levels were observed throughout the facility. Although N-PACs were not detected in area samples, one was detected in personal samples, but at a level below 1 $\mu g/m^3$. Among other organics, toluene was detected in both area and personal samples, but at levels below 0.1 ppm. Phenol was detected only at the solvent recovery area at a concentration of 0.1 ppm.

6.2.4.3 *Medical Surveillance*

A preplacement physical exam was provided and medical histories were obtained for all process workers. Thereafter, exams were given annually. Quarterly skin exams were made by a nurse.

6.2.5 Solvent-Refined Coal I (SRC-I) Pilot Plant, Wilsonville, Alabama

6.2.5.1 *Industrial Hygiene and Medical Surveillance*

In 1979, an industrial hygiene and medical surveillance program was instituted at the SRC-I pilot plant (R. Hickey, U.S. Department of Energy, personal communication). Special change facilities for both male and female workers were provided and included a dirty-clothes area, a shower facility, and a clean change room with lockers. Each employee routinely exposed to coal-derived liquids was provided a daily change of outer clothing. Employees who may have had only incidental exposures were provided coveralls. Overshoes were also provided where necessary. Work practices were reviewed frequently to improve procedures and conditions which may have resulted in unnecessary exposures.

Each employee was given an annual physical exam, and records were monitored to ensure the health of the workforce. Work areas were monitored for physical irritants, noise, gases, vapors, fumes, and dust. Employees were encouraged to use protective creams frequently, to wash and put on fresh clothing immediately after being accidentally sprayed or splashed with coal liquids, and to report any skin abnormalities to their supervisor.

6.2.6 Summary and Conclusions

Coal conversion, whether to coke, gas, liquids, or simply heat, results in products and by-products that may contain materials or compounds which, under certain conditions, can be deleterious to human health. Thus, exposure of workers to objectionable materials in such a production facility is a major concern. The degree of hazard will depend, to some extent, on the specific process and process conditions. However, one chemical class of particular concern, PAHs, is common to all processes. This chemical class can be present in liquid products, as aerosol emissions, in by-products, or associated with particulate matter. Many PAHs are potent carcinogens, hence the concern for occupational exposures to materials containing these compounds. More recently, N-PACs have also been implicated as problematic in certain coal-conversion materials.

As a result, extensive industrial hygiene monitoring and medical surveillance programs have accompanied the operation of coal-conversion facilities, both to develop factual information on occupational exposures to relevant materials and to protect the health of employees. Limited experience has shown that while these are admirable goals, care and judgement must be exercised in the implementation of programs and use of the information generated. Implementing an industrial hygiene and medical surveillance program can be effective in protecting the employee but will only be as good as the management commitment to the program, as perceived by the employee. Unless management is completely and obviously committed to the program, from top management to first-line supervisor, there will not be a serious commitment by the employee. Also, extreme care and good judgement should be exercised in attempting to extrapolate exposure data from pilot plants to demonstration or commercial-scale plants. Pilot plants are not small-scale duplicates of larger-scale demonstration or commercial facilities. Because of their experimental nature, pilot plants are designed to study a variety of conditions and process options. Pilot plants often lack equipment that would be part of the commercial plant and, conversely, may use equipment not anticipated in a commercial plant where a principal goal is economic operation. As a result of frequent on-stream/off stream operations, occupational contact exposures in a pilot plant may be more severe than would be encountered under stable operations in a commercial plant.

With the exception of a coal-hydrogenation plant that operated briefly in Institute, West Virginia, in the early 1950s, coal-liquefaction processes in the United States are essentially new. Although coal liquefaction was done on a commercial scale in Germany during World War II, little information is available on this operation. Thus, much of the concern for occupational health effects in coal liquefaction is based on experiences at the Institute operation and the documented health effects in the coal coking industry. Because of the lack of factual data for coal-liquefaction facilities, pilot-plant operators assumed that repeated exposure of plant personnel to products and by-products of coal liquefaction could cause adverse health effects, particularly skin cancer. To protect employees, extensive industrial hygiene and medical surveillance programs were instituted at each pilot plant. These included the SRC pilot plant, the EDS pilot plant, and the Ashland Synthetic Fuels H-Coal pilot plant. These programs invariably emphasized engineering controls, industrial hygiene monitoring, personal hygiene, work practices, health education, and medical surveillance. These programs appear to have been effective in protecting employees in the short term, since only minor effects were observed. One interesting observation was that the PAHs detected in the plant atmospheres were predominantly of the 2- and 3-ring variety. These would not be expected to be as toxic as the larger condensed aromatics.

The experience at the Institute, West Virginia coal-hydrogenation pilot plant indicated that if worker exposure to these potentially hazardous materials is not properly managed, adverse health effects can occur. Later experiences in the coal-liquefaction pilot plants have shown that when proper management is exercised, available evidence suggests that these effects can be minimized. While these observations are encouraging, results provide information only on the short-term effects of exposures to coal-derived materials. Furthermore, the plants operated for only 7 years or less; the latency period for tumor development is frequently 15–20 years and even as long as 30–40 years. Since most of the pilot plants are no longer operational, the major recourse to obtain information on potential long-term effects from exposure to coal-liquefaction materials is to monitor continually those who had been employed at pilot plants and to collect medical information for as long as is feasible. The SRC and EDS projects include ongoing epidemiological programs to collect and computerize all available medical and exposure data on plant employees for future studies. Although a large database of medical and exposure information is available for workers in the H-Coal project, no epidemiological program was initiated.

6.3 COAL GASIFICATION

The primary occupational health concerns in coal-gasification processes involve by-product tars and oils, carbon monoxide and other gases, and respirable particulates, with their associated organic matter. However, little direct evidence is available on the potential magnitude of this exposure and the degree of hazard. Much has been inferred and extrapolated from related industries, such as coal coking, coal tar, and gas works, where epidemiological studies, discussed earlier, have shown the occupational hazard.

To some extent, materials produced during coal gasification will depend on the process employed (Braunstein et al., 1977; NIOSH, 1978; Munro et al., 1983). For example, because of its lower operating temperature and fixed-bed design, the Lurgi gasifier produces a broader spectrum of products and by-products than entrained-flow, slagging (high-temperature) gasifiers. Tars constitute 15–20% of the total by-product output from Lurgi gasification but are not produced by high-severity gasifiers. Tars and by-product oils constitute an occupational health hazard through potential skin contact and inhalation of aerosols. Also, carbon monoxide may be an inhalation hazard, particularly during start-up and shutdown. Respirable particulates associated with adsorbed organic matter such as PAHs can also pose a problem.

Early coal-gasification plants were primarily small-scale, required only a small workforce, and have yielded no information on industrial hygiene or occupational health practices used. It is assumed that only minimum procedures were used throughout the industry. No epidemiological studies were conducted other than those for related coal industries.

In January 1978, an Interagency Report (NIOSH and EPA, 1978) recommended health and safety guidelines specifically for coal-gasification pilot plants that included various engineering controls and work practices: a health education program, involving an initial meeting followed by periodic meetings and continuous emphasis from management; issuance of clean outer garments and underclothing (including socks) at the beginning of each shift; washing facilities, including a double locker room separated by a large shower facility; availability of ordinary soap and barrier creams to employees on request or under advisement of an occupational physician; a formal respiratory program; designation of regulated areas, including posting; an air-monitoring program, including area and personal monitoring; and a medical surveillance program. The air-monitoring program should measure carbon monoxide as an indicator of potential hazard in most areas and hydrogen sulfide in sulfur recovery areas. The coke-oven standard was recommended for personal monitoring, that is, 0.2 mg/m^3 CTPV (BSF of total particulate matter). The medical surveillance program included a full preemployment physical, history, and laboratory tests; regular checkups, more frequent and detailed for high-risk individuals (those with 5 years of exposure or over 45 years old); long-term follow-up of high-risk individuals after transfer to other job categories or termination; and full recordkeeping, including work history and exposure data, for 40 years.

In September 1978 (NIOSH, 1978), NIOSH published criteria and recommended standards for worker exposures to toxicants and hazardous operating conditions in commercial coal-gasification facilities anticipated in 1985. Epidemiological and toxicological evidence from related processes led NIOSH to conclude that employment in coal-gasification plants might increase the cancer risk in exposed employees. Because of numerous toxicants potentially present in coal-gasification plants, guidelines were presented to monitor carbon monoxide, as an indicator, to allow real-time detection of leaks. Requirements were described for safety procedures, engineering controls, work practices, workplace monitoring (area and personal monitoring for respirable particulates at least monthly in certain process areas), medical surveillance (preplacement exam and annually thereafter), protective clothing and equipment, change rooms (clean and dirty areas), personal hygiene, employee education, and recordkeeping.

In 1982 (Wilzbach and Reilly, 1982), a DOE report concluded that toxicants, including carcinogens, were present in coal-gasification tars and, to a lesser extent, in by-product oils. However, acute toxicity was relatively low, and toxicants were minor components of process streams. Because occupational exposures to by-product tar and particulate and vapor-phase fugitive emissions may be a health concern, proper industrial hygiene procedures should be maintained for plant personnel. However, it was concluded that determining the magnitude of this concern requires more extensive fugitive emission sampling and analysis.

The DOE industrial hygiene program focused on quantifying and controlling worker exposures. Specific objectives of the program were to develop and apply methods to determine area concentrations and personal exposures to specific or surrogate pollutants; to investigate the effectiveness of methods to mitigate exposures; and to ensure, by medical surveillance, that gasification-associated health effects are identified early. Virtually all efforts to date have emphasized the first objective.

Information on industrial hygiene, occupational health practices, and epidemiological studies at gasification plants is summarized below.

6.3.1 Industrial Hygiene and Epidemiology

6.3.1.1 *Combustion Engineering Entrained-Bed Gasifier, Connecticut*

Cubit and Tanita (1982) described the industrial hygiene program at the Combustion Engineering process development gasification unit in Connecticut. The unit is an entrained-bed gasifier producing a low-Btu gas.

Although several sets of protective clothing and equipment (coveralls, shoes, underwear) were given to employees, clean clothing was not provided regularly. Laundering service was provided, and clothing changes were left to the worker's discretion. Shower facilities were available, and the company recommended daily use. Barrier creams and cleansing agents were not generally available. Routine or spot-checking for skin contamination was not done. The educational program involved formal training in the use and care of respirators and preemployment job training by supervisors on an as-needed basis. The medical surveillance program included a preplacement history and physical exam, including blood count, with an annual follow-up. A chest X-ray, pulmonary function test, and EKG were also performed every 3 years. Industrial hygiene monitoring included limited air sampling for benzene solubles. There was no epidemiology program.

An industrial hygiene survey (Cubit and Tanita, 1982) conducted at the plant involved air sampling for PAHs, benzene, toluene, xylene, phenolics, N-PACs, carbon monoxide, and hydrogen sulfides. Operators and maintenance workers showed higher exposures to PAHs than those of laboratory technicians and the combined area sampling. However, this was attributed to repair, cleaning, and dismantling of equipment rather than to fugitive emissions. No significant process emissions of PAHs were detected. The PAHs detected were primarily two- and three-ring compounds. Maximum concentrations of N-PACs were below 0.1 $\mu g/m^3$, while simple organics (benzene, toluene, xylene) were below 0.1 ppm. Emissions of hydrogen sulfide and carbon monoxide were not detected.

6.3.1.2 *Commercial Fixed-Bed Gasifier, York, Pennsylvania*

Cubit and Tanita (1982) also provided similar information on a commercial fixed-bed gasifier at York, Pennsylvania. Because of certain confidentiality requirements, the plant could not be identified.

The personal hygiene and safety program included various protective clothing and equipment. Three sets of coveralls were provided, and changes were left to the worker's discretion. Laundering facilities were available. Shower facilities were also provided, and the plant recommended daily use. Barrier creams and cleansing agents were not generally available. Routine or spot-checking for skin contamination was not done. The educational program included formal training in the use and care of respirators and preemployment job training by supervisors on an as-needed basis. The medical surveillance program included a preplacement history and physical exam with urinalysis, blood chemistries, chest X-ray, and pulmonary function test. Follow-up exams were given annually, and after 5 years semiannually. A skin exam was also included. Industrial hygiene monitoring included a limited air sampling for benzene solubles, with continuous monitoring for carbon monoxide in certain areas; in other areas, readings were taken before entrance. There was no epidemiology program.

Cubit and Tanita (1982) conducted an industrial hygiene survey and calculated geometric mean PAH concentrations that were 10–15 times higher (20–30 $\mu g/m^3$) near individual unit operations than around the plant perimeter. Complete enclosure of the gasification facility explains the high PAH levels in-plant relative to those on the perimeter (mean of 1.8 $\mu g/m^3$). Maximum N-PAC concentrations did not exceed 0.1 $\mu g/m^3$, while simple organics were well below 0.1 ppm. No significant levels (less than 5 ppm) of hydrogen sulfide were found. Carbon monoxide concentrations at this plant have frequently exceeded 50 ppm in certain areas and activated the alarm system. Study results showed basically the same pattern of intermittent carbon monoxide emissions, with highest concentrations measured near the deoilers and detarring unit. Average overall plant levels were below 5 ppm.

6.3.1.3 *Westinghouse Fluidized-Bed Gasifier, Pennsylvania*

Cubit and Tanita (1982) also described the industrial hygiene program at the Westinghouse fluidized-bed gasifier in Pennsylvania. Various protective clothing and equipment were provided to all employees, and clean coveralls were provided daily; laundering service was also provided. Shower facilities were available, and the company recommended daily use. Dual lockers were available in the shower room to separate work and nonwork clothing. Barrier creams and cleansing agents were not generally available. Routine or spot-checking for skin contamination was not done. The educational program included formal training in the use and care of respirators and preemployment job training by supervisors on an as-needed basis. Formal classroom sessions included first aid, accident prevention, and safety rules covering equipment, emergencies, materials, machinery, and fire prevention. Medical surveillance included a preplacement history and physical exam, including urinalysis, complete blood count, serum glucose level, chest X-ray, and pulmonary function. An annual follow-up exam was provided thereafter, with an electrocardiogram (ECG) and audiogram every 2 years. A skin exam was also performed. Industrial hygiene monitoring included limited air sampling for benzene solubles. Hydrogen sulfide, carbon monoxide, and ammonia were measured on each shift; BaP and naphthalene levels were also measured. Cubit and Tanita (1982) stated that most results from this monitoring were considered proprietary information and were therefore unavailable for outside publications. There was no epidemiology program.

During a second plant survey (Cubit and Tanita, 1982), a significant difference was observed between mean PAH concentrations for operators (4.6 $\mu g/m^3$) and combined in-plant process samples (2.0 $\mu g/m^3$). Furthermore, mean PAH concentrations were significantly different among the four unit operations and perimeter samples. The gasifier had the highest recorded mean PAH concentration (5.4 $\mu g/m^3$). In most cases, light aromatics and N-PACs were not detected. Significant hydrogen sulfide and carbon-monoxide exposures occurred at the plant. Concentrations up to 1745 ppm carbon monoxide were measured at the filter unit during filter changing. Employees were sometimes overcome by carbon monoxide during this operation, indicating a serious hazard in this area of the plant. The highest concentrations of hydrogen sulfide, 120 and 160 ppm, were also measured in this area.

6.3.1.4 *SASOL Lurgi Gasifiers, Republic of South Africa*

Only limited information is available on industrial hygiene and occupational health programs at the three SASOL plants in the Republic of South Africa. One plant, SASOL-I, has operated for about 25 years or more.

The SASOL-I industrial complex became operational in 1955. Excluding mine operators, there were 3300 workers at the site during a NIOSH visit in 1976 (Munro et al., 1983).

Carbon monoxide exposure was the primary concern near the Lurgi gasifiers and synthesis gas cleanup units. Tests for leaks were conducted each shift with Draeger tubes. Contamination of equipment by tar was common in the tar-separation plant; older separators were coated with tar overflow. Newer coal-tar separators were relatively clean. Although protective clothing was worn when handling tar, masks were not.

NIOSH investigators were told by the plant physician that skin cancer was not observed among gasifier and cleanup crews nor among workers at the coal-tar separators. However, the plant recordkeeping system did not include chronic diseases, such as cancer, and follow-up studies on workers were not undertaken. Apparently, use of barrier creams was popular among SASOL workers because it helped in washing afterward.

6.3.1.5 *Kosovo Lurgi Gasifiers, Pristina, Yugoslavia*

Little information is available on the industrial hygiene and occupational health practices at the Elektroprivieda Kosovo (EPK), an industrial energy complex including a large commercial gasifier, near Pristina, Yugoslavia. The facility has been in operation about 18 years. The information summarized below was obtained through personal communication with D. Wamback, U.S. Department of Energy.

In 1973, the EPA initiated an environmental and engineering evaluation of the EPK plant. The project was completed successfully and resulted in several publications on process and control equipment used and emissions from the plant (EPA, 1981). In 1977, the DOE, EPA, and NIOSH began investigating the possibility of studying human health effects associated with EPK. Because the general population was exposed to pollutants from several sources, including large electric power-generating plants, the study was restricted to workers at the EPK complex. In 1980, a study protocol was developed that included a retrospective epidemiology study, environmental monitoring to characterize workplace contaminants, job analyses and classification, investigation of a rumored high leukemia increase, initiation of a tumor registry, a prospective epidemiology study, routine monitoring to characterize worker exposure, initiation of indicator compound monitoring, and an annual report, including recommendations for corrective actions.

Initiation of the study was delayed about 1 year. Three preliminary industrial hygiene sampling efforts were conducted between September 1981 and June 1982. The final report on both the epidemiology and industrial hygiene studies was recently completed (Brandt et al., 1986).

A Yugoslav report on the first 2 years of study contains results of three industrial hygiene surveys and a tabulation of medical data from over 300 gasification plant workers with more than 10 years experience. These studies provided data on exposures to ammonia, phenol, hydrogen sulfide, mercaptans, cyanides, methanol, nitrogen dioxide, dust, metals, noise, and heat stress. In general, exposures were reported to be within the Yugoslavian TLV. However, exposures were highly variable, with excursions above standards owing to leaks, repair activities, and adverse weather conditions. Dust levels were very high at some coal-handling operations, and noise levels were commonly above Yugoslavian standards. Sampling and analysis for organic vapors and particulates were unsatisfactory. The large volumes of air collected on silica gel and charcoal tubes resulted in excessive breakthrough. Long delays between sampling and analysis make the few available results questionable.

A site visit in May 1983 by a medical and epidemiology group showed that the medical exams performed were very good and that the medical records were excellent. A retrospective epidemiology study is being performed by Brookhaven National Laboratory under contract to the DOE Office of Health and Environmental Research.

6.3.1.6 *Hygas Fluidized-Bed Gasification Pilot Plant, Chicago, Illinois*

The Hygas plant (Husting, 1978) in Chicago, Illinois, was the first large-scale facility to produce pipeline quality gas from coal. The industrial hygiene and occupational health programs at the pilot plant are extensive (Husting, 1978). Safety eyewear, protective headwear, and protective footwear are mandatory for all process employees. Hard hats and other equipment are mandatory in certain areas. Respirators are available and their use is encouraged. Uniforms, coveralls, and gloves are provided to all employees. Each employee has at least one locker for personal clothing, and a bank of laundry lockers is available for dirty clothes. Employees are urged to change clothes and shower daily, or immediately following accidental contamination by potentially toxic materials.

A first-aid station is provided at the plant, and medical services are provided by a nearby clinic. Preplacement medical exams and chest X-rays are provided for all new employees. A voluntary annual medical exam is offered thereafter and includes a liver profile and chest X-rays. There is no epidemiology program.

6.3.1.7 *University of North Dakota Energy Research Center (UNDERC)**

The GFETC operated a pilot-scale slagging version of a pressurized, fixed-bed gasifier (Wilzbach et al., 1981). The most noxious gasifier waste stream was the gas liquor obtained from condensation of tars, oils, phenols, and water in the spray cooler. In commercial designs, tar and oil are gravity separated from wastewater and recycled to the gasifier or burned to generate steam. Thus, handling these materials and the possibility of spills or leaks are the main industrial hygiene concerns at the pilot plant.

Initially, a program was designed to develop methods to measure and monitor workplace air hazards (Wilzbach et al., 1981). Various toxic organic vapors that might be present could range from the highly volatile light aromatics to the condensable, high-molecular-weight PAHs that include known carcinogens. These compounds may be released to the environment as fugitive emissions or as a result of accidents or process upsets. Three separate samplings were then performed. The initial sampling, in April 1980, was a limited effort designed to determine the types and approximate concentrations of organic compounds in the gasifier building air. The second sampling, in May and June 1980, was more extensive, and vapors and particulates were collected at various times and locations to determine sample reproducibility and temporal and spatial variations in the gasifier building. The third sampling, in October 1981, collected particulate matter to characterize adsorbed organic compounds and for microscopic examinations.

Organic vapors in the gasifier building resembled process oil in composition. Levels of individual compounds were low; for instance, benzene concentrations were well below the TLV. Concentrations of individual compounds decreased with increasing molecular weight and ring size. One- to three-ring aromatic compounds accounted for most (50–70%) of the material, with aliphatic hydrocarbons accounting for most of the remainder (30–40%). Particulate-associated organic compounds were concentrated on particles in the respirable size range. The nature and amount of adsorbed organic compounds were a function of particle size. Particle size and distributions of trace elements were bimodal and resembled data for ambient air. Physical characteristics of particles collected in the gasifier building were similar to those resulting from coal combustion.

*Formerly the Grand Forks Energy Technology Center (GFETC), Grand Forks, North Dakota.

6.3.1.8 *University of Minnesota–Duluth (UMD) Gasifier*

Occupational health is a major risk associated with operation of the UMD gasifier (Cowser, 1978). Because of the severe winter environment in northern Minnesota, the UMD gasifier facility is enclosed, and gasifier workers are in much closer proximity than the general public to process effluents and stresses. Thus, the worker population is the most likely to exhibit acute health effects from gasifier operations. Two complementary industrial hygiene programs have been in effect at the gasifier since 1978 (Cowser, 1982). The industrial hygiene program focused on collecting information on exposure of UMD workers to gases, particulates, tar aerosols, and heat and noise stress.

Fugitive releases from valves, flanges, and seals were especially troublesome during early operations (Cowser, 1982). Modifications and equipment replacement solved many carbon monoxide, tar, and other hazardous exposure problems associated with these sources. Although variable, carbon monoxide levels were generally in the 5–15-ppm range. Thus, no observable health effects were expected to result from routine exposure of workers to measured carbon monoxide levels. Although peak carbon monoxide levels did exceed threshold value for short periods, they did not exceed TLV limits. Exposure to PAHs was measured, using BaP as an indicator. The arithmetic mean of workplace atmospheric BaP concentrations was 14.6 ng/m^3. Estimated worker risk, using this concentration, was low compared to risks associated with other plant operations and to those in other occupations. Thus, atmospheric exposures to PAHs in the UMD gasifier appear well-controlled, and there appear to have been no significant health impacts associated with gasifier operations. The ongoing analysis of worker exposures from valve leaks, tar contact, and so on has resulted in corrective action and significant reductions in potential exposures.

In 1980 and 1981 (Schurosko et al., 1983), two visits were made to the plant during gasifier runs to evaluate several developmental instruments: a derivative ultraviolet adsorption spectrometer (DUVAS), a second-derivative UV spectrometer designed for the real-time monitoring of a number of aromatic vapors; a passive integrating PAH vapor and aerosol monitor that can be worn as a personal exposure meter; Spill Spotter, for detecting low-level surface contamination; and the lightpipe luminoscope for detecting and measuring small amounts of fluorescent contamination on the skin. While all the instruments performed satisfactorily, the DUVAS instrument appeared to be nearest to readiness for routine use as an industrial hygiene monitoring tool. The others required some additional development studies and modifications.

6.3.1.9 *Morgantown Energy Technology Center (METC), Morgantown, West Virginia*

In 1977 (Hanson et al., 1981), an effort was initiated to evaluate potential health risks from the atmospheric-pressure fluidized-bed combustion of coal. The effort involved chemical and physical characterization of potential effluents from the gasification unit operated at METC. Major findings and conclusions were as follows:

1. Fluidized-bed combustion (FBC) particle emissions are comparable to or lower than those from conventional power plants per unit of power produced.
2. FBC trace-element emissions are comparable to those from conventional power plants.
3. FBC generates fewer volatile trace elements in fine particles than has been observed in conventional combustion.
4. FBC organic emissions are primarily vapors.

5. FBC particle-associated organic emissions appear higher than those estimated for conventional power plants.

In June 1980 (Cowser, 1981), measurements were taken at the gasifier with several developmental monitoring instruments, including the luminoscope, passive PAH dosimeter, a DUVAS, and a Spill Spotter. Eleven plant workers were examined with the luminoscope. Coal-tar contamination was detected on the hands of two workers and on the forehead of another. Luminoscope measurements also revealed that cotton gloves provided limited protection. Luminescence of coal-tar contaminants on the hands of one worker were monitored after repeated washing. Significant coal tar was still present after four washings with detergent.

In 1981, four sets of passive dosimeters were used to monitor PAH vapors and aerosols at different locations around the gasifier (Cowser, 1981). The dosimeters were placed at ground level near a tar storage tank and on three different levels of the gas producer unit. Average pyrene concentrations were 6.8, 3.4, and 0.7 ppb at ground level, level 2, and level 3, respectively. Other PAH compounds detected were quinoline, fluoranthene, and phenanthrene. Air sampling with the DUVAS instrument focused on the third gasifier level. Benzene was detected at about 1.9 ppm and naphthalene at about 130 ppb. Samples were also taken near the ash and tar dumping stations. Measured concentrations were lower than those observed on level 3 near the gasifier. The fluorescence Spill Spotter was used to monitor areas frequented by gasifier operating and maintenance personnel. Most measurements were below 0.3 $\mu g/cm^2$ equivalent of BaP.

6.3.1.10 *Pittsburgh Energy Technology Center (PETC), Pittsburgh, Pennsylvania*

The health program at PETC includes personal hygiene and safety, employee education, industrial hygiene monitoring, and medical surveillance (NIOSH, 1979). Protective clothing is required as necessary. If contact with harmful materials occurs, the affected worker is instructed to wash immediately with soap and water. Employees apt to contact coal-derived products are required to shower daily before leaving the plant. Employees are instructed not to wear soiled work clothes or shoes home and to put on clean work clothes daily. Employees are instructed to conduct a daily self-inspection and report any lesions or abnormalities to the medical department.

A comprehensive safety manual has been developed, covering various topics. Supplemental safe-practices bulletins outline emergency first-aid, general housekeeping, personal protective equipment, and details of the occupational medical exam program. Supervisors are required to inform employees regarding carcinogen hazards and to enforce precautionary measures. Employees are trained in the use of self-contained breathing apparatus, first aid, and cardiopulmonary resuscitation.

Except for short-term employees, who are offered a free baseline physical exam at the plant, employees are required to have a preplacement medical exam from a doctor of their choice. Employees over age 45 receive a comprehensive exam annually. Employees under 45 are offered a comprehensive exam every 2 years. A medical and work history is maintained for each employee. Vision, audiometric, and pulmonary function tests are conducted onsite by the PETC nurse. There is no epidemiology program.

Smith (1983) evaluated the respiratory hazards from coal or silica dust and several hydrocarbons at the PETC facility. Results showed that total coal dust and silica were less than 1.7 mg/m^3, with silica less than 1% of the total. Atmospheric dust concentrations ranged from 0 to 1.9 mg/m^3. All values were below the TLV for respirable coal dust (2 mg/m^3) and respirable silica (3 mg/m^3). Hydrocarbons were not detected in most process areas.

A report (EAL Corp., 1983) on the baseline industrial hygiene survey conducted at PETC indicated that atmospheric samples showed relatively low-level worker exposures to coal-conversion products. No exposures approached or exceeded applicable TLV. In most cases, exposure levels to PAHs were in the part-per-trillion range. However, the final report did not conclude that exposures at this facility were irrelevant. During the study, operations activities were being deemphasized; an increase in activities involving coal-conversion processing would increase health-related occupational exposures.

6.3.2 Summary and Conclusions

Health concerns in coal gasification vary among processes. For example, the Lurgi gasifier, because of its lower operating temperature and fixed-bed design, produces a broader spectrum of products and by-products than entrained-flow, slagging (high-temperature) gasifiers. Tars constitute 15–20% of the total by-product output in Lurgi gasification but are not produced by high-severity gasifiers. Toxicants, including carcinogens, are present in coal-gasification by-product tars and oils. These are occupational health hazards through skin contact and possible inhalation of aerosols. Carbon monoxide and other gases and respirable particulates, with their associated organic matter, are health concerns in all coal-gasification processes. However, the information available, although limited for gasification, suggests that suitable engineering controls, combined with good industrial-hygiene practices and medical surveillance, have effectively protected the health of workers. No epidemiological data are available for coal-gasification workers, although such a study has been initiated at the Kosovo Lurgi gasification operation in Yugoslavia.

REFERENCES

Ashland Synthetic Fuels, Inc., H-Coal Pilot Plant, Topical Report, *Industrial Hygiene Personal Sampling Survey During Two Coal Operating Periods at the H-Coal Pilot Plant*, Revised E&H-1, FE-2260, Ashland Synthetic Fuels, Inc., Ashland, KY, 1982a.

Ashland Synthetic Fuels, Inc., H-Coal Pilot Plant, Topical Report, *Radiation Safety*, E&H-2, FE-2260, Ashland Synthetic Fuels, Inc., Ashland, KY, 1982b.

Ashland Synthetic Fuels, Inc., H-Coal Pilot Plant, Topical Report, *Worker Health Protection Training Program at the H-Coal Pilot Plant*, E&H-3, FE-2260, Ashland Synthetic Fuels, Inc., Ashland, KY, 1982c.

Ashland Synthetic Fuels, Inc., H-Coal Pilot Plant, Topical Report, *Evaluation of a Commercial Laundry Process for Cleaning Work Clothing From a Synthetic Fuels Facility*, E&H-12, FE-2260, Ashland Synthetic Fuels, Inc., Ashland, KY, 1983a.

Ashland Synthetic Fuels, Inc., H-Coal Pilot Plant, Topical Report, *Skin Contamination Survey at the H-Coal Pilot Plant*, E&H-13, FE-2260, Ashland Synthetic Fuels, Inc., Ashland, KY, 1983b.

Ashland Synthetic Fuels, Inc., H-Coal Pilot Plant, Topical Report, *Noise Survey at the H-Coal Pilot Plant*, E&H-14, FE-2260, Ashland Synthetic Fuels, Inc., Ashland, KY, 1983c.

Ashland Synthetic Fuels, Inc., H-Coal Pilot Plant, Topical Report, *Industrial Hygiene Air Monitoring at the H-Coal Pilot Plant*, E&H-15, FE-2260, Ashland Synthetic Fuels, Inc., Ashland, KY, 1983d.

Ashland Synthetic Fuels, Inc., H-Coal Pilot Plant Phase III Final Report, Section 7.0, *Health and Safety*, Ashland Synthetic Fuels, Inc., Ashland, KY, 1983e.

Bagchi, N. J. and R. E. Zimmerman, An Industrial Hygiene Evaluation of Chimney Sweeping. *Am. Ind. Hyg. Assoc. J.*, **41,** 297–299 (1980).

Bennett, R. D., C. E. Feigly, E. D. Oswald, and R. H. Hill, The Permeation by Liquefied Coal of Gloves Used in Coal Liquefaction Pilot Plants. *Am. Ind. Hyg. Assoc. J.*, **44,** 447–452 (1983).

Brandt, M. T., J. O. Jackson, C. R. Sutcliff, O. White, E. T. Premuzic, S. C. Morris, M. A. Haxhiu, A. Abazi, M. Jockic, B. Jonuzi, S. Kapor, and V. Sajdel, *Health Effects Study of Kosova Coal Gasification Facility: The Industrial Hygiene Program*, LA-10850-MS, Los Alamos National Laboratory, Los Alamos, NM, 1986.

Braunstein, H. M., E. D. Copenhaver, and H. A. Pfuderer, *Environmental, Health and Control Aspects of Coal Conversion: An Information Overview*, ORNL/EIS-94, Oak Ridge National Laboratory, Oak Ridge, TN, 1977.

Cowser, K. E., *Proposed Environmental and Health Program for the University of Minnesota Gasification Facility*, Oak Ridge National Laboratory, Oak Ridge, TN, 1978.

Cowser, K. E., *Life Sciences Synthetic Fuels Semiannual Progress Report for the Period Ending June 30, 1981*, ORNL/TM-7926, Oak Ridge National Laboratory, Oak Ridge, TN, 1981.

Cowser, K. E., *Life Sciences Synthetic Fuels Semiannual Progress Report for the Period Ending June 30, 1982*, ORNL/TM-7926, Oak Ridge National Laboratory, Oak Ridge, TN, 1982.

Cubit, D. A. and R. K. Tanita, *Industrial Hygiene Assessment of Coal Gasification Plants*, Dynamac Corporation, Rockville, MD. National Institute of Occupational Safety and Health, Cincinnati, OH, 1982.

Doll, R., R. E. W. Fisher, E. J. Gammon, W. Gunn, G. O. Hughes, F. H. Tyrer, and W. Wilson, Mortality of Gasworkers with Special Reference to Cancers of the Lung and Bladder, Chronic Bronchitis, and Pneumoconiosis. *Br. J. Ind. Med.*, **22,** 1–12 (1965).

Doll, R., M. P. Vessey, R. W. R. Beasley, A. R. Buckley, E. C. Fear, R. E. W. Fisher, E. J. Gammon, W. Gunn, G. O. Hughes, K. Lee, and B. Norman-Smith, Mortality of Gasworkers—Final Report of a Prospective Study. *Br. J. Ind. Med.*, **29,** 394–406 (1972).

EAL Corporation, Pittsburgh Energy Technology Center, Task 4, *Industrial Hygiene Survey*, EAL Corporation, Richland, CA, 1983.

EPA, *Coal Tar, Creosote, and Coal Tar Neutral Oil*, Position Document 1, Coal Tar and Coal Tar Products Working Group, Environmental Protection Agency, Washington, DC, undated.

EPA, *Environmental Assessment: Source Test and Evaluation Report—Lurgi (Kosovo) Medium-Btu Gasification, Final Report*, EPA-600/7-81-142, U.S. Environmental Protection Agency, Washington, DC, 1981.

EPA, *Carcinogen Assessment of Coke Oven Emissions*, EPA-600/6-82-003, U.S. Environmental Protection Agency, Washington, DC, 1982.

EPRI, *Combustion Demonstration of SRC II Fuel Oil in a Tangentially Fired Boiler*, FP-1029, Final Report of Projects 1235-5 and 1414-2, Electric Power Research Institute, Palo Alto, CA, 1979.

EPRI, *Combustion of Exxon Donor Solvent Distillate Fuel in a 44.5MW Utility Boiler*, Electric Power Research Institute, Palo Alto, CA, 1981.

EPRI, *Electric Utility Use of Coal-Derived Fuels: Health, Personnel Protection, and Regulatory Considerations*, AP-2288, Final Report of Project 1735-1, Electric Power Research Institute, Palo Alto, CA, 1982.

Evans, J. and B. Pallay, NIOSH Programs for Evaluation and Control of Industrial Hygiene Hazards in Coal Conversion, in *Symposium Proceedings: Environmental Aspects of Fuel Conversion Technology, IV*, Hollywood, FL, April, 1979, EPA-600/7-29-217, Environmental Protection Agency, Washington, DC, 1979.

Exxon, Research & Engineering Company, *EDS Coal Liquefaction Process Development*, Phase V, EDS Environmental Program, Interim Report, FE-2891-79, Exxon Research & Engineering Company, Florham Park, NJ, 1981.

Exxon Research & Engineering Company, *EDS Coal Liquefaction Process Development*, Phase V, ECLP Industrial Hygiene Program Interim Report, FE-2893-96, Exxon Research & Engineering Company, Florham Park, NJ, 1982.

Exxon Research & Engineering Company, *EDS Coal Liquefaction Process Development*, Phase V, ECLP Industrial Hygiene Program—II, Interim Report, FE-2893-96, Exxon Research & Engineering Company, Florham Park, NJ, 1983.

Freudenthal, R. I., G. A. Lutz, and R. I. Mitchell, *Carcinogenic Potential of Coal and Coal Conversion Products*, Battelle Columbus Laboratories, Columbus, OH, 1977.

Gray, R. H., and K. E. Cowser (eds.), *Status of Health and Environmental Research Relative to Direct Coal Liquefaction: 1976 to the Present*, DOE/NBM-1016, PNL-4176, National Technical Information Service, Springfield, VA, 1982.

Hammond, E. C., I. J. Selikoff, P. L. Lawther, and H. Seidman, Inhalation of Benzopyrene and Cancer in Man. *Ann. NY Acad. Sci.*, **271,** 116–124 (1976).

Hangebrauck, R. P., D. J. von Lehmden, and J. E. Meeker, *Sources of Polynuclear Hydrocarbons in the Atmosphere*, Public Health Service Publication 999-AP-33. U.S. Department of Health, Education and Welfare, Cincinnati, OH, 1967.

Hannon, M. A., L. Recio, H. G. Enock, and R. H. Hill, *Development of a Biological Screening Method to Detect Exposure of Synfuel Plant Workers to Carcinogenic Substances*, Appalachian Regional Commission Project ARC 79-207CO-7162-79-I-302-0917, Kentucky Department of Energy, Lexington, KY, 1981.

Hanson, R. L., S. H. Weissman, R. L. Carpenter, G. J. Newton, and S. J. Rothenberg, Sampling and Chemical Characterization of Coal Combustion Effluents from an Experimental Fluidized-Bed Combustor, in D. D. Mahlum, R. H. Gray, and W. D. Felix (eds.), *Coal Conversion and the Environment, Chemical, Biomedical and Ecological Considerations*, CONF-801039, National Technical Information Service, Springfield, VA, 1981, pp. 52–66.

Hill, R. H. and H. G. Enoch, *Industrial Hygiene Air Monitoring Characterization and Correlation Study of Coal Liquefaction Workers Tested for Mutagenic Activity in Urine*, Appalachian Regional Commission Report, Kentucky. Department of Energy, Lexington, KY, 1983.

Husting, E. L., *Occupational Health Programs and Studies Associated with Fossil Energy Development*, TR-13500-2, Flow Resources Corporation, San Rafael, CA, 1978.

Kawai, M., H. Amamoto, and K. Hareda, Epidemiological Study of Occupational Lung Cancer. *Arch. Environ. Health*, **14,** 859–864 (1967).

Keenan, R. R. and S. B. Cole, A Sampling and Analytical Procedure for Skin-Contamination Evaluation. *Am. Ind. Hyg. Assoc. J.*, **43,** 473–476 (1982).

Lloyd, J. W., Long Term Mortality Study of Steelworkers: V. Respiratory Cancer in Coke Plant Workers. *J. Occup. Med.*, **13,** 53–68 (1971).

Montgomery, C. H. Coal Liquefaction Occupational Health Program, Presented at the 65th Annual Meeting of the American Occupational Medical Association, Detroit, MI, 1980.

Munro, N. B., R. J. M. Fry, R. B. Gammage, W. M. Haschek, E. E. Calle, J. A. Klein, and T. W. Schultz, *Indirect Coal Liquefaction: A Review of Potential Health Effects and Worker Exposure During Gasification and Synthesis*, ORNL-5938, Oak Ridge National Laboratory. National Technical Information Service, Springfield, VA, 1983.

NAS, *Particulate Polycyclic Organic Matter*, National Academy of Sciences, Washington, DC, 1972, pp. 13–35.

NIOSH, *Criteria for a Recommended Standard—Occupational Exposure to Coke Oven Emissions*, Publication No. HSM 73-11016, National Institute for Occupational Safety and Health, Cincinnati, OH, 1973.

NIOSH, *Criteria for a Recommended Standard—Occupational Exposure to Coal Tar Products*, DHEW(NIOSH) Publication No. 78-107, National Institute for Occupational Safety and Health, Cincinnati, OH, 1977.

NIOSH, *Criteria for a Recommended Standard—Occupational Exposures in Coal Gasification Plants*, DHEW(NIOSH) Publication No. 78-191, National Institute for Occupational Safety and Health, Cincinnati, OH, 1978.

NIOSH, *Industrial Hygiene Comprehensive Survey, Solvent Refined Coal Plant, Fort Lewis, Washington,* Enviro Control, Inc., Rockville, MD. National Institute for Occupational Safety and Health, Morgantown, WV, 1979.

NIOSH, *Occupational Hazard Assessment, Coal Liquefaction, Volume II—Assessment*, DHHS(NIOSH) Publication No. 81-132, National Institute for Occupational Safety and Health, Cincinnati, OH, 1981a.

NIOSH, *Control Technology Assessment for Coal Gasification and Liquefaction Processes, Solvent Refined Coal Pilot Plant, Fort Lewis, Washington*, Enviro Control Inc., Rockville, MD. National Institute for Occupational Safety and Health, Morgantown, WV, 1981b.

NIOSH, *Industrial Hygiene Survey Protocol, Exxon Coal Liquefaction Pilot Plant, Baytown, Texas*, Enviro Control, Inc., Rockville, MD. National Institute for Occupational Safety and Health, Morgantown, WV, 1981c.

NIOSH, *Final Summary Report, Industrial Hygiene Assessment of Coal Liquefaction Plants*, Dynamac Corporation, Rockville, MD. National Institute for Occupational Safety and Health, Morgantown, WV, 1982.

NIOSH and EPA, *Recommended Health and Safety Guidelines for Coal Gasification Pilot Plants*, DHEW(NIOSH) Publication No. 78-120, EPA-600/778-007, National Technical Information Service, Springfield, VA, 1978.

O'Connor, R. B., Improving the Environmental Health of Coke Oven Workers. *J. Occup. Med.*, **13**, 83–85 (1971).

OSHA, Exposure to Coke Oven Emissions. *Occupational Safety and Health Standards, Fed. Reg.*, **41**, 46742. U.S. Government Printing Office, Washington, DC, 1976.

Palmer, A., Mortality Experience of 50 Workers with Occupational Exposures to the Products of Coal Hydrogenation Processes. *J. Occup. Med.*, **21**, 41–44 (1979).

P&M Co., Solvent Refined Coal (SRC) Process: Health Programs, Research and Development Report No. 53, Interim Report No. 24, Vol. III: Pilot Plant Development Work, Part 4: *Industrial Hygiene, Clinical, and Toxicological Programs*, FE/496-T15, Pittsburg & Midway Coal Mining Co., Denver, CO. National Technical Information Service, Springfield, VA, 1978.

P&M Co., Solvent Refined Coal (SRC) Process: Health Programs, Research and Development Report No. 53, Interim Report No. 28, Vol. III: Pilot Plant Development Work, Part 4: *Industrial Hygiene, Clinical, and Toxicological Programs*, FE/496-T19, Pittsburg & Midway Coal Mining Co., Denver, CO. National Technical Information Service, Springfield, VA, 1979.

P&M Co., Solvent Refined Coal (SRC) Process: Health Programs, Research and Development Report No. 53, Interim Report No. 32, Vol. III: Pilot Plant Development Work, Part 4: *Industrial Hygiene, Clinical, and Toxicological Programs*, FE/496-T23, Pittsburg & Midway Coal Mining Co., Denver, CO. National Technical Information Service, Springfield, VA, 1980.

P&M Co., Solvent Refined Coal (SRC) Process: Final Report, DOE/ET/10104-46, Vol. 2, DE82013554, Pittsburg & Midway Coal Mining Co., Denver, CO. National Technical Information Service, Springfield, VA, 1982.

Pott, P., Cancer Scroti, in Hawes, Clark, and Collins (eds.), *Chirurgical Observations*, London, 1775, pp. 63–67. [Reprinted in *Nat. Cancer Inst. Monogr.* **10,** 7 (1963).]

Redmond, C. K., A. Ciocco, J. W. Lloyd, and H. W. Rush, Long Term Mortality Study of Steelworkers: VI. Mortality from Malignant Neoplasms Among Coke Oven Workers. *J. Occup. Med.*, **14,** 621–629 (1972).

Redmond, C. K., B. R. Strobino, and R. H. Cypess, Cancer Experience Among Coke By-Product Workers. *Ann. N.Y. Acad. Sci.*, **271,** 102–117 (1976).

Schurosko, D. D., A. R. Hawthorne, T. Vo-Dinh, R. B. Gammage, W. G. Dreibelbis, and S. D. Van Hoesen, *Evaluation of Developmental Industrial Hygiene Instrumentation at the University of Minnesota-Duluth Gasifier*, ORNL-5938, Oak Ridge National Laboratory, Oak Ridge, TN, 1983.

Sexton, R. J., N. I. Condra, N. H. Ketcham, and R. W. Norton. The Hazards to Health in the Hydrogenation of Coal. *Arch. Environ. Health*, **1,** 181–231 (1960).

Shepard, H. G., Occupational Health Aspects Related to Coal Liquefaction. *J. Soc. Occup. Med.*, **31,** 9–15 (1981).

Smith, G. F., *Survey Report on Coal and Silica Dust and Organic Solvents in the Air at the DOE/PETC Center*, NUS Corporation, Pittsburgh, PA, 1983.

Tebbens, B. D., J. F. Thomas, and M. Mukai, Aromatic Hydrocarbon Production Related to Incomplete Combustion. *Arch. Ind. Health*, **14,** 112–120 (1956).

Tebbens, B. D., J. F. Thomas, and M. Mukai, Hydrocarbon Synthesis in Combustion. *Arch. Ind. Health*, **17,** 152–165 (1958).

Vo-Dinh, T. and R. B. Gammage, The Lightpipe Luminoscope for Monitoring Occupational Skin Contamination. *Am. Ind. Hyg. Assoc. J.*, **42,** 112–120 (1981).

Wilzbach, K. E. and C. A. Reilly, Jr., (eds.), *Status of Health and Environmental Research Relative to Coal Gasification 1976 to the Present*, DOE/ER-0149, National Technical Information Service, Springfield, VA, 1982.

Wilzbach, K. E., J. R. Stetter, C. A. Reilly, Jr., and W. G. Wilson, *Environmental Research Program for Slagging Fixed-Bed Coal Gasification, Status Report*, ANL/SER-1, Argonne National Laboratory, Argonne, IL, 1981.

Index

For definitions of acronyms and abbreviations, see Glossary starting on page xv.